Group Theory

Special Notation

Commutative Rings and Linear Algebra

A FIRST COURSE
IN ABSTRACT ALGEBRA
with Applications

Third Edition

JOSEPH J. ROTMAN
*University of Illinois
at Urbana-Champaign*

PEARSON
Prentice
Hall

Upper Saddle River, New Jersey 07458

Library of Congress Cataloging-in-Publication Data

Rotman, Joseph J.
A first course in abstract algebra : with applications / Joseph J. Rotman.--3rd ed.
p. cm.
Includes bibliographical references and index.
ISBN 0-13-186267-7
1. Algebra, Abstract 2. Number theory 3. Commutative rings 4. Group Rings I. Title

QA162.R68   2006
512'.02--dc22   2005053492

Editor-in-Chief: *Sally Yagan*
Production Editor: *Raegan Keida*
Senior Managing Editor: *Linda Mihatov Behrens*
Assistant Managing Editor: *Bayani Mendoza de Leon*
Executive Managing Editor: *Kathleen Schiaparelli*
Manufacturing Buyer: *Lisa McDowell*
Marketing Manager: *Halee Dinsey*
Marketing Assistant: *Joon Won Moon*
Cover Designer: *Bruce Kenselaar*
Art Director: *Jayne Conte*
Director of Creative Services: *Paul Belfanti*
Editorial Assistant: *Jennifer Urban*
Cover Image: *The original painting of 10x10 orthogonal Latin squares*
by Emi Kasia hangs in the office of the Mathematics Department
of the University of Illinois at Urbana-Champaign.

© 2006, 2000, 1996 Pearson Education, Inc.
Pearson Prentice Hall
Pearson Education, Inc.
Upper Saddle River, New Jersey 07458

Pearson Prentice Hall™ is a trademark of Pearson Education, Inc.

Printed in the United States of America
10 9 8 7 6 5 4 3 2 1

ISBN 0-13-186267-7

Pearson Education LTD., *London*
Pearson Education Australia PTY, Limited, *Sydney*
Pearson Education Singapore, Pte. Ltd
Pearson Education North Asia Ltd, *Hong Kong*
Pearson Education Canada, Ltd., *Toronto*
Pearson Educacion de Mexico, S.A. de C.V.
Pearson Education - Japan, *Tokyo*
Pearson Education Malaysia, Pte. Ltd

To my two wonderful kids,

Danny and Ella,

whom I love very much

Contents

Preface to the Third Edition

A First Course in Abstract Algebra introduces number theory, groups, and commutative rings. Group theory was invented by Galois in the early 1800s, when he used groups to completely determine those polynomials whose roots can be found with generalizations of the quadratic formula. Nowadays, the language of group theory is the precise way to discuss various types of symmetry, both in geometry and elsewhere. Thus, besides introducing Galois's ideas, we classify certain planar designs called *friezes*, and we also apply group theory to solve some intricate counting problems (how many 6-beaded bracelets are there if each bead is either red, white, or blue?). Commutative rings provide the proper context in which to study number theory as well as many aspects of the theory of polynomials. Ideas such as greatest common divisor of integers and modular arithmetic extend effortlessly to polynomial rings in one variable. There are applications to public access codes, calendars, Latin squares, magic squares, and design of experiments. We then consider vector spaces with scalars in arbitrary fields (not just the reals), and this study allows us to solve the classical Greek problems involving ruler-compass constructions: trisecting an angle; doubling a cube; squaring a circle; constructing regular n-gons. Linear algebra over finite fields is applied to codes, showing how one can decode messages sent over a noisy channel (for example, photographs sent to Earth from other planets). The classical formulas finding the roots of cubic and quartic polynomials are proved, after which both groups and commutative rings are combined in proving Galois's theorem (polynomials whose roots are obtainable by such formulas have solvable Galois groups) and its corollary, Abel's theorem (there are polynomials of degree 5 whose roots cannot be found by a generalization of these formulas). This is only an introduction to Galois theory; readers wishing to learn more of this beautiful subject will have to see a more advanced text. Algebra is fascinating, and I hope that my enthusiasm for it is transmitted to my readers.

To accomodate readers having different backgrounds, this book contains more material than can be covered in a one- or two-semester course. The first four chapters contain all the results usually covered in a first year. But many sections need not be covered in lectures, either because they are well known (induction, binomial theorem, complex numbers, linear algebra), they are not of primary importance, or they will be covered more thoroughly in more advanced courses. However, instructors may assign projects for interested students from these optional sections as well as from later chapters. Those readers whose appetites have been whetted by results in the first chapters may browse in the end of the book, which investigates groups and rings further. The chapter Groups II proves that finite abelian groups are direct products of cyclic groups, gives the existence (and significance) of large p-subgroups of finite groups, and classifies symmetry groups of friezes. The last chapter, an introduction to polynomials in several variables, includes Hilbert's basis theorem, varieties, Hilbert's *Nullstellensatz* for $\mathbb{C}[x_1, \ldots, x_n]$, and algorithmic methods associated with Gröbner bases. Thus, the last two chapters display some directions in which the earlier ideas have developed, and so they can serve as a reference for some algebra beyond the present courses.

Let me mention some new features of this edition.

- I have rewritten the text, making the exposition more smooth.

- In order that the reader know what is essential in the first five chapters, I have inserted a small arrow next to the most important sections, subsections, definitions, theorems, and examples.

- Chapters 2 and 3, which introduce groups and commutative rings, are essentially independent of one another. Thus, with very minor changes, it is possible to study groups first or to study commutative rings first.

- More linear algebra, over arbitrary fields, has been included. This allows me to include a new section on codes, which goes far enough to decode Reed-Solomon codes.

- There is a new section classifying frieze groups in the plane.

- Exercises.

  (i) The previous edition had 414 exercises; this edition has 574 exercises.

  (ii) Each exercise set begins with a multipart true-false question which reviews important items in its section.

  (iii) Every exercise explicitly cited in the text is marked by *; moreover, every citation gives the page number on which the cited exercise appears.

  (iv) Certain exercises, those marked by H, have hints in a section at the end of the book; thus, readers may consider problems on their own before reading the hints.

- One numbering system enumerates all lemmas, theorems, propositions, corollaries, and examples, so that finding back references is easy.

- There are several pages of Special Notation, giving page numbers where notation is introduced.

Today, abstract algebra is viewed as a challenging course; many bright students seem to have inordinate difficulty learning it. Certainly, students must learn to think in a new way. Axiomatic reasoning may be new to some; others may be more visually oriented. Some students have never written proofs; others may have once done so, but their skills have atrophied from lack of use. But none of these obstacles adequately explains the observed difficulties. After all, the same obstacles exist in beginning real analysis courses, but most students in these courses do learn the material, perhaps after some early struggling. However, the difficulty of standard algebra courses persists, whether groups are taught first, whether rings are taught first, or whether texts are changed. I believe that a major factor contributing to the difficulty in learning abstract algebra is that both groups and rings are introduced in the first course; as soon as a student begins to be comfortable with one topic, it is dropped to study the other. Furthermore, leaving group theory or commutative ring theory before significant applications are made gives students the false impression that the theory is either of no real value or, more likely, that it cannot be appreciated until some future indefinite time. Imagine a beginning analysis course in which both real and complex analysis are introduced in the first semester; would there be ample time to prove the intermediate value theorem and Liouville's theorem? If algebra is taught as a one-year (two-semester) course, there is no longer any reason to crowd both topics into the first course, and a truer, more attractive, picture of algebra is presented. This option is more practical today than it was in the past, for the many applications of abstract algebra have increased the numbers of interested students, many of whom work in other disciplines. Therefore, I have rewritten the text for two audiences. On the one hand, this new edition can serve as a text for those who prefer the currently popular arrangement of introducing both groups and rings in the first semester. There is ample material in the book so that it can serve as a text for a sequel course as well. On the other hand, the book can also serve as a text for a one-year course. There are many possible organizations; I suggest covering number theory and commutative rings in the first semester, and linear algebra and group theory in the second. Detailed syllabi for such courses are presented in the next section.

Giving the etymology of mathematical terms is rarely done in mathematics texts. Let me explain, with an analogy, why I have included derivations of many terms. There are many variations of standard poker games and, in my poker group, the dealer announces the game of his choice by naming it. Now some names are better than others. For example, "Little Red" is a game in which one's smallest red card is wild; this is a good name because it reminds the players of its distinctive feature. On the other hand, "Aggravation" is not such a good name, for though it is, indeed, suggestive, the name does not distinguish this particular game from several others. Most terms in mathematics have been well chosen; there are more red names than aggravating ones. An example of a good name is *even* permutation, for a permutation is even if it is a product of an even number of transpositions. Another example of a good term is the *parallelogram law* describing vector addition. But many good names, clear when they were chosen, are now obscure because their roots are either in another language or in another discipline. The trigonometric terms *tangent* and *secant* are good names for those knowing some Latin, but they are obscure otherwise (see a discussion of their etymology on page 32). The term *mathematics* is obscure only because most of us do not know that it comes from the classical Greek word meaning "to learn". The term *corollary* is doubly obscure; it comes from the Latin word meaning "flower", but why should some theorems be called flowers? A plausible explanation is that it was common, in ancient Rome, to give flowers as a gratuity, and so a corollary is a gift bequeathed by a theorem. The term *theorem* comes from the Greek word meaning "to watch" or "to contemplate" (*theatre* has the same root); it was used by Euclid with its present meaning. The term *lemma* comes from the Greek word meaning "taken" or "received"; it is a statement that is taken for granted (for it has already been proved) in the course of proving a theorem. I believe that etymology of terms is worthwhile (and interesting!), for it often aids understanding by removing unnecessary obscurity.

In addition to thanking again those who helped me with the first two editions, I give special thanks to George Bergman for his many suggestions as well as for his generosity in allowing me to use many interesting exercises. I also thank Chris Heil, for pointing out subtle errors I had not discovered, and Iwan Duursma for his help with the new section on coding. Finally, I thank William Chin, Joel S. Foisy, Robert Friedman, Blair F. Goodlin, Zahid Hasan, Ilya Kapovich, Dieter Koller, Fatma Irem Koprulu, Mario Livio, Thomas G. Lucas, Leon McCulloh, Arnold W. Miller, Charles H. Morgan, Jr., Chuang Peng, Eric Schmutz, Brent B. Solie, Paul Weichsel, and John Wetzel.

George Lobell was with Prentice Hall until this edition was essentially complete. He consistently gave me sage advice about its content and style, and my book is significantly better now than it would have been without him. I am happy to thank him for his guidance.

Joseph Rotman
rotman@math.uiuc.edu

Suggested Syllabi

Here are some one-semester courses using this text, where a semester consists of about 45 one-hour lectures (hour lectures are usually 50 minutes in length; Paul Halmos noted that a *microcentury*, one millionth of a century, is about 52.6 minutes). We give five syllabi. The first, Table 1, is a "standard" syllabus designed for the currently popular course organization: a one-semester course which introduces both groups and rings. This syllabus has three topics: Chapter 1: number theory; Chapter 2: groups; Chapter 3: commutative rings. It is possible to invert the order of topics and treat commutative rings before groups, for I have rewritten Chapters 2 and 3 so that they are now essentially independent of one another. As an aside, I disagree with today's received wisdom that expounding groups first is more efficient than doing rings first; in spite of Chapter 3's mentioning almost no group theory, its present version is about the same length as its versions in previous editions.

Either of the second two syllabi, Tables 2 and 3, may be used for a sequel course (there is ample material in the text which can be used to create other sequel courses as well).

My own ideas about teaching abstract algebra have changed. I now think that the best presentation is a year-long two-semester course in which only one of groups or rings is taught in the first semester. Moreover, I recommend such a course whose first semester covers number theory and commutative rings, and whose second semester covers linear algebra and group theory. Tables 4 and 5 are syllabi for such a course. (Of course, I recognize merit in arguments advocated by those who prefer to discuss groups first. A one-year course using this text and organized about this choice should be easy to design.) I think that doing commutative rings first is more natural. As one passes from $\mathbb{Z}$ to $k[x]$, one can watch arithmetic results and proofs generalize to polynomials. If the second semester begins with linear algebra, then the discussion of groups takes on more significance, for matrix groups, with their geometric context, are another source of concrete examples of groups in addition to groups of permutations.

Section	Topics	Hours
1.3	Division algorithm, euclid lemma, euclidean algorithm	3
1.4	Fundamental theorem of arithmetic	1
1.5	Congruences, Fermat, Chinese remainder theorem	3
2.1	Functions	1
2.2	Permutations	4
2.3	Groups and examples	2
2.4	Subgroups and Lagrange's theorem	2
2.5	Homomorphisms	2
2.6	Quotient groups and isomorphism theorems	4
2.7	Group actions	4
2.8	Burnside counting (skim)	1
3.1	Commutative rings and subrings	1
3.2	Fields	1
3.3	Polynomial rings $R[x]$	1
3.4	Homomorphisms	2
3.5	From numbers to polynomials	3
3.6	Unique factorization for polynomials	1
3.7	Irreducibility (skim)	2
3.8	Quotient rings and finite fields	3

Table 1: Standard One-Semester Syllabus: 41 Hours

Section	Topics	Hours
4.1	Vector spaces and dimension	5
4.1	Gaussian elimination	3
4.2	Euclidean constructions	3
4.3	Linear transformations	4
4.4	Determinants and eigenvalues	2
4.5	Coding	6
5.1	Classical formulas	2
5.2	Solvability by radicals	4
5.2	Translation into group theory	4
5.3	Epilog	1
6.1	Finite abelian groups	3
6.2	Sylow theorems	3

Table 2: Second Semester, Syllabus A: 40 Hours

Section	Topics	Hours
4.1	Vector spaces and dimension	5
4.1	Gaussian elimination	3
4.2	Euclidean constructions	3
4.3	Linear transformations	4
4.4	Determinants and eigenvalues	2
6.1	Finite abelian groups	3
6.2	Sylow theorems	3
6.3	Symmetry groups of friezes (skim)	3
7.1	Prime ideals and maximal ideals	1
7.2	Unique factorization	3
7.3	Noetherian rings	2
7.4	Varieties	6

Table 3: Second Semester: Syllabus B: 38 Hours

Section	Topics	Hours
1.3	Division algorithm, euclid lemma, euclidean algorithm	4
1.4	Fundamental theorem of arithmetic	1
1.5	Congruences, Fermat, Chinese remainder theorem	4
2.1	Functions	1
3.1	Commutative rings and subrings	2
3.2	Fields	1
3.3	Polynomial rings $R[x]$	2
5.1	Classical formulas	2
3.4	Homomorphisms	2
3.5	From numbers to polynomials	4
3.6	Unique factorization for polynomials	2
3.7	Irreducibility	3
3.8	Quotient rings and finite fields	4
3.9	Latin squares, magic squares, projective planes (skim)	1
7.1	Prime ideals and maximal ideals	1
7.2	Unique factorization (skim)	1
7.3	Noetherian rings (skim)	1
7.4	Varieties (skim)	3

Table 4: One-Year Version, Semester I: 39 Hours

Section	Topics	Hours
4.1	Vector spaces	5
4.1	Gaussian elimination	3
4.2	Euclidean constructions	3
4.3	Linear transformations	4
4.4	Eigenvalues	2
4.5	Coding (skim)	3
2.2	Permutations	4
2.3	Groups and examples	2
2.4	Subgroups and Lagrange's theorem	2
2.5	Homomorphisms	2
2.6	Quotient groups and isomorphism theorems	4
2.7	Group actions	4
2.8	Burnside counting (skim)	1
6.1	Finite abelian groups (skim)	1
6.2	Sylow theorems (skim)	1
6.3	Ornamental symmetry (skim)	1

Table 5: One-Year Version, Semester II: 42 Hours

To the Reader

The essential sections, subsections, theorems, definitions, and examples in the first five chapters have a small arrow in the margin next to them (some things, though interesting, are not as important as others).

Exercises in a text have two main functions: to reinforce the reader's grasp of the material, and to provide puzzles whose solutions give a certain pleasure. Therefore, the serious reader should attempt *all* the exercises (many are not difficult).

There are two special notations associated to exercises. An asterisk, as in *2.44, means that this exercise is cited elsewhere in the text. For example, the citation reads "Exercise 2.44 on page 146." The letter H, as in H 2.47, means that there is a hint to Exercise 2.47 in the Hints section at the back of the book. Neither of these notations indicates the relative difficulty of an exercise.

Most exercise sets begin with a multipart question labeled "True or false with reasons." If one of the parts is the statement, "The fourth roots of unity are i and $-i$," then the correct answer is, "False; 1 is also a fourth root of unity." The declaration "False" must be supported by a concrete example. If another statement is "$2 + 4 + \cdots + 100 = 50 \times 51$," then the correct response is "True; using Proposition 1.6, we have

$$2 + 4 + \cdots + 100 = 2[1 + 2 + \cdots + 50] = 2\left[\tfrac{1}{2}(50 \times 51)\right] = 50 \times 51.\text{"}$$

The declaration 'True" must be supported either by a "one-line proof" using results proved in the text or by a short argument from first principles.

1

Number Theory

→ 1.1 INDUCTION

There are many styles of proof, and mathematical induction is one of them. We begin by saying what mathematical induction is not. In the natural sciences, *inductive reasoning* is the assertion that a freqently observed phenomenon will always occur. Thus, one says that the Sun will rise tomorrow morning because, from the dawn of time, the Sun has risen every morning. This is not a legitimate kind of proof in mathematics, for even though a phenomenon has been observed many times, it need not occur forever. However, inductive reasoning is still valuable in mathematics, as it is in natural science, because seeing patterns in data often helps in guessing what may be true in general.

On the other hand, a reasonable guess may not be correct. For example, what is the maximum number of regions into which $\mathbb{R}^3$ (3-dimensional space) can be divided by n planes? Two nonparallel planes can divide $\mathbb{R}^3$ into 4 regions, and three planes can divide $\mathbb{R}^3$ into 8 regions (octants). For smaller n, we note that a single plane divides $\mathbb{R}^3$ into 2 regions, while if $n = 0$, then $\mathbb{R}^3$ is not divided at all: there is 1 region. For $n = 0, 1, 2, 3$, the maximum number of regions is thus 1, 2, 4, 8, and it is natural to guess that n planes can be chosen to divide $\mathbb{R}^3$ into 2^n regions. But it turns out that any four chosen planes can divide $\mathbb{R}^3$ into at most 15 regions!

Before proceeding further, let us make sure that we agree on the meaning of some standard terms. An ***integer*** is one of the numbers $0, 1, -1, 2, -2, 3, \ldots$; the set of all the integers is denoted by $\mathbb{Z}$ (from the German *Zahl* meaning *number*):

$$\mathbb{Z} = \{0, 1, -1, 2, -2, 3, \ldots\}.$$

The ***natural numbers*** consists of all those integers n for which $n \geq 0$:

$$\mathbb{N} = \{n \text{ in } \mathbb{Z} : n \geq 0\} = \{0, 1, 2, 3, \ldots\}.$$

1

→ **Definition.**   An integer d is a ***divisor*** of an integer n if $n = da$ for some integer a. A natural number n is called ***prime***[1] if $n \geq 2$ and its only divisors are ± 1 and $\pm n$; a natural number $n \geq 2$ is called ***composite*** if it is not prime.

If a positive integer n is composite, then it has a factorization $n = ab$, where $a < n$ and $b < n$ are positive integers; the inequalities are present to eliminate the uninteresting factorization $n = n \times 1$. The first few primes are 2, 3, 5, 7, 11, 13, 17, 19, 23, 29, 31, 37, 41, ...; that this sequence never ends is proved in Corollary 1.33.

Consider the assertion that

$$f(n) = n^2 - n + 41$$

is prime for every positive integer n. Evaluating $f(n)$ for $n = 1, 2, 3, \ldots, 40$ gives the numbers

$$41, 43, 47, 53, 61, 71, 83, 97, 113, 131,$$
$$151, 173, 197, 223, 251, 281, 313, 347, 383, 421,$$
$$461, 503, 547, 593, 641, 691, 743, 797, 853, 911,$$
$$971, 1033, 1097, 1163, 1231, 1301, 1373, 1447, 1523, 1601.$$

It is tedious, but not very difficult, to show that every one of these numbers is prime (see Proposition 1.3). Inductive reasoning predicts that *all* the numbers of the form $f(n)$ are prime. But the next number, $f(41) = 1681$, is not prime, for $f(41) = 41^2 - 41 + 41 = 41^2$, which is obviously composite. Thus, inductive reasoning is not appropriate for mathematical proofs.

Here is an even more spectacular example (which I first saw in an article by W. Sierpinski). Recall that ***perfect squares*** are numbers of the form n^2, where n is an integer; the first few perfect squares are 0, 1, 4, 9, 16, 25, 36, ... For each $n \geq 1$, consider the statement

$$S(n): 991n^2 + 1 \text{ is not a perfect square.}$$

The nth statement, $S(n)$, is true for many n; in fact, the smallest number n for which $S(n)$ is false is

$$n = 12, 055, 735, 790, 331, 359, 447, 442, 538, 767$$
$$\approx 1.2 \times 10^{28}.$$

The equation $m^2 = 991n^2 + 1$ is an example of ***Pell's equation***—an equation of the form $m^2 = pn^2 + 1$, where p is prime—and there is a way of calculating all possible solutions of it. An even larger example involves the prime $p = 1,000,099$; the smallest

---

[1]One reason the number 1 is not called a prime is that many theorems involving primes would otherwise be more complicated to state.

n for which $1,000,099n^2 + 1$ is a perfect square has 1116 digits. The most generous estimate of the age of the Earth is 10 billion (10,000,000,000) years, or 3.65×10^{12} days, a number insignificant when compared to 1.2×10^{28}, let alone 10^{1115}. If, starting from the Earth's very first day, one verified statement $S(n)$ on the nth day, then there would be today as much evidence of the general truth of these statements as there is that the Sun will rise tomorrow morning. And yet some of the statements $S(n)$ are false!

As a final example, let us consider the following statement, known as **Goldbach's conjecture**: every even number $m \geq 4$ is a sum of two primes. No one has ever found a counterexample to Goldbach's conjecture, but neither has anyone ever proved it. At present, the conjecture has been verified for all even numbers $m < 10^{13}$, and it has been proved by J.-R. Chen that every sufficiently large even number m can be written as $p + q$, where p is prime and q is "almost" a prime; that is, q is either a prime or a product of two primes. Even with all of this positive evidence, however, no mathematician will say that Goldbach's conjecture must, therefore, be true for all even m.

We have seen what (mathematical) induction is not; let us now discuss what induction is. Our discussion is based on the following property of the set of natural numbers (usually called the *Well-Ordering Principle*).

Least Integer Axiom.   There is a smallest integer in every nonempty[2] subset C of the natural numbers $\mathbb{N}$.

Although this axiom cannot be proved (it arises in analyzing what integers are), it is certainly plausible. Consider the following procedure: check whether 0 belongs to C; if it does, then 0 is the smallest integer in C. Otherwise, check whether 1 belongs to C; if it does, then 1 is the smallest integer in C; if not, check 2. Continue this procedure until one bumps into C; this will occur eventually because C is nonempty.

Proposition 1.1 (Least Criminal).   *Let k be a natural number, and let $S(k)$, $S(k+1)$, $\ldots$, $S(n)$, $\ldots$ be a list of statements. If some of these statements are false, then there is a first false statement.*

Proof.   Let C be the set of all those natural numbers $n \geq k$ for which $S(n)$ is false; by hypothesis, C is a nonempty subset of $\mathbb{N}$. The Least Integer Axiom provides a smallest integer m in C, and $S(m)$ is the first false statement.   •

This seemingly innocuous proposition is useful.

→ **Theorem 1.2.**   *Every integer $n \geq 2$ is either a prime or a product of primes.*

---

[2]Saying that C is *nonempty* merely means that there is at least one integer in C.

Proof. Were this not so, there would be "criminals": there are integers $n \geq 2$ which are neither primes nor products of primes; a least criminal m is the smallest such integer. Since m is not a prime, it is composite; there is thus a factorization $m = ab$ with $2 \leq a < m$ and $2 \leq b < m$ (since a is an integer, $1 < a$ implies $2 \leq a$). Since m is the least criminal, both a and b are "honest"; that is,

$$a = pp'p'' \cdots \quad \text{and} \quad b = qq'q'' \cdots ,$$

where the factors $p, p', p'', \ldots$ and $q, q', q'' \ldots$ are primes. Therefore,

$$m = ab = pp'p'' \cdots qq'q'' \cdots$$

is a product of (at least two) primes, which is a contradiction.[3]   •

Proposition 1.3. *If $m \geq 2$ is a positive integer which is not divisible by any prime p with $p \leq \sqrt{m}$, then m is a prime.*

Proof. If m is not prime, then $m = ab$, where $a < m$ and $b < m$ are positive integers. If $a > \sqrt{m}$ and $b > \sqrt{m}$, then $m = ab > \sqrt{m}\sqrt{m} = m$, a contradiction. Therefore, we may assume that $a \leq \sqrt{m}$. By Theorem 1.2, a is either a prime or a product of primes, and any (prime) divisor p of a is also a divisor of m. Thus, if m is not prime, then it has a "small" prime divisor p; that is, $p \leq \sqrt{m}$. The contrapositive says that if m has no small prime divisor, then m is prime.   •

Proposition 1.3 can be used to show that 991 is a prime. It suffices to check whether 991 is divisible by some prime p with $p \leq \sqrt{991} \approx 31.48$; if 991 is not divisible by 2, 3, 5, $\ldots$, or 31, then it is prime. There are 11 such primes, and one checks (by long division) that none of them is a divisor of 991. (One can check that 1,000,099 is a prime in the same way, but it is a longer enterprise because its square root is a bit over 1000.) It is also tedious, but not difficult, to see that the numbers $f(n) = n^2 - n + 41$, for $1 \leq n \leq 40$, are all prime.

Mathematical induction is a version of Least Criminal that is more convenient to use. The key idea is just this: imagine a stairway to the sky. If its bottom step is white and if the next step above a white step is also white, then all the steps of the stairway must be white. (One can trace this idea back to Levi ben Gershon in 1321. There is an explicit description of induction, cited by Pascal, written by Francesco Maurolico

---

[3]The ***contrapositive*** of an implication "P implies Q" is the implication "(not Q) implies (not P)." For example, the contrapositive of "If a series $\sum a_n$ converges, then $\lim_{n \to \infty} a_n = 0$" is "If $\lim_{n \to \infty} a_n \neq 0$, then $\sum a_n$ diverges." If an implication is true, then so is its contrapositive; conversely, if the contrapositive is true, then so is the original implication. The strategy of this proof is to prove the contrapositive of the original implication. Although a statement and its contrapositive are logically equivalent, it is sometimes more convenient to prove the contrapositive. This method is also called ***indirect proof*** or ***proof by contradiction***.

in 1557.) For example, the statement "$2^n > n$ for all $n \geq 1$" can be regarded as an infinite sequence of statements (a stairway to the sky):

$$2^1 > 1; \quad 2^2 > 2; \quad 2^3 > 3; \quad 2^4 > 4; \quad 2^5 > 5; \quad \cdots .$$

Certainly, $2^1 = 2 > 1$. If $2^{100} > 100$, then

$$2^{101} = 2 \times 2^{100} > 2 \times 100 = 100 + 100 > 101.$$

There is nothing magic about the exponent 100; the same idea shows, having reached any stair, that we can climb up to the next one. This argument will be formalized in Proposition 1.5.

$\rightarrow$ **Theorem 1.4 (Mathematical Induction[4]).** *Given statements $S(n)$, one for each natural number $n \geq 1$, suppose that*

(i) ***Base Step***: $S(1)$ *is true*;

(ii) ***Inductive Step***: *if $S(n)$ is true, then $S(n + 1)$ is true.*

Then $S(n)$ is true for all integers $n \geq 1$.

Proof. We must show that the collection C of all those integers $n \geq 1$ for which the statement $S(n)$ is false is empty.

If, on the contrary, C is nonempty, then there is a first false statement $S(m)$. Since $S(1)$ is true, by (i), we must have $m \geq 2$. This implies that $m - 1 \geq 1$, and so there is a statement $S(m - 1)$ [there is no statement $S(0)$]. As m is the least criminal, $m - 1$ must be honest; that is, $S(m - 1)$ is true. But now (ii) says that $S(m) = S([m - 1] + 1)$ is true, and this is a contradiction. We conclude that C is empty and, hence, that all the statements $S(n)$ are true. $\bullet$

We now show how to use induction.

Proposition 1.5. $2^n > n$ *for all integers $n \geq 1$.*

Proof. The nth statement $S(n)$ is

$$S(n) : 2^n > n.$$

Two steps are required for induction, corresponding to the two hypotheses in Theorem 1.4.

Base step. The initial statement

$$S(1) : 2^1 > 1$$

is true, for $2^1 = 2 > 1$.

---

[4]*Induction*, having a Latin root meaning "to lead," came to mean "prevailing upon to do something" or "influencing." This is an apt name here, for the nth statement influences the $(n + 1)$st one.

Inductive step. If $S(n)$ is true, then $S(n + 1)$ is true; that is, using the **inductive hypothesis** $S(n)$, we must prove

$$S(n + 1) : 2^{n+1} > n + 1.$$

If $2^n > n$ is true, then multiply both sides of its inequality by 2; Proposition A.2 in Appendix A gives the inequality:

$$2^{n+1} = 2 \times 2^n > 2n.$$

Now $2n = n + n \geq n + 1$ (because $n \geq 1$), and hence $2^{n+1} > 2n \geq n + 1$, as desired.

Having verified both the base step and the inductive step, we conclude that $2^n > n$ for all $n \geq 1$. •

Induction is plausible in the same sense that the Least Integer Axiom is plausible. Suppose that a given list $S(1)$, $S(2)$, ... of statements has the property that $S(n + 1)$ is true whenever $S(n)$ is true. If, in addition, $S(1)$ is true, then $S(2)$ is true; the truth of $S(2)$ now gives the truth of $S(3)$; the truth of $S(3)$ now gives the truth of $S(4)$; and so forth. Induction replaces the phrase *and so forth* by the inductive step which guarantees, for every n, that there is never an obstruction in the passage from any statement $S(n)$ to the next one, $S(n + 1)$.

Here are two comments before we give more illustrations of induction. First, one must verify both the base step and the inductive step; verification of only one of them is inadequate. For example, consider the statements $S(n) : n^2 = n$. The base step is true, but one cannot prove the inductive step (of course, these statements are false for all $n > 1$). Another example is given by the statements $S(n) : n = n + 1$. It is easy to see that the inductive step is true: if $n = n + 1$, then Proposition A.2 says that adding 1 to both sides gives $n + 1 = (n + 1) + 1 = n + 2$, which is the next statement, $S(n + 1)$. But the base step is false (of course, all these statements are false).

Second, when first seeing induction, many people suspect that the inductive step is circular reasoning: one is using $S(n)$, and this is what one wants to prove! A closer analysis shows that this is not at all what is happening. The inductive step, by itself, does not prove that $S(n + 1)$ is true. Rather, it says that *if $S(n)$ is true, then $S(n + 1)$* is also true. In other words, the inductive step proves that the *implication* "If $S(n)$ is true, then $S(n + 1)$ is true" is correct. The truth of this implication is not the same thing as the truth of its conclusion. For example, consider the two statements: "Your grade on every exam is 100%" and "Your grade in the course is A." The implication "If all your exams are perfect, then you will get the highest grade for the course" is true. Unfortunately, this does not say that it is inevitable that your grade in the course will be A. Our discussion above gives a mathematical example: the implication "If $n = n + 1$, then $n + 1 = n + 2$" is true, but the conclusion "$n + 1 = n + 2$" is false.

Proposition 1.6. $1 + 2 + \cdots + n = \frac{1}{2}n(n + 1)$ *for every integer* $n \geq 1$.

Proof. The proof is by induction on $n \geq 1$.

Base step. If $n = 1$, then the left side is 1 and the right side is $\frac{1}{2}1(1 + 1) = 1$, as desired.

Inductive step. It is always a good idea to write the $(n + 1)$st statement $S(n + 1)$ so one can see what has to be proved. Here, we must prove

$$S(n + 1) : \quad 1 + 2 + \cdots + n + (n + 1) = \tfrac{1}{2}(n + 1)(n + 2).$$

By the inductive hypothesis, i.e., using $S(n)$, the left side is

$$[1 + 2 + \cdots + n] + (n + 1) = \tfrac{1}{2}n(n + 1) + (n + 1),$$

and high school algebra shows that $\frac{1}{2}n(n + 1) + (n + 1) = \frac{1}{2}(n + 1)(n + 2)$. By induction, the formula holds for all $n \geq 1$. •

There is a story (it probably never happened) told about Gauss as a boy. One of his teachers asked the students to add up all the numbers from 1 to 100, thereby hoping to get some time for himself for other tasks. But Gauss quickly volunteered that the answer was 5050. He let s denote the sum of all the numbers from 1 to 100; $s = 1 + 2 + \cdots + 99 + 100$. Of course, $s = 100 + 99 + \cdots + 2 + 1$. Arrange these nicely:

$$s = 1 + 2 + \cdots + 99 + 100$$
$$s = 100 + 99 + \cdots + 2 + 1$$

and add:

$$2s = 101 + 101 + \cdots + 101 + 101,$$

the sum 101 occurring 100 times. We now solve: $s = \frac{1}{2}(100 \times 101) = 5050$. This argument is valid for any number n in place of 100 (and it does not use induction). Not only does this give a new proof of Proposition 1.6, it also shows how the formula could have been discovered.[5]

It is not always the case, in an inductive proof, that the base step is very simple. In fact, all possibilities can occur: both steps can be easy; both can be difficult; one is harder than the other.

---

[5]Actually, this formula goes back at least a thousand years (see Exercise 1.11 on page 15). Alhazen (Ibn al-Haytham) (965–1039), found a geometric way to add

$$1^k + 2^k + \cdots + n^k$$

for any fixed integer $k \geq 1$ (see Exercise 1.12 on page 15).

Proposition 1.7.   *If we assume $(fg)' = f'g + fg'$, the product rule for derivatives, then*

$$(x^n)' = nx^{n-1} \text{ for all integers } n \geq 1.$$

Proof.   We proceed by induction on $n \geq 1$.

Base step.   If $n = 1$, then we ask whether $(x)' = x^0 \equiv 1$, the constant function identically equal to 1. By definition,

$$f'(x) = \lim_{h \to 0} \frac{f(x+h) - f(x)}{h}.$$

When $f(x) = x$, therefore,

$$(x)' = \lim_{h \to 0} \frac{x+h-x}{h} = \lim_{h \to 0} \frac{h}{h} = 1.$$

Inductive step.   We must prove that $(x^{n+1})' = (n+1)x^n$. It is permissible to use the inductive hypothesis, $(x^n)' = nx^{n-1}$, as well as $(x)' \equiv 1$, for the base step has already been proved. Since $x^{n+1} = x^n x$, the product rule gives

$$(x^{n+1})' = (x^n x)' = (x^n)'x + x^n(x)'$$
$$= (nx^{n-1})x + x^n 1 = (n+1)x^n.$$

We conclude that $(x^n)' = nx^{n-1}$ is true for all $n \geq 1$.   •

→ **Remark.**   The Least Integer Axiom is enjoyed not only by $\mathbb{N}$, but also by any of its nonempty subsets Q (indeed, the proof of Proposition 1.1 uses the fact that the axiom holds for $Q = \{n \text{ in } \mathbb{N} : n \geq 2\}$). In terms of induction, this says that the base step can occur at any positive integer k, not necessarily at $k = 1$. The conclusion, then, is that the statements $S(n)$ are true for all $n \geq k$. The Least Integer Axiom is also enjoyed by the larger set $Q_k = \{n \text{ in } \mathbb{Z} : n \geq k\}$, where k is any, possibly negative, integer. If C is a nonempty subset of Q_k and if $C \cap \{k, k+1, \ldots, -1, 0\}^6$ is nonempty, then this finite set contains a smallest integer, which is the smallest integer in C. If $C \cap \{k, k+1, \ldots, -1, 0\}$ is empty, then C is actually a nonempty subset of $\mathbb{N}$, and the original axiom gives a smallest number in C. In terms of induction, this says that the base step can occur at 0 or at any, possibly negative, integer k [assuming, of course, that there is a kth statement $S(k)$]. For example, if one has statements $S(-1), S(0), S(1), \ldots$, then the base step can occur at $n = -1$; the conclusion in this case is that the statements $S(n)$ are true for all $n \geq -1$.   ◄

---

[6]If C and D are subsets of a set X, then their ***intersection***, denoted by $C \cap D$, is the subset consisting of all those x in X lying in both C and D.

Here is an example of an induction whose base step occurs at $n = 5$. Consider the statements

$$S(n) : 2^n > n^2.$$

This is not true for small values of n: if $n = 2$ or 4, then there is equality, not inequality; if $n = 3$, the left side, 8, is smaller than the right side, 9. However, $S(5)$ is true, for $32 > 25$.

Proposition 1.8. $2^n > n^2$ *is true for all integers $n \geq 5$.*

Proof.   We have just checked the base step $S(5)$. In proving

$$S(n + 1) : 2^{n+1} > (n + 1)^2,$$

we are allowed to assume that $n \geq 5$ (actually, we will need only $n \geq 3$ to prove the inductive step) as well as the inductive hypothesis. Multiply both sides of $2^n > n^2$ by 2 to get

$$2^{n+1} = 2 \times 2^n > 2n^2 = n^2 + n^2 = n^2 + nn.$$

Since $n \geq 5$, we have $n \geq 3$, and so

$$nn \geq 3n = 2n + n \geq 2n + 1.$$

Therefore,

$$2^{n+1} > n^2 + nn \geq n^2 + 2n + 1 = (n + 1)^2. \quad \bullet$$

So far, we have used induction to prove some minor results; let us now use it to prove something more substantial. Observe first that if x and y are positive real numbers, then the identity

$$(x + y)^2 = (x - y)^2 + 4xy$$

gives

$$\left[\tfrac{1}{2}(x + y)\right]^2 = xy + \left[\tfrac{1}{2}(x - y)\right]^2.$$

It follows that

$$\tfrac{1}{2}(x + y) \geq \sqrt{xy}, \tag{1}$$

with the term $[\tfrac{1}{2}(x - y)]^2$ showing why, in general, the inequality is not an equality. If equality holds, then $[\tfrac{1}{2}(x - y)]^2 = 0$ and $x = y$; conversely, if $x = y$, then there is equality: $[\tfrac{1}{2}(x + x)]^2 = xx = x^2$, for $[\tfrac{1}{2}(x - x)]^2 = 0$. Here is an application of this observation.

Recall that the **hyperbolic cosine** is defined by $\cosh(x) = \tfrac{1}{2}(e^x + e^{-x})$. Since $e^x e^{-x} = 1$, it follows from inequality (1) that

$$\cosh(x) \geq 1$$

for all x, with equality if and only if $e^x = e^{-x}$; that is, $\cosh(x) = 1$ if and only if $e^{2x} = 1$, so that $\cosh(x) = 1$ if and only if $x = 0$.

Definition.   Given positive numbers $a_1, a_2, \ldots, a_n$, their **arithmetic mean** is their average $(a_1 + a_2 + \cdots + a_n)/n$, and their **geometric mean** is $\sqrt[n]{a_1 a_2 \cdots a_n}$.

We have just shown that the arithmetic mean of two positive numbers a_1, a_2 is larger than their geometric mean, with equality holding precisely when $a_1 = a_2$. We are going to extend this result to several terms, but we begin with an elementary lemma followed by a normalized version of the inequality.

Lemma 1.9.   *If $0 < m < 1 < M$, then $m + M > 1 + mM$.*

Proof.   Since the product of positive numbers is positive,

$$(1 - m)(M - 1) = M - 1 - mM + m$$

is positive. Therefore, $M + m > 1 + mM$, as desired.   •

For example, if θ is an acute angle, i.e., $0° < \theta < 90°$, then $0 < \cos\theta < 1$, and so $1 < 1/\cos\theta = \sec\theta$. Hence, there are inequalities $0 < \sin\theta < 1 < \sec\theta$, and so Lemma 1.9 gives the inequality, whenever θ is acute,

$$\sin\theta + \sec\theta > 1 + \sin\theta\sec\theta = 1 + \tan\theta.$$

Lemma 1.10.   *If $k_1, \ldots, k_n$ are positive numbers with $k_1 \cdots k_n = 1$, then $k_1 + \cdots + k_n \geq n$; moreover, equality holds if and only if $1 = k_1 = \cdots = k_n$.*

Proof.   Clearly, $k_1 + \cdots + k_n = n$ if all $k_i = 1$. Therefore, to prove both statements, it suffices to show that if $k_1 \cdots k_n = 1$ and not all $k_i = 1$, then $k_1 + \cdots + k_n > n$. We prove this by induction on $n \geq 2$.

Base step. Now $k_1 k_2 = 1$. If both k_1 and k_2 are strictly larger than 1, then $k_1 k_2 > 1$; if both are strictly smaller than 1, then $k_1 k_2 < 1$. Hence, we may assume that $0 < k_1 < 1 < k_2$, and Lemma 1.9 gives $k_1 + k_2 > 1 + k_1 k_2 = 2$.

Inductive step. Assume that $k_1 \cdots k_{n+1} = 1$, where $k_1, \cdots, k_{n+1}$ are positive numbers. If all $k_i \geq 1$, then the present assumption that not all $k_i = 1$ gives the contradiction $k_1 \cdots k_{n+1} > 1$. Hence, we may further assume that some $k_i < 1$. For notational convenience, let $k_1 < 1$. A similar argument, with all inequalities reversed, allows us to assume that $k_{n+1} > 1$. Define

$$a = k_1 k_{n+1}.$$

By Lemma 1.9, $k_1 + k_{n+1} > 1 + k_1 k_{n+1} = 1 + a$, so that adding $k_2 + \cdots + k_n$ to both sides of this inequality gives

$$k_1 + k_2 + \cdots + k_n + k_{n+1} > 1 + a + k_2 + \cdots + k_n. \tag{2}$$

It remains to show that $1 + a + k_2 + \cdots + k_n \geq n + 1$ [for we already have strict inequality in (2)]. Note that $a k_2 \cdots k_n = k_1 k_2 \cdots k_{n+1} = 1$. If $a = 1 = k_2 = \cdots = k_n$, then $1 + a + k_2 + \cdots + k_n = n + 1$, and we are done. Otherwise, the inductive hypothesis applies and gives $a + k_2 + \cdots + k_n > n$, and hence $1 + a + k_2 + \cdots + k_n > n + 1$.   •

Theorem 1.11 (Inequality of the Means). *If $a_1, a_2, \cdots, a_n$ are positive numbers, then*

$$(a_1 + a_2 + \cdots + a_n)/n \geq \sqrt[n]{a_1 a_2 \cdots a_n};$$

moreover, equality holds if and only if $a_1 = a_2 = \cdots = a_n$.

Proof. Define $G = \sqrt[n]{a_1 a_2 \cdots a_n}$, and define $k_i = a_i/G$ for all i. It follows that $k_1 k_2 \cdots k_n = a_1 a_2 \cdots a_n / G^n = 1$, and so $k_1 + k_2 + \cdots + k_n \geq n$, by Lemma 1.10; that is, $a_1 + a_2 + \cdots + a_n \geq nG$, or

$$(a_1 + a_2 + \cdots + a_n)/n \geq G = \sqrt[n]{a_1 a_2 \cdots a_n}.$$

Moreover, the lemma states that there is equality if and only if all the $k_i = 1$; that is, equality holds if and only if all the a_i are equal (to G). •

This inequality is used, in Exercise 1.26 on page 17, to prove an isoperimetric inequality: of all the triangles having a given perimeter, the equilateral triangle has the largest area.

There is another version of induction, usually called the *second form of induction*, that is sometimes more convenient to use.

Definition. The *predecessors* of a natural number $n \geq 1$ are the natural numbers k with $k < n$, namely, $0, 1, 2, \ldots, n-1$ (0 has no predecessor).

→ **Theorem 1.12 (Second Form of Induction).** *Let $S(n)$ be a family of statements, one for each integer $n \geq 1$, and suppose that*

(i) $S(1)$ *is true*;

(ii) *if $S(k)$ is true for all predecessors k of n, then $S(n)$ is itself true.*

Then $S(n)$ is true for all integers $n \geq 1$.

Proof. It suffices to show that there are no integers $n \geq 1$ for which $S(n)$ is false; that is, the collection C of all positive integers n for which $S(n)$ is false is empty.

If, on the contrary, C is nonempty, then there is a least criminal m: there is a first false statement $S(m)$. Since $S(1)$ is true, by (i), we must have $m \geq 2$. As m is the *least* criminal, k must be honest for all $k < m$; in other words, $S(k)$ is true for all the predecessors of m. Then, by (ii), $S(m)$ is true, and this is a contradiction. We conclude that C is empty and, hence, that all the statements $S(n)$ are true. •

The second form of induction can be used to give a second proof of Theorem 1.2. As with the first form, the base step need not occur at 1.

→ **Theorem 1.13 (= Theorem 1.2).**   *Every integer $n \geq 2$ is either a prime or a product of primes.*

Proof. [7] *Base step.*   The statement is true when $n = 2$ because 2 is a prime.

Inductive step.   If $n \geq 2$ is a prime, we are done. Otherwise, $n = ab$, where $2 \leq a < n$ and $2 \leq b < n$. As a and b are predecessors of n, each of them is either prime or a product of primes:

$$a = pp'p'' \cdots \quad \text{and} \quad b = qq'q'' \cdots,$$

and so $n = pp'p'' \cdots qq'q'' \cdots$ is a product of (at least two) primes.   •

The reason why the second form of induction is more convenient here is that it is more natural to use $S(a)$ and $S(b)$ than to use $S(n - 1)$; indeed, it is not at all clear how to use $S(n - 1)$.

Here is a notational remark. We can rephrase the inductive step in the first form of induction: if $S(n - 1)$ is true, then $S(n)$ is true (we are still saying that if a statement is true, then so is the next statement). With this rephrasing, we can now compare the inductive steps of the two forms of induction. Each wants to prove $S(n)$: the inductive hypothesis of the first form is $S(n - 1)$; the inductive hypothesis of the second form is any or all of the preceding statements $S(0), S(1), \ldots, S(n - 1)$. Thus, the second form appears to have a stronger inductive hypothesis. In fact, Exercise 1.22 on page 17 asks you to prove that both forms of mathematical induction are equivalent.

The next result says that one can always factor out a largest power of 2 from any integer.

Proposition 1.14.   *Every integer $n \geq 1$ has a unique factorization $n = 2^k m$, where $k \geq 0$ and $m \geq 1$ is odd.*

Proof.   We use the second form of induction on $n \geq 1$ to prove the existence of k and m; the reader should see that it is more appropriate here than the first form.

Base step.   If $n = 1$, take $k = 0$ and $m = 1$.

Inductive step.   If $n \geq 1$, then n is either odd or even. If n is odd, then take $k = 0$ and $m = n$. If n is even, then $n = 2b$. Because $b < n$, it is a predecessor of n, and so the inductive hypothesis allows us to assume $S(b) : b = 2^{\ell} m$, where $\ell \geq 0$ and m is odd. The desired factorization is $n = 2b = 2^{\ell+1} m$.

The word *unique* means "exactly one." We prove uniqueness by showing that if $n = 2^k m = 2^t m'$, where both k and t are nonnegative and both m and m' are odd, then $k = t$ and $m = m'$. We may assume that $k \geq t$. If $k > t$, then canceling 2^t from both sides gives $2^{k-t} m = m'$. Since $k - t > 0$, the left side is even while the right side is odd; this contradiction shows that $k = t$. We may thus cancel 2^k from both sides, leaving $m = m'$.   •

---

[7]The similarity of the proofs of Theorems 1.2 and 1.13 indicates that the second form of induction is merely a variation of Least Criminal.

The ancient Greeks thought that a rectangular figure is most pleasing to the eye if its edges a and b are in the proportion

$$a : b = b : (a + b).$$

In this case, $a(a + b) = b^2$, so that $b^2 - ab - a^2 = 0$; that is, $(b/a)^2 - (b/a) - 1 = 0$. The quadratic formula gives $b/a = \frac{1}{2}(1 \pm \sqrt{5})$. Therefore,

$$b/a = \gamma = \tfrac{1}{2}(1 + \sqrt{5}) \qquad \text{or} \qquad b/a = \delta = \tfrac{1}{2}(1 - \sqrt{5}).$$

The number γ, approximately 1.61803, is called the **golden ratio**. Since γ is a root of $x^2 - x - 1$, as is δ, we have

$$\gamma^2 = \gamma + 1 \qquad \text{and} \qquad \delta^2 = \delta + 1.$$

The reason for discussing the golden ratio is that it is intimately related to the *Fibonacci sequence*.

Definition. The *Fibonacci sequence* $F_0, F_1, F_2, \ldots$ is defined as follows:

$$F_0 = 0, \quad F_1 = 1, \quad \text{and} \quad F_n = F_{n-1} + F_{n-2} \quad \text{for all integers } n \geq 2.$$

The Fibonacci sequence begins: $0, 1, 1, 2, 3, 5, 8, 13, \ldots$

Theorem 1.15. *If F_n denotes the nth term of the Fibonacci sequence, then for all $n \geq 0$,*

$$F_n = \tfrac{1}{\sqrt{5}}(\gamma^n - \delta^n),$$

where $\gamma = \frac{1}{2}(1 + \sqrt{5})$ and $\delta = \frac{1}{2}(1 - \sqrt{5})$.

Proof. We are going to use the second form of induction [the second form is the appropriate induction here, for the equation $F_n = F_{n-1} + F_{n-2}$ suggests that proving $S(n)$ will involve not only $S(n - 1)$ but $S(n - 2)$ as well].

Base step. The formula is true for $n = 0 : \frac{1}{\sqrt{5}}(\gamma^0 - \delta^0) = 0 = F_0$. The formula is also true for $n = 1$:

$$\begin{aligned}
\tfrac{1}{\sqrt{5}}(\gamma^1 - \delta^1) &= \tfrac{1}{\sqrt{5}}(\gamma - \delta) \\
&= \tfrac{1}{\sqrt{5}}\left[\tfrac{1}{2}(1 + \sqrt{5}) - \tfrac{1}{2}(1 - \sqrt{5})\right] \\
&= \tfrac{1}{\sqrt{5}}(\sqrt{5}) = 1 = F_1.
\end{aligned}$$

[We have mentioned both $n = 0$ and $n = 1$ because verifying the inductive hypothesis for F_n requires our using the truth of the statements for both F_{n-1} and F_{n-2}. For example, knowing only that $F_2 = \frac{1}{\sqrt{5}}(\gamma^2 - \delta^2)$ is not enough to prove that the formula for F_3 is true; one also needs the formula for F_1.]

Inductive step. If $n \geq 2$, then

$$
\begin{aligned}
F_n &= F_{n-1} + F_{n-2} \\
&= \tfrac{1}{\sqrt{5}}(\gamma^{n-1} - \delta^{n-1}) + \tfrac{1}{\sqrt{5}}(\gamma^{n-2} - \delta^{n-2}) \\
&= \tfrac{1}{\sqrt{5}}\left[(\gamma^{n-1} + \gamma^{n-2}) - (\delta^{n-1} + \delta^{n-2})\right] \\
&= \tfrac{1}{\sqrt{5}}\left[\gamma^{n-2}(\gamma + 1) - \delta^{n-2}(\delta + 1)\right] \\
&= \tfrac{1}{\sqrt{5}}\left[\gamma^{n-2}(\gamma^2) - \delta^{n-2}(\delta^2)\right] \\
&= \tfrac{1}{\sqrt{5}}(\gamma^n - \delta^n),
\end{aligned}
$$

because $\gamma + 1 = \gamma^2$ and $\delta + 1 = \delta^2$.   •

It is curious that the integers F_n are expressed in terms of the irrational number $\sqrt{5}$.

Corollary 1.16. *If $\gamma = \frac{1}{2}(1 + \sqrt{5})$, then $F_n > \gamma^{n-2}$ for all integers $n \geq 3$.*

Remark.   If $n = 2$, then $F_2 = 1 = \gamma^0$, and so there is equality, not inequality.   ◀

Proof.   Base step. If $n = 3$, then $F_3 = 2 > \gamma$, for $\gamma \approx 1.618$.
Inductive step. We must show that $F_{n+1} > \gamma^{n-1}$. By the inductive hypothesis,

$$
F_{n+1} = F_n + F_{n-1} > \gamma^{n-2} + \gamma^{n-3} = \gamma^{n-3}(\gamma + 1) = \gamma^{n-3}\gamma^2 = \gamma^{n-1}. •
$$

One can also use induction to give definitions. For example, *n factorial*,[8] denoted by $n!$, can be defined by induction on $n \geq 0$. Define $0! = 1$, and if $n!$ is known, then define

$$
(n + 1)! = n!(n + 1).
$$

One reason for defining $0! = 1$ will be apparent in the next section.

EXERCISES

H **1.1** True or false with reasons.
  (**i**)   There is a largest integer in every nonempty set of negative integers.
  (**ii**) There is a sequence of 13 consecutive natural numbers containing exactly 2 primes.

---

[8] The term *factor* comes from the Latin "to make" or "to contribute"; the term *factorial* recalls that $n!$ has many factors.

(iii) There are at least two primes in any sequence of 7 consecutive natural numbers.

(iv) Of all the sequences of consecutive natural numbers not containing 2 primes, there is a sequence of shortest length.

(v) 79 is a prime.

(vi) There exists a sequence of statements $S(1), S(2), \ldots$ with $S(2n)$ true for all $n \geq 1$ and with $S(2n-1)$ false for every $n \geq 1$.

(vii) For all $n \geq 0$, we have $n \leq F_n$, where F_n is the nth Fibonacci number.

(viii) If m and n are natural numbers, then $(mn)! = m!n!$.

*1.2 (i) For any $n \geq 0$ and any $r \neq 1$, prove that

$$1 + r + r^2 + r^3 + \cdots + r^n = (1 - r^{n+1})/(1 - r).$$

H (ii) Prove that

$$1 + 2 + 2^2 + \cdots + 2^n = 2^{n+1} - 1.$$

H **1.3** Show, for all $n \geq 1$, that 10^n leaves remainder 1 after dividing by 9.

1.4 Prove that if $0 \leq a \leq b$, then $a^n \leq b^n$ for all $n \geq 0$.

1.5 Prove that $1^2 + 2^2 + \cdots + n^2 = \frac{1}{6}n(n+1)(2n+1) = \frac{1}{3}n^3 + \frac{1}{2}n^2 + \frac{1}{6}n$.

1.6 Prove that $1^3 + 2^3 + \cdots + n^3 = \frac{1}{4}n^4 + \frac{1}{2}n^3 + \frac{1}{4}n^2$.

1.7 Prove that $1^4 + 2^4 + \cdots + n^4 = \frac{1}{5}n^5 + \frac{1}{2}n^4 + \frac{1}{3}n^3 - \frac{1}{30}n$.

H **1.8** Find a formula for $1 + 3 + 5 + \cdots + (2n - 1)$, and use mathematical induction to prove that your formula is correct. (Inductive reasoning is used in mathematics to help guess what might be true. Once a guess has been made, it must still be proved, perhaps using mathematical induction, perhaps by some other method.)

H **1.9** Find a formula for $1 + \sum_{j=1}^{n} j!j$; use induction to prove that your formula is correct.

1.10 (*M. Barr*) There is a famous anecdote describing a hospital visit of G. H. Hardy to Ramanujan. Hardy mentioned that the number 1729 of the taxi he had taken to the hospital was not an interesting number. Ramanujan disagreed, saying that it is the smallest positive integer that can be written as the sum of two cubes in two different ways.

(i) Prove that Ramanujan's statement is true.

H (ii) Prove that Ramanujan's statement is false.

*H **1.11** Derive the formula for $\sum_{i=1}^{n} i$ by computing the area $(n+1)^2$ of a square with sides of length $n + 1$ using Figure 1.1.

*1.12 H (i) Derive the formula for $\sum_{i=1}^{n} i$ by computing the area $n(n+1)$ of a rectangle with height $n + 1$ and base n, as pictured in Figure 1.2.

H (ii) (*Alhazen*) For fixed $k \geq 1$, use Figure 1.3 to prove

$$(n+1)\sum_{i=1}^{n} i^k = \sum_{i=1}^{n} i^{k+1} + \sum_{i=1}^{n}\left(\sum_{p=1}^{i} p^k\right).$$

H (iii) Given the formula $\sum_{i=1}^{n} i = \frac{1}{2}n(n+1)$, use part (ii) to derive the formula for $\sum_{i=1}^{n} i^2$.

1.13 H (i) Prove that $2^n > n^3$ for all $n \geq 10$.

H (ii) Prove that $2^n > n^4$ for all $n \geq 17$.

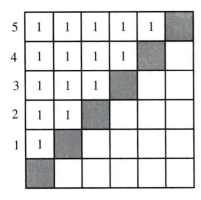

Figure 1.1
$$1 + 2 + \cdots + n = \tfrac{1}{2}(n^2 + n)$$

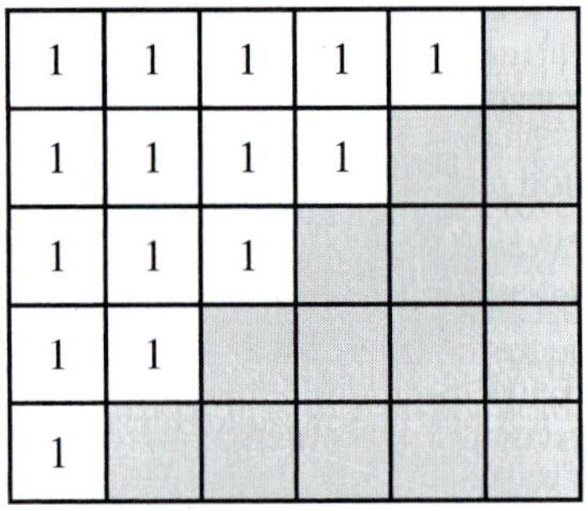

Figure 1.2
$$1 + 2 + \cdots + n = \tfrac{1}{2}n(n + 1)$$

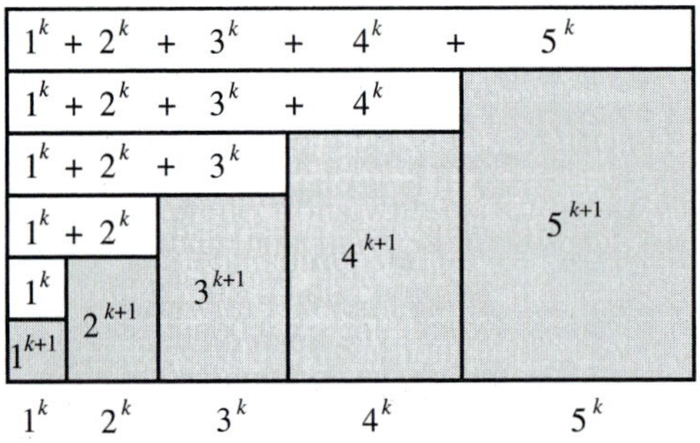

Figure 1.3   Alhazan's Dissection

H **1.14** Around 1350, N. Oresme was able to sum the series $\sum_{n=1}^{\infty} n/2^n$ by dissecting the region in Figure 1.4 in two ways. Let A_n be the vertical rectangle with base $\frac{1}{2^n}$ and height n, so that area(A_n) $= n/2^n$, and let B_n be horizontal rectangle with base $\frac{1}{2^n} + \frac{1}{2^{n+1}} + \cdots$ and height 1. Prove that $\sum_{n=1}^{\infty} n/2^n = 2$.

*H **1.15** Let $g_1(x), \ldots, g_n(x)$ be differentiable functions, and let $f(x)$ be their product: $f(x) = g_1(x) \cdots g_n(x)$. Prove, for all integers $n \geq 2$, that the derivative

$$f'(x) = \sum_{i=1}^{n} g_1(x) \cdots g_{i-1}(x) g_i'(x) g_{i+1}(x) \cdots g_n(x).$$

H **1.16** Prove, for every $n \in \mathbb{N}$, that $(1 + x)^n \geq 1 + nx$ whenever $x \in \mathbb{R}$ and $1 + x > 0$.

H **1.17** Prove that every positive integer a has a unique factorization $a = 3^k m$, where $k \geq 0$ and m is not a multiple of 3.

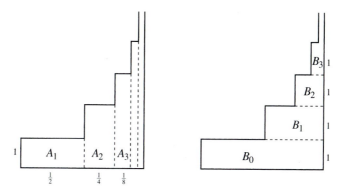

Figure 1.4 Oresme's Dissections

H **1.18** Prove that $F_n < 2^n$ for all $n \geq 0$, where $F_0, F_1, F_2, \ldots$ is the Fibonacci sequence.

H **1.19** If F_n denotes the nth term of the Fibonacci sequence, prove that

$$\sum_{n=1}^{m} F_n = F_{m+2} - 1.$$

H **1.20** Prove that $4^{n+1} + 5^{2n-1}$ is divisible by 21 for all $n \geq 1$.

H **1.21** For any integer $n \geq 2$, prove that there are n consecutive composite numbers. Conclude that the gap between consecutive primes can be arbitrarily large.

***1.22** Prove that the first and second forms of mathematical induction are equivalent; that is, prove that Theorem 1.4 is true if and only if Theorem 1.12 is true.

***1.23** (**Double Induction**) Let $S(m, n)$ be a doubly indexed family of statements, one for each $m \geq 0$ and $n \geq 0$. Suppose that

    (i) $S(0, 0)$ is true;

    (ii) if $S(m, 0)$ is true, then $S(m + 1, 0)$ is true;

    (iii) if $S(m, n)$ is true for all $m \geq 0$, then $S(m, n + 1)$ is true for all $m \geq 0$.

    Prove that $S(m, n)$ is true for all $m \geq 0$ and $n \geq 0$.

1.24 Use double induction to prove that

$$(m + 1)^n > mn$$

for all $m, n \geq 0$.

H **1.25** For every acute angle θ, i.e., $0° < \theta < 90°$, prove that

$$\sin \theta + \cot \theta + \sec \theta \geq 3.$$

***1.26**   H **(i)**   Let p be a positive number. If Δ is an equilateral triangle with perimeter $p = 2s$, prove that area$(\Delta) = s^2/\sqrt{27}$.

         H **(ii)**   Of all the triangles in the plane having perimeter p, prove that the equilateral triangle has the largest area.

H **1.27** Prove that if $a_1, a_2, \ldots, a_n$ are positive numbers, then

$$(a_1 + a_2 + \cdots + a_n)(1/a_1 + 1/a_2 + \cdots + 1/a_n) \geq n^2.$$

→ 1.2   BINOMIAL THEOREM AND COMPLEX NUMBERS

What is the pattern of the coefficients in the formulas for the powers $(1 + x)^n$ of the binomial $1 + x$? The first few such formulas are

$$(1 + x)^0 = 1$$
$$(1 + x)^1 = 1 + 1x$$
$$(1 + x)^2 = 1 + 2x + 1x^2$$
$$(1 + x)^3 = 1 + 3x + 3x^2 + 1x^3$$
$$(1 + x)^4 = 1 + 4x + 6x^2 + 4x^3 + 1x^4.$$

Figure 1.5, called **Pascal's triangle**, after B. Pascal (1623–1662), displays an arrangement of the first few coefficients. Figure 1.6, a picture from China in the year

```
                        1
                     1     1
                  1     2     1
               1     3     3     1
            1     4     6     4     1
         1     5    10    10     5     1
      1     6    15    20    15     6     1
   1     7    21    35    35    21     7     1
```

Figure 1.5

1303, shows that the pattern of coefficients had been recognized long before Pascal was born.

The expansion of $(1 + x)^n$ is an expression of the form

$$c_0 + c_1 x + c_2 x^2 + \cdots + c_n x^n.$$

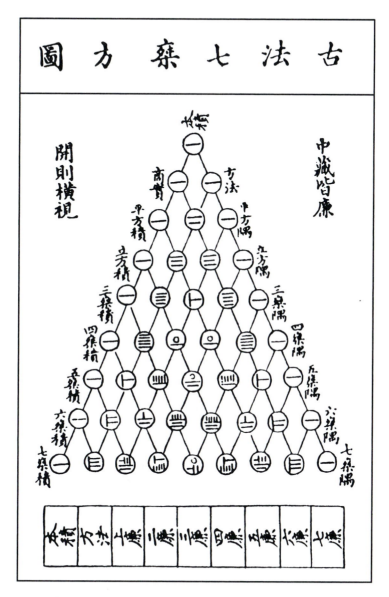

Figure 1.6   Pascal's Triangle, China, ca. 1300

The coefficients c_r are called **binomial coefficients**.[9]   L. Euler (1707–1783) introduced the notation $\left(\frac{n}{r}\right)$ for them; this symbol evolved into $\binom{n}{r}$, which is generally accepted nowadays:

$$\binom{n}{r} = \text{coefficient } c_r \text{ of } x^r \text{ in } (1+x)^n.$$

Hence,

$$(1+x)^n = \sum_{r=0}^{n} \binom{n}{r} x^r.$$

The number $\binom{n}{r}$ is pronounced "n choose r" because it also arises in counting problems, as we shall see later in this section.

Observe, in Figure 1.5, that an inside number (i.e., not a 1 on the border) of the $(n+1)$st row can be computed by going up to the nth row and adding the two neighboring numbers above it. For example, the inside numbers in row 4 can be computed

$$
\begin{array}{ccccccccc}
 & 1 & & 3 & & 3 & & 1 & \\
1 & & 4 & & 6 & & 4 & & 1
\end{array}
$$

from row 3 as follows: $4 = 1 + 3$, $6 = 3 + 3$, and $4 = 3 + 1$. Let us prove that this observation always holds.

→ **Lemma 1.17.**   *For all integers $n \geq 1$ and all r with $0 < r < n + 1$,*

$$\binom{n+1}{r} = \binom{n}{r-1} + \binom{n}{r}.$$

Proof.   We must show, for all $n \geq 1$, that if

$$(1+x)^n = c_0 + c_1 x + c_2 x^2 + \cdots + c_n x^n,$$

---

[9]*Binomial*, coming from the Latin *bi*, meaning "two," and *nomen*, meaning "name" or "term," describes expressions of the form $a + b$. Similarly, *trinomial* describes expressions of the form $a + b + c$, and *monomial* describes expressions with a single term. The word is used here because the binomial coefficients arise when expanding powers of the binomial $1 + x$. The word *polynomial* is a hybrid, coming from the Greek *poly* meaning "many" and the Latin *nomen*; polynomials are certain expressions having many terms.

then the coefficient of x^r in $(1+x)^{n+1}$ is $c_{r-1} + c_r$. Since $c_0 = 1$,

$$
\begin{aligned}
(1+x)^{n+1} &= (1+x)(1+x)^n \\
&= (1+x)^n + x(1+x)^n \\
&= (c_0 + c_1 x + c_2 x^2 + \cdots + c_n x^n) \\
&\quad + x(c_0 + c_1 x + c_2 x^2 + \cdots + c_n x^n) \\
&= (c_0 + c_1 x + c_2 x^2 + \cdots + c_n x^n) \\
&\quad + c_0 x + c_1 x^2 + c_2 x^3 + \cdots + c_n x^{n+1} \\
&= 1 + (c_0 + c_1)x + (c_1 + c_2)x^2 + (c_2 + c_3)x^3 + \cdots .
\end{aligned}
$$

Thus $\binom{n+1}{r}$, the coefficient of x^r in $(1+x)^{n+1}$, is

$$
c_{r-1} + c_r = \binom{n}{r-1} + \binom{n}{r}. \quad \bullet
$$

$\rightarrow$ **Proposition 1.18 (Pascal).** *For all $n \geq 0$ and all r with $0 \leq r \leq n$,*

$$
\binom{n}{r} = \frac{n!}{r!(n-r)!}.
$$

Proof. We prove the proposition by induction on $n \geq 0$.

Base step.[10] If $n = 0$, then $\binom{0}{0} = \dfrac{0!}{0!0!} = 1$.

Inductive step. Assuming the formula for $\binom{n}{r}$ for all r, we must prove

$$
\binom{n+1}{r} = \frac{(n+1)!}{r!(n+1-r)!}.
$$

If $r = 0$, then $\binom{n+1}{0} = 1 = \dfrac{(n+1)!}{0!(n+1-0)!}$; if $r = n+1$, then $\binom{n+1}{n+1} = 1 =$

---

[10]This is one reason why 0! is defined to be 1.

$\dfrac{(n+1)!}{(n+1)!0!}$; if $0 < r < n+1$, we use Lemma 1.17:

$$\binom{n+1}{r} = \binom{n}{r-1} + \binom{n}{r}$$

$$= \frac{n!}{(r-1)!(n-r+1)!} + \frac{n!}{r!(n-r)!}$$

$$= \frac{n!}{(r-1)!(n-r)!}\left(\frac{1}{(n-r+1)} + \frac{1}{r}\right)$$

$$= \frac{n!}{(r-1)!(n-r)!}\left(\frac{r+n-r+1}{r(n-r+1)}\right)$$

$$= \frac{n!}{(r-1)!(n-r)!}\left(\frac{n+1}{r(n-r+1)}\right)$$

$$= \frac{(n+1)!}{r!(n+1-r)!}. \quad \bullet$$

Corollary 1.19. *For any real number x and for all integers $n \geq 0$,*

$$(1+x)^n = \sum_{r=0}^{n}\binom{n}{r}x^r = \sum_{r=0}^{n}\frac{n!}{r!(n-r)!}x^r.$$

Proof. The first equation is the definition of the binomial coefficients, and the second equation replaces $\binom{n}{r}$ by the value given in Pascal's theorem. $\bullet$

→ **Corollary 1.20 (Binomial Theorem).** *For all real numbers a and b and for all integers $n \geq 1$,*

$$(a+b)^n = \sum_{r=0}^{n}\binom{n}{r}a^{n-r}b^r = \sum_{r=0}^{n}\left(\frac{n!}{r!(n-r)!}\right)a^{n-r}b^r.$$

Proof. The result is trivially true when $a = 0$ (if we agree that $0^0 = 1$). If $a \neq 0$, set $x = b/a$ in Corollary 1.19, and observe that

$$\left(1 + \frac{b}{a}\right)^n = \left(\frac{a+b}{a}\right)^n = \frac{(a+b)^n}{a^n}.$$

Therefore,

$$(a+b)^n = a^n\left(1 + \frac{b}{a}\right)^n = a^n\sum_{r=0}^{n}\binom{n}{r}\frac{b^r}{a^r} = \sum_{r=0}^{n}\binom{n}{r}a^{n-r}b^r. \quad \bullet$$

Remark.   The binomial theorem can be proved without first proving Corollary 1.19; just prove the formula for $(a + b)^n$ by induction on $n \geq 0$. We have chosen the proof above for clearer exposition.   ◄

Here is a combinatorial interpretation of the binomial coefficients. Given a set X, an **r-subset** is a subset of X with exactly r elements. If X has n elements, denote the number of its r-subsets by

$$[n, r];$$

that is, $[n, r]$ is the number of ways one can choose r things from a box of n things.

We compute $[n, r]$ by considering a related question. Given an "alphabet" with n (distinct) letters and a number r with $1 \leq r \leq n$, an **r-anagram** is a sequence of r of these letters with no repetitions. For example, the 2-anagrams on the alphabet a, b, c are

$$ab, \quad ba, \quad ac, \quad ca, \quad bc, \quad cb$$

(note that aa, bb, cc are not on this list). How many r-anagrams are there on an alphabet with n letters? We count the number of such anagrams in two ways.

(1) There are n choices for the first letter; since no letter is repeated, there are only $n - 1$ choices for the second letter, only $n - 2$ choices for the third letter, and so forth. Thus, the number of r-anagrams is

$$n(n - 1)(n - 2) \cdots (n - [r - 1]) = n(n - 1)(n - 2) \cdots (n - r + 1).$$

Note the special case $n = r$: the number of n-anagrams on n letters is $n!$.

(2) Here is a second way to count these anagrams. First choose an r-subset of the alphabet (consisting of r letters); there are $[n, r]$ ways to do this, for this is exactly what the symbol $[n, r]$ means. For each chosen r-subset, there are $r!$ ways to arrange the r letters in it (this is the special case of our first count when $n = r$). The number of r-anagrams is thus

$$r![n, r].$$

We conclude that

$$r![n, r] = n(n - 1)(n - 2) \cdots (n - r + 1),$$

from which it follows, by Pascal's formula, that

$$[n, r] = n(n - 1)(n - 2) \cdots (n - r + 1)/r! = \binom{n}{r}.$$

This is why the binomial coefficient $\binom{n}{r}$ is often pronounced as "n choose r."

As an example, how many ways are there to choose 2 hats from a closet containing 14 different hats? (One of my friends does not like the phrasing of this question. After all, one can choose 2 hats with one's left hand, with one's right hand, with one's teeth, . . . ; but I continue the evil tradition.) The answer is $\binom{14}{2}$, and Pascal's formula allows us to compute this as $(14 \times 13)/2 = 91$.

Our first interpretation of the binomial coefficients $\binom{n}{r}$ was *algebraic*; that is, as coefficients of polynomials which can be calculated by Pascal's formula; our second interpretation is *combinatorial*; that is, as n choose r. Quite often, each interpretation can be used to prove a desired result. For example, here is a combinatorial proof of Lemma 1.17. Let X be a set with $n + 1$ elements, and let us color one of its elements red and the other n elements blue. Now $\binom{n+1}{r}$ is the number of r-subsets of X. There are two possibilities for an r-subset Y: either it contains the red element or it is all blue. If Y contains the red element, then Y consists of the red element and $r - 1$ blue elements, and so the number of such Y is the same as the number of all blue $(r - 1)$-subsets, namely, $\binom{n}{r-1}$. The other possibility is that Y is all blue, and there are $\binom{n}{r}$ such r-subsets. Therefore, $\binom{n+1}{r} = \binom{n}{r-1} + \binom{n}{r}$, as desired.

We are now going to apply the binomial theorem to trigonometry, but we begin by reviewing properties of the complex numbers. Recall that the **modulus** $|z|$ of a complex number $z = a + ib$ is defined to be

$$|z| = \sqrt{a^2 + b^2}.$$

If we identify a complex number $z = a + ib$ with the point (a, b) in the plane, then its modulus $|z|$ is the distance from z to the origin. It follows that every complex number z of modulus 1 corresponds to a point P on the unit circle (see Figure 1.7). In the right triangle OPA, we have $\cos\theta = |OA|/|OP| = |OA|$, because $|OP| = 1$, and $\sin\theta = |PA|/|OP| = |PA|$. Therefore, P has coordinates $(\cos\theta, \sin\theta)$.

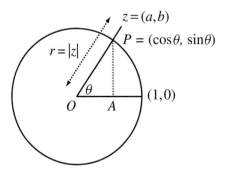

Figure 1.7 $(a, b) = r(\cos\theta + i \sin\theta)$

Here is the simplest way to find the inverse of a nonzero complex number. If $z = a + ib$, where a and b are real numbers, define its **complex conjugate** $\bar{z} = a - ib$. Note that $z\bar{z} = a^2 + b^2$, so that $z \neq 0$ if and only if $z\bar{z} \neq 0$. If $z \neq 0$, then

$$z^{-1} = 1/z = \bar{z}/z\bar{z} = (a/z\bar{z}) - i(b/z\bar{z}).$$

It follows that if z lies on the unit circle, then z^{-1} also lies on the unit circle; moreover, $z^{-1} = \overline{z}$ in this case. The reader may verify the following identities for all complex numbers z and w:

$$\overline{z + w} = \overline{z} + \overline{w}$$
$$\overline{zw} = \overline{z}\,\overline{w}$$
$$\overline{\overline{z}} = z.$$

Moreover, $\overline{z} = z$ if and only if z is a real number.

→ **Proposition 1.21 (Polar Decomposition).**   *For every complex number z, there is a factorization*

$$z = r(\cos\theta + i\sin\theta),$$

where $r = |z| \geq 0$ *and* $0 \leq \theta < 2\pi$.

Proof.   If $z = 0$, then $|z| = 0$ and any choice of θ works. If $z = a + bi \neq 0$, then $|z| \neq 0$. Now $z/|z| = a/|z| + ib/|z|$ has modulus 1, for $(a/|z|)^2 + (b/|z|)^2 = (a^2 + b^2)/|z|^2 = 1$. Therefore, there is an angle θ with

$$\frac{z}{|z|} = \cos\theta + i\sin\theta,$$

and so $z = |z|(\cos\theta + i\sin\theta) = r(\cos\theta + i\sin\theta)$.   •

If $z = a + ib = r(\cos\theta + i\sin\theta)$, then (r, θ) are the ***polar coordinates***[11] of z; this is the reason Proposition 1.21 is called the polar decomposition of z.

The trigonometric addition formulas for $\cos(\theta + \psi)$ and $\sin(\theta + \psi)$ have a lovely translation in the language of complex numbers.

Proposition 1.22 (Addition Theorem).   *If*

$$z = \cos\theta + i\sin\theta \qquad and \qquad w = \cos\psi + i\sin\psi,$$

then

$$zw = \cos(\theta + \psi) + i\sin(\theta + \psi).$$

Proof.

$$zw = (\cos\theta + i\sin\theta)(\cos\psi + i\sin\psi)$$
$$= (\cos\theta\cos\psi - \sin\theta\sin\psi) + i(\sin\theta\cos\psi + \cos\theta\sin\psi).$$

---

[11] A *pole* is an axis about which rotation occurs. For example, the axis of the Earth has endpoints the North and South Poles. Here, we take the pole to be the z-axis (perpendicular to the plane).

The trigonometric addition formulas show that

$$zw = \cos(\theta + \psi) + i \sin(\theta + \psi). \quad \bullet$$

The addition theorem gives a geometric interpretation of complex multiplication: if $z = r(\cos\theta + i \sin\theta)$ and $w = s(\cos\psi + i \sin\psi)$, then

$$zw = rs[\cos(\theta + \psi) + i \sin(\theta + \psi)],$$

and the polar coordinates of zw are

$$(rs, \theta + \psi).$$

$\rightarrow$ **Corollary 1.23.**   *If z and w are complex numbers, then*

$$|zw| = |z|\,|w|.$$

Proof.   If the polar decompositions of z and w are $z = r(\cos\theta + i\sin\theta)$ and $w = s(\cos\psi + i\sin\psi)$, respectively, then we have just seen that $|z| = r$, $|w| = s$, and $|zw| = rs$. $\bullet$

It follows from this corollary that if z and w lie on the unit circle, then their product zw also lies on the unit circle.

In 1707, A. De Moivre (1667–1754) proved the following elegant result.

Theorem 1.24 (De Moivre).   *For every real number x and every positive integer n,*

$$\cos(nx) + i \sin(nx) = (\cos x + i \sin x)^n.$$

Proof.   We prove De Moivre's theorem by induction on $n \geq 1$. The base step $n = 1$ is obviously true. For the inductive step,

$$
\begin{aligned}
(\cos x + i \sin x)^{n+1} &= (\cos x + i \sin x)^n(\cos x + i \sin x) \\
&= [\cos(nx) + i \sin(nx)](\cos x + i \sin x) \\
&\qquad\qquad \text{(inductive hypothesis)} \\
&= \cos(nx + x) + i \sin(nx + x) \\
&\qquad\qquad \text{(addition formula)} \\
&= \cos([n+1]x) + i \sin([n+1]x). \quad \bullet
\end{aligned}
$$

Example 1.25.
Let us find the value of $(\cos 3° + i \sin 3°)^{40}$. By De Moivre's theorem,

$$(\cos 3° + i \sin 3°)^{40} = \cos 120° + i \sin 120° = -\tfrac{1}{2} + i\tfrac{\sqrt{3}}{2}. \quad \blacktriangleleft$$

Here are the double- and triple-angle formulas.

Corollary 1.26.

(i) $\cos(2x) = \cos^2 x - \sin^2 x = 2\cos^2 x - 1$

$\sin(2x) = 2\sin x \cos x.$

(ii) $\cos(3x) = \cos^3 x - 3\cos x \sin^2 x = 4\cos^3 x - 3\cos x$

$\sin(3x) = 3\cos^2 x \sin x - \sin^3 x = 3\sin x - 4\sin^3 x.$

Proof.

(i) De Moivre's theorem gives

$$\cos(2x) + i\sin(2x) = (\cos x + i\sin x)^2$$
$$= \cos^2 x + 2i\sin x \cos x + i^2 \sin^2 x$$
$$= \cos^2 x - \sin^2 x + i(2\sin x \cos x).$$

Equating real and imaginary parts gives both double angle formulas.

(ii) De Moivre's theorem gives

$$\cos(3x) + i\sin(3x) = (\cos x + i\sin x)^3$$
$$= \cos^3 x + 3i\cos^2 x \sin x + 3i^2 \cos x \sin^2 x + i^3 \sin^3 x$$
$$= \cos^3 x - 3\cos x \sin^2 x + i(3\cos^2 x \sin x - \sin^3 x).$$

Equality of the real parts gives $\cos(3x) = \cos^3 x - 3\cos x \sin^2 x$; the second formula for $\cos(3x)$ follows by replacing $\sin^2 x$ by $1 - \cos^2 x$. Equality of the imaginary parts gives $\sin(3x) = 3\cos^2 x \sin x - \sin^3 x = 3\sin x - 4\sin^3 x$; the second formula arises by replacing $\cos^2 x$ by $1 - \sin^2 x$.   •

Corollary 1.26 will be generalized in Proposition 1.27. If $f_2(x) = 2x^2 - 1$, then

$$\cos(2x) = 2\cos^2 x - 1 = f_2(\cos x),$$

and if $f_3(x) = 4x^3 - 3x$, then

$$\cos(3x) = 4\cos^3 x - 3\cos x = f_3(\cos x).$$

Proposition 1.27.   *For all $n \geq 1$, there is a polynomial $f_n(x)$ having all coefficients integers such that*

$$\cos(nx) = f_n(\cos x).$$

Proof.   By De Moivre's theorem,

$$\cos(nx) + i\sin(nx) = (\cos x + i\sin x)^n$$
$$= \sum_{r=0}^{n} \binom{n}{r}(\cos x)^{n-r}(i\sin x)^r.$$

The real part of the left side, $\cos(nx)$, must be equal to the real part of the right side. Now i^r is real if and only if[12] r is even, and so

$$\cos(nx) = \sum_{\substack{r=0 \\ r \text{ even}}}^{n} \binom{n}{r}(\cos x)^{n-r}(i \sin x)^r.$$

If $r = 2k$, then $i^r = i^{2k} = (-1)^k$, and

$$\cos(nx) = \sum_{k=0}^{\lfloor n/2 \rfloor}(-1)^k\binom{n}{2k}(\cos x)^{n-2k}\sin^{2k}x$$

($\lfloor n/2 \rfloor$ denotes the largest integer m with $m \le n/2$).[13] But $\sin^{2k}x = (\sin^2 x)^k = (1 - \cos^2 x)^k$, which is a polynomial in $\cos x$. This completes the proof.   •

It is not difficult to show that $f_n(x)$ begins with $2^{n-1}x^n$. A sine version of Proposition 1.27 can be found in Exercise 1.37 on page 36.

We are now going to present a beautiful formula discovered by Euler, but we begin by recalling some power series formulas from calculus to see how it arises. For every real number x,

$$e^x = 1 + x + \frac{x^2}{2!} + \cdots + \frac{x^n}{n!} + \cdots,$$

$$\cos x = 1 - \frac{x^2}{2!} + \frac{x^4}{4!} - \cdots + \frac{(-1)^n x^{2n}}{(2n)!} + \cdots,$$

and

$$\sin x = x - \frac{x^3}{3!} + \frac{x^5}{5!} - \cdots + \frac{(-1)^n x^{2n+1}}{(2n+1)!} + \cdots.$$

One can define convergence of any power series $\sum_{n=0}^{\infty} c_n z^n$, where z and c_n are complex numbers, and one can show that the series

$$1 + z + \frac{z^2}{2!} + \cdots + \frac{z^n}{n!} + \cdots$$

converges for every complex number z; the **complex exponential** e^z is defined to be the sum of this series.

---

[12] The **converse** of an implication "If P is true, then Q is true" is the implication "If Q is true, then P is true." An implication may be true without its converse being true. For example, "If $a = b$, then $a^2 = b^2$." The phrase **if and only if** means that both the statement and its converse are true.

[13] $\lfloor x \rfloor$, called the **floor** of x or the **greatest integer** in x, is the largest integer m with $m \le x$. For example, $\lfloor 3 \rfloor = 3$ and $\lfloor \pi \rfloor = 3$.

Euler's Theorem. *For all real numbers x,*

$$e^{ix} = \cos x + i \sin x.$$

Proof. (*Sketch*) Now

$$e^{ix} = 1 + ix + \frac{(ix)^2}{2!} + \cdots + \frac{(ix)^n}{n!} + \cdots.$$

As n varies over $0, 1, 2, 3, \ldots$, the powers of i repeat every four steps: that is, i^n takes values

$$1, i, -1, -i, 1, i, -1, -i, 1, \ldots.$$

Thus, the even powers of ix do not involve i, whereas the odd powers do. Collecting terms, one has $e^{ix} = $ even terms $+$ odd terms, where

$$\text{even terms} = 1 + \frac{(ix)^2}{2!} + \frac{(ix)^4}{4!} + \cdots$$

$$= 1 - \frac{x^2}{2!} + \frac{x^4}{4!} - \cdots = \cos x$$

and

$$\text{odd terms} = ix + \frac{(ix)^3}{3!} + \frac{(ix)^5}{5!} + \cdots.$$

$$= i\left(x - \frac{x^3}{3!} + \frac{x^5}{5!} - \cdots\right) = i \sin x.$$

Therefore, $e^{ix} = \cos x + i \sin x$.   •

As a consequence of Euler's theorem, the polar decomposition can be rewritten in exponential form: Every complex number z has a factorization

$$z = re^{i\theta},$$

where $r \geq 0$ and $0 \leq \theta < 2\pi$.

The addition theorem and De Moivre's theorem can be restated in complex exponential form. The first becomes

$$e^{ix}e^{iy} = e^{i(x+y)};$$

the second becomes

$$(e^{ix})^n = e^{inx}.$$

→ **Definition.**   If $n \geq 1$ is an integer, then an ***nth root of unity*** is a complex number ζ with $\zeta^n = 1$.

→ **Corollary 1.28.** *Every nth root of unity ζ is equal to*

$$e^{2\pi ik/n} = \cos(2\pi k/n) + i \sin(2\pi k/n),$$

for some k with $0 \le k \le n - 1$.

Proof. If $\zeta = \cos(2\pi/n) + i \sin(2\pi/n)$, then De Moivre's theorem, Theorem 1.24, gives

$$\begin{aligned}
\zeta^n &= [\cos(2\pi/n) + i \sin(2\pi/n)]^n \\
&= \cos(n2\pi/n) + i \sin(n2\pi/n) \\
&= \cos(2\pi) + i \sin(2\pi) \\
&= 1,
\end{aligned}$$

so that ζ is an nth root of unity. Finally, if k is an integer, then $\zeta^n = 1$ implies $(\zeta^k)^n = (\zeta^n)^k = 1^k = 1$, and so $\zeta^k = \cos(2\pi k/n) + i \sin(2\pi k/n)$ is also an nth root of unity.

Conversely, assume that ζ is an nth root of unity. By the polar decomposition, Proposition 1.21, we have $\zeta = \cos\theta + i\sin\theta$ (because $|\zeta| = 1$). By De Moivre's theorem, $1 = \zeta^n = \cos n\theta + i \sin n\theta$. Since $\cos\theta = 1$ if and only if $\theta = 2k\pi$ for some integer k, we have $n\theta = 2k\pi$; that is, $\zeta = \cos(2k\pi/n) + i \sin(2k\pi/n)$. It is clear that we may choose k so that $0 \le k < n$ because $\cos x$ is periodic with period 2π. •

Here is a more algebraic proof of sufficiency in Corollary 1.28. We will prove (Theorem 3.50) that a polynomial of degree n has at most n roots. Since the n displayed nth roots of unity, namely, $e^{2\pi ik/n}$ for $k - 0, 1, \ldots, n - 1$, are distinct roots of $x^n - 1$, there can be no other nth roots of unity.

Corollary 1.23 states that $|zw| = |z| |w|$ for any complex numbers z and w. It follows that if ζ is an nth root of unity, then $1 = |\zeta^n| = |\zeta|^n$, so that $|\zeta| = 1$ and ζ lies on the unit circle. Given a positive integer n, let $\theta = 2\pi/n$ and let $\zeta = e^{i\theta}$. The polar coordinates of ζ are $(1, \theta)$, the polar coordinates of ζ^2 are $(1, 2\theta)$, the polar coordinates of ζ^3 are $(1, 3\theta), \ldots$, the polar coordinates of ζ^{n-1} are $(1, (n-1)\theta)$, and the polar coordinates of $\zeta^n = 1$ are $(1, n\theta) = (1, 0)$. Thus, the nth roots of unity are evenly spaced around the unit circle. Figure 1.8 shows the 8th roots of unity (here, $\theta = 2\pi/8 = \pi/4$).

Just as there are two square roots of a number a, namely, $\sqrt{a}$ and $-\sqrt{a}$, there are n different nth roots of a, namely, $e^{2\pi ik/n}\sqrt[n]{a}$ for $k = 0, 1, \ldots, n - 1$. For example, the cube roots of unity are 1,

$$\zeta = \cos 120° + i \sin 120° = -\tfrac{1}{2} + i\tfrac{\sqrt{3}}{2}$$

and

$$\zeta^2 = \cos 240° + i \sin 240° = -\tfrac{1}{2} - i\tfrac{\sqrt{3}}{2}.$$

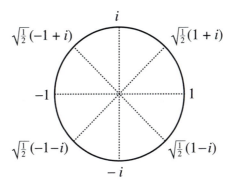

Figure 1.8 8th Roots of Unity

There are 3 cube roots of 2, namely, $\sqrt[3]{2}$, $\zeta \sqrt[3]{2}$, and $\zeta^2 \sqrt[3]{2}$.

Every nth root of unity is, of course, a root of the polynomial $x^n - 1$. Therefore,

$$x^n - 1 = \prod_{\zeta^n = 1} (x - \zeta).$$

If ζ is an nth root of unity, and if n is the smallest positive integer for which $\zeta^n = 1$, we say that ζ is a ***primitive nth root of unity***. For example, $\zeta = e^{2\pi i/n}$ is a primitive nth root of unity. Now i is an 8th root of unity, for $i^8 = 1$; it is not a primitive 8th root of unity, but it is a primitive 4th root of unity.

Lemma 1.29. *Let ζ be a primitive dth root of unity. If $\zeta^n = 1$, then d must be a divisor of n.*

Proof. By long division, $n/d = q + r/d$, where q and r are natural numbers and $0 \le r/d < 1$; that is, $n = qd + r$, where $0 \le r < d$. But

$$1 = \zeta^n = \zeta^{qd+r} = \zeta^{qd} \zeta^r = \zeta^r,$$

because $\zeta^{qd} = (\zeta^d)^q = 1$. If $r \ne 0$, we contradict d being the smallest exponent for which $\zeta^d = 1$. Hence, $n = qd$, as claimed. •

→ **Definition.** If d is a positive integer, then the dth ***cyclotomic***[14] ***polynomial*** is defined by

$$\Phi_d(x) = \prod (x - \zeta),$$

where ζ ranges over all the *primitive dth roots of unity*.

---

[14]The roots of $x^n - 1$ are the nth roots of unity: $1, \zeta, \zeta^2, \dots, \zeta^{n-1}$, where $\zeta = e^{2\pi i/n} = \cos(2\pi/n) + i \sin(2\pi/n)$. Now these roots divide the unit circle $\{\zeta \in \mathbb{C} : |z| = 1\}$ into n equal arcs (see Figure 1.8). This explains the term *cyclotomic*, for its Greek origin means "circle splitting."

In Proposition 3.47, we will prove that all the coefficients of $\Phi_d(x)$ are integers. The following result is almost obvious.

Proposition 1.30. *For every integer $n \geq 1$,*

$$x^n - 1 = \prod_{d \mid n} \Phi_d(x),$$

where d ranges over all the positive divisors d of n [in particular, $\Phi_1(x)$ and $\Phi_n(x)$ occur].

Proof. In light of Corollary 1.28, the proposition follows by collecting, for each divisor d of n, all terms in the equation $x^n - 1 = \prod(x - \zeta)$ with ζ a primitive dth root of unity. •

For example, if p is a prime, then $x^p - 1 = \Phi_1(x)\Phi_p(x)$. Since $\Phi_1(x) = x - 1$, it follows that

$$\Phi_p(x) = x^{p-1} + x^{p-2} + \cdots + x + 1.$$

→ **Definition.** The *Euler ϕ-function* is the degree of the nth cyclotomic polynomial:

$$\phi(n) = \deg(\Phi_n(x)).$$

In Proposition 1.42, we will give another description of the Euler ϕ-function that does not depend on roots of unity.

Corollary 1.31. *For every integer $n \geq 1$, we have*

$$n = \sum_{d \mid n} \phi(d).$$

Proof. Note that $\phi(n)$ is the degree of $\Phi_n(x)$, and use the fact that the degree of a product of polynomials is the sum of the degrees of the factors. •

Where do the names of the trigonometric functions come from? The circle in Figure 1.9 is the unit circle, and the coordinates of the point A are $(\cos\alpha, \sin\alpha)$; that is, $|OD| = \cos\alpha$ and $|AD| = \sin\alpha$. The reader may show that $|BC| = \tan\alpha$ (the Latin word *tangere* means "to touch," and a *tangent* is a line which touches the circle in only one point), and that $|OB| = \sec\alpha$ (the Latin word *secare* means "to cut," and a *secant* is a line that cuts a circle). The *complement* of an acute angle α is $90° - \alpha$, and so the name *cosine* arises from that of sine because of the identity $\cos\alpha = \sin(90° - \alpha)$.

I found the more amusing etymology of the term *sine* in the Oxford English Dictionary. We see in Figure 1.9 that

$$\sin\alpha = |AD| = \tfrac{1}{2}|AE|;$$

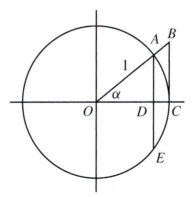

Figure 1.9 Etymology of
Trigonometric Names

that is, $\sin \alpha$ is half the length of the chord AE. The fifth-century Indian mathematician Aryabhata called the sine *ardha-jya* (half chord) in Sanskrit, which was later abbreviated to *jya*. A few centuries later, books in Arabic transliterated *jya* as *jiba*. In Arabic script, there are letters and diacritical marks; roughly speaking, the letters correspond to our consonants, while the diacritical marks correspond to our vowels. It is customary to suppress diacritical marks in writing; for example, the Arabic version of *jiba* is written *jb* (using Arabic characters, of course). Now *jiba*, having no other meaning in Arabic, eventually evolved into *jaib*, which is an Arabic word meaning "bosom of a dress" (a fine word, but having absolutely nothing to do with half-chord). Finally, Gherardo of Cremona, circa 1150, translated *jaib* into its Latin equivalent, *sinus*. And this is why sine is so called, for *sinus* means bosom!

As long as we are discussing etymology, why is a root so called? Just as the Greeks called the bottom side of a triangle its base (as in the area formula $\frac{1}{2}$ altitude $\times$ base), they also called the bottom side of a square its base. A natural question for the Greeks was, given a square of area A, what is the length of its side? Of course, the answer is $\sqrt{A}$. Were we inventing a word for $\sqrt{A}$, we might have called it the *base* of A or the *side* of A. Similarly, consider the analogous three-dimensional question: given a cube of volume V, what is the length of its edge? The answer $\sqrt[3]{V}$ might be called the *cube base* of V, and $\sqrt{A}$ might then be called the *square base* of A. Why, then, do we call these numbers cube *root* and square *root*? What has any of this to do with plants?

Since tracing the etymology of words is not a simple matter, we only suggest the following explanation. Through about the fourth and fifth centuries, most mathematics was written in Greek, but, by the fifth century, India had become a center of mathematics, and important mathematical texts were also written in Sanskrit. The Sanskrit term for square root is *pada*. Both Sanskrit and Greek are Indo-European languages, and the Sanskrit word *pada* is a cognate of the Greek word *podos*; both mean *base* in the sense of the foot of a pillar or, as above, the bottom of a square. In both languages, however, there is a secondary meaning: the root of a plant. In translating from Sanskrit, Arab mathematicians chose the secondary meaning, perhaps in error (Arabic is not an Indo-European language), perhaps for some unknown reason. For example, the influential book by al-Khwarizmi, *Al-jabr w'al muqabala*,[15] which appeared in the year 830, used the Arabic word *jidhr*, meaning root of a plant. (The word *algebra* is a European version of the first word in the title of this book; the author's name has also come into the English language as the word *algorithm*.) This mistranslation has since been handed down through the centuries; the term *jidhr* became standard in Arabic mathematical writings, and European translations from Arabic into Latin used the word *radix* (meaning root, as in *radish* or *radical*). The notation $r2$ for $\sqrt{2}$ occurs in European writings from about the 12th century (but the square root symbol did not arise from the letter r; it evolved from an old dot notation). However, there was a competing notation in use at the same time, for some scholars who translated directly from the Greek denoted $\sqrt{2}$ by $l2$, where l abbreviates the Latin word *latus*, meaning side. Finally, with the invention of logarithms in the 1500s, r won out over l, for the notation $l2$ was then commonly used to denote $\log 2$. The passage from square root to cube root to the root of a polynomial equation other than $x^2 - a$ and $x^3 - a$ is a natural enough generalization. Thus, as pleasant as it would be, there seems to be no botanical connection with roots of equations.

EXERCISES

H **1.28**   True or false with reasons.

    (i)   For all integers r with $0 < r < 7$, the binomial coefficient $\binom{7}{r}$ is a multiple of 7.

    (ii) For any positive integer n and any r with $0 < r < n$, the binomial coefficient $\binom{n}{r}$ is a multiple of n.

    (iii) Let D be a collection of 10 different dogs, and let C be a collection of 10 different cats. There are the same number of quartets of dogs as there are sextets of cats.

    (iv) If q is a rational number, then $e^{2\pi i q}$ is a root of unity.

    (v) Let $f(x) = ax^2 + bx + c$, where a, b, c are real numbers. If z is a root of $f(x)$, then $\overline{z}$ is also a root of $f(x)$.

    (vi) Let $f(x) = ax^2 + bx + c$, where a, b, c are complex numbers. If z is a root of $f(x)$, then $\overline{z}$ is also a root of $f(x)$.

    (vii) The primitive 4th roots of unity are i and $-i$.

---

[15] One can translate this title from Arabic, but the words already had a technical meaning: both *jabr* and *muqabala* refer to certain operations akin to subtracting the same number from both sides of an equation.

H **1.29** Prove that the binomial theorem holds for complex numbers: if u and v are complex numbers, then

$$(u+v)^n = \sum_{r=0}^{n} \binom{n}{r} u^{n-r} v^r.$$

***1.30** Show that the binomial coefficients are "symmetric":

$$\binom{n}{r} = \binom{n}{n-r}$$

for all r with $0 \le r \le n$.

*H **1.31** Show, for every n, that the sum of the binomial coefficients is 2^n:

$$\binom{n}{0} + \binom{n}{1} + \binom{n}{2} + \cdots + \binom{n}{n} = 2^n.$$

1.32 H **(i)** Show, for every $n \ge 1$, that the "alternating sum" of the binomial coefficients is zero:

$$\binom{n}{0} - \binom{n}{1} + \binom{n}{2} - \cdots + (-1)^n \binom{n}{n} = 0.$$

(ii) Use part (i) to prove, for a given n, that the sum of all the binomial coefficients $\binom{n}{r}$ with r even is equal to the sum of all those $\binom{n}{r}$ with r odd.

H **1.33** Prove that if $n \ge 2$, then

$$\sum_{r=1}^{n} (-1)^{r-1} r \binom{n}{r} = 0.$$

***1.34** If $1 \le r \le n$, prove that

$$\binom{n}{r} = \frac{n}{r} \binom{n-1}{r-1}.$$

1.35 Let $\varepsilon_1, \ldots, \varepsilon_n$ be complex numbers with $|\varepsilon_j| = 1$ for all j, where $n \ge 2$.

H **(i)** Prove that

$$\left| \sum_{j=1}^{n} \varepsilon_j \right| \le \sum_{j=1}^{n} |\varepsilon_j| = n.$$

H **(ii)** Prove that there is equality,

$$\left| \sum_{j=1}^{n} \varepsilon_j \right| = n,$$

if and only if all the ε_j are equal.

1.36 (*Star of David*) Prove, for all $n > r \ge 1$, that

$$\binom{n-1}{r-1}\binom{n}{r+1}\binom{n+1}{r} = \binom{n-1}{r}\binom{n}{r-1}\binom{n+1}{r+1}.$$

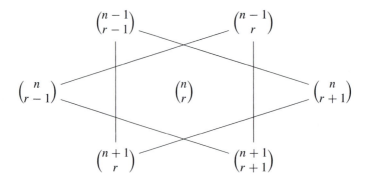

*H **1.37** For all odd $n \geq 1$, prove that there is a polynomial $g_n(x)$, all of whose coefficients are integers, such that

$$\sin(nx) = g_n(\sin x).$$

1.38   **(i)**   What is the coefficient of x^{16} in $(1+x)^{20}$?

H **(ii)**   How many ways are there to choose 4 colors from a palette containing paints of 20 different colors?

1.39 Give at least two different proofs that a set X with n elements has exactly 2^n subsets.

H **1.40** A weekly lottery asks you to select 5 different numbers between 1 and 45. At the week's end, 5 such numbers are drawn at random, and you win the jackpot if all your numbers match the drawn numbers. What is your chance of winning?

Definition.   Define the **n th derivative** $f^{(n)}(x)$ of a function $f(x)$ inductively: set $f^{(0)}(x)$ to be $f(x)$ and, if $n \geq 0$, define $f^{(n+1)}(x) = (f^{(n)})'(x)$.

1.41 Assume that "term-by-term" differentiation holds for power series: if $f(x) = c_0 + c_1 x + c_2 x^2 + \cdots + c_n x^n + \cdots$, then the power series for the derivative $f'(x)$ is

$$f'(x) = c_1 + 2c_2 x + 3c_3 x^2 + \cdots + nc_n x^{n-1} + \cdots .$$

(i)   Prove that $f(0) = c_0$.

(ii)   Prove, for all $n \geq 0$, that

$$f^{(n)}(x) = n!c_n + (n+1)!c_{n+1}x + x^2 g_n(x),$$

where $g_n(x)$ is some power series .

(iii)   Prove that $c_n = f^{(n)}(x)(0)/n!$ for all $n \geq 0$. (Of course, this is Taylor's formula.)

*H **1.42** (**Leibniz**) A function $f : \mathbb{R} \to \mathbb{R}$ is called a **C^∞-function** if it has an nth derivative $f^{(n)}(x)$ for every $n \geq 0$. Prove that if f and g are C^∞-functions, then

$$(fg)^{(n)}(x) = \sum_{k=0}^{n} \binom{n}{k} f^{(k)}(x) \cdot g^{(n-k)}(x).$$

1.43 Find $\sqrt{i}$.

*1.44   **(i)**   If $z = r[\cos\theta + i\sin\theta]$, show that

$$w = \sqrt[n]{r}\,[\cos(\theta/n) + i\sin(\theta/n)]$$

is an nth root of z, where $r \geq 0$.

   (ii)   Show that every nth root of z has the form $\zeta^k w$, where ζ is a primitive nth root of unity and $k = 0, 1, 2, \ldots, n-1$.

1.45   H **(i)**   Find $\sqrt{8 + 15i}$.

   H **(ii)**   Find all the fourth roots of $8 + 15i$.

→ 1.3   GREATEST COMMON DIVISORS

This is an appropriate time to introduce notation for some popular sets of numbers other than $\mathbb{Z}$ (denoting the integers) and $\mathbb{N}$ (denoting the natural numbers).

$\mathbb{Q}$ = the set of all rational numbers (or fractions), that is, all numbers of the form a/b, where a and b are integers and $b \neq 0$ (after the word *quotient*)

$\mathbb{R}$ = the set of all real numbers

$\mathbb{C}$ = the set of all complex numbers

Long division involves dividing an integer b by a nonzero integer a, giving

$$\frac{b}{a} = q + \frac{r}{a},$$

where q is an integer and $0 \leq r/a < 1$. We clear denominators to get a statement wholly in $\mathbb{Z}$.

→ **Theorem 1.32 (Division Algorithm).**   *Given integers a and b with $a \neq 0$, there exist unique integers q and r with*

$$b = qa + r \quad and \quad 0 \leq r < |a|.$$

Proof.   We will prove the theorem in the special case in which $a > 0$ and $b \geq 0$; Exercise 1.47 on page 53 asks the reader to complete the proof. Long division involves finding the largest integer q with $qa \leq b$, which is the same thing as finding the smallest nonnegative integer of the form $b - qa$. We formalize this.

The set C of all nonnegative integers of the form $b - na$, where $n \geq 0$, is not empty because it contains $b = b - 0a$ (we are assuming that $b \geq 0$). By the Least Integer Axiom, C contains a smallest element, say, $r = b - qa$ (for some $q \geq 0$); of course, $r \geq 0$, by its definition. If $r \geq a$, then

$$b - (q+1)a = b - qa - a = r - a \geq 0.$$

Hence, $r - a = b - (q + 1)a$ is an element of C that is smaller than r, contradicting r being the smallest integer in C. Therefore, $0 \leq r < a$.

It remains to prove the uniqueness of q and r. Suppose that $b = qa + r = q'a + r'$, where $0 \leq r, r' < a$, so that

$$(q - q')a = r' - r.$$

We may assume that $r' \geq r$, so that $r' - r \geq 0$ and hence $q - q' \geq 0$. If $q \neq q'$, then $q - q' \geq 1$ (for $q - q'$ is an integer); thus, since $a > 0$,

$$(q - q')a \geq a.$$

On the other hand, since $r' < a$, Proposition A.2 gives

$$r' - r < a - r \leq a.$$

Therefore, $(q - q')a \geq a$ and $r' - r < a$, contradicting the given equation $(q - q')a = r' - r$. We conclude that $q = q'$ and hence $r = r'$.   •

→ **Definition.**   If a and b are integers with $a \neq 0$, then the integers q and r occurring in the division algorithm are called the ***quotient*** and ***remainder*** after dividing b by a.

For example, there are only two possible remainders after dividing by 2, namely, 0 and 1. A number m is even if the remainder is 0; m is odd if the remainder is 1. Thus, either $m = 2q$ or $m = 2q + 1$.

Warning! The division algorithm makes sense, in particular, when b is negative. A careless person may assume that b and $-b$ leave the same remainder after dividing by a, and this is usually false. For example, let us divide 60 and -60 by 7.

$$60 = 7 \cdot 8 + 4 \quad \text{and} \quad -60 = 7 \cdot (-9) + 3.$$

Thus, the remainders after dividing 60 and -60 by 7 are different (see Exercise 1.84 on page 75).

The next result shows that there is no largest prime.

→ **Corollary 1.33.**   *There are infinitely many primes.*

Proof.   (***Euclid***) Suppose, on the contrary, that there are only finitely many primes. If $p_1, p_2, \ldots, p_k$ is the complete list of all the primes, define $M = (p_1 \cdots p_k) + 1$. By Theorem 1.2, M is either a prime or a product of primes. But M is neither a prime ($M > p_i$ for every i) nor does it have any prime divisor p_i, for dividing M by p_i gives remainder 1 and not 0. For example, dividing M by p_1 gives $M = p_1(p_2 \cdots p_k) + 1$, so that the quotient and remainder are $q = p_2 \cdots p_k$ and $r = 1$; dividing M by p_2 gives $M = p_2(p_1 p_3 \cdots p_k) + 1$, so that $q = p_1 p_3 \cdots p_k$ and $r = 1$; and so forth. The assumption that there are only finitely many primes leads to a contradiction, and so there must be an infinite number of them.   •

An *algorithm* solving a problem is a set of directions which gives the correct answer after a finite number of steps, never at any stage leaving the user in doubt as to what to do next. The division algorithm is an algorithm in this sense: one starts with a and b and ends with q and r. Appendix B at the end of the book treats algorithms more formally, using **pseudocodes**, which are general directions that can easily be translated into a programming language. For example, here is a pseudocode for the division algorithm.

$$\text{Input: } b \geq a > 0$$
$$\text{Output: } q, r$$
$$q := 0; \quad r := b$$
$$\text{WHILE } r \geq a \text{ DO}$$
$$r := r - a$$
$$q := q + 1$$
$$\text{END WHILE}$$

→ **Definition.**   If a and b are integers, then a is a **divisor** of b if there is an integer d with $b = ad$ (synonyms are a **divides** b and also b is a **multiple** of a). We denote this by

$$a \mid b.$$

Note that $3 \mid 6$, because $6 = 3 \times 2$, but that $3 \nmid 5$ (that is, 3 does not divide 5): even though $5 = 3 \times \frac{5}{3}$, the fraction $\frac{5}{3}$ is not an integer. The numbers ± 1 and $\pm b$ are divisors of any integer b. We always have $b \mid 0$ (because $0 = b \times 0$); on the other hand, if $0 \mid b$, then $b = 0$ (because there is some d with $b = 0 \times d = 0$).

If a and b are integers with $a \neq 0$, then a is a divisor of b if and only if the remainder r given by the division algorithm is 0. If a is a divisor of b, then the remainder r given by the division algorithm is 0; conversely, if the remainder r is 0, then a is a divisor of b.

→ **Definition.**   A *common divisor* of integers a and b is an integer c with $c \mid a$ and $c \mid b$. The **greatest common divisor** of a and b, denoted by $\gcd(a, b)$ [or, more briefly, by (a, b)], is defined by

$$\gcd(a, b) = \begin{cases} 0 \text{ if } a = 0 = b \\ \text{the largest common divisor of } a \text{ and } b \text{ otherwise.} \end{cases}$$

The notation (a, b) for the gcd is, obviously, the same notation used for the ordered pair, but the reader should have no difficulty understanding the intended meaning from the context in which the symbol occurs.

If a and m are positive integers with $a \mid m$, say, $m = ab$, we claim that $a \leq m$. Since $0 < b$, we have $1 \leq b$, because b is an integer, and so $a \leq ab = m$. It follows that gcd's always exist.

If c is a common divisor of a and b, then so is $-c$. Since one of $\pm c$ is nonnegative, the gcd is always nonnegative. If at least one of a and b is nonzero, then $(a, b) > 0$.

Proposition 1.34.   *If p is a prime and b is any integer, then*

$$\gcd(p, b) = \begin{cases} p \text{ if } p \mid b \\ 1 \text{ otherwise.} \end{cases}$$

Proof.   A common divisor c of p and b is, of course, a divisor of p. But the only positive divisors of p are p and 1, and so $(p, b) = p$ or 1; it is p if $p \mid b$, and it is 1 otherwise.   ●

→ **Definition.**   A *linear combination* of integers a and b is an integer of the form

$$sa + tb,$$

where s and t are integers.

The next result is one of the most useful properties of gcd's.

→ **Theorem 1.35.**   *If a and b are integers, then $\gcd(a, b)$ is a linear combination of a and b.*

Proof.   We may assume that at least one of a and b is not zero (otherwise, the gcd is 0 and the result is obvious). Consider the set I of all the linear combinations:

$$I = \{sa + tb : s, t \text{ in } \mathbb{Z}\}.$$

Both a and b are in I (take $s = 1$ and $t = 0$ or vice versa). It follows that I contains positive integers (if $a \neq 0$, then I contains $\pm a$), and hence the set P of all those positive integers that lie in I is nonempty. By the Least Integer Axiom, P contains a smallest positive integer, say, d, which we claim is the gcd.

Since d is in I, it is a linear combination of a and b: there are integers s and t with

$$d = sa + tb.$$

Let us show that d is a common divisor by trying to divide each of a and b by d. The division algorithm gives $a = qd + r$, where $0 \leq r < d$. If $r > 0$, then

$$r = a - qd = a - q(sa + tb) = (1 - qs)a + (-qt)b \text{ is in } P,$$

contradicting d being the smallest element of P. Hence $r = 0$ and $d \mid a$; a similar argument shows that $d \mid b$.

Finally, if c is a common divisor of a and b, then $a = ca'$ and $b = cb'$, so that c divides d, for $d = sa + tb = c(sa' + tb')$. But if $c \mid d$, then $|c| \leq d$, and so d is the gcd of a and b.   ●

If $d = \gcd(a, b)$ and if c is a common divisor of a and b, then $c \leq d$. The next corollary shows that more is true: $c \mid d$ for every common divisor c.

→ **Corollary 1.36.** *Let a and b be integers. A nonnegative common divisor d is their* gcd *if and only if c | d for every common divisor c.*

Proof. *Necessity* (i.e., the implication ⟹) That every common divisor c of a and b is a divisor of $d = sa+tb$, has already been proved at the end of the proof of Theorem 1.35.

Sufficiency (i.e., the implication ⟸) Let d denote the gcd of a and b, and let d' be a nonnegative common divisor divisible by every common divisor c. Thus, $d' \leq d$, because $c \leq d$ is for every common divisor c. On the other hand, d itself is a common divisor, and so $d \mid d'$, by hypothesis. Hence, $d \leq d'$, and so $d = d'$. •

The proof of Theorem 1.35 contains an idea that will be used again.

→ **Corollary 1.37.** *Let I be a subset of $\mathbb{Z}$ such that*

(i) *0 is in I;*

(ii) *if a and b are in I, then $a - b$ is in I;*

(iii) *if a is in I and q is in $\mathbb{Z}$, then qa is in I.*

Then there is a nonnegative integer d in I with I consisting precisely of all the multiples of d.

Proof. If I consists of only the single integer 0, take $d = 0$. If I contains a nonzero integer a, then $(-1)a = -a$ is in I, by (iii). Thus, I contains $\pm a$, one of which is positive. By the Least Integer Axiom, I contains a smallest positive integer; call it d.

We claim that every element a in I is a multiple of d. The division algorithm gives integers q and r with $a = qd + r$, where $0 \leq r < d$. Since d is in I, so is qd, by (iii), and so (ii) gives $r = a - qd$ in I. But $r < d$, the smallest positive element of I, and so $r = 0$; thus, a is a multiple of d. •

The next result, called *Euclid's lemma*, is of great interest, for it gives one of the most important characterizations of prime numbers. Euclid's lemma is used frequently (at least ten times in this chapter alone), and an analog of it for irreducible polynomials is equally important. Looking further ahead, this lemma motivates the notion of *prime ideal*.

→ **Theorem 1.38 (Euclid's Lemma).** *If p is a prime and $p \mid ab$, then $p \mid a$ or $p \mid b$. More generally, if a prime p divides a product $a_1 a_2 \cdots a_n$, then it must divide at least one of the factors a_i. Conversely, if $m \geq 2$ is an integer such that $m \mid ab$ always implies $m \mid a$ or $m \mid b$, then m is a prime.*

Proof. Assume that $p \nmid a$; that is, p does not divide a; we must show that $p \mid b$. Now the gcd $(p, a) = 1$, by Proposition 1.34. By Theorem 1.35, there are integers s and t with $1 = sp + ta$, and so

$$b = spb + tab.$$

Since $p \mid ab$, we have $ab = pc$ for some integer c, so that $b = spb + tpc = p(sb + tc)$ and $p \mid b$. The second statement now follows easily by induction on $n \geq 2$.

We prove the contrapositive: if m is composite, then there is a product ab divisible by m, yet neither factor is divisible by m. Since m is composite, $m = ab$, where $a < m$ and $b < m$. Thus, m divides ab, but m divides neither factor (if $m \mid a$, then $m \leq a$). •

Here is a concrete illustration showing that Euclid's lemma is not true in general: $6 \mid 12 = 4 \times 3$, but $6 \nmid 4$ and $6 \nmid 3$.

→ **Proposition 1.39.**   *If p is a prime, then $p \mid \binom{p}{j}$ for $0 < j < p$.*

Proof.   Recall that

$$\binom{p}{j} = \frac{p!}{j!(p-j)!} = \frac{p(p-1)\cdots(p-j+1)}{j!}.$$

Cross multiplying gives

$$j!\binom{p}{j} = p(p-1)\cdots(p-j+1),$$

so that $p \mid j!\binom{p}{j}$. If $p \mid j!$, then Euclid's lemma says that p would have to divide some factor $1, 2, \ldots, j$ of $j!$. Since $0 < j < p$, each factor of $j!$ is strictly less than p, and so p is not a divisor of any of them. Therefore, $p \nmid j!$. As $p \mid j!\binom{p}{j}$, Euclid's lemma now shows that p must divide $\binom{p}{j}$. •

Notice that the assumption that p is prime is needed; for example, $\binom{4}{2} = 6$, but $4 \nmid 6$.

→ **Definition.**   Call integers a and b *relatively prime* if their gcd is 1.

Thus, a and b are relatively prime if their only common divisors are ± 1; moreover, 1 is a linear combination of a and b. For example, 2 and 3 are relatively prime, as are 8 and 15.

Here is a generalization of Euclid's lemma having the same proof.

→ **Corollary 1.40.**   *Let a, b, and c be integers. If c and a are relatively prime and if $c \mid ab$, then $c \mid b$.*

Proof.   By hypothesis, $ab = cd$ for some integer d. There are integers s and t with $1 = sc + ta$, and so $b = scb + tab = scb + tcd = c(sb + td)$. •

We see that it is important to know proofs: Corollary 1.40 does not follow from the statement of Euclid's lemma, but it does follow from its proof.

Definition.   An expression a/b for a rational number (where a and b are integers) is in *lowest terms* if a and b are relatively prime.

Lemma 1.41.   *Every nonzero rational number r has an expression in lowest terms.*

Proof.   Since r is rational, $r = a/b$ for integers a and b. If $d = (a, b)$, then $a = a'd$, $b = b'd$, and $a/b = a'd/b'd = a'/b'$. But $(a', b') = 1$, for if $d' > 1$ is a common divisor of a' and b', then $d'd > d$ is a larger common divisor of a and b.   •

Here is a description of the Euler ϕ-function that does not mention cyclotomic polynomials. Recall that $\phi(n)$ was defined as the number of *primitive* nth roots of unity ζ; that is, $\zeta^n = 1$, but $\zeta^d \neq 1$ for $1 \leq d < n$.

→ **Proposition 1.42.**   *If $n \geq 1$ is an integer, then $\phi(n)$ is the number of integers k with $1 \leq k \leq n$ and $(k, n) = 1$.*

Proof.   Since every nth root of unity has the form $\zeta = e^{2\pi ik/n}$, by Corollary 1.28, it suffices to prove that ζ is primitive if and only if $(k, n) = 1$.

If k and n are not relatively prime, then $n = dr$ and $k = ds$, where d, r, and s are integers, and $d > 1$; it follows that $r < n$. Hence, $\frac{k}{n} = \frac{ds}{dr} = \frac{s}{r}$, so that $(e^{2\pi ik/n})^r = (e^{2\pi is/r})^r = 1$, and hence $e^{2\pi ik/n}$ is not a primitive nth root of unity.

Conversely, assume that $(k, n) = 1$. Write $\zeta = e^{2\pi ik/n}$ and $\eta = e^{2\pi i/n}$. There are integers s and t with $sk + tn = 1$. Hence,

$$\eta = e^{2\pi i/n} = e^{2\pi iks/n} e^{2\pi int/n} = e^{2\pi iks/n} = \zeta^s.$$

If there is d with $1 \leq d < n$, then $\zeta^d = 1$ and $\eta^d = 1$, contradicting η being a primitive nth root of unity. Therefore, no such d exists, and ζ is a primitive nth root of unity.   •

Proposition 1.43. $\sqrt{2}$ *is irrational.*

Proof.   Suppose, on the contrary, that $\sqrt{2}$ is rational; that is, $\sqrt{2} = a/b$. We may assume that a/b is in lowest terms; that is, $(a, b) = 1$. Squaring, $a^2 = 2b^2$. By Euclid's lemma,[16] $2 \mid a$, so that $2m = a$, hence $4m^2 = a^2 = 2b^2$, and $2m^2 = b^2$. Euclid's lemma now gives $2 \mid b$, contradicting $(a, b) = 1$.   •

This last result is significant in the history of mathematics. The ancient Greeks defined *number* to mean "positive integer," while (positive) rational numbers were viewed as "ratios" $a : b$ (which we can interpret as fractions a/b). That $\sqrt{2}$ is irrational was a shock to the Pythagoreans (around 600 B.C.), for it told them that $\sqrt{2}$ could not be defined in terms of numbers (positive integers) alone. On the other hand, they knew that the diagonal of a square having sides of length 1 has length $\sqrt{2}$. Thus, there is no

---

[16]This proof can be made more elementary; one needs only Proposition 1.14.

numerical solution to the equation $x^2 = 2$, but there is a geometric solution. By the time of Euclid (around 325 B.C.), this problem was resolved by splitting mathematics into two different disciplines: algebra and geometry. This resolution is probably one of the main reasons that the golden age of classical mathematics declined in Europe after the rise of the Roman Empire. For example, there were geometric ways of viewing addition, subtraction, multiplication, and division of segments (see Theorem 4.47), but it was virtually impossible to do any algebra. A sophisticated geometric argument (due to Eudoxus and given in Euclid's *Elements*) was needed to prove the version of cross-multiplication saying that if $a : b = c : d$, then $a : c = b : d$.

We quote van der Waerden, *Science Awakening*, page 125:

> Nowadays we say that the length of the diagonal is the "irrational number" $\sqrt{2}$, and we feel superior to the poor Greeks who "did not know irrationals." But the Greeks knew irrational ratios very well ... That they did not consider $\sqrt{2}$ as a number was not a result of ignorance, but of strict adherence to the definition of number. *Arithmos* means quantity, therefore whole number. Their logical rigor did not even allow them to admit fractions; they replaced them by ratios of integers.

> For the Babylonians, every segment and every area simply represented a number ... When they could not determine a square root exactly, they calmly accepted an approximation. Engineers and natural scientists have always done this. But the Greeks were concerned with exact knowledge, with "the diagonal itself," as Plato expresses it, not with an acceptable approximation.

> In the domain of numbers (positive integers), the equation $x^2 = 2$ cannot be solved, not even in that of ratios of numbers. But it is solvable in the domain of segments; indeed the diagonal of the unit square is a solution. Consequently, in order to obtain exact solutions of quadratic equations, we have to pass from the domain of numbers (positive integers) to that of geometric magnitudes. Geometric algebra is valid also for irrational segments and is nevertheless an exact science. It is therefore logical necessity, not the mere delight in the visible, which compelled the Pythagoreans to transmute their algebra into a geometric form.

Even though the Greek definition of number is no longer popular, their dichotomy still persists. For example, almost all American high schools teach one year of algebra followed by one year of geometry, instead of two years in which both subjects are developed together. The problem of defining *number* has arisen several times since the classical Greek era. In the 1500s, mathematicians had to deal with negative numbers and with complex numbers (see our discussion of cubic polynomials in Chapter 5); the description of real numbers generally accepted today dates from the late 1800s. There are echos of ancient Athens in our time. L. Kronecker (1823–1891) wrote,

Die ganzen Zahlen hat der liebe Gott gemacht, alles andere ist Menschen-werk. (God created the integers; everything else is the work of Man.)

Even today some logicians argue for a new definition of number.

Our discussion of gcd's is incomplete. What is $\gcd(12327, 2409)$? To ask the question another way, is the expression $2409/12327$ in lowest terms? The next result not only enables one to compute gcd's efficiently, it also allows one to compute integers s and t expressing the gcd as a linear combination.[17] Before giving the theorem, consider the following example. Since $(2, 3) = 1$, there are integers s and t with $1 = 2s + 3t$. A moment's thought gives $s = -1$ and $t = 1$; but another moment's thought gives $s = 2$ and $t = -1$. We conclude that the coefficients s and t expressing the gcd as a linear combination are not uniquely determined. The algorithm below, however, always picks out a particular pair of coefficients.

→ **Theorem 1.44 (Euclidean Algorithm).** *Let a and b be positive integers. There is an algorithm that finds the gcd $d = (a, b)$, and there is an algorithm that finds a pair of integers s and t with $d = sa + tb$.*

Remark. The general case for arbitrary a and b follows from this, for

$$(a, b) = (|a|, |b|). \quad \blacktriangleleft$$

Proof. The idea is to keep repeating the division algorithm (we will show where this idea comes from after the proof is completed). Let us set $b = r_0$ and $a = r_1$. Repeated application of the division algorithm gives integers q_i, positive integers r_i, and equations:

$$
\begin{array}{ll}
b = q_1 a + r_2, & r_2 < a \\
a = r_1 = q_2 r_2 + r_3, & r_3 < r_2 \\
r_2 = q_3 r_3 + r_4, & r_4 < r_3 \\
\quad\vdots & \quad\vdots \\
r_{n-3} = q_{n-2} r_{n-2} + r_{n-1}, & r_{n-1} < r_{n-2} \\
r_{n-2} = q_{n-1} r_{n-1} + r_n, & r_n < r_{n-1} \\
r_{n-1} = q_n r_n &
\end{array}
$$

---

[17]Every positive integer is a product of primes, and this is used, in Proposition 1.55, to compute gcd's. However, finding prime factorizations of large numbers is notoriously difficult; indeed, it is the basic reason why public key cryptography is secure.

(remember that all q_j and r_j are explicitly known from the division algorithm). Notice that there is a last remainder; the procedure stops because the remainders form a strictly decreasing sequence of nonnegative integers (indeed, the number of steps needed is less than a. Proposition 1.46 gives a smaller bound on the number of steps).

We use Corollary 1.36 to show that the last remainder $d = r_n$ is the gcd. Let us rewrite the top equations of the Euclidean algorithm without subscripts.

$$b = qa + r$$
$$a = q'r + s.$$

If c is a common divisor of a and b, then the first equation shows that $c \mid r$. Going down to the second equation, we now know that $c \mid a$ and $c \mid r$, and so $c \mid s$. Continuing down the list, we see that c divides every remainder; in particular, $c \mid d$.

Let us now rewrite the bottom equations of the Euclidean algorithm without subscripts.

$$f = ug + h$$
$$g = u'h + k$$
$$h = u''k + d$$
$$k = vd$$

Going from the bottom up, we have $d \mid k$ and $d \mid d$, so that $d \mid h$; going up again, $d \mid h$ and $d \mid k$ imply $d \mid g$. Working upward ultimately gives $d \mid a$ and $d \mid b$. We conclude that d is a common divisor. But $d = (a, b)$ because we saw, in the preceding paragraph, that if c is any common divisor, then $c \mid d$.

We now find s and t, again working from the bottom up. Rewrite the equation $h = u'k + d$ as $d = h - u'k$, and substitute $k = g - u'h$ from the equation above it:

$$d = h - u''k = h - u''(g - u'h) = (1 + u''u')h - u''g.$$

Thus, d is a linear combination of g and h. Continue this procedure, replacing h by $f - ug$, and so on, until d is written as a linear combination of a and b.   •

We say that n is the **number of steps** in the Euclidean algorithm, for one does not know whether r_n in the $(n-1)$st step

$$r_{n-2} = q_{n-1}r_{n-1} + r_n$$

is the gcd until the division algorithm is applied to r_{n-1} and r_n.

Example 1.45.
Find $(326, 78)$, express it as a linear combination of 326 and 78, and write 78/326 in

lowest terms.

$$\boxed{326} = 4 \times \boxed{78} + \boxed{14} \tag{1}$$

$$\boxed{78} = 5 \times \boxed{14} + \boxed{8} \tag{2}$$

$$\boxed{14} = 1 \times \boxed{8} + \boxed{6} \tag{3}$$

$$\boxed{8} = 1 \times \boxed{6} + \boxed{2} \tag{4}$$

$$\boxed{6} = 3 \times \boxed{2}. \tag{5}$$

The Euclidean algorithm gives $(326, 78) = 2$.

We now express 2 as a linear combination of 326 and 78, working from the bottom up using the equations above.

$$
\begin{aligned}
2 &= \boxed{8} - 1\boxed{6} \quad \text{by Eq. (4)} \\
&= \boxed{8} - 1\left(\boxed{14} - 1\boxed{8}\right) \quad \text{by Eq. (3)} \\
&= 2\boxed{8} - 1\boxed{14} \\
&= 2\left(\boxed{78} - 5\boxed{14}\right) - 1\boxed{14} \quad \text{by Eq. (2)} \\
&= 2\boxed{78} - 11\boxed{14} \\
&= 2\boxed{78} - 11\left(\boxed{326} - 4\boxed{78}\right) \quad \text{by Eq. (1)} \\
&= 46\boxed{78} - 11\boxed{326};
\end{aligned}
$$

thus, $s = 46$ and $t = -11$.

Dividing numerator and denominator by the gcd, namely, 2, gives $78/326 = 39/163$, and the last expression is in lowest terms. ◀

The Greek terms for the Euclidean algorithm are *antanairesis* or *anthyphairesis*, either of which may be freely translated as "back and forth subtraction." Exercise 1.61 on page 54 says that $(b, a) = (b-a, a)$. If $b-a \geq a$, repeat to get $(b, a) = (b-a, a) = (b - 2a, a)$. Keep subtracting until a pair a and $b - qa$ (for some q) is reached with $b - qa < a$. Thus, if $r = b - qa$, where $0 \leq r < a$, then

$$(b, a) = (b - a, a) = (b - 2a, a) = \cdots = (b - qa, a) = (r, a).$$

Now change direction: repeat the procedure beginning with the pair $(r, a) = (a, r)$, for $a > r$; eventually one reaches $(d, 0) = d$.

For example, antanairesis computes the gcd $(326, 78)$ as follows:

$$(326, 78) = (248, 78) = (170, 78) = (92, 78) = (14, 78).$$

So far, we have been subtracting 78 from the other larger numbers. At this point, we now subtract 14 (this is the reciprocal aspect of antanairesis), for $78 > 14$.

$$(78, 14) = (64, 14) = (50, 14) = (36, 14) = (22, 14) = (8, 14).$$

Again we change direction:
$$(14, 8) = (6, 8).$$

Change direction once again to get $(8, 6) = (2, 6)$, and change direction one last time to get
$$(6, 2) = (4, 2) = (2, 2) = (0, 2) = 2.$$

Thus, gcd $(326, 78) = 2$.

The division algorithm (which is just iterated subtraction!) is a more efficient way of performing antanairesis. There are four subtractions in the passage from $(326, 78)$ to $(14, 78)$; the division algorithm expresses this as

$$326 = 4 \times 78 + 14.$$

There are then five subtractions in the passage from $(78, 14)$ to $(8, 14)$; the division algorithm expresses this as
$$78 = 5 \times 14 + 8.$$

There is one subtraction in the passage from $(14, 8)$ to $(6, 8)$:

$$14 = 1 \times 8 + 6.$$

There is one subtraction in the passage from $(8, 6)$ to $(2, 6)$:

$$8 = 1 \times 6 + 2,$$

and there are three subtractions from $(6, 2)$ to $(0, 2) = 2$:

$$6 = 3 \times 2.$$

These are the steps in the Euclidean algorithm.

The Euclidean algorithm was one of the first algorithms for which an explicit bound on the number of its steps in a computation was given. The proof of this involves the Fibonacci sequence

$$F_0 = 0, \qquad F_1 = 1, \qquad F_n = F_{n-1} + F_{n-2} \qquad \text{for all } n \geq 2.$$

Proposition 1.46 (Lamé's[18] Theorem). *Let $b \geq a$ be positive integers, and let $d(a)$ be the number of digits in the decimal expression of a. If n is the number of steps in the Euclidean algorithm computing $\gcd(b, a)$, then*

$$n \leq 5d(a).$$

Proof. Let us denote b by r_0 and a by r_1 in the equations of the Euclidean algorithm on page 45, so that every equation there has the form

$$r_j = r_{j+1}q_{j+1} + r_{j+2}$$

except the last one, which is

$$r_{n-1} = r_n q_n.$$

Note that $q_n \geq 2$: if $q_n \leq 1$, then $r_{n-1} \leq q_n r_n = r_n$, contradicting $r_n < r_{n-1}$. Similarly, all $q_1, q_2, \ldots, q_{n-1} \geq 1$: otherwise $q_j = 0$ for some $j \leq n - 1$, and $r_{j-1} = r_{j+1}$, contradicting the strict inequalities $r_n < r_{n-1} < \cdots < r_1 = b$.
   Now

$$r_n \geq 1 = F_2$$

and, since $q_n \geq 2$,

$$r_{n-1} = r_n q_n \geq 2r_n \geq 2F_2 \geq 2 = F_3.$$

More generally, let us prove, by induction on $j \geq 0$, that

$$r_{n-j} \geq F_{j+2}.$$

The inductive step is

$$r_{n-j-1} = r_{n-j}q_{n-j} + r_{n-j+1}$$
$$\geq r_{n-j} + r_{n-j+1} \qquad \text{(since } q_{n-j} \geq 1\text{)}$$
$$\geq F_{j+2} + F_{j+1} = F_{j+3}.$$

We conclude that $a = r_1 = r_{n-(n-1)} \geq F_{n-1+2} = F_{n+1}$. By Corollary 1.16, $F_{n+1} > \gamma^{n-1}$, where $\gamma = \frac{1}{2}(1 + \sqrt{5})$, and so

$$a > \gamma^{n-1}.$$

Now $\log_{10} \gamma > \log_{10}(1.6) > \frac{1}{5}$, so that

$$\log_{10} a > (n - 1)\log_{10} \gamma > (n - 1)/5.$$

---

[18]This is an example in which a theorem's name is not that of its discoverer. Lamé's proof appeared in 1844. The earliest estimate for the number of steps in the Euclidean algorithm can be found in a rare book by Simon Jacob, published around 1564. There were also estimates by T. F. de Lagny in 1733, A.-A.-L. Reynaud in 1821, E. Léger in 1837, and P.-J.-E. Finck in 1841. [This earlier work is described in articles of P. Shallit (1994) and P. Schreiber (1995), respectively, in the journal *Historia Mathematica*.]

Therefore,

$$n - 1 < 5\log_{10} a < 5d(a),$$

because $d(a) = \lfloor \log_{10} a \rfloor + 1$, and so $n \le 5d(a)$ since $d(a)$, hence $5d(a)$, is an integer.

          •

For example, Lamé's theorem guarantees there are at most 10 steps needed to compute $(326, 78)$, for $d(78) = 2$; actually, there are 5 steps.

The usual notation for the integer 5754 is an abbreviation of

$$5 \times 10^3 + 7 \times 10^2 + 5 \times 10 + 4.$$

The next result shows that there is nothing special about the number 10; any integer $b \ge 2$ can be used instead of 10.

→ **Proposition 1.47.** *If $b \ge 2$ is an integer, then every positive integer m has an expression in **base b**: there are integers d_i with $0 \le d_i < b$ such that*

$$m = d_k b^k + d_{k-1} b^{k-1} + \cdots + d_0;$$

moreover, this expression is unique if $d_k \ne 0$.

Remark. The numbers $d_k, \ldots, d_0$ are called the **b-adic digits** of m. ◄

Proof. We iterate the division algorithm to define integers a_i and d_i as follows.

$$\begin{aligned}
m &= a_0 b + d_0, & 0 \le d_0 < b \\
a_0 &= a_1 b + d_1, & 0 \le d_1 < b \\
a_1 &= a_2 b + d_2, & 0 \le d_2 < b \\
&\vdots & \vdots
\end{aligned}$$

An easy induction shows that $m = b^{i+1}a_i + b^i d_i + b^{i-1}d_{i-1} + \cdots + bd_1 + d_0$. There is an integer k with $b^k \le m < b^{k+1}$; for this k, we have $a_k = 0$ (if $a_k \ne 0$, then $a_k \ge 1$ and $m \ge b^{k+1}a_k \ge b^{k+1}$). Hence

$$m = b^k d_k + b^{k-1}d_{k-1} + b^{k-2}d_{k-2} + \cdots + bd_1 + d_0$$

is an expression for m in base b.

Before proving uniqueness of the digits d_i, we first observe that if $0 \le d_i < b$ for all i, then

$$\sum_{i=0}^{k} d_i b^i \le \sum_{i=0}^{k}(b-1)b^i = \sum_{i=0}^{k} b^{i+1} - \sum_{i=0}^{k} b^i = b^{k+1} - 1 < b^{k+1}. \tag{6}$$

We now prove, by induction on $k \geq 0$, that if $b^k \leq m < b^{k+1}$, then the b-adic digits d_i in the expression $m = \sum_{i=0}^{k} d_i b^i$ are uniquely determined by m. Let $m = \sum_{i=0}^{k} d_i b^i = \sum_{i=0}^{k} c_i b^i$, where $0 \leq d_i < b$ and $0 \leq c_i < b$ for all i. Subtracting, we obtain

$$0 = \sum_{i=0}^{k} (d_i - c_i) b^i.$$

Eliminate any zero coefficients, and transpose all negative coefficients $d_i - c_i$, if any, to obtain an equation in which all coefficients are positive and in which the index sets I and J are disjoint:

$$L = \sum_{i \text{ in } I} (d_i - c_i) b^i = \sum_{j \text{ in } J} (c_j - d_j) b^j = R.$$

Let p be the largest index in I and let q be the largest index in J. Since I and J are disjoint, we may assume that $q < p$. As the left side L involves b^p with a nonzero coefficient, we have $L \geq b^p$; but Eq. (6) shows that the right side $R < b^{q+1} \leq b^p$, a contradiction. Therefore, the b-adic digits are uniquely determined.   ●

Example 1.48.
Let us follow the steps in the proof of Proposition 1.47 to write 12345 in base 7. Repeated use of the division algorithm gives

$$12345 = 1763 \cdot 7 + 4$$
$$1763 = 251 \cdot 7 + 6$$
$$251 = 35 \cdot 7 + 6$$
$$35 = 5 \cdot 7 + 0$$
$$5 = 0 \cdot 7 + 5.$$

The 7-adic digits of 12345 are thus 50664.   ◄

The most popular bases are $b = 10$ (giving everyday *decimal* digits), $b = 2$ (giving *binary* digits, useful because a computer can interpret 1 as "on" and 0 as "off"), and $b = 16$ (*hexadecimal*, also for computers), but let us see that other bases can also be useful.

Example 1.49.
Here is a problem of Bachet de Méziriac from 1624. A merchant had a 40-pound weight that broke into 4 pieces. When the pieces were weighed, it was found that each piece was a whole number of pounds and that the four pieces could be used to weigh every integral weight between 1 and 40 pounds. What were the weights of the pieces?

Weighing means using a balance scale having two pans, with weights being put on either pan. Thus, given weights of 1 and 3 pounds, one can weigh a 2-pound weight □ by putting 1 and □ on one pan and 3 on the other pan.

A solution to Bachet's problem is 1, 3, 9, 27. If □ denotes a given integral weight, let us write the weights on one pan to the left of the semicolon and the weights on the other pan to the right of the semicolon. The number in boldface is the weight of □. The reader should note that Proposition 1.47 gives the uniqueness of the weights used in the pans.

1	1 ; □	**9**	9 ; □
2	3 ; 1, □	**10**	9, 1 ; □
3	3 ; □	**11**	9, 3 ; 1, □
4	3, 1 ; □	**12**	9, 3 ; □
5	9 ; 3, 1, □	**13**	9, 3, 1 ; □
6	9 ; 3, □	**14**	27 ; 9, 3, 1, □
7	9, 1 ; 3, □	**15**	27 ; 9, 3, □
8	9 ; 1, □		

The reader may complete this table for □ ≤ 40.   ◄

Example 1.50.

Given a balance scale, the weight (as an integral number of pounds) of any person weighing at most 364 pounds can be found using only six lead weights.

We begin by proving that every positive integer m can be written

$$m = e_k 3^k + e_{k-1} 3^{k-1} + \cdots + 3e_1 + e_0,$$

where $e_i = -1, 0$, or 1.

The idea is to modify the 3-adic expansion

$$m = d_k 3^k + d_{k-1} 3^{k-1} + \cdots + 3d_1 + d_0.$$

where $d_i = 0, 1, 2$, by "carrying." If $d_0 = 0$ or 1, set $e_0 = d_0$ and leave d_1 alone. If $d_0 = 2$, set $e_0 = -1$, and replace d_1 by $d_1 + 1$ (we have merely substituted $3 - 1$ for 2). Now $1 \leq d_1 + 1 \leq 3$. If $d_1 + 1 = 1$, set $e_1 = 1$, and leave d_2 alone; if $d_1 + 1 = 2$, set $e_1 = -1$, and replace d_2 by $d_2 + 1$; if $d_1 + 1 = 3$, define $e_1 = 0$ and replace d_2 by $d_2 + 1$. Continue in this way (the ultimate expansion of m may begin with either $e_k 3^k$ or $e_{k+1} 3^{k+1}$). Here is a table of the first few numbers in this new expansion (let us write $\bar{1}$ instead of -1).

1	1	**9**	100
2	$1\bar{1}$	**10**	101
3	10	**11**	$11\bar{1}$
4	11	**12**	110
5	$1\bar{1}\bar{1}$	**13**	111
6	$1\bar{1}0$	**14**	$1\bar{1}\bar{1}\bar{1}$
7	$1\bar{1}1$	**15**	$1\bar{1}\bar{1}0$
8	$10\bar{1}$		

The reader should now understand Example 1.49. If $\square$ weighs m pounds, write $m = \sum e_i 3^i$, where $e_i = 1, 0,$ or -1, and then transpose those terms having negative coefficients. Those weights with $e_i = -1$ go on the pan with $\square$, while those weights with $e_i = 1$ go on the other pan.

The solution to the current weighing problem involves choosing as weights 1, 3, 9, 27, 81, and 243 pounds. One can find the weight of anyone under 365 pounds, because $1 + 3 + 9 + 27 + 81 = 364$. ◄

EXERCISES

H **1.46** True or false with reasons.

    (i) $6 \mid 2$.

    (ii) $2 \mid 6$.

    (iii) $6 \mid 0$.

    (iv) $0 \mid 6$.

    (v) $0 \mid 0$.

    (vi) $(n, n+1) = 1$ for every natural number n.

    (vii) $(n, n+2) = 2$ for every natural number n.

    (viii) If b and m are positive integers, then $b \mid m$ if and only if the last b-adic digit d_0 of m is 0.

    (ix) 113 is a sum of distinct powers of 2.

    (x) If a and b are natural numbers, there there are natural numbers s and t with $\gcd(a, b) = sa + tb$.

*H **1.47** Given integers a and b (possibly negative) with $a \neq 0$, prove that there exist unique integers q and r with $b = qa + r$ and $0 \leq r < |a|$.

 1.48 Prove that $\sqrt{2}$ is irrational using Proposition 1.14 instead of Euclid's lemma.

H **1.49** Let $p_1, p_2, p_3, \ldots$ be the list of the primes in ascending order: $p_1 = 2, p_2 = 3, p_3 = 5$, and so forth. Define $f_k = p_1 p_2 \cdots p_k + 1$ for $k \geq 1$. Find the smallest k for which f_k is not a prime.

***1.50** Prove that if d and d' are nonzero integers, each of which divides the other, then $d' = \pm d$.

H **1.51** If ζ is a root of unity, prove that there is a positive integer d with $\zeta^d = 1$ such that whenever $\zeta^k = 1$, then $d \mid k$.

H **1.52** Show that every positive integer m can be written as a sum of distinct powers of 2; show, moreover, that there is only one way in which m can so be written.

1.53 Find the b-adic digits of 1000 for $b = 2, 3, 4, 5,$ and 20.

*__1.54__ H **(i)** Prove that if n is *squarefree* (i.e., $n > 1$ and n is not divisible by the square of any prime), then $\sqrt{n}$ is irrational.

H **(ii)** Prove that $\sqrt[3]{2}$ is irrational.

1.55 **(i)** Find $d = \gcd(12327, 2409)$, find integers s and t with $d = 12327s + 2409t$, and put the fraction $2409/12327$ in lowest terms.

(ii) Find $d = \gcd(7563, 526)$, and express d as a linear combination of 7563 and 526.

(iii) Find $d = \gcd(73122, 7404621)$ and express d as a linear combination of 73122 and 7404621.

*__1.56__ Let a and b be integers, and let $sa + tb = 1$ for s, t in $\mathbb{Z}$. Prove that a and b are relatively prime.

*__1.57__ If $d = (a, b)$, prove that a/d and b/d are relatively prime.

* H **1.58** Prove that if $(r, m) = 1 = (r', m)$, then $(rr', m) = 1$.

H **1.59** Let a, b and d be integers. If $d = sa + tb$, where s and t are integers, find infinitely many pairs of integers (s_k, t_k) with $d = s_k a + t_k b$.

* H **1.60** If a and b are relatively prime and if each divides an integer n, prove that their product ab also divides n.

* H **1.61** Prove, for any (possibly negative) integers a and b, that $(b, a) = (b - a, a)$.

H **1.62** If $a > 0$, prove that $a(b, c) = (ab, ac)$. [One must assume that $a > 0$ lest $a(b, c)$ be negative.]

1.63 Prove that the following pseudocode implements the Euclidean algorithm.

```
Input: a, b
Output: d
d := b;   s := a
WHILE s > 0 DO
    rem := remainder after dividing d by s
    d := s
    s := rem
END WHILE
```

H **1.64** If F_n denotes the nth term of the Fibonacci sequence $0, 1, 1, 2, 3, 5, 8, \ldots$, prove, for all $n \geq 1$, that F_{n+1} and F_n are relatively prime.

Definition. A *common divisor* of integers $a_1, a_2, \ldots, a_n$, where $n \geq 2$, is an integer c with $c \mid a_i$ for all i; the largest of the common divisors, denoted by $(a_1, a_2, \ldots, a_n)$, is called the *greatest common divisor*.

*__1.65__ **(i)** Show that if d is the greatest common divisor of $a_1, a_2, \ldots, a_n$, then $d = \sum t_i a_i$, where t_i is in $\mathbb{Z}$ for all i with $1 \leq i \leq n$.

(ii) Prove that if c is a common divisor of $a_1, a_2, \ldots, a_n$, then $c \mid d$.

*__1.66__ **(i)** Show that (a, b, c), the gcd of a, b, c, is equal to $(a, (b, c))$.

(ii) Compute $(120, 168, 328)$.

*1.67 A *Pythagorean triple* is a triple (a, b, c) of positive integers for which

$$a^2 + b^2 = c^2;$$

it is called *primitive* if the gcd $(a, b, c) = 1$.

(i)   Consider a complex number $z = q + ip$, where $q > p$ are positive integers. Prove that

$$(q^2 - p^2, 2qp, q^2 + p^2)$$

is a Pythagorean triple by showing that $|z^2| = |z|^2$. [One can prove that every *primitive* Pythagorean triple (a, b, c) is of this type.]

(ii)   Show that the Pythagorean triple $(9, 12, 15)$ (which is not primitive) is not of the type given in part (i).

(iii)   Using a calculator which can find square roots but which can display only 8 digits, show that

$$(19597501, 28397460, 34503301)$$

is a Pythagorean triple by finding q and p.

→ **1.4   THE FUNDAMENTAL THEOREM OF ARITHMETIC**

We have already seen, in Theorem 1.2, that every integer $a \geq 2$ is either a prime or a product of primes. We are now going to generalize Proposition 1.14 by showing that the primes in such a factorization and the number of times each of them occurs are uniquely determined by a.

→ **Theorem 1.51 (Fundamental Theorem of Arithmetic).**   *Every integer $a \geq 2$ is a prime or a product of primes. Moreover, if a has factorizations*

$$a = p_1 \cdots p_m \text{ and } a = q_1 \cdots q_n,$$

where the p's and q's are primes, then $n = m$ and the q's may be reindexed so that $q_i = p_i$ for all i.

Proof.   We may assume that $m \geq n$, and the proof is by induction on m.

Base step.   If $m = 1$, then the given equation is $a = p_1 = q_1$, and the result is obvious.

Inductive step.   The equation gives $p_m \mid q_1 \cdots q_n$. By Theorem 1.38, Euclid's lemma, there is some i with $p_m \mid q_i$. But q_i, being a prime, has no positive divisors other than 1 and itself, so that $q_i = p_m$. Reindexing, we may assume that $q_n = p_m$. Canceling, we have $p_1 \cdots p_{m-1} = q_1 \cdots q_{n-1}$. By the inductive hypothesis, $n - 1 = m - 1$ and the q's may be reindexed so that $q_i = p_i$ for all i.   •

$\rightarrow$ **Corollary 1.52.**   *If $a \geq 2$ is an integer, then there are distinct primes p_i, unique up to indexing, and unique integers $e_i > 0$ with*

$$a = p_1^{e_1} \cdots p_n^{e_n}.$$

Proof.   Just collect like terms in a prime factorization.   •

The uniqueness in the Fundamental Theorem of Arithmetic says that the exponents $e_1, \ldots, e_n$ in the prime factorization $a = p_1^{e_1} \cdots p_n^{e_n}$ are well-defined integers determined by a. That is, if, say, $n = p^2 q^5 r^6$ and $n = p^2 q^3 s^8$, where p, q, r, s are distinct primes, then it would not make sense to speak of n's "exponent of q."

It is sometimes convenient to allow factorizations $p_1^{e_1} \cdots p_n^{e_n}$ having some zero exponents, for this device allows us to use the same primes when factoring two given numbers. For example, $168 = 2^3 3^1 7^1$ and $60 = 2^2 3^1 5^1$ may be rewritten as $168 = 2^3 3^1 5^0 7^1$ and $60 = 2^2 3^1 5^1 7^0$.

Corollary 1.53.   *Every positive rational number $r \neq 1$ has a unique factorization*

$$r = p_1^{g_1} \cdots p_n^{g_n}$$

where the p_i are distinct primes and the g_i are nonzero integers. Moreover, r is an integer if and only if $g_i > 0$ for all i.

Proof.   There are positive integers a and b with $r = a/b$. If $a = p_1^{e_1} \cdots p_n^{e_n}$ and $b = p_1^{f_1} \cdots p_n^{f_n}$, then $r = p_1^{g_1} \cdots p_n^{g_n}$, where $g_i = e_i - f_i$ (we may assume that the same primes appear in both factorizations by allowing zero exponents). The desired factorization is obtained if one deletes those factors $p_i^{g_i}$, if any, with $g_i = 0$.

Suppose there were another such factorization

$$r = p_1^{h_1} \cdots p_n^{h_n}$$

(by allowing zero exponents, we may again assume that the same primes occur in each factorization). Suppose that $g_j \neq h_j$ for some j; reindexing if necessary, we may assume that $j = 1$ and that $g_1 > h_1$. Therefore,

$$p_1^{g_1 - h_1} p_2^{g_2} \cdots p_n^{g_n} = p_2^{h_2} \cdots p_n^{h_n}.$$

This is an equation of rational numbers, for some of the exponents may be negative. Cross multiplying gives an equation in $\mathbb{Z}$ whose left side involves the prime p_1 and whose right side does not; this contradicts the fundamental theorem of arithmetic.

If all the exponents in the factorization of r are positive, then r is an integer because it is a product of integers. Conversely, if r is an integer, then it has a prime factorization in which all exponents are positive.   •

Lemma 1.54.   *Let positive integers a and b have prime factorizations*

$$a = p_1^{e_1} \cdots p_n^{e_n} \text{ and } b = p_1^{f_1} \cdots p_n^{f_n},$$

where $p_1, \ldots, p_n$ are distinct primes and $e_i, f_i \geq 0$ for all i. Then $a \mid b$ if and only if $e_i \leq f_i$ for all i.

Proof.   If $e_i \leq f_i$ for all i, then $b = ac$, where $c = p_1^{f_1 - e_1} \cdots p_n^{f_n - e_n}$; Corollary 1.53 shows that c is an integer, since $f_i - e_i \geq 0$ for all i. Therefore, $a \mid b$.

Conversely, if $b = ac$, let the prime factorization of c be $c = p_1^{g_1} \cdots p_n^{g_n}$, where $g_i \geq 0$ for all i. It follows from the Fundamental Theorem of Arithmetic that $e_i + g_i = f_i$ for all i, and so $f_i - e_i = g_i \geq 0$ for all i.   •

→ **Definition.**   A ***common multiple*** of a, b is an integer m with $a \mid m$ and $b \mid m$. The ***least common multiple***, denoted by lcm(a, b) (or, more briefly, by $[a, b]$), is the smallest positive common multiple if all $a, b \neq 0$, and it is 0 otherwise.

More generally, if $n \geq 2$, a ***common multiple*** of $a_1, a_2, \ldots, a_n$ is an integer m with $a_i \mid m$ for all i. The ***least common multiple***, denoted by

$$[a_1, a_2, \ldots, a_n],$$

is the smallest positive common multiple if all $a_i \neq 0$, and it is 0 otherwise.

We can now give a new description of gcd's.

→ **Proposition 1.55.**   *Let $a = p_1^{e_1} \cdots p_n^{e_n}$ and $b = p_1^{f_1} \cdots p_n^{f_n}$, where $p_1, \ldots, p_n$ are distinct primes and $e_i, f_i \geq 0$ for all i; define*

$$m_i = \min\{e_i, f_i\} \quad \text{and} \quad M_i = \max\{e_i, f_i\}.$$

Then

$$\gcd(a, b) = p_1^{m_1} \cdots p_n^{m_n} \quad \text{and} \quad \text{lcm}(a, b) = p_1^{M_1} \cdots p_n^{M_n}.$$

Proof.   Define $d = p_1^{m_1} \cdots p_n^{m_n}$. Lemma 1.54 shows that d is a (positive) common divisor of a and b; moreover, if c is any (positive) common divisor, then $c = p_1^{g_1} \cdots p_n^{g_n}$, where $0 \leq g_i \leq \min\{e_i, f_i\} = m_i$ for all i. Therefore, $c \mid d$.

A similar argument shows that $D = p_1^{M_1} \cdots p_n^{M_n}$ is a common multiple that divides every other such.   •

For small numbers a and b, using their prime factorizations is a more efficient way to compute their gcd than using the Euclidean algorithm. For example, since $168 = 2^3 3^1 5^0 7^1$ and $60 = 2^2 3^1 5^1 7^0$, we have $(168, 60) = 2^2 3^1 5^0 7^0 = 12$ and $[168, 60] = 2^3 3^1 5^1 7^1 = 840$. As we mentioned when we introduced the Euclidean algorithm, finding the prime factorization of a large integer is very inefficient.

Proposition 1.56.   *If a and b are positive integers, then*

$$\text{lcm}(a, b) \gcd(a, b) = ab.$$

Proof.   The result follows from Proposition 1.55 if one uses the identity

$$m_i + M_i = e_i + f_i,$$

where $m_i = \min\{e_i, f_i\}$ and $M_i = \max\{e_i, f_i\}$.   •

Of course, this proposition allows us to compute the lcm as $ab/(a, b)$.

EXERCISES

H **1.68**   True or false with reasons.

    (i) $|2^{19} - 3^{12}| < \frac{1}{2}$.

    (ii)   If $r = p_1^{g_1} \cdots p_n^{g_n}$, where the p_i are distinct primes and the g_i are integers, then r is an integer if and only if all the g_i are nonnegative.

    (iii)   The least common multiple $[2^3 \cdot 3^2 \cdot 5 \cdot 7^2, 3^3 \cdot 5 \cdot 13] = 2^3 \cdot 3^5 \cdot 5^2 \cdot 7^2 \cdot 13/45$.

    (iv)   If a and b are positive integers which are not relatively prime, then there is a prime p with $p \mid a$ and $p \mid b$.

    (v)   If a and b are relatively prime, then $(a^2, b^2) = 1$.

1.69   **(i)**   Find $\gcd(210, 48)$ using factorizations into primes.

    (ii)   Find $\gcd(1234, 5678)$.

*****1.70**   **(i)**   Prove that an integer $m \geq 2$ is a perfect square if and only if each of its prime factors occurs an even number of times.

    H **(ii)**   Prove that if m is a positive integer for which $\sqrt{m}$ is rational, then m is a perfect square. Conclude that if m is not a perfect square, then $\sqrt{m}$ is irrational.

H **1.71**   If a and b are positive integers with $(a, b) = 1$, and if ab is a square, prove that both a and b are squares.

***** H **1.72**   Let $n = p^r m$, where p is a prime not dividing an integer $m \geq 1$. Prove that $p \nmid \binom{n}{p^r}$.

Definition.   If p is a prime, define the **p-adic norm** of a rational number a as follows: if $a \neq 0$, then $a = \pm p^e p_1^{e_1} \cdots p_n^{e_n}$, where $p, p_1, \ldots, p_n$ are distinct primes, and we set $\|a\|_p = p^{-e}$; if $a = 0$, set $\|0\|_p = 0$. Define the **p-adic metric** by $\delta_p(a, b) = \|a - b\|_p$.

*****1.73**   **(i)**   For all rationals a and b, prove that

$$\|ab\|_p = \|a\|_p \|b\|_p \quad \text{and} \quad \|a + b\|_p \leq \max\{\|a\|_p, \|b\|_p\}.$$

    (ii)   For all rationals a, b, prove $\delta_p(a, b) \geq 0$ and $\delta_p(a, b) = 0$ if and only if $a = b$.

    (iii)   For all rationals a, b, prove that $\delta_p(a, b) = \delta_p(b, a)$.

    (iv)   For all rationals a, b, c, prove $\delta_p(a, b) \leq \delta_p(a, c) + \delta_p(c, b)$.

    (v)   If a and b are integers and $p^n \mid (a - b)$, then $\delta_p(a, b) \leq p^{-n}$. (Thus, a and b are "close" if $a - b$ is divisible by a "large" power of p.)

1.74 Let a and b be in $\mathbb{Z}$. Prove that if $\delta_p(a, b) \leq p^{-n}$, then a and b have the same first n p-adic digits, $d_0, \ldots, d_{n-1}$.

1.75 Prove that an integer $M \geq 0$ is the lcm of $a_1, a_2, \ldots, a_n$ if and only if it is a common multiple of $a_1, a_2, \ldots, a_n$ which divides every other common multiple.

*1.76 H **(i)** Give another proof of Proposition 1.56, $[a, b](a, b) = |ab|$, without using the Fundamental Theorem of Arithmetic.

 (ii) Find $[1371, 123]$.

→ **1.5 CONGRUENCES**

When first learning long division, one emphasizes the quotient q; the remainder r is merely the fragment left over. There is now going to be a shift in viewpoint: we are interested in whether or not a given number b is a multiple of a number a, but we are not so interested in which multiple it may be. Hence, from now on, we will emphasize the remainder.

Two integers a and b are said to have the *same parity* if they are both even or both odd. We claim that a and b have the same parity if and only if $a - b$ is even. The claim is surely true if a and b are both even; if a and b are both odd, then $a = 2m + 1$, $b = 2n + 1$, and $a - b = 2(m - n)$ is even. Conversely, if $a - b$ is even, then we cannot have one of them even and the other odd lest $a - b$ be odd. The next definition generalizes this notion of parity, letting any positive integer m play the role of 2.

→ **Definition.** If $m \geq 0$ is fixed, then integers a and b are *congruent modulo m*, denoted by

$$a \equiv b \bmod m,$$

if $m \mid (a - b)$.

Usually, one assumes that the *modulus* $m \geq 2$ because the cases $m = 0$ and $m = 1$ are not very interesting: if a and b are integers, then $a \equiv b \bmod 0$ if and only if $0 \mid (a - b)$, that is, $a = b$, and so congruence mod 0 is ordinary equality. The congruence $a \equiv b \bmod 1$ is true for every pair of integers a and b because $1 \mid (a - b)$ always. Hence, every two integers are congruent mod 1.

The word *modulo* is usually abbreviated to *mod*. The Latin root of this word means "a standard of measure." Thus, the term *modular unit* is used today in architecture: a fixed length m is chosen, say, $m = 1$ foot, and plans are drawn so that the dimensions of every window, door, wall, and so on, are integral multiples of m.

If a and b are positive integers, then $a \equiv b \bmod 10$ if and only if they have the same last digit; more generally, $a \equiv b \bmod 10^n$ if and only if they have same last n digits. For example, $526 \equiv 1926 \bmod 100$.

London time is 6 hours later than Chicago time. What time is it in London if it is 10:00 A.M. in Chicago? Since clocks are set up with 12 hour cycles, this is really a problem about congruence mod 12. To solve it, note that

$$10 + 6 = 16 \equiv 4 \bmod 12,$$

and so it is 4:00 P.M. in London.

The next theorem shows that congruence mod m behaves very much like equality.

→ **Proposition 1.57.**   *If $m \geq 0$ is a fixed integer, then for all integers a, b, c,*

(i) $a \equiv a \bmod m$;

(ii) *if $a \equiv b \bmod m$, then $b \equiv a \bmod m$;*

(iii) *if $a \equiv b \bmod m$ and $b \equiv c \bmod m$, then $a \equiv c \bmod m$.*

Remark.   (i) says that congruence is ***reflexive,*** (ii) says it is ***symmetric,*** and (iii) says it is ***transitive.***   ◄

Proof.
(i) Since $m \mid (a - a) = 0$, we have $a \equiv a \bmod m$.
(ii) If $m \mid (a - b)$, then $m \mid -(a - b) = b - a$ and so $b \equiv a \bmod m$.
(iii) If $m \mid (a - b)$ and $m \mid (b - c)$, then $m \mid [(a - b) + (b - c)] = a - c$, and so $a \equiv c \bmod m$.   •

We now generalize the observation that $a \equiv 0 \bmod m$ if and only if $m \mid a$.

→ **Proposition 1.58.**   *Let $m \geq 0$ be a fixed integer.*

(i) *If $a = qm + r$, then $a \equiv r \bmod m$.*

(ii) *If $0 \leq r' < r < m$, then $r \not\equiv r' \bmod m$; that is, r and r' are not congruent mod m.*

(iii) *$a \equiv b \bmod m$ if and only if a and b leave the same remainder after dividing by m.*

Proof.
(i) The equation $a - r = qm$ shows that $m \mid (a - r)$.
(ii) If $r \equiv r' \bmod m$, then $m \mid (r - r')$ and $m \leq r - r'$. But $r - r' \leq r < m$, a contradiction.
(iii) If $a = qm + r$ and $b = q'm + r'$, where $0 \leq r < m$ and $0 \leq r' < m$, then $a - b = (q - q')m + (r - r')$; that is,

$$a - b \equiv r - r' \bmod m.$$

Therefore, if $a \equiv b \bmod m$, then $a - b \equiv 0 \bmod m$, hence $r - r' \equiv 0 \bmod m$, and $r \equiv r' \bmod m$; by part (ii), $r = r'$.

Conversely, if $r = r'$, then $a = qm + r$ and $b = q'm + r$, so that $a - b = (q' - q)m$ and $a \equiv b \bmod m$. $\bullet$

→ **Corollary 1.59.** *Given $m \geq 2$, every integer a is congruent* mod m *to exactly one of* $0, 1, \ldots, m - 1$.

Proof. The division algorithm says that $a \equiv r \bmod m$, where $0 \leq r < m$; that is, r is an integer on the list $0, 1, \ldots, m - 1$. If a were congruent to two integers on the list, say, r and r', then $r \equiv r' \bmod m$, contradicting part (ii) of Proposition 1.58. Therefore, a is congruent to a unique such r. $\bullet$

We know that every integer a is either even or odd; that is, a has the form $2k$ or $1 + 2k$. We now see that if $m \geq 2$, then every integer a has exactly one of the forms $0 + km, 1 + km, 2 + km, \ldots, (m - 1) + km$; thus, congruence mod m generalizes the even/odd dichotomy from $m = 2$ to $m \geq 2$. Notice how we continue to focus on the remainder in the division algorithm and not upon the quotient.

Congruence is compatible with addition and multiplication.

→ **Proposition 1.60.** *Let $m \geq 0$ be a fixed integer.*

(i) *If $a_i \equiv a_i' \bmod m$ for $i = 1, 2, \ldots, n$, then*

$$a_1 + \cdots + a_n \equiv a_1' + \cdots + a_n' \bmod m.$$

In particular, if $a \equiv a' \bmod m$ and $b \equiv b' \bmod m$, then

$$a + b \equiv a' + b' \bmod m.$$

(ii) *If $a_i \equiv a_i' \bmod m$ for $i = 1, 2, \ldots, n$, then*

$$a_1 \cdots a_n \equiv a_1' \cdots a_n' \bmod m.$$

In particular, if $a \equiv a' \bmod m$ and $b \equiv b' \bmod m$, then

$$ab \equiv a'b' \bmod m.$$

(iii) *If $a \equiv b \bmod m$, then $a^n \equiv b^n \bmod m$ for all $n \geq 1$.*

Proof.
(i) The proof is by induction on $n \geq 2$. For the base step, if $m \mid (a - a')$ and $m \mid (b - b')$, then $m \mid [(a - a') + (b - b')] = (a + b) - (a' + b')$. Therefore, $a + b \equiv a' + b' \bmod m$. The proof of the inductive step is routine.

(ii) The proof is by induction on $n \geq 2$. For the base step, we must show that if $m \mid (a - a')$ and $m \mid (b - b')$, then $m \mid (ab - a'b')$, and this follows from the identity

$$ab - a'b' = (ab - a'b) + (a'b - a'b')$$
$$= (a - a')b + a'(b - b').$$

Therefore, $ab \equiv a'b' \bmod m$. The proof of the inductive step is routine.

(iii) This is the special case of part (ii) when $a_i = a$ and $a'_i = b$ for all i.   •

Let us repeat a warning given on page 38. A number and its negative usually have different remainders after being divided by a number m. For example, $60 = 7 \cdot 8 + 4$ and $-60 = 7 \cdot (-9) + 3$. In terms of congruences,

$$60 \equiv 4 \bmod 7 \qquad \text{while} \qquad -60 \equiv 3 \bmod 7.$$

In light of Proposition 1.58(i), if the remainder after dividing b by m is r and the remainder after dividing $-b$ by m is s, then $b \equiv r \bmod m$ and $-b \equiv s \bmod m$. Therefore, Proposition 1.60(i) gives

$$r + s \equiv b - b \equiv 0 \bmod m.$$

Thus, if b is not a multiple of m, then $r \neq 0$, $s \neq 0$, and $r + s = m$, because $0 \leq r, s < m$. For example, we have just seen that the remainders after dividing 60 and -60 by 7 are 4 and 3, respectively. If both a and $-a$ have the same remainder r after dividing by m, then $-r \equiv r \bmod m$; that is, $2r \equiv 0 \bmod m$. Exercise 1.84 on page 75 asks you to solve this last congruence.

The next example shows how one can use congruences. In each case, the key idea is to solve a problem by replacing numbers by their remainders.

Example 1.61.

(i) If a is in $\mathbb{Z}$, then $a^2 \equiv 0$, 1, or 4 mod 8.

r	0	1	2	3	4	5	6	7
r^2	0	1	4	9	16	25	36	49
$r^2 \bmod 8$	0	1	4	1	0	1	4	1

Table 1.1   Squares mod 8

If a is an integer, then $a \equiv r \bmod 8$, where $0 \leq r \leq 7$; moreover, by Proposition 1.60(iii), $a^2 \equiv r^2 \bmod 8$, and so it suffices to look at the squares of the remainders. We see in Table 1.1 that only 0, 1, or 4 can be a remainder after dividing a perfect square by 8.

(ii) $n = 1003456789$ is not a perfect square.

Since $1000 = 8 \cdot 125$, we have $1000 \equiv 0 \bmod 8$, and so

$$1003456789 = 1003456 \cdot 1000 + 789 \equiv 789 \bmod 8.$$

Dividing 789 by 8 leaves remainder 5; that is, $n \equiv 5 \bmod 8$. But if n were a perfect square, then $n \equiv 0$, 1, or 4 mod 8.

(iii) There are no perfect squares of the form $3^m + 3^n + 1$, where m and n are positive integers.

Again, let us look at remainders mod 8. Now $3^2 = 9 \equiv 1 \bmod 8$, and so we can evaluate $3^m \bmod 8$ as follows: if $m = 2k$, then $3^m = 3^{2k} = 9^k \equiv 1 \bmod 8$; if $m = 2k + 1$, then $3^m = 3^{2k+1} = 9^k \cdot 3 \equiv 3 \bmod 8$. Thus,

$$3^m \equiv \begin{cases} 1 \bmod 8 & \text{if } m \text{ is even;} \\ 3 \bmod 8 & \text{if } m \text{ is odd.} \end{cases}$$

Replacing numbers by their remainders after dividing by 8, we have the following possibilities for the remainder of $3^m + 3^n + 1$, depending on the parities of m and n:

$$3 + 1 + 1 \equiv 5 \bmod 8$$
$$3 + 3 + 1 \equiv 7 \bmod 8$$
$$1 + 1 + 1 \equiv 3 \bmod 8$$
$$1 + 3 + 1 \equiv 5 \bmod 8.$$

In no case is the remainder 0, 1, or 4, and so no number of the form $3^m + 3^n + 1$ can be a perfect square, by part (i). ◄

Every positive integer is congruent to either 0, 1, or 2 mod 3; hence, if $p \neq 3$ is a prime, then either $p \equiv 1 \bmod 3$ or $p \equiv 2 \bmod 3$. For example, 7, 13, and 19 are congruent to 1 mod 3, while 2, 5, 11, and $17 \equiv 2 \bmod 3$. The next theorem is another illustration of the fact that a proof of one theorem may be adapted to prove another theorem.

Proposition 1.62.   *There are infinitely many primes p with $p \equiv 2 \bmod 3$.*

Remark.   This proposition is a special case of a beautiful theorem of Dirichlet about primes in arithmetic progressions: if a, b in $\mathbb{N}$ are relatively prime, then there are infinitely many primes of the form $a + bn$. In this proposition, we show that there are infinitely many primes of the form $2 + 3n$. Even though the proof of this special case is not difficult, the proof of Dirichlet's theorem uses complex analysis and it is deep.   ◄

Proof. We mimic Euclid's proof that there are infinitely many primes. Suppose, on the contrary, that there are only finitely many primes congruent to 2 mod 3; let them be $p_1, \ldots, p_s$. Consider the number

$$m = 1 + p_1^2 \cdots p_s^2.$$

Now $p_i \equiv 2 \bmod 3$ implies $p_i^2 \equiv 4 \equiv 1 \bmod 3$; hence, $p_1^2 \cdots p_s^2 \equiv 1 \bmod 3$, and so $m \equiv 1 + 1 = 2 \bmod 3$. Since $m > p_i$ for all i, the number m is not prime, for it is not one of the p_i. Actually, none of the p_i divide m: if we define $Q_i = p_1^2 \cdots p_{i-1}^2 p_i p_{i+1}^2 \cdots p_s^2$, then the uniqueness part of the division algorithm coupled with the equation $m = p_i Q_i + 1$ shows that m leaves remainder 1 after dividing by p_i. Hence, the prime factorization of m is $m = q_1 \cdots q_t$, where, for each j, either $q_j \equiv 1 \bmod 3$ or $q_j \equiv 0 \bmod 3$; i.e., $q_j = 3$. Thus, $m = q_1 \cdots q_t \equiv 0 \bmod 3$ or $m = q_1 \cdots q_t \equiv 1 \bmod 3$, contradicting $m \equiv 2 \bmod 3$.   •

The next result shows how congruence can simplify complicated expressions.

→ **Proposition 1.63.**   *If p is a prime and a and b are integers, then*

$$(a + b)^p \equiv a^p + b^p \bmod p.$$

Proof.   The binomial theorem gives

$$(a + b)^p = a^p + b^p + \sum_{r=1}^{p-1} \binom{p}{r} a^{p-r} b^r.$$

But Proposition 1.39 gives $\binom{p}{r} \equiv 0 \bmod p$ for $0 < r < p$, and so Proposition 1.60(i) gives $(a + b)^p \equiv a^p + b^p \bmod p$.   •

→ **Theorem 1.64** (*Fermat*).

(i) *If p is a prime, then*

$$a^p \equiv a \bmod p$$

for every a in $\mathbb{Z}$.

(ii) *If p is a prime, then*

$$a^{p^k} \equiv a \bmod p$$

for every a in $\mathbb{Z}$ and every integer $k \geq 1$.

Proof.

(i) Assume first that $a \geq 0$; we proceed by induction on a. The base step $a = 0$ is plainly true. For the inductive step, observe that

$$(a + 1)^p \equiv a^p + 1 \bmod p,$$

by Proposition 1.63. The inductive hypothesis gives $a^p \equiv a \bmod p$, and so $(a+1)^p \equiv a^p + 1 \equiv a + 1 \bmod p$, as desired.

Now consider $-a$, where $a \geq 0$. If $p = 2$, then $-a \equiv a$; hence, $(-a)^2 = a^2 \equiv a \equiv -a \bmod 2$. If p is an odd prime, then $(-a)^p = (-1)^p a^p \equiv (-1)^p a \equiv -a \bmod p$, as desired.

(ii) A straightforward induction on $k \geq 1$; the base step is part (i). •

Corollary 1.65. *A positive integer n is divisible by* 3 *(or by* 9*) if and only if the sum of its (decimal) digits is divisible by* 3 *(or by* 9*).*

Proof. If the decimal notation for n is $d_k \ldots d_1 d_0$, then

$$n = d_k 10^k + \cdots + d_1 10 + d_0.$$

Now $10 \equiv 1 \bmod 3$, so that Proposition 1.60(iii) gives $10^i \equiv 1^i = 1 \bmod 3$ for all i; thus Proposition 1.60(i) gives $n \equiv d_k + \cdots + d_1 + d_0 \bmod 3$. Therefore, n is divisible by 3 if and only if $n \equiv 0 \bmod 3$ if and only if $d_k + \cdots + d_1 + d_0 \equiv 0 \bmod 3$.

Since $10 \equiv 1 \bmod 9$, the same proof gives the result for 9. •

Example 1.66 (Casting Out 9s).

Define two operations on the (decimal) digits of a positive integer n:

(i) delete all 9s or any group of digits whose sum is 9;

(ii) add up all the digits.

For obvious reasons, changing an integer by repeated use of operations (i) and (ii) is called ***casting out 9s***. If n has at least two digits, then either operation replaces n by an integer strictly smaller than it; hence, casting out 9s eventually gives a single digit, say, $r(n)$ with $0 \leq r(n) < 9$ (should the single digit be 9, then the first operation replaces it by 0). A priori, there are many possible values of $r(n)$.

For example, (i) changes 5261934 to 526134 to 2613 (because $5 + 4 = 9$) to 21 (because $6 + 3 = 9$) to $1 + 2 = 3$. Since $2 + 3 + 4 = 9$, we could also have proceeded: $5261934 \rightarrow 526134 \rightarrow 561 \rightarrow 21 \rightarrow 3$. Note that both choices just made give the same value: $r(5261934) = 3$; note also that casting out 9s can be done very quickly.

Corollary 1.65 shows that the sum of the digits of an integer n is congruent to $n \bmod 9$. Thus, applying (ii) to an integer n does not change its remainder mod 9. If some digit is 9, or if some group of digits adds up to 9, then deleting them, as in operation (i), gives an integer congruent to n mod 9. We conclude that $r(n) \equiv n$ mod 9. Moreover, since $0 \leq r(n) < 9$, Corollary 1.59 shows that the integer $r(n)$ does not depend on the choices of operations used. In short, $r(n)$ is the remainder after dividing n by 9.

Here is a bookkeeping[19] trick which checks for arithmetic errors. Let us test the correctness of an alleged equation, $(12345 + 5261944)1776 = 9367119504$, by casting out 9s. Now $r(12345) = 6$, $r(5261934) = 3$, as we saw above, and $r(1776) = 3$. By Proposition 1.60, $r([12345 + 5261944] \times 1776) = r([6 + 3] \times 3) = 0$. Since $r(9367119504) = 0$, both sides have the same remainder, and the computation passes the test of casting out 9s (had the two sides been different, then a mistake would have been made). Unfortunately, this trick does not guarantee correctness of a computation. For example, if n' is obtained from n by interchanging two digits, then $r(n') = r(n)$, and so transposed digits cannot be detected by casting out 9s. ◄

Corollary 1.67.   *Let p be a prime and let n be a positive integer. If $m \geq 0$ and if $\Sigma(m)$ is the sum of the p-adic digits of m, then*

$$n^m \equiv n^{\Sigma(m)} \text{ mod } p.$$

Proof.   Let $m = d_k p^k + \cdots + d_1 p + d_0$ be the expression of m in base p. By Fermat's theorem, Theorem 1.64(ii), $n^{p^i} \equiv n$ mod p for all i; thus, $n^{d_i p^i} = (n^{d_i})^{p^i} \equiv n^{d_i}$ mod p. Therefore,

$$
\begin{aligned}
n^m &= n^{d_k p^k + \cdots + d_1 p + d_0} \\
&= n^{d_k p^k} n^{d_{k-1} p^{k-1}} \cdots n^{d_1 p} n^{d_0} \\
&\equiv n^{d_k} n^{d_{k-1}} \cdots n^{d_1} n^{d_0} \text{ mod } p \\
&\equiv n^{d_k + \cdots + d_1 + d_0} \text{ mod } p \\
&\equiv n^{\Sigma(m)} \text{ mod } p. \quad \bullet
\end{aligned}
$$

Example 1.68.
What is the remainder after dividing 3^{12345} by 7? By Example 1.48, the 7-adic digits of 12345 are 50664. Therefore, $3^{12345} \equiv 3^{21}$ mod 7 (because $5 + 0 + 6 + 6 + 4 = 21$). The 7-adic digits of 21 are 30 (because $21 = 3 \times 7$), and so $3^{21} \equiv 3^3$ mod 7 (because $3 + 0 = 3$). We conclude that $3^{12345} \equiv 3^3 = 27 \equiv 6$ mod 7. ◄

---

[19]The word *bookkeeper* has three consecutive doubled letters: oo, kk, ee. Here is a word with six consecutive doubled letters. Ramon, the raccoon, was a stellar attraction in the Madrid zoo, for Ramon could dance, both classical ballet and flamenco. Because of the crowds he attracted, Ramon was given his own cage. But Ramon needed his own private space to relax from the pressures of performing, a nook where he could hide from his admirers. The zoo hired a special attendant to cater to Ramon's needs; he was a raccoonnookkeeper.

→ **Theorem 1.69.** *If $(a, m) = 1$, then, for every integer b, the congruence*

$$ax \equiv b \bmod m$$

can be solved for x; in fact, $x = sb$, where $sa \equiv 1 \bmod m$. Moreover, any two solutions are congruent mod *m.*

Remark. We consider the case $(a, m) \neq 1$ in Exercise 1.89 on page 75. ◄

Proof. Since $(a, m) = 1$, there is an integer s with $as \equiv 1 \bmod m$ (because there is a linear combination $1 = sa + tm$). It follows that $b = sab + tmb$ and $asb \equiv b \bmod m$, so that $x = sb$ is a solution. [Note that Proposition 1.58(i) allows us to take s with $1 \leq s < m$.]

   If y is another solution, then $ax \equiv ay \bmod m$, and so $m \mid a(x - y)$. Since $(a, m) = 1$, Corollary 1.40 gives $m \mid (x - y)$; that is, $x \equiv y \bmod m$. •

→ **Corollary 1.70.** *If p is prime and $p \nmid a$, then the congruence $ax \equiv b \bmod p$ is always solvable.*

Proof. Since p is a prime, $p \nmid a$ implies $(a, p) = 1$. •

Example 1.71.
When $(a, m) = 1$, Theorem 1.69 says that the solutions to $ax \equiv b \bmod m$ are precisely those integers of the form $sb + km$ for k in $\mathbb{Z}$, where $sa \equiv 1 \bmod m$; that is, where $sa + tm = 1$ for some integer t. Thus, s can always be found by the Euclidean algorithm. However, when m is small, it is easier to find such an integer s by trying each of $ra = 2a, 3a, \ldots, (m - 1)a$ in turn, at each step checking if $ra \equiv 1 \bmod m$.

   For example, let us find all the solutions to

$$2x \equiv 9 \bmod 13.$$

Considering the products $2 \cdot 2, 3 \cdot 2, 4 \cdot 2, \ldots \bmod 13$ quickly leads to $7 \times 2 = 14 \equiv 1 \bmod 13$; that is, $s = 7$ and $x = 7 \cdot 9 = 63 \equiv 11 \bmod 13$. Therefore,

$$x \equiv 11 \bmod 13,$$

and the solutions are $\ldots, -15, -2, 11, 24, \ldots$. ◄

Example 1.72.
Find all the solutions to $51x \equiv 10 \bmod 94$.

   Since 94 is large, seeking an integer s with $51s \equiv 1 \bmod 94$, as in Example 1.71, can be tedious. The Euclidean algorithm gives $1 = -35 \cdot 51 + 19 \cdot 94$, and so $s = 59$, for $59 \equiv -35 \bmod 94$. The solution consists of all integers x with $x \equiv 59 \times 10 \bmod 94$; that is, numbers of the form $590 + 94k$. ◄

There are problems solved in ancient Chinese manuscripts that involve simultaneous congruences with relatively prime moduli.

→ **Theorem 1.73 (Chinese Remainder Theorem).**   *If m and m' are relatively prime, then the two congruences*

$$x \equiv b \bmod m$$
$$x \equiv b' \bmod m'$$

have a common solution, and any two solutions are congruent mod mm'.

Proof.   Every solution of the first congruence has the form $x = b + km$ for some integer k; thus, we must find k with $b + km \equiv b' \bmod m'$; that is, $km \equiv b' - b \bmod m'$. Since $(m, m') = 1$, however, Theorem 1.69 applies at once to show that such an integer k does exist.

If y is another common solution, then both m and m' divide $x - y$; by Exercise 1.60 on page 54, $mm' \mid (x - y)$, and so $x \equiv y \bmod mm'$.   •

Example 1.74.
Find all the solutions to the simultaneous congruences

$$x \equiv 7 \bmod 8$$
$$x \equiv 11 \bmod 15.$$

Every solution to the first congruence has the form

$$x = 7 + 8k,$$

for some integer k. Substituting, $x = 7 + 8k \equiv 11 \bmod 15$, so that

$$8k \equiv 4 \bmod 15.$$

But $2 \cdot 8 = 16 \equiv 1 \bmod 15$, so that multiplying by 2 gives

$$16k \equiv k \equiv 8 \bmod 15.$$

We conclude that $x = 7 + 8 \cdot 8 = 71$ is a solution, and the Chinese Remainder Theorem says that every solution has the form $71 + 120n$ for n in $\mathbb{Z}$.   ◄

Example 1.75.
Solve the simultaneous congruences

$$x \equiv 2 \bmod 5$$
$$3x \equiv 5 \bmod 13.$$

Every solution to the first congruence has the form $x = 5k + 2$ for k in $\mathbb{Z}$. Substituting into the second congruence, we have

$$3(5k + 2) \equiv 5 \bmod 13.$$

Therefore,

$$15k + 6 \equiv 5 \bmod 13$$
$$2k \equiv -1 \bmod 13.$$

Now $7 \times 2 \equiv 1 \bmod 13$, and so multiplying by 7 gives

$$k \equiv -7 \equiv 6 \bmod 13.$$

By the Chinese Remainder Theorem, all the simultaneous solutions x have the form

$$x \equiv 5k + 2 \equiv 5 \cdot 6 + 2 = 32 \bmod 65;$$

that is, the solutions are

$$\ldots, -98, -33, 32, 97, 162, \ldots \ . \quad \blacktriangleleft$$

If we do not assume that the moduli m and m' are relatively prime, then there may be no solutions to a linear system. For example, if $m = m' > 1$, then uniqueness of the remainder in the division algorithm shows that there is no solution to

$$x \equiv 0 \bmod m$$
$$x \equiv 1 \bmod m.$$

Proposition 1.76. *Let $d = (m, m')$. The system*

$$x \equiv b \bmod m$$
$$x \equiv b' \bmod m'$$

has a solution if and only if $b \equiv b' \bmod d$.

Remark. Remember that $b \equiv b' \bmod 1$ is always true. $\blacktriangleleft$

Proof. If $h \equiv b \bmod m$ and $h \equiv b' \bmod m'$, then $m \mid (h - b)$ and $m' \mid (h - b')$. Since d is a common divisor of m and m', we have $d \mid (h - b)$ and $d \mid (h - b')$. Therefore, $d \mid (b - b')$, because $(h - b') - (h - b) = b - b'$, and so $b \equiv b' \bmod d$.

Conversely, assume that $b \equiv b' \bmod d$, so that there is an integer k with $b' = b + kd$. If $m = dc$ and $m' = dc'$, then $(c, c') = 1$, by Exercise 1.57 on page 54. Hence, there are integers s and t with $1 = sc + tc'$. Define $h = b'sc + btc'$. Now

$$\begin{aligned}
h &= b'sc + btc' \\
&= (b + kd)sc + btc' \\
&= b(sc + tc') + kdsc \\
&= b + ksm \\
&\equiv b \bmod m.
\end{aligned}$$

A similar argument, replacing b by $b' - kd$, shows that $h \equiv b' \bmod m'$.   •

Exercise 1.96 on page 76 asks you to prove, given the hypothesis of Proposition 1.76, that any two solutions are congruent mod ℓ, where $\ell = \text{lcm}\{m, m'\}$.

Example 1.77.
Solve the linear system

$$x \equiv 1 \bmod 6$$
$$x \equiv 4 \bmod 15.$$

Here, $m = 6, m' = 15, d = 3, c = 2, c' = 5, s = 3$, and $t = -1$. Proposition 1.76 applies, for $1 \equiv 4 \bmod 3$. Define

$$h = 4 \times 3 \times 2 + 1 \times (-1) \times 5 = 19.$$

We check that $19 \equiv 1 \bmod 6$ and $19 \equiv 4 \bmod 15$. Since $\text{lcm}\{6, 15\} = 30$, the solutions are $\ldots, -41, -11, 19, 49, 79, \ldots$.   ◄

Example 1.78 (A Mayan Calendar).
A congruence arises whenever there is cyclic behavior. For example, suppose we choose some particular Sunday as time zero and enumerate all the days according to the time elapsed since then. Every date now corresponds to some integer, which is negative if it occurred before time zero. Given two dates t_1 and t_2, we ask for the number $x = t_2 - t_1$ of days from one to the other. If, for example, t_1 falls on a Thursday and t_2 falls on a Tuesday, then $t_1 \equiv 4 \bmod 7$ and $t_2 \equiv 2 \bmod 7$, and so $x = t_2 - t_1 = -2 \equiv 5 \bmod 7$. Thus, $x = 7k + 5$ for some k.

About 2500 years ago, the Maya of Central America and Mexico developed three calendars (each having a different use). Their religious calendar, called *tzolkin*, consisted of 20 "months," each having 13 days (so that the tzolkin "year" had 260 days). The months were

1. Imix	6. Cimi	11. Chuen	16. Cib
2. Ik	7. Manik	12. Eb	17. Caban
3. Akbal	8. Lamat	13. Ben	18. Etznab
4. Kan	9. Muluc	14. Ix	19. Cauac
5. Chicchan	10. Oc	15. Men	20. Ahau

Let us describe a tzolkin date by an ordered pair $\{m, d\}$, where $1 \leq m \leq 20$ and $1 \leq d \leq 13$ (thus, m denotes the month and d denotes the day). Instead of enumerating as we do (so that Imix 1 is followed by Imix 2, then by Imix 3, and so forth), the Maya let both month and day cycle simultaneously; that is, the days proceed as follows:

$$\text{Imix 1, Ik 2, Akbal 3,} \ldots, \text{Ben 13, Ix 1, Men 2,} \ldots,$$
$$\text{Cauac 6, Ahau 7, Imix 8, Ik 9,} \ldots.$$

We now ask how many days have elapsed between Oc 11 and Etznab 5. More generally, let x be the number of days from tzolkin $\{m, d\}$ to tzolkin $\{m', d'\}$. As we remarked at the beginning of this example, the cyclic behavior of the days gives the congruence

$$x \equiv d' - d \bmod 13$$

(e.g., there are 13 days between Imix 1 and Ix 1; here, $x \equiv 0 \bmod 13$), while the cyclic behavior of the months gives the congruence

$$x \equiv m' - m \bmod 20$$

(e.g., there are 20 days between Imix 1 and Imix 8; here, $x \equiv 0 \bmod 20$). To answer the original question, Oc 11 corresponds to the ordered pair $\{10, 11\}$ and Etznab 5 corresponds to $\{18, 5\}$ (since $5 - 11 = -6$ and $18 - 10 = 8$). The simultaneous congruences are thus

$$x \equiv -6 \bmod 13$$
$$x \equiv \ \ \ 8 \bmod 20.$$

Now $(13, 20) = 1$, so that we can solve this system as in the proof of the Chinese Remainder Theorem. The first congruence gives

$$x = 13k - 6,$$

and the second gives

$$13k - 6 \equiv 8 \bmod 20;$$

that is,

$$13k \equiv 14 \bmod 20.$$

Since $13 \times 17 = 221 \equiv 1 \bmod 20$,[20] we have $k \equiv 17 \times 14 \bmod 20$, that is,

$$k \equiv 18 \bmod 20,$$

and so the Chinese Remainder Theorem gives

$$x = 13k - 6 \equiv 13 \times 18 - 6 \equiv 228 \bmod 260.$$

It is not clear whether Oc 11 precedes Etznab 5 in a given year (one must look). If it does, then there are 228 days between them; otherwise, there are $32 = 260 - 228$ days between them (the truth is 228). ◄

---

[20]One finds 17 either by trying each number between 1 and 19 or by using the Euclidean algorithm.

→ **Example 1.79 (Public Key Cryptography).**

In a war between A and B, spies for A learn of a surprise attack being planned by B, and so they must send an urgent message back home. If B learns that its plans are known to A, it will, of course, change them, and so A's spies put the message in code before sending it.

It is no problem to convert a message in English into a number. Make a list of the 52 English letters (lower case and upper case) together with a space and the 11 punctuation marks

$$, \quad . \quad ; \quad : \quad ! \quad ? \quad - \quad ' \quad " \quad (\quad)$$

In all, there are 64 symbols. Assign a two-digit number to each symbol. For example,

$$a \mapsto 01, \ldots, z \mapsto 26, A \mapsto 27, \ldots, Z \mapsto 52$$
$$\text{space} \mapsto 53, . \mapsto 54, , \mapsto 55, \ldots, (\mapsto 63,) \mapsto 64.$$

A *cipher* is a code in which distinct letters in the original message are replaced by distinct symbols. It is not difficult to decode any cipher; indeed, many newspapers print daily cryptograms to entertain their readers. In the cipher we have just described, "I love you." is encoded

$$\text{I love you.} = 3553121522055325152154.$$

Notice that each coded message in this cipher has an even number of digits, and so decoding, converting the number into English, is a simple matter. Thus,

$$3553121522055325152154 = (35)(53)(12)(15)(22)(05)(53)(25)(15)(21)(54)$$
$$= \text{I love you.}$$

What makes a good code? If a message is a natural number x (and this is no loss in generality), we need a way to encode x (in a fairly routine way so as to avoid introducing any errors into the coded message), and we need a (fairly routine) method for the recipient to decode the message. Of utmost importance is security: an unauthorized reader of the (coded) message should not be able to decode it. An ingenious way to find a code with these properties, now called *RSA public key cryptography*, was found in 1978 by R. Rivest, A. Shamir, and L. Adleman; they received the 2002 Turing Award for their discovery.

Given natural numbers N, s, and t, suppose that $x^{st} \equiv x \bmod N$ for every natural number x. We can encode any natural number $x < N$ as $[x^s]_N$, the remainder of $x^s \bmod N$, and we can decode this if we know the number t, for

$$(x^s)^t = x^{st} \equiv x \bmod N.$$

It remains to find numbers N, s, and t satisfying the several criteria for a good code.

Ease of Encoding and Decoding

Suppose that N has d (decimal) digits. It is enough to show how to encode a number x with at most d digits, for we can subdivide a longer number into blocks each having at most d digits. An efficient computation of $x^s \bmod N$ is based on the fact that computing $x^2 \bmod N$ is an easy task for a computer. Since computing x^{2^i} is just computing i squares, this, too, is an easy task. Now write the exponent s in base 2, so that computing x^s is the same as multiplying several squares: if $m = 2^i + 2^j + \cdots + 2^z$, then $x^m = x^{2^i + 2^j + \cdots + 2^z} = x^{2^i} x^{2^j} \cdots x^{2^z}$. In short, computers can encode a message in this way with no difficulty.

Decoding involves computing $(x^s)^t \bmod N$, and this is also an easy task (assuming t is known) if, as above, we write t in base 2.

Constructing N and $m = st$

Choose distinct primes p and q, both congruent to 2 mod 3, and define $N = pq$. If $m \geq p$, then

$$x^m = x^{m-p}x^p \equiv x^{m-p}x = x^{m-(p-1)} \bmod p,$$

by Fermat's theorem. If $m - (p-1) \geq p$, we may repeat this, continuing until we have

$$
\begin{aligned}
x^{m-(p-1)} &= x^{m-(p-1)-p}x^p \\
&\equiv x^{m-(p-1)-p}x \\
&= x^{m-2(p-1)} \\
&\ \ \vdots \\
&\equiv x^{m-h(p-1)} \bmod p,
\end{aligned}
$$

where h is the largest integer for which $m - h(p-1) \geq 0$. But this is just the division algorithm: $m = h(p-1) + r$, where r is the remainder after dividing m by $p-1$. Hence, for all x,

$$x^m \equiv x^r \bmod p.$$

Therefore, if $m \equiv 1 \bmod (p-1)$, then

$$x^m \equiv x \bmod p \quad \text{for all } x.$$

Similarly, if $m \equiv 1 \bmod (q-1)$, then $x^m \equiv x \bmod q$ for all x. Therefore, if m is chosen such that

$$m \equiv 1 \bmod (p-1)(q-1),$$

then $m \equiv 1 \bmod (p-1)$ and $m \equiv 1 \bmod (q-1)$. Thus, $x^m \equiv x \bmod p$ and $x^m \equiv x \bmod q$; that is, $p \mid (x^m - x)$ and $q \mid (x^m - x)$. As p and q are distinct primes, they are relatively prime, and so $pq \mid (x^m - x)$, by Exercise 1.60 on page 54; that is, $x^m \equiv x$ mod pq Since $N = pq$, we have shown that if $m \equiv 1$ mod $(p-1)(q-1)$, then

$$x^m \equiv x \text{ mod } N \quad \text{for all } x.$$

It remains, given p and q, to find a number $m \equiv 1$ mod $(p-1)(q-1)$ and a factorization $m = st$. We claim that there is a factorization with $s = 3$. Let us first show that $\big(3, (p-1)(q-1)\big) = 1$. Since $p \equiv 2$ mod 3 and $q \equiv 2$ mod 3, we have $p - 1 \equiv 1$ mod 3 and $q - 1 \equiv 1$ mod 3; hence, $(p-1)(q-1) \equiv 1$ mod 3, so that 3 and $(p-1)(q-1)$ are relatively prime (Proposition 1.34). Thus, there are integers t and u with $1 = 3t + (p-1)(q-1)u$, so that $3t \equiv 1$ mod $(p-1)(q-1)$. To sum up, $x^{3t} \equiv x$ mod N for all x with this choice of t. Choosing $m = 3t$ completes the construction of the ingredients of the code.

Security

Since $3t \equiv 1$ mod $(p-1)(q-1)$, he who knows the factorization $N = pq$ knows the number $(p-1)(q-1)$, and hence he can find t using the Euclidean algorithm. Unauthorized readers may know N, but without knowing its factorization, they do not know t and, hence, they cannot decode. This is why this code is secure today. For example, if both p and q have about 200 digits (and, for technical reasons, they are not too close together), then the fastest existing computers need two or three months to factor N. By Proposition 1.62, there are plenty of primes congruent to 2 mod 3, and so we may choose a different pair of primes p and q every month, say, thereby stymying the enemy. ◄

EXERCISES

H **1.77** True or false with reasons.
   (i)   If a and m are integers with $m > 0$, then $a \equiv i$ mod m for some integer i with $0 \le i \le m - 1$.
   (ii) If a, b and m are integers with $m > 0$, then $a \equiv b$ mod m implies $(a+b)^m \equiv a^m + b^m$ mod m.
   (iii) If a is an integer, then $a^6 \equiv a$ mod 6.
   (iv) If a is an integer, then $a^4 \equiv a$ mod 4.
   (v)   5263980007 is a perfect square.
   (vi) There is an integer n with $n \equiv 1$ mod 100 and $n \equiv 4$ mod 1000.
   (vii) There is an integer n with $n \equiv 1$ mod 100 and $n \equiv 4$ mod 1001.
   (viii) If p is a prime and $m \equiv n$ mod p, then $a^m \equiv a^n$ mod p for every natural number a.

1.78 Find all the integers x which are solutions to each of the following congruences:
   (i) $3x \equiv 2$ mod 5.
   (ii) $7x \equiv 4$ mod 10.

     (iii) $243x + 17 \equiv 101 \bmod 725$.

     (iv) $4x + 3 \equiv 4 \bmod 5$.

     (v) $6x + 3 \equiv 4 \bmod 10$.

     (vi) $6x + 3 \equiv 1 \bmod 10$.

H **1.79** Let m be a positive integer, and let m' be an integer obtained from m by rearranging its (decimal) digits (e.g., take $m = 314159$ and $m' = 539114$). Prove that $m - m'$ is a multiple of 9.

H **1.80** Prove that a positive integer n is divisible by 11 if and only if the alternating sum of its digits is divisible by 11 (if the digits of a are $d_k \ldots d_2 d_1 d_0$, then their **alternating sum** is $d_0 - d_1 + d_2 - \cdots$).

H **1.81** What is the remainder after dividing 10^{100} by 7? (The huge number 10^{100} is called a *googol* [21] in children's stories.)

*$**1.82**$ **(i)**   Prove that $10q + r$ is divisible by 7 if and only if $q - 2r$ is divisible by 7.

       (ii) Given an integer a with decimal digits $d_k d_{k-1} \ldots d_0$, define

$$a' = d_k d_{k-1} \cdots d_1 - 2d_0.$$

          Show that a is divisible by 7 if and only if some one of $a', a'', a''', \ldots$ is divisible by 7. (For example, if $a = 65464$, then $a' = 6546 - 8 = 6538$, $a'' = 653 - 16 = 637$, and $a''' = 63 - 14 = 49$; we conclude that 65464 is divisible by 7.)

*$**1.83**$ **(i)**   Show that $1000 \equiv -1 \bmod 7$.

       (ii) Show that if $a = r_0 + 1000 r_1 + 1000^2 r_2 + \cdots$, then a is divisible by 7 if and only if $r_0 - r_1 + r_2 - \cdots$ is divisible by 7.

Remark. Exercises 1.82 and 1.83 combine to give an efficient way to determine whether large numbers are divisible by 7. If $a = 33456789123987$, for example, then $a \equiv 0 \bmod 7$ if and only if $987 - 123 + 789 - 456 + 33 = 1230 \equiv 0 \bmod 7$. By Exercise 1.82 on page 75, $1230 \equiv 123 \equiv 6 \bmod 7$, and so a is not divisible by 7. ◄

 *$**1.84**$ For a given positive integer m, find all integers r with $0 < r < m$ such that $2r \equiv 0 \bmod m$.

H **1.85** Prove that there are no integers x, y, and z such that $x^2 + y^2 + z^2 = 999$.

H **1.86** Prove that there is no perfect square a^2 whose last two digits are 35.

  1.87 If x is an odd number not divisible by 3, prove that $x^2 \equiv 1 \bmod 24$.

*H **1.88** Prove that if p is a prime and if $a^2 \equiv 1 \bmod p$, then $a \equiv \pm 1 \bmod p$.

 *$**1.89**$ Consider the congruence $ax \equiv b \bmod m$ when $\gcd(a, m) = d$. Show that $ax \equiv b \bmod m$ has a solution if and only if $d \mid b$.

H **1.90** Solve the congruence $x^2 \equiv 1 \bmod 21$.

  1.91 Solve the simultaneous congruences:

     (i) $x \equiv 2 \bmod 5$ and $3x \equiv 1 \bmod 8$;

     (ii) $3x \equiv 2 \bmod 5$ and $2x \equiv 1 \bmod 3$.

H **1.92** Find the smallest positive integer which leaves remainder 4, 3, 1 after dividing by 5, 7, 9, respectively.

---

[21] This word was invented by a 9-year-old boy when his uncle asked him to think up a name for the number 1 followed by a hundred zeros. At the same time, the boy suggested *googolplex* for a 1 followed by a googol zeros.

1.93 How many days are there between Akbal 13 and Muluc 8 in the Mayan tzolkin calendar?

1.94 H **(i)**   Show that $(a + b)^n \equiv a^n + b^n$ mod 2 for all a and b and for all $n \geq 1$.

   (ii)   Show that $(a + b)^2 \not\equiv a^2 + b^2$ mod 3.

1.95 Solve the linear system

$$x \equiv 12 \text{ mod } 25$$
$$x \equiv 2 \text{ mod } 30.$$

*1.96 Let m, m' be positive integers, let $d = (m, m')$, and let $b \equiv b'$ mod d. Prove that any two solutions of the system

$$x \equiv b \text{ mod } m$$
$$x \equiv b' \text{ mod } m'$$

are congruent mod ℓ, where $\ell = \text{lcm}\{m, m'\}$.

H **1.97** On a desert island, five men and a monkey gather coconuts all day, then sleep. The first man awakens and decides to take his share. He divides the coconuts into five equal shares, with one coconut left over. He gives the extra one to the monkey, hides his share, and goes to sleep. Later, the second man awakens and takes his fifth from the remaining pile; he too finds one extra and gives it to the monkey. Each of the remaining three men does likewise in turn. Find the minimum number of coconuts originally present.

1.6   DATES AND DAYS

Congruences can be used to determine on which day of the week a given date falls. For example, on what day of the week was July 4, 1776?

A *year* is the amount of time it takes the Earth to make one complete orbit around the Sun; a *day* is the amount of time it takes the Earth to make a complete rotation about the axis through its north and south poles. There is no reason why the number of days in a year should be an integer, and it is not; a year is approximately 365.2422 days long. In 46 B.C., Julius Caesar (and his scientific advisors) compensated for this by creating the *Julian calendar*, containing a *leap year* every 4 years; that is, every fourth year has an extra day, namely, February 29, and so it contains 366 days (a *common year* is a year that is not a leap year). This would be fine if the year were exactly 365.25 days long, but it has the effect of making the year $365.25 - 365.2422 = .0078$ days (about 11 minutes and 14 seconds) too long. After 128 years, a full day was added to the calendar; that is, the Julian calendar overcounted the number of days. In the year 1582, the vernal equinox (the Spring day on which there are exactly 12 hours of daylight and 12 hours of night) occurred on March 11 instead of on March 21. Pope Gregory XIII (and his scientific advisors) then installed the *Gregorian calendar* by erasing 10 days that year; the day after October 4, 1582 was October 15, 1582, and this caused confusion and fear among the people. The Gregorian calendar modified the Julian calendar as follows. Call a year y ending in 00 a *century year*. If a year y

is not a century year, then it is a leap year if it is divisible by 4; if y is a century year, it is a leap year only if it is divisible by 400. For example, 1900 is not a leap year, but 2000 is a leap year. The Gregorian calendar is the one in common use today, but it was not uniformly adopted throughout Europe. For example, the British did not accept it until 1752, when 11 days were erased, and the Russians did not accept it until 1918, when 13 days were erased (thus, the Russians called their 1917 revolution the October Revolution, even though it occurred in November of the Gregorian calendar).

The true number of days in 400 years is about

$$400 \times 365.2422 = 146096.88 \text{ days.}$$

In this period, the Julian calendar has

$$400 \times 365 + 100 = 146,100 \text{ days,}$$

while the Gregorian calendar has 146,097 days (it eliminated 3 leap years from this time period). Thus, the Julian calendar gains about 3.12 days every 400 years, while the Gregorian calendar gains only 0.12 days (about 2 hours and 53 minutes).

A little arithmetic shows that there are 1628 years from 46 B.C. to 1582. The Julian calendar overcounts one day every 128 years, and so it overcounted 13 days in this period (for $13 \times 128 = 1662$). Why didn't Gregory have to erase 13 days? The Council of Nicaea, meeting in the year 325, defined Easter as the first Sunday strictly after the Paschal full moon, which is the first full moon on or after the vernal equinox. The vernal equinox in 325 fell on March 21, and the Synod of Whitby, in 664, officially defined the vernal equinox to be March 21. The discrepancy observed in 1582 was thus the result of only $1257 = 1582 - 325$ years of the Julian calendar: approximately 10 days.

Let us now seek a calendar formula. For easier calculation, we choose 0000 as our reference year, even though there was no year zero! Assign a number to each day of the week, according to the following scheme:

Sun	Mon	Tues	Wed	Thurs	Fri	Sat
0	1	2	3	4	5	6

In particular, March 1, 0000, has some number a, where $0 \le a \le 6$. In the next year 0001, March 1 has number $a + 1 \pmod 7$, for 365 days have elapsed from March 1, 0000, to March 1, 0001, and

$$365 = 52 \times 7 + 1 \equiv 1 \bmod 7.$$

Similarly, March 1, 0002, has number $a + 2$, and March 1, 0003, has number $a + 3$. However, March 1, 0004, has number $a + 5$, for February 29, 0004, fell between March 1, 0003, and March 1, 0004, and so $366 \equiv 2 \bmod 7$ days had elapsed since the previous March 1. We see, therefore, that every common year adds 1 to the previous number for

March 1, while each leap year adds 2. Thus, if March 1, 0000, has number a, then the number a' of March 1, year y, is

$$a' \equiv a + y + L \bmod 7,$$

where L is the number of leap years from year 0000 to year y. To compute L, count all those years divisible by 4, then throw away all the century years, and then put back those century years that are leap years. Thus,

$$L = \lfloor y/4 \rfloor - \lfloor y/100 \rfloor + \lfloor y/400 \rfloor,$$

where $\lfloor x \rfloor$ denotes the greatest integer in x. Therefore, we have

$$a' \equiv a + y + L$$
$$\equiv a + y + \lfloor y/4 \rfloor - \lfloor y/100 \rfloor + \lfloor y/400 \rfloor \bmod 7.$$

We can actually find a' by looking at a calendar. Since March 1, 1994, fell on a Tuesday,

$$2 \equiv a + 1994 + \lfloor 1994/4 \rfloor - \lfloor 1994/100 \rfloor + \lfloor 1994/400 \rfloor$$
$$\equiv a + 1994 + 498 - 19 + 4 \bmod 7,$$

and so

$$a \equiv -2475 \equiv -4 \equiv 3 \bmod 7$$

(that is, March 1, year 0000, fell on Wednesday). One can now determine the day of the week on which March 1 will fall in any year $y > 0$, namely, the day corresponding to

$$3 + y + \lfloor y/4 \rfloor - \lfloor y/100 \rfloor + \lfloor y/400 \rfloor \bmod 7.$$

There is a reason we have been discussing March 1. Had Julius Caesar decreed that the extra day of a leap year be December 32 instead of February 29, life would have been simpler.[22] Let us now analyze February 28. For example, suppose that February 28, 1600, has number b. As 1600 is a leap year, February 29, 1600, occurs between February 28, 1600, and February 28, 1601; hence, 366 days have elapsed between

---

[22]Actually, March 1 was the first day of the year in the old Roman calendar. This explains why the leap day was added onto February and not onto some other month. It also explains why months 9, 10, 11, and 12, namely, September, October, November, and December, are so named; originally, they were months 7, 8, 9, and 10.

George Washington's birthday, in the Gregorian calendar, is February 22, 1732. But the Gregorian calendar was not introduced in the British colonies until 1752. Thus, his original birthday was February 11. But New Year's Day was also changed from March 1 to January 1, so that February, which had been in 1731, was regarded, after the calendar change, as being in 1732. George Washington used to joke that not only did his birthday change, but so did his birth year. See Exercise 1.102 on page 83.

these two February 28s, so that February 28, 1601, has number $b + 2$. February 28, 1602, has number $b + 3$, February 28, 1603, has number $b + 4$, February 28, 1604, has number $b + 5$, but February 28, 1605, has number $b + 7$ (for there was a February 29 in 1604).

Let us compare the pattern of behavior of February 28, 1600, namely, b, $b + 2$, $b + 3$, $b + 4$, $b + 5$, $b + 7$, ..., with that of some date in 1599. If May 26, 1599, has number c, then May 26, 1600, has number $c + 2$, for February 29, 1600, comes between these two May 26s, and so there are $366 \equiv 2 \bmod 7$ intervening days. The numbers of the next few May 26s, beginning with May 26, 1601, are $c + 3$, $c + 4$, $c + 5$, $c + 7$. We see that the pattern of the days for February 28, starting in 1600, is exactly the same as the pattern of the days for May 26, starting in 1599; indeed, the same is true for any date in January or February. Thus, the pattern of the days for any date in January or February of a year y is the same as the pattern for a date occurring in the preceding year $y - 1$: a year preceding a leap year adds 2 to the number for such a date, whereas all other years add 1. Therefore, we revert to the ancient calendar by making New Year's Day fall on March 1; any date in January or February is treated as if it had occurred in the previous year.

How do we find the day corresponding to a date other than March 1? Since March 1, 0000, has number 3 (as we have seen above), April 1, 0000, has number 6, for March has 31 days and $3 + 31 \equiv 6 \bmod 7$. Since April has 30 days, May 1, 0000, has number $6 + 30 \equiv 1 \bmod 7$. Here is the table giving the number of the first day of each month in year 0000 (see Table 1.2).

Date	Number	Date	Number	Date	Number
March 1	3	July 1	6	November 1	3
April 1	6	August 1	2	December 1	5
May 1	1	September 1	5	January 1	1
June 1	4	October 1	0	February 1	4

Table 1.2 First day of the month.

Remember that we are pretending that March is month 1, April is month 2, and so on. Let us denote these numbers by $1 + j(m)$, where $j(m)$, for $m = 1, 2, \ldots, 12$, is defined by

$$j(m) : 2, 5, 0, 3, 5, 1, 4, 6, 2, 4, 0, 3.$$

It follows that month m, day 1, year y, has number

$$1 + j(m) + g(y) \bmod 7,$$

where

$$g(y) = y + \lfloor y/4 \rfloor - \lfloor y/100 \rfloor + \lfloor y/400 \rfloor.$$

Proposition 1.80 (Calendar[23] Formula).   *The date with month m, day d, year y has number*

$$d + j(m) + g(y) \bmod 7,$$

where

$$j(m) = 2, 5, 0, 3, 5, 1, 4, 6, 2, 4, 0, 3,$$

(March corresponds to m = 1, April to m = 2, ..., February to m = 12) and

$$g(y) = y + \lfloor y/4 \rfloor - \lfloor y/100 \rfloor + \lfloor y/400 \rfloor,$$

provided that dates in January and February are treated as having occurred in the previous year.

Proof.   The number mod 7 corresponding to month m, day 1, year y, is $1 + j(m) + g(y)$. It follows that $2 + j(m) + g(y)$ corresponds to month m, day 2, year y, and, more generally, that $d + j(m) + g(y)$ corresponds to month m, day d, year y.   •

Example 1.81.
Let us use the calendar formula to find the day of the week on which July 4, 1776, fell. Here $m = 5$, $d = 4$, and $y = 1776$. Substituting in the formula, we obtain the number

$$4 + 5 + 1776 + 444 - 17 + 4 = 2216 \equiv 4 \bmod 7;$$

therefore, July 4, 1776, fell on a Thursday.   ◀

Most of us need paper and pencil (or a calculator) to use the calendar formula in the theorem. Here are some ways to simplify the formula so that one can do the calculation in one's head and amaze one's friends.

One mnemonic for $j(m)$ is given by

$$j(m) = \lfloor 2.6m - 0.2 \rfloor, \quad \text{where } 1 \le m \le 12.$$

Another mnemonic for $j(m)$ is the sentence

My Uncle Charles has eaten a cold supper; he eats nothing hot.
2   5   (7 ≡ 0)   3   5   1   4   6   2   4   (7 ≡ 0)   3

Corollary 1.82.   *The date with month m, day d, year y = 100C + N, where 0 ≤ N ≤ 99, has number*

$$d + j(m) + N + \lfloor N/4 \rfloor + \lfloor C/4 \rfloor - 2C \bmod 7,$$

provided that dates in January and February are treated as having occurred in the previous year.

---

[23] The word *calendar* comes from the Greek "to call," which evolved into the Latin for the first day of a month (when accounts were due).

Proof. If we write a year $y = 100C + N$, where $0 \leq N \leq 99$, then

$$y = 100C + N \equiv 2C + N \text{ mod } 7,$$
$$\lfloor y/4 \rfloor = 25C + \lfloor N/4 \rfloor \equiv 4C + \lfloor N/4 \rfloor \text{ mod } 7,$$
$$\lfloor y/100 \rfloor = C, \text{ and } \lfloor y/400 \rfloor = \lfloor C/4 \rfloor.$$

Therefore,

$$y + \lfloor y/4 \rfloor - \lfloor y/100 \rfloor + \lfloor y/400 \rfloor \equiv N + 5C + \lfloor N/4 \rfloor + \lfloor C/4 \rfloor \text{ mod } 7$$
$$\equiv N + \lfloor N/4 \rfloor + \lfloor C/4 \rfloor - 2C \text{ mod } 7. \quad \bullet$$

This formula is simpler than the first one. For example, the number corresponding to July 4, 1776, is now obtained as

$$4 + 5 + 76 + 19 + 4 - 34 = 74 \equiv 4 \text{ mod } 7,$$

agreeing with the calculation in Example 1.81. The reader may now compute the day of his or her birth.

Example 1.83.
The birthday of Amalia, the grandmother of Danny and Ella, is December 5, 1906; on what day of the week was she born?

If A is the number of the day, then

$$A \equiv 5 + 4 + 6 + \lfloor 6/4 \rfloor + \lfloor 19/4 \rfloor - 38$$
$$\equiv -18 \text{ mod } 7$$
$$\equiv \quad 3 \text{ mod } 7.$$

Amalia was born on a Wednesday. ◄

Does every year y contain a Friday 13? We have

$$5 \equiv 13 + j(m) + g(y) \text{ mod } 7.$$

The question is answered positively if the numbers $j(m)$, as m varies from 1 through 12, give all the remainders 0 through 6 mod 7. And this is what happens. The sequence of remainders mod 7 is

$$2, 5, \boxed{0}, \boxed{3}, \boxed{5}, \boxed{1}, \boxed{4}, \boxed{6}, \boxed{2}, 4, 0, 3.$$

Indeed, we see that there must be a Friday 13 occurring between May and November. No number occurs three times on the list, but it is possible that there are three Friday 13's in a year because January and February are viewed as having occurred in the previous year; for example, there were three Friday 13s in 1987 (see Exercise 1.101 on page 83). Of course, we may replace Friday by any other day of the week, and we may replace 13 by any number between 1 and 28.

J. H. Conway has found an even simpler calendar formula. In his system, he calls *doomsday* of a year that day of the week on which the last day of February occurs. We can compute doomsday using Corollary 1.82.

Knowing the doomsday of a century year $100C$, one can find the doomsday of any other year $y = 100C + N$ in that century, as follows. Since $100C$ is a century year, the number of leap years from $100C$ to y does not involve the Gregorian alteration. Thus, if D is doomsday $100C$ (of course, $0 \leq D \leq 6$), then doomsday $100C + N$ is congruent to

$$D + N + \lfloor N/4 \rfloor \bmod 7.$$

For example, since doomsday 1900 is Wednesday = 3, we see that doomsday 1994 is Monday = 1, for

$$3 + 94 + 23 = 120 \equiv 1 \bmod 7.$$

February 29, 1600	2	Tuesday
February 28, 1700	0	Sunday
February 28, 1800	5	Friday
February 28, 1900	3	Wednesday
February 29, 2000	2	Tuesday

Table 1.3   Doomsdays

Proposition 1.84 (*Conway*).   *Let D be doomsday $100C$, and let $0 \leq N \leq 99$. If $N = 12q + r$, where $0 \leq r < 12$, then the formula for doomsday $100C + N$ is*

$$D + q + r + \lfloor r/4 \rfloor \bmod 7.$$

Proof.

$$\begin{aligned} \text{Doomsday } (100C + N) &\equiv D + N + \lfloor N/4 \rfloor \\ &\equiv D + 12q + r + \lfloor (12q + r)/4 \rfloor \\ &\equiv D + 15q + r + \lfloor r/4 \rfloor \\ &\equiv D + q + r + \lfloor r/4 \rfloor \bmod 7 \quad \bullet \end{aligned}$$

For example, $94 = 12 \times 7 + 10$, so that doomsday 1994 is $3 + 7 + 10 + 2 \equiv 1 \bmod 7$; that is, doomsday 1994 is Monday, as we saw above.

Once one knows doomsday of a particular year, one can use various tricks (e.g., my Uncle Charles) to pass from doomsday to any other day in the year. Conway observes that some other dates falling on the same day of the week as the doomsday are

April 4, June 6, August 8, October 10, December 12,

May 9, July 11, September 5, and November 7.

If we return to the usual counting having January as the first month: 1 = January), then it is easier to remember these dates using the notation

4/4, 6/6, 8/8, 10/10, 12/12, and 5/9, 7/11, 9/5, and 11/7,

where m/d denotes month/day. Since doomsday corresponds to the last day of February, we are now within a few weeks of any date in the calendar, and we can easily interpolate to find the desired day.

EXERCISES

H **1.98** A suspect said that he had spent the Easter holiday April 21, 1893, with his ailing mother; Sherlock Holmes challenged his veracity at once. How could the great detective have been so certain?

H **1.99** How many times in 1900 did the first day of a month fall on a Tuesday?

H **1.100** On what day of the week did February 29, 1896 fall? Conclude from your method of solution that no extra fuss is needed to find leap days.

*1.101   **(i)**   Show that 1987 had three Friday 13s.

        (ii)   Show, for any year $y > 0$, that $g(y) - g(y - 1) = 1$ or 2, where $g(y) = y + \lfloor y/4 \rfloor - \lfloor y/100 \rfloor + \lfloor y/400 \rfloor$.

      H **(iii)**   Can there be a year with exactly one Friday 13?

*H **1.102** My Uncle Ben was born in Pogrebishte, a village near Kiev, and he claimed that his birthday was February 29, 1900. I told him that this could not be, for 1900 was not a leap year. Why was I wrong?

2

Groups I

Group theory was invented by E. Galois (1811–1832) in order to solve one of the premiere mathematical problems of his day: when can the roots of a polynomial be found by some generalization of the quadratic formula? Since Galois (who was killed in a duel when he was only 20 years old), group theory has found many other applications. For example, we shall give a new proof of Fermat's theorem (if p is prime, then $a^p \equiv a \bmod p$), and this proof will then be adapted to prove a theorem of Euler: if $m \geq 2$, then $a^{\phi(m)} \equiv 1 \bmod m$, where $\phi(m)$ is the Euler ϕ-function. We will also use groups to solve counting problems such as: how many different bracelets having 10 beads can be assembled from a pile containing 10 red beads, 10 white beads, and 10 blue beads? In Chapter 6, we will illustrate the fact that groups are a precise way to describe symmetry by classifying all possible friezes in the plane.

$\rightarrow$ 2.1 SOME SET THEORY

A *group* is a set whose elements can be "multiplied" and whose multiplication obeys certain rules. Important examples of groups have elements which are permutations, and permutations are certain functions. Moreover, one compares two groups using functions called *homomorphisms*. Thus, this section contains definitions and basic properties of functions. The reader who has seen this material before may skim this section now and return to it later when necessary.

A set X is a collection of elements (numbers, points, herring, etc.); one writes

$$x \in X$$

to denote x belonging to X. The terms *set*, *element*, and *belongs to* are undefined terms (there have to be such in any language), and they are used so that a set is determined by the elements in it.[1] Thus, we define two sets X and Y to be *equal*, denoted by

$$X = Y,$$

if they are comprised of exactly the same elements: for every element x, we have $x \in X$ if and only if $x \in Y$.

A *subset* of a set X is a set S each of whose elements also belongs to X: if $s \in S$, then $s \in X$. One denotes S being a subset of X by

$$S \subseteq X;$$

synonyms for this are S is *contained* in X and S is *included* in X. Note that $X \subseteq X$ is always true; we say that a subset S of X is a *proper subset* of X, denoted by $S \subsetneq X$, if $S \subseteq X$ and $S \neq X$. It follows from these definitions that two sets X and Y are equal if and only if each is a subset of the other:

$$X = Y \qquad \text{if and only if} \qquad X \subseteq Y \text{ and } Y \subseteq X.$$

Because of this remark, many proofs showing that two sets are equal break into two parts, each half showing that one of the sets is a subset of the other. For example, let

$$X = \{a \in \mathbb{R} : a \geq 0\} \qquad \text{and} \qquad Y = \{b \in \mathbb{R} : b = r^2 \text{ for some } r \in \mathbb{R}\}.$$

If $a \in X$, then $a \geq 0$ and $a = r^2$, where $r = \sqrt{a}$; hence, $a \in Y$ and $X \subseteq Y$. For the reverse inclusion, choose $b \in Y$, so that $b = r^2 \in Y$ for some $r \in \mathbb{R}$. If $r \geq 0$, then $r^2 \geq 0$; if $r < 0$, then $r = -s$, where $s > 0$, and $r^2 = (-1)^2 s^2 = s^2 \geq 0$. In either case, $b = r^2 \geq 0$ and $b \in X$. Therefore, $Y \subseteq X$, and so $X = Y$.

→ **Definition.**   The *empty set* is the set $\varnothing$ having no elements.

We claim, for every set X, that $\varnothing \subseteq X$. The negation of "If $s \in \varnothing$, then $s \in X$" is "There exists $s \in \varnothing$ with $s \notin X$"; as there is no $s \in \varnothing$, however, this cannot be true. It follows that there is a unique empty set, for if $\varnothing_1$ were a second such, then $\varnothing \subseteq \varnothing_1$ and, similarly, $\varnothing_1 \subseteq \varnothing$. Therefore, $\varnothing = \varnothing_1$.

Here are some ways to create new sets from old; see Figure 2.1.

→ **Definition.**   If X and Y are subsets of a set Z, then their *intersection* is the set

$$X \cap Y = \{z \in Z : z \in X \text{ and } z \in Y\}.$$

More generally, if $\{A_i : i \in I\}$ is any, possibly infinite, family of subsets of a set Z, then their *intersection* is

$$\bigcap_{i \in I} A_i = \{z \in Z : z \in A_i \text{ for all } i \in I\}.$$

---

[1] There are some rules governing the usage of $\in$; for example, $x \in a \in x$ is always a false statement.

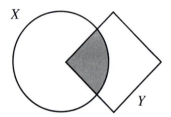 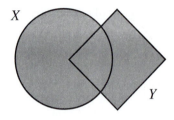

Figure 2.1 $X \cap Y$ $X \cup Y$

It is clear that $X \cap Y \subseteq X$ and $X \cap Y \subseteq Y$. In fact, the intersection is the largest such subset: if $S \subseteq X$ and $S \subseteq Y$, then $S \subseteq X \cap Y$. Similarly, $\bigcap_{i \in I} A_i \subseteq A_j$ for all $j \in I$.

→ **Definition.**   If X and Y are subsets of a set Z, then their ***union*** is the set

$$X \cup Y = \{z \in Z : z \in X \text{ or } z \in Y\}.$$

More generally, if $\{A_i : i \in I\}$ is any, possibly infinite, family of subsets of a set Z, then their ***union*** is

$$\bigcup_{i \in I} A_i = \{z \in Z : z \in A_i \text{ for some } i \in I\}.$$

It is clear that $X \subseteq X \cup Y$ and $Y \subseteq X \cup Y$. In fact, the union is the smallest such subset: if $X \subseteq S$ and $Y \subseteq S$, then $X \cup Y \subseteq S$. Similarly, $A_j \subseteq \bigcup_{i \in I} A_i$ for all $j \in I$.

→ **Definition.**   If X and Y are sets, then their ***difference*** is the set

$$X - Y = \{x \in X : x \notin Y\}.$$

The difference $Y - X$ has a similar definition and, of course, $Y - X$ and $X - Y$ have no elements in common: $(Y - X) \cap (X - Y) = \varnothing$ (see Figures 2.2 and 2.3).

In particular, if X is a subset of a set Z, then its ***complement*** in Z is the set

$$X' = Z - X = \{z \in Z : z \notin X\}.$$

It is clear that X' is ***disjoint*** from X; that is, there is no element $z \in Z$ lying in both X and X', so that $X \cap X' = \varnothing$. (Thus, the empty set $\varnothing$ is needed to guarantee that the intersection of two subsets A and B always be a subset; that is, $A \cap B$ should always be defined.) In fact, X' is the largest subset of Z disjoint from X: if $S \subseteq Z$ and $S \cap X = \varnothing$, then $S \subseteq X'$.

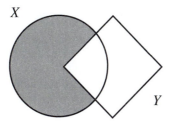

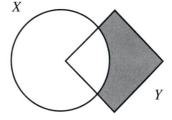

Figure 2.2 $X - Y$ **Figure 2.3** $Y - X$

→ **Functions**

The idea of a *function* occurs in calculus (and earlier); examples are x^2, $\sin x$, $\sqrt{x}$, $1/x$, $x + 1$, e^x, etc. Calculus books define a function $f(x)$ as a "rule" that assigns, to each number a, exactly one number, namely, $f(a)$. Thus, the squaring function assigns the number 81 to the number 9; the square root function assigns the number 3 to the number 9. Notice that there are two candidates for $\sqrt{9}$, namely, 3 and -3. In order that there be exactly one number assigned to 9, one must select one of the two possible values ± 3; everyone has agreed that $\sqrt{x} \geq 0$ whenever $x \geq 0$, and so this agreement implies that $\sqrt{x}$ is a function.

The calculus definition of function is in the right spirit, but it has a defect: what is a rule? To ask this question another way, when are two rules the same? For example, consider the functions

$$f(x) = (x + 1)^2 \qquad \text{and} \qquad g(x) = x^2 + 2x + 1.$$

Is $f(x) = g(x)$? The evaluation procedures are certainly different: for example, $f(6) = (6 + 1)^2 = 7^2$, while $g(6) = 6^2 + 2 \cdot 6 + 1 = 36 + 12 + 1$. Since the term *rule* has not been defined, it is ambiguous, and our question cannot be answered. Surely the calculus description is inadequate if one cannot decide whether these two functions are equal.

To find a reasonable definition, let us return to examples of what we seek to define. Each of the functions x^2, $\sin x$, and so on, has a *graph*, namely, the subset of the plane consisting of all those points of the form $(a, f(a))$. For example, the graph of $f(x) = x^2$ is the parabola consisting of all the points of the form (a, a^2).

A graph is a concrete thing, and the upcoming formal definition of a function amounts to saying that a function *is* its graph. The informal calculus definition of a function as a rule remains, but we will have avoided the problem of saying what a rule is. In order to give the definition, we first need an analog of the plane (for we will want to use functions $f(x)$ whose argument x does not vary over numbers).

→ **Definition.**   If X and Y are (not necessarily distinct) sets, then their ***cartesian***[2] ***product*** $X \times Y$ is the set of all ordered pairs (x, y), where $x \in X$ and $y \in Y$.

The plane is $\mathbb{R} \times \mathbb{R}$.
The only thing one needs to know about ordered pairs is that

$$(x, y) = (x', y') \quad \text{if and only if} \quad x = x' \text{ and } y = y'$$

(see Exercise 2.5 on page 104).

Observe that if X and Y are finite sets, say, $|X| = m$ and $|Y| = n$ (we denote the number of elements in a finite set X by $|X|$), then $|X \times Y| = mn$.

→ **Definition.**   Let X and Y be (not necessarily distinct) sets. A ***function*** f from X to Y, denoted by

$$f : X \to Y,$$

is a subset $f \subseteq X \times Y$ such that, for each $a \in X$, there is a unique $b \in Y$ with $(a, b) \in f$.

For each $a \in X$, the unique element $b \in Y$ for which $(a, b) \in f$ is called the ***value*** of f at a, and b is denoted by $f(a)$. Thus, f consists of all those points in $X \times Y$ of the form $(a, f(a))$. When $f : \mathbb{R} \to \mathbb{R}$, then f *is the graph of* $f(x)$.

→ **Example 2.1.**

(i) If X is a set, then the ***identity function*** on X, denoted by $1_X : X \to X$, is defined by $1_X(x) = x$ for every $x \in X$ [when $X = \mathbb{R}$, the graph of the identity function is the 45° line through the origin consisting of all those points in the plane of the form (a, a)].

(ii) ***Constant functions***: If $y_0 \in Y$, then $f(x) = y_0$ for all $x \in X$ (when $X = \mathbb{R} = Y$, then the graph of a constant function is a horizontal line).   ◄

From now on, we depart from the calculus notation; we denote a function by f and not by $f(x)$; the latter notation is reserved for the value of f at an element x (there are a few exceptions; we will continue to write the familiar functions, e.g., polynomials, $\sin x$, e^x, $\sqrt{x}$, $\log x$, as usual). Here are some more words. If $f : X \to Y$, call X the ***domain*** of f, call Y the ***target*** (or *codomain*) of f, and define the ***image*** (or *range*) of f, denoted by im f, to be the subset of Y consisting of all the values of f. When we say that X is the domain of a function $f : X \to Y$, we mean that $f(x)$ is defined for every $x \in X$. For example, the domain of $\sin x$ is $\mathbb{R}$, its target is usually $\mathbb{R}$, and its image is $[-1, 1]$. The domain of $1/x$ is the set of all nonzero reals and its image is also the nonzero reals; the domain of the square root function is the set $\mathbb{R}^{\geq} = \{x \in \mathbb{R} : x \geq 0\}$ of all nonnegative reals and its image is also $\mathbb{R}^{\geq}$.

---

[2]This term honors R. Descartes, one of the founders of analytic geometry.

→ **Definition.**   Functions $f \colon X \to Y$ and $g \colon X' \to Y'$ are *equal* if $X = X'$, $Y = Y'$, and the subsets $f \subseteq X \times Y$ and $g \subseteq X' \times Y'$ are equal.

A function $f \colon X \to Y$ has three ingredients–its domain X, its target Y, and its graph–and we are saying that two functions are equal if and only if they have the same domains, the same targets, and the same graphs. It is plain that the domain and the graph are essential parts of a function, and some reasons for caring about the target are given in a remark at the end of this section.

→ **Definition.**   If $f \colon X \to Y$ is a function, and if S is a subset of X, then the *restriction* of f to S is the function $f|S \colon S \to Y$ defined by $(f|S)(s) = f(s)$ for all $s \in S$.

If S is a subset of a set X, define the *inclusion* $i \colon S \to X$ to be the function defined by $i(s) = s$ for all $s \in S$.

If S is a proper subset of X, then the inclusion i is not the identity function 1_S because its target is X, not S; it is not the identity function 1_X because its domain is S, not X. If S is a proper subset of X, then $f|S \neq f$ because they have different domains.

→ **Proposition 2.2.**   *Let $f \colon X \to Y$ and $g \colon X' \to Y'$ be functions. Then $f = g$ if and only if $X = X'$, $Y = Y'$, and $f(a) = g(a)$ for every $a \in X$.*

Remark.   This proposition resolves the problem raised by the ambiguous term *rule*. If $f, g \colon \mathbb{R} \to \mathbb{R}$ are given by $f(x) = (x+1)^2$ and $g(x) = x^2 + 2x + 1$, then $f = g$ because $f(a) = g(a)$ for every number a.   ◄

Proof.   Assume that $f = g$. Functions are subsets of $X \times Y$, and so $f = g$ means that each of f and g is a subset of the other (informally, we are saying that f and g have the same graph). If $a \in X$ and $(a, f(a)) \in f = g$, then $(a, f(a)) \in g$. But there is only one ordered pair in g with first coordinate a, namely, $(a, g(a))$ [because the definition of function says that g gives a unique value to a]. Therefore, $(a, f(a)) = (a, g(a))$, and equality of ordered pairs gives $f(a) = g(a)$, as desired.

Conversely, assume that $f(a) = g(a)$ for every $a \in X$. To see that $f = g$, it suffices to show that $f \subseteq g$ and $g \subseteq f$. Each element of f has the form $(a, f(a))$. Since $f(a) = g(a)$, we have $(a, f(a)) = (a, g(a))$, and hence $(a, f(a)) \in g$. Therefore, $f \subseteq g$. The reverse inclusion $g \subseteq f$ is proved similarly.   ●

Let us make the contrapositive explicit: if $f, g \colon X \to Y$ are functions that disagree at even one point, that is, if there is some $a \in X$ with $f(a) \neq g(a)$, then $f \neq g$.

We continue to regard a function f as a rule sending $x \in X$ to $f(x) \in Y$, but the precise definition is now available whenever we need it, as in Proposition 2.2. However, to reinforce our wanting to regard functions $f \colon X \to Y$ as dynamic things sending points in X to points in Y, we often write

$$f \colon x \mapsto y$$

instead of $f(x) = y$. For example, we may write $f : x \mapsto x^2$ instead of $f(x) = x^2$, and we may describe the identity function by $x \mapsto x$ for all x.

Example 2.3.

Our definitions allow us to treat a degenerate case. If X is a set, what are the functions $X \to \varnothing$? Note first that an element of $X \times \varnothing$ is an ordered pair (x, y) with $x \in X$ and $y \in \varnothing$; since there is no $y \in \varnothing$, there are no such ordered pairs, and so $X \times \varnothing = \varnothing$. Now a function $X \to \varnothing$ is a subset of $X \times \varnothing$ of a certain type; but $X \times \varnothing = \varnothing$, so there is only one subset, namely $\varnothing$, and hence at most one function, namely, $f = \varnothing$. The definition of function $X \to \varnothing$ says, for each $x \in X$, that there exists a unique $y \in \varnothing$ with $(x, y) \in f$. If $X \neq \varnothing$, then there exists $x \in X$ for which no such y exists (there are no elements y at all in $\varnothing$), and so f is not a function. Thus, if $X \neq \varnothing$, there are no functions from X to $\varnothing$. On the other hand, if $X = \varnothing$, we claim that $f = \varnothing$ is a function. Otherwise, the negation of the statement "f is a function" would be true: "there exists $x \in \varnothing$, etc." We need not go on; since $\varnothing$ has no elements in it, there is no way to complete the sentence so that it is a true statement. We conclude that $f = \varnothing$ is a function $\varnothing \to \varnothing$, and we declare it to be the identity function $1_\varnothing$.   ◀

There is a name for functions whose image is equal to the whole target.

→ **Definition.**   A function $f : X \to Y$ is *surjective* (or *onto*) if

$$\operatorname{im} f = Y.$$

Thus, f is surjective if, for each $y \in Y$, there is some $x \in X$ (probably depending on y) with $y = f(x)$.

Example 2.4.

(i) Of course, identity functions are surjections.

(ii) The sine function $\mathbb{R} \to \mathbb{R}$ is not surjective, for its image is $[-1, 1]$ which is a proper subset of its target $\mathbb{R}$.

(iii) The functions $x^2 : \mathbb{R} \to \mathbb{R}$ and $e^x : \mathbb{R} \to \mathbb{R}$ have target $\mathbb{R}$. Now $\operatorname{im} x^2$ consists of the nonnegative reals and $\operatorname{im} e^x$ consists of the positive reals, so that neither x^2 nor e^x is surjective.

(iv) Let $f : \mathbb{R} \to \mathbb{R}$ be defined by

$$f(a) = 6a + 4.$$

To see whether f is a surjection, we ask whether every $b \in \mathbb{R}$ has the form $b = f(a)$ for some a; that is, given b, can one find a so that

$$6a + 4 = b?$$

One can always solve this equation for a, obtaining $a = \frac{1}{6}(b - 4)$. Therefore, f is a surjection.

(v) Let $f : \mathbb{R} - \{\frac{3}{2}\} \to \mathbb{R}$ be defined by

$$f(a) = \frac{6a + 4}{2a - 3}.$$

To see whether f is a surjection, we seek a solution a for a given b: can we always solve

$$\frac{6a + 4}{2a - 3} = b?$$

This leads to the equation $a(6 - 2b) = -3b - 4$, which can be solved for a if $6 - 2b \neq 0$ [note that $(-3b - 4)/(6 - 2b) \neq 3/2$]. On the other hand, it suggests that there is no solution when $b = 3$ and, indeed, there is not: if $(6a+4)/(2a-3) = 3$, cross multiplying gives the false equation $6a+4 = 6a-9$. Thus, $3 \notin \operatorname{im} f$, and f is not a surjection. ◀

Instead of saying that the values of a function f are unique, one sometimes says that f is **single-valued**. For example, if $\mathbb{R}^{\geq}$ denotes the set of nonnegative reals, then $\sqrt{\ } : \mathbb{R}^{\geq} \to \mathbb{R}^{\geq}$ is a function because we have agreed that $\sqrt{a} \geq 0$ for every positive number a. On the other hand, $f(a) = \pm\sqrt{a}$ is not single-valued, and hence it is not a function.

The simplest way to verify whether an alleged function f is single-valued is to phrase uniqueness of values as an implication:

$$\text{if } a = a', \text{ then } f(a) = f(a').$$

Does the formula $g\left(\frac{a}{b}\right) = ab$ define a function $g : \mathbb{Q} \to \mathbb{Q}$? There are many ways to write a fraction; since $\frac{1}{2} = \frac{3}{6}$, we see that $g\left(\frac{1}{2}\right) = 1 \cdot 2 \neq 3 \cdot 6 = g\left(\frac{3}{6}\right)$, and so g is not a function. Had we said that the formula $g\left(\frac{a}{b}\right) = ab$ holds whenever $\frac{a}{b}$ is in lowest terms, then g would be a function.

The formula $f\left(\frac{a}{b}\right) = 3 \cdot \frac{a}{b}$ does define a function $f : \mathbb{Q} \to \mathbb{Q}$, for it is single-valued: if $\frac{a}{b} = \frac{a'}{b'}$, then $f\left(\frac{a}{b}\right) = f\left(\frac{a'}{b'}\right)$. To see this, note that $\frac{a}{b} = \frac{a'}{b'}$ gives $ab' = a'b$; hence, $3ab' = 3a'b$ and $3 \cdot \frac{a}{b} = 3 \cdot \frac{a'}{b'}$. Thus, f is a bona fide function.

The following definition gives another important property a function may have.

→ **Definition.** A function $f : X \to Y$ is **injective** (or *one-to-one*) if, whenever a and a' are distinct elements of X, then $f(a) \neq f(a')$. Equivalently, (the contrapositive states that) f is injective if, for every pair $a, a' \in X$, we have

$$f(a) = f(a') \text{ implies } a = a'.$$

The reader should note that being injective is the converse of being single-valued: f is single-valued if $a = a'$ implies $f(a) = f(a')$; f is injective if $f(a) = f(a')$ implies $a = a'$.

Most functions are neither injective nor surjective. For example, the squaring function $f : \mathbb{R} \to \mathbb{R}$, defined by $f(x) = x^2$, is neither.

Example 2.5.

(i) Identity functions 1_X are injective.

(ii) Let $f : \mathbb{R} - \{\frac{3}{2}\} \to \mathbb{R}$ be defined by

$$f(a) = \frac{6a + 4}{2a - 3}.$$

To check whether f is injective, suppose that $f(a) = f(b)$:

$$\frac{6a + 4}{2a - 3} = \frac{6b + 4}{2b - 3}.$$

Cross multiplying yields

$$12ab + 8b - 18a - 12 = 12ab + 8a - 18b - 12,$$

which simplifies to $26a = 26b$ and hence $a = b$. We conclude that f is injective. [We saw, in Example 2.4(v), that f is not surjective.]

(iii) Consider $f : \mathbb{R} \to \mathbb{R}$ given by $f(x) = x^2 - 2x - 3$. If we try to check whether f is an injection by looking at the consequences of $f(a) = f(b)$, as in part (ii), we arrive at the equation $a^2 - 2a = b^2 - 2b$; it is not instantly clear whether this forces $a = b$. Instead, we seek the roots of $f(x)$, which are 3 and -1. It follows that f is not injective, for $f(3) = 0 = f(-1)$; that is, there are two distinct numbers having the same value. ◀

Sometimes there is a way of combining two functions to form another function, their *composite*.

→ **Definition.**   If $f : X \to Y$ and $g : Y \to Z$ are functions (the target of f is the domain of g), then their **composite**, denoted by $g \circ f$, is the function $X \to Z$ given by

$$g \circ f : x \mapsto g(f(x));$$

that is, first evaluate f on x and then evaluate g on $f(x)$.

Composition is thus a two-step process: $x \mapsto f(x) \mapsto g(f(x))$. For example, the function $h : \mathbb{R} \to \mathbb{R}$, defined by $h(x) = e^{\cos x}$, is the composite $g \circ f$, where $f(x) = \cos x$ and $g(x) = e^x$. This factorization is plain as soon as one tries to evaluate, say, $h(\pi)$; one must first evaluate $f(\pi) = \cos\pi = -1$ and then evaluate $g(f(\pi)) = g(-1) = e^{-1}$ in order to evaluate $h(\pi)$. The chain rule in calculus is a formula for computing the derivative $(g \circ f)'$ in terms of g' and f':

$$(g \circ f)'(x) = g'(f(x)) \cdot f'(x).$$

If $f : \mathbb{N} \to \mathbb{N}$ and $g : \mathbb{N} \to \mathbb{R}$ are functions, then $g \circ f : \mathbb{N} \to \mathbb{R}$ is defined, but $f \circ g$ is not defined [for target$(g) = \mathbb{R} \neq \mathbb{N} = \text{domain}(f)$]. Even when $f : X \to Y$ and $g : Y \to X$, so that both composites $g \circ f$ and $f \circ g$ are defined, these composites need not be equal. For example, define $f, g : \mathbb{N} \to \mathbb{N}$ by $f : n \mapsto n^2$ and $g : n \mapsto 3n$; then $g \circ f : 2 \mapsto g(4) = 12$ and $f \circ g : 2 \mapsto f(6) = 36$. Hence, $g \circ f \neq f \circ g$.
    Given a set X, let

$$\mathcal{F}(X) = \{\text{all functions } X \to X\}.$$

The composite of two functions in $\mathcal{F}(X)$ is always defined, and it is, again, a function in $\mathcal{F}(X)$. As we have just seen, composition is not ***commutative***; that is, $f \circ g$ and $g \circ f$ need not be equal. Let us now show that composition is always *associative*.

→ **Lemma 2.6.**   *Composition of functions is associative: if*

$$f : X \to Y, \quad g : Y \to Z, \quad and \quad h : Z \to W$$

are functions, then

$$h \circ (g \circ f) = (h \circ g) \circ f.$$

Proof.   We show that the value of either composite on an element $a \in X$ is just $w = h(g(f(a)))$. If $x \in X$, then

$$h \circ (g \circ f) : x \mapsto (g \circ f)(x) = g(f(x)) \mapsto h(g(f(x))) = w,$$

and

$$(h \circ g) \circ f : x \mapsto f(x) \mapsto (h \circ g)(f(x)) = h(g(f(x))) = w.$$

It follows from Proposition 2.2 that the composites are equal.   •

    In light of this lemma, we need not write parentheses: the notation $h \circ g \circ f$ is unambiguous. Suppose that $f : X \to Y$ and $g : Z \to W$ are functions. If $Y \subseteq Z$, then some authors define the composite $h : X \to W$ by $h(x) = g(f(x))$. We do not allow composition if $Y \neq Z$. However, we can define h as the composite $h = g \circ i \circ f$, where $i : Y \to Z$ is the inclusion.
    The next result implies that the identity function 1_X behaves for composition in $\mathcal{F}(X)$ just as the number one does for multiplication of numbers.

→ **Lemma 2.7.** *If* $f: X \rightarrow Y$, *then* $1_Y \circ f = f = f \circ 1_X$.

Proof.   If $x \in X$, then

$$1_Y \circ f : x \mapsto f(x) \mapsto f(x)$$

and

$$f \circ 1_X : x \mapsto x \mapsto f(x). \quad \bullet$$

Are there "reciprocals" in $\mathcal{F}(X)$; that is, are there any functions f for which there is $g \in \mathcal{F}(X)$ with $f \circ g = 1_X$ and $g \circ f = 1_X$? The following discussion will allow us to answer this question.

→ **Definition.**   A function $f: X \rightarrow Y$ is **bijective** (or is a *one-one correspondence*) if it is both injective and surjective.

Example 2.8.

  (i) Identity functions are always bijections.

  (ii) Let $X = \{1, 2, 3\}$ and define $f: X \rightarrow X$ by

$$f(1) = 2, \quad f(2) = 3, \quad f(3) = 1.$$

It is easy to see that f is a bijection.   ◄

We can draw a picture of a function in the special case when X and Y are finite sets. Let $X = \{1, 2, 3, 4, 5\}$, let $Y = \{a, b, c, d, e\}$, and define $f: X \rightarrow Y$ by

$$f(1) = b; \qquad f(2) = e; \qquad f(3) = a; \qquad f(4) = b; \qquad f(5) = c.$$

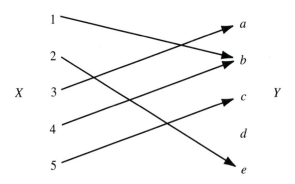

Figure 2.4   A Function

We see that f is not injective because $f(1) = b = f(4)$, and f is not surjective because there is no $x \in X$ with $f(x) = d$. Can one reverse the arrows to get a function

$g \colon Y \to X$? There are two reasons why one cannot. First, there is no arrow going to d, and so $g(d)$ is not defined. Second, what is $g(b)$? Is it 1 or 4? The first problem is that the domain of g is not all of Y, and it arises because f is not surjective; the second problem is that g is not single-valued, and it arises because f is not injective (this reflects the fact that being single-valued is the converse of being injective). Neither problem arises when f is a bijection.

→ **Definition.**   A function $f \colon X \to Y$ has an *inverse* if there exists a function $g \colon Y \to X$ with both composites $g \circ f$ and $f \circ g$ being identity functions.

We do not say that every function f has an inverse; on the contrary, we have just given some reasons why a function may not have an inverse. Notice that if an inverse function g does exist, then it "reverses the arrows" in Figure 2.4. If $f(a) = y$, then there is an arrow from a to y. Now $g \circ f$ being the identity says that $a = (g \circ f)(a) = g(f(a)) = g(y)$; therefore $g \colon y \mapsto a$, and so the picture of g is obtained from the picture of f by reversing arrows. If f twists something, then its inverse g untwists it.

Lemma 2.9.   *If $f \colon X \to Y$ and $g \colon Y \to X$ are functions such that $g \circ f = 1_X$, then f is injective and g is surjective.*

Proof.   Suppose that $f(a) = f(a')$; apply g to obtain $g(f(a)) = g(f(a'))$; that is, $a = a'$ [because $g(f(a)) = a$], and so f is injective. If $x \in X$, then $x = g(f(x))$, so that $x \in \operatorname{im} g$; hence g is surjective.   •

→ **Proposition 2.10.**   *A function $f \colon X \to Y$ has an inverse $g \colon Y \to X$ if and only if it is a bijection.*

Proof.   If f has an inverse g, then Lemma 2.9 shows that f is injective and surjective, for both composites $g \circ f$ and $f \circ g$ are identities.

Assume that f is a bijection. Let $y \in Y$. Since f is surjective, there is some $a \in X$ with $f(a) = y$; since f is injective, this element a is unique. Defining $g(y) = a$ thus gives a (single-valued) function whose domain is Y [g merely "reverses arrows:" since $f(a) = y$, there is an arrow from a to y, and the reversed arrow goes from y to a]. It is plain that g is the inverse of f; that is, $f(g(y)) = f(a) = y$ for all $y \in Y$ and $g(f(a)) = g(y) = a$ for all $a \in X$.   •

Notation.   The inverse of a bijection f is denoted by f^{-1} (Exercise 2.9 on page 105 says that a function cannot have two inverses). This is the same notation used for inverse trigonometric functions in calculus; for example, $\sin^{-1} x = \arcsin x$ satisfies $\sin(\arcsin(x)) = x$ and $\arcsin(\sin(x)) = x$. Of course, $\sin^{-1}$ does *not* denote the reciprocal $1/\sin x$, which is $\csc x$.

Example 2.11.
Here is an example of two functions f and g whose composite $g \circ f$ is the identity but whose composite $f \circ g$ is not the identity; thus, f and g are not inverse functions.
Define $f, g \colon \mathbb{N} \to \mathbb{N}$ as follows:

$$f(n) = n + 1;$$

$$g(n) = \begin{cases} 0 & \text{if } n = 0 \\ n - 1 & \text{if } n \geq 1. \end{cases}$$

The composite $g \circ f = 1_{\mathbb{N}}$, for $g(f(n)) = g(n+1) = n$ (because $n + 1 \geq 1$). On the other hand, $f \circ g \neq 1_{\mathbb{N}}$ because $f(g(0)) = f(0) = 1 \neq 0$. ◀

Example 2.12.
If a is a real number, then ***multiplication by a*** is the function $\mu_a \colon \mathbb{R} \to \mathbb{R}$ defined by $r \mapsto ar$ for all $r \in \mathbb{R}$. If $a \neq 0$, then μ_a is a bijection; its inverse function is ***division by a***, namely, $\delta_a \colon \mathbb{R} \to \mathbb{R}$, defined by $r \mapsto \frac{1}{a}r$; of course, $\delta_a = \mu_{1/a}$. If $a = 0$, however, then $\mu_a = \mu_0$ is the constant function $\mu_0 \colon r \mapsto 0$ for all $r \in \mathbb{R}$, which has no inverse function because it is not a bijection. ◀

Two strategies are now available to determine whether a given function is a bijection: use the definitions of injective and surjective, or find an inverse. For example, if $\mathbb{R}^+$ denotes the positive real numbers, let us show that the exponential function $f \colon \mathbb{R} \to \mathbb{R}^+$, defined by $f(x) = e^x = \sum x^n/n!$, is a bijection. A direct proof that f is an injection would require showing that if $e^a = e^b$, then $a = b$; a direct proof showing that f is surjective would involve showing that every positive real number c has the form e^a for some a. It is simplest to prove these statements using the (natural) logarithm $g(y) = \log y$. The usual formulas $e^{\log y} = y$ and $\log e^x = x$ say that both composites $f \circ g$ and $g \circ f$ are identities, and so f and g are inverse functions. Therefore, f is a bijection, for it has an inverse.

Let us summarize the results of this section.

→ **Proposition 2.13.**   *If the set of all the bijections from a set X to itself is denoted by S_X, then composition of functions satisfies the following properties:*

(i) *if $f, g \in S_X$, then $f \circ g \in S_X$;*

(ii) $h \circ (g \circ f) = (h \circ g) \circ f$ *for all $f, g, h \in S_X$;*

(iii) *the identity 1_X lies in S_X, and $1_X \circ f = f = f \circ 1_X$ for every $f \in S_X$;*

(iv) *for every $f \in S_X$, there is $g \in S_X$ with $g \circ f = 1_X = f \circ g$.*

Proof.   We have merely restated results of Exercise 2.14(ii) on page 105, Lemma 2.6, 2.7, and Proposition 2.10. •

Here is one interesting use of bijections. It is easy to prove (see Exercise 2.12 on page 105) that two finite sets X and Y have the same number of elements if and only if there is a bijection $f: X \to Y$. This suggests the following definition, due to G. Cantor (1845–1918).

Definition.   Two (possibly infinite) sets X and Y have the *same number of elements*, denoted by $|X| = |Y|$, if there exists a bijection $f: X \to Y$.

For example, a set X is called *countable* if either X is finite or X has the same number of elements as the natural numbers $\mathbb{N}$. If X is infinite and countable, then there is a bijection $f: \mathbb{N} \to X$; that is, there is a list $x_0, x_1, x_2, \ldots$, with no repetitions, of all the elements of X, where $x_n = f(n)$ for all $n \in \mathbb{N}$. Cantor proved that $\mathbb{R}$ is *uncountable*; that is, $\mathbb{R}$ is not countable. Thus, there are different sizes of infinity (in fact, there are infinitely many different sizes of infinity). The difference in size can be useful. For example, one calls a real number z *algebraic* if it is a root of some polynomial $f(x) = q_0 + q_1x + \cdots + q_nx^n$, all of whose coefficients $q_0, q_1, \ldots, q_n$ are rational; one calls z *transcendental* if it is not algebraic. Of course, every rational r is algebraic, for it is a root of $x - r$. But irrational algebraic numbers do exist; for example, $\sqrt{2}$ is algebraic, being a root of $x^2 - 2$. Are there any transcendental numbers? One can prove that there are only countably many algebraic numbers, and so it follows from Cantor's theorem, the uncountability of $\mathbb{R}$, that there exist (uncountably many) transcendental numbers.

Remark.   Why should we care about the target of a function when its image is more important? As a practical matter, when first defining a function, one usually does not know its image. For example, let $f: \mathbb{R} \to \mathbb{R}$ be defined by

$$f(x) = |x|^{e^{-x}} \sqrt[5]{x^2 + \sin^2 x}.$$

We must analyze f to find its image, and this is no small task. But if targets have to be images, then we could not even write down $f: X \to Y$ without having first found the image of f. Thus, targets are convenient to use.

Part of the definition of equality of functions is that their targets are equal; changing the target changes the function. Suppose we do not do this. Consider a function $f: X \to Y$ that is not surjective, let $Y' = \operatorname{im} f$, and define $g: X \to Y'$ by $g(x) = f(x)$ for all $x \in X$. The functions f and g have the same domain and the same values (i.e., the same graph); they differ only in their targets. Now g is surjective. Had we decided that targets are not a necessary ingredient in the definition of a function, then we would not be able to distinguish between f, which is not surjective, and g, which is. It would then follow that every function is a surjection (this would not shake the foundations of mathematics, but it would force us into using cumbersome circumlocutions). See Exercise 4.36 on page 382 for a more practical reason for using targets. ◄

If X and Y are sets, then a function $f : X \to Y$ defines a "forward motion" carrying the subsets of X into subsets of Y: if $S \subseteq X$, then

$$f(S) = \{y \in Y : y = f(s) \text{ for some } s \in S\}.$$

One calls $f(S)$ the ***direct image*** of S. A function f also defines a "backward motion" carrying the subsets of Y into subsets of X: if $W \subseteq Y$, then

$$f^{-1}(W) = \{x \in X : f(x) \in W\}.$$

We are not assuming that f is a bijection, and so f^{-1} does *not* mean the inverse function in this context. Here, $f^{-1}(W)$ means the set of all those elements in X, if any, which f sends into W. One calls $f^{-1}(W)$ the ***inverse image*** of W.

In Exercise 2.16 on page 105, it is shown that direct image preserves unions: if $f : X \to Y$ and if $\{S_i : i \in I\}$ is a family of subsets of X, then $f(\bigcup_{i \in I} S_i) = \bigcup_{i \in I} f(S_i)$. On the other hand, $f(S_1 \cap S_2) \neq f(S_1) \cap f(S_2)$ is possible. Exercise 2.17 on page 105 shows that inverse image is better behaved than direct image.

Proposition 2.14. *Let X and Y be sets, and let $f : X \to Y$ be a function.*

(i) *If $T \subseteq S$ are subsets of X, then $f(T) \subseteq f(S)$, and if $U \subseteq V$ are subsets of Y, then $f^{-1}(U) \subseteq f^{-1}(V)$.*

(ii) *If $U \subseteq Y$, then $ff^{-1}(U) \subseteq U$; if f is a surjection, then $ff^{-1}(U) = U$.*

(iii) *If $T \subseteq X$, then $T \subseteq f^{-1}f(T)$; if f is an injection, then $T \subseteq X$, then $W = f^{-1}f(T)$.*

Proof.
(i) If $y \in f(T)$, then $y = f(t)$ for some $t \in T$. But $t \in S$, because $T \subseteq S$, and so $f(t) \in f(S)$. Therefore, $f(T) \subseteq f(S)$. The other inclusion is proved just as easily.
(ii) If $a \in ff^{-1}(U)$, then $a = f(x')$ for some $x' \in f^{-1}(U)$; that is, $a = f(x') \in U$. We prove the reverse inclusion when f is surjective. If $u \in U$, then there is $x \in X$ with $f(x) = u$; hence, $x \in f^{-1}(U)$, and so $u = f(x) \in ff^{-1}(U)$.
(iii) If $t \in T$, then $f(t) \in f(T)$, and so $t \in f^{-1}f(t) \subseteq f^{-1}(T)$. We prove the reverse inclusion when f is injective. If $x \in f^{-1}f(T)$, then $f(x) \in f(T)$; hence, there is $t \in T$ with $f(x) = f(t)$. Since f is an injection, $x = t$ and $x \in T$.   •

Strict inequality is possible in Proposition 2.14(ii). If $f : \mathbb{Z} \to \mathbb{Q}$ is the inclusion, then

$$ff^{-1}(\{\tfrac{1}{2}\}) = f(\varnothing) = \varnothing \subsetneq \{\tfrac{1}{2}\}.$$

Strict inequality is also possible in Proposition 2.14(iii). Let $f : \mathbb{R} \to S^1$, where S^1 is the unit circle, be defined by $f(x) = e^{2\pi i x}$. If $A = \{0\}$, then $f(A) = \{1\}$ and

$$f^{-1}f(A) = f^{-1}f(\{0\}) = f^{-1}(\{1\}) = \mathbb{Z} \supsetneq A.$$

Corollary 2.15. *If $f: X \to Y$ is a surjection, then $B \mapsto f^{-1}(B)$ is an injection $\mathcal{P}(Y) \to \mathcal{P}(X)$, where $\mathcal{P}(Y)$ denotes the family of all the subsets of Y.*

Proof. If $B, C \subseteq Y$ and $f^{-1}(B) = f^{-1}(C)$, then Proposition 2.14(ii) gives

$$B = ff^{-1}(B) = ff^{-1}(C) = C. \quad \bullet$$

$\to$ Equivalence Relations

We are going to define the important notion of *equivalence relation*, but we begin with the general notion of *relation*.

$\to$ **Definition.** Given sets X and Y, a **relation from X to Y** is a subset R of $X \times Y$; if $X = Y$, then we say that R is a **relation on** X. One usually writes xRy instead of $(x, y) \in R$.

Here is a concrete example. Certainly, $\leq$ should be a relation on $\mathbb{R}$; to see that it is, define the relation

$$R = \{(x, y) \in \mathbb{R} \times \mathbb{R} : (x, y) \text{ lies on or above the line } y = x\}.$$

The reader should check that $x \leq y$ if and only if $(x, y) \in R$.

Example 2.16.

(i) Every function $f: X \to Y$ is a relation from X to Y.

(ii) Equality is a relation on any set X.

(iii) Congruence mod m is a relation on $\mathbb{Z}$. $\blacktriangleleft$

$\to$ **Definition.** A relation $x \equiv y$ on a set X is

reflexive if $x \equiv x$ for all $x \in X$;
symmetric if $x \equiv y$ implies $y \equiv x$ for all $x, y \in X$;
transitive if $x \equiv y$ and $y \equiv z$ imply $x \equiv z$ for all $x, y, z \in X$.

A relation on X that has all three properties–reflexivity, symmetry, and transitivity–is called an **equivalence relation** on X.

Example 2.17.

(i) Ordinary equality is an equivalence relation on any set.

(ii) If $m \geq 0$, then Proposition 1.57 says that $x \equiv y$ mod m is an equivalence relation on $X = \mathbb{Z}$.

(iii) Let $X = \{(a, b) \in \mathbb{Z} \times \mathbb{Z} : b \neq 0\}$, and define a relation $\equiv$ on X by cross multiplication:

$$(a, b) \equiv (c, d) \quad \text{if} \quad ad = bc.$$

We claim that $\equiv$ is an equivalence relation. Verification of reflexivity and symmetry is easy. For transitivity, assume that $(a, b) \equiv (c, d)$ and $(c, d) \equiv (e, f)$. Now $ad = bc$ gives $adf = bcf$, and $cf = de$ gives $bcf = bde$; thus, $adf = bde$. We may cancel the nonzero integer d to get $af = be$; that is, $(a, b) \equiv (e, f)$.

(iv) In calculus, equivalence relations are implicit in the discussion of vectors. An *arrow* from a point P to a point Q can be denoted by the ordered pair (P, Q); call P its *foot* and Q its *head*. An equivalence relation on arrows can be defined by saying that $(P, Q) \equiv (P', Q')$ if these arrows have the same length and the same direction. More precisely, $(P, Q) \equiv (P', Q')$ if the quadrilateral obtained by joining P to P' and Q to Q' is a parallelogram [this definition is incomplete, for one must also relate collinear arrows as well as "degenerate" arrows (P, P)]. Note that the direction of an arrow from P to Q is important; if $P \neq Q$, then $(P, Q) \not\equiv (Q, P)$. ◀

An equivalence relation on a set X yields a family of subsets of X.

→ **Definition.** Let $\equiv$ be an equivalence relation on a set X. If $a \in X$, the *equivalence class* of a, denoted by $[a]$, is defined by

$$[a] = \{x \in X : x \equiv a\} \subseteq X.$$

We now display the equivalence classes arising from the equivalence relations given above.

→ **Example 2.18.**

(i) Let $\equiv$ be equality on a set X. If $a \in X$, then $[a] = \{a\}$, the subset having only one element, namely, a. After all, if $x = a$, then x and a are equal!

(ii) Consider the relation of congruence mod m on $\mathbb{Z}$, and let $a \in \mathbb{Z}$. The *congruence class* of a is defined by

$$\{x \in \mathbb{Z} : x = a + km \text{ where } k \in \mathbb{Z}\}.$$

On the other hand, the equivalence class of a is, by definition,

$$\{x \in \mathbb{Z} : x \equiv a \bmod m\}.$$

Since $x \equiv a \bmod m$ if and only if $x = a + km$ for some $k \in \mathbb{Z}$, these two subsets coincide; that is, the equivalence class $[a]$ is the congruence class.

(iii) The equivalence class of (a, b) under cross multiplication, where $a, b \in \mathbb{Z}$ and $b \neq 0$, is

$$[(a, b)] = \{(c, d) : ad = bc\}.$$

If we denote $[(a, b)]$ by a/b, then this equivalence class *is* precisely the fraction usually denoted by a/b. After all, it is plain that $(1, 2) \neq (2, 4)$, but $[(1, 2)] = [(2, 4)]$; that is, $1/2 = 2/4$.

(iv) An equivalence class $[(P, Q)]$ of arrows, as in Example 2.17(iv), is called a *vector*; we denote it by $[(P, Q)] = \overrightarrow{PQ}$. ◄

It is instructive to compare rational numbers and vectors, for both are defined as equivalence classes. Every rational a/b has a "favorite" name–its expression in lowest terms; every vector has a favorite name–an arrow (O, Q) with its foot at the origin. Working with fractions in lowest terms is not always convenient; for example, even if both a/b and c/d are in lowest terms, their sum $(ad + bc)/bd$ may not be in lowest terms. Vector addition is defined by the parallelogram law (see Figure 2.5): $\overrightarrow{OP} + \overrightarrow{OQ} = \overrightarrow{OR}$, where O, P, Q, and R are the vertices of a parallelogram. But $\overrightarrow{OQ} = \overrightarrow{PR}$, because $(O, Q) \equiv (P, R)$, and it is more natural to write $\overrightarrow{OP} + \overrightarrow{OQ} = \overrightarrow{OP} + \overrightarrow{PR} = \overrightarrow{OR}$.

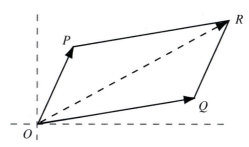

Figure 2.5 Parallelogram Law

→ **Lemma 2.19.** *If $\equiv$ is an equivalence relation on a set X, then $x \equiv y$ if and only if $[x] = [y]$.*

Proof. Assume that $x \equiv y$. If $z \in [x]$, then $z \equiv x$, and so transitivity gives $z \equiv y$; hence $[x] \subseteq [y]$. By symmetry, $y \equiv x$, and this gives the reverse inclusion $[y] \subseteq [x]$. Thus, $[x] = [y]$.

Conversely, if $[x] = [y]$, then $x \in [x]$, by reflexivity, and so $x \in [x] = [y]$. Therefore, $x \equiv y$.   •

In words, this lemma says that one can replace equivalence by honest equality at the cost of replacing elements by their equivalence classes.

Here is a set-theoretic idea that we shall show is intimately involved with equivalence relations.

→ **Definition.**   A family $\mathcal{P}$ of subsets of a set X is called *pairwise disjoint* if, for all $A, B \in \mathcal{P}$, either $A = B$ or $A \cap B = \varnothing$.

A *partition* of a set X is a family of nonempty pairwise disjoint subsets, called *blocks*, whose union is all of X.

Notice that if X is a finite set and $A_1, A_2, \ldots, A_n$ is a partition of X, then

$$|X| = |A_1| + |A_2| + \cdots + |A_n|.$$

We are now going to prove that equivalence relations and partitions are merely different views of the same thing.

→ **Proposition 2.20.**   *If $\equiv$ is an equivalence relation on a set X, then the equivalence classes form a partition of X. Conversely, given a partition $\mathcal{P}$ of X, there is an equivalence relation on X whose equivalence classes are the blocks in $\mathcal{P}$.*

Proof.   Assume that an equivalence relation $\equiv$ on X is given. Each $x \in X$ lies in the equivalence class $[x]$ because $\equiv$ is reflexive; it follows that the equivalence classes are nonempty subsets whose union is X. To prove pairwise disjointness, assume that $a \in [x] \cap [y]$, so that $a \equiv x$ and $a \equiv y$. By symmetry, $x \equiv a$, and so transitivity gives $x \equiv y$. Therefore, $[x] = [y]$, by Lemma 2.19, and so the equivalence classes form a partition of X.

Conversely, let $\mathcal{P}$ be a partition of X. If $x, y \in X$, define $x \equiv y$ if there is $A \in \mathcal{P}$ with $x \in A$ and $y \in A$. It is plain that $\equiv$ is reflexive and symmetric. To see that $\equiv$ is transitive, assume that $x \equiv y$ and $y \equiv z$; that is, there are $A, B \in \mathcal{P}$ with $x, y \in A$ and $y, z \in B$. Since $y \in A \cap B$, pairwise disjointness gives $A = B$ and so $x, z \in A$; that is, $x \equiv z$. We have shown that $\equiv$ is an equivalence relation.

It remains to show that the equivalence classes are the subsets in $\mathcal{P}$. If $x \in X$, then $x \in A$ for some $A \in \mathcal{P}$. By definition of $\equiv$, if $y \in A$, then $y \equiv x$ and $y \in [x]$; hence, $A \subseteq [x]$. For the reverse inclusion, let $z \in [x]$, so that $z \equiv x$. There is some B with $x \in B$ and $z \in B$; thus, $x \in A \cap B$. By pairwise disjointness, $A = B$, so that $z \in A$, and $[x] \subseteq A$. Hence, $[x] = A$.   •

Example 2.21.

(i) If $\equiv$ is the identity relation on a set X, then the blocks are the 1-point subsets of X.

(ii) Let $X = [0, 2\pi]$, and define the partition of X whose blocks are $\{0, 2\pi\}$ and the singletons $\{x\}$, where $0 < x < 2\pi$. This partition identifies the endpoints of the interval (and nothing else), and so we may regard this as a construction of the unit circle. ◄

Given an equivalence relation on a set X, it is a common practice to construct the set $\widetilde{X}$ whose elements are the equivalence classes $[x]$ of elements $x \in X$. For example, in Example 2.17(iii), we have $X = \{(a, b) \in \mathbb{Z} \times \mathbb{Z} : b \neq 0\}$ and $\widetilde{X} = \mathbb{Q}$. Whenever one defines a function, one must be certain that it is single-valued; in particular, this is so when the domain is $\widetilde{X}$. We discussed this on page 91; we saw there that the formula $f(a/b) = ab$ does not define a function $\mathbb{Q} \to \mathbb{Z}$ because its values depend on a choice of representative (a, b) in the equivalence class $[(a, b)] = a/b$. In contrast, addition of rationals, $\alpha \colon \mathbb{Q} \times \mathbb{Q} \to \mathbb{Q}$, given by $(a/b) + (c/d) = (ad + bd)/bd$, does define a function. Even though the formula for α appears to depend on the choices of name for a/b and c/d, it is actually independent of such choices. The reader can prove that if $a/b = a'/b'$ and $c/d = c'/d'$, then $(ad + bc)/bd = (a'd' + b'c')/b'd'$. When the values of a supposed function f appear to depend on choices, then one is obliged to prove independence of choices before declaring that f is a (single-valued) function. Checking whether an alleged function f with domain $\widetilde{X}$ is single-valued is often described as checking that f is **well-defined**.

EXERCISES

H **2.1** True or false with reasons.

(i) If $S \subseteq T$ and $T \subseteq X$, then $S \subseteq X$.

(ii) Any two functions $f \colon X \to Y$ and $g \colon Y \to Z$ have a composite $f \circ g \colon X \to Z$.

(iii) Any two functions $f \colon X \to Y$ and $g \colon Y \to Z$ have a composite $g \circ f \colon X \to Z$.

(iv) For every set X, we have $X \times \varnothing = \varnothing$.

(v) If $f \colon X \to Y$ and $j \colon \operatorname{im} f \to Y$ is the inclusion, then there is a surjection $g \colon X \to \operatorname{im} f$ with $f = j \circ g$.

(vi) If $f \colon X \to Y$ is a function for which there is a function $g \colon Y \to X$ with $f \circ g = 1_Y$, then f is a bijection.

(vii) The formula $f(\frac{a}{b}) = (a + b)(a - b)$ is a well-defined function $\mathbb{Q} \to \mathbb{Z}$.

(viii) If $f \colon \mathbb{N} \to \mathbb{N}$ is given by $f(n) = n + 1$ and $g \colon \mathbb{N} \to \mathbb{N}$ is given by $g(n) = n^2$, then the composite $g \circ f$ is $n \mapsto n^2(n + 1)$.

(ix) Complex conjugation $z = a + ib \mapsto \bar{z} = a - ib$ is a bijection $\mathbb{C} \to \mathbb{C}$.

2.2 If A and B are subsets of a set X, prove that $A - B = A \cap B'$, where $B' = X - B$ is the complement of B.

*2.3 Let A and B be subsets of a set X. Prove the **de Morgan laws**

$$(A \cup B)' = A' \cap B' \quad \text{and} \quad (A \cap B)' = A' \cup B',$$

where $A' = X - A$ denotes the complement of A.

*2.4 If A and B are subsets of a set X, define their **symmetric difference** (see Figure 2.6) by

$$A + B = (A - B) \cup (B - A).$$

   (i)   Prove that $A + B = (A \cup B) - (A \cap B)$.
   (ii)   Prove that $A + A = \varnothing$.
   (iii) Prove that $A + \varnothing = A$.
H (iv)   Prove that $A + (B + C) = (A + B) + C$ (see Figure 2.7).
   (v)   Prove that $A \cap (B + C) = (A \cap B) + (A \cap C)$.

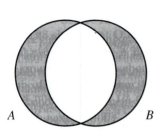

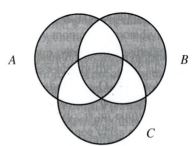

Figure 2.6   Symmetric Difference   **Figure 2.7**   Associativity

*H 2.5 Let A and B be sets, and let $a \in A$ and $b \in B$. Define their **ordered pair** as follows:

$$(a, b) = \{a, \{a, b\}\}.$$

If $a' \in A$ and $b' \in B$, prove that $(a', b') = (a, b)$ if and only if $a' = a$ and $b' = b$.

2.6 Let $\Delta = \{(x, x) : x \in \mathbb{R}\}$; thus, Δ is the line in the plane which passes through the origin and which makes an angle of $45°$ with the x-axis.

   H (i)   If $P = (a, b)$ is a point in the plane with $a \neq b$, prove that Δ is the perpendicular bisector of the segment PP' having endpoints $P = (a, b)$ and $P' = (b, a)$.

   (ii)   If $f : \mathbb{R} \to \mathbb{R}$ is a bijection whose graph consists of certain points (a, b) [of course, $b = f(a)$], prove that the graph of f^{-1} is

$$\{(b, a) : (a, b) \in f\}.$$

*2.7 Let X and Y be sets, and let $f : X \to Y$ be a function.

   H (i)   If S is a subset of X, prove that the restriction $f|S$ is equal to the composite $f \circ i$, where $i : S \to X$ is the inclusion map.

   (ii)   If $\operatorname{im} f = A \subseteq Y$, prove that there exists a surjection $f' : X \to A$ with $f = j \circ f'$, where $j : A \to Y$ is the inclusion.

H **2.8** If $f: X \to Y$ has an inverse g, show that g is a bijection.

***2.9** Show that if $f: X \to Y$ is a bijection, then it has exactly one inverse.

H **2.10** Show that $f: \mathbb{R} \to \mathbb{R}$, defined by $f(x) = 3x + 5$, is a bijection, and find its inverse.

H **2.11** Determine whether $f: \mathbb{Q} \times \mathbb{Q} \to \mathbb{Q}$, given by

$$f(a/b, c/d) = (a + c)/(b + d)$$

is a function.

*H **2.12** Let $X = \{x_1, \ldots, x_m\}$ and $Y = \{y_1, \ldots, y_n\}$ be finite sets, where the x_i are distinct and the y_j are distinct. Show that there is a bijection $f: X \to Y$ if and only if $|X| = |Y|$; that is, $m = n$.

2.13** (Pigeonhole Principle***)

    H **(i)**   If X and Y are finite sets with the same number of elements, show that the following conditions are equivalent for a function $f: X \to Y$:

        (i) f is injective;

        (ii) f is bijective;

        (iii) f is surjective.

    (ii)   Suppose there are 11 pigeons, each sitting in some pigeonhole. If there are only 10 pigeonholes, prove that there is a hole containing more than one pigeon.

***2.14** Let $f: X \to Y$ and $g: Y \to Z$ be functions.

    (i)   If both f and g are injective, prove that $g \circ f$ is injective.

    (ii)   If both f and g are surjective, prove that $g \circ f$ is surjective.

    (iii)   If both f and g are bijective, prove that $g \circ f$ is bijective.

    (iv)   If $g \circ f$ is a bijection, prove that f is an injection and g is a surjection.

  2.15 H **(i)**   If $f: (-\pi/2, \pi/2) \to \mathbb{R}$ is defined by $a \mapsto \tan a$, then f has an inverse function g; indeed, $g = \arctan$.

    (ii)   Show that each of $\arcsin x$ and $\arccos x$ is an inverse function (of $\sin x$ and $\cos x$, respectively) as defined in this section. (Domains and targets must be chosen with care.)

***2.16** **(i)**   Let $f: X \to Y$ be a function, and let $\{S_i : i \in I\}$ be a family of subsets of X. Prove that

$$f\left(\bigcup_{i \in I} S_i\right) = \bigcup_{i \in I} f(S_i).$$

    (ii)   If S_1 and S_2 are subsets of a set X, and if $f: X \to Y$ is a function, prove that $f(S_1 \cap S_2) \subseteq f(S_1) \cap f(S_2)$. Give an example in which $f(S_1 \cap S_2) \neq f(S_1) \cap f(S_2)$.

    (iii)   If S_1 and S_2 are subsets of a set X, and if $f: X \to Y$ is an injection, prove that $f(S_1 \cap S_2) = f(S_1) \cap f(S_2)$.

***2.17** Let $f: X \to Y$ be a function.

    (i)   If $B_i \subseteq Y$ is a family of subsets of Y, prove that

$$f^{-1}\left(\bigcup_i B_i\right) = \bigcup_i f^{-1}(B_i) \quad \text{and} \quad f^{-1}\left(\bigcap_i B_i\right) = \bigcap_i f^{-1}(B_i).$$

(ii) If $B \subseteq Y$, prove that $f^{-1}(B') = f^{-1}(B)'$, where B' denotes the complement of B.

2.18 Let $f: X \to Y$ be a function. Define a relation on X by $x \equiv x'$ if $f(x) = f(x')$. Prove that $\equiv$ is an equivalence relation. If $x \in X$ and $f(x) = y$, the equivalence class $[x]$ is denoted by $f^{-1}(y)$; it is called the **fiber** over y.

2.19 Let $X = \{$rock, paper, scissors$\}$. Recall the game whose rules are: paper dominates rock, rock dominates scissors, and scissors dominates paper. Draw a subset of $X \times X$ showing that domination is a relation on X.

2.20   H **(i)**   Find the error in the following argument which claims to prove that a symmetric and transitive relation R on a set X must be reflexive; that is, R is an equivalence relation on X. If $x \in X$ and $x \, R \, y$, then symmetry gives $y \, R \, x$ and transitivity gives $x \, R \, x$.

(ii)   Give an example of a symmetric and transitive relation on the closed unit interval $X = [0, 1]$ which is not reflexive.

→ 2.2   PERMUTATIONS

In high school mathematics, the words *permutation* and *arrangement* are used interchangeably, if the word *arrangement* is used at all. We draw a distinction between them.

→ **Definition.**   If X is a set, then a **list in** X is a function $f: \{1, 2, \ldots, n\} \to X$. If a list f in X is a bijection (so that X is now a finite set with $|X| = n$), then f is called an **arrangement of** X.

If f is a list, denote its values $f(i)$ by x_i, where $1 \le i \le n$. Thus, a list in X is merely an n-tuple $(x_1, x_2, \ldots, x_n)$. To say that a list f is injective is to say that there are no repeated coordinates [if $i \ne j$, then $x_i = f(i) \ne f(j) = x_j$]; to say that f is surjective is to say that every $x \in X$ occurs as some coordinate. Thus, an arrangement of X is an n-tuple $(x_1, x_2, \ldots, x_n)$ of all the elements of X with no repetitions. We often omit parentheses and write a list as $x_1, x_2, \ldots, x_n$. For example, there are 27 lists in $X = \{a, b, c\}$ and 6 arrangements:

$$abc; \quad acb; \quad bac; \quad bca; \quad cab; \quad cba.$$

All we can do with such lists is count the number of them; there are exactly n^n lists and $n!$ arrangements of an n-element set X.

→ **Definition.**   A **permutation** of a possibly infinite set X is a bijection $\alpha: X \to X$.

Given a finite set X with $|X| = n$, let $\varphi: \{1, 2, \ldots, n\} \to X$ be an arrangement chosen once for all; of course, φ is a bijection. If $f: \{1, 2, \ldots, n\} \to X$ is an arrangement of X, then $f \circ \varphi^{-1}: X \to X$ is a permutation of X. Conversely, if $\alpha: X \to X$ is a permutation of X, then $\alpha \circ \varphi: \{1, 2, \ldots, n\} \to X$ is an arrangement of X. Thus, arrangements and permutations are simply different ways of describing the same thing. The advantage of using permutations, instead of arrangements, is that permutations can be composed and, by Exercise 2.14(ii) on page 105, their composite is also a permutation.

If $X = \{1, 2, \ldots, n\}$, then we may use a two-rowed notation to denote a permutation α:

$$\alpha = \begin{pmatrix} 1 & 2 & \cdots & j & \cdots & n \\ \alpha(1) & \alpha(2) & \cdots & \alpha(j) & \cdots & \alpha(n) \end{pmatrix}.$$

Thus, the bottom row is the arrangement $\alpha(1), \alpha(2), \ldots, \alpha(n)$.

Most of the results in this section first appeared in an article of Cauchy in 1815 (see Figure 2.9).

→ **Definition.** The family of all the permutations of a set X, denoted by S_X, is called the *symmetric group* on X. When $X = \{1, 2, \ldots, n\}$, S_X is usually denoted by S_n, and it is called the *symmetric group on n letters*.

Notice that composition in S_3 is not commutative. If

$$\alpha = \begin{pmatrix} 1 & 2 & 3 \\ 2 & 3 & 1 \end{pmatrix} \qquad \text{and} \qquad \beta = \begin{pmatrix} 1 & 2 & 3 \\ 2 & 1 & 3 \end{pmatrix},$$

then their composites[3] are

$$\alpha \circ \beta = \begin{pmatrix} 1 & 2 & 3 \\ 3 & 2 & 1 \end{pmatrix} \qquad \text{and} \qquad \beta \circ \alpha = \begin{pmatrix} 1 & 2 & 3 \\ 1 & 3 & 2 \end{pmatrix}.$$

Thus, $\alpha \circ \beta \colon 1 \mapsto \alpha(\beta(1)) = \alpha(2) = 3$ while $\beta \circ \alpha \colon 1 \mapsto 2 \mapsto 1$, and so $\alpha \circ \beta \neq \beta \circ \alpha$.
On the other hand, some permutations do commute; for example,

$$\gamma = \begin{pmatrix} 1 & 2 & 3 & 4 \\ 2 & 1 & 3 & 4 \end{pmatrix} \qquad \text{and} \qquad \delta = \begin{pmatrix} 1 & 2 & 3 & 4 \\ 1 & 2 & 4 & 3 \end{pmatrix}$$

commute, as the reader may check.

Composition in S_X satisfies the *cancellation law*:

$$\text{if } \gamma \circ \alpha = \gamma \circ \beta, \text{ then } \alpha = \beta.$$

---

[3]There are authors who multiply permutations differently; their $\alpha \circ \beta$ is our $\beta \circ \alpha$. In more detail, let $\alpha, \beta \colon X \to X$. Since we write the value of α on $i \in X$ as $\alpha(i)$, then the composite in which we first apply α and then β sends $i \mapsto \alpha(i) \mapsto \beta(\alpha(i))$; thus, it is natural for us to denote the composite α followed by β as $\beta \circ \alpha$. But some other authors use a right-handed notation; they denote the value of α on i by $(i)\alpha$. For them, α followed by β sends $i \mapsto (i)\alpha \mapsto ((i)\alpha)\beta$, and so they denote this composite by $\alpha \circ \beta$; that is, our $\beta \circ \alpha$ is their $\alpha \circ \beta$. We will always write $\beta \circ \alpha$ to mean α followed by β, but the reader should be aware that other books may use the right-handed notation.

To see this,

$$\alpha = 1_X \circ \alpha$$
$$= (\gamma^{-1} \circ \gamma) \circ \alpha$$
$$= \gamma^{-1} \circ (\gamma \circ \alpha)$$
$$= \gamma^{-1} \circ (\gamma \circ \beta)$$
$$= (\gamma^{-1} \circ \gamma) \circ \beta$$
$$= 1_X \circ \beta = \beta.$$

A similar argument shows that

$$\alpha \circ \gamma = \beta \circ \gamma \text{ implies } \alpha = \beta.$$

Aside from being cumbersome, there is a major problem with the two-rowed notation for permutations. It hides the answers to such elementary questions as: Is the square of a permutation the identity? What is the smallest positive integer m so that the mth power of a permutation is the identity? Can one factor a permutation into simpler permutations? The special permutations introduced below will remedy this problem.

Let us first simplify notation by writing $\beta\alpha$ instead of $\beta \circ \alpha$ and (1) instead of 1_X.

→ **Definition.**   If $\alpha \in S_n$ and $i \in \{1, 2, \ldots, n\}$, then α *fixes* i if $\alpha(i) = i$, and α *moves* i if $\alpha(i) \neq i$.

→ **Definition.**   Let $i_1, i_2, \ldots, i_r$ be distinct integers in $\{1, 2, \ldots, n\}$. If $\alpha \in S_n$ fixes the other integers (if any) and if

$$\alpha(i_1) = i_2, \quad \alpha(i_2) = i_3, \quad \ldots, \quad \alpha(i_{r-1}) = i_r, \quad \alpha(i_r) = i_1,$$

then α is called an *r-cycle*. One also says that α is a cycle of *length* r.

A 2-cycle interchanges i_1 and i_2 and fixes everything else; 2-cycles are also called *transpositions*. A 1-cycle is the identity, for it fixes every i; thus, all 1-cycles are equal: $(i) = (1)$ for all i.

Consider the permutation

$$\alpha = \begin{pmatrix} 1 & 2 & 3 & 4 & 5 \\ 4 & 3 & 1 & 5 & 2 \end{pmatrix}.$$

The two-rowed notation does not help us recognize that α is, in fact, a 5-cycle: $\alpha(1) = 4$, $\alpha(4) = 5$, $\alpha(5) = 2$, $\alpha(2) = 3$, and $\alpha(3) = 1$. We now introduce new notation: an r-cycle α, as in the definition, shall be denoted by

$$\alpha = (i_1 \ i_2 \ \ldots \ i_r).$$

For example, the preceding 5-cycle α will be written $\alpha = (1\ 4\ 5\ 2\ 3)$. The reader may check that

$$\begin{pmatrix} 1 & 2 & 3 & 4 \\ 2 & 3 & 4 & 1 \end{pmatrix} = (1\ 2\ 3\ 4),$$

$$\begin{pmatrix} 1 & 2 & 3 & 4 & 5 \\ 5 & 1 & 4 & 2 & 3 \end{pmatrix} = (1\ 5\ 3\ 4\ 2),$$

and

$$\begin{pmatrix} 1 & 2 & 3 & 4 & 5 \\ 2 & 3 & 1 & 4 & 5 \end{pmatrix} = (1\ 2\ 3).$$

Notice that

$$\beta = \begin{pmatrix} 1 & 2 & 3 & 4 \\ 2 & 1 & 4 & 3 \end{pmatrix}$$

is not a cycle; in fact, $\beta = (1\ 2)(3\ 4)$. The term *cycle* comes from the Greek word

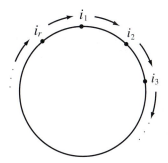

Figure 2.8 A Cycle as a Rotation

for circle. Picture the cycle $(i_1\ i_2\ \ldots\ i_r)$ as a clockwise rotation of the circle, as in Figure 2.8. Any i_j can be taken as the "starting point," and so there are r different cycle notations for any r-cycle:

$$(i_1\ i_2\ \ldots\ i_r) = (i_2\ i_3\ \ldots\ i_r\ i_1) = \cdots = (i_r\ i_1\ i_2\ \ldots\ i_{r-1}).$$

Figure 2.9 is a page from Cauchy's 1815 paper in which he introduces the calculus of permutations. Notice that his notation for a cycle is a circle.

Let us now give an ***algorithm*** to factor a permutation into a product of cycles. For example, take

$$\alpha = \begin{pmatrix} 1 & 2 & 3 & 4 & 5 & 6 & 7 & 8 & 9 \\ 6 & 4 & 7 & 2 & 5 & 1 & 8 & 9 & 3 \end{pmatrix}.$$

Nous observerons d'abord que, si dans la substitution $\begin{pmatrix} A_s \\ A_t \end{pmatrix}$ formée par deux permutations prises à volonté dans la suite

$$A_1, \quad A_2, \quad A_3, \quad \ldots, \quad A_N,$$

les deux termes A_s, A_t renferment des indices correspondants qui soient respectivement égaux, on pourra, sans inconvénient, supprimer les mêmes indices pour ne conserver que ceux des indices correspondants qui sont respectivement inégaux. Ainsi, par exemple, si l'on fait $n = 5$, les deux substitutions

$$\begin{pmatrix} 1.2.3.4.5 \\ 2.3.1.4.5 \end{pmatrix} \quad \text{et} \quad \begin{pmatrix} 1.2.3 \\ 2.3.1 \end{pmatrix}$$

seront équivalentes entre elles. Je dirai qu'une substitution aura été réduite à sa plus simple expression lorsqu'on aura supprimé, dans les deux termes, tous les indices correspondants égaux.

Soient maintenant $\alpha, \beta, \gamma, \ldots, \zeta, \eta$ plusieurs des indices $1, 2, 3, \ldots, n$ en nombre égal à p, et supposons que la substitution $\begin{pmatrix} A_s \\ A_t \end{pmatrix}$ réduite à sa plus simple expression prenne la forme

$$\begin{pmatrix} \alpha & \beta & \gamma & \ldots & \zeta & \eta \\ \beta & \gamma & \delta & \ldots & \eta & \alpha \end{pmatrix},$$

en sorte que, pour déduire le second terme du premier, il suffise de ranger en cercle, ou plutôt en polygone régulier, les indices $\alpha, \beta, \gamma, \delta, \ldots, \zeta, \eta$ de la manière suivante :

et de remplacer ensuite chaque indice par celui qui, le premier, vient prendre sa place lorsqu'on fait tourner d'orient en occident le polygone

Figure 2.9

A. Cauchy, *Mémoire sur le nombre des valeurs qu'une fonction peut acquérir lorsqu'on y permute de toutes les manières possibles les quantités qu'elle renferme*
J. de l'École Poly., XVIIe Cahier, Tome X (1815), pp. 1–28
From *Oeuvres Completes d'Augustin Cauchy*, II Serie, Tome I, Gauthier-Villars, Paris, 1905.

Begin by writing "(1." Now $\alpha: 1 \mapsto 6$, so write "(1 6." Next, $\alpha: 6 \mapsto 1$, and so the parentheses close: α begins "(1 6)." The first number not having appeared is 2, and so we write "(1 6)(2." Now $\alpha: 2 \mapsto 4$, so we write "(1 6)(2 4." Since $\alpha: 4 \mapsto 2$, the parentheses close once again, and we write "(1 6)(2 4)." The smallest remaining number is 3; now $3 \mapsto 7, 7 \mapsto 8, 8 \mapsto 9$, and $9 \mapsto 3$; this gives the 4-cycle (3 7 8 9). Finally, $\alpha(5) = 5$; we claim that

$$\alpha = (1\ 6)(2\ 4)(3\ 7\ 8\ 9)(5).$$

Since multiplication in S_n is composition of functions, our claim is that

$$\alpha(i) = [(1\ 6)(2\ 4)(3\ 7\ 8\ 9)(5)](i)$$

for every i between 1 and n (after all, two functions f and g are equal if and only if $f(i) = g(i)$ for every i in their common domain). The right side is the composite $\beta\gamma\delta$, where $\beta = (1\ 6)$, $\gamma = (2\ 4)$, and $\delta = (3\ 7\ 8\ 9)$ (actually, there is also the 1-cycle (5), which we may ignore when we are evaluating, for any 1-cycle is the identity function). Now $\alpha(1) = 6$; multiplication of permutations views the permutations as functions and then takes their composite. For example, if $i = 1$, then

$$\begin{aligned}
\beta\gamma\delta(1) &= \beta(\gamma(\delta(1))) \\
&= \beta(\gamma(1)) \quad \delta \text{ fixes } 1 \\
&= \beta(1) \quad\quad \gamma \text{ fixes } 1 \\
&= 6.
\end{aligned}$$

In Proposition 2.24, we will give a more satisfactory proof that α has been factored as a product of cycles.

Factorizations into cycles are very convenient for multiplication of permutations. For example, in S_5, let us simplify the product

$$\sigma = (1\ 2)(1\ 3\ 4\ 2\ 5)(2\ 5\ 1\ 3)$$

by displaying the "partial outputs" of the algorithm: $\sigma: 1 \mapsto 3 \mapsto 4 \mapsto 4$, so that σ begins (1 4. Next, $\sigma: 4 \mapsto 4 \mapsto 2 \mapsto 1$; hence, σ begins (1 4). The smallest number not yet considered is 2, and $\sigma: 2 \mapsto 5 \mapsto 1 \mapsto 2$; thus, σ fixes 2, and σ begins (1 4)(2). The smallest number not yet considered is 3, and $\sigma: 3 \mapsto 2 \mapsto 5 \mapsto 5$. Finally, $\sigma: 5 \mapsto 1 \mapsto 3 \mapsto 3$, and we conclude that

$$\sigma = (1\ 4)(2)(3\ 5).$$

In the factorization of a permutation into cycles, given by the algorithm above, one notes that the family of cycles is *disjoint* in the following sense.

→ **Definition.**   Two permutations $\alpha, \beta \in S_n$ are ***disjoint*** if every i moved by one is fixed by the other: if $\alpha(i) \neq i$, then $\beta(i) = i$, and if $\beta(j) \neq j$, then $\alpha(j) = j$. A family $\beta_1 \ldots, \beta_t$ of permutations is ***disjoint*** if each pair of them is disjoint.

Consider the special case of cycles. If $\alpha = (i_1 \ i_2 \ \ldots \ i_r)$ and $\beta = (j_1 \ j_2 \ \ldots \ j_s)$, then any k in the intersection $\{i_1, i_2, \ldots, i_r\} \cap \{j_1, j_2, \ldots, j_s\}$ is moved by both α and β. Thus, it is easy to see that two cycles are disjoint if and only if $\{i_1, i_2, \ldots, i_r\} \cap \{j_1, j_2, \ldots, j_s\} = \varnothing$; that is, $\{i_1, i_2, \ldots, i_r\}$ and $\{j_1, j_2, \ldots, j_s\}$ are disjoint sets.

When permutations α and β are disjoint, there are exactly three distinct possibilities for a number i: it is moved by α, it is moved by β, or it is moved by neither (that is, it is fixed by both).

→ **Lemma 2.22.**   *Disjoint permutations $\alpha, \beta \in S_n$ commute.*

Proof.   It suffices to prove that if $1 \leq i \leq n$, then $\alpha\beta(i) = \beta\alpha(i)$. If β moves i, say, $\beta(i) = j \neq i$, then β also moves j [otherwise, $\beta(j) = j$ and $\beta(i) = j$ contradicts β's being an injection]; since α and β are disjoint, $\alpha(i) = i$ and $\alpha(j) = j$. Hence $\beta\alpha(i) = j = \alpha\beta(i)$. A similar argument shows that $\alpha\beta(i) = \beta\alpha(i)$ if α moves i. The last possibility is that neither α nor β moves i; in this case, $\alpha\beta(i) = i = \beta\alpha(i)$. Therefore, $\alpha\beta = \beta\alpha$, by Proposition 2.2.   •

In particular, disjoint cycles commute.

Permutations which are not disjoint may commute; for example, the reader may check that $(1\ 2\ 3)(4\ 5)$ and $(1\ 3\ 2)(6\ 7)$ do commute. An even simpler example arises from a permutation commuting with itself.

Lemma 2.23.   *Let $X = \{1, 2, \ldots, n\}$, let $\alpha \in S_X = S_n$, and, if $i_1 \in X$, define i_j for all $j \geq 1$ by induction: $i_{j+1} = \alpha(i_j)$. Write $Y = \{i_j : j \geq 1\}$, and let Y' be the complement of Y.*

(i) *If α moves i_1, then there is $r > 1$ with $i_1, \ldots, i_r$ all distinct and with $i_{r+1} = \alpha(i_r) = i_1$.*

(ii) *$\alpha(Y) = Y$ and $\alpha(Y') = Y'$.*

Proof.
(i) Since X is finite, there is a smallest $r > 1$ with $i_1, \ldots, i_r$ all distinct, but with $i_{r+1} = \alpha(i_r) \in \{i_1, \ldots, i_r\}$; that is, $\alpha(i_r) = i_j$ for $1 \leq j \leq r$. If $j > 1$, then $\alpha(i_r) = i_j = \alpha(i_{j-1})$. But α is an injection, so that $i_r = i_{j-1}$, contradicting $i_1, \ldots, i_r$ all being distinct. Therefore, $\alpha(i_r) = i_1$.
(ii) It is obvious that $\alpha(Y) \subseteq Y$, for if $i_j \in Y$, then $\alpha(i_j) = i_{j+1} \in Y$. If $k \in Y'$, then either $\alpha(k) \in Y$ or $\alpha(k) \in Y'$, for Y' is the complement of Y, and so $X = Y \cup Y'$. If $\alpha(k) \in Y$, then $\alpha(k) = i_j = \alpha(i_{j-1})$ for some j (by part (i), this is even true for $i_j = i_1$). Since α is injective, $k = i_{j-1} \in Y$, contradicting $Y \cap Y' = \varnothing$. Therefore, $\alpha(Y') \subseteq Y'$.

We now show that the inclusions $\alpha(Y) \subseteq Y$ and $\alpha(Y') \subseteq Y'$ are actually equalities. Now $\alpha(X) = \alpha(Y \cup Y') = \alpha(Y) \cup \alpha(Y')$, and this is a disjoint union because α is an injection. But $\alpha(Y) \subseteq Y$ gives $|\alpha(Y)| \leq |Y|$, and $\alpha(Y') \subseteq Y'$ gives $|\alpha(Y')| \leq |Y'|$. If either of these inequalities is strict, then $|\alpha(X)| < |X|$. But $\alpha(X) = X$, because α is a surjection, and this is a contradiction. $\bullet$

The argument in the proof of Lemma 2.23(i) will be used again.

→ **Proposition 2.24.**   *Every permutation $\alpha \in S_n$ is either a cycle or a product of disjoint cycles.*

Proof.   The proof is by induction on the number $k \geq 0$ of points moved by α. The base step $k = 0$ is true, for α is now the identity, which is a 1-cycle.

If $k > 0$, there is a point, say i_1, moved by α. As in Lemma 2.23, define $Y = \{i_1, \ldots, i_r\}$, where $i_1, \ldots, i_r$ are all distinct, $\alpha(i_j) = i_{j+1}$ for $j < r$, and $\alpha(i_r) = i_1$. Let $\sigma \in S_X$ be the r-cycle $(i_1 \; i_2 \; i_3 \; \ldots \; i_r)$, so that σ fixes each point, if any, in the complement Y' of Y. If $r = n$, then $\alpha = \sigma$. If $r < n$, then $\alpha(Y') = Y'$, as in the lemma. Define $\alpha' = \alpha\sigma^{-1}$; we claim that α' and σ are disjoint. If σ moves i, then $i = i_j \in Y$. But $\alpha'(i_j) = \alpha\sigma^{-1}(i_j) = \alpha(i_{j-1}) = i_j$; that is, α' fixes i_j. Suppose that α' moves some point i'. We have just seen that $i' \notin Y$, so that we may assume that $i' \in Y'$. By definition, σ fixes every point in Y'; hence, σ fixes i'. Therefore, $\alpha = \alpha'\sigma$ is a factorization into disjoint permutations. The number of points moved by α' is $k - r < k$, and so the inductive hypothesis gives $\alpha' = \beta_1 \cdots \beta_t$, where $\beta_1, \ldots, \beta_t$ are disjoint cycles. Therefore, $\alpha = \alpha'\sigma = \beta_1 \cdots \beta_t \sigma$ is a product of disjoint cycles, as desired. $\bullet$

We have just proved that the output of the algorithm on page 109 is always a product of disjoint cycles.

Usually one suppresses the 1-cycles in this factorization [for 1-cycles equal the identity (1)]. However, a factorization of α containing one 1-cycle for each i fixed by α, if any, will arise several times in the sequel.

→ **Definition.**   A *complete factorization* of a permutation α is a factorization of α into disjoint cycles that contains one 1-cycle (i) for every i fixed by α.

The factorization algorithm always yields a complete factorization. For example, if

$$\alpha = \begin{pmatrix} 1 & 2 & 3 & 4 & 5 \\ 1 & 3 & 4 & 2 & 5 \end{pmatrix},$$

then the algorithm gives $\alpha = (1)(2 \; 3 \; 4)(5)$, which is a complete factorization. However, if one suppresses 1-cycles, the factorizations

$$\alpha = (2 \; 3 \; 4) = (1)(2 \; 3 \; 4) = (2 \; 3 \; 4)(5)$$

are not complete factorizations. In a complete factorization $\alpha = \beta_1 \cdots \beta_t$, every symbol i between 1 and n occurs in exactly one of the β's.

If $\beta \in S_n$ and $k \geq 0$, define the **powers** of β inductively: $\beta^0 = (1)$ and $\beta^{k+1} = \beta\beta^k$. Thus, β^k is the composite of β with itself k times. There is a relation between an r-cycle β and its powers. We modify notation a bit for the next observation; write $\beta = (i_0 \ i_1 \ \ldots \ i_{r-1})$. Note that $i_1 = \beta(i_0)$, $i_2 = \beta(i_1) = \beta(\beta(i_0)) = \beta^2(i_0)$, $i_3 = \beta(i_2) = \beta(\beta^2(i_0)) = \beta^3(i_0)$, and, for all $k \leq r - 1$,

$$i_k = \beta^k(i_0). \tag{1}$$

Since $\beta(i_{r-1}) = i_0$, it is easy to see that the equation $i_k = \beta^k(i_0)$ holds if subscripts j in the notation i_j are taken mod r.

Lemma 2.25.

   (i) *Let $\alpha = \beta\delta$ be a factorization into disjoint permutations. If β moves i, then $\alpha^k(i) = \beta^k(i)$ for all $k \geq 1$.*

   (ii) *If β and γ are cycles both of which move $i = i_0$, and if $\beta^k(i) = \gamma^k(i)$ for all $k \geq 1$, then $\beta = \gamma$.*

Remark.   The hypothesis in part (ii) does not assume that the cycles β and γ have the same length, but this is part of the conclusion.   ◄

Proof.
(i) Since β moves i, disjointness implies that δ fixes i; indeed, every power of δ fixes i. Now β and δ commute, by Lemma 2.22, and so Exercise 2.30(i) on page 124 gives $(\beta\delta)^k(i) = \beta^k(\delta^k(i)) = \beta^k(i)$, as desired.
(ii) By Eq. (1), if $\beta = (i_0 \ i_1 \ \ldots \ i_{r-1})$, then $i_k = \beta^k(i_0)$ for all $k < r - 1$. Similarly, if $\gamma = (i_0 \ j_1 \ldots j_{s-1})$, then $j_k = \gamma^k(i_0)$ for $k < s - 1$. We may assume that $r \leq s$, so that $i_1 = j_1, \ldots, i_{r-1} = j_{r-1}$. Since $j_r = \gamma^r(i_0) = \beta^r(i_0) = i_0$, it follows that $s - 1 = r - 1$ and $j_k = i_k$ for all k. Therefore, $\beta = (i_0 \ i_1 \ \ldots \ i_{r-1}) = \gamma$.   •

The next theorem is an analog of the fundamental theorem of arithmetic.

→ **Theorem 2.26.**   *Let $\alpha \in S_n$ and let $\alpha = \beta_1 \cdots \beta_t$ be a complete factorization into disjoint cycles. This factorization is unique except for the order in which the cycles occur.*

Proof.   Let $\alpha = \gamma_1 \cdots \gamma_s$ be a second complete factorization of α into disjoint cycles. Since every complete factorization of α has exactly one 1-cycle for each i fixed by α, it suffices to prove, by induction on ℓ, the larger of t and s, that the cycles of length > 1 are uniquely determined by α.

The base step is true, for when $\ell = 1$, the hypothesis is $\beta_1 = \alpha = \gamma_1$.

To prove the inductive step, note first that if β_t moves $i = i_0$, then $\beta_t^k(i_0) = \alpha^k(i_0)$ for all $k \geq 1$, by Lemma 2.25(i). Now some γ_j must move i_0; since disjoint cycles commute, we may re-index so that γ_s moves i_0. As in the first paragraph, $\gamma_s^k(i_0) = \alpha^k(i_0)$ for all k. It follows from Lemma 2.25(ii) that $\beta_t = \gamma_s$, and the cancellation law on page 107 gives $\beta_1 \cdots \beta_{t-1} = \gamma_1 \cdots \gamma_{s-1}$. By the inductive hypothesis, $s = t$ and the γ's can be reindexed so that $\gamma_1 = \beta_1, \ldots, \gamma_{t-1} = \beta_{t-1}$. •

Every permutation is a bijection; how do we find its inverse? In Figure 2.8 on page 109, the pictorial representation of a cycle β as a clockwise rotation of a circle, the inverse β^{-1} is just a counterclockwise rotation.

Proposition 2.27.

(i) *The inverse of the cycle* $\alpha = (i_1 \ i_2 \ \ldots \ i_r)$ *is the cycle* $(i_r \ i_{r-1} \ \ldots \ i_1)$:

$$(i_1 \ i_2 \ \ldots \ i_r)^{-1} = (i_r \ i_{r-1} \ \ldots \ i_1).$$

(ii) *If* $\gamma \in S_n$ *and* $\gamma = \beta_1 \cdots \beta_k$, *then*

$$\gamma^{-1} = \beta_k^{-1} \cdots \beta_1^{-1}$$

(note that the order of the factors in γ^{-1} *has been reversed).*

Proof.
(i) If $\alpha \in S_n$, we show that both composites of the displayed cycles are equal to (1). Now the composite $(i_1 \ i_2 \ \ldots \ i_r)(i_r \ i_{r-1} \ \ldots \ i_1)$ fixes each integer between 1 and n, if any, other than $i_1, \ldots, i_r$. The composite also sends $i_1 \mapsto i_r \mapsto i_1$ while it acts on i_j, for $j \geq 2$, by $i_j \mapsto i_{j-1} \mapsto i_j$. Thus, each integer between 1 and n is fixed by the composite, and so it is (1). A similar argument proves that the composite in the other order is also equal to (1), from which it follows that

$$(i_1 \ i_2 \ \ldots \ i_r)^{-1} = (i_r \ i_{r-1} \ \ldots \ i_1).$$

(ii) The proof is by induction on $k \geq 2$. For the base step $k = 2$, we have

$$(\beta_1 \beta_2)(\beta_2^{-1} \beta_1^{-1}) = \beta_1(\beta_2 \beta_2^{-1})\beta_1^{-1} = \beta_1 \beta_1^{-1} = (1).$$

Similarly, $(\beta_2^{-1} \beta_1^{-1})(\beta_1 \beta_2) = (1)$.

For the inductive step, let $\delta = \beta_1 \cdots \beta_k$, so that $\beta_1 \cdots \beta_k \beta_{k+1} = \delta \beta_{k+1}$. Then

$$(\beta_1 \cdots \beta_k \beta_{k+1})^{-1} = (\delta \beta_{k+1})^{-1}$$
$$= \beta_{k+1}^{-1} \delta^{-1}$$
$$= \beta_{k+1}^{-1}(\beta_1 \cdots \beta_k)^{-1}$$
$$= \beta_{k+1}^{-1} \beta_k^{-1} \cdots \beta_1^{-1}. \quad •$$

Thus, $(1\ 2\ 3\ 4)^{-1} = (4\ 3\ 2\ 1) = (1\ 4\ 3\ 2)$ and $(1\ 2)^{-1} = (2\ 1) = (1\ 2)$ (every transposition is equal to its own inverse).

Example 2.28.
The result in Proposition 2.27 holds, in particular, if the factors are disjoint cycles (in which case the reversal of the order of the factors is unnecessary because they commute with one another, by Lemma 2.22). Thus, if

$$\alpha = \begin{pmatrix} 1 & 2 & 3 & 4 & 5 & 6 & 7 & 8 & 9 \\ 6 & 4 & 7 & 2 & 5 & 1 & 8 & 9 & 3 \end{pmatrix},$$

then $\alpha = (1\ 6)(2\ 4)(3\ 7\ 8\ 9)(5)$ and

$$\alpha^{-1} = (5)(9\ 8\ 7\ 3)(4\ 2)(6\ 1)$$
$$= (1\ 6)(2\ 4)(3\ 9\ 8\ 7). \quad \blacktriangleleft$$

→ **Definition.**   Two permutations $\alpha, \beta \in S_n$ have the **same cycle structure** if their complete factorizations have the same number of r-cycles for each $r \geq 1$.

According to Exercise 2.24 on page 124, there are

$$(1/r)[n(n-1)\cdots(n-r+1)]$$

r-cycles in S_n. This formula can be used to count the number of permutations having any given cycle structure if one is careful about factorizations having several cycles of the same length. For example, the number of permutations in S_4 with cycle structure $(a\ b)(c\ d)$ is

$$\tfrac{1}{2}\left[\tfrac{1}{2}(4\times3)\right]\times[\tfrac{1}{2}(2\times1)] = 3,$$

the extra factor $\tfrac{1}{2}$ occurring so that we do not count $(a\ b)(c\ d) = (c\ d)(a\ b)$ twice. Similarly, the number of permutations in S_n of the form $(a\ b)(c\ d)(e\ f)$ is

$$\frac{1}{3!2^3}[n(n-1)(n-2)(n-3)(n-4)(n-5)]$$

(see Exercise 2.24 on page 124).

Example 2.29.

Cycle Structure	Number
(1)	1
(1 2)	6
(1 2 3)	8
(1 2 3 4)	6
(1 2)(3 4)	3
	$\overline{24}$

Table 2.1   Permutations in S_4 ◄

Example 2.30.

Cycle Structure	Number
(1)	1
(1 2)	10
(1 2 3)	20
(1 2 3 4)	30
(1 2 3 4 5)	24
(1 2)(3 4 5)	20
(1 2)(3 4)	15
	$\overline{120}$

Table 2.2 Permutations in S_5 ◄

We present a computational aid after the next lemma.

Lemma 2.31. *Let $\alpha, \gamma \in S_n$. For all i, if $\gamma : i \to j$, then $\alpha\gamma\alpha^{-1} : \alpha(i) \to \alpha(j)$.*

Proof.
$$\alpha\gamma\alpha^{-1}(\alpha(i)) = \alpha\gamma(i) = \alpha(j). \quad \bullet$$

Proposition 2.32. *If $\gamma, \alpha \in S_n$, then $\alpha\gamma\alpha^{-1}$ has the same cycle structure as γ. In more detail, if the complete factorization of γ is*

$$\gamma = \beta_1\beta_2 \cdots (i \ j \ \ldots) \cdots \beta_t,$$

then $\alpha\gamma\alpha^{-1}$ is the permutation σ which is obtained from γ by applying α to the symbols in the cycles of γ.

Remark. For example, if $\gamma = (1\ 3)(2\ 4\ 7)(5)(6)$ and $\alpha = (2\ 5\ 6)(1\ 4\ 3)$, then

$$\alpha\gamma\alpha^{-1} = (\alpha 1 \ \alpha 3)(\alpha 2 \ \alpha 4 \ \alpha 7)(\alpha 5)(\alpha 6) = (4\ 1)(5\ 3\ 7)(6)(2). \quad ◄$$

Proof. If γ fixes i, then Lemma 2.31 shows that σ fixes $\alpha(i)$. Assume that γ moves a symbol i, say, $\gamma(i) = j$, so that one of the cycles in the complete factorization of γ is

$$(i \ j \ \ldots).$$

By the definition of σ, one of its cycles is

$$(\alpha(i) \ \alpha(j) \ \ldots);$$

that is, $\sigma : \alpha(i) \mapsto \alpha(j)$. But Lemma 2.31 says that $\alpha\gamma\alpha^{-1} : \alpha(i) \mapsto \alpha(j)$, so that σ and $\alpha\gamma\alpha^{-1}$ agree on all numbers of the form $\alpha(i)$. But every $k \in X$ has the form $k = \alpha(i)$, because $\alpha : X \to X$ is a surjection, and so $\sigma = \alpha\gamma\alpha^{-1}$. $\quad \bullet$

→ **Proposition 2.33.** *If $\gamma, \gamma' \in S_n$, then γ and γ' have the same cycle structure if and only only if there exists $\alpha \in S_n$ with $\gamma' = \alpha \gamma \alpha^{-1}$.*

Proof. Sufficiency has just been proved, in Proposition 2.32.

Conversely, assume that γ and γ' have the same cycle structure; that is, $\gamma = \beta_1 \cdots \beta_t$ and $\gamma' = \sigma_1 \cdots \sigma_t$ are complete factorizations with β_λ and σ_λ having the same length for all $\lambda \le t$. Let $\beta_\lambda = (i_1^\lambda, \ldots, i_{r(\lambda)}^\lambda)$ and $\sigma_\lambda = (j_1^\lambda, \ldots, j_{r(\lambda)}^\lambda)$. Define

$$\alpha(i_1^\lambda) = j_1^\lambda, \quad \alpha(i_2^\lambda) = j_2^\lambda, \quad \ldots, \quad \alpha(i_{r(\lambda)}^\lambda) = j_{r(\lambda)}^\lambda,$$

for all λ. Since $\beta_1 \cdots \beta_t$ is a complete factorization, every $i \in X = \{1, \ldots, n\}$ occurs in exactly one β_λ; hence, $\alpha(i)$ is defined for every $i \in X$, and $\alpha \colon X \to X$ is a (single-valued) function. Since every $j \in X$ occurs in some σ_λ, because $\sigma_1 \cdots \sigma_t$ is a complete factorization, it follows that α is surjective. By Exercise 2.13 on page 105, α is a bijection, and so $\alpha \in S_n$. Proposition 2.32 says that $\alpha \gamma \alpha^{-1}$ has the same cycle structure as γ and the λth cycle, for each λ, is

$$\left(\alpha(i_1^\lambda) \; \alpha(i_2^\lambda) \; \ldots \; \alpha(i_{r(\lambda)}^\lambda) \right) = \sigma_\lambda.$$

Therefore, $\alpha \gamma \alpha^{-1} = \gamma'$.   •

Example 2.34.

If

$$\gamma = (1\ 2\ 3)(4\ 5)(6) \quad \text{and} \quad \gamma' = (2\ 5\ 6)(3\ 1)(4),$$

then $\gamma' = \alpha \gamma \alpha^{-1}$, where

$$\alpha = \begin{pmatrix} 1 & 2 & 3 & 4 & 5 & 6 \\ 2 & 5 & 6 & 3 & 1 & 4 \end{pmatrix} = (1\ 2\ 5)(3\ 6\ 4).$$

Note that there are other choices for α as well.   ◀

Here is another useful factorization of a permutation.

→ **Proposition 2.35.** *If $n \ge 2$, then every $\alpha \in S_n$ is a product of transpositions.*

Proof. By Proposition 2.24, it suffices to factor an r-cycle β into a product of transpositions. This is done as follows. If $r = 1$, then β is the identity, and $\beta = (1\ 2)(1\ 2)$. If $r \ge 2$, then

$$\beta = (1\ 2\ \ldots\ r) = (1\ r)(1\ r-1) \cdots (1\ 3)(1\ 2).$$

[One checks that this is an equality by evaluating each side. For example, the left side β sends $1 \mapsto 2$; each of $(1\ r), (1\ r-1), \ldots, (1\ 3)$ fixes 2, and so the right side also sends $1 \mapsto 2$.]   •

Every permutation can thus be realized as a sequence of interchanges. Such a factorization is not as nice as the factorization into disjoint cycles. First of all, the transpositions occurring need not commute: $(1\ 2\ 3) = (1\ 3)(1\ 2) \neq (1\ 2)(1\ 3)$; second, neither the factors themselves nor the number of factors are uniquely determined. For example, here are some factorizations of $(1\ 2\ 3)$ in S_4:

$$
\begin{aligned}
(1\ 2\ 3) &= (1\ 3)(1\ 2) \\
&= (2\ 3)(1\ 3) \\
&= (1\ 3)(4\ 2)(1\ 2)(1\ 4) \\
&= (1\ 3)(4\ 2)(1\ 2)(1\ 4)(2\ 3)(2\ 3).
\end{aligned}
$$

Is there any uniqueness at all in such a factorization? We are going to prove that the *parity* of the number of factors is the same for all factorizations of a permutation α; that is, the number of transpositions is always even or always odd [as is suggested by the factorizations of $\alpha = (1\ 2\ 3)$ displayed above].

Example 2.36.
The ***15-puzzle*** consists of a ***starting position***, which is a 4×4 array of the numbers between 1 and 15 and a symbol # (which we interpret as "blank"), and *simple moves*. For example, consider the starting position shown below.

3	15	4	12
10	11	1	8
2	5	13	9
6	7	14	#

A ***simple move*** interchanges the blank with a symbol adjacent to it; for example, there are two beginning simple moves for this starting position: either interchange # and 14 or interchange # and 9. One wins the game if, after a sequence of simple moves, the starting position is transformed into the standard array 1, 2, 3, ..., 15, #.

To analyze this game, note that the given array is really a permutation $\alpha \in S_{16}$. More precisely, α permutes $\{1, 2, \ldots, 15, \#\}$: if the spaces are labeled 1 through 15, #, then let $\alpha(i)$ be the symbol occupying the ith square. For example, the starting position given above is

$$
\begin{pmatrix}
1 & 2 & 3 & 4 & 5 & 6 & 7 & 8 & 9 & 10 & 11 & 12 & 13 & 14 & 15 & \# \\
3 & 15 & 4 & 12 & 10 & 11 & 1 & 8 & 2 & 5 & 13 & 9 & 6 & 7 & 14 & \#
\end{pmatrix}.
$$

Each simple move is a special transposition, namely, one that moves #. Moreover, performing a simple move (corresponding to a special transposition τ) from a position (corresponding to a permutation β) yields a new position which corresponds to the permutation $\tau\beta$. For example, if α is the position above and τ is the transposition interchanging 14 and #, then $\tau\alpha(\#) = \tau(\#) = 14$ and $\tau\alpha(15) = \tau(14) = \#$, while $\tau\alpha(i) = i$ for all other i. That is, the new configuration has all the numbers in their original positions except for 14 and # being interchanged. To win the game, one needs special transpositions $\tau_1, \tau_2, \ldots, \tau_m$ such that

$$\tau_m \cdots \tau_2\tau_1\alpha = (1).$$

It turns out that there are some choices of α for which the game can be won, but there are others for which it cannot be won, as we shall see in Example 2.42. ◄

The following discussion will enable us to analyze the 15-game further.

Lemma 2.37. *If $k, \ell \geq 0$ and the letters a, b, c_i, d_j are all distinct, then*

$$(a\ b)(a\ c_1\ \ldots\ c_k\ b\ d_1\ \ldots\ d_\ell) = (a\ c_1\ \ldots\ c_k)(b\ d_1\ \ldots\ d_\ell)$$

and

$$(a\ b)(a\ c_1\ \ldots\ c_k)(b\ d_1\ \ldots\ d_\ell) = (a\ c_1\ \ldots\ c_k\ b\ d_1\ \ldots\ d_\ell).$$

Proof.   The left side of the first asserted equation sends

$$
\begin{aligned}
a &\mapsto c_1 \mapsto c_1; \\
c_i &\mapsto c_{i+1} \mapsto c_{i+1} \quad \text{if } i < k; \\
c_k &\mapsto b \mapsto a; \\
b &\mapsto d_1 \mapsto d_1; \\
d_j &\mapsto d_{j+1} \mapsto d_{j+1} \quad \text{if } j < \ell; \\
d_\ell &\mapsto a \mapsto b.
\end{aligned}
$$

Similar evaluation of the right side shows that both permutations agree on a, b, and all c_i, d_j. Since each side fixes all other numbers in $\{1, 2, \ldots, n\}$, if any, both sides are equal.

For the second equation, reverse the first equation,

$$(a\ c_1\ \ldots\ c_k)(b\ d_1\ \ldots\ d_\ell) = (a\ b)(a\ c_1\ \ldots\ c_k\ b\ d_1\ \ldots\ d_\ell),$$

and multiply both sides on the left by $(a\ b)$:

$$
\begin{aligned}
(a\ b)(a\ c_1\ \ldots\ c_k)(b\ d_1\ \ldots\ d_\ell) &= (a\ b)(a\ b))(a\ c_1\ \ldots\ c_k\ b\ d_1\ \ldots\ d_\ell) \\
&= (a\ c_1\ \ldots\ c_k\ b\ d_1\ \ldots\ d_\ell). \quad \bullet
\end{aligned}
$$

An illustration of the lemma is

$$(1\ 2)(1\ 3\ 4\ 2\ 5\ 6\ 7) = (1\ 3\ 4)(2\ 5\ 6\ 7).$$

→ **Definition.** If $\alpha \in S_n$ and $\alpha = \beta_1 \cdots \beta_t$ is a complete factorization into t disjoint cycles, then ***signum***[4] α is defined by

$$\text{sgn}(\alpha) = (-1)^{n-t}.$$

Theorem 2.26 shows that sgn is a (single-valued) function, for t, the number of cycles, is uniquely determined by α. If ε is a 1-cycle, then $\text{sgn}(\varepsilon) = 1$, for $t = n$ and so $\text{sgn}(\varepsilon) = (-1)^0 = 1$. If τ is a transposition, then it moves two numbers, and it fixes each of the $n - 2$ other numbers; therefore, $t = 1 + (n - 2) = n - 1$, and so $\text{sgn}(\tau) = (-1)^{n-(n-1)} = -1$.

Lemma 2.38. *If $\alpha, \tau \in S_n$, where τ is a transposition, then*

$$\text{sgn}(\tau\alpha) = -\text{sgn}(\alpha).$$

Proof. Let $\alpha = \beta_1 \cdots \beta_t$ be a complete factorization of α into disjoint cycles, and let $\tau = (a\ b)$. If a and b occur in the same β, say, in β_1, then $\beta_1 = (a\ c_1 \ldots c_k\ b\ d_1 \ldots d_\ell)$, where $k, \ell \geq 0$. By Lemma 2.37,

$$\tau\beta_1 = (a\ c_1 \ldots c_k)(b\ d_1 \ldots d_\ell).$$

This is a complete factorization of $\tau\alpha = (\tau\beta_1)\beta_2 \cdots \beta_t$, for the cycles in it are pairwise disjoint and every number in $\{1, 2, \ldots, n\}$ occurs in exactly one cycle. Thus, $\tau\beta$ has $t + 1$ cycles, for $\tau\beta_1$ splits into two disjoint cycles. Therefore, $\text{sgn}(\tau\alpha) = (-1)^{n-(t+1)} = -\text{sgn}(\alpha)$.

The other possibility is that a and b occur in different cycles, say, $\beta_1 = (a\ c_1 \ldots c_k)$ and $\beta_2 = (b\ d_1 \ldots d_\ell)$, where $k, \ell \geq 0$. But $\tau\alpha = (\tau\beta_1\beta_2)\beta_3 \cdots \beta_t$, and Lemma 2.37 gives

$$\tau\beta_1\beta_2 = (a\ c_1 \ldots c_k\ b\ d_1 \ldots d_\ell).$$

Therefore, $\text{sgn}(\tau\alpha) = (-1)^{n-(t-1)} = -\text{sgn}(\alpha)$, for $\tau\alpha$ has a complete factorization with $t - 1$ cycles. •

→ **Theorem 2.39.** *For all $\alpha, \beta \in S_n$,*

$$\text{sgn}(\alpha\beta) = \text{sgn}(\alpha)\,\text{sgn}(\beta).$$

Proof. Assume that $\alpha \in S_n$ is given and that α has a factorization as a product of m transpositions: $\alpha = \tau_1 \cdots \tau_m$. We prove, by induction on m, that $\text{sgn}(\alpha\beta) = \text{sgn}(\alpha)\,\text{sgn}(\beta)$ for every $\beta \in S_n$. The base step $m = 1$ is precisely Lemma 2.38,

---

[4]*Signum* is the Latin word for "mark" or "token"; of course, it has become the word *sign*.

for $m = 1$ says that α is a transposition. If $m > 1$, then the inductive hypothesis applies to $\tau_2 \cdots \tau_m$, and so

$$
\begin{aligned}
\mathrm{sgn}(\alpha\beta) &= \mathrm{sgn}(\tau_1 \cdots \tau_m \beta) \\
&= -\,\mathrm{sgn}(\tau_2 \cdots \tau_m \beta) && \text{(Lemma 2.38)} \\
&= -\,\mathrm{sgn}(\tau_2 \cdots \tau_m)\,\mathrm{sgn}(\beta) && \text{(by induction)} \\
&= \mathrm{sgn}(\tau_1 \cdots \tau_m)\,\mathrm{sgn}(\beta) && \text{(Lemma 2.38)} \\
&= \mathrm{sgn}(\alpha)\,\mathrm{sgn}(\beta). \quad \bullet
\end{aligned}
$$

It follows by induction on $k \geq 2$ that

$$
\mathrm{sgn}(\alpha_1\alpha_2 \cdots \alpha_k) = \mathrm{sgn}(\alpha_1)\,\mathrm{sgn}(\alpha_2) \cdots \mathrm{sgn}(\alpha_k).
$$

→ **Definition.**   A permutation $\alpha \in S_n$ is **even** if $\mathrm{sgn}(\alpha) = 1$, and α is **odd** if $\mathrm{sgn}(\alpha) = -1$. We say that α and β have the **same parity** if both are even or both are odd.

Let us return to factorizations of a permutation into a product of transpositions. We saw on page 119 that there are many such factorizations of a permutation, and the only common feature of these different factorizations appeared to be the parity of the number of factors. To prove this, one must show that a permutation cannot be both a product of an even number of transpositions and a product of an odd number of transpositions.

→ **Theorem 2.40.**

   (i) *Let $\alpha \in S_n$. If α is even, then α is a product of an even number of transpositions, and if α is odd, then α is a product of an odd number of transpositions.*

   (ii) *If $\alpha = \tau_1 \cdots \tau_q = \tau_1' \cdots \tau_p'$ are factorizations into transpositions, then q and p have the same parity.*

Proof.
(i) If $\alpha = \tau_1 \cdots \tau_q$ is a factorization of α into transpositions, then Theorem 2.39 gives $\mathrm{sgn}(\alpha) = \mathrm{sgn}(\tau_1) \cdots \mathrm{sgn}(\tau_q) = (-1)^q$, for we know that every transposition is odd. Therefore, if α is even; that is, if $\mathrm{sgn}(\alpha) = 1$, then q is even, while if α is odd; that is, if $\mathrm{sgn}(\alpha) = -1$, then q is odd.
(ii) If there were two factorizations of α, one into an odd number of transpositions and the other into an even number of transpositions, then $\mathrm{sgn}(\alpha)$ would have two different values. $\bullet$

Corollary 2.41.   *Let $\alpha, \beta \in S_n$. If α and β have the same parity, then $\alpha\beta$ is even, while if α and β have distinct parity, then $\alpha\beta$ is odd.*

Proof. If $\text{sgn}(\alpha) = (-1)^q$ and $\text{sgn}(\beta) = (-1)^p$, then Theorem 2.39 gives $\text{sgn}(\alpha\beta) = (-1)^{q+p}$, and the result follows. $\bullet$

We return to the 15-game.

Example 2.42.
Further analysis of the 15-puzzle in Example 2.36 shows that if $\alpha \in S_{16}$ is the starting position, then the game can be won if and only if α is an even permutation that fixes #. For a proof of this, we refer the reader to McCoy and Janusz, *Introduction to Modern Algebra*. The proof in one direction is fairly clear, however. The blank # starts in position 16. Each simple move takes # up, down, left, or right. Thus, the total number m of moves is $u + d + l + r$, where u is the number of up moves, and so on. If # is to return home, each one of these must be undone: there must be the same number of up moves as down moves, i.e., $u = d$, and the same number of left moves as right moves, i.e., $r = l$. Thus, the total number of moves is even: $m = 2u + 2r$. That is, if $\tau_m \cdots \tau_1 \alpha = (1)$, then m is even; hence, $\alpha = \tau_1 \cdots \tau_m$ (because $\tau^{-1} = \tau$ for every transposition τ), and so α is an even permutation. Armed with this theorem, we examine the complete factorization of the starting position α in Example 2.36:

$$\alpha = (1\ 3\ 4\ 12\ 9\ 2\ 15\ 14\ 7)(5\ 10)(6\ 11\ 13)(8)(\#),$$

where (8) and (#) are 1-cycles. Now $\text{sgn}(\alpha) = (-1)^{16-5} = -1$, so that α is an odd permutation; therefore, the game starting with α cannot be won. ◄

EXERCISES

H **2.21** True or false with reasons.

    (i) The symmetric group on n letters is a set of n elements.

    (ii) If $\sigma \in S_6$, then $\sigma^n = 1$ for some $n \geq 1$.

    (iii) If $\alpha, \beta \in S_n$, then $\alpha\beta$ is an abbreviation for $\alpha \circ \beta$.

    (iv) If α, β are cycles in S_n, then $\alpha\beta = \beta\alpha$.

    (v) If σ, τ are r-cycles in S_n, then $\sigma\tau$ is an r-cycle.

    (vi) If $\sigma \in S_n$ is an r-cycle, then $\alpha\sigma\alpha^{-1}$ is an r-cycle for every $\alpha \in S_n$.

    (vii) Every transposition is an even permutation.

    (viii) If a permutation α is a product of 3 transpositions, then it cannot be a product of 4 transpositions.

    (ix) If a permutation α is a product of 3 transpositions, then it cannot be a product of 5 transpositions.

    (x) If $\sigma\alpha\sigma^{-1} = \omega\alpha\omega^{-1}$, then $\sigma = \omega$.

****2.22** Find $\text{sgn}(\alpha)$ and α^{-1}, where

$$\alpha = \begin{pmatrix} 1 & 2 & 3 & 4 & 5 & 6 & 7 & 8 & 9 \\ 9 & 8 & 7 & 6 & 5 & 4 & 3 & 2 & 1 \end{pmatrix}.$$

H **2.23** If $\sigma \in S_n$ fixes some j, where $1 \le j \le n$ (that is, $\sigma(j) = j$), define $\sigma' \in S_X$, where $X = \{1, \ldots, \widehat{j}, \ldots, n\}$, by $\sigma'(i) = \sigma(i)$ for all $i \ne j$. Prove that

$$\mathrm{sgn}(\sigma') = \mathrm{sgn}(\sigma).$$

*2.24   H **(i)**   If $1 < r \le n$, prove that there are

$$\tfrac{1}{r}[n(n-1)\cdots(n-r+1)]$$

r-cycles in S_n.

(ii)   If $kr \le n$, where $1 < r \le n$, prove that the number of permutations $\alpha \in S_n$, where α is a product of k disjoint r-cycles, is

$$\tfrac{1}{k!}\tfrac{1}{r^k}[n(n-1)\cdots(n-kr+1).]$$

*2.25   H **(i)**   If α is an r-cycle, show that $\alpha^r = (1)$.

H **(ii)**   If α is an r-cycle, show that r is the least positive integer k such that $\alpha^k = (1)$.

2.26 Show that an r-cycle is an even permutation if and only if r is odd.

H **2.27** Given $X = \{1, 2, \ldots, n\}$, let us call a permutation τ of X an **adjacency** if it is a transposition of the form $(i \ i+1)$ for $i < n$. If $i < j$, prove that $(i \ j)$ is a product of an odd number of adjacencies.

*2.28 Define $f\colon \{0, 1, 2, \ldots, 10\} \to \{0, 1, 2, \ldots, 10\}$ by

$$f(n) = \text{the remainder after dividing } 4n^2 - 3n^7 \text{ by } 11.$$

(i)   Show that f is a permutation.

(ii)   Compute the parity of f.

(iii)   Compute the inverse of f.

2.29   H **(i)**   A permutation $\alpha \in S_n$ is **regular** if either α has no fixed points and it is the product of disjoint cycles of the same length, or $\alpha = (1)$. Prove that α is regular if and only if α is a power of an n-cycle.

(ii)   Prove that if α is an r-cycle, then α^k is a product of (r, k) disjoint cycles, each of length $r/(r, k)$.

(iii)   If p is a prime, prove that every power of a p-cycle is either a p-cycle or (1).

(iv)   How many regular permutations are there in S_5? How many regular permutations are there in S_8?

*2.30   H **(i)**   Prove that if α and β are (not necessarily disjoint) permutations that commute, then $(\alpha\beta)^k = \alpha^k \beta^k$ for all $k \ge 1$.

(ii)   Give an example of two permutations α and β for which $(\alpha\beta)^2 \ne \alpha^2 \beta^2$.

*2.31   **(i)**   Prove, for all i, that $\alpha \in S_n$ moves i if and only if α^{-1} moves i.

(ii)   Prove that if $\alpha, \beta \in S_n$ are disjoint and if $\alpha\beta = (1)$, then $\alpha = (1)$ and $\beta = (1)$.

* H **2.32** If $n \ge 2$, prove that the number of even permutations in S_n is $\tfrac{1}{2}n!$.

2.33 Give an example of $\alpha, \beta, \gamma \in S_5$, none of which is the identity (1), with $\alpha\beta = \beta\alpha$ and $\alpha\gamma = \gamma\alpha$, but with $\beta\gamma \ne \gamma\beta$.

*2.34 If $n \ge 3$, show that if $\alpha \in S_n$ commutes with every $\beta \in S_n$, then $\alpha = (1)$.

H **2.35** Can the following 15-puzzle be won?

4	10	9	1
8	2	15	6
12	5	11	3
7	14	13	#

→ **2.3   GROUPS**

Generalizations of the quadratic formula for finding the roots of cubic and quartic polynomials were discovered in the early 1500s. Over the next three centuries, many tried to find analogous formulas for the roots of higher-degree polynomials, but in 1824, N. H. Abel (1802–1829) proved that there is no such formula giving the roots of the general polynomial of degree 5. In 1831, E. Galois (1811–1832) completely solved this problem by finding precisely which polynomials, of arbitrary degree, admit such a formula for their roots. His fundamental idea involved his invention of the idea of *group*. Since Galois's time, groups have arisen in many other areas of mathematics, for they are also the way to describe the notion of symmetry, as we will see later in this section and also in Chapter 6.

The essence of a "product" is that two things are combined to form a third thing of the same kind. For example, ordinary multiplication, addition, and subtraction combine two numbers to give another number, while composition combines two permutations to give another permutation.

→ **Definition.**   A (binary) *operation* on a set G is a function

$$*: G \times G \to G.$$

In more detail, an operation assigns an element $*(x, y)$ in G to each ordered pair (x, y) of elements in G. It is more natural to write $x * y$ instead of $*(x, y)$; thus, composition of functions is the function $(f, g) \mapsto g \circ f$, while multiplication, addition, and subtraction are, respectively, the functions $(x, y) \mapsto xy$, $(x, y) \mapsto x + y$, and $(x, y) \mapsto x - y$. The examples of composition and subtraction show why we want ordered pairs, for $x * y$ and $y * x$ may be distinct. As any function, an operation is single-valued; when one says this explicitly, it is usually called the ***law of substitution***:

$$\text{if } x = x' \text{ and } y = y', \text{ then } x * y = x' * y'.$$

→ **Definition.**   A *group* is a set G equipped with an operation $*$ and a special element $e \in G$, called the *identity*, such that

(i) the *associative law* holds: for every $a, b, c \in G$,

$$a * (b * c) = (a * b) * c;$$

(ii) $e * a = a$ for all $a \in G$;

(iii) for every $a \in G$, there is $a' \in G$ with $a' * a = e$.

By Proposition 2.13, the set S_X of all permutations of a set X, with composition as the operation and 1_X as the identity, is a group (the *symmetric group* on X).

We are now at the precise point when algebra becomes *abstract* algebra. In contrast to the concrete group S_n consisting of all the permutations of the set $X = \{1, 2, \ldots, n\}$ under composition, we will be proving general results about groups without specifying either their elements or their operation. Thus, products of elements are not explicitly computable but are, instead, merely subject to certain rules. It will be seen that this approach is quite fruitful, for theorems now apply to many different groups, and it is more efficient to prove theorems once for all instead of proving them anew for each group encountered. For example, the next proposition and three lemmas give properties that hold in every group G. In addition to this obvious economy, it is often simpler to work with the "abstract" viewpoint even when dealing with a particular concrete group. For example, we will see that certain properties of S_n are simpler to treat without recognizing that the elements in question are permutations (see Example 2.52).

→ **Definition.**   A group G is called *abelian*[5] if it satisfies the *commutative law*: $x * y = y * x$ holds for every $x, y \in G$.

The groups S_n, for $n \geq 3$, are not abelian because (1 2) and (1 3) are elements of S_n that do not commute: (1 2)(1 3) = (1 3 2) and (1 3)(1 2) = (1 2 3).

We prove some basic facts before giving more examples of groups.

How does one multiply three numbers? Given the expression $2 \times 3 \times 4$, for example, one can first multiply $2 \times 3 = 6$, and then multiply $6 \times 4 = 24$. Alternatively, one can first multiply $3 \times 4 = 12$ and then multiply $2 \times 12 = 24$; of course, the two answers agree because multiplication of numbers is associative. Not all operations are associative, however. For example, subtraction is not associative: if $c \neq 0$, then

$$a - (b - c) \neq (a - b) - c.$$

More generally, how does one multiply three elements $a * b * c$? Since one can only multiply two elements, there is a choice: multiply $b * c$ to get a new element of G,

---

[5]This term honors N. H. Abel who proved a theorem, in 1827, equivalent to there being a formula for the roots of a polynomial if its Galois group is commutative. This theorem is virtually forgotten today, because it was superseded by a theorem of Galois around 1830.

and now multiply this new element by a to obtain $a * (b * c)$; or, one can multiply $a * b$ and then multiply this new element by c to obtain $(a * b) * c$. Associativity says that both products are the same, $a * (b * c) = (a * b) * c$, and so it is unambiguous to write $a * b * c$ without parentheses. The next lemma shows that some associativity carries over to products with four factors (that associativity allows us to dispense with parentheses for all products having $n \geq 3$ factors is proved in Theorem 2.49).

Lemma 2.43. *If $*$ is an associative operation on a set G, then*

$$(a * b) * (c * d) = [a * (b * c)] * d$$

for all $a, b, c, d \in G$

Proof. If we write $g = a * b$, then $(a * b) * (c * d) = g * (c * d) = (g * c) * d = [(a * b) * c] * d = [a * (b * c)] * d$. $\bullet$

Lemma 2.44. *If G is a group and $a \in G$ satisfies $a * a = a$, then $a = e$.*

Proof. There is $a' \in G$ with $a' * a = e$. Multiplying both sides on the left by a' gives $a' * (a * a) = a' * a$. The right side is e, and the left side is $a' * (a * a) = (a' * a) * a = e * a = a$, and so $a = e$. $\bullet$

→ **Proposition 2.45.** *Let G be a group with operation $*$ and identity e.*

  (i) *$a * a' = e$ for all $a \in G$.*

  (ii) *$a * e = a$ for all $a \in G$.*

  (iii) *If $e_0 \in G$ satisfies $e_0 * a = a$ for all $a \in G$, then $e_0 = e$.*

  (iv) *Let $a \in G$. If $b \in G$ satisfies $b * a = e$, then $b = a'$.*

Proof.
(i) We know that $a' * a = e$, and we now show that $a * a' = e$. By Lemma 2.43,

$$\begin{aligned}
(a * a') * (a * a') &= [a * (a' * a)] * a' \\
&= (a * e) * a' \\
&= a * (e * a') \\
&= a * a'.
\end{aligned}$$

By Lemma 2.44, $a * a' = e$.
(ii) We use part (i).

$$a * e = a * (a' * a) = (a * a') * a = e * a = a.$$

Therefore, $a * e = a$.

(iii) We now prove that a group has a unique identity element; that is, no other element in G shares its defining property $e * a = a$ for all $a \in G$. If $e_0 * a = a$ for all $a \in G$, then we have, in particular, $e_0 * e_0 = e_0$. By Lemma 2.44, $e_0 = e$.

(iv) In part (i), we proved that if $a' * a = e$, then $a * a' = e$. Now

$$b = b * e = b * (a * a') = (b * a) * a' = e * a' = a'. \quad \bullet$$

In light of part (iii) of the proposition, for each $a \in G$, there is exactly one element $a' \in G$ with $a' * a = e$.

→ **Definition.**   If G is a group and $a \in G$, then the unique element $a' \in G$ such that $a' * a = e$ is called the **inverse of** a, and it is denoted by a^{-1}.

Here are three more properties holding in all groups.

→ **Lemma 2.46.**   *Let G be a group.*

  (i) *The **cancellation laws** hold: if $a, b, x \in G$, and either $x * a = x * b$ or $a * x = b * x$, then $a = b$.*

 (ii) $(a^{-1})^{-1} = a$ *for all $a \in G$.*

(iii) *If $a, b \in G$, then*
$$(a * b)^{-1} = b^{-1} * a^{-1}.$$

Proof.
(i)

$$a = e * a = (x^{-1} * x) * a = x^{-1} * (x * a)$$
$$= x^{-1} * (x * b) = (x^{-1} * x) * b = e * b = b.$$

A similar proof, using $x * x^{-1} = e$, works when x is on the right.

(ii) By Proposition 2.45(i), we have $a * a^{-1} = e$. But uniqueness of inverses, Proposition 2.45(iv), says that $(a^{-1})^{-1}$ is the unique $x \in G$ such that $x * a^{-1} = e$. Therefore, $(a^{-1})^{-1} = a$.

(iii) By Lemma 2.43,

$$(a * b) * (b^{-1} * a^{-1}) = [a * (b * b^{-1})] * a^{-1} = (a * e) * a^{-1} = a * a^{-1} = e.$$

Hence, $(a * b)^{-1} = b^{-1} * a^{-1}$, by Proposition 2.45(iv). $\bullet$

   In the proofs just given, we have been very careful about justifying every step and displaying all parentheses, for we are only beginning to learn the ideas of group theory. As one becomes more adept, however, the need for explicitly writing all such details lessens. This does not mean that one is allowed to become careless; it only means that one is growing. Of course, you must always be prepared to supply omitted details if your proof is challenged.

From now on, we will usually denote the product $a * b$ in a group by ab (we have already abbreviated $\alpha \circ \beta$ to $\alpha\beta$ in symmetric groups), and we will denote the identity e by 1. When a group is abelian, however, we will often use ***additive notation***. Here is the definition of group written in additive notation.

An ***additive group*** is a set G equipped with an operation $+$ and an identity element $0 \in G$ such that

(i) $a + (b + c) = (a + b) + c$ for every $a, b, c \in G$;

(ii) $0 + a = a$ for all $a \in G$;

(iii) for every $a \in G$, there is $-a \in G$ with $(-a) + a = 0$.

Note that the inverse of a, in additive notation, is written $-a$ instead of a^{-1}.

We now give too many examples of groups (and there are more!). Glance over the list and choose several that look interesting to you.

$\rightarrow$ **Example 2.47.**

(i) We remind the reader that S_X, the set of all permutations of a set X, is a group under composition. In particular, S_n, the set of all permutations of $X = \{1, 2, \ldots, n\}$, is a group.

(ii) The set $\mathbb{Z}$ of all integers is an additive abelian group with $a * b = a + b$, with identity $e = 0$, and with the inverse of an integer n being $-n$. Similarly, one can see that $\mathbb{Q}$, $\mathbb{R}$, and $\mathbb{C}$ are additive abelian groups.

(iii) The set $\mathbb{Q}^\times$ of all nonzero rationals is an abelian group, where $*$ is ordinary multiplication, the number 1 is the identity, and the inverse of $r \in \mathbb{Q}^\times$ is $1/r$. Similarly, $\mathbb{R}^\times$ and $\mathbb{C}^\times$ are multiplicative abelian groups.

Note that $\mathbb{Z}^\times$ is not a group, for none of its elements (aside from ± 1) has a multiplicative inverse in $\mathbb{Z}^\times$.

(iv) The circle S^1 of radius 1 with center the origin can be made into a multiplicative abelian group if we regard its points as complex numbers of modulus 1. The ***circle group*** is defined by

$$S^1 = \{z \in \mathbb{C} : |z| = 1\},$$

where the operation is multiplication of complex numbers; that this is an operation on S^1 follows from Corollary 1.23. Of course, complex multiplication is associative, the identity is 1 (which has modulus 1), and the inverse of any complex number of modulus 1 is its complex conjugate, which also has modulus 1. Therefore, S^1 is a group. Even though S^1 is an abelian group, we still write it multiplicatively, for it would be confusing to write it additively.

(v) For any positive integer n, let

$$\Gamma_n = \left\{ \zeta^k : 0 \le k < n \right\}$$

be the set of all the nth roots of unity, where

$$\zeta = e^{2\pi i/n} = \cos(2\pi/n) + i \sin(2\pi/n).$$

The reader may use De Moivre's theorem to see that Γ_n is an abelian group with operation multiplication of complex numbers; moreover, the inverse of any root of unity is its complex conjugate.

(vi) The plane $\mathbb{R} \times \mathbb{R}$ is an additive abelian group with operation vector addition; that is, if $\mathbf{v} = (x, y)$ and $\mathbf{v'} = (x', y')$, then $\mathbf{v} + \mathbf{v'} = (x + x', y + y')$. The identity is the origin $O = (0, 0)$, and the inverse of $\mathbf{v} = (x, y)$ is $-\mathbf{v} = (-x, -y)$.

(vii) The **parity group** $\mathcal{P}$ has two elements, the words *even* and *odd*, with operation

$$\text{even} + \text{even} = \text{even} = \text{odd} + \text{odd}$$

and

$$\text{even} + \text{odd} = \text{odd} = \text{odd} + \text{even}.$$

The reader may show that $\mathcal{P}$ is an abelian group.

(viii) Let X be a set. Recall that if A and B are subsets of X, then their *symmetric difference* is $A + B = (A - B) \cup (B - A)$ (symmetric difference is pictured in Figure 2.6 on page 104). The **Boolean group** $\mathcal{B}(X)$ [named after the logician G. Boole (1815–1864)] is the family of all the subsets of X equipped with addition given by symmetric difference.

It is plain that $A + B = B + A$, so that symmetric difference is commutative. The identity is $\varnothing$, the empty set, and the inverse of A is A itself, for $A + A = \varnothing$ (see Exercise 2.4 on page 104). Thus, $\mathcal{B}(X)$ is an abelian group. ◀

$\rightarrow$ **Example 2.48.**

(i) A (2×2 real) **matrix**[6] A is $\left[\begin{smallmatrix} a & c \\ b & d \end{smallmatrix} \right]$, where $a, b, c, d \in \mathbb{R}$. If $B = \left[\begin{smallmatrix} w & y \\ x & z \end{smallmatrix} \right]$, then the **product** AB is defined by

$$AB = \begin{bmatrix} a & c \\ b & d \end{bmatrix} \begin{bmatrix} w & y \\ x & z \end{bmatrix} = \begin{bmatrix} aw + cx & ay + cz \\ bw + dx & by + dz \end{bmatrix}.$$

---

[6]The word **matrix** (derived from the word meaning "mother") means "womb" in Latin; more generally, it means something that contains the essence of a thing. Its mathematical usage arises because a 2×2 matrix, which is an array of four numbers, completely describes a certain type of function $\mathbb{R}^2 \to \mathbb{R}^2$ called a linear transformation (more generally, larger matrices contain the essence of linear transformations between higher-dimensional spaces).

The elements a, b, c, d are called the *entries* of A. Call (a, c) the first *row* of A and call (b, d) the second row; call (a, b) the first *column* of A and call (c, d) the second column. Thus, each entry of the product AB is a dot product of a row of A with a column of B. The ***determinant*** of A, denoted by $\det(A)$, is the number $ad - bc$, and a matrix A is called ***nonsingular*** if $\det(A) \neq 0$. The reader may calculate that

$$\det(AB) = \det(A)\det(B),$$

from which it follows that the product of nonsingular matrices is itself nonsingular. The set GL(2, $\mathbb{R}$) of all nonsingular matrices, with operation matrix multiplication, is a (nonabelian) group, called the 2×2 ***real general linear group***: the identity is the *identity matrix*

$$I = \begin{bmatrix} 1 & 0 \\ 0 & 1 \end{bmatrix}$$

and the inverse of a nonsingular matrix A is

$$A^{-1} = \begin{bmatrix} d/\Delta & -c/\Delta \\ -b/\Delta & a/\Delta \end{bmatrix},$$

where $\Delta = ad - bc = \det(A)$. (The proof of associativity is routine, though tedious; a "clean" proof of associativity can be given once one knows the relation between matrices and linear transformations [see Corollary 4.72].)

(ii) The previous example can be modified in two ways. First, we may allow the entries to lie in $\mathbb{Q}$ or in $\mathbb{C}$, giving the groups GL(2, $\mathbb{Q}$) or GL(2, $\mathbb{C}$). We may even allow the entries to be in $\mathbb{Z}$, in which case GL(2, $\mathbb{Z}$) is defined to be the set of all such matrices with determinant ± 1 (one wants all the entries of A^{-1} to be in $\mathbb{Z}$). For readers familiar with linear algebra, all nonsingular $n \times n$ matrices form a group GL(n, $\mathbb{R}$) under multiplication.

(iii) All ***special***[7] ***orthogonal*** matrices, that is, all matrices of the form

$$A = \begin{bmatrix} \cos\alpha & -\sin\alpha \\ \sin\alpha & \cos\alpha \end{bmatrix},$$

form a group denoted by $SO(2, \mathbb{R})$, called the 2×2 ***special orthogonal group***. Let us show that matrix multiplication is an operation on $SO(2, \mathbb{R})$. The product

$$\begin{bmatrix} \cos\alpha & -\sin\alpha \\ \sin\alpha & \cos\alpha \end{bmatrix} \begin{bmatrix} \cos\beta & -\sin\beta \\ \sin\beta & \cos\beta \end{bmatrix}$$

is

$$\begin{bmatrix} \cos\alpha\cos\beta - \sin\alpha\sin\beta & -[\cos\alpha\sin\beta + \sin\alpha\cos\beta] \\ \sin\alpha\cos\beta + \cos\alpha\sin\beta & \cos\alpha\cos\beta - \sin\alpha\sin\beta \end{bmatrix}.$$

---

[7]The adjective *special* applied to a matrix usually means that its determinant is 1.

The addition theorem for sine and cosine shows that this product is again a special orthogonal matrix, for it is

$$
\begin{bmatrix}
\cos(\alpha + \beta) & -\sin(\alpha + \beta) \\
\sin(\alpha + \beta) & \cos(\alpha + \beta)
\end{bmatrix}.
$$

In fact, this calculation shows that $SO(2, \mathbb{R})$ is abelian. It is clear that the identity matrix is special orthogonal, and we let the reader check that the inverse of a special orthogonal matrix (which exists because special orthogonal matrices have determinant 1) is also special orthogonal.

In Exercise 2.77 on page 170, we will see that $SO(2, \mathbb{R})$ is a disguised version of the circle group S^1, and that this group consists of all the rotations of the plane about the origin.

(iv) The **affine** [8] **group** Aff$(1, \mathbb{R})$ consists of all functions $\mathbb{R} \to \mathbb{R}$ (called *affine maps*) of the form

$$
f_{a,b}(x) = ax + b,
$$

where a and b are fixed real numbers with $a \neq 0$. Let us check that Aff$(1, \mathbb{R})$ is a group under composition. If $f_{c,d}(x) = cx + d$, then

$$
\begin{aligned}
f_{a,b} f_{c,d}(x) &= f_{a,b}(cx + d) \\
&= a(cx + d) + b \\
&= acx + (ad + b) \\
&= f_{ac, ad+b}(x).
\end{aligned}
$$

Since $ac \neq 0$, the composite is an affine map. The identity function $1_\mathbb{R} \colon \mathbb{R} \to \mathbb{R}$ is an affine map ($1_\mathbb{R} = f_{1,0}$), while the inverse of $f_{a,b}$ is easily seen to be $f_{a^{-1}, -a^{-1}b}$. The reader should note that this composition is reminiscent of matrix multiplication.

$$
\begin{bmatrix}
a & b \\
0 & 1
\end{bmatrix}
\begin{bmatrix}
c & d \\
0 & 1
\end{bmatrix}
=
\begin{bmatrix}
ac & ad + b \\
0 & 1
\end{bmatrix}.
$$

Similarly, replacing $\mathbb{R}$ by $\mathbb{Q}$ gives the group Aff$(1, \mathbb{Q})$, and replacing $\mathbb{R}$ by $\mathbb{C}$ gives the group Aff$(1, \mathbb{C})$. ◀

The following discussion is technical, and it can be skipped as long as the reader is aware of the statement of Theorem 2.49. Informally, this theorem says that if an operation is associative, then no parentheses are needed in products involving $n \geq 3$ factors.

---

[8]Projective geometry involves enlarging the plane (and higher-dimensional spaces) by adjoining "points at infinity." The enlarged plane is called the projective plane, and the original plane is called an *affine* plane. Affine functions are special functions between affine planes.

Let G be a set with a binary operation. An n-tuple $(a_1, a_2, \ldots, a_n) \in G \times \cdots \times G$ (n factors) is called an *n-expression*; it yields many elements of G by the following procedure. Choose two adjacent a's, multiply them, and obtain an $(n-1)$-expression: the new product just formed and $n-2$ original a's. In this shorter new expression, choose two adjacent factors (either an original pair or an original one together with the new product from the first step) and multiply them. Repeat until a 2-expression (W, X) is reached; now multiply and obtain the element WX in G. Call WX an *ultimate product* derived from the original expression. For example, consider the 4-expression (a, b, c, d). Let us multiply ab, obtaining the 3-expression (ab, c, d). We may now choose either adjacent pair ab, c or c, d; in either case, multiply these and obtain 2-expressions $((ab)c, d)$ or (ab, cd). The elements in either of these last expressions can now be multiplied to give the ultimate products $[(ab)c]d$ or $(ab)(cd)$. Other ultimate products derived from (a, b, c, d) arise by multiplying bc or cd as the first step, yielding (a, bc, d) or (a, b, cd). To say that an operation is associative is to say that the two ultimate products arising from 3-expressions (a, b, c) are equal. It is not obvious, even when an operation is associative, whether all the ultimate products derived from a longer expression are equal.

Definition. An n-expression $(a_1, a_2, \ldots, a_n)$ *needs no parentheses* if all ultimate products derived from it are equal; that is, no matter what choices are made of adjacent factors to multiply, all the resulting products in G are equal.

Theorem 2.49 (Generalized Associativity). *If $n \geq 3$, then every n-expression $(a_1, a_2, \ldots, a_n)$ in a group G needs no parentheses.*

Remark. Note that neither the identity element nor inverses will be used in the proof. Thus, the hypothesis of the theorem can be weakened by assuming that G is only a *semigroup*; that is, G is a nonempty set equipped with an associative binary operation. ◀

Proof. The proof is by (the second form of) induction. The base step $n = 3$ is associativity. For the inductive step, consider 2-expressions of G obtained from an n-expression $(a_1, a_2, \ldots, a_n)$ after two series of choices:

$$(W, X) = (a_1 \cdots a_i, a_{i+1} \cdots a_n) \quad \text{and} \quad (Y, Z) = (a_1 \cdots a_j, a_{j+1} \cdots a_n).$$

We must prove that $WX = YZ$ in G. By induction, each of the elements $W = a_1 \cdots a_i$, $X = a_{i+1} \cdots a_n$, $Y = a_1 \cdots a_j$, and $Z = a_{j+1} \cdots a_n$, is the (one and only!) ultimate product from m-expressions with $m < n$. Without loss of generality, we may assume that $i \leq j$. If $i = j$, then the inductive hypothesis gives $W = Y$ and $X = Z$ in G, and so $WX = YZ$, as desired.

We may now assume that $i < j$. Let A be an ultimate product of the i-expression $(a_1, \ldots, a_i)$, let B be an ultimate product from the expression $(a_{i+1}, \ldots, a_j)$, and let C be an ultimate product from the expression $a_{j+1} \cdots a_n$. The group elements A, B, and C are unambiguously defined, for the inductive hypothesis says that each of the shorter expressions yields only one ultimate product. Now $W = A$, for both are ultimate products from the i-expression $(a_1, \ldots, a_i)$, $Z = C$ [both are ultimate products from the $(n - j)$-expression $(a_{j+1}, \ldots, a_n)$], $X = BC$ [both are ultimate products from the $(n - i)$-expression $(a_{i+1}, \ldots, a_n)$], and $Y = AB$ [both are ultimate products from the j-expression $(a_1, \ldots, a_j)$]. We conclude that $WX = A(BC)$ and $YZ = (AB)C$, and so associativity, the base step $n = 3$, gives $WX = YZ$, as desired. •

→ **Definition.** If G is a group and if $a \in G$, define the **powers**[9] a^n, for $n \geq 1$, inductively:

$$a^1 = a \quad \text{and} \quad a^{n+1} = aa^n.$$

Define $a^0 = 1$ and, if n is a positive integer, define

$$a^{-n} = (a^{-1})^n.$$

We let the reader prove that $(a^{-1})^n = (a^n)^{-1}$; this is a special case of the equation in Lemma 2.46(iii).

There is a hidden complication here. The first and second powers are fine: $a^1 = a$ and $a^2 = aa$. There are two possible cubes: we have defined $a^3 = aa^2 = a(aa)$, but there is another reasonable contender: $(aa)a = a^2a$. If one assumes associativity, then these are equal:

$$a^3 = aa^2 = a(aa) = (aa)a = a^2a.$$

Generalized associativity shows that all powers of an elements are unambiguously defined.

Corollary 2.50. *If G is a group, if $a \in G$, and if m, $n \geq 1$, then*

$$a^{m+n} = a^m a^n \quad \text{and} \quad (a^m)^n = a^{mn}.$$

Proof. Both a^{m+n} and $a^m a^n$ arise from the expression having $m+n$ factors each equal to a; in the second instance, both $(a^m)^n$ and a^{mn} arise from the expression having mn factors each equal to a. •

---

[9]The terminology x square and x cube for x^2 and x^3 is, of course, geometric in origin. Usage of the word *power* in this context goes back to Euclid, who wrote, "The power of a line is the square of the same line" (from the first English translation of Euclid, in 1570, by H. Billingsley). "Power" was the standard European rendition of the Greek *dunamis* (from which dynamo derives). However, contemporaries of Euclid, such as Aristotle and Plato, often used *dunamis* to mean amplification, and this seems to be a more appropriate translation, for Euclid was probably thinking of a 1-dimensional line sweeping out a 2-dimensional square. (I thank Donna Shalev for informing me of the classical usage of *dunamis*.)

It follows that any two powers of an element a in a group commute:

$$a^m a^n = a^{m+n} = a^{n+m} = a^n a^m.$$

The proofs of the various statements in the next proposition, while straightforward, are not short.

→ **Proposition 2.51 (Laws of Exponents).**   *Let G be a group, let $a, b \in G$, and let m and n be (not necessarily positive) integers.*

(i) *If a and b commute, then $(ab)^n = a^n b^n$.*

(ii) *$(a^n)^m = a^{mn}$.*

(iii) *$a^m a^n = a^{m+n}$.*

Proof.   Exercises for the reader.   •

The notation a^n is the natural way to denote $a * a * \cdots * a$ if a appears n times. However, if the operation is $+$, then it is more natural to denote $a + a + \cdots + a$ by na. Let G be a group written additively; if $a, b \in G$ and m and n are (not necessarily positive) integers, then Proposition 2.51 is usually rewritten:

(i) $n(a + b) = na + nb$.

(ii) $m(na) = (mn)a$.

(iii) $ma + na = (m + n)a$.

→ **Example 2.52.**
Suppose a deck of cards is shuffled, so that the order of the cards has changed from $1, 2, 3, 4, \ldots, 52$ to $2, 1, 4, 3, \ldots, 52, 51$. If we shuffle again in the same way, then the cards return to their original order. But a similar thing happens for any permutation α of the 52 cards: if one repeats α sufficiently often, the deck is eventually restored to its original order. One way to see this uses our knowledge of permutations. Write α as a product of disjoint cycles, say, $\alpha = \beta_1 \beta_2 \cdots \beta_t$, where β_i is an r_i-cycle. Now $\beta_i^{r_i} = (1)$ for every i, by Exercise 2.25 on page 124, and so $\beta_i^k = (1)$, where $k = r_1 \cdots r_t$. Since disjoint cycles commute, Exercise 2.30 on page 124 gives

$$\alpha^k = (\beta_1 \cdots \beta_t)^k = \beta_1^k \cdots \beta_t^k = (1).$$

Here is a more general result with a simpler proof (abstract algebra can be easier than algebra): if G is a finite group and $a \in G$, then $a^k = 1$ for some $k \geq 1$. We use the argument in Lemma 2.23(i). Consider the sequence

$$1, a, a^2, \ldots, a^n, \ldots.$$

Since G is finite, there must be a repetition occurring in this sequence: there are integers $m > n$ with $a^m = a^n$, and hence $1 = a^m a^{-n} = a^{m-n}$. We have shown that there is some positive power of a equal to 1. Our original argument that $\alpha^k = (1)$ for a permutation α of 52 cards is not worthless, for Proposition 2.55 will show that we may choose k to be the $\mathrm{lcm}(r_1, \dots, r_t)$. ◀

→ **Definition.**   Let G be a group and let $a \in G$. If $a^k = 1$ for some $k \geq 1$, then the smallest such exponent $k \geq 1$ is called the *order* of a; if no such power exists, then one says that a has *infinite order*.

The argument given in Example 2.52 shows that every element in a finite group has finite order. In any group G, the identity has order 1, and it is the only element in G of order 1; an element has order 2 if and only if it is not the identity and it is equal to its own inverse. The matrix $A = \begin{bmatrix} 1 & 1 \\ 0 & 1 \end{bmatrix}$ in the group $\mathrm{GL}(2, \mathbb{R})$ has infinite order, for $A^k = \begin{bmatrix} 1 & k \\ 0 & 1 \end{bmatrix} \neq \begin{bmatrix} 1 & 0 \\ 0 & 1 \end{bmatrix}$ for all $k \geq 1$.

Lemma 2.53.   *Let G be a group and assume that $x \in G$ has finite order k. If $x^n = 1$, then $k \mid n$.*

Proof.   By the division algorithm, $n = qk + r$, where $0 \leq r < k$. Hence,

$$1 = x^n = x^{qk+r} = (x^k)^q x^r = x^r.$$

Since $x^r = 1$ and $r < k$, we must have $r = 0$; that is, $k \mid n$. •

Here is alternative proof of Lemma 2.53 which shows that there is a connection between the order of a group element and *ideals*: subsets of $\mathbb{Z}$ that arose in the proof that gcd's are linear combinations.

→ **Proposition 2.54.**   *Let G be a group and assume that $x \in G$ has finite order k.*

(i) *$I = \{n \in \mathbb{Z} : x^n = 1\}$ is the set of all the multiples of k.*

(ii) *If $x^n = 1$, then $k \mid n$.*

Proof.
(i) We show that I satisfies the hypotheses of Corollary 1.37 on page 41:
(*a*): $0 \in I$ because $x^0 = 1$.
(*b*): If $n, m \in I$, then $x^n = 1$ and $x^m = 1$, so that $x^{n-m} = x^n x^{-m} = 1$; hence, $n - m \in I$.
(*c*): If $n \in I$ and $q \in \mathbb{Z}$, then $x^n = 1$ and $x^{qn} = (x^n)^q = 1$; hence, $qn \in I$.

Therefore, I consists of all the multiples of d, where d is the smallest positive integer in I. But the order k of x is, by definition, the smallest such integer, and so $d = k$.
(ii) If $x^n = 1$, then $n \in I = (k)$, and so $k \mid n$. •

What is the order of a permutation in S_n?

→ **Proposition 2.55.** *Let $\alpha \in S_n$.*

(i) *If α is an r-cycle, then α has order r.*

(ii) *If $\alpha = \beta_1 \cdots \beta_t$ is a product of disjoint r_i-cycles β_i, then α has order $m = \text{lcm}(r_1, \ldots, r_t)$.*

(iii) *If p is a prime, then α has order p if and only if it is a p-cycle or a product of disjoint p-cycles.*

Proof.
(i) This is Exercise 2.25(i) on page 124.
(ii) Each β_i has order r_i, by (i). Suppose that $\alpha^M = (1)$. Since the β_i commute, $(1) = \alpha^M = (\beta_1 \cdots \beta_t)^M = \beta_1^M \cdots \beta_t^M$. By Exercise 2.31(ii) on page 124, disjointness of the β's implies that $\beta_i^M = (1)$ for each i, so that Lemma 2.53 gives $r_i \mid M$ for all i; that is, M is a common multiple of $r_1, \ldots, r_t$. But if $m = \text{lcm}(r_1, \ldots, r_t)$, then it is easy to see that $\alpha^m = (1)$. Hence, α has order m.
(iii) Write α as a product of disjoint cycles and use part (ii). •

For example, a permutation in S_n has order 2 if and only if it is either a transposition or a product of disjoint transpositions.

We can now augment Table 2.2 in Example 2.30.

Cycle Structure	Number	Order	Parity
(1)	1	1	Even
(1 2)	10	2	Odd
(1 2 3)	20	3	Even
(1 2 3 4)	30	4	Odd
(1 2 3 4 5)	24	5	Even
(1 2)(3 4 5)	20	6	Odd
(1 2)(3 4)	15	2	Even
	120		

Table 2.3 Permutations in S_5

Symmetry

We now present a connection between groups and geometric symmetry (and so we will be using some linear algebra over $\mathbb{R}$). What do we mean when we say that an isosceles triangle Δ is symmetric? Figure 2.10 shows $\Delta = \Delta ABC$ with its base AB

on the x-axis and with the y-axis being the perpendicular-bisector of AB. Close your eyes; let Δ be reflected in the y-axis (so that the vertices A and B are interchanged); open your eyes. You cannot tell that Δ has been reflected; that is, Δ is symmetric about the y-axis. On the other hand, if Δ were reflected in the x-axis, then it would be obvious, once your eyes are reopened, that a reflection had taken place; that is, Δ is not symmetric about the x-axis. Reflection is a special kind of *isometry*.

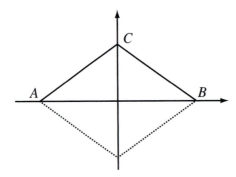

Figure 2.10   Isosceles Triangle

→ **Definition.**   An *isometry* of the plane is a function $\varphi \colon \mathbb{R}^2 \to \mathbb{R}^2$ that is *distance preserving*: for all points $P = (a, b)$ and $Q = (c, d)$ in $\mathbb{R}^2$,

$$\|\varphi(P) - \varphi(Q)\| = \|P - Q\|,$$

where $\|P - Q\| = \sqrt{(a-c)^2 + (b-d)^2}$ is the distance from P to Q.

Let $P \cdot Q$ denote the *dot product*:

$$P \cdot Q = (a, b) \cdot (c, d) = ac + bd.$$

Now

$$
\begin{aligned}
(P - Q) \cdot (P - Q) &= P \cdot P - 2(P \cdot Q) + Q \cdot Q \\
&= (a^2 + b^2) - 2(ac + bd) + (c^2 + d^2) \\
&= (a^2 - 2ac + c^2) + (b^2 - 2bd + d^2) \\
&= (a - c)^2 + (b - d)^2 \\
&= \|P - Q\|^2.
\end{aligned}
$$

Lemma 2.56. *Let φ be an isometry of the plane. Then φ preserves dot products [that is, $\varphi(P) \cdot \varphi(Q) = P \cdot Q$ for all points P and Q] if and only if $\varphi(O) = O$.*

Proof. If $\varphi(P)\cdot\varphi(Q) = P\cdot Q$ for all points P and Q, then $\varphi(O)\cdot\varphi(O) = O\cdot O = 0$. It follows that φ fixes the origin, for if $\varphi(O) \neq O$, then $\varphi(O) \cdot \varphi(O) = \|\varphi(O)\|^2 \neq 0$.

Conversely, if $\varphi(O) = O$, then

$$\|P\| = \|P - O\| = \|\varphi(P) - \varphi(O)\| = \|\varphi(P)\|$$

for all P, because φ is an isometry. Hence, for all points P and Q,

$$\begin{aligned}\|\varphi(P)\|^2 + \|\varphi(Q)\|^2 - 2\varphi(P)\varphi(Q) &= [\varphi(P) - \varphi(Q)] \cdot [\varphi(P) - \varphi(Q)] \\ &= \|\varphi(P) - \varphi(Q)\|^2 \\ &= \|P - Q\|^2 \\ &= (P - Q) \cdot (P - Q) \\ &= \|P\|^2 + \|Q\|^2 - 2P \cdot Q.\end{aligned}$$

Therefore, $\varphi(P) \cdot \varphi(Q) = P \cdot Q$. •

Recall the formula giving the geometric interpretation of the dot product:

$$P \cdot Q = \|P\| \|Q\| \cos\theta,$$

where θ is the angle between P and Q. It follows that every isometry preserves angles. In particular, P and Q are orthogonal if and only if $P \cdot Q = 0$, and so isometries preserve perpendicularity. Conversely, if $\varphi(P) \cdot \varphi(Q) = P \cdot Q$, that is, if φ preserves dot products, then the formula $(P - Q) \cdot (P - Q) = \|P - Q\|^2$ shows that φ is an isometry.

We denote the set of all isometries of the plane by **Isom**$(\mathbb{R}^2)$; its subset consisting of all those isometries φ with $\varphi(O) = O$ is called the ***orthogonal group of the plane***, and it is denoted by $O(2, \mathbb{R})$. We will see, in Proposition 2.61, that both **Isom**$(\mathbb{R}^2)$ and $O(2, \mathbb{R})$ are groups under composition.

We introduce some notation to help us analyze isometries.

Notation. If P and Q are distinct points in the plane, let $L[P, Q]$ denote the line they determine, and let PQ denote the line segment with endpoints P and Q.

Here are some examples of isometries.

Example 2.57.

(i) Given an angle θ, ***rotation*** R_θ about the origin O is defined as follows: $R_\theta(O) = O$; if $P \neq O$, draw the line segment PO in Figure 2.11, rotate it θ (counterclockwise if θ is positive, clockwise if θ is negative) to OP', and define $R_\theta(P) = P'$. Of course, one can rotate about any point in the plane.

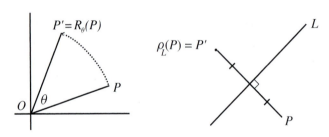

Figure 2.11   Rotation   **Figure 2.12**   Reflection

(ii) **Reflection** ρ_L in a line L, called its **axis**, fixes each point in L; if $P \notin L$, then $\rho_L(P) = P'$, as in Figure 2.12 (L is the perpendicular-bisector of PP'). If one pretends that the axis L is a mirror, then P' is the mirror image of P. Now $\rho_L \in \mathbf{Isom}(\mathbb{R}^2)$; if L passes through the origin, then $\rho_L \in O(2, \mathbb{R})$.

(iii) Given a point V, **translation**[10] **by** V is the function $\tau_V : \mathbb{R}^2 \to \mathbb{R}^2$ defined by $\tau_V(U) = U + V$. Translations lie in $\mathbf{Isom}(\mathbb{R}^2)$; a translation τ_V fixes the origin if and only if $V = O$, so that the identity is the only translation which is also a rotation. ◄

Proposition 2.58.   *If φ is an isometry of the plane, then distinct points P, Q, R in $\mathbb{R}^2$ are collinear if and only if $\varphi(P), \varphi(Q), \varphi(R)$ are collinear. Hence, if L is a line, then $\varphi(L)$ is also a line.*

Proof.   Suppose that P, Q, R are collinear. Choose notation so that R is between P and Q; hence, $\|P - Q\| = \|P - R\| + \|R - Q\|$. If $\varphi(P), \varphi(Q), \varphi(R)$ are not collinear, then they are the vertices of a triangle. The triangle inequality gives

$$\|\varphi(P) - \varphi(Q)\| < \|\varphi(P) - \varphi(R)\| + \|\varphi(R) - \varphi(Q)\|,$$

contradicting φ preserving distance. A similar argument proves the converse. If P, Q, R are not collinear, then they are the vertices of a triangle. If $\varphi(P), \varphi(Q), \varphi(R)$ are collinear, then the strict inequality displayed above now becomes an equality, contradicting φ preserving distance.

If $\varphi(L)$ is not a line, then it contains 3 noncollinear points $\varphi(P), \varphi(Q), \varphi(R)$, where P, Q, R lie on L, a contradiction. •

Every isometry φ is an injection. If $P \neq Q$, then $\|P - Q\| \neq 0$, so that $\|\varphi(P) - \varphi(Q)\| = \|P - Q\| \neq 0$; hence, $\varphi(P) \neq \varphi(Q)$. It is less obvious that isometries are surjections, but we will soon see that they are.

---

[10]The word *translation* comes from the Latin word meaning "to transfer." It usually means passing from one language to another, but here it means a special way of moving each point to another.

Proposition 2.59. *Every isometry φ of $\mathbb{R}^2$ fixing the origin is a linear transformation.*

Proof. Let $C_d = \{Q \in \mathbb{R}^2 : \|Q - O\| = d\}$ be the circle of radius $d > 0$ having center O. We claim that $\varphi(C_d) \subseteq C_d$. If $P \in C_d$, then $\|P - O\| = d$; since φ preserves distance, $d = \|\varphi(P) - \varphi(O)\| = \|\varphi(P) - O\|$; thus, $\varphi(P) \in C_d$.

Let $P \neq O$ be a point in $\mathbb{R}^2$, and let $r \in \mathbb{R}$. If $\|P - O\| = p$, then $\|rP - O\| = |r|p$. Hence, $rP \in L[O, P] \cap C_{|r|p}$, where $C_{|r|p}$ is the circle with center O and radius $|r|p$. Since φ preserves collinearity, by Lemma 2.58, $\varphi(L[O, P] \cap C_{|r|p}) \subseteq L[O, \varphi(P)] \cap C_{|r|p}$; that is, $\varphi(rP) = \pm r\varphi(P)$ (for a line intersects a circle in at most two points).

If we eliminate the possibility $\varphi(rP) = -r\varphi(P)$, then we can conclude that $\varphi(rP) = r\varphi(P)$. In case $r > 0$, the origin O lies between $-rP$ and P, and so the distance from $-rP$ to P is $rp + p$. On the other hand, the distance from rP to P is $|rp - p|$ (if $r > 1$, the distance is $rp - p$; if $0 < r < 1$, then the distance is $p - pr$). But $r + rp \neq |rp - p|$, and so $\varphi(rP) \neq -r\varphi(P)$ (because φ preserves distance). A similar argument works in case $r < 0$.

It remains to prove that $\varphi(P + Q) = \varphi(P) + \varphi(Q)$. If O, P, Q are collinear, then choose a point U on the line $L[O, P]$ whose distance to the origin is 1. Thus, $P = pU$, $Q = qU$, and $P + Q = (p + q)U$. The points $O = \varphi(O), \varphi(U), \varphi(P), \varphi(Q)$ are collinear. Since φ preserves scalar multiplication, we have

$$
\begin{aligned}
\varphi(P) + \varphi(Q) &= \varphi(pU) + \varphi(qU) \\
&= p\varphi(U) + q\varphi(U) \\
&= (p + q)\varphi(U) \\
&= \varphi((p + q)U) \\
&= \varphi(P + Q).
\end{aligned}
$$

If O, P, Q are not collinear, then $P + Q$ is given by the parallelogram law: $P + Q$ is the point S such that O, P, Q, S are the vertices of a parallelogram. Since φ preserves distance, the points $O = \varphi(U), \varphi(P), \varphi(Q), \varphi(S)$ are the vertices of a parallelogram, and so $\varphi(S) = \varphi(P) + \varphi(Q)$. But $S = P + Q$, and so $\varphi(P + Q) = \varphi(P) + \varphi(Q)$, as desired. •

Corollary 2.60. *Every isometry $\varphi \colon \mathbb{R}^2 \to \mathbb{R}^2$ is a bijection, and every isometry fixing 0 is a nonsingular linear transformation.*

Proof. Let us first assume that φ fixes the origin: $\varphi(0) = 0$. By Proposition 2.59, φ is a linear transformation. Since φ is injective, $P = \varphi(e_1)$, $Q = \varphi(e_2)$ is a basis of $\mathbb{R}^2$, where $e_1 = (1, 0)$, $e_2 = (0, 1)$ is the standard basis of $\mathbb{R}^2$. It follows that the function $\psi \colon \mathbb{R}^2 \to \mathbb{R}^2$, defined by $\psi \colon aP + bQ \mapsto ae_1 + be_2$, is a (single-valued) function, and that ψ and φ are inverse functions. Therefore, φ is a bijection, and hence it is nonsingular.

Suppose that φ is any isometry, so that $\varphi(0) = U$. Now $\tau_{-U} \circ \varphi \colon 0 \mapsto U \mapsto 0$, so that $\tau_{-U} \circ \varphi = \theta$, where θ is a nonsingular linear transformation. Therefore, $\varphi = \tau_U \circ \theta$ is a bijection, being the composite of bijections.   •

We will study $\mathbf{Isom}(\mathbb{R}^2)$ more carefully in Chapter 6. In particular, we will see that all isometries are either rotations, reflections, translations, or a fourth type, *glide reflections*.

→ **Definition.**   The *orthogonal group* $O(2, \mathbb{R})$ is the set of all isometries of the plane which fix the origin.

Proposition 2.61.   *Both* $\mathbf{Isom}(\mathbb{R}^2)$ *and* $O(2, \mathbb{R})$ *are groups under composition.*

Proof.   We show that $\mathbf{Isom}(\mathbb{R}^2)$ is a group. Clearly, $1_{\mathbb{R}^2}$ is an isometry, so that $1_{\mathbb{R}^2} \in \mathbf{Isom}(\mathbb{R}^2)$. Let φ' and φ be isometries. For all points P and Q, we have

$$\|(\varphi'\varphi)(P)) - (\varphi'\varphi)(Q))\| = \|\varphi'(\varphi(P)) - \varphi'(\varphi(Q))\|$$
$$= \|\varphi(P) - \varphi(Q)\|$$
$$= \|P - Q\|,$$

and so $\varphi'\varphi$ is also an isometry; that is, composition is an operation on $\mathbf{Isom}(\mathbb{R}^2)$. If $\varphi \in \mathbf{Isom}(\mathbb{R}^2)$, then φ is a bijection, by Corollary 2.60, and so it has an inverse φ^{-1}. Now φ^{-1} is also an isometry:

$$\|P - Q\| = \|\varphi(\varphi^{-1}(P)) - \varphi(\varphi^{-1}(Q))\| = \|\varphi^{-1}(P) - \varphi^{-1}(Q)\|.$$

Therefore, $\mathbf{Isom}(\mathbb{R}^2)$ is a group, for composition of functions is always associative, by Lemma 2.6.

The reader may adapt this proof to show that $O(2, \mathbb{R})$ is also a group.   •

Corollary 2.62.   *If* O, P, Q *are noncollinear points, and if* φ *and* ψ *are isometries of the plane such that* $\varphi(P) = \psi(P)$ *and* $\varphi(Q) = \psi(Q)$, *then* $\varphi = \psi$.

Proof.   Since O, P, Q are noncollinear points, the list P, Q is linearly independent in the vector space $\mathbb{R}^2$. Since $\dim(\mathbb{R}^2) = 2$, this is a basis [see Corollary 4.24(ii)], and any two linear transformations that agree on a basis are equal (see Corollary 4.63).   •

Let us return to symmetry.

Example 2.63.
If Δ is a triangle with vertices P, Q, U and if φ is an isometry, then $\varphi(\Delta)$ is the triangle with vertices $\varphi(P), \varphi(Q), \varphi(U)$, by Proposition 2.58. If we assume further that $\varphi(\Delta) = \Delta$, then φ permutes the vertices P, Q, U (see Figure 2.13). Assume that

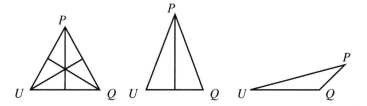

Figure 2.13 Equilateral, Isosceles, Scalene

the center of Δ is O. If Δ is isosceles (with equal sides PQ and PU), and if ρ_ℓ is the reflection with axis $\ell = L[O, P]$, then $\rho_\ell(\Delta) = \Delta$ (we can thus describe ρ_ℓ by the transposition $(Q\ U)$, for it fixes P and interchanges Q and U); on the other hand, if Δ is not isosceles, then $\rho_{\ell'}(\Delta) \neq \Delta$, where $\ell' = L[O, Q]$. If Δ is equilateral, then $\rho_{\ell''}(\Delta) = \Delta$, where $\ell'' = L[O, U]$ and $\rho_{\ell''}(\Delta) = \Delta$ [we can describe these reflections by the transpositions $(P\ U)$ and $(P\ Q)$, respectively]; these reflections do not carry Δ into itself when Δ is only isosceles. Moreover, the rotation about O by 120° and 240° also carry Δ into itself [these rotations can be described by the 3-cycles $(P\ Q\ U)$ and $(P\ U\ Q)$]. We see that an equilateral triangle is "more symmetric" than an isosceles triangle, and that an isosceles triangle is "more symmetric" than a triangle Δ that is not even isosceles [for such a triangle, $\varphi(\Delta) = \Delta$ implies that $\varphi = 1$]. ◀

→ **Definition.** The *symmetry group* $\Sigma(\Omega)$ of a figure Ω in the plane is the set of all isometries φ of the plane with $\varphi(\Omega) = \Omega$. The elements of $\Sigma(\Omega)$ are called *symmetries* of Ω.

It is straightforward to see that $\Sigma(\Omega)$ is always a group.

Example 2.64.

(i) A regular 3-gon π_3 is an equilateral triangle, and $|\Sigma(\pi_3)| = 6$, as we saw in Example 2.63.

(ii) Let π_4 be a square (a regular 4-gon) having vertices $\{v_0, v_1, v_2, v_3\}$; draw π_4 in the plane so that its center is at the origin O and its sides are parallel to the axes. It is easy to see that every $\varphi \in \Sigma(\pi_4)$ permutes the vertices; indeed, a symmetry φ of π_4 is determined by $\{\varphi(v_i) : 0 \leq i \leq 3\}$, and so there are at most $24 = 4!$ possible symmetries. Not every permutation in S_4 arises from a symmetry of π_4, however. If v_i and v_j are adjacent, then $\|v_i - v_j\| = 1$, but $\|v_0 - v_2\| = \sqrt{2} = \|v_1 - v_3\|$; it follows that φ must preserve adjacency (for isometries preserve distance). There are only eight symmetries of π_4 (this is proved in Theorem 2.65). Aside from the identity and the three rotations about

O by 90°, 180°, and 270°, there are four reflections, respectively, with axes $L[v_0, v_2]$, $L[v_1, v_3]$, the x-axis, and the y-axis. The group $\Sigma(\pi_4)$ is called a ***dihedral group*** with eight elements, and it is denoted by D_8.

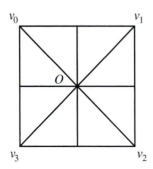

Figure 2.14   Symmetries of π_4 **Figure 2.15**   Symmetries of π_5

(iii) The symmetry group $\Sigma(\pi_5)$ of a regular pentagon π_5 having vertices $v_0, \ldots, v_4$ and center O has 10 elements: the rotations about the origin by $(72j)°$, where $0 \leq j \leq 4$, as well as the reflections with axes $L[O, v_k]$ for $0 \leq k \leq 4$ (Theorem 2.65 shows that there are no other symmetries). The symmetry group $\Sigma(\pi_5)$ is called a ***dihedral group*** with 10 elements, and it is denoted by D_{10}.   ◄

The symmetry group $\Sigma(\pi_n)$ of a regular polygon π_n with center at O and vertices $v_0, v_1, \ldots, v_{n-1}$ is called a *dihedral* [11] *group* D_{2n}. However, we give a definition that does not depend on geometry.

→ **Definition.**   A group D_{2n} with exactly $2n$ elements is called a ***dihedral group*** if it contains an element a of order n and an element b of order 2 such that $bab = a^{-1}$.

If $n = 2$, then a dihedral group D_4 is abelian; if $n \geq 3$, then D_{2n} is not abelian. Exercise 2.72 on page 169 shows that there is essentially only one dihedral group with $2n$ elements (more precisely, any two such are isomorphic).

→ **Theorem 2.65.**   *The symmetry group $\Sigma(\pi_n)$ of a regular n-gon π_n is a dihedral group with $2n$ elements.*

---

[11] F. Klein was investigating those finite groups occurring as subgroups of the group of isometries of $\mathbb{R}^3$. Some of these occur as symmetry groups of regular polyhedra [from the Greek *poly* meaning "many" and *hedron* meaning "two-dimensional side." He invented a degenerate polyhedron that he called a *dihedron*, from the Greek word *di* meaning "two," which consists of two congruent regular polygons of zero thickness pasted together. The symmetry group of a dihedron is thus called a *dihedral group*.

Proof.   Let π_n have vertices $v_0, \ldots, v_{n-1}$ and center O. Define a to be the rotation about O by $(360/n)°$:

$$a(v_i) = \begin{cases} v_{i+1} & \text{if } 0 \leq i < n - 1 \\ v_0 & \text{if } i = n - 1. \end{cases}$$

It is clear that a has order n. Define b to be the reflection with axis $L[O, v_0]$; thus,

$$b(v_i) = \begin{cases} v_0 & \text{if } i = 0 \\ v_{n-i} & \text{if } 1 \leq i \leq n - 1. \end{cases}$$

It is clear that b has order 2. There are n distinct symmetries $1, a, a^2, \ldots, a^{n-1}$ (because a has order n), and $b, ab, a^2b, \ldots, a^{n-1}b$ are all distinct as well (by the cancellation law). If $a^s = a^r b$, where $0 \leq r \leq n - 1$ and $s = 0, 1$, then $a^s(v_i) = a^r b(v_i)$ for all i. Now $a^s(v_0) = v_s$ while $a^r b(v_0) = v_{r-1}$; hence, $s = r - 1$. If $i = 1$, then $a^{r-1}(v_1) = v_{r+1}$ while $a^r b(v_1) = v_{r-1}$. Therefore, $a^s \neq a^r b$ for all r, s, and we have exhibited $2n$ distinct symmetries in $\Sigma(\pi_n)$.

We now show that there are no other symmetries of π_n. We may assume that π_n has its center O at the origin, and so every symmetry φ fixes O; that is, φ is a linear transformation (by Proposition 2.59). The vertices adjacent to v_0, namely, v_1 and v_{n-1}, are the closest vertices to v_0; that is, if $2 \leq i \leq n - 2$, then $\|v_i - v_0\| > \|v_1 - v_0\|$. Therefore, if $\varphi(v_0) = v_j$, then $\varphi(v_1) = v_{j+1}$ or $\varphi(v_1) = v_{j-1}$. In the first case, $a^j(v_0) = \varphi(v_0)$ and $a^j(v_1) = \varphi(v_1)$, so that Corollary 2.62 gives $\varphi = a^j$. In the second case, $a^j b(v_0) = v_j$ and $a^j b(v_1) = v_{j-1}$, and this corollary gives $\varphi = a^j b$. Therefore, $|\Sigma(\pi_n)| = 2n$.

We have shown that $\Sigma(\pi_n)$ is a group with exactly $2n$ elements and which contains elements a and b of orders n and 2, respectively. It remains to show that $bab = a^{-1}$. By Corollary 2.62, it suffices to evaluate each of these on v_0 and v_1. But $bab(v_0) = v_{n-1} = a^{-1}(v_0)$ and $bab(v_1) = v_0 = a^{-1}(v_1)$.   •

Symmetry arises in calculus when describing figures in the plane. We quote from Edwards and Penny, *Calculus and Analytic Geometry*, 3d ed., 1990, p. 456, as they describe different kinds of symmetry that might be enjoyed by a curve with equation $f(x, y) = 0$.

(i)   *Symmetry about the x-axis*: the equation of the curve is unaltered when y is replaced by $-y$.

(ii)   *Symmetry about the y-axis*: the equation of the curve is unaltered when x is replaced by $-x$.

(iii)   *Symmetry with respect to the origin*: the equation of the curve is unaltered when x is replaced by $-x$ and y is replaced by $-y$.

(iv)   *Symmetry about the 45° line $y = x$*: the equation is unaltered when x and y are interchanged.

In our language, their first symmetry is ρ_x, reflection with axis x-axis, the second is ρ_y, reflection with axis y-axis, the third is R_{180}, rotation by $180°$, and the fourth is ρ_L, where L is the $45°$ line. One can now say when a function of two variables has symmetry. For example, a function $f(x, y)$ has the first type of symmetry if $f(x, y) = f(x, -y)$. In this case, the graph Γ of the equation $f(x, y) = 0$ [consisting of all the points (a, b) for which $f(a, b) = 0$] is symmetric about the x-axis, for $(a, b) \in \Gamma$ implies $(a, -b) \in \Gamma$.

EXERCISES

H **2.36** True or false with reasons.

    (i)   The function $e \colon \mathbb{N} \times \mathbb{N} \to \mathbb{N}$, defined by $e(m, n) = m^n$, is an associative operation.

    (ii)   Every group is abelian.

    (iii)   The set of all positive real numbers is a group under multiplication.

    (iv)   The set of all positive real numbers is a group under addition.

    (v)   For all $a, b \in G$, where G is a group, $aba^{-1}b^{-1} = 1$.

    (vi)   Every permutation of the vertices v_1, v_2, v_3 of an equilateral triangle π_3 is the restriction of a symmetry of π_3.

    (vii)   Every permutation of the vertices v_1, v_2, v_3, v_4 of a square π_4 is the restriction of a symmetry of π_4.

    (viii)   If $a, b \in G$, where G is a group, then $(ab)^n = a^n b^n$ for all $n \in \mathbb{N}$.

    (ix)   Every infinite group contains an element of infinite order.

    (x)   Complex conjugation permutes the roots of every polynomial having real coefficients.

2.37 If $a_1, a_2 \ldots, a_n$ are (not necessarily distinct) elements in a group G, prove that

$$(a_1 a_2 \cdots a_n)^{-1} = a_n^{-1} \cdots a_2^{-1} a_1^{-1}.$$

2.38   **(i)**   Compute the order, inverse, and parity of

$$\alpha = (1\ 2)(4\ 3)(1\ 3\ 5\ 4\ 2)(1\ 5)(1\ 3)(2\ 3).$$

    (ii)   What are the respective orders of the permutations in Exercises 2.22 and 2.28 on page 124?

2.39 H **(i)**   How many elements of order 2 are there in S_5 and in S_6?

    H **(ii)**   How many elements of order 2 are there in S_n?

*H **2.40** Let G be a group, and let $y \in G$ have of order m. If $m = dt$ for some $d \geq 1$, prove that y^t has order d.

*2.41 Let G be a group and let $a \in G$ have order dk, where $d, k > 1$. Prove that if there is $x \in G$ with $x^d = a$, then the order of x is $d^2 k$. Conclude that the order of x is larger than the order of a.

2.42 Let $G = \text{GL}(2, \mathbb{Q})$, and let $A = \begin{bmatrix} 0 & -1 \\ 1 & 0 \end{bmatrix}$ and $B = \begin{bmatrix} 0 & 1 \\ -1 & 1 \end{bmatrix}$. Show that $A^4 = I = B^6$, but that $(AB)^n \neq I$ for all $n > 0$. Conclude that AB can have infinite order even though both factors A and B have finite order (this cannot happen in a finite group).

2.43 H **(i)** Prove, by induction on $k \geq 1$, that

$$\begin{bmatrix} \cos\theta & -\sin\theta \\ \sin\theta & \cos\theta \end{bmatrix}^k = \begin{bmatrix} \cos k\theta & -\sin k\theta \\ \sin k\theta & \cos k\theta \end{bmatrix}.$$

(ii) Find all the elements of finite order in $SO(2, \mathbb{R})$, the special orthogonal group [see Example 2.48(iii)].

*2.44 If G is a group in which $x^2 = 1$ for every $x \in G$, prove that G must be abelian. [The Boolean groups $\mathcal{B}(X)$ of Example 2.47(viii) are such groups.]

*H 2.45 Let G be a finite group in which every element has a square root; that is, for each $x \in G$, there exists $y \in G$ with $y^2 = x$. Prove that every element in G has a unique square root.

*H 2.46 If G is a group with an even number of elements, prove that the number of elements in G of order 2 is odd. In particular, G must contain an element of order 2.

H 2.47 What is the largest order of an element in S_n, where $n = 1, 2, \ldots, 10$?

*2.48 The **stochastic**[12] **group** $\Sigma(2, \mathbb{R})$ consists of all those matrices in $\text{GL}(2, \mathbb{R})$ whose column sums are 1; that is, $\Sigma(2, \mathbb{R})$ consists of all the nonsingular matrices $\begin{bmatrix} a & c \\ b & d \end{bmatrix}$ with $a + b = 1 = c + d$. [There are also stochastic groups $\Sigma(2, \mathbb{Q})$ and $\Sigma(2, \mathbb{C})$.]

Prove that the product of two stochastic matrices is again stochastic, and that the inverse of a stochastic matrix is stochastic.

2.49 Show that the symmetry group $\Sigma(C)$ of a circle C is infinite.

*2.50 Prove that every element in a dihedral group D_{2n} has a unique factorization of the form $a^i b^j$, where $0 \leq i < n$ and $j = 0$ or 1.

2.51 Let $e_1 = (1, 0)$ and $e_2 = (0, 1)$, If φ is an isometry of the plane fixing O, let $\varphi(e_1) = (a, b)$, $\varphi(e_2) = (c, d)$, and let $A = \begin{bmatrix} a & c \\ b & d \end{bmatrix}$. Prove that $\det(A) = \pm 1$.

→ 2.4 SUBGROUPS AND LAGRANGE'S THEOREM

A *subgroup* of a group G is a subset which is a group under the same operation as on G. The following definition makes this last phrase precise.

→ **Definition.** Let $*$ be an operation on a set G, and let $S \subseteq G$ be a subset. We say that S is **closed under** $*$ if $x * y \in S$ for all $x, y \in S$.

The operation on a group G is a function $*: G \times G \to G$. If $S \subseteq G$, then $S \times S \subseteq G \times G$, and to say that S is closed under the operation $*$ means that $*(S \times S) \subseteq S$. For example, the subset $\mathbb{Z}$ of the additive group $\mathbb{Q}$ of rational numbers is closed under $+$. However, if $\mathbb{Q}^\times$ is the multiplicative group of nonzero rational numbers, then $\mathbb{Q}^\times$ is closed under multiplication, but it is not closed under $+$ (for example, 2 and -2 lie in $\mathbb{Q}^\times$, but their sum $-2 + 2 = 0 \notin \mathbb{Q}^\times$).

---

[12]The term *stochastic* comes from the Greek word meaning "to guess." Its mathematical usage occurs in statistics, and stochastic matrices first arose in the study of certain statistical problems.

→ **Definition.**   A subset H of a group G is a **subgroup** if

   (i) $1 \in H$;

   (ii) if $x, y \in H$, then $xy \in H$; that is, H is closed under $*$;

   (iii) if $x \in H$, then $x^{-1} \in H$.

We write $H \leq G$ to denote H being a subgroup of a group G. Observe that $\{1\}$ and G are always subgroups of a group G, where $\{1\}$ denotes the subset consisting of the single element 1. We call a subgroup H of G **proper** if $H \neq G$, and we write $H < G$. We call a subgroup H of G **nontrivial** if $H \neq \{1\}$. More interesting examples of subgroups will be given below.

→ **Proposition 2.66.**   *Every subgroup $H \leq G$ of a group G is itself a group.*

Proof.   Axiom (ii) (in the definition of subgroup) shows that H is closed under the operation of G; that is, H has an operation (essentially, the restriction of the operation $*: G \times G \to G$ to $H \times H \subseteq G \times G$). This operation is associative: since the equation $(xy)z = x(yz)$ holds for all $x, y, z \in G$, it holds, in particular, for all $x, y, z \in H$. Finally, axiom (i) gives the identity, and axiom (iii) gives inverses.   •

It is quicker to check that a subset H of a group G is a subgroup (and hence that it is a group in its own right) than to verify the group axioms for H, for associativity is inherited from the operation on G and hence it need not be verified again.

→ **Example 2.67.**

   (i) Recall that **Isom**$(\mathbb{R}^2)$ is the group of all isometries of the plane. The subset $O(2, \mathbb{R})$, consisting of all isometries fixing the origin, is a subgroup of **Isom**$(\mathbb{R}^2)$. If $\Omega \subseteq \mathbb{R}^2$, then the symmetry group $\Sigma(\Omega)$ is also a subgroup of **Isom**$(\mathbb{R}^2)$. If the center of gravity of Ω exists and is at the origin, then $\Sigma(\Omega) \leq O(2, \mathbb{R})$.

   (ii) The four permutations

$$\mathbf{V} = \big\{(1), (1\ 2)(3\ 4), (1\ 3)(2\ 4), (1\ 4)(2\ 3)\big\}$$

form a group, because $\mathbf{V}$ is a subgroup of $S_4 : (1) \in \mathbf{V}$; $\alpha^2 = (1)$ for each $\alpha \in \mathbf{V}$, and so $\alpha^{-1} = \alpha \in \mathbf{V}$; the product of any two distinct permutations in $\mathbf{V} - \{(1)\}$ is the third one. One calls $\mathbf{V}$ the **four-group** (or the **Klein group**) ($\mathbf{V}$ abbreviates the original German term *Vierergruppe*).

Consider what verifying associativity $a(bc) = (ab)c$ would involve: there are 4 choices for each of a, b, and c, and so there are $4^3 = 64$ equations to be checked. Of course, we may assume that none is (1), leaving us with only $3^3 = 27$ equations but, obviously, proving $\mathbf{V}$ is a group by showing it is a subgroup of S_4 is the best way to proceed.

(iii) If $\mathbb{R}^2$ is the plane considered as an (additive) abelian group, then any line L through the origin is a subgroup. The easiest way to see this is to choose a nonzero point (a, b) on L and then note that L consists of all the scalar multiples (ra, rb). The reader may now verify that the axioms in the definition of subgroup do hold for L. ◀

One can shorten the list of items needed to verify that a subset is, in fact, a subgroup.

→ **Proposition 2.68.** *A subset H of a group G is a subgroup if and only if H is nonempty and, whenever $x, y \in H$, then $xy^{-1} \in H$.*

Proof. If H is a subgroup, then it is nonempty, for $1 \in H$. If $x, y \in H$, then $y^{-1} \in H$, by axiom (iii) of the definition of subgroup, and so $xy^{-1} \in H$, by axiom (ii).

Conversely, assume that H is a subset satisfying the new condition. Since H is nonempty, it contains some element, say, h. Taking $x = h = y$, we see that $1 = hh^{-1} \in H$, and so axiom (i) holds. If $y \in H$, then set $x = 1$ (which we can now do because $1 \in H$), giving $y^{-1} = 1y^{-1} \in H$, and so axiom (iii) of the definition holds. Finally, we know that $(y^{-1})^{-1} = y$, by Lemma 2.46. Hence, if $x, y \in H$, then $y^{-1} \in H$, and so $xy = x(y^{-1})^{-1} \in H$. Therefore, H is a subgroup of G. •

Since every subgroup contains 1, one may replace the hypothesis "H is nonempty" in Proposition 2.68 by "$1 \in H$."

Note that if the operation in G is addition, then the condition in the proposition is that H is a nonempty subset such that $x, y \in H$ implies $x - y \in H$.

For Galois, a group was just a subset H of S_n that is closed under composition; that is, if $\alpha, \beta \in H$, then $\alpha\beta \in H$. A. Cayley, in 1854, was the first to define an abstract group, mentioning associativity, inverses, and identity explicitly.

→ **Proposition 2.69.** *A nonempty subset H of a finite group G is a subgroup if and only if H is closed under the operation of G; that is, if $a, b \in H$, then $ab \in H$. In particular, a nonempty subset of S_n is a subgroup if and only if it is closed under composition.*

Proof. Every subgroup is nonempty, by axiom (i) in the definition of subgroup, and it is closed, by axiom (ii).

Conversely, assume that H is a nonempty subset of G closed under the operation on G; thus, axiom (ii) holds. It follows that H contains all the powers of its elements. In particular, there is some element $a \in H$, because H is nonempty, and $a^n \in H$ for all $n \geq 1$. As we saw in Example 2.52, every element in G has finite order: there is an integer m with $a^m = 1$; hence $1 \in H$ and axiom (i) holds. Finally, if $h \in H$ and $h^m = 1$, then $h^{-1} = h^{m-1}$ (for $hh^{m-1} = 1 = h^{m-1}h$), so that $h^{-1} \in H$ and axiom (iii) holds. Therefore, H is a subgroup of G. •

This last proposition can be false when G is an infinite group. For example, the subset $\mathbb{N}$ of the additive group $\mathbb{Z}$ is closed under +, but it is not a subgroup of $\mathbb{Z}$.

→ **Example 2.70.**
The subset A_n of S_n, consisting of all the even permutations, is a subgroup because it is closed under multiplication: even $\circ$ even = even. This subgroup of S_n is called the *alternating*[13] *group* on n letters, and it is denoted by A_n. ◄

→ **Definition.**   If G is a group and $a \in G$, write

$$\langle a \rangle = \{a^n : n \in \mathbb{Z}\} = \{\text{all powers of } a\};$$

$\langle a \rangle$ is called the *cyclic subgroup* of G *generated* by a.

A group G is called *cyclic* if there is some $a \in G$ with $G = \langle a \rangle$; in this case a is called a *generator* of G.

It is easy to see that $\langle a \rangle$ is, in fact, a subgroup: $1 = a^0 \in \langle a \rangle$; $a^n a^m = a^{n+m} \in \langle a \rangle$; $a^{-1} \in \langle a \rangle$. Example 2.47(v) shows, for every $n \geq 1$, that the multiplicative group Γ_n of all nth roots of unity is a cyclic group with the primitive nth root of unity $\zeta = e^{2\pi i/n}$ as a generator.

A cyclic group can have several different generators. For example, $\langle a \rangle = \langle a^{-1} \rangle$. If $(k, n) = 1$, then the primitive nth root $e^{2\pi i k/n}$ is also a generator of Γ_n.

→ **Proposition 2.71.**   *If $G = \langle a \rangle$ is a cyclic group of order n, then a^k is a generator of G if and only if $\gcd(k, n) = 1$.*

Proof.   If a^k is a generator, then $a \in \langle a^k \rangle$, so there is s with $a = a^{ks}$. Hence, $a^{ks-1} = 1$, so that Lemma 2.53 shows that $n \mid (ks - 1)$; that is, there is an integer t with $ks - 1 = tn$, or $sk - tn = 1$. Hence, $(k, n) = 1$, by Exercise 1.56 on page 54.

Conversely, since $\gcd(k, n) = 1$, there are integers s and t with $1 = sk + tn$. Hence, $a = a^{sk+tn} = a^{sk}$ (because $a^{tn} = 1$), and so $a \in \langle a^k \rangle$. Therefore, $G = \langle a \rangle \leq \langle a^k \rangle$, and so $G = \langle a^k \rangle$. •

→ **Corollary 2.72.**   *The number of generators of a cyclic group of order n is $\phi(n)$.*

Proof.   This follows at once from the Propositions 2.71 and 1.42. •

---

[13]The *alternating group* first arose in studying polynomials. If

$$f(x) = (x - u_1)(x - u_2) \cdots (x - u_n),$$

then the number $D = \prod_{i<j}(u_i - u_j)$ changes sign if one permutes the roots: if α is a permutation of $\{u_1, u_2, \ldots, u_n\}$, then it is easy to see that $\prod_{i<j}[\alpha(u_i) - \alpha(u_j)] = \pm D$. Thus, the sign of the product alternates as various permutations α are applied to its factors. The sign does not change for those α in the alternating group.

→ **Proposition 2.73.** *Every subgroup S of a cyclic group $G = \langle a \rangle$ is itself cyclic. In fact, a^k is a generator of S, where k is the smallest positive integer m with $a^m \in S$.*

Proof. We may assume that S is nontrivial; that is, $S \neq \{1\}$, for the proposition is obviously true when $S = \{1\}$. Let $I = \{m \in \mathbb{Z} : a^m \in S\}$; we check that I satisfies the three conditions in Corollary 1.37. First, $0 \in I$, for $a^0 = 1 \in S$. Second, if $m, n \in I$, then $a^m, a^n \in S$, and so $a^m a^{-n} = a^{m-n} \in S$; hence, $m - n \in I$. Third, if $m \in I$ and $i \in \mathbb{Z}$, then $a^m \in S$, and so $(a^m)^i = a^{im} \in S$; hence, $im \in I$. Since $S \neq \{1\}$, there is some $a^q \in S$ with $a^q \neq 1$; thus, $q \in I$ and $I \neq \{0\}$. By Corollary 1.37, $I = (k)$, where k is the smallest positive integer in I, and so $k \mid m$ for every $m \in I$. We claim that $\langle a^k \rangle = S$. Clearly, $\langle a^k \rangle \leq S$. For the reverse inclusion, take $s \in S$. Now $s = a^m$ for some m, so that $m \in I$ and $m = k\ell$ for some ℓ. Therefore, $s = a^m = a^{k\ell} \in \langle a^k \rangle$.
●

Proposition 2.80 will give a number-theoretic interpretation of this last result.

→ **Proposition 2.74.** *Let G be a finite group and let $a \in G$. Then the order of a is the number of elements in $\langle a \rangle$.*

Proof. We will use the idea in Lemma 2.23. Since G is finite, there is an integer $k \geq 1$ with $1, a, a^2, \ldots, a^{k-1}$ consisting of k distinct elements, while $1, a, a^2, \ldots, a^k$ has a repetition; hence $a^k \in \{1, a, a^2, \ldots, a^{k-1}\}$; that is, $a^k = a^i$ for some i with $0 \leq i < k$. If $i \geq 1$, then $a^{k-i} = 1$, contradicting the original list having no repetitions. Therefore, $a^k = a^0 = 1$, and k is the order of a (being the smallest positive such k).

If $H = \{1, a, a^2, \ldots, a^{k-1}\}$, then $|H| = k$; it suffices to show that $H = \langle a \rangle$. Clearly, $H \subseteq \langle a \rangle$. For the reverse inclusion, take $a^i \in \langle a \rangle$. By the division algorithm, $i = qk + r$, where $0 \leq r < k$. Hence $a^i = a^{qk+r} = a^{qk} a^r = (a^k)^q a^r = a^r \in H$; this gives $\langle a \rangle \subseteq H$, and so $\langle a \rangle = H$. ●

→ **Definition.** If G is a finite group, then the number $|G|$ of elements in G is called the *order* of G.

The word *order* has two different meanings in group theory: the order of an *element* $a \in G$ and the order $|G|$ of a *group* G. Proposition 2.74 shows that the order of a group element a is equal to $|\langle a \rangle|$.

The following characterization of finite cyclic groups will be used to prove Theorem 3.55 showing that the multiplicative group of a finite field is cyclic.

→ **Proposition 2.75.** *Let G be a group of order n. If G is cyclic, then G has a unique subgroup of order d for each divisor d of n. Conversely, if there is at most one cyclic subgroup of order d, where $d \mid n$, then G is cyclic.*

Proof. Suppose that $G = \langle a \rangle$ is a cyclic group of order n. We claim that $\langle a^{n/d} \rangle$ has order d. Clearly, $(a^{n/d})^d = a^n = 1$, and it suffices to show that d is the smallest such positive integer. If $(a^{n/d})^r = 1$, then $n \mid (n/d)r$, by Lemma 2.53. Hence, there is some integer s with $(n/d)r = ns$, so that $r = ds$ and $r \geq d$.

To prove uniqueness, let C be a subgroup of G of order d; by Proposition 2.73, the subgroup C is cyclic, say, $C = \langle x \rangle$. Now $x = a^m$ has order d, so that $1 = (x^m)^d$. Hence, $n \mid md$, by Lemma 2.53, and so $md = nk$ for some integer k. Therefore, $x = a^m = (a^{n/d})^k$, so that $C = \langle x \rangle \subseteq \langle a^{n/d} \rangle$. Since both subgroups have the same order, however, it follows that $C = \langle a^{n/d} \rangle$.

Conversely, define a relation on a group G by $a \equiv b$ if $\langle a \rangle = \langle b \rangle$. It is easy to see that this is an equivalence relation and that the equivalence class $[a]$ of $a \in G$ consists of all the generators of $C = \langle a \rangle$. Thus, we denote $[a]$ by $\mathrm{gen}(C)$, and

$$G = \bigcup_{C \text{ cyclic}} \mathrm{gen}(C).$$

Hence, $n = |G| = \sum_C |\mathrm{gen}(C)|$, where the sum is over all the cyclic subgroups of G. But Corollary 2.72 gives $|\mathrm{gen}(C)| = \phi(|C|)$. By hypothesis, G has at most one (cyclic) subgroup of any order, so that

$$n = \sum_C |\mathrm{gen}(C)| \leq \sum_{d \mid n} \phi(d) = n,$$

the last equality being Corollary 1.31. Therefore, for each divisor d of n, there must be a cyclic subgroup C of order d contributing $\phi(d)$ to $\sum_C |\mathrm{gen}(C)|$. In particular, there must be a cyclic subgroup C of order n, and so G is cyclic.   •

Here is a way of constructing a new subgroup from given ones.

→ **Proposition 2.76.**   *The intersection $\bigcap_{i \in I} H_i$ of any family of subgroups of a group G is again a subgroup of G. In particular, if H and K are subgroups of G, then $H \cap K$ is a subgroup of G.*

Proof.   Let $D = \bigcap_{i \in I} H_i$; we prove that D is a subgroup by verifying each of the parts in the definition. Note first that $D \neq \varnothing$, because $1 \in H_i$ for all i implies $1 \in D$. If $x \in D$, then x got into D by being in each H_i; as each H_i is a subgroup, $x^{-1} \in H_i$ for all i, and so $x^{-1} \in D$. Finally, if $x, y \in D$, then both x and y lie in every H_i, hence their product xy lies in every H_i, and so $xy \in D$.   •

→ **Corollary 2.77.**   *If X is a subset of a group G, then there is a subgroup $\langle X \rangle$ of G containing X that is **smallest** in the sense that $\langle X \rangle \leq H$ for every subgroup H of G which contains X.*

Proof. First of all, note that there exist subgroups of G which contain X; for example, G itself contains X. Define $\langle X \rangle = \bigcap_{X \subseteq H} H$, the intersection of all the subgroups H of G which contain X. By Proposition 2.76, $\langle X \rangle$ is a subgroup of G; of course, $\langle X \rangle$ contains X because every H contains X. Finally, if H is any subgroup containing X, then H is one of the subgroups whose intersection is $\langle X \rangle$; that is, $\langle X \rangle \leq H$. $\bullet$

Note that there is no restriction on the subset X in the last corollary; in particular, $X = \varnothing$ is allowed. Since the empty set is a subset of every set, we have $\varnothing \subseteq H$ for every subgroup H of G. Thus, $\langle \varnothing \rangle$ is the intersection of *all* the subgroups of G, one of which is $\{1\}$, and so $\langle \varnothing \rangle = \{1\}$.

$\rightarrow$ **Definition.** If X is a subset of a group G, then $\langle X \rangle$ is called the ***subgroup generated by*** X.

Example 2.78.

(i) If $G = \langle a \rangle$ is a cyclic group with generator a, then G is generated by the subset $X = \{a\}$. However, we always write $\langle a \rangle$ instead of $\langle \{a\} \rangle$.

(ii) The symmetry group $\Sigma(\pi_n)$ of a regular n-gon π_n is generated by a, b, where a is a rotation about the origin by $(360/n)°$ and b is a reflection (see Theorem 2.65). These generators satisfy relations $a^n = 1$, $b^2 = 1$, and $bab = a^{-1}$, and $\Sigma(\pi_n)$ is a dihedral group D_{2n}. $\blacktriangleleft$

The next proposition gives a more concrete description of the subgroup generated by a subset.

Definition. Let X be a subset of a group G. Then a ***word on*** X is the identity element or an element of G of the form $w = x_1^{e_1} x_2^{e_2} \cdots x_n^{e_n}$, where $n \geq 1$, $x_i \in X$ for all i, and $e_i = \pm 1$ for all i.

Proposition 2.79. *If X is a subset of a group G, then $\langle X \rangle$ is the set of all the words on X.*

Proof. We begin by showing that W, the set of all words on X, is a subgroup of G. By definition, $1 \in W$, even if $X = \varnothing$. If $w, w' \in W$, then $w = x_1^{e_1} x_2^{e_2} \cdots x_n^{e_n}$ and $w' = y_1^{f_1} y_2^{f_2} \cdots y_m^{f_m}$, where $x_i, y_j \in X$ and $e_i, f_j = \pm 1$. Hence, $ww' = x_1^{e_1} x_2^{e_2} \cdots x_n^{e_n} y_1^{f_1} y_2^{f_2} \cdots y_m^{f_m}$, which is a word on X, and so $ww' \in W$. Finally, $(w)^{-1} = x_n^{-e_n} x_{n-1}^{-e_{n-1}} \cdots x_1^{-e_1} \in W$. Thus, W is a subgroup of G, and it clearly contains every element of X. We conclude that $\langle X \rangle \leq W$. For the reverse inequality, we show that if S is any subgroup of G containing X, then S contains every word on X. But this is obvious: since S is a subgroup, it contain x^e whenever $x \in X$ and $e = \pm 1$, and it contains all possible products of such elements. Therefore, $W \leq S$ for all such S, and so $W \leq \bigcap S = \langle X \rangle$. $\bullet$

$\rightarrow$ **Proposition 2.80.**   *Let a and b be integers and let $A = \langle a \rangle$ and $B = \langle b \rangle$ be the cyclic subgroups of $\mathbb{Z}$ they generate.*

(i) *If $A + B$ is defined to be $\{sa + tb : s, t \in \mathbb{Z}\}$, then $A + B = \langle d \rangle$, where $d = \gcd(a, b)$.*

(ii) *$A \cap B = \langle m \rangle$, where $m = \mathrm{lcm}(a, b)$.*

Proof.
(i) In additive notation, a word is just a linear combination, and so Proposition 2.79 says that $A + B$ is the subgroup of $\mathbb{Z}$ generated by $A \cup B$. By Proposition 2.73, $A + B$ is cyclic, so that $A + B = \langle d \rangle$, where d is the smallest non-negative linear combination of a and b. Thus, the proof of Theorem 1.35 gives $d = \gcd(a, b)$.
(ii) If $c \in A \cap B$, then $c \in A$ and $a \mid c$; similarly, if $c \in A \cap B$, then $c \in B$ and $b \mid c$. Thus, every element in $A \cap B$ is a common multiple of a and b. Conversely, every common multiple lies in the intersection. By Proposition 2.73, the subgroup $A \cap B$ is cyclic: $A \cap B = \langle m \rangle$, where m can be chosen to be the smallest non-negative number in $A \cap B$. Therefore, m is the smallest common multiple; that is, $m = \mathrm{lcm}(a, b)$. $\bullet$

Perhaps the most fundamental fact about subgroups H of a finite group G is that their orders are constrained. Certainly, we have $|H| \leq |G|$, but it turns out that $|H|$ must be a divisor of $|G|$. To prove this, we introduce the notion of coset.

$\rightarrow$ **Definition.**   If H is a subgroup of a group G and $a \in G$, then the ***coset***[14] aH is the subset aH of G, where

$$aH = \{ah : h \in H\}.$$

Of course, $a = a1 \in aH$. Cosets are usually not subgroups. For example, if $a \notin H$, then $1 \notin aH$ (otherwise $1 = ah$ for some $h \in H$, and this gives the contradiction $a = h^{-1} \in H$).

If we use the $*$ notation for the operation in a group G, then we denote the coset aH by $a * H$, where

$$a * H = \{a * h : h \in H.\}$$

In particular, if the operation is addition, then the coset is denoted by

$$a + H = \{a + h : h \in H\}.$$

---

[14]The cosets defined here are often called ***left cosets***; there are also ***right cosets*** of H, namely, subsets of the form $Ha = \{ha : h \in H\}$; these arise in further study of groups, but we shall work almost exclusively with (left) cosets.

→ **Example 2.81.**

(i) Consider the plane $\mathbb{R}^2$ as an (additive) abelian group and let ℓ be a line through the origin O (see Figure 2.16); as in Example 2.67(iii), the line ℓ is a subgroup of $\mathbb{R}^2$. If $\beta \in \mathbb{R}^2$, then the coset $\beta + \ell$ is the line ℓ' containing β which is parallel to ℓ, for if $r\alpha \in \ell$, then the parallelogram law gives $\beta + r\alpha \in \ell'$.

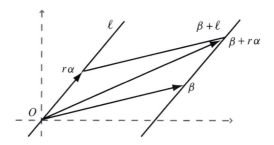

Figure 2.16 The Coset $\beta + \ell$

(ii) If $G = S_3$ and $H = \langle (1\ 2) \rangle$, there are exactly three cosets of H, namely

$$
\begin{aligned}
H = &\quad \{(1), (1\ 2)\} = (1\ 2)H, \\
(1\ 3)H = &\ \{(1\ 3), (1\ 2\ 3)\} = (1\ 2\ 3)H, \\
(2\ 3)H = &\ \{(2\ 3), (1\ 3\ 2)\} = (1\ 3\ 2)H,
\end{aligned}
$$

each of which has size 2. ◄

Observe, in our examples, that different cosets of a given subgroup do not overlap. If H is a subgroup of a group G, then the relation on G, defined by

$$a \equiv b \qquad \text{if} \qquad a^{-1}b \in H,$$

is an equivalence relation on G. If $a \in G$, then $a^{-1}a = 1 \in H$, and $a \equiv a$; hence, $\equiv$ is reflexive. If $a \equiv b$, then $a^{-1}b \in H$; since subgroups are closed under inverses, $(a^{-1}b)^{-1} = b^{-1}a \in H$ and $b \equiv a$; hence $\equiv$ is symmetric. If $a \equiv b$ and $b \equiv c$, then $a^{-1}b, b^{-1}c \in H$; since subgroups are closed under multiplication, $(a^{-1}b)(b^{-1}c) = a^{-1}c \in H$, and $a \equiv c$. Therefore, $\equiv$ is transitive, and hence it is an equivalence relation.

We claim that the equivalence class of $a \in H$ is the coset aH. If $x \equiv a$, then there is $h \in H$ with $a^{-1}x = h$; hence, $x = ah \in aH$, and $[a] \subseteq aH$. For the reverse inclusion, it is easy to see that if $x = ah \in aH$, then $x^{-1}a = (ah)^{-1}a = h^{-1}a^{-1}a = h \in H$, so that $x \equiv a$ and $x \in [a]$. Hence, $aH \subseteq [a]$, and so $[a] = aH$.

Lemma 2.82.   *Let H be a subgroup of a group G, and let $a, b \in G$.*

(i) *$aH = bH$ if and only if $b^{-1}a \in H$. In particular, $aH = H$ if and only if $a \in H$.*

(ii) *If $aH \cap bH \neq \varnothing$, then $aH = bH$.*

(iii) *$|aH| = |H|$ for all $a \in G$.*

Proof.
(i) This is a special case of Lemma 2.19, for cosets are equivalence classes. Since $H = 1H$, we have $aH = H = 1H$ if and only if $a = 1^{-1}a \in H$.
(ii) This is a special case of Proposition 2.20, for the equivalence classes comprise a partition of X.
(iii) The function $f \colon H \to aH$, given by $f(h) = ah$, is easily seen to be a bijection [its inverse $aH \to H$ is given by $ah \mapsto a^{-1}(ah) = h$]. Therefore, H and aH have the same number of elements, by Exercise 2.12 on page 105.   •

$\rightarrow$ **Theorem 2.83 (Lagrange's Theorem).**   *If H is a subgroup of a finite group G, then $|H|$ is a divisor of $|G|$.*

Proof.   Let $\{a_1H, a_2H, \dots, a_tH\}$ be the family of all the distinct cosets of H in G. The cosets of H partition G because they are equivalence classes. Hence, G is the disjoint union

$$G = a_1H \cup a_2H \cup \cdots \cup a_tH.$$

(Exercise 2.53 on page 158 asks you to prove that G is the disjoint union of the cosets of H without using equivalence relations.) It follows that

$$|G| = |a_1H| + |a_2H| + \cdots + |a_tH|.$$

But $|a_iH| = |H|$ for all i, by Lemma 2.82(iii), so that $|G| = t|H|$.   •

$\rightarrow$ **Definition.**   The *index* of a subgroup H in G, denoted by $[G : H]$, is the number of cosets of H in G.

When G is finite, the index $[G : H]$ is the number t in the formula $|G| = t|H|$ in the proof of Lagrange's theorem, so that

$$|G| = [G : H]|H|.$$

This formula shows that the index $[G : H]$ is also a divisor of $|G|$.

$\rightarrow$ **Corollary 2.84.**   *If H is a subgroup of a finite group G, then*

$$[G : H] = |G|/|H|.$$

Proof.   This follows at once from Lagrange's theorem.   •

Recall Theorem 2.65: the symmetry group $\Sigma(\pi_n)$ of a regular n-gon is a dihedral group of order $2n$. It contains a cyclic subgroup of order n, generated by a rotation a, and the subgroup $\langle a \rangle$ has index $[\Sigma(\pi_n) : \langle a \rangle] = 2$. Thus, there are two cosets: $\langle a \rangle$ and $b\langle a \rangle$, where b is any symmetry outside of $\langle a \rangle$.

We now see why the orders of elements in S_5, displayed in Table 2.3 on page 137, are divisors of 120. Corollary 2.146 will explain why the number of permutations in S_5 of any given cycle structure is a divisor of 120.

→ **Corollary 2.85.** *If G is a finite group and $a \in G$, then the order of a divides $|G|$.*

Proof. By Proposition 2.74, the order of the element a is equal to the order of the subgroup $H = \langle a \rangle$. •

→ **Corollary 2.86.** *If a finite group G has order m, then $a^m = 1$ for all $a \in G$.*

Proof. By Corollary 2.85, a has order d, where $d \mid m$; that is, $m = dk$ for some integer k. Thus, $a^m = a^{dk} = (a^d)^k = 1$. •

→ **Corollary 2.87.** *If p is a prime, then every group G of order p is cyclic.*

Proof. Choose $a \in G$ with $a \neq 1$, and let $H = \langle a \rangle$ be the cyclic subgroup generated by a. By Lagrange's theorem, $|H|$ is a divisor of $|G| = p$. Since p is a prime and $|H| > 1$, it follows that $|H| = p = |G|$, and so $H = G$. •

Lagrange's theorem says that the order of a subgroup of a finite group G is a divisor of $|G|$. Is the "converse" of Lagrange's theorem true? That is, if d is a divisor of $|G|$, must there exist a subgroup of G having order d? The answer is "no"; Proposition 2.99 will show that the alternating group A_4 is a group of order 12 which has no subgroup of order 6.

EXERCISES

H **2.52** True or false with reasons. Here, G is always a group.
   (i)   If H is a subgroup of K and K is a subgroup of G, then H is a subgroup of G.
   (ii) G is a subgroup of itself.
   (iii) The empty set $\varnothing$ is a subgroup of G.
   (iv) If G is a finite group and m is a divisor of $|G|$, then G contains an element of order m.
   (v)   Every subgroup of S_n has order dividing $n!$.
   (vi) If H is a subgroup of G, then the intersection of two (left) cosets of H is a (left) coset of H.
   (vii) The intersection of two cyclic subgroups of G is a cyclic subgroup.

(viii) If X is a finite subset of G, then $\langle X \rangle$ is a finite subgroup.

(ix) If X is an infinite set, then

$$F = \{\sigma \in S_X : \sigma \text{ moves only finitely many elements of } X\}$$

is a subgroup of S_X.

(x)   Every proper subgroup of S_3 is cyclic.

(xi)   Every proper subgroup of S_4 is cyclic.

*2.53 Let H be a subgroup of a finite group G, and let $a_1 H, \ldots, a_t H$ be a list of all the distinct cosets of H in G. Prove the following statements without using the equivalence relation on G defined by $a \equiv b$ if $b^{-1}a \in H$.

    (i) Prove that each $g \in G$ lies in the coset gH, and that $gH = a_i H$ for some i. Conclude that $G = a_1 H \cup \cdots \cup a_t H$.

    (ii) If $a, b \in G$ and $aH \cap bH \neq \varnothing$, prove that $aH = bH$. Conclude that if $i \neq j$, then $a_i H \cap a_j H = \varnothing$.

2.54 **(i)** Define the *special linear group* by

$$\text{SL}(2, \mathbb{R}) = \{A \in \text{GL}(2, \mathbb{R}) : \det(A) = 1\}.$$

    Prove that $\text{SL}(2, \mathbb{R})$ is a subgroup of $\text{GL}(2, \mathbb{R})$.

    (ii) Prove that $\text{GL}(2, \mathbb{Q})$ is a subgroup of $\text{GL}(2, \mathbb{R})$.

*H 2.55 Give an example of two subgroups H and K of a group G whose union $H \cup K$ is not a subgroup of G.

*2.56 Let G be a finite group with subgroups H and K. If $H \leq K$, prove that

$$[G : H] = [G : K][K : H].$$

H 2.57 If H and K are subgroups of a group G and if $|H|$ and $|K|$ are relatively prime, prove that $H \cap K = \{1\}$.

H 2.58 Prove that every infinite group contains infinitely many subgroups.

*2.59 Let G be a group of order 4. Prove that either G is cyclic or $x^2 = 1$ for every $x \in G$. Conclude, using Exercise 2.44 on page 147, that G must be abelian.

2.60 **(i)** Prove that the stochastic group $\Sigma(2, \mathbb{R})$, the set of all nonsingular 2×2 matrices whose row sums are 1, is a subgroup of $\text{GL}(2, \mathbb{R})$ (see Exercise 2.48 on page 147).

    (ii) Define $\Sigma'(2, \mathbb{R})$ to be the set of all nonsingular *doubly stochastic* matrices (all row sums are 1 and all column sums are 1). Prove that $\Sigma'(2, \mathbb{R})$ is a subgroup of $\text{GL}(2, \mathbb{R})$.

*H 2.61 Let G be a finite group, and let S and T be (not necessarily distinct) nonempty subsets. Prove that either $G = ST$ or $|G| \geq |S| + |T|$.

2.62 **(i)** If $\{S_i : i \in I\}$ is a family of subgroups of a group G, prove that an intersection of cosets $\bigcap_{i \in I} x_i S_i$ is either empty or a coset of $\bigcap_{i \in I} S_i$.

    H (ii) (***B. H. Neumann.***)   If a group G is the set-theoretic union of finitely many cosets,

$$G = x_1 S_1 \cup \cdots \cup x_n S_n,$$

prove that at least one of the subgroups S_i has finite index in G.

2.63   (i)   Show that a left coset of $\langle (1\ 2) \rangle$ in S_3 may not be equal to a right coset of $\langle (1\ 2) \rangle$ in S_3; that is, there is $\alpha \in S_3$ with $\alpha \langle (1\ 2) \rangle \neq \langle (1\ 2) \rangle \alpha$.

H **(ii)**   Let G be a finite group and let $H \leq G$ be a subgroup. Prove that the number of left cosets of H in G is equal to the number of right cosets of H in G.

→ **2.5   HOMOMORPHISMS**

An important problem is determining whether two given groups G and H are somehow the same. For example, we have investigated S_3, the group of all permutations of $X = \{1, 2, 3\}$. The group S_Y of all the permutations of $Y = \{a, b, c\}$ is a group different from S_3 because permutations of $\{1, 2, 3\}$ are different than permutations of $\{a, b, c\}$. But even though S_3 and S_Y are different, they surely bear a strong resemblance to each other (see Example 2.88). The notions of homomorphism and isomorphism allow one to compare different groups, as we shall see.

→ **Definition.**   If $(G, *)$ and $(H, \circ)$ are groups (we have displayed the operation in each), then a function $f : G \to H$ is a **homomorphism**[15] if

$$f(x * y) = f(x) \circ f(y)$$

for all $x, y \in G$. If f is also a bijection, then f is called an **isomorphism**. We say that G and H are **isomorphic**, denoted by $G \cong H$, if there exists an isomorphism $f : G \to H$.

In Exercise 2.67 on page 169, we will see that isomorphism is an equivalence relation on any family of groups. In particular, if $G \cong H$, then $H \cong G$.

Two obvious examples of homomorphisms are the identity $1_G : G \to G$, which is an isomorphism, and the trivial homomorphism $f : G \to H$, defined by $f(a) = 1$ for all $a \in G$.

Here are more interesting examples. Let $\mathbb{R}$ be the group of all real numbers with operation addition, and let $\mathbb{R}^>$ be the group of all positive real numbers with operation multiplication. The function $f : \mathbb{R} \to \mathbb{R}^>$, defined by $f(x) = e^x$, is a homomorphism, for if $x, y \in \mathbb{R}$, then

$$f(x + y) = e^{x+y} = e^x e^y = f(x) f(y).$$

---

[15]The word *homomorphism* comes from the Greek *homo* meaning "same" and *morph* meaning "shape" or "form." Thus, a homomorphism carries a group to another group (its image) of similar form. The word *isomorphism* involves the Greek *iso* meaning "equal," and isomorphic groups have identical form.

Now f is an isomorphism, for its inverse function $g: \mathbb{R}^> \to \mathbb{R}$ is $\log(x)$. Therefore, the additive group $\mathbb{R}$ is isomorphic to the multiplicative group $\mathbb{R}^>$. Note that the inverse function g is also an isomorphism:

$$g(xy) = \log(xy) = \log(x) + \log(y) = g(x) + g(y).$$

As a second example, we claim that the additive group $\mathbb{C}$ of complex numbers is isomorphic to the additive group $\mathbb{R}^2$ [see Example 2.47(vi)]. Define $f: \mathbb{C} \to \mathbb{R}^2$ by

$$f: a + ib \mapsto (a, b).$$

It is easy to check that f is a bijection; f is a homomorphism because

$$
\begin{aligned}
f([a + ib] + [a' + ib']) &= f([a + a'] + i[b + b']) \\
&= (a + a', b + b') \\
&= (a, b) + (a', b') \\
&= f(a + ib) + f(a' + ib').
\end{aligned}
$$

→ **Definition.**   Let $a_1, a_2, \ldots, a_n$ be a list with no repetitions of all the elements of a finite group G of order n. A ***multiplication table*** for G is an $n \times n$ matrix whose ij entry is $a_i a_j$.

G	a_1	$\cdots$	a_j	$\cdots$	a_n
a_1	$a_1 a_1$	$\cdots$	$a_1 a_j$	$\cdots$	$a_1 a_n$
a_i	$a_i a_1$	$\cdots$	$a_i a_j$	$\cdots$	$a_i a_n$
a_n	$a_n a_1$	$\cdots$	$a_n a_j$	$\cdots$	$a_n a_n$

Let us agree, when writing a multiplication table, that the identity element is listed first; that is, $a_1 = 1$. In this case, the first row and first column of the table are redundant, and we usually omit them.

Consider two almost trivial examples of groups: let Γ_2 denote the multiplicative group $\{1, -1\}$, and let $\mathcal{P}$ denote the parity group [Example 2.47(vii)]. Here are their multiplication tables:

$\Gamma_2:$

1	-1
-1	1

$\mathcal{P}:$

even	odd
odd	even

It is clear that Γ_2 and $\mathcal{P}$ are distinct groups; it is equally clear that there is no significant difference between them. The notion of isomorphism formalizes this idea; Γ_2 and $\mathcal{P}$ are isomorphic, for the function $f: \Gamma_2 \to \mathcal{P}$, defined by $f(1) = $ even and $f(-1) = $ odd, is an isomorphism, as the reader can quickly check.

There are many multiplication tables for a group G of order n, one for each of the $n!$ arrangements of its elements. If $a_1, a_2, \ldots, a_n$ is a list of all the elements of G with no repetitions, and if $f : G \to H$ is a bijection, then $f(a_1), f(a_2), \ldots, f(a_n)$ is a list of all the elements of H with no repetitions, and so this latter list determines a multiplication table for H. That f is an isomorphism says that if we superimpose the multiplication table for G (determined by $a_1, a_2, \ldots, a_n$) upon the multiplication table for H [determined by $f(a_1), f(a_2), \ldots, f(a_n)$], then the tables match: if $a_i a_j$ is the ij entry in the given multiplication table of G, then $f(a_i)f(a_j) = f(a_i a_j)$ is the ij entry of the multiplication table of H. In this sense, isomorphic groups have the same multiplication table. Thus, isomorphic groups are essentially the same, differing only in the notation for the elements and the operations.

Example 2.88.
Here is an algorithm to check whether a given bijection $f : G \to H$ between a pair of groups is actually an isomorphism: enumerate the elements $a_1, \ldots, a_n$ of G, form the multiplication table of G arising from this list, form the multiplication table for H from the list $f(a_1), \ldots, f(a_n)$, and compare the n^2 entries of the two tables one row at a time.

We illustrate this for $G = S_3$, the symmetric group permuting $\{1, 2, 3\}$, and $H = S_Y$, the symmetric group of all the permutations of $Y = \{a, b, c\}$. First, enumerate G:

$$(1), \quad (1\ 2), \quad (1\ 3), \quad (2\ 3), \quad (1\ 2\ 3), \quad (1\ 3\ 2).$$

We define the obvious function $\varphi : S_3 \to S_Y$ that replaces numbers by letters:

$$(1), \quad (a\ b), \quad (a\ c), \quad (b\ c), \quad (a\ b\ c), \quad (a\ c\ b).$$

Compare the multiplication table for S_3 arising from this list of its elements with the multiplication table for S_Y arising from the corresponding list of its elements. The reader should write out the 6×6 tables of each and superimpose one on the other to see that they match. We will illustrate this by checking the 4,5 entry. The 4,5 position in the table for S_3 is the product $(2\ 3)(1\ 2\ 3) = (1\ 3)$, while the 4,5 position in the table for S_Y is the product $(b\ c)(a\ b\ c) = (a\ c)$.

This result is generalized in Exercise 2.65 on page 169. ◄

We now turn from isomorphisms to more general homomorphisms.

Lemma 2.89. *Let $f : G \to H$ be a homomorphism.*

(i) $f(1) = 1$;

(ii) $f(x^{-1}) = f(x)^{-1}$;

(iii) $f(x^n) = f(x)^n$ *for all $n \in \mathbb{Z}$.*

Proof.

(i) Applying f to the equation $1 \cdot 1 = 1$ in G gives the equation $f(1)f(1) = f(1)$ in H, and multiplying both sides by $f(1)^{-1}$ gives $f(1) = 1$.

(ii) Apply f to the equation $x^{-1}x = 1$ in G to obtain the equation $f(x^{-1})f(x) = 1$ in H. Proposition 2.45(iv), uniqueness of the inverse, gives $f(x^{-1}) = f(x)^{-1}$.

(iii) It is routine to prove by induction that $f(x^n) = f(x)^n$ for all $n \geq 0$. For negative exponents, we have $(y^{-1})^n = y^{-n}$ for all y in a group, and so

$$f(x^{-n}) = f((x^{-1})^n) = f((x^{-1}))^n = (f(x)^{-1})^n = f(x)^{-n}. \quad \bullet$$

→ **Example 2.90.**

We show that any two finite cyclic groups G and H of the same order m are isomorphic. It will then follow from Corollary 2.87 that any two groups of prime order p are isomorphic.

Suppose that $G = \langle x \rangle$ and $H = \langle y \rangle$. Define $f : G \to H$ by $f(x^i) = y^i$ for $0 \leq i < m$. Now $G = \{1, x, x^2, \ldots, x^{m-1}\}$ and $H = \{1, y, y^2, \ldots, y^{m-1}\}$, and so it follows that f is a bijection. To see that f is a homomorphism (and hence an isomorphism), we must show that $f(x^i x^j) = f(x^i)f(x^j)$ for all i and j with $0 \leq i$, $j < m$. The desired equation clearly holds if $i + j < m$, for $f(x^{i+j}) = y^{i+j}$, and so

$$f(x^i x^j) = f(x^{i+j}) = y^{i+j} = y^i y^j = f(x^i)f(x^j).$$

If $i + j \geq m$, then $i + j = m + r$, where $0 \leq r < m$, so that

$$x^{i+j} = x^{m+r} = x^m x^r = x^r$$

(because $x^m = 1$); similarly, $y^{i+j} = y^r$ (because $y^m = 1$). Hence

$$f(x^i x^j) = f(x^{i+j}) = f(x^r)$$
$$= y^r = y^{i+j} = y^i y^j = f(x^i)f(x^j).$$

Therefore, f is an isomorphism and $G \cong H$. (See Example 2.117 for a nicer proof of this.) ◀

A property of a group G that is shared by every other group isomorphic to it is called an *invariant* of G. For example, the order, $|G|$, is an invariant of G, for isomorphic groups have the same order. Being abelian is an invariant [if a and b commute, then $ab = ba$ and

$$f(a)f(b) = f(ab) = f(ba) = f(b)f(a);$$

hence, $f(a)$ and $f(b)$ commute]. Thus, $\mathbb{R}$ and $\mathrm{GL}(2, \mathbb{R})$ are not isomorphic, for $\mathbb{R}$ is abelian and $\mathrm{GL}(2, \mathbb{R})$ is not. There are other invariants of a group (see Exercise 2.69 on page 169); for example, the number of elements in it of any given order r, or whether or not the group is cyclic. In general, however, it is a challenge to decide whether two given groups are isomorphic.

→ **Example 2.91.**

We present two nonisomorphic groups of the same order.

As in Example 2.67(ii), let $\mathbf{V}$ be the four-group consisting of the following four permutations:

$$\mathbf{V} = \{(1),\ (1\ 2)(3\ 4),\ (1\ 3)(2\ 4),\ (1\ 4)(2\ 3)\},$$

and let $\Gamma_4 = \langle i \rangle = \{1, i, -1, -i\}$ be the multiplicative cyclic group of fourth roots of unity, where $i^2 = -1$. If there were an isomorphism $f: \mathbf{V} \to \Gamma_4$, then surjectivity of f would provide some $x \in \mathbf{V}$ with $i = f(x)$. But $x^2 = (1)$ for all $x \in \mathbf{V}$, so that $i^2 = f(x)^2 = f(x^2) = f((1)) = 1$, contradicting $i^2 = -1$. Therefore, $\mathbf{V}$ and Γ_4 are not isomorphic.

There are other ways to prove this result. For example, Γ_4 is cyclic and $\mathbf{V}$ is not, or Γ_4 has an element of order 4 and $\mathbf{V}$ does not, or Γ_4 has a unique element of order 2, but $\mathbf{V}$ has 3 elements of order 2. At this stage, you should really believe that Γ_4 and $\mathbf{V}$ are not isomorphic! ◄

→ **Definition.**   If $f: G \to H$ is a homomorphism, define

$$\mathbf{\textit{kernel}}^{16} f = \{x \in G : f(x) = 1\}$$

and

$$\mathbf{\textit{image}}\ f = \{h \in H : h = f(x) \text{ for some } x \in G\}.$$

We usually abbreviate kernel f to ker f and image f to im f.

Example 2.92.

(i) If $\Gamma_n = \langle \zeta \rangle$, where $\zeta = e^{2\pi i/n}$ is a primitive nth root of unity, then $f: \mathbb{Z} \to \Gamma_n$, given by $f(m) = \zeta^m$, is a surjective homomorphism with ker f all the multiples of n.

(ii) If Γ_2 is the multiplicative group $\Gamma_2 = \{\pm 1\}$, then Theorem 2.39 says that sgn: $S_n \to \Gamma_2$ is a homomorphism. The image of sgn $= \{\pm 1\}$, that is, sgn is surjective, because sgn$(\tau) = -1$ for a transposition τ; the kernel of sgn is the alternating group A_n, the set of all even permutations.

(iii) Determinant is a homomorphism det: $\mathrm{GL}(2, \mathbb{R}) \to \mathbb{R}^{\times}$, the multiplicative group of nonzero reals. Now im det $= \mathbb{R}^{\times}$, that is, det is surjective, because if $r \in \mathbb{R}^{\times}$, then $r = \det(\left[\begin{smallmatrix} r & 0 \\ 0 & 1 \end{smallmatrix}\right])$. The kernel of det is the special linear group $\mathrm{SL}(2, \mathbb{R})$. [This example can be extended to $\mathrm{GL}(n, \mathbb{R})$; see Example 2.48(ii).]

---

[16]*Kernel* comes from the German word meaning "grain" or "seed" (*corn* comes from the same word). Its usage here indicates an important ingredient of a homomorphism.

(iv) Let $f : G \to H$ be a homomorphism with ker $f = K$. Recall the definition of the *inverse image* of a function: if $f : X \to Y$ is a function and if $B \subseteq Y$ is a subset, then

$$f^{-1}(B) = \{x \in X : f(x) \in B\}.$$

If $f : G \to H$ is a homomorphism and if $B \leq H$ is a subgroup of H, then we show that $f^{-1}(B)$ is a subgroup of G. Now $1 \in f^{-1}(B)$, for $f(1) = 1 \in B \leq H$. If $x, y \in f^{-1}(B)$, then $f(x), f(y) \in B$ and so $f(x)f(y) \in B$; hence, $f(xy) = f(x)f(y) \in B$, and $xy \in f^{-1}(B)$. Finally, if $x \in f^{-1}(B)$, then $f(x) \in B$; hence, $f(x^{-1}) = f(x)^{-1} \in B$ and $x^{-1} \in f^{-1}(B)$. In particular, if $B = \{1\}$, then $f^{-1}(B) = f^{-1}(1) = \ker f$. It follows that if $f : G \to H$ is a homomorphism and B is a subgroup of H, then $f^{-1}(B)$ is a subgroup of G containing ker f. ◀

→ **Proposition 2.93.**   *Let $f : G \to H$ be a homomorphism.*

  (i) ker f *is a subgroup of G and* im f *is a subgroup of H.*

 (ii) *If $x \in$ ker f and if $a \in G$, then $axa^{-1} \in$ ker f.*

(iii) *f is an injection if and only if* ker $f = \{1\}$.

Proof.
(i) Lemma 2.89 shows that $1 \in \ker f$, for $f(1) = 1$. Next, if $x, y \in \ker f$, then $f(x) = 1 = f(y)$; hence, $f(xy) = f(x)f(y) = 1 \cdot 1 = 1$, and so $xy \in \ker f$. Finally, if $x \in \ker f$, then $f(x) = 1$ and so $f(x^{-1}) = f(x)^{-1} = 1^{-1} = 1$; thus, $x^{-1} \in \ker f$, and ker f is a subgroup of G.
  We now show that im f is a subgroup of H. First, $1 = f(1) \in$ im f. Next, if $h = f(x) \in$ im f, then $h^{-1} = f(x)^{-1} = f(x^{-1}) \in$ im f. Finally, if $k = f(y) \in$ im f, then $hk = f(x)f(y) = f(xy) \in$ im f. Hence, im f is a subgroup of H.
(ii) If $x \in \ker f$, then $f(x) = 1$ and

$$f(axa^{-1}) = f(a)f(x)f(a)^{-1} = f(a)1f(a)^{-1} = f(a)f(a)^{-1} = 1;$$

therefore, $axa^{-1} \in \ker f$.
(iii) If f is an injection, then $x \neq 1$ implies $f(x) \neq f(1) = 1$, and so $x \notin \ker f$. Conversely, assume that ker $f = \{1\}$ and that $f(x) = f(y)$. Then $1 = f(x)f(y)^{-1} = f(xy^{-1})$, so that $xy^{-1} \in \ker f = 1$; therefore, $xy^{-1} = 1$, $x = y$, and f is an injection.   ●

→ **Definition.**   A subgroup K of a group G is called a ***normal subgroup*** if $k \in K$ and $g \in G$ imply $gkg^{-1} \in K$. If K is a normal subgroup of G, one writes

$$K \lhd G.$$

The proposition says that the kernel of a homomorphism is always a normal subgroup. If G is an abelian group, then every subgroup K is normal, for if $k \in K$ and $g \in G$, then $gkg^{-1} = kgg^{-1} = k \in K$.

The cyclic subgroup $H = \langle(1\ 2)\rangle$ of S_3, consisting of the two elements (1) and $(1\ 2)$, is not a normal subgroup of S_3: if $\alpha = (1\ 2\ 3)$, then $\alpha^{-1} = (3\ 2\ 1)$, and

$$\alpha(1\ 2)\alpha^{-1} = (1\ 2\ 3)(1\ 2)(3\ 2\ 1) = (2\ 3) \notin H.$$

On the other hand, the cyclic subgroup $K = \langle(1\ 2\ 3)\rangle$ of S_3 is a normal subgroup, as the reader should verify.

It follows from Examples 2.92(ii) and 2.92(iii) that A_n is a normal subgroup of S_n and $SL(2, \mathbb{R})$ is a normal subgroup of $GL(2, \mathbb{R})$ (however, it is also easy to prove these facts directly).

→ **Definition.** If G is a group and $a \in G$, then a ***conjugate*** of a is an element in G of the form

$$gag^{-1},$$

where $g \in G$.

It is clear that a subgroup $K \leq G$ is a normal subgroup if and only if K contains all the conjugates of its elements: if $k \in K$, then $gkg^{-1} \in K$ for all $g \in G$. In Proposition 2.33, we showed that $\alpha, \beta \in S_n$ are conjugate in S_n if and only if they have the same cycle structure.

If $H \leq S_n$, then $\alpha, \beta \in H$ being conjugate in S_n (that is, α and β have the same cycle structure) does not imply that α and β are conjugate in H. For example, $(1\ 2)(3\ 4)$ and $(1\ 3)(2\ 4)$ are conjugate in S_4, for they have the same cycle structure, but they are not conjugate in $\mathbf{V}$ because the four-group $\mathbf{V}$ is abelian.

Remark. In linear algebra, a linear transformation $T: V \to V$, where V is an n-dimensional vector space over $\mathbb{R}$, determines an $n \times n$ matrix A if one uses a basis of V; if one uses another basis, then T determines another matrix B. It turns out that A and B are *similar*; that is, there is a nonsingular matrix P with $B = PAP^{-1}$. Thus, conjugacy in $GL(n, \mathbb{R})$ is similarity. ◄

→ **Definition.** If G is a group and $g \in G$, define ***conjugation*** $\gamma_g: G \to G$ by

$$\gamma_g(a) = gag^{-1}$$

for all $a \in G$.

Proposition 2.94.

(i) *If G is a group and $g \in G$, then conjugation $\gamma_g : G \to G$ is an isomorphism.*

(ii) *Conjugate elements have the same order.*

Proof.
(i) If $g, h \in G$, then

$$(\gamma_g \circ \gamma_h)(a) = \gamma_g(hah^{-1}) = g(hah^{-1})g^{-1} = (gh)a(gh)^{-1} = \gamma_{gh}(a);$$

that is,

$$\gamma_g \circ \gamma_h = \gamma_{gh}.$$

It follows that each γ_g is a bijection, for $\gamma_g \circ \gamma_{g^{-1}} = \gamma_1 = 1 = \gamma_{g^{-1}} \circ \gamma_g$. We now show that γ_g is an isomorphism: if $a, b \in G$,

$$\gamma_g(ab) = g(ab)g^{-1} = (gag^{-1})(gbg^{-1}) = \gamma_g(a)\gamma_g(b).$$

(ii) To say that a and b are conjugate is to say that there is $g \in G$ with $b = gag^{-1}$; that is, $b = \gamma_g(a)$. But γ_g is an isomorphism, and so Exercise 2.69(ii) on page 169 shows that a and $b = \gamma_g(a)$ have the same order.   ●

→ **Example 2.95.**
Define the ***center*** of a group G, denoted by $Z(G)$, to be

$$Z(G) = \{z \in G : zg = gz \text{ for all } g \in G\};$$

that is, $Z(G)$ consists of all elements commuting with everything in G. (Note that the equation $zg = gz$ can be rewritten as $z = gzg^{-1}$, so that no other elements in G are conjugate to z.)

Let us show that $Z(G)$ is a subgroup of G. Clearly $1 \in Z(G)$, for 1 commutes with everything. If $y, z \in Z(G)$, then $yg = gy$ and $zg = gz$ for all $g \in G$. Therefore, $(yz)g = y(zg) = y(gz) = (yg)z = g(yz)$, so that yz commutes with everything, and $yz \in Z(G)$. Finally, if $z \in Z(G)$, then $zg = gz$ for all $g \in G$; in particular, $zg^{-1} = g^{-1}z$. Therefore,

$$gz^{-1} = (zg^{-1})^{-1} = (g^{-1}z)^{-1} = z^{-1}g$$

[we are using Lemma 2.46: $(ab)^{-1} = b^{-1}a^{-1}$ and $(a^{-1})^{-1} = a$].

The center $Z(G)$ is a normal subgroup: if $z \in Z(G)$ and $g \in G$, then

$$gzg^{-1} = zgg^{-1} = z \in Z(G).$$

A group G is abelian if and only if $Z(G) = G$. At the other extreme are groups G for which $Z(G) = \{1\}$; such groups are called ***centerless***. For example, it is easy to see that $Z(S_3) = \{1\}$; indeed, all large symmetric groups are centerless, for Exercise 2.34 on page 124 shows that $Z(S_n) = \{1\}$ for all $n \geq 3$.   ◀

→ **Example 2.96.**

The four-group **V** is a normal subgroup of S_4. Recall that the elements of **V** are

$$\mathbf{V} = \{(1), (1\ 2)(3\ 4), (1\ 3)(2\ 4), (1\ 4)(2\ 3)\}.$$

By Proposition 2.32, every conjugate of a product of two transpositions is another such. But we saw, in Example 2.29, that only 3 permutations in S_4 have this cycle structure, and so **V** is a normal subgroup of S_4. ◄

→ **Proposition 2.97.**

(i) *If H is a subgroup of index 2 in a group G, then $g^2 \in H$ for every $g \in G$.*

(ii) *If H is a subgroup of index 2 in a group G, then H is a normal subgroup of G.*

Proof.

(i) Since H has index 2, there are exactly two cosets, namely, H and aH, where $a \notin H$. Thus, G is the disjoint union $G = H \cup aH$; that is, aH is the (set-theoretic) complement of H. Take $g \in G$ with $g \notin H$, so that $g \in aH$; that is, $g = ah$ for some $h \in H$. Similarly, if $g^2 \notin H$, then $g^2 = ah'$, where $h' \in H$. Hence,

$$g = g^{-1}g^2 = (ah)^{-1}ah' = h^{-1}a^{-1}ah' = h^{-1}h' \in H,$$

and this is a contradiction.

(ii) It suffices to prove that if $h \in H$, then the conjugate $ghg^{-1} \in H$ for every $g \in G$. As we mentioned in part (i), that H has index 2 says that aH is the complement of H. Now, either $g \in H$ or $g \in aH$. If $g \in H$, then $ghg^{-1} \in H$, because H is a subgroup. In the second case, write $g = ax$, where $x \in H$. Then $ghg^{-1} = a(xhx^{-1})a^{-1} = ah'a^{-1}$, where $h' = xhx^{-1} \in H$ (for h' is a product of three elements in H). If $ghg^{-1} \notin H$, then $ghg^{-1} = ah'a^{-1} \in aH$; that is, $ah'a^{-1} = ay$ for some $y \in H$. Canceling a, we have $h'a^{-1} = y$, which gives the contradiction $a = y^{-1}h' \in H$. Therefore, if $h \in H$, every conjugate of h also lies in H; that is, H is a normal subgroup of G. •

→ **Definition.** The group of *quaternions*[17] is the group **Q** of order 8 consisting of the matrices in GL(2, $\mathbb{C}$)

$$\mathbf{Q} = \{I, A, A^2, A^3, B, BA, BA^2, BA^3\},$$

where I is the identity matrix, $A = \begin{bmatrix} 0 & 1 \\ -1 & 0 \end{bmatrix}$, and $B = \begin{bmatrix} 0 & i \\ i & 0 \end{bmatrix}$.

---

[17]The operations of addition, subtraction, multiplication, and division (by nonzero numbers) can be extended from $\mathbb{R}$ to the plane in such a way that all the usual laws of arithmetic hold; of course, the plane is usually called the complex numbers $\mathbb{C}$ in this context. W. R. Hamilton invented a way of extending all these operations from $\mathbb{C}$ to four-dimensional space in such a way that all the usual laws of arithmetic still hold (except for commutativity of multiplication); he called the new "numbers" *quaternions* (from the Latin word meaning "four"). The multiplication is determined by knowing

The reader should note that the element $A \in \mathbf{Q}$ has order 4, so that $\langle A \rangle$ is a subgroup of order 4 and hence of index 2; the other coset is $B\langle A \rangle = \{B, BA, BA^2, BA^3\}$.

Example 2.98.
In Exercise 2.86 on page 170, the reader will check that $\mathbf{Q}$ is a nonabelian group of order 8. We claim that every subgroup of $\mathbf{Q}$ is normal. Lagrange's theorem says that every subgroup of Q has order a divisor of 8, and so the only possible orders of subgroups are 1, 2, 4, or 8. Clearly, the subgroup $\{1\}$ and the subgroup of order 8 (namely, $\mathbf{Q}$ itself) are normal subgroups. By Proposition 2.97(ii), any subgroup of order 4 must be normal, for it has index 2. Finally, the only element in $\mathbf{Q}$ having order 2 is $-I$, as the reader may quickly check, and so $\langle -I \rangle$ is the only subgroup of order 2. But this subgroup is normal, for if M is any matrix, then $M(-I) = (-I)M$, so that $M(-I)M^{-1} = (-I)MM^{-1} = -I \in \langle -I \rangle$. [Exercise 2.86 asks you to prove that $\langle -I \rangle = Z(\mathbf{Q})$.] ◄

Example 2.98 shows that $\mathbf{Q}$ is a nonabelian group which is like abelian groups in the sense that every subgroup is normal. This is essentially the only such example: every finite group with every subgroup normal has the form $\mathbf{Q} \times A$, where A is an abelian group of a special form: $A = B \times C$, where every nonidentity element in B has order 2 and every element in C has odd order (*direct products* $A \times B$ will be introduced in the next section).

Lagrange's theorem states that the order of a subgroup of a finite group G must be a divisor of $|G|$. This suggests the question, given some divisor d of $|G|$, whether G must contain a subgroup of order d. The next result shows that there need not be such a subgroup.

→ **Proposition 2.99.**   *The alternating group A_4 is a group of order* 12 *having no subgroup of order* 6.

Proof.   First of all, $|A_4| = 12$, by Exercise 2.32 on page 124. If A_4 contains a subgroup H of order 6, then H has index 2, and so $\alpha^2 \in H$ for every $\alpha \in A_4$, by Proposition 2.97(i). If α is a 3-cycle, however, then α has order 3, so that $\alpha = \alpha^4 = (\alpha^2)^2$. Thus, H contains every 3-cycle. This is a contradiction, for there are 8 3-cycles in A_4.   •

---

how to multiply the particular 4-tuples 1, $\mathbf{i}$, $\mathbf{j}$, and $\mathbf{k}$:

$$\mathbf{i}^2 = -1 = \mathbf{j}^2 = \mathbf{k}^2;$$

$$\mathbf{ij} = \mathbf{k}; \quad \mathbf{ji} = -\mathbf{k}; \quad \mathbf{jk} = \mathbf{i}; \quad \mathbf{kj} = -\mathbf{i}; \quad \mathbf{ki} = \mathbf{j}; \mathbf{jk} = -\mathbf{j}.$$

All the nonzero quaternions form a multiplicative group, and the group of quaternions is isomorphic to the smallest subgroup (it has order 8) containing these four elements.

Proposition 2.124 will show that if G is an *abelian* group of order n, then G does have a subgroup of order d for every divisor d of n.

EXERCISES

H **2.64** True or false with reasons.

    (i)   If G and H are additive groups, then every homomorphism $f : G \to H$ satisfies $f(x + y) = f(x) + f(y)$ for all $x, y \in G$.

    (ii)   A function $f : \mathbb{R} \to \mathbb{R}^{\times}$ is a homomorphism if and only if $f(x + y) = f(x) + f(y)$ for all $x, y \in \mathbb{R}$.

    (iii) The inclusion $\mathbb{Z} \to \mathbb{R}$ is a homomorphism of additive groups.

    (iv) The subgroup $\{0\}$ of $\mathbb{Z}$ is isomorphic to the subgroup $\{(1)\}$ of S_5.

    (v)   Any two finite groups of the same order are isomorphic.

    (vi) If p is a prime, any two groups of order p are isomorphic.

    (vii) The subgroup $\langle (1\ 2) \rangle$ is a normal subgroup of S_3.

    (viii) The subgroup $\langle (1\ 2\ 3) \rangle$ is a normal subgroup of S_3.

    (ix) If G is a group, then $Z(G) = G$ if and only if G is abelian.

    (x)   The 3-cycles $(7\ 6\ 5)$ and $(5\ 26\ 34)$ are conjugate in S_{100}.

* H **2.65** If there is a bijection $f : X \to Y$ (that is, if X and Y have the same number of elements), prove that there is an isomorphism $\varphi : S_X \to S_Y$.

2.66 Let G be a group, let X be a set, and let $\varphi : G \to X$ be a bijection. Prove that there is an operation on X which makes X into a group such that $\varphi : G \to X$ is an isomorphism.

2.67   **(i)**   Prove that the composite of homomorphisms is itself a homomorphism.

    (ii) Prove that the inverse of an isomorphism is an isomorphism.

    (iii) Prove that isomorphism is an equivalence relation on any family of groups.

    (iv) Prove that two groups that are isomorphic to a third group are isomorphic to each other.

2.68 Prove that a group G is abelian if and only if the function $f : G \to G$, given by $f(a) = a^{-1}$, is a homomorphism.

*2.69 This exercise gives some invariants of a group G. Let $f : G \to H$ be an isomorphism.

    (i)   Prove that if $a \in G$ has infinite order, then so does $f(a)$, and if a has finite order n, then so does $f(a)$. Conclude that if G has an element of some order n and H does not, then $G \not\cong H$.

    (ii) Prove that if $G \cong H$, then, for every divisor k of $|G|$, both G and H have the same number of elements of order k.

2.70   **(i)**   Show that every group G with $|G| < 6$ is abelian.

    (ii) Find two nonisomorphic groups of order 6.

2.71 Prove that a dihedral group of order 4 is isomorphic to $\mathbf{V}$, the 4-group, and a dihedral group of order 6 is isomorphic to S_3.

*2.72 Prove that any two dihedral groups of order $2n$ are isomorphic.

*2.73 This exercise is for readers familiar with $n \times n$ matrices (see Example 4.66). Define a function $f: S_n \to \mathrm{GL}(n, \mathbb{R})$ by $f: \sigma \mapsto P_\sigma$, where P_σ is the matrix obtained from the $n \times n$ identity matrix I by permuting its columns by σ (the matrix P_σ is called a **permutation matrix**). Prove that f is an isomorphism from S_n to a subgroup of $\mathrm{GL}(n, \mathbb{R})$.

2.74   **(i)**   Find a subgroup $H \leq S_4$ with $H \cong \mathbf{V}$ but with $H \neq \mathbf{V}$.

  (ii)   Prove that the subgroup H in part (i) is not a normal subgroup.

H **2.75** If G is a group and $a, b \in G$, prove that ab and ba have the same order.

2.76   **(i)**   If $f: G \to H$ is a homomorphism and $x \in G$ has order k, prove that $f(x) \in H$ has order m, where $m \mid k$.

  (ii)   If $f: G \to H$ is a homomorphism and if $(|G|, |H|) = 1$, prove that $f(x) = 1$ for all $x \in G$.

*2.77 H **(i)**   Prove that the special orthogonal group $SO(2, \mathbb{R})$ is isomorphic to the circle group S^1.

  (ii)   Prove that all the rotations of the plane about the origin form a group under composition which is isomorphic to $SO(2, \mathbb{R})$.

H **2.78** Let G be the additive group of all polynomials in x with coefficients in $\mathbb{Z}$, and let H be the multiplicative group of all positive rationals. Prove that $G \cong H$.

*2.79 Show that if H is a subgroup with $bH = Hb = \{hb : h \in H\}$ for every $b \in G$, then H must be a normal subgroup. (The converse is proved in Lemma 2.112.)

2.80 Prove that the intersection of any family of normal subgroups of a group G is itself a normal subgroup of G.

2.81 Define $W = \langle (1\ 2)(3\ 4) \rangle$, the cyclic subgroup of S_4 generated by $(1\ 2)(3\ 4)$. Show that W is a normal subgroup of $\mathbf{V}$, but that W is not a normal subgroup of S_4. Conclude that normality is not transitive: $K \lhd H$ and $H \lhd G$ need not imply $K \lhd G$.

* H **2.82** Let G be a finite group written multiplicatively. Prove that if $|G|$ is odd, then every $x \in G$ has a square root. Conclude, using Exercise 2.45 on page 147, that there exists exactly one $g \in G$ with $g^2 = x$.

H **2.83** Give an example of a group G, a subgroup $H \leq G$, and an element $g \in G$ with $[G : H] = 3$ and $g^3 \notin H$.

* H **2.84** Show that the center of $\mathrm{GL}(2, \mathbb{R})$ is the set of all *scalar matrices* $\begin{bmatrix} a & 0 \\ 0 & a \end{bmatrix}$ with $a \neq 0$.

*2.85 Let $\zeta = e^{2\pi i/n}$ be a primitive nth root of unity, and define $A = \begin{bmatrix} \zeta & 0 \\ 0 & \zeta^{-1} \end{bmatrix}$ and $B = \begin{bmatrix} 0 & 1 \\ 1 & 0 \end{bmatrix}$.

  (i)   Prove that A has order n and that B has order 2.

  (ii)   Prove that $BAB = A^{-1}$.

  H **(iii)**   Prove that the matrices of the form A^i and BA^i, for $0 \leq i < n$, form a multiplicative subgroup $G \leq \mathrm{GL}(2, \mathbb{C})$.

  (iv)   Prove that each matrix in G has a unique expression of the form $B^i A^j$, where $i = 0, 1$ and $0 \leq j < n$. Conclude that $|G| = 2n$ and that $G \cong D_{2n}$.

*2.86 Recall that the group of quaternions $\mathbf{Q}$ (defined in Example 2.98) consists of the 8 matrices in $\mathrm{GL}(2, \mathbb{C})$,

$$\mathbf{Q} = \{I, A, A^2, A^3, B, BA, BA^2, BA^3\},$$

where $A = \begin{bmatrix} 0 & 1 \\ -1 & 0 \end{bmatrix}$ and $B = \begin{bmatrix} 0 & i \\ i & 0 \end{bmatrix}$.

  H **(i)**   Prove that $\mathbf{Q}$ is a nonabelian group with operation matrix multiplication.

  (ii)   Prove that $-I$ is the only element in $\mathbf{Q}$ of order 2, and that all other elements $M \neq I$ satisfy $M^2 = -I$.

      (iii) Show that **Q** has a unique subgroup of order 2, and it is the center of **Q**.

      (iv) Prove that $\langle -I \rangle$ is the center $Z(\mathbf{Q})$.

*H **2.87** Prove that the quaternions **Q** and the dihedral group D_8 are nonisomorphic groups of order 8.

  2.88 If G is a finite group generated by two elements of order 2, prove that $G \cong D_{2n}$ for some $n \geq 2$.

*2.89   **(i)** Prove that A_3 is the only subgroup of S_3 of order 3.

   H **(ii)** Prove that A_4 is the only subgroup of S_4 of order 12. (In Exercise 2.135 on page 208, this will be generalized from S_4 and A_4 to S_n and A_n for all $n \geq 3$.)

*2.90   **(i)** Let $\mathcal{A}$ be the set of all 2×2 matrices of the form $A = \begin{bmatrix} a & b \\ 0 & 1 \end{bmatrix}$, where $a \neq 0$. Prove that $\mathcal{A}$ is a subgroup of GL$(2, \mathbb{R})$.

      (ii) Prove that $\psi \colon \text{Aff}(1, \mathbb{R}) \to \mathcal{A}$, defined by $f \mapsto A$, is an isomorphism, where $f(x) = ax + b$ [see Example 2.48(iv)].

   H **(iii)** Prove that the stochastic group $\Sigma(2, \mathbb{R})$ [see Exercise 2.48 on page 147] is isomorphic to the affine group Aff$(1, \mathbb{R})$ by showing that $\varphi \colon \Sigma(2, \mathbb{R}) \to \mathcal{A} \cong$ Aff$(1, \mathbb{R})$, given by $\varphi(M) = QMQ^{-1}$, is an isomorphism, where $Q = \begin{bmatrix} 1 & 0 \\ 1 & 1 \end{bmatrix}$ and $Q^{-1} = \begin{bmatrix} 1 & 0 \\ -1 & 1 \end{bmatrix}$.

 H **2.91** Prove that the symmetry group $\Sigma(\pi_n)$, where π_n is a regular polygon with n vertices, is isomorphic to a subgroup of S_n.

  2.92 An *automorphism* of a group G is an isomorphism $G \to G$.

      (i) Prove that Aut(G), the set of all the automorphisms of a group G, is a group under composition.

      (ii) Prove that $\gamma \colon G \to \text{Aut}(G)$, defined by $g \mapsto \gamma_g$ (conjugation by g), is a homomorphism.

    (iii) Prove that ker $\gamma = Z(G)$.

    (iv) Prove that im $\gamma \lhd \text{Aut}(G)$.

  2.93 If G is a group, prove that Aut$(G) = \{1\}$ if and only if $|G| \leq 2$.

  2.94 If C is a finite cyclic group of order n, prove that $|\text{Aut}(C)| = \phi(n)$, where $\phi(n)$ is the Euler ϕ-function.

→ 2.6   Quotient Groups

We are now going to construct a group using congruence mod m. Once this is done, we will be able to give a proof of Fermat's theorem using group theory. This construction is the prototype of a more general way of building new groups from given groups, called *quotient groups*.

    Recall, given $m \geq 2$ and $a \in \mathbb{Z}$, that the *congruence class* of a mod m is the subset $[a]$ of $\mathbb{Z}$:

$$[a] = \{b \in \mathbb{Z} : b \equiv a \bmod m\}$$
$$= \{a + km : k \in \mathbb{Z}\}$$
$$= \{\ldots, a - 2m, a - m, a, a + m, a + 2m, \ldots\}.$$

→ **Definition.**   The *integers mod m*, denoted by $\mathbb{I}_m$,[18] is the family of all congruence classes mod m.

For example, if $m = 2$, then $[0] = \{b \in \mathbb{Z} : b \equiv 0 \bmod 2\}$ is the set of all the even integers and $[1] = \{b \in \mathbb{Z} : b \equiv 1 \bmod 2\}$ is the set of all the odd integers. Notice that $[2] = \{2 + 2k : k \in \mathbb{Z}\}$ is also the set of all even integers, so that $[2] = [0]$; indeed, $[0] = [2] = [-2] = [4] = [-4] = [6] = [-6] = \cdots$.

Remark.   Given m, we may form the cyclic subgroup $\langle m \rangle$ of $\mathbb{Z}$ generated by m. In Example 2.18(ii), we saw that the congruence class $[a]$ is precisely the coset $a + \langle m \rangle$.
◀

The notation $[a]$ is incomplete in that it does not mention the modulus m: for example, $[1]$ in $\mathbb{I}_2$ is not the same as $[1]$ in $\mathbb{I}_3$ (the former is the set of all odd numbers and the latter is $\{1 + 3k : k \in \mathbb{Z}\} = \{\ldots, -2, 1, 4, 7, \ldots\}$). This will not cause problems, for, almost always, one works with only one $\mathbb{I}_m$ at a time. However, if there is a danger of confusion, as in Theorem 2.128, we will denote the congruence class of a in $\mathbb{I}_m$ by $[a]_m$. The next proposition is a special case of Lemma 2.19 about equivalence classes.

→ **Proposition 2.100.** $[a] = [b]$ *in* $\mathbb{I}_m$ *if and only if* $a \equiv b \bmod m$.

Proof.   If $[a] = [b]$, then $a \in [a]$, by reflexivity, and so $a \in [a] = [b]$. Therefore, $a \equiv b \bmod m$.

Conversely, if $c \in [a]$, then $c \equiv a \bmod m$, and so transitivity gives $c \equiv b \bmod m$; hence $[a] \subseteq [b]$. By symmetry, $b \equiv a \bmod m$, and this gives the reverse inclusion $[b] \subseteq [a]$. Thus, $[a] = [b]$.   •

In words, Proposition 2.100 says that congruence mod m between numbers can be converted into equality at the cost of replacing numbers by congruence classes.

In particular, $[a] = [0]$ in $\mathbb{I}_m$ if and only if $a \equiv 0 \bmod m$; that is, $[a] = [0]$ in $\mathbb{I}_m$ if and only if m is a divisor of a.

Proposition 2.101.   *Let* $m \geq 2$ *be given.*

(i) *If* $a \in \mathbb{Z}$, *then* $[a] = [r]$ *for some* r *with* $0 \leq r < m$.

---

[18]Nowadays, $\mathbb{Z}$ is widely accepted as the notation for the set of all integers, and the two most popular notations for the integers mod m are $\mathbb{Z}/m\mathbb{Z}$ and $\mathbb{Z}_m$. Both notations are good ones: the first reminds us that the group is a quotient group of $\mathbb{Z}$, but the notation is cumbersome; the second notation is compact, but it causes confusion because it is also used by number theorists, when m is a prime p, to denote all the rational numbers whose denominator is prime to p (the ring of p-adic fractions). In fact, many number theorists denote the ring of p-adic integers by $\mathbb{Z}_p$. To avoid possible confusion, I am introducing the notation $\mathbb{I}_m$.

(ii) *If $0 \le r' < r < m$, then $[r'] \ne [r]$.*

(iii) $\mathbb{I}_m$ *has exactly m elements, namely,* $[0], [1], \ldots, [m-1]$.

Proof.
(i) For each $a \in \mathbb{Z}$, the division algorithm gives $a = qm + r$, where $0 \le r < m$; hence $a - r = qm$ and $a \equiv r \bmod m$. Therefore, $[a] = [r]$, where r is the remainder after dividing a by m.
(ii) Proposition 1.58(ii) gives $r' \not\equiv r \bmod m$.
(iii) Part (i) shows that every $[a]$ in $\mathbb{I}_m$ occurs on the list $[0], [1], [2], \ldots, [m-1]$; part (ii) shows that this list of m items has no repetitions. ●

We are now going to make $\mathbb{I}_m$ into an abelian group by equipping it with an addition. Now Proposition 2.100 says that $[a] = [b]$ in $\mathbb{I}_m$ if and only if $a \equiv b \bmod m$, so that each $[a] \in \mathbb{I}_m$ has many names. The operation we propose to define on $\mathbb{I}_m$ will appear to depend on choices of names, and so we will be obliged to prove that the operation is well-defined.

Lemma 2.102.   *If $m \ge 2$, then the function $\alpha: \mathbb{I}_m \times \mathbb{I}_m \to \mathbb{I}_m$, given by*

$$\alpha([a], [b]) = [a + b],$$

is an operation on $\mathbb{I}_m$.

Proof.   To see that α is a (well-defined) function, we must show that if $[a] = [a']$ and $[b] = [b']$, then $\alpha([a], [b]) = \alpha([a'], [b'])$, that is, $[a + b] = [a' + b']$. But this is precisely Proposition 1.60(i). ●

→ **Proposition 2.103.**   *If $m \ge 2$, the set $\mathbb{I}_m$ of integers* mod *m is an additive cyclic group of order m with generator* $[1]$.

Proof.   In this proof only, we shall write $\boxplus$ for addition of congruence classes:

$$\alpha([a], [b]) = [a] \boxplus [b] = [a + b].$$

Associativity of the operation $\boxplus$ follows from associativity of ordinary addition:

$$
\begin{aligned}
[a] \boxplus \big([b] \boxplus [c]\big) &= [a] \boxplus [b + c] \\
&= [a + (b + c)] \\
&= [(a + b) + c] \\
&= [a + b] \boxplus [c] \\
&= \big([a] \boxplus [b]\big) \boxplus [c].
\end{aligned}
$$

Commutativity of the operation $\boxplus$ follows from commutativity of ordinary addition:

$$[a] \boxplus [b] = [a + b] = [b + a] = [b] \boxplus [a].$$

The identity element is [0]: since 0 is the (additive) identity in $\mathbb{Z}$,

$$[0] \boxplus [a] = [0 + a] = [a].$$

The inverse of $[a]$ is $[-a]$; since $-a$ is the additive inverse of a in $\mathbb{Z}$,

$$[-a] \boxplus [a] = [-a + a] = [0].$$

Therefore, $\mathbb{I}_m$ is an abelian group of order m; it is cyclic with generator [1], for if $0 \leq r < m$, then $[r] = [1] + \cdots + [1]$ is the congruence class obtained by adding [1] to itself r times.   •

The reader should notice that the group axioms in $\mathbb{I}_m$ are "inherited" from the group axioms in $\mathbb{Z}$.

Here is an alternative construction of the group $\mathbb{I}_m$. Define G_m to be the set $\{0, 1, \ldots, m - 1\}$, and define an operation on G_m by

$$a \boxplus b = \begin{cases} a + b & \text{if } a + b \leq m - 1; \\ a + b - m & \text{if } a + b > m - 1. \end{cases}$$

Although this definition is simpler than what we have just done, proving associativity is now is very tedious. It is also more awkward to use, for proofs usually require case analyses (for example, see Example 2.90).

We now drop the notation $\boxplus$; henceforth, we shall write

$$[a] + [b] = [a + b]$$

for the sum of congruence classes in $\mathbb{I}_m$.

→ **Corollary 2.104.**   *Every cyclic group of order $m \geq 2$ is isomorphic to $\mathbb{I}_m$.*

Proof.   We have already seen, in Example 2.90, that any two finite cyclic groups of the same order are isomorphic.   •

We now focus on multiplication.

→ **Proposition 2.105.**   *The function $\mu \colon \mathbb{I}_m \times \mathbb{I}_m \to \mathbb{I}_m$, given by*

$$\mu([a], [b]) = [ab],$$

is an operation on $\mathbb{I}_m$. This operation is associative and commutative, and [1] is an identity element.

Proof. To see that μ is a (well-defined) function, we must show that if $[a] = [a']$ and $[b] = [b']$, then $\mu([a], [b]) = \mu([a'], [b'])$, that is, $[ab] = [a'b']$. But this is precisely Proposition 1.60(ii).

In this proof only, we are going to write $\boxtimes$ for multiplication of congruence classes:

$$\mu([a], [b]) = [a] \boxtimes [b] = [ab].$$

Associativity of $\boxtimes$ follows from associativity of ordinary multiplication:

$$\begin{aligned}
[a] \boxtimes \big([b] \boxtimes [c]\big) &= [a] \boxtimes [bc] \\
&= [a(bc)] \\
&= [(ab)c] \\
&= [ab] \boxtimes [c] \\
&= \big([a] \boxtimes [b]\big) \boxtimes [c].
\end{aligned}$$

Commutativity of $\boxtimes$ follows from commutativity of ordinary multiplication:

$$[a] \boxtimes [b] = [ab] = [ba] = [b] \boxtimes [a].$$

The identity element is $[1]$, because

$$[1] \boxtimes [a] = [1a] = [a]$$

for all $a \in \mathbb{Z}$. •

We now drop the notation $\boxtimes$; henceforth, we shall write

$$[a][b] = [ab]$$

for the product of congruence classes in $\mathbb{I}_m$ instead of $[a] \boxtimes [b]$. Note that $\mathbb{I}_m$ is *not* a group under multiplication because some elements, such as $[0]$, do not have inverses.

→ **Proposition 2.106.**

(i) *If $(a, m) = 1$, then $[a][x] = [b]$ can be solved for $[x]$ in $\mathbb{I}_m$.*

(ii) *If p is a prime, then $\mathbb{I}_p^\times$, the set of nonzero elements in $\mathbb{I}_p$, is a multiplicative abelian group of order $p - 1$.*

Proof.
(i) By Theorem 1.69, the congruence $ax \equiv b \bmod m$ can be solved for x if $(a, m) = 1$; that is, $[a][x] = [b]$ can be solved for $[x]$ in $\mathbb{I}_m$ when a and m are relatively prime. (Recall that if $sa + tm = 1$, then $[x] = [sb]$.)
(ii) Assume that $m = p$ is prime; if $0 < a < p$, then $(a, p) = 1$ and the equation $[a][x] = [1]$ can be solved in $\mathbb{I}_p$, by part (i); that is, $[a]$ has an inverse in $\mathbb{I}_p$. We have proved that $\mathbb{I}_p^\times$ is an abelian group; its order is $p - 1$ because, as a set, it is obtained from $\mathbb{I}_p$ by throwing away one element, namely, $[0]$. •

In Theorem 3.55 we will prove, for every prime p, that $\mathbb{I}_p^\times$ is a cyclic group.

We now give a new proof of Fermat's theorem which is entirely different from our earlier proof, Theorem 1.64.

→ **Corollary 2.107** (*Fermat*).   *If p is a prime and $a \in \mathbb{Z}$, then*

$$a^p \equiv a \bmod p.$$

Proof.   By Proposition 2.100, it suffices to show that $[a^p] = [a]$ in $\mathbb{I}_p$. If $[a] = [0]$, then Proposition 2.105 gives $[a^p] = [a]^p = [0]^p = [0] = [a]$. If $[a] \neq [0]$, then $[a] \in \mathbb{I}_p^\times$, the multiplicative group of nonzero elements in $\mathbb{I}_p$. By Corollary 2.86 to Lagrange's theorem, $[a]^{p-1} = [1]$, because $|\mathbb{I}_p^\times| = p - 1$. Multiplying by $[a]$ gives the desired result $[a^p] = [a]^p = [a]$. Therefore, $a^p \equiv a \bmod p$.   •

Note that if $m \geq 2$ is not a prime, then $\mathbb{I}_m^\times$ is not a group: if $m = ab$, where $1 < a, b < m$, then $[a], [b] \in \mathbb{I}_m^\times$, but their product $[a][b] = [ab] = [m] = [0] \notin \mathbb{I}_m^\times$. We are now going to define an analog of $\mathbb{I}_p^\times$ that can be used to generalize Fermat's theorem.

→ **Definition.**   Let $U(\mathbb{I}_m)$ be the set of all those congruence classes in $\mathbb{I}_m$ having an inverse; that is, $[a] \in U(\mathbb{I}_m)$ if there is $[s] \in \mathbb{I}_m$ with $[s][a] = [1]$.

→ **Lemma 2.108.**

(i) $$U(I_m) = \{\, [r] \in \mathbb{I}_m : (r, m) = 1 \}.$$

(ii) $U(I_m)$ *is a multiplicative abelian group of order $\phi(m)$, the Euler ϕ-function.*

Proof.
(i) Let $E = \{\, [r] \in \mathbb{I}_m : (r, m) = 1 \}$. If $[r] \in E$, then $(r, m) = 1$, so there are integers s and t with $sr + tm = 1$. Hence, $sr \equiv 1 \bmod m$. Therefore, $[sr] = [s][r] = [1]$, and so $[r] \in U(I_m)$. For the reverse inclusion, assume that $[r] \in U(I_m)$; that is, there is $[s] \in U(I_m)$ with $[s][r] = [1]$. But $[s][r] = [sr] = [1]$, so that $m \mid (sr - 1)$; that is, there is an integer t with $tm = sr - 1$. By Exercise 1.56 on page 54, $(r, m) = 1$, and so $[r] \in E$.
(ii) By Exercise 1.58 on page 54, $(r, m) = 1 = (r', m)$ implies $(rr', m) = 1$. Hence, $[r]$ and $[r']$ in $U(I_m)$ imply $[r][r'] = [rr'] \in U(I_m)$, so that multiplication is an operation on $U(\mathbb{I}_m)$. Proposition 2.105 shows that multiplication is associative and commutative, and that $[1]$ is the identity. By Proposition 2.106(i), the equation $[r][x] = [1]$ can be solved for $[x] \in \mathbb{I}_m$. That is, each $[r] \in U(I_m)$ has an inverse. Therefore, $U(I_m)$ is an abelian group and, by Proposition 1.42, its order is $|U(\mathbb{I}_m)| = \phi(m)$.   •

If p is a prime, then $\phi(p) = p - 1$, and $U(I_p) = I_p^\times$.

→ **Theorem 2.109 (*Euler*).**   *If $(r, m) = 1$, then*

$$r^{\phi(m)} \equiv 1 \bmod m.$$

Proof.   If G is a finite group of order n, then Corollary 2.86 to Lagrange's theorem gives $x^n = 1$ for all $x \in G$. Here, if $[r] \in U(I_m)$, then $[r]^{\phi(m)} = [1]$, by Lemma 2.108. In congruence notation, this says that if $(r, m) = 1$, then $r^{\phi(m)} \equiv 1 \bmod m$.   •

Example 2.110.
It is straightforward to see that

$$U(\mathbb{I}_8) = \{ [1], [3], [5], [7] \} \cong \mathbf{V},$$

for $[3]^2 = [9] = [1]$, $[5]^2 = [25] = [1]$, and $[7]^2 = [49] = [1]$.
    Moreover,

$$U(\mathbb{I}_{10}) = \{ [1], [3], [7], [9] \} \cong \mathbb{I}_4,$$

for $[3]^4 = [81] = [1]$, while $[3]^2 = [9] = [-1] \neq [1]$.   ◄

→ **Theorem 2.111 (Wilson's Theorem).**   *An integer p is a prime if and only if*

$$(p - 1)! \equiv -1 \bmod p.$$

Proof.   Let p be a prime; we may assume that $p \geq 3$, for $1 \equiv -1 \bmod 2$. If $a_1, a_2, \ldots, a_n$ is a list of all the elements of a finite abelian group G, then the product $a_1 a_2 \ldots a_n$ is the same as the product of all elements a with $a^2 = 1$, for any other element cancels against its inverse. Since $p \geq 3$ is prime, Exercise 1.88 on page 75 implies that $\mathbb{I}_p^{\times}$ has only one element of order 2, namely, $[-1]$. It follows that the product of all the elements in $\mathbb{I}_p^{\times}$, namely, $[(p - 1)!]$, is equal to $[-1]$; therefore, $(p - 1)! \equiv -1 \bmod p$.
    Conversely, if $(m - 1)! \equiv -1 \bmod m$, then $(m, (m - 1)!) = 1$. If m is composite, then there is an integer $a \mid m$ with $1 < a \leq m - 1$. Now $a \mid a!$ implies $a \mid (m - 1)!$. Thus, $a > 1$ is a common divisor of m and $(m - 1)!$, a contradiction. Therefore, m is prime.   •

Remark.   One can generalize Wilson's theorem in the same way that Euler's theorem generalizes Fermat's theorem: replace $U(\mathbb{I}_p)$ by $U(\mathbb{I}_n)$. For example, one can prove, for all $m \geq 3$, that $U(\mathbb{I}_{2^m})$ has exactly 3 elements of order 2, namely, $[-1]$, $[1 + 2^{m-1}]$, and $[-(1 + 2^{m-1})]$. It now follows that the product of all the odd numbers r, where $1 \leq r < 2^m$ is congruent to 1 mod 2^m, because

$$(-1)(1 + 2^{m-1})(-1 - 2^{m-1}) = (1 + 2^{m-1})^2$$

$$= 1 + 2^m + 2^{2m-2} \equiv 1 \bmod 2^m.   ◄$$

The homomorphism $\pi : \mathbb{Z} \to \mathbb{I}_m$, defined by $\pi : a \mapsto [a]$, is surjective, so that $\mathbb{I}_m$ is equal to $\mathrm{im}\, \pi$. Thus, every element of $\mathbb{I}_m$ has the form $\pi(a)$ for some $a \in \mathbb{Z}$, and $\pi(a) + \pi(b) = \pi(a + b)$. This description of the additive group $\mathbb{I}_m$ in terms of the additive group $\mathbb{Z}$ can be generalized to arbitrary, not necessarily abelian, groups. Suppose that $f : G \to H$ is a surjective homomorphism between groups G and H. Since f is surjective, each element of H has the form $f(a)$ for some $a \in G$, and the operation in H is given by $f(a)f(b) = f(ab)$, where $a, b \in G$. Now $K = \ker f$ is a normal subgroup of G, and we are going to reconstruct $H = \mathrm{im}\, f$ from G and K alone.

We begin by introducing an operation on the set

$$\mathcal{S}(G)$$

of all nonempty subsets of a group G. If $X, Y \in \mathcal{S}(G)$, define

$$XY = \{xy : x \in X \text{ and } y \in Y\}.$$

This multiplication is associative: $X(YZ)$ is the set of all $x(yz)$, where $x \in X$, $y \in Y$, and $z \in Z$, $(XY)Z$ is the set of all such $(xy)z$, and these subsets are the same because associativity in G says that their elements are the same.

An instance of this multiplication is the product of a one-point subset $\{a\}$ and a subgroup $H \leq G$, which is the coset aH.

As a second example, we show that if H is any subgroup of G, then

$$HH = H.$$

If $h, h' \in H$, then $hh' \in H$, because subgroups are closed under multiplication, and so $HH \subseteq H$. For the reverse inclusion, if $h \in H$, then $h = h1 \in HH$ (because $1 \in H$), and so $H \subseteq HH$.

It is possible for two subsets X and Y in $\mathcal{S}(G)$ to commute even though their constituent elements do not commute. One example has just been given; take $X = Y = H$, where H is a nonabelian subgroup of G. Here is a more interesting example: let $G = S_3$ and $K = \langle (1\ 2\ 3) \rangle$. Now $(1\ 2)$ does not commute with $(1\ 2\ 3) \in K$, but we claim that $(1\ 2)K = K(1\ 2)$.

$\to$ **Lemma 2.112.** *A subgroup K of a group G is a normal subgroup if and only if $bK = Kb$ for every $b \in G$.*

Proof. Let $bk \in bK$. Since K is normal, $bkb^{-1} \in K$, say $bkb^{-1} = k' \in K$, so that $bk = (bkb^{-1})b = k'b \in Kb$, and so $bK \subseteq Kb$. For the reverse inclusion, let $kb \in Kb$. Since K is normal, $(b^{-1})k(b^{-1})^{-1} = b^{-1}kb \in K$, say $b^{-1}kb = k'' \in K$. Hence, $kb = b(b^{-1}kb) = bk'' \in bK$ and $Kb \subseteq bK$. Therefore, $bK = Kb$ when $K \lhd G$.

Even though sufficiency has already appeared as Exercise 2.79 on page 170, we prove it here. Assume that $bK = Kb$ for all $b \in G$. If $x \in K$, then $bx \in bK = Kb$; hence, there is $x' \in K$ with $bx = x'b$, so that $bxb^{-1} = x' \in K$. Therefore, $K \lhd G$. •

It follows from Lemma 2.112 that if $K \lhd G$, then every left coset of K in G is a right coset of K; the converse is Exercise 2.107 on page 191.

Here is a fundamental construction of a new group from a given group.

→ **Theorem 2.113.** *Let G/K denote the family of all the cosets of a subgroup K of G. If K is a normal subgroup, then*

$$aKbK = abK$$

for all $a, b \in G$, and G/K is a group under this operation.

Remark. The group G/K is called the **quotient group** G mod K; when G is finite, its order $|G/K|$ is the index $[G : K] = |G|/|K|$ (presumably, this is the reason *quotient groups* are so called). ◄

Proof. The product of two cosets $(aK)(bK)$ can also be viewed as the product of 4 elements in $\mathcal{S}(G)$. Hence, by Theorem 2.49, associativity in the semigroup $\mathcal{S}(G)$ gives generalized associativity:

$$(aK)(bK) = a(Kb)K = a(bK)K = abKK = abK,$$

for normality of K gives $Kb = bK$ for all $b \in K$, by Lemma 2.112, while $KK = K$ because K is a subgroup. Thus, the product of two cosets of K is again a coset of K, and so an operation on G/K has been defined. Because multiplication in $\mathcal{S}(G)$ is associative, equality $X(YZ) = (XY)Z$ holds, in particular, when X, Y, and Z are cosets of K, so that the operation on G/K is associative. The identity is the coset $K = 1K$, for $(1K)(bK) = 1bK = bK$, and the inverse of aK is $a^{-1}K$, for $(a^{-1}K)(aK) = a^{-1}aK = K$. Therefore, G/K is a group. •

→ **Example 2.114.**

We show that the quotient group $\mathbb{Z}/\langle m \rangle$ is precisely $\mathbb{I}_m$, where $\langle m \rangle$ is the (cyclic) subgroup consisting of all the multiples of a positive integer m. Since $\mathbb{Z}$ is abelian, $\langle m \rangle$ is necessarily a normal subgroup. The sets $\mathbb{Z}/\langle m \rangle$ and $\mathbb{I}_m$ coincide because they are comprised of the same elements: the coset $a + \langle m \rangle$ is the congruence class $[a]$:

$$a + \langle m \rangle = \{a + km : k \in \mathbb{Z}\} = [a].$$

The operations also coincide: addition in $\mathbb{Z}/\langle m \rangle$ is given by

$$(a + \langle m \rangle) + (b + \langle m \rangle) = (a + b) + \langle m \rangle;$$

since $a + \langle m \rangle = [a]$, this last equation is just $[a] + [b] = [a + b]$, which is the sum in $\mathbb{I}_m$. Therefore, $\mathbb{I}_m$ is equal to the quotient group $\mathbb{Z}/\langle m \rangle$. ◄

Here is the converse of Proposition 2.93(ii). Recall Lemma 2.82(i): if K is a subgroup of G, then two cosets aK and bK are equal if and only if $b^{-1}a \in K$. In particular, if $b = 1$, then $aK = K$ if and only if $a \in K$.

→ **Corollary 2.115.**   *Every normal subgroup is the kernel of some homomorphism.*

Proof.   If $K \lhd G$, define the ***natural map*** $\pi : G \to G/K$ by $\pi(a) = aK$. With this notation, the formula $aKbK = abK$ can be rewritten as $\pi(a)\pi(b) = \pi(ab)$; thus, π is a (surjective) homomorphism. Since K is the identity element in G/K,

$$\ker \pi = \{a \in G : \pi(a) = K\} = \{a \in G : aK = K\} = K,$$

by Lemma 2.82(i).   •

The next theorem shows that every homomorphism gives rise to an isomorphism, and that quotient groups are merely constructions of homomorphic images. It was E. Noether (1882–1935) who emphasized the fundamental importance of this fact.

→ **Theorem 2.116 (First Isomorphism Theorem).**   *If $f : G \to H$ is a homomorphism, then*

$$\ker f \lhd G \quad and \quad G/\ker f \cong \operatorname{im} f.$$

In more detail, if $\ker f = K$, then the function $\varphi : G/K \to \operatorname{im} f \leq H$, given by $\varphi : aK \mapsto f(a)$, is an isomorphism.

Proof.   We have already seen, in Proposition 2.93(ii), that $K = \ker f$ is a normal subgroup of G. Now φ is well-defined: if $aK = bK$, then $a = bk$ for some $k \in K$, and so $f(a) = f(bk) = f(b)f(k) = f(b)$, because $f(k) = 1$.

Let us now see that φ is a homomorphism. Since f is a homomorphism and $\varphi(aK) = f(a)$,

$$\varphi(aKbK) = \varphi(abK) = f(ab) = f(a)f(b) = \varphi(aK)\varphi(bK).$$

It is clear that $\operatorname{im} \varphi \leq \operatorname{im} f$. For the reverse inclusion, note that if $y \in \operatorname{im} f$, then $y = f(a)$ for some $a \in G$, and so $y = f(a) = \varphi(aK)$. Thus, φ is surjective.

Finally, we show that φ is injective. If $\varphi(aK) = \varphi(bK)$, then $f(a) = f(b)$. Thus, $1 = f(b)^{-1}f(a) = f(b^{-1}a)$, so that $b^{-1}a \in \ker f = K$. Thus, $aK = bK$, by Lemma 2.82(i), and so φ is injective. Hence, $\varphi : G/K \to \operatorname{im} f$ is an isomorphism.   •

Remark.   The following diagram describes the proof of the first isomorphism theorem, where $\pi : G \to G/K$ is the natural map $\pi : a \mapsto aK$.

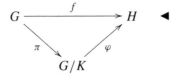

Given any homomorphism $f : G \to H$, one should immediately ask for its kernel and its image; the first isomorphism theorem will then provide an isomorphism

$G/\ker f \cong \operatorname{im} f$. Since there is no significant difference between isomorphic groups, the first isomorphism theorem also says that there is no significant difference between quotient groups and homomorphic images.

Example 2.117.
Let us revisit Example 2.90, which showed that any two cyclic groups of order m are isomorphic. If $G = \langle a \rangle$ is a cyclic group of order m, define a homomorphism $f \colon \mathbb{Z} \to G$ by $f(n) = a^n$ for all $n \in \mathbb{Z}$. Now f is surjective (because a is a generator of G), while $\ker f = \{n \in \mathbb{Z} : a^n = 1\} = \langle m \rangle$, by Lemma 2.53. The first isomorphism theorem gives an isomorphism $\mathbb{Z}/\langle m \rangle \cong G$. We have shown that every cyclic group of order m is isomorphic to $\mathbb{Z}/\langle m \rangle$, and hence that any two cyclic groups of order m are isomorphic to each other. Now Example 2.114 shows that $\mathbb{Z}/\langle m \rangle = \mathbb{I}_m$, so that every cyclic group of order m is isomorphic to $\mathbb{I}_m$. ◀

Example 2.118.
What is the quotient group $\mathbb{R}/\mathbb{Z}$? Define $f \colon \mathbb{R} \to S^1$, where S^1 is the circle group, by

$$f \colon x \mapsto e^{2\pi i x}.$$

Now f is a homomorphism; that is, $f(x+y) = f(x)f(y)$, by the addition formulas for sine and cosine. The map f is surjective, and $\ker f$ consists of all $x \in \mathbb{R}$ for which $1 = e^{2\pi i x} = \cos 2\pi x + i \sin 2\pi x$. Clearly, $\mathbb{Z} \subseteq \ker f$, for if $n \in \mathbb{Z}$, then $f(n) = e^{2\pi i n} = 1$; for the reverse inclusion, if $1 = f(x) = e^{2\pi i x}$, then $\cos 2\pi x = 0 = \sin 2\pi x$ forces x to be an integer. Therefore, $\ker f = \mathbb{Z}$, and the first isomorphism theorem now gives

$$\mathbb{R}/\mathbb{Z} \cong S^1. \quad ◀$$

A natural question is whether HK is a subgroup when both H and K are subgroups. In general, HK need not be a subgroup. For example, let $G = S_3$, let $H = \langle (1\ 2) \rangle$, and let $K = \langle (1\ 3) \rangle$. Then

$$HK = \{(1), (1\ 2), (1\ 3), (1\ 3\ 2)\}$$

is not a subgroup lest we contradict Lagrange's theorem. Exercise 2.106 on page 191 gives a necessary and sufficient condition for the product HK of subgroups H and K to be a subgroup.

Proposition 2.119.

(i) *If H and K are subgroups of a group G, and if one of them is a normal subgroup, then HK is a subgroup of G; moreover, $HK = KH$ in this case.*

(ii) *If both H and K are normal subgroups, then HK is a normal subgroup.*

Proof.
(i) Assume that $K \lhd G$. We claim that $HK = KH$. If $hk \in HK$, then $k' = hkh^{-1} \in K$, because $K \lhd G$, and
$$hk = hkh^{-1}h = k'h \in KH.$$

Hence, $HK \subseteq KH$. For the reverse inclusion, write $kh = hh^{-1}kh = hk'' \in HK$. (Note that the same argument shows that $HK = KH$ if $H \lhd G$.)

We now show that HK is a subgroup. Since $1 \in H$ and $1 \in K$, we have $1 = 1 \cdot 1 \in HK$; if $hk \in HK$, then $(hk)^{-1} = k^{-1}h^{-1} \in KH = HK$; if $hk, h_1k_1 \in HK$, then $h_1^{-1}kh_1 = k' \in K$ and

$$hkh_1k_1 = hh_1(h_1^{-1}kh_1)k_1 = (hh_i)(k'k_1) \in HK.$$

Therefore, HK is a subgroup of G.
(ii) If $g \in G$, then
$$ghkg^{-1} = (ghg^{-1})(gkg^{-1}) \in HK.$$

Therefore, $HK \lhd G$ in this case. •

Here is a useful counting result.

Proposition 2.120 (Product Formula). *If H and K are subgroups of a finite group G, then*
$$|HK||H \cap K| = |H||K|,$$

where $HK = \{hk : h \in H \text{ and } k \in K\}$.

Remark. Since we are not assuming that either H or K is a normal subgroup, the subset HK need not be a subgroup. ◄

Proof. Define a function $f : H \times K \to HK$ by $f : (h, k) \mapsto hk$. Clearly, f is a surjection. It suffices to show, for every $x \in HK$, that $|f^{-1}(x)| = |H \cap K|$, where $f^{-1}(x) = \{(h, k) \in H \times K : f(h, k) = x\}$ [because $H \times K$, which has order $|H \times K| = |H||K|$, is the disjoint union $\bigcup_{x \in HK} f^{-1}(x)$, which here has order $|HK||H \cap K|$].

We claim that if $x = hk$, then
$$f^{-1}(x) = \{(hd, d^{-1}k) : d \in H \cap K\}.$$

Each $(hd, d^{-1}k) \in f^{-1}(x)$, for $f(hd, d^{-1}k) = hdd^{-1}k = hk = x$. For the reverse inclusion, let $(h', k') \in f^{-1}(x)$, so that $h'k' = hk$. Then $h^{-1}h' = kk'^{-1} \in H \cap K$; call this element d. Then $h' = hd$ and $k' = d^{-1}k$, and so (h', k') lies in the right side. Therefore,
$$|f^{-1}(x)| = |\{(hd, d^{-1}k) : d \in H \cap K\}| = |H \cap K|,$$

because $d \mapsto (hd, d^{-1}k)$ is a bijection. •

The next two results are variants of the first isomorphism theorem.

→ **Theorem 2.121 (Second Isomorphism Theorem).** *If H and K are subgroups of a group G with $H \lhd G$, then HK is a subgroup, $H \cap K \lhd K$, and*

$$K/(H \cap K) \cong HK/H.$$

Proof. We begin by showing first that HK/H makes sense and then describing its elements. Since $H \lhd G$, Proposition 2.119 shows that HK is a subgroup. Normality of H in HK follows from a more general fact: if $H \leq S \leq G$ and if H is normal in G, then H is normal in S (if $ghg^{-1} \in H$ for every $g \in G$, then, in particular, $ghg^{-1} \in H$ for every $g \in S$).

We now show that each coset $xH \in HK/H$ has the form kH for some $k \in K$. Of course, $xH = hkH$, where $h \in H$ and $k \in K$. But $hk = k(k^{-1}hk) = kh'$ for some $h' \in H$, so that $hkH = kh'H = kH$.

It follows that the function $f : K \to HK/H$, given by $f : k \mapsto kH$, is surjective. Now f is a homomorphism, for it is the restriction of the natural map $\pi : G \to G/H$. Since $\ker \pi = H$, it follows that $\ker f = H \cap K$, and so $H \cap K$ is a normal subgroup of K. The first isomorphism theorem now gives $K/(H \cap K) \cong HK/H$. •

The second isomorphism theorem gives the product formula in the special case when one of the subgroups is normal: if $K/(H \cap K) \cong HK/H$, then $|K/(H \cap K)| = |HK/H|$, and so $|HK||H \cap K| = |H||K|$.

→ **Theorem 2.122 (Third Isomorphism Theorem).** *If H and K are normal subgroups of a group G with $K \leq H$, then $H/K \lhd G/K$ and*

$$(G/K)/(H/K) \cong G/H.$$

Proof. Define $f : G/K \to G/H$ by $f : aK \mapsto aH$. Note that f is a (well-defined) function, for if $a' \in G$ and $a'K = aK$, then $a^{-1}a' \in K \leq H$, and so $aH = a'H$. It is easy to see that f is a surjective homomorphism.

Now $\ker f = H/K$, for $aK = H$ if and only if $a \in H$, and so H/K is a normal subgroup of G/K. Since f is surjective, the first isomorphism theorem gives $(G/K)/(H/K) \cong G/H$. •

The third isomorphism theorem is easy to remember: the K's in the fraction $(G/K)/(H/K)$ can be canceled. One can better appreciate the first isomorphism theorem after having proved the third one. The elements of $(G/K)/(H/K)$ are cosets of H/K whose representatives are themselves cosets (of G/K). A direct proof of the third isomorphism theorem could be nasty.

The next result, which describes the subgroups of a quotient group G/K, can be regarded as a fourth isomorphism theorem. Recall that a function $f : X \to Y$ sets up a correspondence, using direct and inverse images, between subsets of X and subsets of Y. We now adapt this viewpoint to the special case when $f : G \to H$ is a homomorphism.

If G is a group and $K \lhd G$, let **Sub**$(G; K)$ denote the family of all those subgroups S of G containing K, and let **Sub**(G/K) denote the family of all the subgroups of G/K.

→ **Proposition 2.123 (Correspondence Theorem).**   *Let G be a group and let $K \lhd G$ be a normal subgroup.*

(i) *The function $S \mapsto S/K$ is a bijection* **Sub**$(G; K) \to$ **Sub**(G/K).

(ii) *Denoting S/K by S^*, we have $T \leq S \leq G$ in* **Sub**$(G; K)$ *if and only if $T^* \leq S^*$ in* **Sub**(G/K), *in which case $[S : T] = [S^* : T^*]$;*

(iii) *$T \lhd S$ in* **Sub**$(G; K)$ *if and only if $T^* \lhd S^*$ in* **Sub**(G/K), *in which case $S/T \cong S^*/T^*$.*

Proof.

(i) Let $\Phi \colon$ **Sub**$(G; K) \to$ **Sub**(G/K) denote the function $\Phi \colon S \mapsto S/K$ (it is routine to check that if S is subgroup of G containing K, then S/K is a subgroup of G/K).

To see that Φ is injective, we first show that if $K \leq S \leq G$, then $\pi^{-1}\pi(S) = S$, where $\pi \colon G \to G/K$ is the natural map. As always, $S \leq \pi^{-1}\pi(S)$, by Proposition 2.14(iii). For the reverse inclusion, let $a \in \pi^{-1}\pi(S)$, so that $\pi(a) = \pi(s)$ for some $s \in S$. It follows that $as^{-1} \in \ker \pi = K$, so that $a = sk$ for some $k \in K$. But $K \leq S$, and so $a = sk \in S$, as desired.

Assume now that $\pi(S) = \pi(S')$, where S and S' are subgroups of G containing K (note that $\pi(S) = S/K$). Then $\pi^{-1}\pi(S) = \pi^{-1}\pi(S')$, as we have just proved in the preceding paragraph, and so $S = S'$; hence, Φ is injective.

To see that Φ is surjective, let U be a subgroup of G/K. By Example 2.92(iv) $\pi^{-1}(U)$ is a subgroup of G containing $K = \pi^{-1}(\{1\})$, and $\pi(\pi^{-1}(U)) = U$, by Proposition 2.14(ii).

(ii) Proposition 2.14(i) shows that $T \leq S \leq G$ implies $T/K = \pi(T) \leq \pi(S) = S/K$. Conversely, assume that $T/K \leq S/K$. If $t \in T$, then $tK \in T/K \leq S/K$ and so $tK = sK$ for some $s \in S$. Hence, $t = sk$ for some $k \in K \leq S$, and so $t \in S$.

In the important special case when G is finite, we prove $[S : T] = [S^* : T^*]$ as follows:

$$
\begin{aligned}
[S^* : T^*] &= |S^*|/|T^*| \\
&= |S/K|/|T/K| \\
&= (|S|/|K|)\,/\,(|T|/|K|) \\
&= |S|/|T| \\
&= [S : T].
\end{aligned}
$$

To prove that $[S : T] = [S^* : T^*]$ in the general case, it suffices to show that there is a bijection from the family of all cosets of the form sT, where $s \in S$, and the family of all cosets of the form s^*T^*, where $s^* \in S^*$, and the reader may check that $sT \mapsto \pi(s)T^*$ is such a bijection.

(iii) The third isomorphism theorem shows that if $T \lhd S$, then $T/K \lhd S/K$ and $(S/K)/(T/K) \cong S/T$; that is, $S^*/T^* \cong S/T$. It remains to show that $T \lhd S$ if $T^* \lhd S^*$; that is, if $t \in T$ and $s \in S$, then $sts^{-1} \in T$. Now

$$\pi(sts^{-1}) = \pi(s)\pi(t)\pi(s)^{-1} \in \pi(s)T^*\pi(s)^{-1} = T^*,$$

so that $sts^{-1} \in \pi^{-1}(T^*) = T$. $\bullet$

When dealing with quotient groups, one usually says, without mentioning the correspondence theorem explicitly, that every subgroup of G/K has the form S/K for a unique subgroup $S \leq G$ containing K.

$\rightarrow$ **Proposition 2.124.**

(i) *If G is a finite abelian group and p is a prime divisor of $|G|$, then G contains an element of order p.*

(ii) *If G is a finite abelian group, then G has a subgroup of order d for every divisor d of $|G|$.*

Proof.
(i) We prove, by induction on $n = |G|$, that if p is a prime divisor of $|G|$, then there is an element of order p in G. The base step $n = 1$ is true, for there are no prime divisors of 1. For the inductive step, choose $a \in G$ of order $k > 1$. If $p \mid k$, say $k = p\ell$, then Exercise 2.40 on page 146 says that a^ℓ has order p. If $p \nmid k$, consider the cyclic subgroup $H = \langle a \rangle$. Now $H \lhd G$, because G is abelian, and so the quotient group G/H exists. Note that $|G/H| = n/k$ is divisible by p, and so the inductive hypothesis gives an element $bH \in G/H$ of order p. If b has order m, then $(bH)^m = b^m H = H$ in G/H, and so Lemma 2.53 gives $p \mid m$. We have returned to the first case.

(ii) We prove the general result by induction on $d \geq 1$. The base step $d = 1$ is obviously true, and so we may assume that $d > 1$; that is, we may assume that d has a prime divisor, say, p. By induction, G contains a subgroup H of order p. Since G is abelian, $H \lhd G$, and so the quotient group G/H is defined. Moreover, $|G/H| = |G|/p$, so that $(d/p) \mid |G/H|$. The inductive hypothesis gives a subgroup $S^* \leq G/H$ with $|S^*| = d/p$. By the correspondence theorem, there is an intermediate subgroup S (i.e., $H \leq S \leq G$) with $S^* = S/H$. Therefore, $|S| = p|S^*| = p \cdot (d/p) = d$. $\bullet$

Part (i) of Proposition 2.124 generalizes: Cauchy's theorem, Theorem 2.147, says that if p is a prime divisor of $|G|$, where G is any finite, not necessarily abelian, group, then G has an element of order p. However, part (ii) is not true in general: Proposition 2.99 shows that A_4 is a group of order 12 having no subgroup of order 6.

Here is another construction of a new group from two given groups.

→ **Definition.**   If H and K are groups, then their ***direct product***, denoted by $H \times K$, is the set of all ordered pairs (h, k) with $h \in H$ and $k \in K$ equipped with the operation

$$(h, k)(h', k') = (hh', kk').$$

It is routine to check that $H \times K$ is a group [the identity is $(1, 1)$ and $(h, k)^{-1} = (h^{-1}, k^{-1})$]. Note that $H \times K$ is abelian if and only if both H and K are abelian.

→ **Example 2.125.**
The four-group $\mathbf{V}$ is isomorphic to $\mathbb{I}_2 \times \mathbb{I}_2$. The reader may check that the function $f : \mathbf{V} \to \mathbb{I}_2 \times \mathbb{I}_2$, defined by

$$f : (1) \mapsto ([0], [0]),$$
$$f : (1\ 2)(3\ 4) \mapsto ([1], [0]),$$
$$f : (1\ 3)(2\ 4) \mapsto ([0], [1]),$$
$$f : (1\ 4)(2\ 3) \mapsto ([1], [1]),$$

is an isomorphism.   ◄

We now apply the first isomorphism theorem to direct products.

→ **Proposition 2.126.**   *Let G and G' be groups, and let $K \lhd G$ and $K' \lhd G'$ be normal subgroups. Then $K \times K'$ is a normal subgroup of $G \times G'$, and there is an isomorphism*

$$(G \times G')/(K \times K') \cong (G/K) \times (G'/K').$$

Proof.   Let $\pi : G \to G/K$ and $\pi' : G' \to G'/K'$ be the natural maps. The reader may check that $f : G \times G' \to (G/K) \times (G'/K')$, given by

$$f : (g, g') \mapsto (\pi(g), \pi'(g')) = (gK, g'K')$$

is a surjective homomorphism with $\ker f = K \times K'$. The first isomorphism theorem now gives the desired isomorphism.   •

Here is a characterization of direct products.

→ **Proposition 2.127.**   *If G is a group containing normal subgroups H and K with $H \cap K = \{1\}$ and $HK = G$, then $G \cong H \times K$.*

Proof.   If $g \in G = HK$, then $g = hk$, where $h \in H$ and $k \in K$. We show first that if $g \in G$, then the factorization $g = hk$ is unique. If $hk = h'k'$, then $h'^{-1}h = k'k^{-1} \in H \cap K = \{1\}$. Therefore, $h' = h$ and $k' = k$. We may now define a function $\varphi : G \to H \times K$ by $\varphi(g) = (h, k)$, where $g = hk, h \in H$, and $k \in K$. To see whether φ is a homomorphism, let $g' = h'k'$, so that $gg' = hkh'k' = hh'kk'$. Hence,

$\varphi(gg') = \varphi(hkh'k')$, which is not in the proper form for evaluation. If we knew that if $h \in H$ and $k \in K$, then $hk = kh$, then we could continue:

$$\varphi(hkh'k') = \varphi(hh'kk')$$
$$= (hh', kk')$$
$$= (h, k)(h', k')$$
$$= \varphi(g)\varphi(g').$$

Let $h \in H$ and $k \in K$. Since $K \lhd G$, we have $hkh^{-1} \in K$, and so $(hkh^{-1})k^{-1} \in K$. Since $H \lhd G$, we have $kh^{-1}k^{-1} \in H$, and so $h(kh^{-1}k^{-1}) \in H$. But $H \cap K = \{1\}$, so that $hkh^{-1}k^{-1} = 1$ and $hk = kh$.

Now φ is surjective, for if $(h, k) \in H \times K$, then the element $g = hk \in G$ satisfies $\varphi(g) = (h, k)$. Finally, if $\varphi(g) = (1, 1)$, then $g = hk$, where $h = 1 = k$, so that $g = 1$. Hence, $\ker \varphi = \{1\}$, φ is injective, and φ is an isomorphism.   •

All the hypotheses in Proposition 2.127 are needed. For example, let $G = S_3$, $H = \langle (1\ 2\ 3) \rangle$, and $K = \langle (1\ 2) \rangle$. Now $S_3 = HK$, $\{1\} = H \cap K$, and $H \lhd S_3$, but K is not a normal subgroup. It is not true that $S_3 \cong H \times K$, for S_3 is not abelian, while the direct product $H \times K$ of abelian groups is abelian.

→ **Theorem 2.128.** *If m and n are relatively prime, then*

$$\mathbb{I}_{mn} \cong \mathbb{I}_m \times \mathbb{I}_n.$$

Proof. Let us write the elements of $\mathbb{I}_m$ and $\mathbb{I}_n$ as $[a]_m$ and $[a]_n$, respectively. It is easy to check that $f : \mathbb{Z} \to \mathbb{I}_m \times \mathbb{I}_n$, defined by $f(a) = ([a]_m, [a]_n)$, is a homomorphism. We claim that f is a surjection; if $([b]_m, [c]_n) \in \mathbb{I}_m \times \mathbb{I}_n$, then the Chinese remainder theorem applies (since m and n are relatively prime,), and there is an integer a with $([b]_m, [c]_n) = ([a]_m, [a]_n) = f(a)$. Now $a \in \ker f$ if and only if $a \in \langle m \rangle \cap \langle n \rangle$. But Proposition 2.80 says that $\langle m \rangle \cap \langle n \rangle = \langle \ell \rangle$, where $\ell = \text{lcm}\{m, n\}$. That m and n are relatively prime gives $\text{lcm}\{m, n\} = mn$, by Proposition 1.56, and so $\ker f = \langle mn \rangle$. By the first isomorphism theorem, the function $g : \mathbb{Z}/\langle mn \rangle \to (\mathbb{Z}/\langle m \rangle) \times (\mathbb{Z}/\langle n \rangle)$, given by $g : [a]_{mn} \mapsto f(a) = ([a]_m, [a]_n)$, is an isomorphism. Therefore, $\mathbb{I}_{mn} \cong \mathbb{I}_m \times \mathbb{I}_n$.   •

For example, it follows that $\mathbb{I}_6 \cong \mathbb{I}_2 \times \mathbb{I}_3$. Note that there is no isomorphism if m and n are not relatively prime. Example 2.125 shows that $\mathbf{V} \cong \mathbb{I}_2 \times \mathbb{I}_2$, which is not isomorphic to $\mathbb{I}_4$ because $\mathbf{V}$ has no element of order 4.

In light of Proposition 2.74, an element $a \in G$ has order n if $\langle a \rangle \cong \mathbb{I}_n$. Theorem 2.128 can now be interpreted as saying that if a and b are commuting elements having relatively prime orders m and n, then ab has order mn. Let us give a direct proof of this result.

Proposition 2.129.   *Let G be a group, and let $a, b \in G$ be commuting elements of orders m and n, respectively. If $(m, n) = 1$, then ab has order mn.*

Proof.   Since a and b commute, we have $(ab)^r = a^r b^r$ for all r, so that $(ab)^{mn} = a^{mn} b^{mn} = 1$. It suffices to prove that if $(ab)^k = 1$, then $mn \mid k$. If $1 = (ab)^k = a^k b^k$, then $a^k = b^{-k}$. Since a has order m, we have $1 = a^{mk} = b^{-mk}$. Since b has order n, Lemma 2.53 gives $n \mid mk$. As $(m, n) = 1$, however, Corollary 1.40 gives $n \mid k$; a similar argument gives $m \mid k$. Finally, Exercise 1.60 on page 54 shows that $mn \mid k$. Therefore, $mn \leq k$, and mn is the order of ab.   •

Here is a variant of Proposition 2.75.

→ **Proposition 2.130.**   *If G is a finite abelian group having a unique subgroup of order p for every prime divisor p of $|G|$, then G is cyclic.*

Proof.   Choose $a \in G$ of largest order, say, n. If p is a prime divisor of $|G|$, let $C = C_p$ be the unique subgroup of G having order p; the subgroup C must be cyclic, say $C = \langle c \rangle$. We show that $p \mid n$ by showing that $c \in \langle a \rangle$ (and hence $C \leq \langle a \rangle$). If $(p, n) = 1$, then ca has order $pn > n$, by Proposition 2.129, contradicting a being an element of largest order. If $p \mid n$, say, $n = pq$, then a^q has order p, and hence it lies in the unique subgroup $\langle c \rangle$ of order p. Thus, $a^q = c^i$ for some i. Now $(i, p) = 1$, so there are integers u and v with $1 = ui + vp$; hence, $c = c^{ui+vp} = c^{ui} c^{vp} = c^{ui}$. Therefore, $a^{qu} = c^{ui} = c$, so that $c \in \langle a \rangle$, as desired. It follows that $\langle a \rangle$ contains every element $x \in G$ with $x^p = 1$ for some prime p.

If $\langle a \rangle = G$, we are finished. Therefore, we may assume that there is $b \in G$ with $b \notin \langle a \rangle$. Now $b^{|G|} = 1 \in \langle a \rangle$; let k be the smallest positive integer with $b^k \in \langle a \rangle$:

$$b^k = a^q.$$

Note that $k \mid |G|$ because k is the order of $b \langle a \rangle$ in $G / \langle a \rangle$. Of course, $k \neq 1$, and so there is a factorization $k = pm$, where p is prime. There are now two possibilities. If $p \mid q$, then $q = pu$ and

$$b^{pm} = b^k = a^q = a^{pu}.$$

Hence, $(b^m a^{-u})^p = 1$, and so $b^m a^{-u} \in \langle a \rangle$. Thus, $b^m \in \langle a \rangle$, and this contradicts k being the smallest exponent with this property. The second possibility is that $p \nmid q$, in which case $(p, q) = 1$. There are integers s and t with $1 = sp + tq$, and so

$$a = a^{sp+tq} = a^{sp} a^{tq} = a^{sp} b^{pmt} = (a^s b^{mt})^p.$$

Therefore, $a = x^p$, where $x = a^s b^{mt}$, and Exercise 2.41 on page 146, which applies because $p \mid n$, says that the order of x is greater than that of a, a contradiction. We conclude that $G = \langle a \rangle$.   •

Proposition 2.130 is false for nonabelian groups, for the group of quaternions **Q** is a counterexample; it is a noncyclic group of order 8 having a unique subgroup of order 2.

Here is a number-theoretic application of direct products.

→ **Corollary 2.131.**   *If $(m, n) = 1$, then $\phi(mn) = \phi(m)\phi(n)$, where ϕ is the Euler ϕ-function.*

Proof. [19]   Since m and n are relatively prime, the proof of Theorem 2.128 shows that $g \colon \mathbb{Z}/\langle mn \rangle \to (\mathbb{Z}/\langle m \rangle) \times (\mathbb{Z}/\langle n \rangle)$, given by $g \colon [a]_{mn} \mapsto ([a]_m, [a]_n)$, is an isomorphism. If $U(\mathbb{I}_m) = \{[r] \in \mathbb{I}_m : (r, m) = 1\}$, then Lemma 2.108 says that $|U(\mathbb{I}_m)| = \phi(m)$. Thus, if we prove that $g(U(\mathbb{I}_{mn})) = U(\mathbb{I}_m) \times U(\mathbb{I}_n)$, then the result will follow:

$$\phi(mn) = |U(\mathbb{I}_{mn})| = |g(U(\mathbb{I}_{mn}))|$$
$$= |U(\mathbb{I}_m) \times U(\mathbb{I}_n)| = |U(\mathbb{I}_m)| \cdot |U(\mathbb{I}_n)| = \phi(m)\phi(n).$$

We claim that $g(U(\mathbb{I}_{mn})) = U(\mathbb{I}_m) \times U(\mathbb{I}_n)$. If $[a]_{mn} \in U(\mathbb{I}_{mn})$, then $[a]_{mn}[b]_{mn} = [1]_{mn}$ for some $[b]_{mn} \in \mathbb{I}_{mn}$, and

$$g([ab]_{mn}) = ([ab]_m, [ab]_n) = ([a]_m[b]_m, [a]_n[b]_n)$$
$$= ([a]_m, [a]_n)([b]_m, [b]_n) = ([1]_m, [1]_n).$$

Hence, $[1]_m = [a]_m[b]_m$ and $[1]_n = [a]_n[b]_n$, so that $g([a]_{mn}) = ([a]_m, [a]_n) \in U(\mathbb{I}_m) \times U(\mathbb{I}_n)$, and $g(U(\mathbb{I}_{mn})) \leq U(\mathbb{I}_m) \times U(\mathbb{I}_n)$.

For the reverse inclusion, if $g([c]_{mn}) = ([c]_m, [c]_n) \in U(\mathbb{I}_m) \times U(\mathbb{I}_n)$, then we must show that $[c]_{mn} \in U(\mathbb{I}_{mn})$. There is $[d]_m \in \mathbb{I}_m$ with $[c]_m[d]_m = [1]_m$, and there is $[e]_n \in \mathbb{I}_n$ with $[c]_n[e]_n = [1]_n$. Since g is surjective, there is $b \in \mathbb{Z}$ with $([b]_m, [b]_n) = ([d]_m, [e]_n)$, so that

$$g([1]_{mn}) = ([1]_m, [1]_n) = ([c]_m[b]_m, [c]_n[b]_n) = g([c]_{mn}[b]_{mn}).$$

Since g is an injection, $[1]_{mn} = [c]_{mn}[b]_{mn}$ and $[c]_{mn} \in U(\mathbb{I}_{mn})$.   •

Direct products of several groups can be defined.

→ **Definition.**   If $H_1, \ldots, H_n$ are groups, then their ***direct product***

$$H_1 \times \cdots \times H_n$$

is the set of all n-tuples $(h_1, \ldots, h_n)$, where $h_i \in H_i$ for all i, with coordinatewise multiplication:

$$(h_1, \ldots, h_n)(h'_1, \ldots, h'_n) = (h_1 h'_1, \ldots, h_n h'_n).$$

The basis theorem, Theorem 6.11, says that every finite abelian group is a direct product of cyclic groups.

---

[19]See Exercise 3.54(iii) on page 252 for a less computational proof.

EXERCISES

H **2.95** True or false with reasons.

  (i)   If $[a] = [b]$ in $\mathbb{I}_m$, then $a = b$ in $\mathbb{Z}$.

  (ii)   There is a homomorphism $\mathbb{I}_m \to \mathbb{Z}$ defined by $[a] \mapsto a$.

  (iii) If $a = b$ in $\mathbb{Z}$, then $[a] = [b]$ in $\mathbb{I}_m$.

  (iv)   If G is a group and $K \lhd G$, then there is a homomorphism $G \to G/K$ having kernel K.

  (v)   If G is a group and $K \lhd G$, then every homomorphism $G \to G/K$ has kernel K.

  (vi)   Every quotient group of an abelian group is abelian.

  (vii) If G and H are abelian groups, then $G \times H$ is an abelian group.

  (viii) If G and H are cyclic groups, then $G \times H$ is a cyclic group.

  (ix)   If every subgroup of a group G is a normal subgroup, then G is abelian.

  (x)   If G is a group, then $\{1\} \lhd G$ and $G/\{1\} \cong G$.

2.96 Prove that $U(\mathbb{I}_9) \cong \mathbb{I}_6$ and $U(\mathbb{I}_{15}) \cong \mathbb{I}_4 \times \mathbb{I}_2$.

2.97 (i)   If H and K are groups, prove, without using the first isomorphism theorem, that $H^* = \{(h, 1) : h \in H\}$ and $K^* = \{(1, k) : k \in K\}$ are normal subgroups of $H \times K$ with $H \cong H^*$ and $K \cong K^*$.

  (ii)   Prove that $f \colon H \to (H \times K)/K^*$, defined by $f(h) = (h, 1)K^*$, is an isomorphism without using the first isomorphism theorem.

  H (iii) Prove $K^* \lhd (H \times K)$ and $(H \times K)/K^* \cong H$ using the first isomorphism theorem.

* H **2.98** If G is a group and $G/Z(G)$ is cyclic, where $Z(G)$ denotes the center of G, prove that G is abelian; that is, $G = Z(G)$. Conclude that if G is not abelian, then $G/Z(G)$ is never cyclic.

* H **2.99** Let G be a finite group, let p be a prime, and let H be a normal subgroup of G. Prove that if both $|H|$ and $|G/H|$ are powers of p, then $|G|$ is a power of p.

H **2.100** Call a group G *finitely generated* if there is a finite subset $X \subseteq G$ with $G = \langle X \rangle$. Prove that every subgroup S of a finitely generated *abelian* group G is itself finitely generated. (This can be false if G is not abelian.)

***2.101** (i)   Let $\pi \colon G \to H$ be a surjective homomorphism with $\ker \pi = T$. Let $H = \langle X \rangle$, and, for each $x \in X$, choose an element $g_x \in G$ with $\pi(g_x) = x$. Prove that G is generated by $T \cup \{g_x : x \in X\}$.

  (ii)   Let G be a group and let $T \lhd G$. If both T and G/T are finitely generated, prove that G is finitely generated.

***2.102** Let A, B and C be groups, and let α, β and γ be homomorphisms with $\gamma \circ \alpha = \beta$.

If α is surjective, prove that $\ker \gamma = \alpha(\ker \beta)$.

2.103 Let A and B be groups, let $A' \lhd A$ and $B' \lhd B$ be normal subgroups, and let $\alpha \colon A \to B$ be a homomorphism with $\alpha(A') \le B'$.

   (i)   Prove that there is a (well-defined) homomorphism $\alpha_* \colon A/A' \to B/B'$ given by $\alpha_* \colon aA' \mapsto \alpha(a)B'$.

   (ii)   Prove that if α is surjective, then α_* is surjective.

   (iii)   Give an example in which α is injective and α_* is not injective.

2.104 **(i)**   Prove that $\mathbf{Q}/Z(\mathbf{Q}) \cong \mathbf{V}$, where $\mathbf{Q}$ is the group of quaternions and $\mathbf{V}$ is the four-group. Conclude that the quotient of a nonabelian group by its center can be abelian.

   (ii)   Prove that $\mathbf{Q}$ has no subgroup isomorphic to $\mathbf{V}$. Conclude that the quotient $\mathbf{Q}/Z(\mathbf{Q})$ is not isomorphic to a subgroup of $\mathbf{Q}$.

H **2.105** Let G be a finite group with $K \lhd G$. If $(|K|, [G : K]) = 1$, prove that K is the unique subgroup of G having order $|K|$.

* **2.106** Let H and K be subgroups of a group G.

   H **(i)**   Prove that HK is a subgroup of G if and only if $HK = KH$. In particular, the condition holds if $hk = kh$ for all $h \in H$ and $k \in K$.

   (ii)   If $HK = KH$ and $H \cap K = \{1\}$, prove that $HK \cong H \times K$.

* **2.107** Prove the converse of Lemma 2.112: if K is a subgroup of a group G, and if every left coset aK is equal to a right coset Kb, then $K \lhd G$.

2.108 Let G be a group and regard $G \times G$ as the direct product of G with itself. If the multiplication $\mu \colon G \times G \to G$ is a group homomorphism, prove that G must be abelian.

* **2.109** Generalize Theorem 2.128 as follows. Let G be a finite (additive) abelian group of order mn, where $(m, n) = 1$. Define

$$G_m = \{g \in G : \text{order } (g) \mid m\} \quad \text{and} \quad G_n = \{h \in G : \text{order } (h) \mid n\}.$$

   (i)   Prove that G_m and G_n are subgroups with $G_m \cap G_n = \{0\}$.

   (ii)   Prove that $G = G_m + G_n = \{g + h : g \in G_m \text{ and } h \in G_n\}$.

   (iii)   Prove that $G \cong G_m \times G_n$.

* **2.110** **(i)**   Generalize Theorem 2.128 by proving that if the prime factorization of an integer m is $m = p_1^{e_1} \cdots p_n^{e_n}$, then

$$\mathbb{I}_m \cong \mathbb{I}_{p_1^{e_1}} \times \cdots \times \mathbb{I}_{p_n^{e_n}}.$$

   (ii)   Generalize Corollary 2.131 by proving that if the prime factorization of an integer m is $m = p_1^{e_1} \cdots p_n^{e_n}$, then

$$U(\mathbb{I}_m) \cong U(\mathbb{I}_{p_1^{e_1}}) \times \cdots \times U(\mathbb{I}_{p_n^{e_n}}).$$

2.111 **(i)**   If p is a prime, prove that $\phi(p^k) = p^k(1 - \frac{1}{p})$.

   H **(ii)**   If the distinct prime divisors of a positive integer h are $p_1, p_2, \ldots, p_n$, prove that

$$\phi(h) = h(1 - \tfrac{1}{p_1})(1 - \tfrac{1}{p_2}) \cdots (1 - \tfrac{1}{p_n}).$$

H **2.112** Let p be an odd prime, and assume that $a_i \equiv i \bmod p$ for $1 \le i \le p - 1$. Prove that there exist a pair of distinct integers i and j with $ia_i \equiv ja_j \bmod p$.

*2.113 If G is a group and $x, y \in G$, define their **commutator** to be $xyx^{-1}y^{-1}$, and define the **commutator subgroup** G' to be the subgroup generated by all the commutators (the product of two commutators need not be a commutator).

   (i)   Prove that $G' \lhd G$.
   (ii) Prove that G/G' is abelian.
   (iii) If $\varphi \colon G \to A$ is a homomorphism, where A is an abelian group, prove that $G' \le \ker \varphi$. Conversely, if $G' \le \ker \varphi$, prove that $\operatorname{im} \varphi$ is abelian.
   (iv) If $G' \le H \le G$, prove that $H \lhd G$.

$\rightarrow$ 2.7   GROUP ACTIONS

Groups of permutations led us to abstract groups; the next result, due to A. Cayley (1821–1895), shows that abstract groups are not so far removed from permutations.

$\rightarrow$ **Theorem 2.132 (Cayley).**   *Every group G is (isomorphic to) a subgroup of the symmetric group S_G. In particular, if G is a finite group of order n, then G is isomorphic to a subgroup of S_n.*

Proof.   For each $a \in G$, define "translation" $\tau_a \colon G \to G$ by $\tau_a(x) = ax$ for every $x \in G$ (if $a \ne 1$, then τ_a is not a homomorphism). For $a, b \in G$, $(\tau_a \circ \tau_b)(x) = \tau_a(\tau_b(x)) = \tau_a(bx) = a(bx) = (ab)x = \tau_{ab}(x)$, by associativity, so that

$$\tau_a \tau_b = \tau_{ab}.$$

It follows that each τ_a is a bijection, for its inverse is $\tau_{a^{-1}}$:

$$\tau_a \tau_{a^{-1}} = \tau_{aa^{-1}} = \tau_1 = 1_G,$$

and so $\tau_a \in S_G$.
   Define $\varphi \colon G \to S_G$ by $\varphi(a) = \tau_a$. Rewriting,

$$\varphi(a)\varphi(b) = \tau_a \tau_b = \tau_{ab} = \varphi(ab),$$

so that φ is a homomorphism. Finally, φ is an injection. If $\varphi(a) = \varphi(b)$, then $\tau_a = \tau_b$, and hence $\tau_a(x) = \tau_b(x)$ for all $x \in G$; in particular, when $x = 1$, this gives $a = b$, as desired.
   The last statement follows from Exercise 2.65 on page 169, which says that if X is a set with $|X| = n$, then $S_X \cong S_n$.   •

The reader may note, in the proof of Cayley's theorem, that the permutation τ_a is just the ath row of the multiplication table of G.
   To tell the truth, Cayley's theorem itself is only mildly interesting. However, the identical proof works in a larger setting which is more useful.

→ **Theorem 2.133 (Representation on Cosets).**   *Let G be a group, and let H be a subgroup of G having finite index n. Then there exists a homomorphism $\varphi \colon G \to S_n$ with $\ker \varphi \leq H$.*

Proof.   Even though H may not be a normal subgroup, we will denote the family of all the cosets of H in G by G/H in this proof.

For each $a \in G$, define "translation" $\tau_a \colon G/H \to G/H$ by $\tau_a(xH) = axH$ for every $x \in G$. For $a, b \in G$,

$$(\tau_a \circ \tau_b)(xH) = \tau_a(\tau_b(xH)) = \tau_a(bxH) = a(bxH) = (ab)xH = \tau_{ab}(aH),$$

by associativity, so that

$$\tau_a \tau_b = \tau_{ab}.$$

It follows that each τ_a is a bijection, for its inverse is $\tau_{a^{-1}}$:

$$\tau_a \tau_{a^{-1}} = \tau_{aa^{-1}} = \tau_1 = 1_{G/H},$$

and so $\tau_a \in S_{G/H}$. Define $\varphi \colon G \to S_{G/H}$ by $\varphi(a) = \tau_a$. Rewriting,

$$\varphi(a)\varphi(b) = \tau_a \tau_b = \tau_{ab} = \varphi(ab),$$

so that φ is a homomorphism. Finally, if $a \in \ker \varphi$, then $\varphi(a) = 1_{G/H}$, so that $\tau_a(xH) = xH$ for all $x \in G$; in particular, when $x = 1$, this gives $aH = H$, and $a \in H$, by Lemma 2.82(i). The result follows from Exercise 2.65 on page 169, for $|G/H| = n$, and so $S_{G/H} \cong S_n$.   •

When $H = \{1\}$, this is the Cayley theorem.

We are now going to classify all the groups of order up to 7. By Example 2.90, every group of prime order p is isomorphic to $\mathbb{I}_p$, and so, up to isomorphism, there is just one group of order p. Of the possible orders through 7, four of them, 2, 3, 5, and 7, are primes, and so we need look only at orders 4 and 6.

→ **Proposition 2.134.**   *Every group G of order 4 is isomorphic to either $\mathbb{I}_4$ or the four-group $\mathbf{V}$. Moreover, $\mathbb{I}_4$ and $\mathbf{V}$ are not isomorphic.*

Proof.   By Lagrange's theorem, every element in G, other than 1, has order either 2 or 4. If there is an element of order 4, then G is cyclic. Otherwise, $x^2 = 1$ for all $x \in G$, so that Exercise 2.44 on page 147 shows that G is abelian.

If distinct elements x and y in G are chosen, neither being 1, then one quickly checks that $xy \notin \{1, x, y\}$; hence,

$$G = \{1, x, y, xy\}.$$

It is easy to see that the bijection $f \colon G \to \mathbf{V}$, defined by $f(1) = 1$, $f(x) = (1\ 2)(3\ 4)$, $f(y) = (1\ 3)(2\ 4)$, and $f(xy) = (1\ 4)(2\ 3)$, is an isomorphism, for the product of any two elements of order 2 here is the other element of order 2.

We have already seen, in Example 2.91, that $\mathbb{I}_4 \not\cong \mathbf{V}$.   •

→ **Proposition 2.135.** *If G is a group of order 6, then G is isomorphic to either $\mathbb{I}_6$ or S_3.*[20] *Moreover, $\mathbb{I}_6$ and S_3 are not isomorphic.*

Proof. By Lagrange's theorem, the only possible orders of nonidentity elements are 2, 3, and 6. Of course, $G \cong \mathbb{I}_6$ if G has an element of order 6. Now Exercise 2.46 on page 147 shows that G must contain an element of order 2, say, t. Let $T = \langle t \rangle$.

Since $[G : T] = 3$, the representation on the cosets of T is a homomorphism $\rho \colon G \to S_{G/T} \cong S_3$ with $\ker \rho \leq T$. Thus, $\ker \rho = \{1\}$ or $\ker \rho = T$. In the first case, ρ is an injection, and hence it is an isomorphism, for $|G| = 6 = |S_3|$. In the second case, $\ker \rho = T$, so that $T \lhd G$ and the quotient group G/T is defined. We will see that G is cyclic in this second case. Now G/T is cyclic, for $|G/T| = 3$, so there is $a \in G$ with $G/T = \{T, aT, a^2 T\}$. Moreover, ρ_t is the permutation

$$\rho_t = \begin{pmatrix} T & aT & a^2 T \\ tT & taT & ta^2 T \end{pmatrix}.$$

Since $t \in T = \ker \rho$, we have ρ_t the identity. In particular, $aT = \rho_t(aT) = taT$, so that $a^{-1}ta \in T = \{1, t\}$, by Lemma 2.82(i). But $a^{-1}ta \neq 1$, so that $a^{-1}ta = t$; that is, $ta = at$. Now a has order 3 or order 6 (for $a \neq 1$ and $a^2 \neq 1$). We claim, in either case, that G has an element of order 6. If a has order 3, then at has order 6, by Proposition 2.129 [alternatively, just note that $(at)^6 = 1$ and that $(at)^i \neq 1$ for $i < 6$]. Therefore, G is cyclic of order 6, and $G \cong \mathbb{I}_6$.

It is clear that $\mathbb{I}_6$ and S_3 are not isomorphic, for one is abelian and the other is not.

•

One consequence of this result is another proof that $\mathbb{I}_6 \cong \mathbb{I}_2 \times \mathbb{I}_3$ (see Theorem 2.128).

Classifying groups of order 8 is more difficult, for we have not yet developed enough theory (see my book, *Advanced Modern Algebra*, Theorem 5.83). It turns out that there are only 5 nonisomorphic groups of order 8; three are abelian, namely, $\mathbb{I}_8, \mathbb{I}_4 \times \mathbb{I}_2$, and $\mathbb{I}_2 \times \mathbb{I}_2 \times \mathbb{I}_2$; two are nonabelian, namely, D_8 and $\mathbf{Q}$.

One can continue this discussion for larger orders, but things soon get out of hand, as Table 2.4 shows (the calculation of the numbers in the table is very sophisticated). The number of nonisomorphic groups having order ≤ 2000 was found by E. O'Brien,

---

[20]Cayley states this proposition in an article he wrote in 1854. However, in 1878, in the *American Journal of Mathematics*, he wrote, "The general problem is to find all groups of a given order n; ... if $n = 6$, there are three groups; a group

$$1, \alpha, \alpha^2, \alpha^3, \alpha^4, \alpha^5 \quad (\alpha^6 = 1),$$

and two more groups

$$1, \beta, \beta^2, \alpha, \alpha\beta, \alpha\beta^2 \quad (\alpha^2 = 1, \beta^3 = 1),$$

viz., in the first of these $\alpha\beta = \beta\alpha$ while in the other of them, we have $\alpha\beta = \beta^2\alpha, \alpha\beta^2 = \beta\alpha$." Cayley's list is $\mathbb{I}_6, \mathbb{I}_2 \times \mathbb{I}_3$, and S_3. Of course, $\mathbb{I}_2 \times \mathbb{I}_3 \cong \mathbb{I}_6$. Even Homer nods.

Order of Group	Number of Groups
2	1
4	2
8	5
16	14
32	51
64	267
128	2, 328
256	56, 092
512	10, 494, 213
1024	49, 487, 365, 422

Table 2.4

but focusing on the numbers in Table 2.4 is more dramatic. A. McIver and P. M. Neumann proved, for large n, that the number of nonisomorphic groups of order n is about $n^{\mu^2+\mu+2}$, where $\mu(n)$ is the largest exponent occurring in the prime factorization of n. Obviously, making a telephone directory of groups is not the way to study them.

Groups arose by abstracting the fundamental properties enjoyed by permutations. But there is an important feature of permutations that the axioms do not mention: permutations are functions. We shall see that there are interesting consequences when this feature is restored.

Let us agree on some notation before giving the next definition. A function of two variables, $\alpha \colon X \times Y \to Z$, can be regarded as a one-parameter family of functions of one variable: each $x \in X$ gives a function $\alpha_x \colon Y \to Z$, namely, $\alpha_x(y) = \alpha(x, y)$.

→ **Definition.** If X is a set and G is a group, then G *acts* on X[21] if there exists a function $\alpha \colon G \times X \to X$, called an *action*, such that

(i) for $g, h \in G$, $\alpha_g \circ \alpha_h = \alpha_{gh}$;

(ii) $\alpha_1 = 1_X$, the identity function.

If G acts on X, we shall usually write gx instead of $\alpha_g(x)$. In this notation, axiom (i) reads $g(hx) = (gh)x$.

Of course, every subgroup $G \leq S_X$ acts on X. More generally, actions of a group G on a set X correspond to homomorphisms $G \to S_X$.

→ **Proposition 2.136.** *If $\alpha \colon G \times X \to X$ is an action of a group G on a set X, then $g \mapsto \alpha_g$ defines a homomorphism $G \to S_X$. Conversely, if $B \colon G \to S_X$ is a homomorphism,*

---

[21] If G acts on X, then one often calls X a *G-set*.

then $\beta \colon G \times X \to X$, defined by $\beta(g, x) = B(g)(x)$, is an action.

Proof. If $\alpha \colon G \times X \to X$ is an action, then we claim that each α_g is a permutation of X. Indeed, its inverse is $\alpha_{g^{-1}}$, because $\alpha_g \alpha_{g^{-1}} = \alpha_{gg^{-1}} = \alpha_1 = 1_X$. It follows that $A \colon G \to S_X$, defined by $A(g) = \alpha_g$, is a function with the stated target. That A is a homomorphism follows from axiom (i):

$$A(gh) = \alpha_{gh} = \alpha_g \circ \alpha_h = A(g) \circ A(h).$$

Conversely, given a homomorphism $B \colon G \to S_X$, the function $\beta \colon G \times X \to X$, given by $\beta(g, x) = B(g)(x)$, is an action. According to our notational agreement, $\beta_g = B(g)$. Thus, axiom (i) merely says that $B(g) \circ B(h) = B(gh)$, which is true because B is a homomorphism, while axiom (ii), $B(1) = 1_X$, holds because every homomorphism takes the identity to the identity. $\bullet$

Cayley's theorem says that a group G acts on itself by (left) translation, and its generalization, the representation on cosets (Theorem 2.133), shows that G also acts on the family of cosets of a subgroup H by (left) translation.

→ **Example 2.137.**
We show that G ***acts on itself by conjugation***. For each $g \in G$, define $\alpha_g \colon G \to G$ by

$$\alpha_g(x) = gxg^{-1}.$$

To verify axiom (i), note that for each $x \in G$,

$$\begin{aligned}
(\alpha_g \circ \alpha_h)(x) &= \alpha_g(\alpha_h(x)) \\
&= \alpha_g(hxh^{-1}) \\
&= g(hxh^{-1})g^{-1} \\
&= (gh)x(gh)^{-1} \\
&= \alpha_{gh}(x).
\end{aligned}$$

Therefore, $\alpha_g \circ \alpha_h = \alpha_{gh}$.
   To prove axiom (ii), note that for each $x \in G$,

$$\alpha_1(x) = 1x1^{-1} = x,$$

and so $\alpha_1 = 1_G$. ◄

The following two definitions are fundamental.

→ **Definition.** If G acts on X and $x \in X$, then the *orbit* of x, denoted by $\mathcal{O}(x)$, is the subset of X:

$$\mathcal{O}(x) = \{gx : g \in G\} \subseteq X;$$

the *stabilizer* of x, denoted by G_x, is the subgroup of G:

$$G_x = \{g \in G : gx = x\} \leq G.$$

It is easy to check that the stabilizer G_x of a point x is a subgroup of G.
Let us find orbits and stabilizers in the examples above.

Example 2.138.

(i) Cayley's theorem says that G acts on itself by translations: $\tau_a : x \mapsto ax$. If $x \in G$, then the orbit $\mathcal{O}(x) = G$, for if $g \in G$, then $g = (gx^{-1})x$. The stabilizer G_x of x is $\{1\}$, for if $x = \tau_a(x) = ax$, then $a = 1$. One says that G acts *transitively* on X when there is some $x \in X$ with $\mathcal{O}(x) = X$.

(ii) When G acts on G/H [the family of cosets of a (not necessarily normal) subgroup H] by translations $\tau_a : xH \mapsto axH$, then the orbit $\mathcal{O}(xH) = G/H$, for if $g \in G$ and $a = gx^{-1}$, then $\tau_a : xH \mapsto gH$. Thus, G acts transitively on G/H. The stabilizer G_{xH} of xH is xHx^{-1}, for $axH = xH$ if and only if $x^{-1}ax \in H$ if and only if $a \in xHx^{-1}$. ◄

Example 2.139.
Let $X = \{v_0, v_1, v_2, v_3\}$ be the set of vertices of a square, and let G be the dihedral group D_8 acting on X, as in Figure 2.17 (for clarity, the vertices in the figure are labeled 0, 1, 2, 3 instead of v_0, v_1, v_2, v_3).

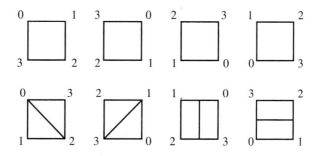

Figure 2.17 Dihedral Group D_8

$$G = \{\text{rotations} : (1), (v_0\ v_1\ v_2\ v_3), (v_0\ v_2)(v_1\ v_3), (v_0\ v_3\ v_2\ v_1)\} \cup$$
$$\{\text{reflections} : (v_1\ v_3), (v_0\ v_2), (v_0\ v_1)(v_2\ v_3), (v_0\ v_3)(v_1\ v_2)\}.$$

For each vertex $v_i \in X$, there is some $g \in G$ with $g v_0 = v_i$; therefore, $\mathcal{O}(v_0) = X$ and D_8 acts transitively.

What is the stabilizer G_{v_0} of v_0? Aside from the identity, there is only one $g \in D_8$ fixing v_0, namely, $g = (v_1\ v_3)$; therefore G_{v_0} is a subgroup of order 2. (This example can be generalized to the dihedral group D_{2n} acting on a regular n-gon.) ◄

$\rightarrow$ **Example 2.140.**
Let a group G act on itself by conjugation. If $x \in G$, then
$$\mathcal{O}(x) = \{y \in G : y = axa^{-1} \text{ for some } a \in G\};$$
$\mathcal{O}(x)$ is called the ***conjugacy class*** of x, and it is often denoted by x^G. For example, Proposition 2.33 shows that if $\alpha \in S_n$, then the conjugacy class of α consists of all the permutations in S_n having the same cycle structure as α.

If $x \in G$, then the stabilizer G_x of x is
$$C_G(x) = \{g \in G : gxg^{-1} = x\}.$$

This subgroup of G, consisting of all $g \in G$ that commute with x, is called the ***centralizer*** of x in G. ◄

Example 2.141.
Let $X = \{1, 2, \ldots, n\}$, let $\sigma \in S_n$, and regard the cyclic group $G = \langle \sigma \rangle$ as acting on X. If $i \in X$, then
$$\mathcal{O}(i) = \{\sigma^k(i) : k \in \mathbb{Z}\}.$$
Let $\sigma = \beta_1 \cdots \beta_{t(\sigma)}$ be the complete factorization of σ, and let $i = i_0$ be moved by σ. If the cycle involving i_0 is $\beta_j = (i_0\ i_1\ \ldots\ i_{r-1})$, then the proof of Theorem 2.26 shows that $i_k = \sigma^k(i_0)$ for all $k < r - 1$. Therefore,
$$\mathcal{O}(i) = \{i_0, i_1, \ldots, i_{r-1}\},$$
where $i = i_0$. It follows that $|\mathcal{O}(i)| = r$. The stabilizer G_ℓ of a symbol ℓ is G if σ fixes ℓ, and it is a proper subgroup of G if σ moves ℓ. ◄

A group G acting on a set X gives an equivalence relation on X. Define
$$x \equiv y \text{ if there exists } g \in G \text{ with } y = gx.$$
If $x \in X$, then $1x = x$, where $1 \in G$, and so $x \equiv x$; hence, $\equiv$ is reflexive. If $x \equiv y$, so that $y = gx$, then
$$g^{-1}y = g^{-1}(gx) = (g^{-1}g)x = 1x = x,$$
so that $x = g^{-1}y$ and $y \equiv x$; hence, $\equiv$ is symmetric. If $x \equiv y$ and $y \equiv z$, there are $g, h \in G$ with $y = gx$ and $z = hy$, so that $z = hy = h(gx) = (hg)x$, and $x \equiv z$. Therefore, $\equiv$ is transitive, and hence it is an equivalence relation. Now the equivalence class of $x \in X$ is its orbit, for
$$[x] = \{y \in X : y \equiv x\} = \{gx : g \in G\} = \mathcal{O}(x).$$

→ **Proposition 2.142.** *If G acts on a set X, then X is the disjoint union of the orbits. If X is finite, then*

$$|X| = \sum_i |\mathcal{O}(x_i)|,$$

where one x_i is chosen from each orbit.

Proof. This follows from Proposition 2.20, for the orbits form a partition of X. The count given in the second statement is correct: the orbits are disjoint, and so no element in X is counted twice. •

Here is the connection between orbits and stabilizers.

→ **Theorem 2.143.** *If G acts on a set X and $x \in X$, then*

$$|\mathcal{O}(x)| = [G : G_x];$$

the size of the orbit $\mathcal{O}(x)$ is the index of the stabilizer G_x in G.

Proof. Let G/G_x denote the family of all the cosets of G_x in G (we are not assuming that G_x is a normal subgroup, nor are we assuming that G/G_x is a group). We will exhibit a bijection $\varphi \colon \mathcal{O}(x) \to G/G_x$; this will give the result, since $|G/G_x| = [G : G_x]$, by Corollary 2.84 of Lagrange's theorem. If $y \in \mathcal{O}(x)$, then $y = gx$ for some $g \in G$; define $\varphi(y) = gG_x$. Now φ is well-defined: if $y = hx$ for some $h \in G$, then $h^{-1}gx = x$ and $h^{-1}g \in G_x$; hence $hG_x = gG_x$. To see that φ is injective, suppose that $\varphi(y) = \varphi(z)$; then there are $g, h \in G$ with $y = gx$, $z = hx$, and $gG_x = hG_x$; that is, $h^{-1}g \in G_x$. It follows that $h^{-1}gx = x$, and so $y = gx = hx = z$. Finally, φ is a surjection: if $gG_x \in G/G_x$, then let $y = gx \in \mathcal{O}(x)$, and note that $\varphi(y) = gG_x$. •

Example 2.139 describes D_8 acting on the four corners of a square; we saw there that $|\mathcal{O}(v_0)| = 4$, $|G_{v_0}| = 2$, and $[G : G_{v_0}] = 8/2 = 4$. Example 2.141 describes $G = \langle \sigma \rangle \leq S_n$ acting on $X = \{1, 2, \ldots, n\}$. We saw there that if, in the complete factorization of σ into disjoint cycles $\sigma = \beta_1 \cdots \beta_{t(\sigma)}$, the r-cycle β_j moves ℓ, then $r = |\mathcal{O}(\ell)|$ for any ℓ occurring in β_j. Theorem 2.143 says that r is a divisor of the order k of σ. (But Theorem 2.55 tells us more: k is the lcm of the lengths of the cycles occurring in the factorization.)

→ **Corollary 2.144.** *If a finite group G acts on a set X, then the number of elements in any orbit is a divisor of $|G|$.*

Proof. This follows at once from Theorem 2.143 and Lagrange's theorem. •

→ **Corollary 2.145.** *If x lies in a finite group G, then the number of conjugates of x is the index of its centralizer:*

$$|x^G| = [G : C_G(x)],$$

and hence it is a divisor of $|G|$.

Proof. As in Example 2.140, the orbit of x is its conjugacy class x^G, and the stabilizer G_x is the centralizer $C_G(x)$. •

Table 2.1 in Example 2.29 displays the number of permutations in S_4 of each cycle structure; these numbers are 1, 6, 8, 6, 3. Note that each of these numbers is a divisor of $|S_4| = 24$. Table 2.2 in Example 2.30 displays the corresponding numbers for S_5; they are 1, 10, 20, 30, 24, 20, and 15, all of which are divisors of $|S_5| = 120$. We now recognize these subsets as being conjugacy classes, and the next corollary explains why these numbers divide the group order.

→ **Corollary 2.146.** *If $\alpha \in S_n$, then the number of permutations in S_n having the same cycle structure as α is a divisor of $n!$.*

Proof. This follows at once from Corollary 2.145 once one recalls Proposition 2.33 which says that two permutations in S_n are conjugate in S_n if and only if they have the same cycle structure. •

When we began classifying groups of order 6, it would have been helpful to be able to assert that any such group has an element of order 3 (we were able to use an earlier exercise to assert the existence of an element of order 2). We now prove that every finite group G contains an element of prime order p for every $p \mid |G|$ [Proposition 2.124(i) is the special case of this when G is abelian].

If the conjugacy class x^G of an element x in a group G consists of x alone, then x commutes with every $g \in G$, for $gxg^{-1} = x$; that is, $x \in Z(G)$. Conversely, if $x \in Z(G)$, then $x^G = \{x\}$. Thus, the center $Z(G)$ consists of all those elements in G whose conjugacy class has exactly one element.

→ **Theorem 2.147 (Cauchy).** *If G is a finite group whose order is divisible by a prime p, then G contains an element of order p.*

Proof. We prove the theorem by induction on $|G|$; the base step $|G| = 1$ is vacuously true, for there are no prime divisors of 1. If $x \in G$, then the number of conjugates of x is $|x^G| = [G : C_G(x)]$, where $C_G(x)$ is the centralizer of x in G. As noted above, if $x \notin Z(G)$, then x^G has more than one element, and so $|C_G(x)| < |G|$. If $p \mid |C_G(x)|$ for some noncentral x, then the inductive hypothesis says there is an element of order p in $C_G(x) \leq G$, and we are done. Therefore, we may assume that $p \nmid |C_G(x)|$ for all noncentral $x \in G$. Better, since $|G| = [G : C_G(x)]|C_G(x)|$, Euclid's lemma gives

$$p \mid [G : C_G(x)].$$

After recalling that $Z(G)$ consists of all those elements $x \in G$ with $|x^G| = 1$, we may use Proposition 2.142 to see

$$|G| = |Z(G)| + \sum_i [G : C_G(x_i)],$$

where one x_i is selected from each conjugacy class having more than one element. Since $|G|$ and all $[G : C_G(x_i)]$ are divisible by p, it follows that $|Z(G)|$ is divisible by p. But $Z(G)$ is abelian, and so Proposition 2.124(i) says that $Z(G)$, and hence G, contains an element of order p.   •

→ **Definition.**   The *class equation* of a finite group G is

$$|G| = |Z(G)| + \sum_i [G : C_G(x_i)],$$

where one x_i is selected from each conjugacy class having more than one element.

→ **Definition.**   If p is a prime, then a *p-group* is a group of order p^n for some $n \geq 0$.

Exercise 2.117 on page 207 shows that a finite group G is a p-group if and only if every element in G has order some power of p.

There are groups whose center is trivial; for example, $Z(S_3) = \{1\}$. For p-groups with more than one element, however, this is never true.

→ **Theorem 2.148.**   *If p is a prime and G is a p-group with more than one element, then $Z(G) \neq \{1\}$.*

Proof.   We may assume that G is not abelian, for the theorem is obviously true when G is abelian. In the class equation $|G| = |Z(G)| + \sum_i [G : C_G(x_i)]$, each $C_G(x_i)$ is a proper subgroup of G, for $x_i \notin Z(G)$. Since G is a p-group, $[G : C_G(x_i)]$ is a divisor of $|G|$, hence is itself a power of p. Thus, p divides each of the terms in the class equation other than $|Z(G)|$, and so $p \mid |Z(G)|$ as well. Therefore, $Z(G) \neq \{1\}$. •

→ **Corollary 2.149.**   *If p is a prime, then every group G of order p^2 is abelian.*

Proof.   If a group G is not abelian, then its center $Z(G)$ is a proper subgroup, so that $|Z(G)| = 1$ or p, by Lagrange's theorem. But Theorem 2.148 says that $Z(G) \neq \{1\}$, and so $|Z(G)| = p$. The center is always a normal subgroup, so that the quotient $G/Z(G)$ is defined; it has order p, and hence $G/Z(G)$ is cyclic. This contradicts Exercise 2.98 on page 190.   •

→ **Example 2.150.**

For every prime p, we exhibit a nonabelian group of order p^3. Define $UT(3, \mathbb{I}_p)$ to be the subgroup of $GL(3, \mathbb{I}_p)$ consisting of all upper triangular matrices having 1's on the diagonal; that is,

$$UT(3, \mathbb{I}_p) = \left\{ \begin{bmatrix} 1 & a & b \\ 0 & 1 & c \\ 0 & 0 & 1 \end{bmatrix} : a, b, c \in \mathbb{I}_p \right\}.$$

It is easy to see that $\mathrm{UT}(3, \mathbb{I}_p)$ is a subgroup of $\mathrm{GL}(3, \mathbb{I}_p)$; it has order p^3 because there are p choices for each of a, b, c. The reader will have no difficulty finding two matrices in $\mathrm{UT}(3, \mathbb{I}_p)$ that do not commute. Exercise 2.123 on page 207 says that $\mathrm{UT}(3, \mathbb{I}_2) \cong D_8$, and Proposition 6.29 shows, for every odd prime p, that $A^p = I$ for all $A \in \mathrm{UT}(3, \mathbb{I}_p)$. ◄

$\rightarrow$ **Example 2.151.**
Who would have guessed that Cauchy's theorem and Fermat's theorem are special cases of some common theorem?[22] The elementary yet ingenious proof of this is due to J. H. McKay (as A. Mann has shown me). If G is a finite group and p is a prime, denote the cartesian product of p copies of G by G^p, and define

$$X = \{(a_1, a_2, \ldots, a_p) \in G^p : a_1 a_2 \ldots a_p = 1\}.$$

Note that $|X| = |G|^{p-1}$, for having chosen the first $p - 1$ entries arbitrarily, the pth entry must equal $(a_1 a_2 \cdots a_{p-1})^{-1}$. Now make X into an $\mathbb{I}_p$-set by defining, for $0 \leq i \leq p - 1$,

$$[i](a_1, a_2, \ldots, a_p) = (a_{i+1}, a_{i+2}, \ldots, a_p, a_1, a_2, \ldots, a_i).$$

The product of the entries in the new p-tuple is a conjugate of $a_1 a_2 \cdots a_p$:

$$a_{i+1} a_{i+2} \cdots a_p a_1 a_2 \cdots a_i = (a_1 a_2 \cdots a_i)^{-1}(a_1 a_2 \cdots a_p)(a_1 a_2 \cdots a_i).$$

This conjugate is 1 (for $g^{-1} 1 g = 1$), and so $[i](a_1, a_2, \ldots, a_p) \in X$. By Corollary 2.144, the size of every orbit of X is a divisor of $|\mathbb{I}_p| = p$; since p is prime, these sizes are either 1 or p. Now orbits with just one element consist of a p-tuple all of whose entries a_i are equal, for all cyclic permutations of the p-tuple are the same. In other words, such an orbit corresponds to an element $a \in G$ with $a^p = 1$. Clearly, $(1, 1, \ldots, 1)$ is such an orbit; if it were the only such, then we would have

$$|G|^{p-1} = |X| = 1 + kp$$

for some $k \geq 0$; that is, $|G|^{p-1} \equiv 1 \bmod p$. If p is a divisor of $|G|$, then we have a contradiction, for $|G|^{p-1} \equiv 0 \bmod p$. We have thus proved Cauchy's theorem: if a prime p is a divisor of $|G|$, then G has an element of order p.

Suppose now that G is a group of order n and that p is not a divisor of n; for example, let $G = \mathbb{I}_n$. By Lagrange's theorem, G has no elements of order p, so that if $a \in G$ and $a^p = 1$, then $a = 1$. Therefore, the only orbit in G^p of size 1 is $(1, 1, \ldots, 1)$, and so

$$n^{p-1} = |G|^{p-1} = |X| = 1 + kp;$$

that is, if p is not a divisor of n, then $n^{p-1} \equiv 1 \bmod p$. Multiplying both sides by n, we have $n^p \equiv n \bmod p$. This congruence also holds when p is a divisor of n, and this is Fermat's theorem. ◄

---

[22]If G is a group of order n and p is a prime, then the number of solutions $x \in G$ of the equation $x^p = 1$ is congruent to $n^{p-1} \bmod p$.

We have seen, in Proposition 2.99, that A_4 is a group of order 12 having no subgroup of order 6. Thus, the assertion that if d is a divisor of $|G|$, then G must have a subgroup of order d, is false. However, this assertion is true when G is a p-group. Indeed, more is true; G must have a normal subgroup of order d.

→ **Proposition 2.152.**   *If G is a group of order $|G| = p^e$, then G has a normal subgroup of order p^k for every $k \leq e$.*

Proof.   We prove the result by induction on $e \geq 0$. The base step is obviously true, and so we proceed to the inductive step. By Theorem 2.148, the center of G is nontrivial: $Z(G) \neq \{1\}$. If $Z(G) = G$, then G is abelian, and we have already proved the result in Proposition 2.124. Therefore, we may assume that $Z(G)$ is a proper subgroup of G. Since $Z(G) \lhd G$, we have $G/Z(G)$ a p-group of order strictly smaller than $|G|$. Assume that $|Z(G)| = p^c$. If $k \leq c$, then $Z(G)$ and, hence G, contains a normal subgroup of order p^k, because $Z(G)$ is abelian. If $k > c$, then $G/Z(G)$ contains a normal subgroup S^* of order p^{k-c}, by induction. The correspondence theorem gives a normal subgroup S of G with

$$Z(G) \leq S \leq G$$

such that $S/Z(G) \cong S^*$. By Corollary 2.84 to Lagrange's theorem,

$$|S| = |S^*||Z(G)| = p^{k-c} \cdot p^c = p^k. \quad \bullet$$

Abelian groups (and the quaternions) have the property that every subgroup is normal. At the opposite pole are groups having no normal subgroups other than the two obvious ones: $\{1\}$ and G.

→ **Definition.**   A group G is called *simple* if $G \neq \{1\}$ and G has no normal subgroups other than $\{1\}$ and G itself.

→ **Proposition 2.153.**   *An abelian group G is simple if and only if it is finite and of prime order.*

Proof.   If G is finite of prime order p, then G has no subgroups H other than $\{1\}$ and G, otherwise Lagrange's theorem would show that $|H|$ is a divisor of p. Therefore, G is simple.

Conversely, assume that G is simple. Since G is abelian, every subgroup is normal, and so G has no subgroups other than $\{1\}$ and G. Choose $x \in G$ with $x \neq 1$. Since $\langle x \rangle$ is a subgroup, we have $\langle x \rangle = G$. If x has infinite order, then all the powers of x are distinct, and so $\langle x^2 \rangle < \langle x \rangle$ is a forbidden subgroup of $\langle x \rangle$, a contradiction. Therefore, every $x \in G$ has finite order, say, m. If m is composite, then $m = k\ell$ and $\langle x^k \rangle$ is a proper nontrivial subgroup of $\langle x \rangle$, a contradiction. Therefore, $G = \langle x \rangle$ has prime order. $\bullet$

We are now going to show that A_5 is a nonabelian simple group (indeed, it is the smallest such; there is no nonabelian simple group of order less than 60).

Suppose that an element $x \in G$ has k conjugates; that is,

$$|x^G| = |\{gxg^{-1} : g \in G\}| = k.$$

If there is a subgroup $H \leq G$ with $x \in H \leq G$, how many conjugates does x have in H? Since

$$x^H = \{hxh^{-1} : h \in H\} \subseteq \{gxg^{-1} : g \in G\} = x^G,$$

we have $|x^H| \leq |x^G|$. It is possible that there is strict inequality $|x^H| < |x^G|$. For example, take $G = S_3$, $x = (1\ 2)$, and $H = \langle x \rangle$. We know that $|x^G| = 3$ (because all transpositions are conjugate), whereas $|x^H| = 1$ (because H is abelian).

Now let us consider this question, in particular, for $G = S_5$, $x = (1\ 2\ 3)$, and $H = A_5$.

Lemma 2.154. *All 3-cycles are conjugate in A_5.*

Proof. Let $G = S_5$, $\alpha = (1\ 2\ 3)$, and $H = A_5$. We know that $|\alpha^{S_5}| = 20$, for there are 20 3-cycles in S_5 (as we saw in Example 2.30). Therefore, $20 = |S_5|/|C_{S_5}(\alpha)| = 120/|C_{S_5}(\alpha)|$, by Corollary 2.145, so that $|C_{S_5}(\alpha)| = 6$; that is, there are exactly six permutations in S_5 that commute with α. Here they are:

$$(1), \ (1\ 2\ 3), \ (1\ 3\ 2), \ (4\ 5), \ (4\ 5)(1\ 2\ 3), \ (4\ 5)(1\ 3\ 2).$$

The last three of these are odd permutations, so that $|C_{A_5}(\alpha)| = 3$. We conclude that

$$|\alpha^{A_5}| = |A_5|/|C_{A_5}(\alpha)| = 60/3 = 20;$$

that is, all 3-cycles are conjugate to $\alpha = (1\ 2\ 3)$ in A_5.   •

This lemma, which says that A_5 is generated by the 3-cycles, can be generalized from A_5 to A_n for all $n \geq 5$; see Exercise 2.127 on page 207.

Lemma 2.155. *Every element in A_5 is a 3-cycle or a product of 3-cycles.*

Proof. If $\alpha \in A_5$, then α is a product of an even number of transpositions: $\alpha = \tau_1\tau_2\cdots\tau_{2n-1}\tau_{2n}$. As the transpositions may be grouped in pairs $\tau_{2i-1}\tau_{2i}$, it suffices to consider products $\tau\tau'$, where τ and τ' are transpositions. If τ and τ' are not disjoint, then $\tau = (i\ j)$, $\tau' = (i\ k)$, and $\tau\tau' = (i\ k\ j)$; if τ and τ' are disjoint, then $\tau\tau' = (i\ j)(k\ \ell) = (i\ j)(j\ k)(j\ k)(k\ \ell) = (i\ j\ k)(j\ k\ \ell)$.   •

→ **Theorem 2.156.** *A_5 is a simple group.*

Proof. We shall show that if H is a normal subgroup of A_5 and $H \neq \{(1)\}$, then $H = A_5$. Now if H contains a 3-cycle, then normality forces H to contain all its conjugates. By Lemma 2.154, H contains every 3-cycle, and by Lemma 2.155, $H = A_5$. Therefore, it suffices to prove that H contains a 3-cycle.

As $H \neq \{(1)\}$, it contains some $\sigma \neq (1)$. We may assume, after a harmless relabeling, that either $\sigma = (1\ 2\ 3)$, $\sigma = (1\ 2)(3\ 4)$, or $\sigma = (1\ 2\ 3\ 4\ 5)$. As we have just remarked, we are done if σ is a 3-cycle.

If $\sigma = (1\ 2)(3\ 4) \in H$, use Proposition 2.32: conjugate σ by $\beta = (3\ 4\ 5)$ to have $\beta\sigma\beta^{-1} = \sigma' = (1\ 2)(4\ 5) \in H$ (because $\beta \in A_5$ and $H \lhd S_5$). Hence, $\sigma\sigma' = (3\ 4\ 5) \in H$.

If $\sigma = (1\ 2\ 3\ 4\ 5) \in H$, use Proposition 2.32: conjugate σ by $\gamma = (1\ 2\ 3)$ to have $\gamma\sigma\gamma^{-1} = \sigma'' = (2\ 3\ 1\ 4\ 5) \in H$ (because $\gamma \in A_5$ and $H \lhd S_5$). Hence, $\sigma''\sigma^{-1} = (2\ 3\ 1\ 4\ 5)(5\ 4\ 3\ 2\ 1) = (1\ 2\ 4) \in H$. We should say how this last equation arose. If $\sigma \in H$ and γ is a 3-cycle, then $\gamma\sigma\gamma^{-1} \in H$, and so $(\gamma\sigma\gamma^{-1})\sigma^{-1} \in H$. Reassociating, $\gamma(\sigma\gamma^{-1}\sigma^{-1}) \in H$. But $\sigma\gamma^{-1}\sigma^{-1}$ is a 3-cycle, so that H contains a product of two 3-cycles. We have chosen γ more carefully to force this product of two 3-cycles to be a 3-cycle.

We have shown, in all cases, that H contains a 3-cycle. Therefore, the only normal subgroups in A_5 are $\{(1)\}$ and A_5 itself, and so A_5 is simple. •

As we shall see in Chapter 5, Theorem 2.156 turns out to be the basic reason why the quadratic formula has no generalization giving the roots of polynomials of degree 5 or higher.

Without much more effort, we can prove that the alternating groups A_n are simple for all $n \geq 5$. Observe that A_4 is not simple, for the four-group **V** is a normal subgroup of A_4.

Lemma 2.157. *A_6 is a simple group.*

Proof. Let $H \neq \{(1)\}$ be a normal subgroup of A_6; we must show that $H = A_6$. Assume that there is some $\alpha \in H$ with $\alpha \neq (1)$ which fixes some i, where $1 \leq i \leq 6$. Define

$$F = \{\sigma \in A_6 : \sigma(i) = i\}.$$

Now $\alpha \in H \cap F$, so that $H \cap F \neq \{(1)\}$. The second isomorphism theorem gives $H \cap F \lhd F$. But F is simple, for $F \cong A_5$, by Exercise 2.130 on page 207, and so the only normal subgroups in F are $\{(1)\}$ and F. Since $H \cap F \neq \{(1)\}$, we have $H \cap F = F$; that is, $F \leq H$. It follows that H contains a 3-cycle, and so $H = A_6$, by Exercise 2.127 on page 207.

We may now assume that there is no $\alpha \in H$ with $\alpha \neq (1)$ which fixes some i with $1 \leq i \leq 6$. If one considers the cycle structures of permutations in A_6, however, any such α must have cycle structure (1 2)(3 4 5 6) or (1 2 3)(4 5 6). In the first case, $\alpha^2 \in H$ is a nontrivial permutation which fixes 1 (and also 2), a contradiction. In the second case, H contains $\alpha(\beta\alpha^{-1}\beta^{-1})$, where $\beta = (2\ 3\ 4)$, and it is easily checked that this is a nontrivial element in H which fixes 6, another contradiction. Therefore, no such normal subgroup H can exist, and so A_6 is a simple group.   •

→ **Theorem 2.158.**   *A_n is a simple group for all $n \geq 5$.*

Proof. If H is a nontrivial normal subgroup of A_n [that is, $H \neq (1)$], then we must show that $H = A_n$; by Exercise 2.127 on page 207, it suffices to prove that H contains a 3-cycle. If $\beta \in H$ is nontrivial, then there exists some i that β moves; say, $\beta(i) = j \neq i$. Choose a 3-cycle α which fixes i and moves j. The permutations α and β do not commute: $\beta\alpha(i) = \beta(i) = j$, while $\alpha\beta(i) = \alpha(j) \neq j$. It follows that $\gamma = (\alpha\beta\alpha^{-1})\beta^{-1}$ is a nontrivial element of H. But $\beta\alpha^{-1}\beta^{-1}$ is a 3-cycle, by Proposition 2.32, and so $\gamma = \alpha(\beta\alpha^{-1}\beta^{-1})$ is a product of two 3-cycles. Hence, γ moves at most 6 symbols, say, $i_1, \ldots, i_6$ (if γ moves fewer than 6 symbols, just adjoin others so we have a list of 6). Define

$$F = \{\sigma \in A_n : \sigma \text{ fixes all } i \neq i_1, \ldots, i_6\}.$$

Now $F \cong A_6$, by Exercise 2.130 on page 207, and $\gamma \in H \cap F$. Hence, $H \cap F$ is a nontrivial normal subgroup of F. But F is simple, being isomorphic to A_6, and so $H \cap F = F$; that is, $F \leq H$. Therefore, H contains a 3-cycle, and so $H = A_n$; the proof is complete.   •

EXERCISES

H **2.114** True or false with reasons.

(i) Every group G is isomorphic to the symmetric group S_G.

(ii) Every group of order 4 is abelian.

(iii) Every group of order 6 is abelian.

(iv) If a group G acts on a set X, then X is a group.

(v) If a group G acts on a set X, and if $g, h \in G$ satisfy $gx = hx$ for some $x \in X$, then $g = h$.

(vi) If a group G acts on a set X, and if $x, y \in X$, then there exists $g \in G$ with $y = gx$.

(vii) If $g \in G$, where G is a finite group, then the number of conjugates of g is a divisor of $|G|$.

(viii) Every group of order 100 contains an element of order 5.

(ix) Every group of order 100 contains an element of order 4.

(x) Every group of order 5^8 contains a normal subgroup of order 5^6.

(xi) If G is a simple group of order p^n, where p is a prime, then $n = 1$.

(xii) The alternating group A_4 is simple.

(xiii) The alternating group A_5 is simple.

(xiv) The symmetric group S_6 is simple.

2.115 Prove that every translation $\tau_a \in S_G$, where $\tau_a \colon g \mapsto ag$, is a regular permutation (see Exercise 2.29 on page 124). The homomorphism $\varphi \colon G \to S_G$, defined by $\varphi(a) = \tau_a$, is often called the *regular representation* of G.

2.116 Prove that no pair of the following groups of order 8,

$$\mathbb{I}_8; \quad \mathbb{I}_4 \times \mathbb{I}_2; \quad \mathbb{I}_2 \times \mathbb{I}_2 \times \mathbb{I}_2; \quad D_8; \quad \mathbf{Q},$$

are isomorphic.

*H **2.117** If p is a prime and G is a finite group in which every element has order a power of p, prove that G is a p-group.

***2.118** Prove that a finite p-group G is simple if and only if $|G| = p$.

***2.119** Show that S_4 has a subgroup isomorphic to D_8.

*H **2.120** Prove that $S_4/\mathbf{V} \cong S_3$.

2.121 H **(i)**   Prove that $A_4 \not\cong D_{12}$.

   H **(ii)** Prove that $D_{12} \cong S_3 \times \mathbb{I}_2$.

***2.122**   **(i)**   If H is a subgroup of G and if $x \in H$, prove that

$$C_H(x) = H \cap C_G(x).$$

   H **(ii)** If H is a subgroup of index 2 in a finite group G and if $x \in H$, prove that either $|x^H| = |x^G|$ or $|x^H| = \frac{1}{2}|x^G|$, where x^H is the conjugacy class of x in H.

*H **2.123** Prove that the group $\mathrm{UT}(3, \mathbb{I}_2)$ in Example 2.150 is isomorphic to D_8.

2.124 H **(i)**   How many permutations in S_5 commute with $(1\ 2)(3\ 4)$, and how many *even* permutations in S_5 commute with $(1\ 2)(3\ 4)$.

   (ii)   How many permutations in S_7 commute with $(1\ 2)(3\ 4\ 5)$?

   (iii) Exhibit all the permutations in S_7 commuting with $(1\ 2)(3\ 4\ 5)$.

2.125 H **(i)**   Show that there are two conjugacy classes of 5-cycles in A_5, each of which has 12 elements.

   (ii)   Prove that the conjugacy classes in A_5 have sizes 1, 12, 12, 15, and 20.

   (iii) Prove that every normal subgroup H of a group G is a union of conjugacy classes of G, one of which is $\{1\}$.

   (iv)   Use parts (ii) and (iii) to give a second proof of the simplicity of A_5.

***2.126** If $\sigma, \tau \in S_5$, where σ is a 5-cycle and τ is a transposition, prove that $\langle \sigma, \tau \rangle = S_5$.

***2.127** H **(i)**   For all $n \geq 3$, prove that every $\alpha \in A_n$ is a product of 3-cycles.

   (ii)   Prove that if a normal subgroup $H \lhd A_n$ contains a 3-cycle, where $n \geq 5$, then $H = A_n$. (*Remark.* See Lemmas 2.155 and 2.155.)

2.128 Prove that $(A_{10})' = A_{10}$, where G' denotes the commutator subgroup of a group G. (See Exercise 2.113 on page 192.)

H **2.129** Prove that the only normal subgroups of S_4 are $\{(1)\}$, $\mathbf{V}$, A_4, and S_4.

***2.130** Let $\{i_1, \ldots, i_r\} \subseteq \{1, 2, \ldots n\}$, and let

$$F = \{\sigma \in A_n : \sigma \text{ fixes all } i \text{ with } i \neq i_1, \ldots, i_r\}.$$

Prove that $F \cong A_r$.

H **2.131** Prove that A_5 is a group of order 60 having no subgroup of order 30.

2.132 Let $X = \{1, 2, 3, \ldots\}$ be the set of all positive integers, and let S_X be the symmetric group on X.

>   **(i)**   Prove that $F_\infty = \{\sigma \in S_X : \sigma$ moves only finitely many $n \in X\}$ is a subgroup of S_X.
>
>   **(ii)** Define A_∞ to be the subgroup of F_∞ generated by the 3-cycles. Prove that A_∞ is an infinite simple group.

2.133   H **(i)**   Prove that if a simple group G has a subgroup of index n, then G is isomorphic to a subgroup of S_n.

>   H **(ii)** Prove that an infinite simple group has no subgroups of finite index $n > 1$.

* H **2.134** Let G be a group with $|G| = mp$, where p is a prime and $1 < m < p$. Prove that G is not simple.

>   **Remark.**   Of all the numbers smaller than 60, we can now show that all but 11 are not orders of nonabelian simple groups (namely, 12, 18, 24, 30, 36, 40, 45, 48, 50, 54, 56). Theorem 2.148 eliminates all prime powers (for the center is always a normal subgroup), and Exercise 2.134 eliminates all numbers of the form mp, where p is a prime and $m < p$. (We will complete the proof that there are no nonabelian simple groups of order less than 60 in Theorem 6.25.)   ◄

* H **2.135** If $n \geq 3$, prove that A_n is the only subgroup of S_n of order $\frac{1}{2}n!$.

* **2.136** Prove that A_6 has no subgroups of prime index.

2.8   COUNTING WITH GROUPS

We are now going to use group theory to do some fancy counting.

Lemma 2.159.

>   (i) *Let a group G act on a set X. If $x \in X$ and $\sigma \in G$, then $G_{\sigma x} = \sigma G_x \sigma^{-1}$.*
>
>   (ii) *If a finite group G acts on a finite set X and if x and y lie in the same orbit, then $|G_y| = |G_x|$.*

Proof.
(i) If $\tau \in G_x$, then $\tau x = x$. If $\sigma x = y$, we have

$$\sigma \tau \sigma^{-1} y = \sigma \tau \sigma^{-1} \sigma x = \sigma \tau x = \sigma x = y.$$

Therefore, $\sigma \tau \sigma^{-1}$ fixes y, and so $\sigma G_x \sigma^{-1} \leq G_y$. The reverse inclusion is proved in the same way, for $x = \sigma^{-1} y$.
(ii) If x and y are in the same orbit, then there is $\sigma \in G$ with $y = \sigma x$, and so $|G_y| = |G_{\sigma x}| = |\sigma G_x \sigma^{-1}| = |G_x|$.   •

Theorem 2.160 (Burnside's Lemma).[23] *Let G act on a finite set X. If N is the number of orbits, then*

$$N = \frac{1}{|G|} \sum_{\tau \in G} F(\tau),$$

where $F(\tau)$ is the number of $x \in X$ fixed by τ.

Proof.   List the elements of X as follows: choose $x_1 \in X$, and then list all the elements in the orbit $\mathcal{O}(x_1)$; say, $\mathcal{O}(x_1) = \{x_1, x_2, \ldots, x_r\}$; then choose $x_{r+1} \notin \mathcal{O}(x_1)$, and list the elements of $\mathcal{O}(x_{r+1})$ as $x_{r+1}, x_{r+2}, \ldots$; continue this procedure until all the elements of X are listed. Now list the elements $\tau_1, \tau_2, \ldots, \tau_n$ of G, and form the following array of 0s and 1s, where

$$f_{i,j} = \begin{cases} 1 & \text{if } \tau_i \text{ fixes } x_j \\ 0 & \text{if } \tau_i \text{ moves } x_j. \end{cases}$$

	x_1	$\cdots$	x_r	x_{r+1}	$\cdots$	x_j	$\cdots$
τ_1	$f_{1,1}$	$\cdots$	$f_{1,r}$	$f_{1,r+1}$	$\cdots$	$f_{1,j}$	$\cdots$
τ_2	$f_{2,1}$	$\cdots$	$f_{2,r}$	$f_{2,r+1}$	$\cdots$	$f_{2,j}$	$\cdots$
τ_i	$f_{i,1}$	$\cdots$	$f_{i,r}$	$f_{i,r+1}$	$\cdots$	$f_{i,j}$	$\cdots$
τ_n	$f_{n,1}$	$\cdots$	$f_{n,r}$	$f_{n,r+1}$	$\cdots$	$f_{n,j}$	$\cdots$

Now $F(\tau_i)$, the number of x fixed by τ_i, is the number of 1s in the ith row of the array; therefore, $\sum_{\tau \in G} F(\tau)$ is the total number of 1s in the array. Let us now look at the columns. The number of 1s in the first column is the number of τ_i that fix x_1; by definition, these τ_i comprise G_{x_1}. Thus, the number of 1s in column 1 is $|G_{x_1}|$. Similarly, the number of 1s in column 2 is $|G_{x_2}|$. By Lemma 2.159(ii), $|G_{x_1}| = |G_{x_2}|$. By Theorem 2.143, the number of 1s in the r columns labeled by the $x_i \in \mathcal{O}(x_1)$ is thus

$$r|G_{x_1}| = |\mathcal{O}(x_1)| \cdot |G_{x_1}| = \big(|G|/|G_{x_1}|\big) |G_{x_1}| = |G|.$$

---

[23] Burnside's influential book *The Theory of Groups of Finite Order* had two editions. In the first edition (1897), he attributed this theorem to G. Frobenius; in the second edition (1911), he gave no attribution at all. However, the commonly accepted name of this theorem is *Burnside's lemma*. To avoid the confusion that would be caused by changing a popular name, P. M. Neumann suggested that it be called "not-Burnside's lemma." Burnside was a fine mathematician, and there do exist theorems properly attributed to him. For example, Burnside proved that if p and q are primes, then there are no simple groups of order $p^m q^n$.

The same is true for any other orbit: its columns contain exactly $|G|$ 1s. Therefore, if there are N orbits, there are $N|G|$ 1s in the array. We conclude that

$$\sum_{\tau \in G} F(\tau) = N|G|. \quad \bullet$$

We are going to use Burnside's lemma to solve problems of the following sort. How many striped flags are there having six stripes (of equal width) each of which can be colored red, white, or blue? Clearly, the two flags in Figure 2.18 are the same: the bottom flag is just the reverse of the top one (the flag may be viewed by standing in front of it or by standing in back of it).

r	w	b	r	w	b

b	w	r	b	w	r

Figure 2.18   A Flag

Let X be the set of all 6-tuples of colors; if $x \in X$, then

$$x = (c_1, c_2, c_3, c_4, c_5, c_6),$$

where each c_i denotes either red, white, or blue. Let τ be the permutation that reverses all the indices:

$$\tau = \begin{pmatrix} 1 & 2 & 3 & 4 & 5 & 6 \\ 6 & 5 & 4 & 3 & 2 & 1 \end{pmatrix} = (1\ 6)(2\ 5)(3\ 4)$$

(thus, τ "turns over" each 6-tuple x of colored stripes). The cyclic group $G = \langle \tau \rangle$ acts on X; since $|G| = 2$, the orbit of any 6-tuple x consists of either 1 or 2 elements: either τ fixes x or it does not. Since a flag is unchanged by turning it over, it is reasonable to identify a flag with an orbit of a 6-tuple. For example, the orbit consisting of the 6-tuples

$$(r, w, b, r, w, b) \quad \text{and} \quad (b, w, r, b, w, r)$$

describes the flag in Figure 2.18. The number of flags is thus the number N of orbits; by Burnside's lemma, $N = \frac{1}{2}[F((1)) + F(\tau)]$. The identity permutation (1) fixes every $x \in X$, and so $F((1)) = 3^6$ (there are 3 colors). Now τ fixes a 6-tuple x if and only if x is a "palindrome," that is, if the colors in x read the same forward as backward. For example,

$$x = (r, r, w, w, r, r)$$

is fixed by τ. Conversely, if

$$x = (c_1, c_2, c_3, c_4, c_5, c_6)$$

is fixed by $\tau = (1\ 6)(2\ 5)(3\ 4)$, then $c_1 = c_6$, $c_2 = c_5$, and $c_3 = c_4$; that is, x is a palindrome. It follows that $F(\tau) = 3^3$, for there are 3 choices for each of c_1, c_2, and c_3. The number of flags is thus

$$N = \tfrac{1}{2}(3^6 + 3^3) = 378.$$

Let us make the notion of coloring more precise.

Definition.   Given an action of a group G on $X = \{1, \ldots, n\}$ and a set $\mathcal{C}$ of q *colors*, then G acts on the set $\mathcal{C}^n$ of all n-tuples of colors by

$$\tau(c_1, \ldots, c_n) = (c_{\tau 1}, \ldots, c_{\tau n}) \quad \text{for all } \tau \in G.$$

An orbit of $(c_1, \ldots, c_n) \in \mathcal{C}^n$ is called a *(q, G)-coloring* of X.

Example 2.161.
Color each square in a 4×4 grid red or black (adjacent squares may have the same color; indeed, one possibility is that all the squares have the same color).

1	2	3	4
5	6	7	8
9	10	11	12
13	14	15	16

13	9	5	1
14	10	6	2
15	11	7	3
16	12	8	4

Figure 2.19   Chessboard

If X consists of the 16 squares in the grid and if $\mathcal{C}$ consists of the two colors red and black, then the cyclic group $G = \langle R \rangle$ of order 4 acts on X, where R is clockwise rotation by $90°$; Figure 2.19 shows how R acts: the right square is R's action on the left square. In cycle notation,

$$R = (1,\ 4,\ 16,\ 13)(2,\ 8,\ 15,\ 9)(3,\ 12,\ 14,\ 5)(6,\ 7,\ 11,\ 10),$$
$$R^2 = (1,\ 16)(4,\ 13)(2,\ 15)(8,\ 9)(3,\ 14)(12,\ 5)(6,\ 11)(7,\ 10),$$
$$R^3 = (1,\ 13,\ 16,\ 4)(2,\ 9,\ 15,\ 8)(3,\ 5,\ 14,\ 12)(6,\ 10,\ 11,\ 7).$$

A red-and-black chessboard does not change when it is rotated; it is merely viewed from a different position. Thus, we may regard a chessboard as a $(2, G)$-coloring of X; the orbit of a 16-tuple corresponds to the four ways of viewing the board.

By Burnside's lemma, the number of chessboards is

$$\tfrac{1}{4}\Big[F((1)) + F(R) + F(R^2) + F(R^3)\Big].$$

Now $F((1)) = 2^{16}$, for every 16-tuple is fixed by the identity. To compute $F(R)$, note that squares $1, 4, 16, 13$ must all have the same color in a 16-tuple fixed by R. Similarly, squares $2, 8, 15, 9$ must have the same color, squares $3, 12, 14, 5$ must have the same color, and squares $6, 7, 11, 10$ must have the same color. We conclude that $F(R) = 2^4$; note that the exponent 4 is the number of cycles in the complete factorization of R. A similar analysis shows that $F(R^2) = 2^8$, for the complete factorization of R^2 has 8 cycles, and $F(R^3) = 2^4$, because the cycle structure of R^3 is the same as that of R. Therefore, the number N of chessboards is

$$N = \tfrac{1}{4}\Big[2^{16} + 2^4 + 2^8 + 2^4\Big] = 16,456.$$

Doing this count without group theory is more difficult because of the danger of counting the same chessboard more than once. ◄

We now show that the cycle structure of a permutation τ allows one to calculate $F(\tau)$.

Theorem 2.162.   *Let C be a set of q colors, and let $\tau \in S_n$.*

(i) *If $F(\tau)$ is the number of $x \in C^n$ fixed by τ, and if $t(\tau)$ is the number of cycles in the complete factorization of τ, then*

$$F(\tau) = q^{t(\tau)}.$$

(ii) *If a finite group G acts on $X = \{1, \ldots, n\}$, then the number N of (q, G)-colorings of X is*

$$N = \frac{1}{|G|} \sum_{\tau \in G} q^{t(\tau)},$$

where $t(\tau)$ is the number of cycles in the complete factorization of τ.

Proof.
(i) Let $\tau \in S_n$ and let $\tau = \beta_1 \cdots \beta_t$ be a complete factorization, where each β_j is an r_j-cycle. If $i_1, \ldots, i_{r_j}$ are the symbols moved by β_j, then $i_{k+1} = \tau^k i_1$ for $k < r_j$. Since τ fixes $x = (c_1, \ldots, c_n)$, we have $\tau(c_1, \ldots, c_n) = (c_{\tau 1}, \ldots, c_{\tau n}) = (c_1, \ldots, c_n)$; thus, $c_{\tau i_1} = c_{i_1}$, so that $c_{\tau i_1}$ and c_{i_1} have the same color. But $\tau^2 i_1$ also has the same color as i_1; in fact, $\tau^k i_1$ has the same color as i_1 for all k. Now there is another way to view these coordinates. By Example 2.141, the coordinates $\tau^k i_1$ are precisely the symbols moved by β_j; that is, $\beta_j = (i_1, i_2, \ldots, i_{r_j})$. Thus, $(c_1, \ldots, c_n)$ is fixed by τ if, for each j, all the symbols c_k for k moved by β_j must have the same color. As there are q colors and $t(\tau)$ β_j's, there are $q^{t(\tau)} n$-tuples fixed by τ.
(ii) Substitute $q^{t(\tau)}$ for $F(\tau)$ into the formula in Burnside's lemma. ●

Example 2.163.
We can now simplify the computations in Example 2.161. The group G acting on the set X of all 4×4 grids consists of the 4 elements $1, R, R^2, R^3$. The complete factorizations of these elements were given in the example, from which we see that

$$t((1)) = 16, \quad t(R) = 4 = t(R^3), \quad t(R^2) = 8.$$

It follows from Theorem 2.162 that

$$N = \tfrac{1}{4}\left[2^{16} + 2 \cdot 2^4 + 2^8\right]. \quad \blacktriangleleft$$

We introduce a polynomial in several variables to allow us to state a more delicate counting result due to Pólya.

Definition.   If the complete factorization of $\tau \in S_n$ has $e_r(\tau) \geq 0$ r-cycles, then the *index* of τ is the monomial

$$\operatorname{ind}(\tau) = x_1^{e_1(\tau)} x_2^{e_2(\tau)} \cdots x_n^{e_n(\tau)}.$$

If G is a subgroup of S_n, then the *cycle index* of G is the polynomial in n variables with coefficients in $\mathbb{Q}$:

$$P_G(x_1, \ldots, x_n) = \frac{1}{|G|} \sum_{\tau \in G} \operatorname{ind}(\tau).$$

In our earlier discussion of striped flags, the group G was a cyclic group of order 2 with generator $\tau = (1\ 6)(2\ 5)(3\ 4)$. Thus, $\operatorname{ind}((1)) = x_1^6$, $\operatorname{ind}(\tau) = x_2^3$, and

$$P_G(x_1, \ldots, x_6) = \tfrac{1}{2}(x_1^6 + x_2^3).$$

As a second example, consider all possible blue-and-white flags having 9 stripes. Here $|X| = 9$ and $G = \langle \tau \rangle \leq S_9$, where

$$\tau = (1\ 9)(2\ 8)(3\ 7)(4\ 6)(5).$$

Now, $\operatorname{ind}((1)) = x_1^9$, $\operatorname{ind}(\tau) = x_1 x_2^4$, and the cycle index of $G = \langle \tau \rangle$ is thus

$$P_G(x_1, \ldots, x_9) = \tfrac{1}{2}(x_1^9 + x_1 x_2^4).$$

In Example 2.161, we saw that the cyclic group $G = \langle R \rangle$ of order 4 acts on a grid with 16 squares, and

$$\operatorname{ind}((1)) = x_1^{16}; \quad \operatorname{ind}(R) = x_4^4; \quad \operatorname{ind}(R^2) = x_2^8; \quad \operatorname{ind}(R^3) = x_4^4.$$

The cycle index is thus

$$P_G(x_1, \ldots, x_{16}) = \tfrac{1}{4}(x_1^{16} + x_2^8 + 2x_4^4).$$

Proposition 2.164. *If $|X| = n$ and G is a subgroup of S_n, then the number of (q, G)-colorings of X is $P_G(q, \ldots, q)$, where $P_G(x_1, \ldots, x_n)$ is the cycle index.*

Proof. By Theorem 2.162, the number of (q, G)-colorings of X is

$$\frac{1}{|G|} \sum_{\tau \in G} q^{t(\tau)},$$

where $t(\tau)$ is the number of cycles in the complete factorization of τ. On the other hand,

$$P_G(x_1, \ldots, x_n) = \frac{1}{|G|} \sum_{\tau \in G} \mathrm{ind}(\tau)$$

$$= \frac{1}{|G|} \sum_{\tau \in G} x_1^{e_1(\tau)} x_2^{e_2(\tau)} \cdots x_n^{e_n(\tau)},$$

and so

$$P_G(q, \ldots, q) = \frac{1}{|G|} \sum_{\tau \in G} q^{e_1(\tau) + e_2(\tau) + \cdots + e_n(\tau)}$$

$$= \frac{1}{|G|} \sum_{\tau \in G} q^{t(\tau)}. \quad \bullet$$

Let us count once again the number of red-and-black chessboards with sixteen squares in Example 2.161. Here,

$$P_G(x_1, \ldots, x_{16}) = \tfrac{1}{4}(x_1^{16} + x_2^8 + 2x_4^4).$$

and so the number of chessboards is

$$P_G(2, \ldots, 2) = \tfrac{1}{4}(2^{16} + 2^8 + 2 \cdot 2^4).$$

The reason for introducing the cycle index is to allow us to state Pólya's generalization of Burnside's lemma, which solves the following sort of problem. How many blue-and-white flags with 9 stripes have 4 blue stripes and 5 white stripes? More generally, we want to count the number of orbits in which we prescribe the number of "stripes" of any given color.

Theorem 2.165 (Pólya). *Let $G \leq S_X$, where $|X| = n$, let $|\mathcal{C}| = q$, and, for each $i \geq 1$, define $\sigma_i = c_1^i + \cdots + c_q^i$. Then the number of (q, G)-colorings of X having f_r elements of color c_r, for every r, is the coefficient of $c_1^{f_1} c_2^{f_2} \cdots c_q^{f_q}$ in $P_G(\sigma_1, \ldots, \sigma_n)$.*

Proofs of Pólya's theorem can be found in combinatorics books (for example, Biggs, *Discrete Mathematics* or Tucker, *Applied Combinatorics*). To solve the flag problem posed above, first note that the cycle index for blue-and-white flags having 9 stripes is

$$P_G(x_1, \ldots, x_9) = \tfrac{1}{2}(x_1^9 + x_1 x_2^4).$$

and so the number of flags is $P_G(2, \ldots, 2) = \tfrac{1}{2}(2^9 + 2^5) = 272$. Using Pólya's theorem, the number of flags with 4 blue stripes and 5 white ones is the coefficient of $b^4 w^5$ in

$$P_G(\sigma_1, \ldots, \sigma_9) = \tfrac{1}{2} \left[(b + w)^9 + (b + w)(b^2 + w^2)^4 \right].$$

A calculation using the binomial theorem shows that the coefficient of $b^4 w^5$ is 66.

EXERCISES

H **2.137** True or false with reasons.

    (i) If a finite group G acts on a set X, then X must be finite.

    (ii) If a group G acts on a finite set X, then G must be finite.

    (iii) If a group G acts on a set X, and if $x, y \in X$, then $G_x \cong G_y$.

    (iv) If a group G acts on a set X, and if $x, y \in X$ lie in the same orbit, then $G_x \cong G_y$.

    (v) If D_{10} acts on a bracelet with 5 beads, then the cycle structure of $\tau \in D_{10}$ is (1), (1 2)(3 4), or (1 2 3 4 5).

    (vi) If D_{10} acts on a bracelet with 5 beads, and if τ is the reflection about the axis passing through one bead and perpendicular to the opposite side, then the cycle index of τ is $x_1 x_2^2$.

H **2.138** How many flags are there with n stripes each of which can be colored any one of q given colors?

2.139 Let X be the squares in an $n \times n$ grid, and let ρ be a rotation by $90°$. Define a ***chessboard*** to be a (q, G)-coloring, where the cyclic group $G = \langle \rho \rangle$ of order 4 is acting. Show that the number of chessboards is

$$\tfrac{1}{4} \left(q^{n^2} + q^{\lfloor (n^2+1)/2 \rfloor} + 2q^{\lfloor (n^2+3)/4 \rfloor} \right),$$

where $\lfloor x \rfloor$ is the greatest integer in the number x.

2.140 Let X be a disk divided into n congruent circular sectors, and let ρ be a rotation by $(360/n)°$. Define a ***roulette wheel*** to be a (q, G)-coloring, where the cyclic group $G = \langle \rho \rangle$ of order n is acting. Prove that if $n = 6$, then there are $\tfrac{1}{6}(2q + 2q^2 + q^3 + q^6)$ roulette wheels having 6 sectors.

    [The formula for the number of roulette wheels with n sectors is

$$\frac{1}{n} \sum_{d \mid n} \phi(n/d) q^d,$$

where ϕ is the Euler ϕ-function.]

2.141 Let X be the vertices of a regular n-gon, and let the dihedral group $G = D_{2n}$ act (as the usual group of symmetries [see Example 2.64]). Define a ***bracelet*** to be a (q, G)-coloring of a regular n-gon, and call each of its vertices a ***bead***. (Not only can one rotate a bracelet; one can also flip it.)

H **(i)**   How many bracelets are there having 5 beads, each of which can be colored any one of q available colors?

H **(ii)**   How many bracelets are there having 6 beads, each of which can be colored any one of q available colors?

(iii)   How many bracelets are there with exactly 6 beads having 1 red bead, 2 white beads, and 3 blue beads?

3

Commutative Rings I

→ **3.1 FIRST PROPERTIES**

Often, in high school algebra classes, one is presented with a list of "rules" for ordinary addition and multiplication of real numbers; these lists[1] are often quite long, having perhaps 20 or more items. For example, one rule is the ***additive cancellation law***:

$$\text{if } a + c = b + c, \text{ then } a = b.$$

Some rules, as this one, follow from properties of subtraction–just subtract c from both sides–but there are also rules involving two operations. One such is the ***distributive law***:

$$(a + b)c = ac + bc;$$

when read from left to right, it says that c can be "multiplied through" $a + b$; when read from right to left, it says that c can be "factored out" of $ac + bc$. There is also the "mysterious" rule:

$$(-1) \times (-1) = 1. \tag{M}$$

Lists of rules can be shrunk by deleting redundant items, but there is a good reason for so shrinking them aside from the obvious economy provided by a shorter list: a short list makes it easier to see analogies between numbers and other objects, such as polynomials or congruence classes, which can be added and multiplied. Before exploring such analogies, let us dispel the mystery of (M).

Lemma 3.1. $0 \cdot a = 0$ *for every number a.*

---

[1]For example, see H. S. Hall and S. R. Knight, *Algebra for Colleges and Schools*, Macmillan, 1923, or J. C. Stone and V. S. Mallory, *A Second Course in Algebra*, Sanborn, 1937.

Proof.   Since $0 = 0 + 0$, the distributive law gives

$$0 \cdot a = (0 + 0) \cdot a = (0 \cdot a) + (0 \cdot a).$$

Now subtract $0 \cdot a$ from both sides (that is, use the additive cancellation law) to get $0 = 0 \cdot a$.   •

We can now see why dividing by 0 is forbidden: given a number b, its reciprocal $1/b$ must satisfy $b(1/b) = 1$. In particular, $1/0$ would be a number satisfying $0 \cdot (1/0) = 1$. But Lemma 3.1 gives $0 \cdot (1/0) = 0$, contradicting $1 \neq 0$.

Lemma 3.2.   *If $-a$ is that number which, when added to a, gives 0, then*

$$(-1)(-a) = a.$$

Proof.   The distributive law and Lemma 3.1 give

$$0 = 0 \cdot (-a) = (-1 + 1)(-a) = (-1)(-a) + (-a);$$

now add a to both sides (the additive cancellation law again) to get $a = (-1)(-a)$.   •

Setting $a = 1$ gives the (no longer) mysterious (M).

While we are proving elementary properties, let us show that, fortunately, the product $(-1)a$ is the same as $-a$.

Corollary 3.3. $(-1)a = -a$ *for every number a.*

Proof.   By Lemma 3.2, $(-1)(-a) = a$. Multiplying both sides by -1 gives

$$(-1)(-1)(-a) = (-1)a.$$

But Lemma 3.2 gives $(-1)(-1) = 1$, so that $-a = (-1)a$.   •

Mathematical objects other than numbers can be added and multiplied. For example, one adds and multiplies functions in calculus. Now the constant function $\varepsilon(x) \equiv 1$ behaves just like the number 1 under multiplication; that is, $\varepsilon f = f$. Is the analog of Lemma 3.2 true; is $[-\varepsilon(x)][-f(x)] = f(x)$? The answer is yes, and the proof of this fact is exactly the same as the proof just given for numbers: just replace every occurrence of the letter a by $f(x)$ and the numeral 1 by ε.

We now focus on certain simple properties enjoyed by ordinary addition and multiplication, elevating them to the status of axioms. In essence, we are describing more general realms in which we shall be working.

→ **Definition.**   A *commutative ring*[2] R is a set with two operations, addition and multiplication, such that

(i) $a + b = b + a$ for all $a, b \in R$;

(ii) $a + (b + c) = (a + b) + c$ for all $a, b, c \in R$;

(iii) there is an element $0 \in R$ with $0 + a = a$ for all $a \in R$;

(iv) for each $a \in R$, there is $a' \in R$ with $a' + a = 0$;

(v) $ab = ba$ for all $a, b \in R$;

(vi) $a(bc) = (ab)c$ for every $a, b, c \in R$;

(vii) there is an element $1 \in R$, called *one* (or *the unit*),[3] with $1a = a$ for every $a \in R$;

(viii) $a(b + c) = ab + ac$ for every $a, b, c \in R$.

Of course, axioms (i) through (iv) say that R is an abelian group under addition. Addition and multiplication in a commutative ring R are operations, so there are functions

$$\alpha: R \times R \to R \qquad \text{with} \qquad \alpha(r, r') = r + r' \in R$$

and

$$\mu: R \times R \to R \qquad \text{with} \qquad \mu(r, r') = rr' \in R$$

for all $r, r' \in R$. Since α and μ are functions, they are single-valued, so that the *law of substitution* holds: if $r = r'$ and $s = s'$, then $r + s = r' + s'$ and $rs = r's'$. For example, the proof of Lemma 3.1 begins with $\mu(0, a) = \mu(0 + 0, a)$, and the proof of Lemma 3.2 begins with $\alpha(0, -a) = \alpha(-1 + 1, -a)$.

Example 3.4.

(i) The reader may assume that $\mathbb{Z}$, $\mathbb{Q}$, $\mathbb{R}$, and $\mathbb{C}$ are commutative rings with the usual addition and multiplication (the ring axioms are verified in courses in the foundations of mathematics).

(ii) Let $\mathbb{Z}[i]$ be the set of all complex numbers of the form $a + bi$, where $a, b \in \mathbb{Z}$ and $i^2 = -1$. It is a boring exercise to check that $\mathbb{Z}[i]$ is, in fact, a commutative ring (this exercise will be significantly shortened once the notion of *subring* has been introduced). $\mathbb{Z}[i]$ is called the ring of *Gaussian integers*.

---

[2]This term was probably coined by D. Hilbert, in 1897, when he wrote *Zahlring*. One of the meanings of the word *ring*, in German as in English, is collection, as in the phrase "a ring of thieves." (It has also been suggested that Hilbert used this term because, for a ring of algebraic integers, an appropriate power of each element "cycles back" to being a linear combination of lower powers.)

[3]Some authors do not demand that commutative rings have 1. For them, the set of all even integers is a commutative ring, but we do not recognize it as such. They refer to our rings as *commutative rings with one*.

(iii) Consider the set R of all real numbers x of the form

$$x = a + b\omega,$$

where $a, b \in \mathbb{Q}$ and $\omega = \sqrt[3]{2}$. It is easy to see that R is closed under ordinary addition, but we claim that R is not closed under multiplication. If $\omega^2 \in R$, then there are rationals a and b with

$$\omega^2 = a + b\omega.$$

Multiplying both sides by ω gives the equations

$$\begin{aligned}
2 &= a\omega + b\omega^2 \\
&= a\omega + b(a + b\omega) \\
&= a\omega + ab + b^2\omega \\
&= ab + (a + b^2)\omega.
\end{aligned}$$

If $a + b^2 = 0$, then $a = -b^2$, and the last equation gives $2 = ab$; hence, $2 = (-b^2)b = -b^3$. But this says that the cube root of 2 is rational, contradicting Exercise 1.54(ii) on page 54. Therefore, $a + b^2 \neq 0$ and $\omega = (2 - ab)/(a + b^2)$. Since a and b are rational, we have ω rational, again contradicting Exercise 1.54(ii). Therefore, R is not closed under multiplication, and so R is not a commutative ring. ◀

→ **Remark.** In the term *commutative ring*, the adjective modifies the operation of multiplication, for commutativity of addition is part of the general concept of ring. There are noncommutative rings; that is, there are sets with addition and multiplication satisfying all the axioms of a commutative ring except the commutativity axiom: $ab = ba$. [Actually, the definition replaces the axiom $1a = a$ by $1a = a = a1$, and it replaces the distributive law by two distributive laws, one on either side: $a(b + c) = ab + ac$ and $(b + c)a = ba + ca$.] For example, let M denote the set of all 2×2 matrices with real entries. Example 2.48(i) defines multiplication of matrices, and we now define addition by

$$\begin{bmatrix} a & b \\ c & d \end{bmatrix} + \begin{bmatrix} a' & b' \\ c' & d' \end{bmatrix} = \begin{bmatrix} a + a' & b + b' \\ c + c' & d + d' \end{bmatrix}.$$

It is easy to see that M, equipped with this addition and multiplication, satisfies all the new ring axioms, but M is not a commutative ring.

Even though there are interesting examples of noncommutative rings, we shall consider only commutative rings in this book. ◀

Proposition 3.5. Lemma 3.1, Lemma 3.2, *and* Corollary 3.3 *hold for every commutative ring.*

Proof. Each of these results can be proved using only the defining axioms of a commutative ring. To illustrate, here is a very fussy proof of Lemma 3.1: if R is a commutative ring and $a \in R$, then $0 \cdot a = 0$.

Since $0 = 0 + 0$, the distributive law gives

$$0 \cdot a = (0 + 0) \cdot a = (0 \cdot a) + (0 \cdot a).$$

Now add $-(0 \cdot a)$ to both sides:

$$-(0 \cdot a) + (0 \cdot a) = -(0 \cdot a) + [(0 \cdot a) + (0 \cdot a)].$$

The defining property of $-(0 \cdot a)$ gives the left side $-(0 \cdot a) + (0 \cdot a) = 0$, and so

$$0 = -(0 \cdot a) + [(0 \cdot a) + (0 \cdot a)].$$

We use associativity to simplify the right side.

$$\begin{aligned}
0 &= -(0 \cdot a) + [(0 \cdot a) + (0 \cdot a)] \\
&= [-(0 \cdot a) + (0 \cdot a)] + (0 \cdot a) \\
&= 0 + (0 \cdot a) \\
&= 0 \cdot a. \quad \bullet
\end{aligned}$$

It is unusual to give such a detailed proof, for it tends to make a simple idea look difficult. You should regard a proof as an explanation why a statement is true. But an explanation depends on whom you are talking to: you would probably give one explanation to a beginning high school student, another to one of your classmates, and yet another to your professor. As a rule of thumb, your proofs should be directed toward your peers, one of whom is yourself. Make your proof as clear as possible, not too long, not too short. If your proof is challenged, you must be prepared to explain further, so try to anticipate challenges by giving enough details in your original proof.

What have we shown? Formulas such as $(-1)(-a) = a$ hold, not because of the nature of the numbers a and 1, not because of the particular definitions of the operations of addition and multiplication, but merely as consequences of the axioms for addition and multiplication stated in the definition of commutative ring. For example, we shall see, in Proposition 3.6, that the binomial theorem holds in every commutative ring. Once we see that all functions $\mathbb{R} \to \mathbb{R}$ form a commutative ring [Example 3.10], it will then follow that the binomial theorem $(f + g)^n = \sum \binom{n}{i} f^i g^{n-i}$ holds for all functions $f, g \colon \mathbb{R} \to \mathbb{R}$. Thus, a theorem about commutative rings applies not only to numbers but to other realms as well, thereby proving many theorems all at once instead of one at a time. The abstract approach allows us to be more efficient; the same result need not be proved over and over again. There is a second advantage of abstraction.

The things one adds and multiplies may be very complicated, but many properties may be consequences of the rules of manipulating them and not of their intrinsic structure. Thus, as we have seen when we studied groups, the abstract approach allows us to focus on the essential parts of a problem; we need not be distracted by any features irrelevant to it.

→ **Definition.**   If R is a commutative ring and $a, b \in R$, then **subtraction** is defined by

$$a - b = a + (-b).$$

In light of Corollary 3.3,

$$a - b = a + (-1)b.$$

Here is one more ultrafussy proof (we shall not be so fussy again!): the distributive law $ca - cb = c(a - b)$ holds for subtraction.

$$
\begin{aligned}
a(b - c) = a[b + (-1)c] &= ab + a[(-1)c] \\
&= ab + [a(-1)]c = ab + [(-1)a]c \\
&= ab + (-1)(ac) = ab - ac.
\end{aligned}
$$

It R is a commutative ring and $r \in R$, it is natural to denote rr as r^2 and rrr as r^3. Similarly, it is natural to denote $r + r$ as $2r$ and $r + r + r$ as $3r$. Here is the formal definition.

→ **Definition.**   Let R be a commutative ring, let $a \in R$, and let $n \in \mathbb{N}$. Define $0a = 0$ (the 0 on the left is the number zero, while the 0 on the right is the zero element of R), and define $(n + 1)a = na + a$. Define $(-n)a = -(na)$.

Thus, if $n \in \mathbb{N}$ and $a \in R$, we have $na = a + a + \cdots + a$, where there are n summands. It is easy to see that $(-n)a = -(na) = n(-a)$. The element $n^* = n1 = 1 + \cdots + 1$, where 1 is the one element in R, has the property that na, as defined above, is equal to n^*a. Thus, na, the product of a natural number and a ring element, can also be viewed as the product of two ring elements.

Proposition 3.6 (Binomial Theorem).   *If $a, b \in R$, where R is a commutative ring, then for all $n \geq 0$,*

$$(a + b)^n = \sum_{r=0}^{n} \binom{n}{r} a^r b^{n-r}.$$

Proof.   Adapt the proof of Proposition 1.18, the binomial theorem in $\mathbb{Z}$. In particular, define $a^0 = 1$ for every $a \in R$, even for $a = 0$.   •

The definition of commutative ring does not require that $1 \neq 0$.

→ **Proposition 3.7.** *If R is a commutative ring in which* $1 = 0$, *then R has only one element*: $R = \{0\}$. *One calls R the **zero ring**.*

Proof. If $r \in R$, then $r = 1r = 0r = 0$, by Proposition 3.5. •

The zero ring arises occasionally, but we agree that it is not very interesting.

→ **Definition.** An ***integral domain*** is a commutative ring R with $1 \neq 0$ which satisfies an extra axiom, the ***cancellation law*** for multiplication:

$$\text{if } ca = cb \text{ and } c \neq 0, \text{ then } a = b.$$

We will consistently abbreviate this term to ***domain*** (unless it occurs in a context in which it might be confused with the domain of some function).

The familiar examples of commutative rings: $\mathbb{Z}$, $\mathbb{Q}$, $\mathbb{R}$, $\mathbb{C}$, are domains, but we shall soon exhibit honest examples of commutative rings that are not domains.

→ **Proposition 3.8.** *A nonzero commutative ring R is a domain if and only if the product of any two nonzero elements of R is nonzero.*

Proof. Assume that R is a domain, so that the cancellation law holds. Suppose, by way of contradiction, that there are nonzero elements $a, b \in R$ with $ab = 0$. Proposition 3.5 gives $0 \cdot b = 0$, so that $ab = 0 \cdot b$. The cancellation law now gives $a = 0$ (for $b \neq 0$), and this is a contradiction.

Conversely, assume that the product of nonzero elements in R is always nonzero. If $ca = cb$ with $c \neq 0$, then $0 = ca - cb = c(a - b)$. Since $c \neq 0$, the hypothesis that the product of nonzero elements is nonzero forces $a - b = 0$. Therefore, $a = b$, as desired. •

Informally, a *subring* of a ring R is a subset which is a ring having the same addition and multiplication as in R.

→ **Definition.** A subset S of a commutative ring R is a ***subring*** of R if

(i) $1 \in S$;[4]

(ii) if $a, b \in S$, then $a - b \in S$;

(iii) if $a, b \in S$, then $ab \in S$.

Just as a subgroup is a group in its own right, so is a subring of a commutative ring a commutative ring in its own right.

---

[4] The even integers do *not* form a subring of $\mathbb{Z}$ because 1 is not even. Their special structure will be recognized when *ideals* are introduced.

→ **Proposition 3.9.**   *A subring S of a commutative ring R is itself a commutative ring.*

Proof.   By hypothesis, $1 \in S$; since $1r = r$ for all $r \in R$, we have, in particular, that $1s = s$ for all $s \in S$. We now show that S is closed under addition; that is, if $s, s' \in S$, then $s + s' \in S$. Axiom (ii) in the definition of subring gives $0 = 1 - 1 \in S$. Another application of this axiom shows that if $b \in S$, then $0 - b = -b \in S$; finally, if $a, b \in S$, then Lemma 3.3 shows that S contains

$$a - (-b) = a + (-1)(-b) = a + (-1)(-1)b = a + b.$$

Thus, S is closed under addition and multiplication. It contains 1 and 0 and, for each $s \in S$, it contains $-s$. All the other axioms in the definition of commutative ring are inherited by S from their holding in the commutative ring R. For example, we know that the distributive law $a(b+c) = ab + ac$ holds for all $a, b, c \in R$. In particular, this equation holds for all $a, b, c \in S \subseteq R$, and so the distributive law holds in S.   •

To verify that a set S is a commutative ring requires checking ten items: closure under addition and multiplication and eight axioms; to verify that a subset S of a commutative ring is a subring requires checking only three items, which is obviously more economical. For example, it is simpler to show that the ring of Gaussian integers,

$$\mathbb{Z}[i] = \{z \in \mathbb{C} : z = a + ib : a, b \in \mathbb{Z}\},$$

is a subring of $\mathbb{C}$ than to verify all the axioms in the definition of a commutative ring. Of course, one must first have shown that $\mathbb{C}$ is a commutative ring.

Example 3.10.
If $n \geq 3$ is an integer, let $\zeta_n = e^{2\pi i/n}$ be a primitive nth root of unity, and define

$$\mathbb{Z}[\zeta_n] = \{z \in \mathbb{C} : z = a_0 + a_1\zeta_n + a_2\zeta_n^2 + \cdots + a_{n-1}\zeta_n^{n-1}, \text{ all } a_i \in \mathbb{Z}\}.$$

When $n = 4$, then $\mathbb{Z}[\zeta_4]$ is the Gaussian integers $\mathbb{Z}[i]$. It is easy to check that $\mathbb{Z}[\zeta_n]$ is a subring of $\mathbb{C}$; to prove that $\mathbb{Z}[\zeta_n]$ is closed under multiplication, note that if $m \geq n$, then $m = qn + r$, where $0 \leq r < n$, and $\zeta_n^m = \zeta_n^r$.   ◀

Here is an example of a commutative ring that is not a domain.

Example 3.11.

(i) Let $\mathcal{F}(\mathbb{R})$ be the set of all the functions $\mathbb{R} \to \mathbb{R}$ equipped with the operations of ***pointwise addition*** and ***pointwise multiplication***: for functions $f, g \in \mathcal{F}(\mathbb{R})$, define new functions $f + g$ and fg by

$$f + g : a \mapsto f(a) + g(a) \qquad \text{and} \qquad fg : a \mapsto f(a)g(a)$$

(notice that fg is *not* their composite).

Pointwise addition and pointwise multiplication are precisely those operations on functions that occur in calculus. For example, recall the product rule for derivatives:

$$(fg)' = f'g + fg'.$$

The $+$ in the sum $f'g + fg'$ is pointwise addition, and fg is the pointwise product of f and g.

We claim that $\mathcal{F}(\mathbb{R})$ with these operations is a commutative ring. Verification of the axioms is left to the reader with the following hint: the zero in $\mathcal{F}(\mathbb{R})$ is the constant function z with $z(a) = 0$ for all $a \in \mathbb{R}$, and the one is the constant function ε with $\varepsilon(a) = 1$ for all $a \in \mathbb{R}$.

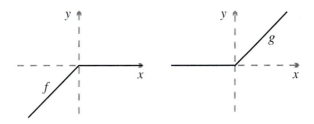

Figure 3.1 $\mathcal{F}(\mathbb{R})$ Is Not a Domain

We now show that $\mathcal{F}(\mathbb{R})$ is not a domain. Define f and g by

$$f(a) = \begin{cases} a & \text{if } a \leq 0 \\ 0 & \text{if } a \geq 0; \end{cases} \qquad g(a) = \begin{cases} 0 & \text{if } a \leq 0 \\ a & \text{if } a \geq 0. \end{cases}$$

Clearly, neither f nor g is zero (i.e., $f \neq z$ and $g \neq z$). On the other hand, for each $a \in \mathbb{R}$, $fg \colon a \mapsto f(a)g(a) = 0$, because at least one of the factors $f(a)$ or $g(a)$ is the number zero. Therefore, $fg = z$, by Proposition 2.2, and $\mathcal{F}(\mathbb{R})$ is not a domain.

(ii) Recall that a function $f \colon \mathbb{R} \to \mathbb{R}$ is *differentiable* if $f'(a)$ exists for all $a \in \mathbb{R}$. Let $\mathcal{D}(\mathbb{R}) = \{$all differentiable functions $f \colon \mathbb{R} \to \mathbb{R}\}$ We claim that $\mathcal{D}(\mathbb{R})$ is a subring of $\mathcal{F}(\mathbb{R})$. Now ε lies in $\mathcal{D}(\mathbb{R})$, for $\varepsilon' = z$. If $f, g \in \mathcal{D}(\mathbb{R})$, then $f + g \in \mathcal{D}(\mathbb{R})$, for $(f + g)' = f' + g'$, while $(fg)'$ exists, by the product rule. Therefore, $\mathcal{D}(\mathbb{R})$ is a subring of $\mathcal{F}(\mathbb{R})$, and so $\mathcal{D}(\mathbb{R})$ is a ring in its own right, by Proposition 3.9. ◀

→ **Proposition 3.12.**

(i) *If $m \geq 2$, then $\mathbb{I}_m$, the integers* mod *m, is a commutative ring.*

(ii) *The commutative ring $\mathbb{I}_m$ is a domain if and only if m is a prime.*

Proof.

(i) In Theorem 2.103, we proved that $[a + b] = [a] + [b]$ defines an addition on $\mathbb{I}_m$ satisfying axioms (i) through (iv) in the definition of commutative ring ($[a]$ is the congruence class $[a] = \{b \in \mathbb{Z} : b \equiv a \bmod m\}$). In Theorem 2.105, we proved that there is a multiplication defined on $\mathbb{I}_m$, namely, $[a][b] = [ab]$, which satisfies axioms (v) through (vii). Only the distributive law needs checking. Since distributivity does hold in $\mathbb{Z}$, we have

$$\begin{aligned}
[a]\big([b] + [c]\big) &= [a][b + c] \\
&= [a(b + c)] \\
&= [ab + ac] \\
&= [ab] + [ac] \\
&= [a][b] + [a][c].
\end{aligned}$$

Therefore, $\mathbb{I}_m$ is a commutative ring. (We remark that if $m = 0$, then $\mathbb{I}_m = \mathbb{Z}$, and if $m = 1$, then $\mathbb{I}_m$ is the zero ring.)

(ii) If m is not a prime, then $m = ab$, where $0 < a, b < m$. Now both $[a]$ and $[b]$ are not $[0]$ in $\mathbb{I}_m$, because m divides neither a nor b, but $[a][b] = [m] = [0]$. Thus, $\mathbb{I}_m$ is not a domain.

Conversely, suppose that m is prime. Since $m \geq 2$, we have $[1] \neq [0]$. If $[a][b] = [0]$, then $ab \equiv 0 \bmod m$, that is, $m \mid ab$. Since m is a prime, Euclid's lemma gives $m \mid a$ or $m \mid b$; that is, $a \equiv 0 \bmod m$ or $b \equiv 0 \bmod m$; that is, $[a] = [0]$ or $[b] = [0]$. Therefore, $\mathbb{I}_m$ is a domain.   •

The ring $\mathbb{I}_6$ is not a domain because $[2] \neq 0$ and $[3] \neq 0$, yet $[2][3] = [6] = [0]$.

Many theorems of ordinary arithmetic, that is, properties of the commutative ring $\mathbb{Z}$, hold in more generality. We now generalize some familiar definitions from $\mathbb{Z}$ to arbitrary commutative rings.

→ **Definition.**   Let a and b be elements of a commutative ring R. Then a ***divides*** b ***in*** R (or a is a ***divisor*** of b or b is a ***multiple*** of a), denoted[5] by

$$a \mid b,$$

if there exists an element $c \in R$ with $b = ca$.

As an extreme example, if $0 \mid a$, then $a = 0 \cdot b$ for some $b \in R$. Since $0 \cdot b = 0$, however, we must have $a = 0$. Thus, $0 \mid a$ if and only if $a = 0$.

Notice that whether $a \mid b$ depends not only on the elements a and b but on the commutative ring R as well. For example, 3 does divide 2 in $\mathbb{Q}$, for $2 = 3 \times \frac{2}{3}$, and

---

[5]Do not confuse the notations $a \mid b$ and a/b. The first notation denotes the *statement* "a is a divisor of b," whereas the second denotes an *element* $c \in R$ with $bc = a$.

$\frac{2}{3} \in \mathbb{Q}$; on the other hand, 3 does not divide 2 in $\mathbb{Z}$, because there is no *integer c* with $3c = 2$.

The reader can quickly check each of the following facts. For every $a \in R$, we have $a \mid a$, $1 \mid a$, $-a \mid a$, $-1 \mid a$, and $a \mid 0$.

Lemma 3.13. *Let R be a commutative ring, and let a, b, c be elements of R.*

(i) *If $a \mid b$ and $b \mid c$, then $a \mid c$.*

(ii) *If $a \mid b$ and $a \mid c$, then a divides every element of R of the form $sb + tc$, where s, $t \in R$.*

Proof. Exercises for the reader. •

→ **Definition.** If R is a commutative ring and $a, b \in R$, then a *linear combination* of them is an element of R of the form $sa + tb$, where $s, t \in R$.

Thus, Lemma 3.13 says that any common divisor of elements $a, b \in R$ must also divide every linear combination of a and b.

→ **Definition.** An element u in a commutative ring R is called a *unit* if $u \mid 1$ in R, that is, if there exists $v \in R$ with $uv = 1$; the element v is called the *inverse* of u (uniqueness of the inverse is Exercise 3.3 on page 229), and v is often denoted by u^{-1}. An element $a \in R$ is an *associate* of an element $r \in R$ if there is a unit $u \in R$ with $a = ur$.

→ **Example 3.14.**
The only units in $\mathbb{Z}$ are ± 1, and the associates of $n \in \mathbb{Z}$ are $\pm n$. ◀

Units are of interest because one can always divide by them. If u is a unit in R, then there is $v \in R$ with $uv = 1$, and if $a \in R$, then $u \mid a$ because

$$a = u(va)$$

is a factorization of a in R. Thus, it is reasonable to define the quotient a/u as $a/u = va = u^{-1}a$. (Recall that this last equation is the reason why zero is never a unit; that is, why dividing by zero is forbidden.)

Just as divisibility depends on the commutative ring R, so does the question whether an element $u \in R$ is a unit depend on R (for it is a question whether $u \mid 1$ in R). For example, the number 2 is a unit in $\mathbb{Q}$, for $\frac{1}{2}$ lies in $\mathbb{Q}$ and $2 \times \frac{1}{2} = 1$, but 2 is not a unit in $\mathbb{Z}$, because there is no *integer v* with $2v = 1$.

The following theorem generalizes Exercise 1.50 on page 53.

Proposition 3.15.   *Let R be a domain, and let $a, b \in R$ be nonzero. Then $a \mid b$ and $b \mid a$ if and only if $b = ua$ for some unit $u \in R$.*

Proof.   If $a \mid b$ and $b \mid a$, there are elements $u, v \in R$ with $b = ua$ and $a = vb$. Substituting, $b = ua = uvb$. Since $b = 1b$ and $b \neq 0$, the cancellation law in the domain R gives $1 = uv$, and so u is a unit.

Conversely, assume that $b = ua$, where u is a unit in R. Plainly, $a \mid b$. If $v \in R$ satisfies $uv = 1$, then $vb = vua = a$, and so $b \mid a$.   •

There exist examples of commutative rings R in which the conclusion of Proposition 3.15 is false, and so the hypothesis in this proposition that R be a domain is needed.

What are the units in $\mathbb{I}_m$?

Proposition 3.16.   *If a is an integer, then $[a]$ is a unit in $\mathbb{I}_m$ if and only if a and m are relatively prime. In fact, if $sa + tm = 1$, then $[a]^{-1} = [s]$.*

Proof.   If $[a]$ is a unit in $\mathbb{I}_m$, then there is $[s] \in \mathbb{I}_m$ with $[s][a] = [1]$. Therefore, $sa \equiv 1 \bmod m$, and so there is an integer t with $sa - 1 = tm$; hence, $1 = sa - tm$. By Exercise 1.56 on page 54, a and m are relatively prime.

Conversely, if a and m are relatively prime, there are integers s and t with $1 = sa + tm$. Hence, $sa - 1 = -tm$ and so $sa \equiv 1 \bmod m$. Thus, $[s][a] = [1]$, and $[a]$ is a unit in $\mathbb{I}_m$.   •

Corollary 3.17.   *If p is a prime, then every nonzero $[a]$ in $\mathbb{I}_p$ is a unit.*

Proof.   If $[a] \neq [0]$, then $a \not\equiv 0 \bmod p$, and hence $p \nmid a$. Therefore, a and p are relatively prime because p is prime.   •

→ **Notation.**   If R is a commutative ring, then the subset of all its units is denoted by

$$U(R) = \{\text{all units in } R\}.$$

It is easy to check that $U(R)$ is a multiplicative group, and we call $U(R)$ the ***group of units*** of R. Corollary 3.17 says that $U(\mathbb{I}_m) = \{[a] \in \mathbb{I}_m : (a, m) = 1\}$ [we have already met $U(\mathbb{I}_m)$ in the proof of Theorem 2.109].

The introduction of the commutative ring $\mathbb{I}_m$ makes the solution of congruence problems much more natural. A congruence $ax \equiv b \bmod m$ in $\mathbb{Z}$ becomes an equation $[a][x] = [b]$ in $\mathbb{I}_m$. If $[a]$ is a unit in $\mathbb{I}_m$, that is, if $(a, m) = 1$, then it has an inverse $[a]^{-1} = [s]$, and we can divide by it; the solution is $[x] = [a]^{-1}[b] = [s][b] = [sb]$. In other words, congruences are solved just as ordinary linear equations $\alpha x = \beta$ are solved over $\mathbb{R}$; that is, $x = \alpha^{-1}\beta$.

Exercises

H **3.1** True or false with reasons.

    (i)   The subset $\{r + s\pi : r, s \in \mathbb{Q}\}$ is a subring of $\mathbb{R}$.

    (ii)   Every subring of a domain is a domain.

    (iii) The zero ring is a subring of $\mathbb{Z}$.

    (iv) There are infinitely many positive integers m for which $\mathbb{I}_m$ is a domain.

    (v)   If S is a subring of a commutative ring R, then $U(S)$ is a subgroup of $U(R)$.

    (vi) If S is a subring of a commutative ring R, then $U(S) = U(R) \cap S$.

    (vii) If R is an infinite commutative ring, then $U(R)$ is infinite.

    (viii) If X is an infinite set, then the family of all finite subsets of X forms a subring of the Boolean ring $\mathcal{B}(X)$.

3.2 Prove that a commutative ring R has a unique one 1; that is, if $e \in R$ satisfies $er = r$ for all $r \in R$, then $e = 1$.

3.3 Let R be a commutative ring.

    (i)   Prove the additive cancellation law.

    (ii)   Prove that every $a \in R$ has a unique additive inverse: if $a + b = 0$ and $a + c = 0$, then $b = c$.

    (iii) If $u \in R$ is a unit, prove that its inverse is unique: if $ub = 1$ and $uc = 1$, then $b = c$.

3.4   **(i)**   Prove that subtraction in $\mathbb{Z}$ is not an associative operation.

    (ii)   Give an example of a commutative ring R in which subtraction is associative.

3.5 Assume that S is a subset of a commutative ring R such that

    (i) $1 \in S$;

    (ii) if $a, b \in S$, then $a + b \in S$;

    (iii) if $a, b \in S$, then $ab \in S$.

(In contrast to the definition of subring, we are now assuming $a + b \in S$ instead of $a - b \in S$.) Give an example of a commutative ring R containing such a subset S which is not a subring of R.

3.6 Find the multiplicative inverses of the nonzero elements in $\mathbb{I}_{11}$.

3.7 H **(i)**   If X is a set, prove that the Boolean group $\mathcal{B}(X)$ [see Example 2.47(viii)] with elements the subsets of X and addition given by symmetric difference, $U + V = (U - V) \cup (V - U)$, is a commutative ring if one defines multiplication by $UV = U \cap V$. One calls $\mathcal{B}(X)$ a ***Boolean ring***.

    (ii)   Prove that $\mathcal{B}(X)$ contains exactly one unit.

    (iii) If Y is a proper subset of X, show that the one in $\mathcal{B}(Y)$ is distinct from the one in $\mathcal{B}(X)$. Conclude that $\mathcal{B}(Y)$ is *not* a subring of $\mathcal{B}(X)$.

    (iv) Prove that every element $U \in \mathcal{B}(X)$ satisfies $U^2 = U$.

3.8   **(i)**   If R is a domain and $a \in R$ satisfies $a^2 = a$, prove that either $a = 0$ or $a = 1$.

    (ii)   Show that the commutative ring $\mathcal{F}(\mathbb{R})$ in Example 3.11(i) contains elements $f \neq 0, 1$ with $f^2 = f$.

3.9 Find all the units in the commutative ring $\mathcal{F}(\mathbb{R})$ defined in Example 3.11(i).

*3.10 Generalize the construction of $\mathcal{F}(\mathbb{R})$ to a set X and an arbitrary commutative ring R: let $\mathcal{F}(X, R)$ be the set of all functions from X to R, with pointwise addition $f + g : x \mapsto f(x) + g(x)$ and pointwise multiplication $fg : x \mapsto f(x)g(x)$ for $x \in X$.

   (i)   Show that $\mathcal{F}(X, R)$ is a commutative ring.
   (ii) Show that if X has at least two elements, then $\mathcal{F}(X, R)$ is not a domain.
   (iii) If R is a commutative ring, denote $\mathcal{F}(R, R)$ by $\mathcal{F}(R)$:

$$\mathcal{F}(R) = \{\text{all functions } R \to R\}.$$

   Show that $\mathcal{F}(\mathbb{I}_2)$ has exactly four elements, and that $f + f = 0$ for every $f \in \mathcal{F}(\mathbb{I}_2)$.

*3.11 H **(i)** Prove that the commutative ring $\mathbb{C}$ is a domain.
   (ii) Prove that $\mathbb{Z}$, $\mathbb{Q}$, and $\mathbb{R}$ are domains.
   (iii) Prove that the ring of Gaussian integers is a domain.

*3.12 Prove that the intersection of any family of subrings of a commutative ring R is a subring of R.

H **3.13** Prove that the only subring of $\mathbb{Z}$ is $\mathbb{Z}$ itself.

H **3.14** Let a and m be relatively prime integers. Prove that if $sa + tm = 1 = s'a + t'm$, then $s \equiv s' \bmod m$. See Exercise 1.56 on page 54.

3.15 H **(i)** Is $R = \{a + b\sqrt{2} : a, b \in \mathbb{Z}\}$ a domain?
   H **(ii)** Is $R = \{\frac{1}{2}(a + b\sqrt{2}) : a, b \in \mathbb{Z}\}$ a domain?
   (iii) Using the fact that $\alpha = \frac{1}{2}(1 + \sqrt{-19})$ is a root of $x^2 - x + 5$, prove that $R = \{a + b\alpha : a, b \in \mathbb{Z}\}$ is a domain.

3.16 Prove that the set of all C^∞-functions is a subring of $\mathcal{F}(\mathbb{R})$. (See Exercise 1.42 on page 36.)

$\to$ 3.2 FIELDS

There is an obvious difference between $\mathbb{Q}$ and $\mathbb{Z}$: every nonzero element of $\mathbb{Q}$ is a unit.

$\to$ **Definition.**   A *field* [6] F is a commutative ring with $1 \neq 0$ in which every nonzero element a is a unit; that is, there is $a^{-1} \in F$ with $a^{-1}a = 1$.

The first examples of fields are $\mathbb{Q}$, $\mathbb{R}$, and $\mathbb{C}$.

The definition of field can be restated in terms of the group of units; a commutative ring R is a field if and only if $U(R)$ is the set $R^\times$ of nonzero elements in R. To say this another way, R is a field if and only if $R^\times$ is a multiplicative group.

---

[6]The derivation of the mathematical usage of the English term *field* (first used by E. H. Moore in 1893 in his article classifying the finite fields) as well as the German term *Körper* and the French term *corps* is probably similar to the derivation of the words *group* and *ring*: each word denotes a "realm," a "body" of things, or a "collection of things." The word *domain* abbreviates the usual English translation *integral domain* of the German word *Integretätsbereich*, a collection of things analogous to integers.

Proposition 3.18. *Every field F is a domain.*

Proof. Assume that $ab = ac$, where $a \neq 0$. Multiplying both sides by a^{-1} gives $a^{-1}ab = a^{-1}ac$, and so $b = c$. •

Of course, the converse of this proposition is false, for $\mathbb{Z}$ is a domain that is not a field.

→ **Proposition 3.19.** *The commutative ring $\mathbb{I}_m$ is a field if and only if m is prime.*

Proof. If m is prime, then Corollary 3.17 shows that $\mathbb{I}_m$ is a field.

Conversely, if m is composite, then Proposition 3.12 shows that $\mathbb{I}_m$ is not a domain. By Proposition 3.18, $\mathbb{I}_m$ is not a field. •

Notation. When p is a prime, we will usually denote the field $\mathbb{I}_p$ by

$$\mathbb{F}_p.$$

At the end of this chapter (see Theorem 3.120), we shall prove that there are finite fields other than $\mathbb{F}_p$ for p prime (a field with four elements is constructed in Exercise 3.19 on page 234).

When I was a graduate student, one of my fellow students was hired to tutor a mathematically gifted 10-year-old boy. To illustrate how gifted the boy was, the tutor described the session in which he introduced 2×2 matrices and matrix multiplication to the boy. The boy's eyes lit up when he was shown multiplication by the identity matrix, and he immediately went off in a corner by himself. In a few minutes, he told his tutor that a matrix $\left[\begin{smallmatrix} a & b \\ c & d \end{smallmatrix}\right]$ has a multiplicative inverse if and only if $ad - bc \neq 0$!

In another session, the boy was shown the definition of a field. He was quite content as the familiar examples of the rationals, reals, and complex numbers were displayed. But when he was shown a field with 2 elements, he became very agitated. After carefully checking that every axiom really does hold, he exploded in a rage. I tell this story to illustrate how truly surprising and unexpected are the finite fields.

In Chapter 2 we introduced $GL(2, \mathbb{R})$, the group of nonsingular matrices with entries in $\mathbb{R}$. Afterward, we observed that replacing $\mathbb{R}$ by $\mathbb{Q}$ or by $\mathbb{C}$ also gives a group. We now observe that $\mathbb{R}$ can be replaced by any field k: $GL(2, k)$ is a group for every field k. In particular, $GL(2, \mathbb{F}_p)$ is a finite group for every prime p.

It was shown in Exercise 3.11 on page 230 that every subring of a domain is itself a domain. Since fields are domains, it follows that every subring of a field is a domain. The converse of this exercise is true, and it is much more interesting: every domain is a subring of a field.

Given four elements a, b, c, and d in a field F with $b \neq 0$ and $d \neq 0$, assume that $ab^{-1} = cd^{-1}$. Multiply both sides by bd to obtain $ad = bc$. In other words, were ab^{-1} written as a/b, then we have just shown that $a/b = c/d$ implies $ad = bc$; that is, "cross multiplication" is valid. Conversely, if $ad = bc$ and both b and d are nonzero, then multiplication by $b^{-1}d^{-1}$ gives $ab^{-1} = cd^{-1}$, that is, $a/b = c/d$. We now generalize Example 2.17(iii), which shows that cross multiplication is an equivalence relation on $\{(a, b) \in \mathbb{Z} \times \mathbb{Z} : b \neq 0\}$.

Lemma 3.20.   *If R is a domain and $X = \{(a, b) \in R \times R : b \neq 0\}$, then the relation on X defined by cross multiplication, $(a, b) \equiv (c, d)$ if $ad = bc$, is an equivalence relation.*

Proof.   Verifications of reflexivity and of symmetry are easy. For transitivity, assume that $(a, b) \equiv (c, d)$ and $(c, d) \equiv (e, f)$. Now $ad = bc$ gives $adf = bcf$, and $cf = de$ gives $bcf = bde$; thus, $adf = bde$. Since R is a domain, we may cancel the nonzero d to get $af = be$; that is, $(a, b) \equiv (e, f)$.   •

The proof of the next theorem is a straightforward generalization of the standard construction of the field of rational numbers $\mathbb{Q}$ from the domain of integers $\mathbb{Z}$.

$\rightarrow$ **Theorem 3.21.**   *If R is a domain, then there is a field F containing R as a subring. Moreover, F can be chosen so that each $f \in F$ has a factorization $f = ab^{-1}$ with a, $b \in R$ and $b \neq 0$.*

Proof.   Cross multiplication, $(a, b) \equiv (c, d)$ if $ad = bc$, is an equivalence relation on $X = R \times R^\times$, by Lemma 3.20. Denote the equivalence class of $(a, b) \in X$ by $[a, b]$, and define F to be the set of all equivalence classes. Equip F with the following addition and multiplication (if we pretend that $[a, b]$ is the fraction a/b, then these are just the usual formulas):

$$[a, b] + [c, d] = [ad + bc, bd]$$

and

$$[a, b][c, d] = [ac, bd].$$

Notice that the symbols on the right make sense, for $b \neq 0$ and $d \neq 0$ imply $bd \neq 0$ because R is a domain. The proof that F is a field is now a series of routine steps.

Addition $F \times F \rightarrow F$ is well-defined: if $[a, b] = [a', b']$ and $[c, d] = [c', d']$, then $[ad + bc, bd] = [a'd' + b'c', b'd']$. We are told that $ab' = a'b$ and $cd' = c'd$. Hence,

$$(ad + bc)b'd' = adb'd' + bcb'd' = (ab')dd' + bb'(cd')$$
$$= a'bdd' + bb'c'd = (a'd' + b'c')bd;$$

that is, $(ad + bc, bd) \equiv (a'd' + b'c', b'd')$, as desired. A similar computation shows that multiplication $F \times F \rightarrow F$ is well-defined.

The verification that F is a commutative ring is also routine, and it is left to the reader, with the remark that the zero element is $[0, 1]$, the one is $[1, 1]$, and the negative of $[a, b]$ is $[-a, b]$. If we identify $a \in R$ with $[a, 1] \in F$, then it is easy to see that the family R' of all such elements is a subring of F:

$$[1, 1] \in R';$$
$$[a, 1] - [c, 1] = [a, 1] + [-c, 1] = [a - c, 1] \in R';$$
$$[a, 1][c, 1] = [ac, 1] \in R'.$$

To see that F is a field, observe first that if $[a, b] \neq 0$, then $a \neq 0$ (for the zero element of F is $[0, 1] = [0, b]$). The inverse of $[a, b]$ is $[b, a]$, for $[a, b][b, a] = [ab, ab] = [1, 1]$.

Finally, if $b \neq 0$, then $[1, b] = [b, 1]^{-1}$ (as we have just seen). Therefore, if $[a, b] \in F$, then $[a, b] = [a, 1][1, b] = [a, 1][b, 1]^{-1}$. This completes the proof, for $[a, 1]$ and $[b, 1]$ are in R'.   •

The statement of Theorem 3.21 is not quite accurate; the field F does not contain R as a subring, for R is not even a subset of F. Instead, we proved that F contains a subring, namely, $R' = \{[a, 1] : a \in R\}$, having the desired properties. Now R' strongly resembles R, and we will be able to identify R' with R once the notion of *isomorphism* is introduced (see Example 3.31).

→ **Definition.** The field F just constructed from a domain R in Theorem 3.21 is called the *fraction field* of R; we denote it by

$$\text{Frac}(R),$$

and we denote the element $[a, b] \in \text{Frac}(R)$ by a/b. In particular, the elements $[a, 1]$ of R' are denoted by $a/1$ or, more simply, by a.

Notice that the fraction field of $\mathbb{Z}$ is $\mathbb{Q}$; that is, $\text{Frac}(\mathbb{Z}) = \mathbb{Q}$. In the next section, we will see that if k is a field, then the set of all the polynomials $f(x)$ whose coefficients lie in k forms a domain, denoted by $k[x]$. The elements $f(x)/g(x)$ of $\text{Frac}(k[x])$ are usually called *rational functions*.

→ **Definition.** A *subfield* of a field K is a subring k of K which is also a field.

→ **Proposition 3.22.**

(i) *A subset k of a field K is a subfield if and only if it is a subring that is closed under inverses; that is, if $a \in k$ and $a \neq 0$, then $a^{-1} \in k$.*

(ii) *If $\{F_i : i \in I\}$ is any (possibly infinite) family of subfields of a field K, then $k = \bigcap_{i \in I} F_i$ is a subfield of K.*

Proof.

(i) If a subset k of a field K is a subfield, then it obviously contains inverses of its non-zero elements. Conversely, if k is a subring that contains inverses of nonzero elements, then it is a field, and hence it is a subfield of K.

(ii) We use part (i). Since any intersection of subrings is itself a subring, by Exercise 3.12 on page 230, k is a subring of K. If $a \in k$ is nonzero, then it got into k by being in every F_i. But since F_i is a subfield, $a^{-1} \in F_i$, and so $a^{-1} \in \bigcap_i F_i = k$. Therefore, k is a subfield of K.   ●

→ **Definition.**   If K is a field, the intersection k of all the subfields of K is called the *prime field* of k.

Of course, every field has a unique prime field. In Proposition 3.110, we will see that every prime field is essentially $\mathbb{Q}$ or $\mathbb{F}_p$ for some prime p.

EXERCISES

H **3.17** True or false with reasons.

    (i)   Every field is a domain.

    (ii)   There is a finite field with more than 10^{100} elements.

    (iii)   If R is a domain, then there is a unique field containing R.

    (iv)   Every commutative ring is a subring of some field.

    (v)   The subset $R = \mathbb{Q}[i] = \{a + bi : a, b \in \mathbb{Q}\}$ is a subfield of $\mathbb{C}$.

    (vi)   The prime field of $\mathbb{Q}[i] = \{a + bi : a, b \in \mathbb{Q}\}$ is $\mathbb{Q}$.

    (vii)   If $R = \mathbb{Q}[\sqrt{2}]$, then $\mathrm{Frac}(R) = \mathbb{R}$.

3.18   **(i)**   If R is a commutative ring, define the *circle operation* $a \circ b$ by

$$a \circ b = a + b - ab.$$

      Prove that the circle operation is associative and that $0 \circ a = a$ for all $a \in R$.

    (ii)   Prove that a commutative ring R is a field if and only if the set

$$R^{\#} = \{r \in R : r \neq 1\}$$

      is an abelian group under the circle operation.

*3.19 (R. A. Dean) Define $\mathbb{F}_4$ to be the set of all 2×2 matrices

$$\mathbb{F}_4 = \left\{ \begin{bmatrix} a & b \\ b & a+b \end{bmatrix} : a, b \in \mathbb{F}_2 \right\}.$$

    (i)   Prove that $\mathbb{F}_4$ is a commutative ring whose operations are matrix addition and matrix multiplication.

    (ii)   Prove that $\mathbb{F}_4$ is a field having exactly 4 elements.

    (iii)   Show that $\mathbb{I}_4$ is not a field.

H **3.20** Prove that every domain R with a finite number of elements must be a field. Using Proposition 3.12, this gives a new proof of sufficiency in Proposition 3.19.

*H **3.21** Find all the units in the ring $\mathbb{Z}[i]$ of Gaussian integers.

3.22 Show that $F = \{a + b\sqrt{2} : a, b \in \mathbb{Q}\}$ is a field.

*3.23 **(i)** Show that $F = \{a + bi : a, b \in \mathbb{Q}\}$ is a field.

(ii) Show that every $u \in F$ has a factorization $u = \alpha\beta^{-1}$, where $\alpha, \beta \in \mathbb{Z}[i]$. (See Exercise 3.50(ii) on page 252.)

*3.24 If R is a commutative ring, define a relation $\equiv$ on R by $a \equiv b$ if there is a unit $u \in R$ with $b = ua$.

(i) Prove that $\equiv$ is an equivalence relation.

(ii) If $a \equiv b$, prove that $(a) = (b)$, where $(a) = \{ra : r \in R\}$. Conversely, prove that if R is a domain, then $(a) = (b)$ implies $a \equiv b$.

3.25 If R is a domain, prove that there is no subfield K of $\mathrm{Frac}(R)$ such that

$$R \subseteq K \subsetneq \mathrm{Frac}(R).$$

*3.26 Let k be a field, and let R be the subring

$$R = \{n \cdot 1 : n \in \mathbb{Z}\} \subseteq k.$$

(i) If F is a subfield of k, prove that $R \subseteq F$.

(ii) Prove that a subfield F of k is the prime field of k if and only if it is the *smallest* subfield of k containing R; that is, there is no subfield F' with $R \subseteq F' \subsetneq F$.

(iii) If R is a subfield of k, prove that R is the prime field of k.

3.27 **(i)** Show that every subfield of $\mathbb{C}$ contains $\mathbb{Q}$.

(ii) Show that the prime field of $\mathbb{R}$ is $\mathbb{Q}$.

(iii) Show that the prime field of $\mathbb{C}$ is $\mathbb{Q}$.

*3.28 H **(i)** For any field F, prove that $\Sigma(2, F) \cong \mathrm{Aff}(1, F)$, where $\Sigma(2, F)$ denotes the stochastic group (defined in Exercise 2.48 on page 147).

(ii) If F is a finite field with q elements, prove that $|\Sigma(2, F)| = q(q - 1)$.

(iii) Prove that $\Sigma(2, \mathbb{F}_3) \cong S_3$.

$\rightarrow$ **3.3 POLYNOMIALS**

Even though the reader is familiar with polynomials, we now introduce them carefully. One modest consequence is that the mystery surrounding the "unknown" x will vanish.

Informally, a polynomial is an "expression" $s_0 + s_1 x + s_2 x^2 + \cdots + s_n x^n$. The key observation is that one should pay attention to where the coefficients $s_0, s_1, s_2, \ldots, s_n$ of polynomials live.

Definition. If R is a commutative ring, then a *formal power series over* R is a function $\sigma : \mathbb{N} \to R$. As any function, a formal power series $\sigma : \mathbb{N} \to R$ is determined by its values; for each $i \in \mathbb{N}$, write $\sigma(i) = s_i \in R$, so that

$$\sigma = (s_0, s_1, s_2, \ldots, s_i, \ldots).$$

The values $s_i \in R$ are called the **coefficients**[7] of the formal[8] power series.

By Proposition 2.2, two formal power series σ and τ over R are equal if and only if their coefficients match; that is, $\sigma(i) = \tau(i)$ for all $i \geq 0$.

$\to$ **Definition.** A formal power series $\sigma = (s_0, s_1, \ldots, s_i, \ldots)$ over a commutative ring R is called a **polynomial over** R if there is some integer $n \geq 0$ with $s_i = 0$ for all $i > n$; that is,

$$\sigma = (s_0, s_1, \ldots, s_n, 0, 0, \ldots).$$

Thus, a polynomial has only finitely many nonzero coefficients, while a formal power series may have infinitely many nonzero coefficients.

$\to$ **Definition.** The **zero polynomial** 0 is the polynomial $(0, 0, 0, \ldots)$ having all coefficients 0. If $\sigma = (s_0, s_1, s_2, \ldots, s_n, 0, 0, \ldots)$ is not the zero polynomial, then there is a natural number n with $s_n \neq 0$ and $s_i = 0$ for all $i > n$. One calls s_n the **leading coefficient** of σ, one calls n the **degree**[9] of σ, and one denotes the degree of σ by $\deg(\sigma)$.

The zero polynomial 0 does not have a degree because it has no nonzero coefficients; every other polynomial does have a degree.

Notation. If R is a commutative ring, then the set of all polynomials with coefficients in R is denoted by $R[x]$.

Equip $R[x]$ with the following operations. Define

$$\sigma + \tau = (s_0 + t_0, s_1 + t_1, \ldots, s_i + t_i, \ldots)$$

and

$$\sigma\tau = (a_0, a_1, \ldots, a_k, \ldots),$$

---

[7] The term *coefficient* means "acting together to some single end." Here, coefficients collectively give one formal power series.

[8] One usually denotes a formal power series $\sigma = (s_0, s_1, s_2, \ldots, s_i \ldots)$ by $s_0 + s_1 x + s_2 x^2 + \cdots = \sum_{i=0}^{\infty} s_i x^i$. The adjective *formal* is used because there is no notion of convergence here. Indeed, even if $R = \mathbb{R}$, so that convergence does make sense, the set of all convergent power series is a proper subset of the set of all formal power series over $\mathbb{R}$.

[9] The word *degree* comes from the Latin word meaning "step."

where $a_k = \sum_{i+j=k} s_i t_j = \sum_{i=0}^{k} s_i t_{k-i}$; thus,

$$\sigma \tau = (s_0 t_0, \, s_0 t_1 + s_1 t_0, \, s_0 t_2 + s_1 t_1 + s_2 t_0, \dots).$$

The next proposition shows where the formula for multiplication comes from.

Proposition 3.23. *If R is a commutative ring and r, s_i, $t_j \in R$ for $i \geq 0$ and $j \geq 0$, then*

$$(s_0 + s_1 r + \cdots)(t_0 + t_1 r + \cdots) = a_0 + a_1 r + \cdots + a_k r^k + \cdots,$$

where $a_k = \sum_{i+j=k} s_i t_j$ for all $k \geq 0$.

Remark. This proof should be an induction on $k \geq 0$, but we give an informal argument because it is clearer. ◀

Proof. Write $\sum_i s_i r^i = f(r)$ and $\sum_j t_j r^j = g(r)$. Then

$$
\begin{aligned}
f(r)g(r) &= (s_0 + s_1 r + s_2 r^2 + \cdots)g(r) \\
&= s_0 g(r) + s_1 r g(r) + s_2 r^2 g(r) + \cdots \\
&= s_0(t_0 + t_1 r + \cdots) + s_1 r(t_0 + t_1 r + \cdots) \\
&\quad + s_2 r^2(t_0 + t_1 r + \cdots) + \cdots \\
&= s_0 t_0 + (s_1 t_0 + s_0 t_1)r + (s_2 t_0 + s_1 t_1 + s_0 t_2)r^2 + \\
&\quad (s_0 t_3 + s_1 t_2 + s_2 t_1 + s_3 t_0)r^3 + \cdots. \quad \bullet
\end{aligned}
$$

→ **Lemma 3.24.** *Let R be a commutative ring and let σ, $\tau \in R[x]$ be nonzero polynomials.*

(i) *Either $\sigma\tau = 0$ or $\deg(\sigma\tau) \leq \deg(\sigma) + \deg(\tau)$.*

(ii) *If R is a domain, then $\sigma\tau \neq 0$ and*

$$\deg(\sigma\tau) = \deg(\sigma) + \deg(\tau).$$

Proof.
(i) Let $\sigma = (s_0, s_1, \dots)$ have degree m, let $\tau = (t_0, t_1, \dots)$ have degree n, and let $\sigma\tau = (a_0, a_1, \dots)$. It suffices to prove that $a_k = 0$ for all $k > m + n$. By definition,

$$a_k = \sum_{i+j=k} s_i t_j.$$

If $i \leq m$, then $j = k - i \geq k - m > n$ (because $k > m + n$), and so $t_j = 0$ (because τ has degree n); if $i > m$, then $s_i = 0$ because σ has degree m. In either case, each term $s_i t_j = 0$, and so $a_k = \sum_{i+j=k} s_i t_j = 0$.

(ii) If $k = m + n$, then the same calculation as in part (i) shows, with the possible exception of $s_m t_n$ (the product of the leading coefficients of σ and τ), that each term $s_i t_j$ in a_{m+n} is 0:

$$s_0 t_{m+n} = \cdots = s_{n-1} t_{m+1} = 0 = s_{n+1} t_{m-1} = \cdots = s_{m+n} t_0.$$

If $i < m$, then $m - i > 0$, hence $j = m - i + n > n$, and so $t_j = 0$; if $i > m$, then $s_i = 0$. Hence

$$a_{m+n} = s_m t_n.$$

Since R is a domain, $s_m \neq 0$ and $t_n \neq 0$ imply $s_m t_n \neq 0$; hence, $\sigma \tau \neq 0$ and $\deg(\sigma \tau) = m + n = \deg(\sigma) + \deg(\tau)$.   •

→ **Proposition 3.25.**

(i) *If R is a commutative ring, then $R[x]$ is a commutative ring that contains R as a subring.*

(ii) *If R is a domain, then $R[x]$ is a domain.*

Proof.
(i) Addition and multiplication are operations on $R[x]$: the sum of two polynomials σ and τ is a polynomial [indeed, either $\sigma + \tau = 0$ or $\deg(\sigma + \tau) \leq \max\{\deg(\sigma), \deg(\tau)\}$], while Lemma 3.24 shows that the product of two polynomials is a polynomial as well. Verifications of the axioms for a commutative ring are again routine, and they are left to the reader. Note that the zero is the zero polynomial, the one is the polynomial $(1, 0, 0, \dots)$, and the negative of $(s_0, s_1, \dots, s_i, \dots)$ is $(-s_0, -s_1, \dots, -s_i, \dots)$. The only possible problem is proving associativity of multiplication; we give the hint that if $\rho = (r_0, r_1, \dots, r_i, \dots)$, then the ℓth coordinate of the polynomial $\rho(\sigma \tau)$ turns out to be $\sum_{i+j+k=\ell} r_i(s_j t_k)$, while the ℓth coordinate of the polynomial $(\rho \sigma) \tau$ turns out to be $\sum_{i+j+k=\ell} (r_i s_j) t_k$; these are equal because of associativity of the multiplication in R.

It is easy to check that $R' = \{(r, 0, 0, \dots) : r \in R\}$ is a subring of $R[x]$, and we identify R' with R by identifying $r \in R$ with $(r, 0, 0, \dots)$.

(ii) If R is a domain and if σ and τ are nonzero polynomials, then Lemma 3.24 shows that $\sigma \tau \neq 0$. Therefore, $R[x]$ is a domain.   •

→ **Definition.**   If R is a commutative ring, then $R[x]$ is called the ***ring of polynomials over R***.

Just as our assertion (in Theorem 3.21) that a domain is a subring of its fraction field was not quite true, so, too, our assertion here that a commutative ring R is a subring of $R[x]$ is not quite correct. We have proved that $R[x]$ does contain a subring, namely, $R' = \{(r, 0, 0, \dots) : r \in R\}$, which strongly resembles R, and we will identify R' with R once the notion of *isomorphism* is introduced (see Example 3.31).

→ **Definition.**   Define the *indeterminate* to be the element

$$x = (0, 1, 0, 0, \dots) \in R[x].$$

Even though x is neither "the unknown" nor a variable, we call it the *indeterminate* to recall one's first encounter with it in high school (see the discussion of polynomial functions on page 240). The indeterminate x is a specific element in the ring $R[x]$, namely, the polynomial $(s_0, s_1, s_2, \dots)$ with $s_1 = 1$ and all other $s_i = 0$. One reason we insist that a commutative ring have a unit is to enable us to make this definition; if the set E of even integers were a commutative ring, then $E[x]$ would not contain x (it would contain $2x$, however). Note that if R is the zero ring, then $R[x]$ is also the zero ring.

Lemma 3.26.

   (i) *If $\sigma = (s_0, s_1, \dots, s_j, \dots)$, then*

$$x\sigma = (0, s_0, s_1, \dots, s_j, \dots);$$

   that is, multiplying by x shifts each coefficient one step to the right.

   (ii) *If $n \geq 1$, then x^n is the polynomial having 0 everywhere except for 1 in the nth coordinate.*

   (iii) *If $r \in R$, then*

$$(r, 0, 0, \dots)(s_0, s_1, \dots, s_j, \dots) = (rs_0, rs_1, \dots, rs_j, \dots).$$

Proof.
(i) Write $x = (t_0, t_1, \dots, t_i, \dots)$, where $t_1 = 1$ and all other $t_i = 0$, and let $x\sigma = (a_0, a_1, \dots, a_k, \dots)$. Now $a_0 = t_0 s_0 = 0$ because $t_0 = 0$. If $k \geq 1$, then the only nonzero term in the sum $a_k = \sum_{i+j=k} s_i t_j$ is $s_{k-1}t_1 = s_{k-1}$, because $t_1 = 1$ and $t_i = 0$ for $i \neq 1$; thus, for $k \geq 1$, the kth coordinate a_k of $x\sigma$ is s_{k-1}, and $x\sigma = (0, s_0, s_1, \dots, s_i, \dots)$.
(ii) An easy induction, using part (i).
(iii) This follows easily from the definition of multiplication.   •

If we identify $(r, 0, 0, \dots)$ with r, then Lemma 3.26(iii) reads

$$r(s_0, s_1, \dots, s_i, \dots) = (rs_0, rs_1, \dots, rs_i, \dots).$$

We can now recapture the usual notation.

→ **Proposition 3.27.**   *If $\sigma = (s_0, s_1, \dots, s_n, 0, 0, \dots)$, then*

$$\sigma = s_0 + s_1 x + s_2 x^2 + \cdots + s_n x^n,$$

where each element $s \in R$ is identified with the polynomial $(s, 0, 0, \dots)$.

Proof.

$$\sigma = (s_0, s_1, \ldots, s_n, 0, 0, \ldots)$$
$$= (s_0, 0, 0, \ldots) + (0, s_1, 0, \ldots) + \cdots + (0, 0, \ldots, s_n, 0, \ldots)$$
$$= s_0(1, 0, 0, \ldots) + s_1(0, 1, 0, \ldots) + \cdots + s_n(0, 0, \ldots, 1, 0, \ldots)$$
$$= s_0 + s_1 x + s_2 x^2 + \cdots + s_n x^n \quad \bullet$$

We shall use this familiar (and standard) notation from now on. As is customary, we shall write

$$f(x) = s_0 + s_1 x + s_2 x^2 + \cdots + s_n x^n$$

instead of $\sigma = (s_0, s_1, \ldots, s_n, 0, 0, \ldots)$.

Here is some standard vocabulary associated with polynomials. If $f(x) = s_0 + s_1 x + s_2 x^2 + \cdots + s_n x^n$, where $s_n \neq 0$, then s_0 is called its **constant term** and, as we have already said, s_n is called its *leading coefficient*. If its leading coefficient $s_n = 1$, then $f(x)$ is called **monic**. Every polynomial other than the zero polynomial 0 (having all coefficients 0) has a degree. A **constant polynomial** is either the zero polynomial or a polynomial of degree 0. Polynomials of degree 1, namely, $a + bx$ with $b \neq 0$, are called **linear**, polynomials of degree 2 are **quadratic**,[10] degree 3's are **cubic**, then **quartic**, **quintic**, and so on.

→ **Corollary 3.28.**   *Polynomials $f(x) = s_0 + s_1 x + s_2 x^2 + \cdots + s_n x^n$ and $g(x) = t_0 + t_1 x + t_2 x^2 + \cdots + t_m x^m$ are equal if and only if $s_i = t_i$ for all $i \in \mathbb{N}$.*

Proof.   We have merely restated the definition of equality of polynomials in terms of the familiar notation. $\bullet$

We can now describe the usual role of the indeterminate x as a variable. If R is a commutative ring, each polynomial $f(x) = s_0 + s_1 x + s_2 x^2 + \cdots + s_n x^n \in R[x]$ defines a **polynomial function** $f^\flat \colon R \to R$ by evaluation: if $r \in R$, define $f^\flat(r) = s_0 + s_1 r + s_2 r^2 + \cdots + s_n r^n \in R$ [usually, one is not so fussy, and one writes $f(r)$ instead of $f^\flat(r)$]. The reader should realize that polynomials and polynomial functions are distinct objects. For example, if R is a finite ring (e.g., $\mathbb{I}_m$), then there are only finitely many functions from R to itself; a fortiori, there are only finitely many polynomial functions. On the other hand, if R is not the zero ring, there are infinitely many polynomials. For example, all the powers $1, x, x^2, \ldots, x^n, \ldots$ are distinct, by Corollary 3.28.

---

[10]Quadratic polynomials are so called because the particular quadratic x^2 gives the area of a square (*quadratic* comes from the Latin word meaning "four," which is to remind one of the 4-sided figure); similarly, *cubic* polynomials are so called because x^3 gives the volume of a cube. *Linear* polynomials are so called because the graph of a linear polynomial in $\mathbb{R}[x]$ is a line.

→ **Definition.**   If k is a field, then the fraction field Frac($k[x]$) of $k[x]$, denoted by $k(x)$, is called the ***function field over*** k. The elements of $k(x)$ are called ***rational functions over*** k.

Proposition 3.29.   *The elements of the function field* $k(x)$ *have the form* $f(x)/g(x)$, *where* $f(x), g(x) \in k[x]$ *and* $g(x) \neq 0$.

Proof.   Since the function field is a fraction field, Theorem 3.21 shows that elements in $k(x)$ have the form $f(x)g(x)^{-1}$.   •

Proposition 3.30.   *If* p *is a prime, then the function field* $\mathbb{F}_p(x)$ *is an infinite field whose prime field is* $\mathbb{F}_p$.

Proof.   By Proposition 3.25, $\mathbb{F}_p[x]$ is a domain. Its fraction field $\mathbb{F}_p(x)$ is a field containing $\mathbb{F}_p[x]$ as a subring, while $\mathbb{F}_p[x]$ contains $\mathbb{F}_p$ as a subring, by Proposition 3.25. That $\mathbb{F}_p$ is the prime field follows from Exercise 3.26 on page 235.   •

In spite of the difference between polynomials and polynomial functions (we shall see, in Corollary 3.52, that these objects coincide when the coefficient ring R is an infinite field), one often calls $R[x]$ the ring of all *polynomials over R in one variable* (or *polynomials over R in one indeterminate*). If we write $A = R[x]$, then the polynomial ring $A[y]$ is called the ring of all *polynomials over R in two variables x and y* (or *indeterminates*), and it is denoted by $R[x, y]$. For example, the quadratic polynomial $ax^2 + bxy + cy^2 + dx + ey + f$ can be written $cy^2 + (bx + e)y + (ax^2 + dx + f)$, a polynomial in y with coefficients in $R[x]$. By induction, one can form the commutative ring $R[x_1, x_2, \dots, x_n]$ of all ***polynomials in n variables*** (or *indeterminates*) with coefficients in R. Proposition 3.25 can ~~~~~~~ ~~~~~~~~ alized, by induction on n, to say that if R is a domain, ~~~~~~~~~~~~~~~~~~~~]. Moreover, when F is a field, we can describe Frac(~~~~~~~~~~~~~~~~~~~~~~~~) by $F(x_1, x_2, \dots, x_n)$, as all rational functions in n varia~~~~~~~~~~~~~~~~~~~~~~~~~ elements have the form $f(x_1, x_2, \dots, x_n)/g(x_1, x_2, \dots,$ ~~~~~~~~~~ $x_1, x_2, \dots, x_n]$.

EXERCISES

H **3.29** True or false with reasons.

    (i)   The sequence notation for $x^3 - 2x + 5$ is $(5, -2, 0, 1, 0, \cdots)$.

    (ii)   If R is a domain, then $R[x]$ is a domain.

    (iii) $\mathbb{Q}[x]$ is a field.

    (iv)   If k is a field, then the prime field of $k[x]$ is k.

    (v)   If R is a domain and $f(x), g(x) \in R[x]$ are nonzero, then $\deg(fg) = \deg(f) + \deg(g)$.

(vi) If R is a domain and $f(x)$, $g(x) \in R[x]$ are nonzero, then either $f(x) + g(x) = 0$ or $\deg(f + g) \leq \max\{\deg(f), \deg(g)\}$.

(vii) If k is a field, then $k[x] = k(x)$.

H **3.30** Show that if R is a nonzero commutative ring, then $R[x]$ is never a field.

*$**3.31**$ Let k be a field and let A be an $n \times n$ matrix with entries in k (so that the powers A^i are defined). If $f(x) = c_0 + c_1 x + \cdots + c_m x^m \in k[x]$, define

$$f(A) = c_0 I + c_1 A + \cdots + c_m A^m.$$

(i) Prove that $k[A]$, defined by $k[A] = \{f(A) : f(x) \in k[x]\}$, is a commutative ring under matrix addition and matrix multiplication.

(ii) If $f(x) = p(x)q(x) \in k[x]$ and if A is an $n \times n$ matrix over k, prove that $f(A) = p(A)q(A)$.

(iii) Give examples of $n \times n$ matrices A and B such that $k[A]$ is a domain and $k[B]$ is not a domain.

*$**3.32**$ H (i) Let R be a domain. Prove that a polynomial $f(x)$ is a unit in $R[x]$ if and only if $f(x)$ is a nonzero constant which is a unit in R.

(ii) Show that $([2]x + [1])^2 = [1]$ in $\mathbb{I}_4[x]$. Conclude that the statement in part (i) may be false for commutative rings that are not domains. [An element $z \in R$ is called **nilpotent** if $z^m = 0$ for some integer $m \geq 1$. For any commutative ring R, it can be proved that a polynomial $f(x) = a_0 + a_1 x + \cdots + a_n x^n \in R[x]$ is a unit in $R[x]$ if and only if a_0 is a unit in R and a_i is nilpotent for all $i \geq 1$.]

*H **3.33** Show that if $f(x) = x^p - x \in \mathbb{F}_p[x]$, then its polynomial function $f^\flat : \mathbb{F}_p \to \mathbb{F}_p$ is identically zero.

3.34 H (i) If p is a prime and $m, n \in \mathbb{N}$, prove that $\binom{pm}{pn} \equiv \binom{m}{n}$ mod p.

(ii) Prove that $\binom{p^r m}{p^r n} \equiv \binom{m}{n}$ mod p for all $r \geq 0$.

(iii) Give another proof of Exercise 1.72: if p is a prime not dividing an integer $m \geq 1$, then $p \nmid \binom{p^r m}{p^r}$.

*$**3.35**$ Let $\alpha \in \mathbb{C}$, and let $\mathbb{Z}[\alpha]$ be the smallest subring of $\mathbb{C}$ containing α; that is, $\mathbb{Z}[\alpha] = \bigcap S$, where S ranges over all those subrings of $\mathbb{C}$ containing α. Prove that

$$\mathbb{Z}[\alpha] = \{f(\alpha) : f(x) \in \mathbb{Z}[x]\}.$$

H **3.36** If R is a commutative ring and $f(x) = \sum_{i=0}^{n} a_i x^i \in R[x]$ has degree $n \geq 1$, define its **derivative** $f'(x) \in R[x]$ by

$$f'(x) = a_1 + 2a_2 x + 3a_3 x^2 + \cdots + n a_n x^{n-1};$$

if $f(x)$ is a constant polynomial, define its derivative to be the zero polynomial. Prove that the usual rules of calculus hold for this definition of derivative; that is,

$$(f + g)' = f' + g';$$
$$(rf)' = rf' \quad \text{if } r \in R;$$
$$(fg)' = fg' + f'g;$$
$$(f^n)' = n f^{n-1} f' \quad \text{for all } n \geq 1.$$

*3.37 Assume that $(x - a) \mid f(x)$ in $R[x]$. Prove that $(x - a)^2 \mid f(x)$ if and only if $(x - a) \mid f'$ in $R[x]$.

3.38   **(i)**   If $f(x) = ax^{2p} + bx^p + c \in \mathbb{F}_p[x]$, prove that $f'(x) = 0$.

   H **(ii)**   State and prove a necessary and sufficient condition that a polynomial $f(x) \in \mathbb{F}_p[x]$ have $f'(x) = 0$.

*3.39 If R is a commutative ring, define $R[[x]]$ to be the set of all formal power series over R.

   H **(i)**   Show that the formulas defining addition and multiplication on $R[x]$ make sense for $R[[x]]$, and prove that $R[[x]]$ is a commutative ring under these operations.

   (ii)   Prove that $R[x]$ is a subring of $R[[x]]$.

   (iii)   Denote a formal power series $\sigma = (s_0, s_1, s_2, \ldots, s_n, \ldots)$ by

$$\sigma = s_0 + s_1 x + s_2 x^2 + \cdots.$$

Prove that if $\sigma = 1 + x + x^2 + \cdots$, then $\sigma = 1/(1 - x)$ is in $R[[x]]$.

*3.40 If $\sigma = (s_0, s_1, s_2, \ldots, s_n, \ldots)$ is a nonzero formal power series, define $\mathrm{ord}(\sigma) = m$, where m is the smallest natural number for which $s_m \neq 0$. Note that $\sigma \neq 0$ if and only if it has an order.

   H **(i)**   Prove that if R is a domain, then $R[[x]]$ is a domain.

   (ii)   Prove that if k is a field, then a nonzero formal power series $\sigma \in k[[x]]$ is a unit if and only if $\mathrm{ord}(\sigma) = 0$; that is, if its constant term is nonzero.

   (iii)   Prove that if $\sigma \in k[[x]]$ and $\mathrm{ord}(\sigma) = n$, then

$$\sigma = x^n u,$$

where u is a unit in $k[[x]]$.

→ **3.4 HOMOMORPHISMS**

Just as one can use homomorphisms to compare groups, so too can one use homomorphisms to compare commutative rings.

→ **Definition.**   If A and R are commutative rings, a (*ring*) *homomorphism* is a function $f : A \to R$ such that

(i) $f(1) = 1$;

(ii) $f(a + a') = f(a) + f(a')$ for all $a, a' \in A$;

(iii) $f(aa') = f(a)f(a')$ for all $a, a' \in A$.

A homomorphism that is also a bijection is called an *isomorphism*. Commutative rings A and R are called *isomorphic*, denoted by $A \cong R$, if there is an isomorphism $f : A \to R$.

$\rightarrow$ **Example 3.31.**

(i) Let R be a domain and let $F = \text{Frac}(R)$ denote its fraction field. In Theorem 3.21 we said that R is a subring of F, but that is not the truth; R is not even a subset of F. We did find a subring R' of F, however, that has a very strong resemblance to R, namely, $R' = \{[a, 1] : a \in R\} \subseteq F$. The function $f : R \rightarrow R'$, given by $f(a) = [a, 1]$, is easily seen to be an isomorphism. Henceforth, we will regard a domain R as a subring of $\text{Frac}(R)$, assuming that this identification has been made.

(ii) We implied that a commutative ring R is a subring of $R[x]$ when we "identified" elements $r \in R$ with constant polynomials $(r, 0, 0, \dots)$ [see Lemma 3.26(iii)]. The subset $R' = \{(r, 0, 0, \dots) : r \in R\}$ is a subring of $R[x]$, and it is easy to see that the function $f : R \rightarrow R'$, defined by $f(r) = (r, 0, 0, \dots)$, is an isomorphism. Henceforth, we will regard R as a subring of $R[x]$, assuming that this identification has been made. $\blacktriangleleft$

Example 3.32.

(i) Complex conjugation, defined by $z = a + ib \mapsto \bar{z} = a - ib$, is a homomorphism $\mathbb{C} \rightarrow \mathbb{C}$ because $\bar{1} = 1, \overline{z + w} = \bar{z} + \bar{w}$, and $\overline{zw} = \bar{z}\,\bar{w}$. The formula $\bar{\bar{z}} = z$ shows that conjugation is an isomorphism, for it is its own inverse.

(ii) Here are two examples of a homomorphism of rings that are not isomorphisms. If R is a commutative ring, then the (inclusion) homomorphism $R \rightarrow R[x]$ is not surjective (it is injective). If $m \geq 2$, then the map $f : \mathbb{Z} \rightarrow \mathbb{I}_m$, given by $f(n) = [n]$, is not injective (it is surjective).

(iii) The preceding example can be generalized. If R is a commutative ring whose unit element is denoted by 1, then the function $\chi : \mathbb{Z} \rightarrow R$, defined by $\chi(n) = n \cdot 1$, is a ring homomorphism. $\blacktriangleleft$

$\rightarrow$ **Theorem 3.33.** *Let R and S be commutative rings, and let $\varphi : R \rightarrow S$ be a homomorphism. If $s_1, \dots, s_n \in S$, then there exists a unique homomorphism*

$$\widetilde{\varphi} : R[x_1, \dots, x_n] \rightarrow S$$

with $\widetilde{\varphi}(x_i) = s_i$ for all i and $\widetilde{\varphi}(r) = \varphi(r)$ for all $r \in R$.

Proof. The proof is by induction on $n \geq 1$. If $n = 1$, denote x_1 by x and s_1 by s. Define $\widetilde{\varphi} : R[x] \rightarrow S$ as follows: if $f(x) = \sum_i r_i x^i$, then

$$\widetilde{\varphi} : r_0 + r_1 x + \cdots + r_n x^n \mapsto \varphi(r_0) + \varphi(r_1)s + \cdots + \varphi(r_n)s^n = \widetilde{\varphi}(f).$$

This formula shows that $\widetilde{\varphi}(x) = s$ and $\widetilde{\varphi}(r) = \varphi(r)$ for all $r \in R$.

It remains to prove that $\widetilde{\varphi}$ is a homomorphism. First, $\widetilde{\varphi}(1) = \varphi(1) = 1$, because φ is a homomorphism. Second, if $g(x) = a_0 + a_1 x + \cdots + a_m x^m$, then

$$\widetilde{\varphi}(f + g) = \widetilde{\varphi}\left(\sum_i (r_i + a_i)x^i\right)$$

$$= \sum_i \varphi(r_i + a_i)s^i$$

$$= \sum_i (\varphi(r_i) + \varphi(a_i))s^i$$

$$= \sum_i \varphi(r_i)s^i + \sum_i \varphi(a_i)s^i$$

$$= \widetilde{\varphi}(f) + \widetilde{\varphi}(g).$$

Third, let $f(x)g(x) = \sum_k c_k x^k$, where $c_k = \sum_{i+j=k} r_i a_j$. Then

$$\widetilde{\varphi}(fg) = \widetilde{\varphi}\left(\sum_k c_k x^k\right)$$

$$= \sum_k \varphi(c_k)s^k$$

$$= \sum_k \varphi\left(\sum_{i+j=k} r_i a_j\right)s^k$$

$$= \sum_k \left(\sum_{i+j=k} \varphi(r_i)\varphi(a_j)\right)s^k.$$

On the other hand,

$$\widetilde{\varphi}(f)\widetilde{\varphi}(g) = \left(\sum_i \varphi(r_i)s^i\right)\left(\sum_j \varphi(a_j)s^j\right)$$

$$= \sum_k \left(\sum_{i+j=k} \varphi(r_i)\varphi(a_j)\right)s^k.$$

We let the reader show uniqueness of $\widetilde{\varphi}$ by proving, by induction on $d \geq 0$, that when $\theta\colon R[x] \to S$ is a homomorphism with $\theta(x) = s$ and $\theta(r) = \varphi(r)$ for all $r \in R$, then

$$\theta(r_0 + r_x + \cdots + r_d x^d) = \varphi(r_0) + \varphi(r_1)s + \cdots + \varphi(r_d)s^d.$$

For the inductive step, we need a homomorphism $\widetilde{\varphi}\colon R[x_1, \ldots, x_{n+1}] \to S$ with $\widetilde{\varphi}(x_i) = s_i$ for all $i \leq n + 1$ and $\widetilde{\varphi}(r) = \varphi(r)$ for all $r \in R$. If we define $A = R[x_1, \ldots, x_n]$, then the inductive hypothesis gives a homomorphism $\psi\colon A \to S$ with $\psi(x_i) = s_i$ for all $i \leq n$ and $\psi(r) = \varphi(r)$ for all $r \in R$. The base step gives a homomorphism $\widetilde{\psi}\colon A[x_{n+1}] \to S$ with $\widetilde{\psi}(x_{n+1}) = s_{n+1}$ and $\widetilde{\psi}(a) = \psi(a)$ for all

$a \in A$. The result now follows from the fact that $R[x_1, \ldots, x_{n+1}] = A[x_{n+1}]$, $\widetilde{\psi}(x_i) = \psi(x_i) = s_i$ for all $i \leq n$, $\widetilde{\psi}(x_{n+1}) = \psi(x_{n+1}) = s_{n+1}$, and $\widetilde{\psi}(r) = \psi(r) = \varphi(r)$ for all $r \in R$.   •

→ **Definition.**   If R is a commutative ring and $a \in R$, then *evaluation at* a is the function $e_a \colon R[x] \to R$, defined by $e_a(f(x)) = f(a)$; that is, $e_a(\sum_i r_i x^i) = \sum_i r_i a^i$.

→ **Corollary 3.34.**   *If R is a commutative ring and $a \in R$, then the evaluation map $e_a \colon R[x] \to R$ is a homomorphism.*

Proof.   If we set $R = S$ and $\varphi = 1_R$ in Theorem 3.33, then $\widetilde{\varphi} = e_a$.   •

Certain properties of a ring homomorphism $f \colon A \to R$ follow from its being a homomorphism between the additive groups A and R. For example: $f(0) = 0$, $f(-a) = -f(a)$, and $f(na) = nf(a)$ for all $n \in \mathbb{Z}$. For readers not familiar with groups, we prove these statements. Since $0 + 0 = 0$, we have $f(0) = f(0) + f(0)$, so that subtracting $f(0)$ from each side gives $0 = f(0)$. Since $-a + a = 0$, we have $f(-a) + f(a) = f(0) = 0$; subtracting $f(a)$ from both sides gives $f(-a) = -f(a)$. The statement $f(na) = nf(a)$ for all $n \geq 0$ and all $a \in R$ is proved by induction for all $n \geq 0$; finally, if $n < 0$, then the result follows by replacing a by $-a$. That a homomorphism preserves multiplication has similar consequences.

Lemma 3.35.   *If $f \colon A \to R$ is a ring homomorphism, then, for all $a \in A$,*

(i) $f(a^n) = f(a)^n$ *for all $n \geq 0$;*

(ii) *if a is a unit, then $f(a)$ is a unit and $f(a^{-1}) = f(a)^{-1}$;*

(iii) *if a is a unit, then $f(a^{-n}) = f(a)^{-n}$ for all $n \geq 1$.*

Proof.
(i) If $n = 0$, then $f(a^0) = 1 = (f(a))^0$; this follows from our convention that $r^0 = 1$ for any ring element r, together with the property $f(1) = 1$ satisfied by every ring homomorphism. The statement for positive n is proved by induction on $n \geq 1$.
(ii) Applying f to the equation $a^{-1}a = 1$ shows that $f(a)$ is a unit with inverse $f(a^{-1})$.
(iii) Recall that $a^{-n} = (a^{-1})^n$, and invoke parts (i) and (ii).   •

Corollary 3.36.   *If $f \colon A \to R$ is a ring homomorphism, then*

$$f(U(A)) \subseteq U(R),$$

where $U(A)$ is the group of units of A; if f is an isomorphism, then there is a group isomorphism

$$U(A) \cong U(R).$$

Proof. The first statement is just a rephrasing of part (ii) of Lemma 3.35: if a is a unit in A, then $f(a)$ is a unit in R.

If f is an isomorphism, then its inverse $f^{-1}: R \to A$ is also a ring homomorphism, by Exercise 3.44(i) on page 251; hence, if r is a unit in R, then $f^{-1}(r)$ is a unit in A. It is now easy to check that $\varphi: U(A) \to U(R)$, defined by $a \mapsto f(a)$, is a (group) isomorphism, for its inverse $\psi: U(R) \to U(A)$ is given by $r \mapsto f^{-1}(r)$. $\bullet$

Example 3.37.

If X is a nonempty set, define a ***bitstring*** on X to be a function $\beta: X \to \mathbb{F}_2$, and denote the set of all bitstrings on X by $b(X)$. When X is finite, say, $X = \{x_1, \dots, x_n\}$, then a bitstring is just a sequence of 0s and 1s of length n.

Define binary operations on $b(X)$: if $\beta, \gamma \in b(X)$, define

$$\beta\gamma: x \mapsto \beta(x)\gamma(x)$$

and

$$\beta + \gamma: x \mapsto \beta(x) + \gamma(x).$$

That $b(X)$ is a commutative ring under these operations is the special case of Exercise 3.10 on page 230 with $R = \mathbb{F}_2$.

Recall the Boolean ring $\mathcal{B}(X)$ [see Exercise 3.7(i) on page 229] whose elements are the subsets of X, with multiplication defined as their intersection: $AB = A \cap B$, and with addition defined as their symmetric difference: $A + B = (A - B) \cup (B - A)$. We now show that $\mathcal{B}(X) \cong b(X)$.

If A is a subset of a set X, define its ***characteristic function*** $\chi_A: X \to \mathbb{F}_2$ by

$$\chi_A(x) = \begin{cases} 1 & \text{if } x \in A \\ 0 & \text{if } x \notin A. \end{cases}$$

For example, $\chi_\varnothing$ is the constant function $\chi_\varnothing(x) = 0$ for all $x \in X$, while χ_X is the constant function $\chi_X(x) = 1$ for all $x \in X$.

Define $\varphi: \mathcal{B}(X) \to b(X)$ by $\varphi(A) = \chi_A$, its characteristic function. If $x \in X$, then $x \in A$ if and only if $\chi_A(x) = 1$. Hence, if $\chi_A = \chi_B$, then $x \in A$ if and only if $x \in B$; that is, $A = B$. It follows that φ is an injection. In fact, φ is a bijection, for if $f: X \to \mathbb{F}_2$ is a bitstring, then $\varphi(A) = f$, where $A = \{x \in X : f(x) = 1\}$.

We now show that φ is a ring isomorphism. The identity in $\mathcal{B}(X)$ is X, and $\varphi(X) = \chi_X$, the constant function at 1, which is the identity in $b(X)$. Let A and B be subsets of X. If $x \in X$, then

$$(\chi_A\chi_B)(x) = 1 \quad \text{if and only if} \quad x \in A \text{ and } x \in B;$$

that is, $\chi_A\chi_B = \chi_{A \cap B}$. Hence, $\varphi(AB) = \varphi(A)\varphi(B)$. Also,

$$(\chi_A + \chi_B)(x) = 1 \quad \text{if and only if} \quad x \in A \text{ or } x \in B \text{ but not both;}$$

that is, $\chi_A + \chi_B = \chi_{(A\cup B)-(A\cap B)} = \chi_{A+B}$ [recall Exercise 2.4 on page 104: $A + B = (A\cup B) - (A\cap B)$]. Hence, $\varphi(A+B) = \varphi(A)+\varphi(B)$. Therefore, φ is an isomorphism. ◀

→ **Definition.**   If $f: A \to R$ is a ring homomorphism, then its **kernel** is

$$ker\, f = \{a \in A \text{ with } f(a) = 0\},$$

and its **image** is

$$im\, f = \{r \in R : r = f(a) \text{ for some } a \in A\}.$$

Notice that if we forget their multiplications, then the rings A and R are additive abelian groups and these definitions coincide with the group-theoretic ones.

Let k be a field, let $a \in k$ and, as in Corollary 3.34, consider the evaluation homomorphism $e_a: k[x] \to k$ sending $f(x) \mapsto f(a)$. Now e_a is always surjective, for if $b \in k$, then $b = e_a(f)$, where $f(x) = x - a + b$. Thus, im $e_a = k$. By definition, ker e_a consists of all those polynomials $g(x)$ for which $g(a) = 0$; that is, ker e_a consists of all those polynomials in $k[x]$ having a as a root.

Proposition 3.38.   *If $f: A \to R$ is a ring homomorphism, where R is a nonzero ring, then* im f *is a subring of R and* ker f *is a proper subset of A satisfying the conditions*

  (i) $0 \in$ ker f;

 (ii) $x, y \in$ ker f *implies* $x + y \in$ ker f;

(iii) $x \in$ ker f *and* $a \in A$ *imply* $ax \in$ ker f.

Proof.   If $r, r' \in$ im f, then $r = f(a)$ and $r' = f(a')$ for some $a, a' \in A$. Hence, $r - r' = f(a) - f(a') = f(a - a') \in$ im f and $rr' = f(a)f(a') = f(aa') \in$ im f. Since $f(1) = 1$, by the definition of a homomorphism, im f is a subring of R.

We observed on page 246 that $f(0) = 0$, so that $0 \in$ ker f. If $x, y \in$ ker f, then $f(x + y) = f(x) + f(y) = 0 + 0 = 0$, so that $x + y \in$ ker f. If $x \in$ ker f and $a \in A$, then $f(ax) = f(a)f(x) = f(a)0 = 0$, and so $ax \in$ ker f. Note that ker f is a proper subset of A, for $f(1) = 1 \neq 0$, and so $1 \notin$ ker f.   •

The kernel of a group homomorphism $G \to H$ is not merely a subgroup; it is a *normal* subgroup: it is closed under conjugation by any element in the group G. Similarly, if $f: A \to R$ is a ring homomorphism, then ker f is almost[11] a subring because it is closed under addition and multiplication. But ker f is also closed under multiplication by any element in the commutative ring A.

---

[11]If $f: A \to R$ and neither A nor R is the zero ring, then ker f is not a subring because it does not contain 1; if 1 is the identity in A, then $1 \neq 0$, and $f(1) = 1 \neq 0$ in R.

→ **Definition.** An *ideal* in a commutative ring R is a subset I of R such that

(i) $0 \in I$;

(ii) if $a, b \in I$, then $a + b \in I$;

(iii) if $a \in I$ and $r \in R$, then $ra \in I$.

An ideal $I \neq R$ is called a ***proper ideal***.

Proposition 3.38 can be restated. If $f \colon A \to R$ is a ring homomorphism, where R is a nonzero ring, then im f is a subring of R and ker f is a proper ideal in A.

There are two obvious examples of ideals in every nonzero commutative ring R: the ring R itself and the subset $\{0\}$ consisting of 0 alone. In Proposition 3.43, we will see that a commutative ring having only these ideals must be a field.

→ **Example 3.39.**
If $b_1, b_2, \ldots, b_n$ lie in R, then the set of all their linear combinations,

$$I = \{r_1 b_1 + r_2 b_2 + \cdots + r_n b_n : r_i \in R \text{ for all } i\},$$

is an ideal in R. One writes $I = (b_1, b_2, \ldots, b_n)$ in this case. In particular, if $n = 1$, then

$$I = (b) = \{rb : r \in R\}$$

is the ideal in R consisting of all the multiples of b; it is called the ***principal ideal*** generated by b.

Notice that R and $\{0\}$ are always principal ideals: $R = (1)$ and $\{0\} = (0)$. In $\mathbb{Z}$, the even integers form the principal ideal (2). ◄

→ **Theorem 3.40.** *Every ideal in $\mathbb{Z}$ is a principal ideal.*

Proof. This is just a restatement of Corollary 1.37. •

Can two elements generate the same principal ideal?

Proposition 3.41. *If R is a commutative ring and $a = ub$ for some unit $u \in R$, then $(a) = (b)$. Conversely, if R is a domain, then $(a) = (b)$ implies $a = ub$ for some unit $u \in R$.*

Proof. Suppose that $a = ub$ for some unit $u \in R$. If $x \in (a)$, then $x = ra = rub \in (b)$ for some $r \in R$, so that $(a) \subseteq (b)$. For the reverse inclusion, if $y \in (b)$, then $y = sb$ for some $s \in R$. Hence, $y = sb = su^{-1}a \in (a)$, so that $(b) \subseteq (a)$ and $(a) = (b)$.

Conversely, if $(a) = (b)$, then $a \in (a) = (b)$ says that $a = rb$ for some $r \in R$; that is, $b \mid a$; similarly, $b \in (b) = (a)$ implies $a \mid b$. Since R is a domain, Proposition 3.15 applies to show that there is a unit $u \in R$ with $a = ub$. •

Example 3.42.

If an ideal I in a commutative ring R contains 1, then $I = R$, because I contains $r = r1$ for every $r \in R$. Indeed, an ideal I contains a unit u if and only if $I = R$. Sufficiency is obvious: if $I = R$, then I contains a unit, namely, 1. Conversely, if $u \in I$ for some unit u, then I contains $u^{-1}u = 1$, and so I contains $r = r1$ for every $r \in R$. ◄

→ **Proposition 3.43.**   *A nonzero commutative ring R is a field if and only if its only ideals are $\{0\}$ and R itself.*

Proof.   Assume that R is a field. If $I \neq \{0\}$, it contains some nonzero element, and every nonzero element in a field is a unit. Therefore, $I = R$, by Example 3.42.

Conversely, assume that R is a commutative ring whose only ideals are $\{0\}$ and R itself. If $a \in R$ and $a \neq 0$, then the principal ideal $(a) = R$, for $(a) \neq 0$, and so $1 \in R = (a)$. There is thus $r \in R$ with $1 = ra$; that is, a has an inverse in R, and so R is a field. •

Proposition 3.44.   *A ring homomorphism $f \colon A \to R$ is an injection if and only if* $\ker f = \{0\}$.

Proof.   If f is an injection, then $a \neq 0$ implies $f(a) \neq f(0) = 0$, and so $a \notin \ker f$. Therefore, $\ker f = \{0\}$. Conversely, if $\ker f = \{0\}$ and $f(a) = f(a')$, then $0 = f(a) - f(a') = f(a - a')$. Hence, $a - a' \in \ker f = \{0\}$ and $a = a'$; that is, f is an injection. (The reader may check that the proposition is true if either A or R is the zero ring.) •

Corollary 3.45.   *If $f \colon k \to R$ is a ring homomorphism, where R is not the zero ring, and if k is a field, then f is an injection.*

Proof.   By the proposition, it suffices to prove that $\ker f = \{0\}$. But $\ker f$ is a proper ideal in k, by Proposition 3.38, and Proposition 3.43 shows that k has only two ideals: k and $\{0\}$. Now $\ker f \neq k$, because $f(1) = 1 \neq 0$ (R is not the zero ring). Therefore, $\ker f = \{0\}$ and f is an injection. •

EXERCISES

H **3.41** True or false with reasons.
  (i)   If R and S are commutative rings and $f \colon R \to S$ is a ring homomorphism, then f is also a homomorphism from the additive group of R to the additive group of S.
  (ii)   If R and S are commutative rings and if f is a homomorphism from the additive group of R to the additive group of S with $f(1) = 1$, then f is a ring homomorphism.

(iii) If R and S are isomorphic commutative rings, then any ring homomorphism $f: R \to S$ is an isomorphism.

(iv) If $f: R \to S$ is a ring homomorphism, where S is a nonzero ring, then $\ker f$ is a proper ideal in R.

(v) If I and J are ideals in a commutative ring R, then $I \cap J$ and $I \cup J$ are also ideals in R.

(vi) If $\varphi: R \to S$ is a ring homomorphism and if I is an ideal in R, then $\varphi(I)$ is an ideal in S.

(vii) If $\varphi: R \to S$ is a ring homomorphism and if J is an ideal in S, then the inverse image $\varphi^{-1}(J)$ is an ideal in R.

(viii) If R and S are commutative rings, then the projection $(r, s) \mapsto r$ is a ring homomorphism $R \times S \to R$.

(ix) If k is a field and $f: k \to R$ is a surjective ring homomorphism, then R is a field.

(x) If $f(x) = e^x$, then f is a unit in $\mathcal{F}(\mathbb{R})$.

3.42 Let A be a commutative ring. Prove that a subset J of A is an ideal if and only if $0 \in J$, $u, v \in J$ implies $u - v \in J$, and $u \in J$, $a \in A$ imply $au \in J$. (In order that J be an ideal, $u, v \in J$ should imply $u + v \in J$ instead of $u - v \in J$.)

3.43 (i) Prove that a field with 4 elements (see Exercise 3.19 on page 234) and $\mathbb{I}_4$ are not isomorphic commutative rings.

H (ii) Prove that any two fields having exactly four elements are isomorphic.

*3.44 (i) Let $\varphi: A \to R$ be an isomorphism, and let $\psi: R \to A$ be its inverse. Show that ψ is an isomorphism.

(ii) Show that the composite of two homomorphisms (or two isomorphisms) is again a homomorphism (or an isomorphism).

(iii) Show that $A \cong R$ defines an equivalence relation on any family of commutative rings.

3.45 Let R be a commutative ring and let $\mathcal{F}(R)$ be the commutative ring of all functions $f: R \to R$ (see Exercise 3.10 on page 230).

(i) Show that R is isomorphic to the subring of $\mathcal{F}(R)$ consisting of all the constant functions.

(ii) If $f(x) = a_0 + a_1 x + \cdots + a_n x^n \in R[x]$, let $f^\flat: R \to R$ be defined by $f^\flat(r) = a_0 + a_1 r + \cdots + a_n r^n$ [thus, $f^\flat$ is the polynomial function associated to $f(x)$]. Show that the function $\varphi: R[x] \to \mathcal{F}(R)$, defined by $\varphi: f(x) \mapsto f^\flat$, is a ring homomorphism.

(iii) Show that if $R = \mathbb{F}_p$, where p is a prime, then $x^p - x \in \ker \varphi$. (It will be shown, in Theorem 3.50, that φ is injective when R is an infinite field.)

3.46 Let R be a commutative ring. Show that the function $\eta: R[x] \to R$, defined by

$$\eta: a_0 + a_1 x + a_2 x^2 + \cdots + a_n x^n \mapsto a_0,$$

is a homomorphism. Describe $\ker \eta$ in terms of roots of polynomials.

*3.47 Let $\psi: R \to S$ be a homomorphism, where R and S are commutative rings, and let $\ker \psi = I$. If J is an ideal in S, prove that $\psi^{-1}(J)$ is an ideal in R which contains I.

H **3.48** If R is a commutative ring and $c \in R$, prove that the function $\varphi: R[x] \to R[x]$, defined by $f(x) \mapsto f(x + c)$, is an isomorphism. In more detail, $\varphi(\sum_i s_i x^i) = \sum_i s_i (x + c)^i$.

3.49 If R is a field, show that $R \cong \mathrm{Frac}(R)$. More precisely, show that the homomorphism $f \colon R \to \mathrm{Frac}(R)$ in Example 3.31, namely, $r \mapsto [r, 1]$, is an isomorphism.

*3.50 Let R be a domain and let F be a field containing R as a subring.

    (i) Prove that $E = \{uv^{-1} : u, v \in R \text{ and } v \neq 0\}$ is a subfield of F containing R as a subring.

    (ii) Prove that $\mathrm{Frac}(R) \cong E$, where E is the subfield of F defined in part (i). (See Exercise 3.23 on page 235.)

*3.51 H **(i)** If A and R are domains and $\varphi \colon A \to R$ is a ring isomorphism, then $[a, b] \mapsto [\varphi(a), \varphi(b)]$ is a ring isomorphism $\mathrm{Frac}(A) \to \mathrm{Frac}(R)$.

    (ii) Show that a field k containing an isomorphic copy of $\mathbb{Z}$ as a subring must contain an isomorphic copy of $\mathbb{Q}$.

3.52 Let R be a domain with fraction field $F = \mathrm{Frac}(R)$.

    (i) Prove that $\mathrm{Frac}(R[x]) \cong F(x)$.

    (ii) Prove that $\mathrm{Frac}(R[x_1, x_2, \ldots, x_n]) \cong F(x_1, x_2, \ldots, x_n)$.

3.53 **(i)** If R and S are commutative rings, show that their **_direct product_** $R \times S$ is also a commutative ring, where addition and multiplication in $R \times S$ are defined "coordinatewise":

$$(r, s) + (r', s') = (r + r', s + s') \quad \text{and} \quad (r, s)(r', s') = (rr', ss').$$

    (ii) Show that $R \times \{0\}$ is an ideal in $R \times S$.

    (iii) Show that $R \times S$ is not a domain if neither R nor S is the zero ring.

*3.54 H **(i)** If R and S are commutative rings, prove that

$$U(R \times S) = U(R) \times U(S),$$

where $U(R)$ is the group of units of R.

  H **(ii)** Show that if m and n are relatively prime, then $\mathbb{I}_{mn} \cong \mathbb{I}_m \times \mathbb{I}_n$ as rings.

    (iii) Use part (ii) to give a new proof of Corollary 2.131: if $(m, n) = 1$, then $\phi(mn) = \phi(m)\phi(n)$, where ϕ is the Euler ϕ-function.

3.55 **(i)** Prove that the set F of all 2×2 real matrices of the form $A = \left[\begin{smallmatrix} a & b \\ -b & a \end{smallmatrix}\right]$ is a field with operations matrix addition and matrix multiplication.

  H **(ii)** Prove that F is isomorphic to $\mathbb{C}$.

→ 3.5   FROM NUMBERS TO POLYNOMIALS

We are now going to see that virtually all the theorems proved for $\mathbb{Z}$ in Chapter 1 have polynomial analogs in $k[x]$, where k is a field; we shall also see that the proofs there can be translated into proofs here.

The *division algorithm* for polynomials with coefficients in a field says that long division is possible.

$$
\begin{array}{r}
s_n^{-1} t_m x^{m-n} + \cdots \\
s_n x^n + s_{n-1} x^{n-1} + \cdots \,\big|\, \overline{t_m x^m + t_{m-1} x^{m-1} + \cdots}
\end{array}
$$

→ **Definition.** If $f(x) = s_n x^n + \cdots + s_1 x + s_0$ is a polynomial of degree n, then its *leading term* is

$$\mathrm{LT}(f) = s_n x^n.$$

Recall that the *leading coefficient* of $f(x)$ is s_n.

Let k be a field and let $f(x) = s_n x^n + \cdots + s_1 x + s_0$ and $g(x) = t_m x^m + \cdots + t_1 x + t_0$ be polynomials in $k[x]$ with $\deg(f) \le \deg(g)$; that is, $n \le m$. Then $s_n^{-1} \in k$, because k is a field, and

$$\frac{\mathrm{LT}(g)}{\mathrm{LT}(f)} = s_n^{-1} t_m x^{m-n} \in k[x];$$

thus, $\mathrm{LT}(f) \mid \mathrm{LT}(g)$. More generally, we have $\mathrm{LT}(f) \mid \mathrm{LT}(g)$ in $k[x]$ for any commutative ring k if s_n is a unit in k.

→ **Theorem 3.46 (Division Algorithm).** *Let R be a commutative ring, let $f(x)$, $g(x) \in R[x]$, and let the leading coefficient of $f(x)$ be a unit in R.*

(i) *There are polynomials $q(x), r(x) \in R[x]$ with*

$$g(x) = q(x)f(x) + r(x),$$

where either $r(x) = 0$ or $\deg(r) < \deg(f)$.

(ii) *If R is a domain, then the polynomials $q(x)$ and $r(x)$ in part* (i) *are unique.*

Remark. If R is a field, the hypothesis that the leading coefficient of $f(x)$ is a unit is equivalent to $f(x) \ne 0$. ◀

Proof.
(i) We prove the existence of $q(x), r(x) \in R[x]$ as in the statement. If $f \mid g$, then $g = qf$ for some q; define the remainder $r = 0$, and we are done. If $f \nmid g$, then consider all (necessarily nonzero) polynomials of the form $g - qf$ as q varies over $R[x]$. The least integer axiom provides a polynomial $r = g - qf$ having least degree among all such polynomials. Since $g = qf + r$, it suffices to show that $\deg(r) < \deg(f)$. Write $f(x) = s_n x^n + \cdots + s_1 x + s_0$ and $r(x) = t_m x^m + \cdots + t_1 x + t_0$. By hypothesis, s_n is a unit, and so s_n^{-1} exists in k. If $\deg(r) \ge \deg(f)$, define $h(x)$ by

$$h(x) = r(x) - t_m s_n^{-1} x^{m-n} f(x);$$

that is, $h = r - [\mathrm{LT}(r)/\mathrm{LT}(f)]\, f$; note that $h = 0$ or $\deg(h) < \deg(r)$. If $h = 0$, then $r = [\mathrm{LT}(r)/\mathrm{LT}(f)]f$ and

$$g = qf + r = qf + \frac{\mathrm{LT}(r)}{\mathrm{LT}(f)} f = \left[q + \frac{\mathrm{LT}(r)}{\mathrm{LT}(f)} \right] f,$$

contradicting $f \nmid g$. If $h \neq 0$, then $\deg(h) < \deg(r)$ and

$$g - qf = r = h + \frac{LT(r)}{LT(f)} f.$$

Thus, $g - [q + LT(r)/LT(f)]f = h$, contradicting r being a polynomial of least degree having this form. Therefore, $\deg(r) < \deg(f)$.

(ii) To prove uniqueness of $q(x)$ and $r(x)$, assume that $g = q'f + r'$, where $\deg(r') < \deg(f)$. Then

$$(q - q')f = r' - r.$$

If $r' \neq r$, then each side has a degree. But $\deg((q - q')f) = \deg(q - q') + \deg(f) \geq \deg(f)$, while $\deg(r' - r) \leq \max\{\deg(r'), \deg(r)\} < \deg(f)$, a contradiction. Hence, $r' = r$ and $(q - q')f = 0$. Now $R[x]$ is a domain because R is a domain. It follows that $q - q' = 0$ and $q = q'$.   •

Our proof of the division algorithm for polynomials is written as an indirect proof, but the proof can be recast so that it is a true algorithm, as is the division algorithm for integers. Here is a pseudocode implementing it.

> Input: g, f
> Output: q, r
> $q := 0$; $r := g$
> WHILE $r \neq 0$ AND $LT(f) \mid LT(r)$ DO
> $q := q + \left[LT(r)/LT(f)\right] x^{\deg(r) - \deg(f)}$
> $r := r - \left[LT(q)/LT(f)\right] f$
> END WHILE

→ **Definition.**   If $f(x)$ and $g(x)$ are polynomials in $k[x]$, where k is a field, then the polynomials $q(x)$ and $r(x)$ occurring in the division algorithm are called the **quotient** and **remainder** after dividing $g(x)$ by $f(x)$.

The next theorem uses the division algorithm to divide by a monic polynomial in $\mathbb{Z}[x]$. Recall that cyclotomic polynomials were defined on page 31.

Proposition 3.47.   *For every positive integer n, the cyclotomic polynomial $\Phi_n(x)$ is a monic polynomial all of whose coefficients are integers.*

Proof.   The proof is by induction on $n \geq 1$. The base step is true because $\Phi_1(x) = x - 1$. For the inductive step, we assume that $\Phi_d(x)$ is a monic polynomial with integer coefficients. From the equation $x^n - 1 = \prod_d \Phi_d(x)$ (see Proposition 1.30), we have

$$x^n - 1 = \Phi_n(x) f(x),$$

where $f(x)$ is the product of all $\Phi_d(x)$, where d is a proper divisor of n (i.e., $d \mid n$ and $d < n$). By the inductive hypothesis, $f(x)$ is a monic polynomial with integer coefficients. Because $f(x)$ is monic, the division algorithm for monic polynomials in $\mathbb{Z}[x]$ shows that $(x^n - 1)/f(x) \in \mathbb{Z}[x]$ and, hence, all the coefficients of $\Phi_n(x) = (x^n - 1)/f(x)$ are integers, as desired. $\bullet$

We now turn our attention to roots of polynomials.

$\rightarrow$ **Definition.** If $f(x) \in k[x]$, where k is a field, then a **root** of $f(x)$ **in** k is an element $a \in k$ with $f(a) = 0$.

Remark. The polynomial $f(x) = x^2 - 2$ has its coefficients in $\mathbb{Q}$, but we usually say that $\sqrt{2}$ is a root of $f(x)$ even though $\sqrt{2} \notin \mathbb{Q}$. We shall see later, in Theorem 3.118, that for every polynomial $f(x) \in k[x]$, where k is any field, there is a larger field E which contains k as a subfield and which "contains all the roots" of $f(x)$. $\blacktriangleleft$

Lemma 3.48. *Let* $f(x) \in k[x]$, *where* k *is a field, and let* $a \in k$. *Then there is* $q(x) \in k[x]$ *with*

$$f(x) = q(x)(x - a) + f(a).$$

Proof. Use the division algorithm to obtain

$$f(x) = q(x)(x - a) + r;$$

the remainder r is a constant because $x - a$ has degree 1. By Corollary 3.34, evaluation at a is a ring homomorphism $e_a : k[x] \to k$:

$$e_a(f) = e_a(q)e_a(x - a) + e_a(r).$$

Hence, $f(a) = q(a)(a - a) + r$, and so $r = f(a)$. $\bullet$

There is a connection between roots and factoring.

$\rightarrow$ **Proposition 3.49.** *If* $f(x) \in k[x]$, *where* k *is a field, then* $a \in k$ *is a root of* $f(x)$ *in* k *if and only if* $x - a$ *divides* $f(x)$ *in* $k[x]$.

Proof. If a is a root of $f(x)$ in k, then $f(a) = 0$ and the Lemma 3.48 gives $f(x) = q(x)(x - a)$. Conversely, if $f(x) = q(x)(x - a)$, then evaluating at a (i.e., applying e_a) gives $f(a) = q(a)(a - a) = 0$. $\bullet$

$\rightarrow$ **Theorem 3.50.** *If* k *is a field and* $f(x) \in k[x]$ *has degree* n, *then* $f(x)$ *has at most* n *roots in* k.

Proof. We prove the statement by induction on $n \geq 0$. If $n = 0$, then $f(x)$ is a nonzero constant, and so the number of its roots in k is zero. Now let $n > 0$. If $f(x)$ has no roots in k, then we are done, for $0 \leq n$. Otherwise, we may assume that there is $a \in k$ with a a root of $f(x)$; hence, by Proposition 3.49,

$$f(x) = q(x)(x - a),$$

and $q(x) \in k[x]$ has degree $n - 1$. If there is a root $b \in k$ with $b \neq a$, then

$$0 = f(b) = q(b)(b - a).$$

Since $b - a \neq 0$ and k is a field (and, hence, a domain), we have $q(b) = 0$, so that b is a root of $q(x)$. But $\deg(q) = n - 1$, so that the inductive hypothesis says that $q(x)$ has at most $n - 1$ distinct roots in k. Therefore, $f(x)$ has at most n distinct roots in k.   •

→ **Example 3.51.**
Theorem 3.50 is false for arbitrary commutative rings. The polynomial $x^2 - [1] \in \mathbb{I}_8[x]$ has 4 distinct roots in $\mathbb{I}_8$, namely, $[1]$, $[3]$, $[5]$, and $[7]$.   ◄

Recall that every polynomial $f(x) = c_n x^n + c_{n-1} x^{n-1} + \cdots + c_0 \in k[x]$ determines the polynomial function $f^\flat : k \to k$ with $f^\flat(a) = c_n a^n + c_{n-1} a^{n-1} + \cdots + c_0$ for all $a \in k$. In Exercise 3.33 on page 242, however, we saw that a nonzero polynomial in $\mathbb{F}_p[x]$, e.g., $x^p - x$, can determine the constant function zero. This pathology vanishes when the field k is infinite. See Exercise 3.57 on page 273 for a fancy restatement of the next corollary.

→ **Corollary 3.52.** *Let k be an infinite field and let $f(x)$ and $g(x)$ be polynomials in $k[x]$. If $f(x)$ and $g(x)$ determine the same polynomial function, i.e., if $f^\flat(a) = g^\flat(a)$ for all $a \in k$, then $f(x) = g(x)$.*

Proof. If $f(x) \neq g(x)$, then the polynomial $f(x) - g(x)$ is nonzero, so that it has some degree, say, n. Now every element of k is a root of $f(x) - g(x)$, for $f^\flat(a) = g^\flat(a)$ for all $a \in k$. Since k is infinite, this polynomial of degree n has more than n roots, contradicting Theorem 3.50.   •

The last proof gives a bit more.

Corollary 3.53. *Let k be any (possibly finite) field, let $f(x)$, $g(x) \in k[x]$, and let $n = \max\{\deg(f), \deg(g)\}$. If there are $n+1$ distinct elements $a \in k$ with $f(a) = g(a)$, then $f(x) = g(x)$.*

Proof. If $f(x) \neq g(x)$, then $h(x) = f(x) - g(x) \neq 0$, and

$$\deg(h) \leq \max\{\deg(f), \deg(g)\} = n.$$

By hypothesis, there are $n + 1$ elements $a \in k$ with $h(a) = f(a) - g(a) = 0$, contradicting Theorem 3.50(i). Therefore, $h(x) = 0$ and $f(x) = g(x)$.   •

Notation. If $a_1, a_2, \ldots, a_n$ is a list, then $a_1, \ldots, \widehat{a_i}, \ldots, a_n$ denotes the sublist obtained from the original list by deleting a_i.

Corollary 3.54 (Lagrange Interpolation).   *Let k be a field, and let $u_0, \ldots, u_n$ be distinct elements of k. Given any list $y_0, \ldots, y_n$ in k, there exists a unique $f(x) \in k[x]$ of degree $\leq n$ with $f(u_i) = y_i$ for all $i = 0, \ldots, n$. In fact,*

$$f(x) = \sum_{i=0}^{n} y_i \frac{(x - u_0) \cdots \widehat{(x - u_i)} \cdots (x - u_n)}{(u_i - u_0) \cdots \widehat{(u_i - u_i)} \cdots (u_i - u_n)}.$$

Proof.   The polynomial $f(x)$ displayed in the formula has degree at most n and $f(u_i) = y_i$ for all i. Uniqueness follows from Corollary 3.53.   •

Remark.   The Lagrange interpolation formula can be shortened using Exercise 1.15 on page 16: if $f = g_1 \cdots g_n$, then $f' = \sum_{i=1}^{n} d_i f$, where $d_i f = g_1 \cdots g_i' \cdots g_n$. If $g_i(x) = x - u_i$, then $d_i f = (x - u_1) \cdots \widehat{(x - u_i)} \cdots (x - u_n)$. Therefore,

$$f(x) = \sum_{i=1}^{n} y_i \frac{d_i f(x)}{d_i f(u_i)}. \quad \blacktriangleleft$$

→ **Theorem 3.55.**   *If k is a field and G is a finite subgroup of the multiplicative group $k^\times$, then G is cyclic. In particular, if k itself is finite (e.g., $k = \mathbb{F}_p$), then $k^\times$ is cyclic.*

Proof.   Let d be a divisor of $|G|$. If there are two subgroups of G of order d, say, S and T, then $|S \cup T| > d$. But each $a \in S \cup T$ satisfies $a^d = 1$, and hence it is a root of $x^d - 1$. This contradicts Theorem 3.50(i), for $x^d - 1$ now has too many roots in k. Thus, G is cyclic, by Proposition 2.75.   •

Although the multiplicative groups $\mathbb{F}_p^\times$ are cyclic, no explicit formula giving generators of each of them is known; that is, if $[s(p)]$ is a generator of $\mathbb{F}_p^\times$, no efficient algorithm is known that computes $s(p)$ for every prime p.

The definition of greatest common divisor in $k[x]$, where k is a field, is essentially the same as the corresponding definition for integers. We shall define greatest common divisors in general domains on page 259.

→ **Definition.**   If k is a field and $f(x), g(x) \in k[x]$, then a **common divisor** is a polynomial $c(x) \in k[x]$ with $c(x) \mid f(x)$ and $c(x) \mid g(x)$.

If at least one of $f(x), g(x)$ is nonzero, then the **greatest common divisor**, abbreviated gcd and denoted by (f, g), is a monic common divisor $d(x) \in k[x]$ with $\deg(c) \leq \deg(d)$ for every common divisor $c(x)$. If $f(x) = 0 = g(x)$, then the **greatest common divisor** is defined to be 0.

→ **Proposition 3.56.** *If k is a field, then every pair* $f(x)$, $g(x) \in k[x]$ *has a* gcd.

Proof. Since the theorem is obvious if both $f(x)$ and $g(x)$ are 0, we may assume that $f(x) \neq 0$. If $h(x)$ is a divisor of $f(x)$, then $\deg(h) \leq \deg(f)$; a fortiori, $\deg(f)$ is an upper bound for the degrees of the common divisors of $f(x)$ and $g(x)$. Let $d(x)$ be a common divisor of largest degree; since k is a field, we may assume that $d(x)$ is monic [if $d(x) = a_m x^m + a_{m-1} x^{m-1} + \cdots + a_0$, then $a_m^{-1} d(x)$ is a monic common divisor having $m = \deg(d)$]. Therefore, $d(x)$ is a gcd.   •

Here is the analog of Theorem 1.35; we will soon use it to prove uniqueness of gcd's.

→ **Theorem 3.57.** *If k is a field and* $f(x)$, $g(x) \in k[x]$, *then any* gcd *of them is a linear combination of* $f(x)$ *and* $g(x)$.

Remark. By *linear combination*, we now mean $sf + tg$, where both $s = s(x)$ and $t = t(x)$ are polynomials in $k[x]$.   ◄

Proof. We may assume that at least one of f and g is not zero (for the gcd is 0 otherwise). Consider the set I of all the linear combinations:

$$I = \{s(x)f(x) + t(x)g(x) : s(x), t(x) \in k[x]\} .$$

Now f and g are in I (take $s = 1$ and $t = 0$ or vice versa). It follows that if $N = \{n \in \mathbb{N} : n = \deg(h), \text{ where } h(x) \in I\}$, then N is nonempty. By the least integer axiom, N contains a smallest integer, say, n and there is some $d(x) \in I$ of degree n; replacing $d(x)$ by $a_n^{-1} d(x)$ if necessary, where a_n is the leading coefficient of $d(x)$, we may assume that $d(x)$ is monic. We claim that $d(x)$ is a gcd of f and g.

Since $d \in I$, it is a linear combination of f and g:

$$d = sf + tg.$$

Let us show that d is a common divisor by trying to divide each of f and g by d. The division algorithm gives $f = qd + r$, where $r = 0$ or $\deg(r) < \deg(d)$. If $r \neq 0$, then

$$r = f - qd = f - q(sf + tg) = (1 - qs)f - qtg \in I,$$

contradicting d having smallest degree among all linear combinations of f and g. Hence, $r = 0$ and $d \mid f$; a similar argument shows that $d \mid g$.

Finally, if c is a common divisor of f and g, then c divides $d = sf + tg$. But $c \mid d$ implies $\deg(c) \leq \deg(d)$. Therefore, d is a gcd of f and g.   •

→ **Corollary 3.58.**   *Let k be a field and let $f(x), g(x) \in k[x]$.*

    (i) *A monic common divisor $d(x)$ is a gcd if and only if $d(x)$ is divisible by every common divisor.*

    (ii) *Every two polynomials $f(x)$ and $g(x)$ have a unique gcd.*

Proof.   Necessity has already been proved (in the last paragraph of the proof of Theorem 3.57): if $d(x)$ is a gcd, then every common divisor c of f and g is a divisor of $d = sf + tg$.

Conversely, let d denote a gcd of f and g, and let d' be a common divisor divisible by every common divisor c; thus, $d \mid d'$. On the other hand, d is divisible by every common divisor (by necessity just discussed), so that $d' \mid d$. By Proposition 3.15 (which applies because $k[x]$ is a domain), there is a unit $u(x) \in k[x]$ with $d'(x) = u(x)d(x)$. By Exercise 3.32 on page 242, $u(x)$ is a nonzero constant; call it u. Hence, $d'(x) = ud(x)$; if the leading coefficients of d' and d are s' and s, respectively, then $s' = us$. But both $d(x)$ and $d'(x)$ are monic, so that $u = 1$ and $d'(x) = d(x)$. This last argument also proves uniqueness of the gcd.   •

Greatest common divisors can be defined in any domain R once one takes note of Corollaries 1.36 and 3.58.

→ **Definition.**   If R is a domain and $a, b \in R$, then a ***greatest common divisor***, denoted by (a, b), is either 0 (if $a = 0 = b$) or, if at least one of a, b is nonzero, it is a common divisor $d \in R$ with $c \mid d$ for every common divisor $c \in R$.

Exercise 3.81 on page 275 gives an example of a domain R containing a pair of elements having no gcd.

The definition of gcd just given is not quite the same as our definition of gcd in $\mathbb{Z}$ (in Chapter 1) because it does not insist that gcd's be nonnegative. For example, both 2 and -2 are gcd's of 4 and 6 in the new sense; this explains why we wrote "a gcd" in the latest definition instead of "the gcd." Similarly, this new definition of gcd in $k[x]$ differs from our earlier definition on page 257 because gcd's in the present sense need not be monic. In order to make them unique, gcd's in $\mathbb{Z}$ are defined to be nonnegative and gcd's in $k[x]$ are defined to be monic. In more general domains, however, gcd's are not unique. If both d and d' are gcd's of a, b in a domain R, then there is a unit $u \in R$ with $d' = ud$ (for each divides the other). In $\mathbb{Z}$, gcd's are unique to sign (because the only units in $\mathbb{Z}$ are ± 1), and we choose the positive gcd as our favorite; in $k[x]$, where k is a field, any two gcd's differ by a constant nonzero multiple (for the only units are the nonzero constants), and we choose the monic polynomial as our favorite. But, in general domains, there is no obvious choice of a "favorite," and so we cannot single out one of the gcd's as being "the gcd." By Proposition 3.41, the principal ideals generated by two gcd's d and d' of a and b coincide: $(d') = (d)$. Thus, even though gcd's of elements a and b are not unique, if they exist, they all determine the same principal ideal.

Theorem 3.40 says that every ideal in $\mathbb{Z}$ is a principal ideal; its analog below has essentially the same proof.

→ **Theorem 3.59.** *If k is a field, then every ideal I in $k[x]$ is a principal ideal. Moreover, if $I \neq \{0\}$, there is a unique monic polynomial that generates I.*

Proof. If I consists of 0 alone, take $d = 0$. If there are nonzero polynomials in I, the Least Integer Axiom allows us to choose a polynomial $d(x) \in I$ of least degree. Replacing $d(x)$ by $a^{-1}d(x)$ if necessary, where a is the leading coefficient of $d(x)$, we may assume that $d(x)$ is monic.

We claim that every f in I is a multiple of d. The division algorithm gives polynomials q and r with $f = qd + r$, where either $r = 0$ or $\deg(r) < \deg(d)$. Now $d \in I$ gives $qd \in I$, by part (iii) in the definition of ideal; hence part (ii) gives $r = f - qd \in I$. If $r \neq 0$, then it has a degree and $\deg(r) < \deg(d)$, contradicting d having least degree among all the polynomials in I. Therefore, $r = 0$ and f is a multiple of d.

Finally, $d(x)$ is unique, for if $d'(x)$ is another monic polynomial with $(d) = (d')$, then $d = d'$, by Proposition 3.41. •

The proof of Theorem 3.57 identifies the gcd of $f(x)$ and $g(x)$ in $k[x]$ (when at least one of them is not the zero polynomial) as the monic generator of the ideal $I = (f(x), g(x))$ consisting of all the linear combinations of $f(x)$ and $g(x)$ Recall the notation introduced in Example 3.39: if $b_1, \ldots, b_n \in R$, then the ideal consisting of all their linear combinations is denoted by $(b_1, \ldots, b_n)$. This explains the notation (f, g) for the gcd. Notice that this interpretation in terms of ideals makes sense even when both $f(x)$ and $g(x)$ are the zero polynomial: the gcd is 0, which is the generator of the ideal $(0, 0) = \{0\}$.

→ **Definition.**   A commutative ring R is a **PID** (or *principal ideal domain*) if it is a domain in which every ideal is a principal ideal.

→ **Example 3.60.**

(i) The ring $\mathbb{Z}$ of integers is a PID, by Theorem 3.40.

(ii) Every field is a PID, by Proposition 3.43.

(iii) If k is a field, then the polynomial ring $k[x]$ is a PID, by Theorem 3.59.

(iv) If k is a field, then Exercise 3.72 on page 274 shows that the ring $k[[x]]$ of formal power series is a PID.

(v) There are rings other than $\mathbb{Z}$ and $k[x]$ where k is a field, that have a division algorithm; the ring of Gaussian integers $\mathbb{Z}[i]$ is an example of such a ring. These rings are called *Euclidean rings*, and they are PIDs (see Proposition 3.78). ◄

There are commutative rings having ideals which are not principal.

→ **Example 3.61.**

(i) Let $R = \mathbb{Z}[x]$, the commutative ring of all polynomials over $\mathbb{Z}$. It is easy to see that the set I of all polynomials having even constant term is an ideal in $\mathbb{Z}[x]$. We show that I is *not* a principal ideal.

Suppose there is $d(x) \in \mathbb{Z}[x]$ with $I = (d(x))$. The constant $2 \in I$, so that there is $f(x) \in \mathbb{Z}[x]$ with $2 = d(x)f(x)$. Since the degree of a product is the sum of the degrees of the factors, $0 = \deg(2) = \deg(d) + \deg(f)$. Since degrees are nonnegative, it follows that $\deg(d) = 0$; that is, $d(x)$ is a nonzero constant. As constants here are integers, the candidates for $d(x)$ are ± 1 and ± 2. Of course, neither 1 nor -1 is even, so that $d(x) = \pm 2$, because $d(x) \in I$. Since $x \in I$, there is $g(x) \in \mathbb{Z}[x]$ with $x = d(x)g(x) = \pm 2g(x)$. But every coefficient on the right side is even, while the coefficient of x on the left side is 1. This contradiction shows that no such $d(x)$ exists; that is, the ideal I is not a principal ideal.

(ii) Exercise 3.75 on page 275 shows that $k[x, y]$, where k is a field, is not a PID. More precisely, it asks you to prove that the ideal (x, y) is not a principal ideal. ◀

→ **Example 3.62.**

If I and J are ideals in a commutative ring R, we now show that $I \cap J$ is also an ideal in R. Since $0 \in I$ and $0 \in J$, we have $0 \in I \cap J$. If $a, b \in I \cap J$, then $a - b \in I$ and $a - b \in J$, for each is an ideal, and so $a - b \in I \cap J$. If $a \in I \cap J$ and $r \in R$, then $ra \in I$ and $ra \in J$, hence $ra \in I \cap J$. Therefore, $I \cap J$ is an ideal. With minor alterations, this argument also proves that the intersection of any, possibly infinite, family of ideals in R is also an ideal in R. ◀

→ **Definition.** If $f(x), g(x) \in k[x]$, where k is a field, then a *common multiple* is a polynomial $m(x) \in k[x]$ with $f(x) \mid m(x)$ and $g(x) \mid m(x)$.

Given polynomials $f(x)$ and $g(x)$ in $k[x]$, both nonzero, define their *least common multiple*, abbreviated lcm, to be a monic common multiple of them having smallest degree. If $f(x) = 0$ or $g(x) = 0$, define their lcm $= 0$. A lcm of $f(x)$ and $g(x)$ is often denoted by

$$[f(x), g(x)].$$

Proposition 3.63. *Assume that k is a field and that $f(x), g(x) \in k[x]$ are nonzero.*

(i) *$[f(x), g(x)]$ is the monic generator of $(f(x)) \cap (g(x))$.*

(ii) *If $m(x)$ is a monic common multiple of $f(x)$ and $g(x)$, then $m(x) = [f(x), g(x)]$ if and only if $m(x)$ divides every common multiple of $f(x)$ and $g(x)$.*

(iii) *Every pair of polynomials $f(x)$ and $g(x)$ has a unique* lcm.

Proof.
(i) Since $f(x) \neq 0$ and $g(x) \neq 0$, we have $(f) \cap (g) \neq 0$, because $0 \neq fg \in (f) \cap (g)$. By Theorem 3.59, $(f) \cap (g) = (m)$, where m is the monic polynomial of least degree in $(f) \cap (g)$. As $m \in (f)$, $m = qf$ for some $q(x) \in k[x]$, and so $f \mid m$; similarly, $g \mid m$, so that m is a common multiple of f and g. If M is another common multiple, then $M \in (f)$ and $M \in (g)$, hence $M \in (f) \cap (g) = (m)$, and so $m \mid M$. Therefore, $\deg(m) \leq \deg(M)$, and $m = [f, g]$.
(ii) We have just seen that $[f, g]$ divides every common multiple M of f and g. Conversely, assume that m' is a monic common multiple that divides every other common multiple. Now $m' \mid [f, g]$, because $[f, g]$ is a common multiple, while $[f, g] \mid m'$, by part (i). Proposition 3.15 provides a unit $u(x) \in k[x]$ with $m'(x) = u(x)m(x)$; by Exercise 3.32 on page 242, $u(x)$ is a nonzero constant. Since both $m(x)$ and $m'(x)$ are monic, it follows that $m(x) = m'(x)$.
(iii) If ℓ and ℓ' are both lcm's of $f(x)$ and $g(x)$, then part (ii) shows that each divides the other. Uniqueness now follows from Proposition 3.15, for both ℓ and ℓ' are monic. •

Here is a generalization of prime numbers.

→ **Definition.**   An element p in a commutative ring R is ***irreducible*** if p is neither 0 nor a unit and if, in any factorization $p = ab$ in R, either a or b is a unit.

The irreducible elements in $\mathbb{Z}$ are $\pm p$, where p is a prime. The next proposition describes irreducible polynomials in $k[x]$, where k is a field.

→ **Proposition 3.64.**   *If k is a field, then a nonconstant polynomial $p(x) \in k[x]$ is irreducible in $k[x]$ if and only if $p(x)$ has no factorization in $k[x]$ of the form $p(x) = f(x)g(x)$ in which both factors have degree smaller than $\deg(p)$.*

Proof.   If $p(x)$ is irreducible, it must be nonconstant. If $p(x) = f(x)g(x)$ in $k[x]$, where both factors have degree smaller than $\deg(p)$, then neither $\deg(f)$ nor $\deg(g)$ is 0, and so neither factor is a unit in $k[x]$. This is a contradiction.
Conversely, if $p(x)$ cannot be factored into polynomials of smaller degree, then its only factors have the form a or $ap(x)$, where a is a nonzero constant. Since k is a field, nonzero constants are units, and so $p(x)$ is irreducible.   •

This characterization of irreducible polynomials does not apply to polynomial rings $R[x]$ when R not a field. In $\mathbb{Q}[x]$, the polynomial $f(x) = 2x + 2 = 2(x + 1)$ is irreducible, because every linear polynomial over a field is irreducible; here, 2 is a unit in $\mathbb{Q}[x]$. However, $f(x)$ is not irreducible in $\mathbb{Z}[x]$ because neither 2 nor $x + 1$ is a unit in $\mathbb{Z}[x]$.

A linear polynomial $f(x) \in k[x]$, where k is a field, is always irreducible [if $f = gh$, then $1 = \deg(f) = \deg(g) + \deg(h)$, and so one of g or h must have degree 0 while the other has degree $1 = \deg(f)$]. There are polynomial rings over fields in which linear polynomials are the only irreducible polynomials. For example, the fundamental theorem of algebra says that $\mathbb{C}[x]$ is such a ring.

Just as the definition of divisibility in a commutative ring R depends on R, so, too, does irreducibility of a polynomial $p(x) \in k[x]$ depend on the commutative ring $k[x]$ and hence on the field k. For example, $p(x) = x^2 - 2$ is irreducible in $\mathbb{Q}[x]$, but it factors as $(x + \sqrt{2})(x - \sqrt{2})$ in $\mathbb{R}[x]$.

→ **Proposition 3.65.** *Let k be a field and let $f(x) \in k[x]$ be a quadratic or cubic polynomial. Then $f(x)$ is irreducible in $k[x]$ if and only if $f(x)$ does not have a root in k.*

Proof. If $f(x)$ has a root a in k, then Proposition 3.49 shows that $f(x)$ has an honest factorization, and so it is not irreducible.

Conversely, assume that $f(x)$ is not irreducible, i.e., there is a factorization $f(x) = g(x)h(x)$ in $k[x]$ with $\deg(g) < \deg(f)$ and $\deg(h) < \deg(f)$. By Lemma 3.18, $\deg(f) = \deg(g) + \deg(h)$. Since $\deg(f) = 2$ or 3, one of $\deg(g)$, $\deg(h)$ must be 1, and Proposition 3.49 says that $f(x)$ has a root in k. •

This corollary is false for larger degrees. For example,

$$x^4 + 2x^2 + 1 = (x^2 + 1)^2$$

obviously factors in $\mathbb{R}[x]$, but it has no real roots.

→ **Proposition 3.66.** *If k is a field, then every nonconstant polynomial $f(x) \in k[x]$ has a factorization*

$$f(x) = ap_1(x) \cdots p_t(x),$$

where a is a nonzero constant and the $p_i(x)$ are monic irreducible polynomials.

Proof. We prove the proposition for a polynomial $f(x)$ by (the second form of) induction on $\deg(f) \geq 1$. If $\deg(f) = 1$, then $f(x) = ax + c = a(x + a^{-1}c)$; as every linear polynomial, $x + a^{-1}c$ is irreducible, and so it is a product of irreducibles (our usage of the word *product* allows only one factor). Assume now that $\deg(f) \geq 1$. If $f(x)$ is irreducible and its leading coefficient is a, write $f(x) = a(a^{-1}f(x))$; we are done, for $a^{-1}f(x)$ is monic. If $f(x)$ is not irreducible, then $f(x) = g(x)h(x)$, where $\deg(g) < \deg(f)$ and $\deg(h) < \deg(f)$. By the inductive hypothesis, there are factorizations $g(x) = bp_1(x) \cdots p_m(x)$ and $h(x) = cq_1(x) \cdots q_n(x)$, where the p's and q's are monic irreducibles. It follows that $f(x) = (bc)p_1(x) \cdots p_m(x)q_1(x) \cdots q_n(x)$, as desired. •

The analog of the Fundamental Theorem of Arithmetic–uniqueness of the factors–will be proved in the next section.

If k is a field, it is easy to see that if $p(x), q(x) \in k[x]$ are irreducible polynomials, then $p(x) \mid q(x)$ if and only if there is a unit u with $q(x) = up(x)$. If, in addition, both $p(x)$ and $q(x)$ are monic, then $p(x) \mid q(x)$ implies $p(x) = q(x)$. Here is the analog of Proposition 1.34.

Lemma 3.67. *Let k be a field, let $p(x), f(x) \in k[x]$, and let $d(x) = (p, f)$ be their gcd. If $p(x)$ is a monic irreducible polynomial, then*

$$d(x) = \begin{cases} 1 & \text{if } p(x) \nmid f(x) \\ p(x) & \text{if } p(x) \mid f(x). \end{cases}$$

Proof. The only monic divisors of $p(x)$ are 1 and $p(x)$. If $p(x) \mid f(x)$, then $d(x) = p(x)$, for $p(x)$ is monic. If $p(x) \nmid f(x)$, then the only monic common divisor is 1, and so $d(x) = 1$. •

→ **Theorem 3.68 (Euclid's Lemma).** *Let k be a field and let $f(x), g(x) \in k[x]$. If $p(x)$ is an irreducible polynomial in $k[x]$, and if $p(x) \mid f(x)g(x)$, then $p(x) \mid f(x)$ or $p(x) \mid g(x)$. More generally, if $p(x) \mid f_1(x) \cdots f_n(x)$, where $n \geq 2$, then $p(x) \mid f_i(x)$ for some i.*

Proof. If $p \mid f$, we are done. If $p \nmid f$, then the lemma says that $\gcd(p, f) = 1$. There are thus polynomials $s(x)$ and $t(x)$ with $1 = sp + tf$, and so

$$g = spg + tfg.$$

Since $p \mid fg$, it follows that $p \mid g$, as desired. The second statement follows by induction on $n \geq 2$. •

→ **Definition.** Two polynomials $f(x), g(x) \in k[x]$, where k is a field, are called ***relatively prime*** if their gcd is 1.

Corollary 3.69. *Let $f(x), g(x), h(x) \in k[x]$, where k is a field, and let $h(x)$ and $f(x)$ be relatively prime. If $h(x) \mid f(x)g(x)$, then $h(x) \mid g(x)$.*

Proof. By hypothesis, $fg = hq$ for some $q(x) \in k[x]$. There are polynomials s and t with $1 = sf + th$, and so $g = sfg + thg = shq + thg = h(sq + tg)$; that is, $h \mid g$.
  •

Definition. If k is a field, then a rational function $f(x)/g(x) \in k(x)$ is in ***lowest terms*** if $f(x)$ and $g(x)$ are relatively prime.

Proposition 3.70. *If k is a field, every nonzero $f(x)/g(x) \in k(x)$ can be put in lowest terms.*

Proof. If $d = (f, g)$, then $f = df'$ and $g = dg'$ in $k[x]$. Moreover, f' and g' are relatively prime, for if h were a nonconstant common divisor of f' and g', then hd would be a common divisor of f and g of degree greater than that of d. Now $f/g = df'/dg' = f'/g'$, and the latter is in lowest terms. •

The same complaint about computing gcd's that arose in $\mathbb{Z}$ arises here, and it has the same resolution.

→ **Theorem 3.71 (Euclidean Algorithm).** *If k is a field and $f(x), g(x) \in k[x]$, then there is an algorithm computing the gcd $(f(x), g(x))$, and there is an algorithm finding a pair of polynomials $s(x)$ and $t(x)$ with $(f, g) = sf + tg$.*

Proof. The proof is just a repetition of the proof of the Euclidean algorithm in $\mathbb{Z}$: iterated application of the division algorithm.

$$
\begin{array}{ll}
g = q_0 f + r_1 & \deg(r_1) < \deg(f) \\
f = q_1 r_1 + r_2 & \deg(r_2) < \deg(r_1) \\
r_1 = q_2 r_2 + r_3 & \deg(r_3) < \deg(r_2) \\
r_2 = q_3 r_3 + r_4 & \deg(r_4) < \deg(r_3) \\
\quad \vdots & \quad \vdots
\end{array}
$$

As in the proof of Theorem 1.44, the last nonzero remainder is a common divisor which is divisible by every common divisor. Since the remainder may not be monic (even if f and g are both monic, the remainder $r = g - qf$ may not be monic), one must make it monic by multiplying it by the inverse of its leading coefficient. •

Example 3.72.
Use the Euclidean algorithm to find the gcd $(x^5 + 1, x^3 + 1)$ in $\mathbb{Q}[x]$.

$$
\begin{aligned}
x^5 + 1 &= x^2(x^3 + 1) + (-x^2 + 1) \\
x^3 + 1 &= (-x)(-x^2 + 1) + (x + 1) \\
-x^2 + 1 &= (-x + 1)(x + 1)
\end{aligned}
$$

We conclude that $x + 1$ is the gcd. ◄

Example 3.73.
Find the gcd in $\mathbb{Q}[x]$ of

$$
f(x) = x^3 - x^2 - x + 1 \quad \text{and} \quad g(x) = x^3 + 4x^2 + x - 6.
$$

Note that $f(x), g(x) \in \mathbb{Z}[x]$, and $\mathbb{Z}$ is not a field. As we proceed, rational numbers may enter, for $\mathbb{Q}$ is the smallest field containing $\mathbb{Z}$. Here are the equations:

$$g = 1 \cdot f + (5x^2 + 2x - 7)$$
$$f = (\tfrac{1}{5}x - \tfrac{7}{25})(5x^2 + 2x - 7) + \left(\tfrac{24}{25}x - \tfrac{24}{25}\right)$$
$$5x^2 + 2x - 7 = \left(\tfrac{25}{24}5x + \tfrac{175}{24}\right)\left(\tfrac{24}{25}x - \tfrac{24}{25}\right).$$

We conclude that the gcd is $x - 1$ (which is $\tfrac{24}{25}x - \tfrac{24}{25}$ made monic). The reader should find $s(x), t(x)$ expressing $x - 1$ as a linear combination (as in arithmetic, work from the bottom up).

As a computational aid, one can clear denominators at any stage. For example, one can replace the second equation above by

$$(5x - 7)(5x^2 + 2x - 7) + (24x - 24);$$

after all, we ultimately multiply by a unit to obtain a monic gcd.   ◀

Example 3.74.
Find the gcd in $\mathbb{F}_5[x]$ of

$$f(x) = x^3 - x^2 - x + 1 \quad \text{and} \quad g(x) = x^3 + 4x^2 + x - 6.$$

The terms in the Euclidean algorithm simplify considerably.

$$g = 1 \cdot f + (2x + 3)$$
$$f = (3x^2 + 2)(2x + 3)$$

The gcd is $x - 1$ (which is $2x + 3$ made monic).   ◀

Here are factorizations of the polynomials $f(x)$ and $g(x)$ in Example 3.73:

$$f(x) = x^3 - x^2 - x + 1 = (x - 1)^2(x + 1)$$

and

$$g(x) = x^3 + 4x^2 + x - 6 = (x - 1)(x + 2)(x + 3);$$

one could have seen that $x - 1$ is the gcd, had these factorizations been known at the outset. This suggests that an analog of the Fundamental Theorem of Arithmetic could provide another way to compute gcd's. Such an analog does exist (see Proposition 3.86). As a practical matter, however, factoring polynomials is a very difficult task, and the Euclidean algorithm is the best way to compute gcd's.

Here is an unexpected bonus from the Euclidean algorithm.

→ **Corollary 3.75.**   *Let k be a subfield of a field K, so that k[x] is a subring of K[x]. If f(x), g(x) ∈ k[x], then their* gcd *in k[x] is equal to their* gcd *in K[x].*

Proof.   The division algorithm in $K[x]$ gives

$$g(x) = Q(x)f(x) + R(x),$$

where $Q(x), R(x) \in K[x]$ and either $R(x) = 0$ or deg$(R) <$ deg(f); since $f(x), g(x) \in k[x]$, the division algorithm in $k[x]$ gives

$$g(x) = q(x)f(x) + r(x),$$

where $q(x), r(x) \in k[x]$ and either $r(x) = 0$ or deg$(r) <$ deg(f). But the equation $g(x) = q(x)f(x) + r(x)$ also holds in $K[x]$ because $k[x] \subseteq K[x]$, so that the uniqueness of quotient and remainder in the division algorithm in $K[x]$ gives $Q(x) = q(x) \in k[x]$ and $R(x) = r(x) \in k[x]$. Therefore, the list of equations occurring in the Euclidean algorithm in $K[x]$ is exactly the same list occurring in the Euclidean algorithm in the smaller ring $k[x]$, and so the same gcd is obtained in both polynomial rings.   •

Even though there are more divisors in $\mathbb{C}[x]$, the gcd of $x^3 - x^2 + x - 1$ and $x^4 - 1$ is $x^2 + 1$, whether computed in $\mathbb{R}[x]$ or in $\mathbb{C}[x]$.

We have seen, when k is a field, that there are many analogs for $k[x]$ of theorems proved for $\mathbb{Z}$. The essential reason for this is that both rings are PIDs.

Euclidean Rings

There are rings other than $\mathbb{Z}$ and $k[x]$, where k is a field, that have a division algorithm. In particular, we present an example of such a ring in which the quotient and remainder are not unique. We begin by generalizing a property shared by both $\mathbb{Z}$ and $k[x]$.

Definition.   A commutative ring R is a **Euclidean ring** if it is a domain and there is a function

$$\partial \colon R^\times \to \mathbb{N}$$

(where $R^\times$ denotes the nonzero elements of R), called a **degree function**, such that

(i) $\partial(f) \leq \partial(fg)$ for all $f, g \in R^\times$;

(ii) for all $f, g \in R$ with $f \in R^\times$, there exist $q, r \in R$ with

$$g = qf + r,$$

and either $r = 0$ or $\partial(r) < \partial(f)$.

Example 3.76.

(i) Every field R is a Euclidean ring with degree function ∂ identically 0: if $g \in R$ with $f \in R^\times$, set $q = f^{-1}$ and $r = 0$. Conversely, if R is a domain for which the zero function $\partial \colon R^\times \to \mathbb{N}$ is a degree function, then R is a field. If $f \in R^\times$, then there are $q, r \in R$ with $1 = qf + r$. If $r \neq 0$, then $\partial(r) < \partial(f) = 0$, a contradiction. Hence, $r = 0$ and $1 = qf$, so that f is a unit; therefore, R is a field.

(ii) The domain $\mathbb{Z}$ is a Euclidean ring with degree function $\partial(m) = |m|$. In $\mathbb{Z}$, we have
$$\partial(mn) = |mn| = |m||n| = \partial(m)\partial(n).$$

(iii) When k is a field, the domain $k[x]$ is a Euclidean ring with degree function the usual degree of a nonzero polynomial. In $k[x]$, we have
$$
\begin{aligned}
\partial(fg) &= \deg(fg) \\
&= \deg(f) + \deg(g) \\
&= \partial(f) + \partial(g) \\
&\geq \partial(f).
\end{aligned}
$$

Certain properties of a particular degree function may not be enjoyed by all degree functions. For example, the degree function in $\mathbb{Z}$ in part (ii) is multiplicative: $\partial(mn) = \partial(m)\partial(n)$, while the degree function on $k[x]$ here is not multiplicative. If a degree function ∂ is multiplicative, that is, if
$$\partial(fg) = \partial(f)\partial(g),$$
then ∂ is called a **Euclidean norm**. ◀

Example 3.77.
The Gaussian[12] integers $\mathbb{Z}[i]$ form a Euclidean ring whose degree function
$$\partial(a + bi) = a^2 + b^2$$
is a a Euclidean norm.

To see that ∂ is a multiplicative degree function, note that if $\alpha = a + bi$, then
$$\partial(\alpha) = \alpha\overline{\alpha},$$
where $\overline{\alpha} = a - bi$ is the complex conjugate of α. It follows that $\partial(\alpha\beta) = \partial(\alpha)\partial(\beta)$ for all $\alpha, \beta \in \mathbb{Z}[i]$, because
$$\partial(\alpha\beta) = \alpha\beta\overline{\alpha\beta} = \alpha\beta\overline{\alpha}\overline{\beta} = \alpha\overline{\alpha}\beta\overline{\beta} = \partial(\alpha)\partial(\beta);$$
indeed, this is even true for all $\alpha, \beta \in \mathbb{Q}[i] = \{x + yi : x, y \in \mathbb{Q}\}$, by Corollary 1.23.

---

[12]The Gaussian integers are so called because Gauss tacitly used $\mathbb{Z}[i]$ and its Euclidean norm ∂ to investigate biquadratic residues.

We now show that ∂ satisfies the first property of a degree function. If $\beta = c + id \in \mathbb{Z}[i]$ and $\beta \neq 0$, then

$$1 \leq \partial(\beta),$$

for $\partial(\beta) = c^2 + d^2$ is a positive integer; it follows that if $\alpha, \beta \in \mathbb{Z}[i]$ and $\beta \neq 0$, then

$$\partial(\alpha) \leq \partial(\alpha)\partial(\beta) = \partial(\alpha\beta).$$

Let us show that ∂ also satisfies the second desired property. Given $\alpha, \beta \in \mathbb{Z}[i]$ with $\beta \neq 0$, regard α/β as an element of $\mathbb{C}$. Rationalizing the denominator gives $\alpha/\beta = \alpha\overline{\beta}/\beta\overline{\beta} = \alpha\overline{\beta}/\partial(\beta)$, so that

$$\alpha/\beta = x + yi,$$

where $x, y \in \mathbb{Q}$. Write $x = m + u$ and $y = n + v$, where $m, n \in \mathbb{Z}$ are integers closest to x and y, respectively; thus, $|u|, |v| \leq \frac{1}{2}$. [If x or y has the form $m + \frac{1}{2}$, where m is an integer, then there is a choice of nearest integer: $x = m + \frac{1}{2}$ or $x = (m+1) - \frac{1}{2}$; a similar choice arises if x or y has the form $m - \frac{1}{2}$.] It follows that

$$\alpha = \beta(m + ni) + \beta(u + vi).$$

Notice that $\beta(u + vi) \in \mathbb{Z}[i]$, for it is equal to $\alpha - \beta(m + ni)$. Finally, we have $\partial\bigl(\beta(u + vi)\bigr) = \partial(\beta)\partial(u + vi)$, and so ∂ will be a degree function if $\partial(u + vi) < 1$. And this is so, for the inequalities $|u| \leq \frac{1}{2}$ and $|v| \leq \frac{1}{2}$ give $u^2 \leq \frac{1}{4}$ and $v^2 \leq \frac{1}{4}$, and hence $\partial(u + vi) = u^2 + v^2 \leq \frac{1}{4} + \frac{1}{4} = \frac{1}{2} < 1$. Therefore, $\partial(\beta(u + vi)) < \partial(\beta)$, and so $\mathbb{Z}[i]$ is a Euclidean ring whose degree function is a Euclidean norm.

The ring $\mathbb{Z}[i]$ of Gaussian integers is a Euclidean ring in which quotients and remainders may not be unique.[13] For example, let $\alpha = 3 + 5i$ and $\beta = 2$. Then $\alpha/\beta = \frac{3}{2} + \frac{5}{2}i$; the choices are

$$m = 1 \text{ and } u = \tfrac{1}{2} \quad \text{or} \quad m = 2 \text{ and } u = -\tfrac{1}{2};$$
$$n = 2 \text{ and } v = \tfrac{1}{2} \quad \text{or} \quad n = 3 \text{ and } v = -\tfrac{1}{2}.$$

There are four quotients after dividing $3 + 5i$ by 2 in $\mathbb{Z}[i]$, and each of the remainders (e.g., $1 + i$) has degree $2 < 4 = \partial(2)$:

$$3 + 5i = 2(1 + 2i) + (1 + i);$$
$$= 2(1 + 3i) + (1 - i);$$

---

[13]Note the equations in $\mathbb{Z}$:

$$3 = 1 \cdot 2 + 1;$$
$$3 = 2 \cdot 2 - 1.$$

Now $|-1| = |1| < |2|$, so that quotients and remainders in $\mathbb{Z}$ are not unique! In Theorem 1.32, we forced uniqueness by demanding that remainders be nonnegative.

$$= 2(2 + 2i) + (-1 + i);$$
$$= 2(2 + 3i) + (-1 - i). \quad \blacktriangleleft$$

Proposition 3.78. *Every Euclidean ring R is a PID. In particular, the ring $\mathbb{Z}[i]$ of Gaussian integers is a PID.*

Proof. The reader can adapt the proof of Proposition 3.59: if I is a nonzero ideal in R, then $I = (d)$, where d is an element in I of least degree. $\quad \bullet$

The converse of Proposition 3.78 is false: there are PIDs that are not Euclidean rings, as we will see in the next example.

Example 3.79.
It is shown in algebraic number theory (see the remark on page 276) that the ring

$$\mathbb{Z}[\alpha] = \{a + b\alpha : a, b \in \mathbb{Z}\},$$

where $\alpha = \frac{1}{2}(1 + \sqrt{-19})$, is a PID [$\alpha$ is a root of $x^2 - x + 5$, and $\mathbb{Z}[\alpha]$ is the ring of algebraic integers in the quadratic number field $\mathbb{Q}(\alpha)$]. In 1949, T. S. Motzkin showed that $\mathbb{Z}[\alpha]$ is not a Euclidean ring. To do this, he found the following property of Euclidean rings that does not mention its degree function.

Definition. An element u in a domain R is a ***universal side divisor*** if u is not a unit and, for every $x \in R$, either $u \mid x$ or there is a unit $z \in R$ with $u \mid (x + z)$.

Proposition 3.80. *If R is a Euclidean ring that is not a field, then R has a universal side divisor.*

Proof. Define
$$S = \{\partial(v) : v \neq 0 \text{ and } v \text{ is not a unit}\},$$

where ∂ is a degree function on R. Since R is not a field, there is some $v \in R^\times$ which is not a unit, and so S is a nonempty subset of the natural numbers. By the Least Integer Axiom, there is a nonunit $u \in R^\times$ with $\partial(u)$ the smallest element of S. We claim that u is a universal side divisor. If $x \in R$, there are elements q and r with $x = qu + r$, where either $r = 0$ or $\partial(r) < \partial(u)$. If $r = 0$, then $u \mid x$; if $r \neq 0$, then r must be a unit, otherwise its existence contradicts $\partial(u)$ being the smallest number in S. We have shown that u is a universal side divisor. $\quad \bullet$

Motzkin then showed that if $\alpha = \frac{1}{2}(1 + \sqrt{-19})$, then the ring

$$\mathbb{Z}[\alpha] = \{a + b\alpha : a, b \in \mathbb{Z}\}$$

has no universal side divisors, concluding that the PID $\mathbb{Z}[\alpha]$ is not a Euclidean ring (see Exercise 3.70 on page 274). For more details, we refer the reader to K. S. Williams, "Note on Non-Euclidean Principal Ideal Domains," *Math. Mag.* 48 (1975), 176–177. $\quad \blacktriangleleft$

The following result holds for every PID; we will soon apply it to $\mathbb{Z}[i]$.

Proposition 3.81. *Let R be a PID.*

(i) *Each $\alpha, \beta \in R$ has a gcd, δ, which is a linear combination of α and β: there are $\sigma, \tau \in R$ such that*

$$\delta = \sigma\alpha + \tau\beta.$$

(ii) *If an irreducible element $\pi \in R$ divides a product $\alpha\beta$, then either $\pi \mid \alpha$ or $\pi \mid \beta$.*

Proof.
(i) We may assume that at least one of α and β is not zero (otherwise, the gcd is 0 and the result is obvious). Consider the set I of all the linear combinations:

$$I = \{\sigma\alpha + \tau\beta : \sigma, \tau \text{ in } R\}.$$

Now α and β are in I (take $\sigma = 1$ and $\tau = 0$ or vice versa). It is easy to check that I is an ideal in R, and so there is $\delta \in I$ with $I = (\delta)$, because R is a PID; we claim that δ is a gcd of α and β.

Since $\alpha \in I = (\delta)$, we have $\alpha = \rho\delta$ for some $\rho \in R$; that is, δ is a divisor of α; similarly, δ is a divisor of β, and so δ is a common divisor of α and β.

Since $\delta \in I$, it is a linear combination of α and β: there are $\sigma, \tau \in R$ with

$$\delta = \sigma\alpha + \tau\beta.$$

Finally, if γ is any common divisor of α and β, then $\alpha = \gamma\alpha'$ and $\beta = \gamma\beta'$, so that γ divides δ, for $\delta = \sigma\alpha + \tau\beta = \gamma(\sigma\alpha' + \tau\beta')$. We conclude that δ is a gcd.

(ii) If $\pi \mid \alpha$, we are done. If $\pi \nmid \alpha$, then Exercise 3.77 on page 275 says that 1 is a gcd of π and α. There are thus $\sigma, \tau \in R$ with $1 = \sigma\pi + \tau\alpha$, and so

$$\beta = \sigma\pi\beta + \tau\alpha\beta.$$

Since $\pi \mid \alpha\beta$, it follows that $\pi \mid \beta$, as desired. $\bullet$

If n is an odd number, then either $n \equiv 1 \bmod 4$ or $n \equiv 3 \bmod 4$. In particular, the odd prime numbers are divided into two classes. Thus, 5, 13, 17 are congruent to 1 mod 4, for example, while 3, 7, 11 are congruent to 3 mod 4.

There are number-theoretic proofs of the following lemma, but we are going to use the algebra we have already developed.

Lemma 3.82. *If p is a prime and $p \equiv 1 \bmod 4$, then there is an integer m with*

$$m^2 \equiv -1 \bmod p.$$

Proof. The multiplicative group $\mathbb{F}_p^\times$ is a cyclic group, by Theorem 3.55, and so Proposition 2.75 says that it has a unique subgroup of order d for every divisor d of

$|\mathbb{F}_p^\times| = p - 1$. But $p - 1 \equiv 0 \bmod 4$, so that $\mathbb{F}_p^\times$ contains a subgroup S of order 4. Since every subgroup of a cyclic group is cyclic, by Proposition 2.73, there is an integer m with $S = \langle [m] \rangle$. As $[m]$ has order 4, the element $[m^2]$ has order 2; that is, $[m^2] = [-1]$ (for $[-1]$ is the unique element of order 2 in $\mathbb{F}_p^\times$). Therefore, $m^2 \equiv -1 \bmod p$.   •

Theorem 3.83 (Fermat's Two-Squares Theorem).[14] *An odd prime p is a sum of two squares,*

$$p = a^2 + b^2,$$

where a and b are integers, if and only if $p \equiv 1 \bmod 4$.

Proof. For any integer a, we have $a \equiv r \bmod 4$, where $r = 0, 1, 2$ or 3, and so $a^2 \equiv r^2 \bmod 4$. But, mod 4,

$$0^2 \equiv 0, \quad 1^2 \equiv 1, \quad 2^2 = 4 \equiv 0, \quad \text{and} \quad 3^2 = 9 \equiv 1,$$

so that $a^2 \equiv 0$ or $1 \bmod 4$. It follows, for any integers a and b, that $a^2 + b^2 \not\equiv 3 \bmod 4$. Therefore, if $p = a^2 + b^2$, where a and b are integers, then $p \not\equiv 3 \bmod 4$. Since p is odd, either $p \equiv 1 \bmod 4$ or $p \equiv 3 \bmod 4$. We have just ruled out the latter possibility, and so $p \equiv 1 \bmod 4$.

Conversely, assume that $p \equiv 1 \bmod 4$. By the lemma, there is an integer m such that

$$p \mid (m^2 + 1).$$

In $\mathbb{Z}[i]$, there is a factorization $m^2 + 1 = (m + i)(m - i)$, and so

$$p \mid (m + i)(m - i) \quad \text{in} \quad \mathbb{Z}[i].$$

If $p \mid (m + i)$ in $\mathbb{Z}[i]$, then there are integers u and v with $m + i = p(u + iv)$. Taking complex conjugates, we have $m - i = p(u - iv)$, so that $p \mid (m - i)$ as well. Therefore, p divides $(m + i) - (m - i) = 2i$, which is a contradiction, for $\partial(p) = p^2 > \partial(2i) = 4$. We conclude that p is not an irreducible element, for it does not satisfy the analog of Euclid's lemma in Proposition 3.81. Since $\mathbb{Z}[i]$ is a PID, there is a factorization

$$p = \alpha\beta \quad \text{in} \quad \mathbb{Z}[i]$$

in which neither $\alpha = a + ib$ nor $\beta = c + id$ is a unit; thus, $\partial(\alpha) = a^2 + b^2 \neq 1$ and $\partial(\beta) = c^2 + d^2 \neq 1$. Therefore, taking norms gives an equation in $\mathbb{Z}$:

$$\begin{aligned} p^2 &= \partial(p) \\ &= \partial(\alpha\beta) \\ &= \partial(\alpha)\partial(\beta) \\ &= (a^2 + b^2)(c^2 + d^2). \end{aligned}$$

---

[14]Fermat was the first to state this theorem, but the first published proof of it is due to Euler. Gauss proved that there is only one pair of numbers a and b with $p = a^2 + b^2$.

Since $a^2 + b^2 \neq 1$ and $c^2 + d^2 \neq 1$, the Fundamental Theorem of Arithmetic gives $p = a^2 + b^2$ (and $p = c^2 + d^2$). •

EXERCISES

H **3.56** True or false with reasons.

    (i) If $a(x), b(x) \in \mathbb{F}_5[x]$ with $b(x) \neq 0$, then there exist $c(x), d(x) \in \mathbb{F}_5[x]$ with $a(x) = b(x)c(x) + d(x)$, where either $d(x) = 0$ or $\deg(d) < \deg(b)$.

    (ii) If $g(x), f(x) \in \mathbb{Z}[x]$ with $f(x) \neq 0$, then there exist $q(x), r(x) \in \mathbb{Z}[x]$ with $g(x) = f(x)q(x) + r(x)$, where either $r(x) = 0$ or $\deg(r) < \deg(f)$.

    (iii) The gcd of $2x^2 + 4x + 2$ and $4x^2 + 12x + 8$ in $\mathbb{Q}[x]$ is $2x + 2$.

    (iv) If R is a domain, then every unit in $R[x]$ has degree 0.

    (v) If k is a field and $p(x) \in k[x]$ is a nonconstant polynomial having no roots in k, then $p(x)$ is irreducible in $k[x]$.

    (vi) For every quadratic $s(x) \in \mathbb{C}[x]$, there are $a, b \in \mathbb{C}$ and $q(x) \in \mathbb{C}[x]$ with $(x + 1)^{1000} = s(x)q(x) + ax + b$.

    (vii) If $k = \mathbb{F}_p(x)$, where p is a prime, and if $f(x), g(x) \in k[x]$ satisfy $f(a) = g(a)$ for all $a \in k$, then $f(x) = g(x)$.

    (viii) If k is a field, then $k[x]$ is a PID.

    (ix) $\mathbb{Z}$ is a Euclidean ring.

    (x) There is an integer m such that $m^2 \equiv -1 \bmod 89$.

*3.57 Given a commutative ring R, we saw, in Exercise 3.10 on page 230, that $\mathcal{F}(R) = \{\text{all functions } R \rightarrow R\}$ is a commutative ring under pointwise operations.

    (i) If R is a commutative ring, prove that $\varphi \colon R[x] \rightarrow \mathcal{F}(R)$, given by $\varphi \colon f(x) \mapsto f^\flat$, is a homomorphism.

    H **(ii)** If k is an infinite field, prove that φ is an injection.

H **3.58** Find the gcd of $x^2 - x - 2$ and $x^3 - 7x + 6$ in $\mathbb{F}_5[x]$, and express it as a linear combination of them.

*3.59 Let k be a field, let $f(x) \in k[x]$ be nonzero, and let $a_1, a_2, \ldots, a_t$ be some distinct roots of $f(x)$ in k. Prove that $f(x) = (x - a_1)(x - a_2) \cdots (x - a_t)g(x)$ for some $g(x) \in k[x]$.

H **3.60** If R is a domain and $f(x) \in R[x]$ has degree n, show that $f(x)$ has at most n roots in R.

3.61 Let R be an arbitrary commutative ring. If $f(x) \in R[x]$ and if $a \in R$ is a root of $f(x)$, that is, $f(a) = 0$, prove that there is a factorization $f(x) = (x - a)g(x)$ in $R[x]$.

3.62 Prove the converse of Euclid's lemma. Let k be a field and let $f(x) \in k[x]$ be a polynomial of degree ≥ 1; if, whenever $f(x)$ divides a product of two polynomials, it necessarily divides one of the factors, then $f(x)$ is irreducible.

H **3.63** Let $f(x), g(x) \in R[x]$, where R is a domain. If the leading coefficient of $f(x)$ is a unit in R, then the division algorithm gives a quotient $q(x)$ and a remainder $r(x)$ after dividing $g(x)$ by $f(x)$. Prove that $q(x)$ and $r(x)$ are uniquely determined by $g(x)$ and $f(x)$.

*H **3.64** Let k be a field, and let $f(x), g(x) \in k[x]$ be relatively prime. If $h(x) \in k[x]$, prove that $f(x) \mid h(x)$ and $g(x) \mid h(x)$ imply $f(x)g(x) \mid h(x)$.

3.65 (i)   Show that the following pseudocode implements the Euclidean algorithm finding gcd $f(x)$ and $g(x)$ in $k[x]$, where k is a field.

>Input: g, f
>Output: d
>$d := g;$ $s := f$
>WHILE $s \neq 0$ DO
>   rem := remainder(d, s)
> $d := s$
> $s :=$ rem
>END WHILE
>$a :=$ leading coefficient of d
>$d := a^{-1}d$

  (ii)   Find (f, g), where $f(x) = x^2 + 1$, $g(x) = x^3 + x + 1 \in \mathbb{I}_3[x]$.

* H **3.66** If k is a field in which $1 + 1 \neq 0$, prove that $\sqrt{1 - x^2} \notin k(x)$, where $k(x)$ is the field of rational functions.

* H **3.67** Let $f(x) = (x - a_1) \cdots (x - a_n) \in R[x]$, where R is a commutative ring. Show that $f(x)$ has **no repeated roots** (that is, all the a_i are distinct) if and only if the gcd $(f, f') = 1$, where f' is the derivative of f.

3.68 Let ∂ be the degree function of a Euclidean ring R. If $m, n \in \mathbb{N}$ and $m \geq 1$, prove that ∂' is also a degree function on R, where $\partial'(x) = m\partial(x) + n$ for all $x \in R$. Conclude that a Euclidean ring may have no elements of degree 0 or degree 1.

3.69 Let R be a Euclidean ring with degree function ∂.
  (i)   Prove that $\partial(1) \leq \partial(a)$ for all nonzero $a \in R$.
  H (ii)   Prove that a nonzero $u \in R$ is a unit if and only if $\partial(u) = \partial(1)$.

* **3.70** Let $\alpha = \frac{1}{2}\left(1 + \sqrt{-19}\right)$, and let $R = \mathbb{Z}[\alpha]$.
  (i)   Prove that $N \colon R^\times \to \mathbb{N}$, defined by $N(m + n\alpha) = m^2 - mn + 5n^2$, is multiplicative: $N(uv) = N(u)N(v)$.
  (ii)   Prove that the only units in R are ± 1.
  (iii)   Prove that there is no surjective ring homomorphism $R \to \mathbb{I}_2$ or $R \to \mathbb{I}_3$.
  (iv)   Assume that R has a degree function $\partial \colon R^\times \to \mathbb{N}$. Choose $u \in R - \{0, 1, -1\}$ with $\partial(u)$ minimal. Prove, for all $r \in R$ that there exists $d \in \{0, 1, -1\}$ such that $r - d \in (u)$.
  (v)   Use the ring $R/(u)$ to prove that R is not a Euclidean ring.

H **3.71** Let R be a Euclidean ring with degree function ∂, and assume that $b \in R$ is neither zero nor a unit. Prove, for every $i \geq 0$, that $\partial(b^i) < \partial(b^{i+1})$.

* H **3.72** If k is a field, prove that the ring of formal power series $k[[x]]$ is a PID.

3.73 Let k be a field, and let polynomials $a_1(x), a_2(x), \ldots, a_n(x)$ in $k[x]$ be given.
  H (i)   Show that the greatest common divisor $d(x)$ of these polynomials has the form $\sum t_i(x)a_i(x)$, where $t_i(x) \in k[x]$ for $1 \leq i \leq n$.
  (ii)   Prove that if $c(x)$ is a monic common divisor of these polynomials, then $c(x)$ is a divisor of $d(x)$.

H **3.74** Let $[f(x), g(x)]$ denote the lcm of $f(x), g(x) \in k[x]$, where k is a field. Show that if $f(x)g(x)$ is monic, then
$$[f, g](f, g) = fg.$$

*3.75 If k is a field, show that the ideal (x, y) in $k[x, y]$ is not a principal ideal.

3.76 For every $m \geq 1$, prove that every ideal in $\mathbb{I}_m$ is a principal ideal. (If m is composite, then $\mathbb{I}_m$ is not a PID because it is not a domain.)

*H 3.77 Let R be a PID and let $\pi \in R$ be an irreducible element. If $\beta \in R$ and $\pi \nmid \beta$, prove that π and β are relatively prime.

3.78 H **(i)** Show that $x, y \in k[x, y]$ are relatively prime but that 1 is not a linear combination of them; that is, there do not exist $s(x, y), t(x, y) \in k[x, y]$ with $1 = xs(x, y) + yt(x, y)$.

(ii) Show that 2 and x are relatively prime in $\mathbb{Z}[x]$, but that 1 is not a linear combination of them; that is, there do not exist $s(x), t(x) \in \mathbb{Z}[x]$ with $1 = 2s(x) + xt(x)$.

H 3.79 Because $x - 1 = (\sqrt{x} + 1)(\sqrt{x} - 1)$, a student claims that $x - 1$ is not irreducible. Explain the error of his ways.

3.80 **(i)** Factor each of the integers 5 and 13 as a product of two nonunits in $\mathbb{Z}[x]$, and factor 65 as a product of four nonunits in $\mathbb{Z}[x]$.

(ii) Find two different ways of grouping the factors of 65 in part (i) as a product $\alpha\overline{\alpha}$ of two conjugate factors. Use these to get two different expressions for 65 as a sum of two squares in $\mathbb{Z}$.

*H 3.81 Prove that there are domains R containing a pair of elements having no gcd. (See the definition of gcd in general domains on page 259.)

→ **3.6 Unique Factorization**

Here is the analog for polynomials of the fundamental theorem of arithmetic; it shows that irreducible polynomials are "building blocks" of arbitrary polynomials in the same sense that primes are building blocks of arbitrary integers. To avoid long sentences, let us agree that a "product" may have only one factor. Thus, when we say that a polynomial $f(x)$ is a product of irreducibles, we allow the possibility that the product has only one factor; that is, $f(x)$ itself is irreducible.

→ **Theorem 3.84 (Unique Factorization).** *If k is a field, then every polynomial $f(x) \in k[x]$ of degree ≥ 1 is a product of a nonzero constant and monic irreducibles. Moreover, if*

$$f(x) = ap_1(x) \cdots p_m(x) \qquad and \qquad f(x) = bq_1(x) \cdots q_n(x),$$

where a and b are nonzero constants and the p's and q's are monic irreducibles, then $a = b$, $m = n$, and the q's may be reindexed so that $q_i = p_i$ for all i.

Proof. The existence of a factorization of a polynomial $f(x) \in k[x]$ into irreducibles was proved in Proposition 3.66, and so only the uniqueness assertion needs proof.

An equation $f(x) = ap_1(x) \cdots p_m(x)$ gives a the leading coefficient of $f(x)$, since a product of monic polynomials is monic. Hence, two factorizations of $f(x)$

give $a = b$, for each is equal to the leading coefficient of $f(x)$. It now suffices to prove uniqueness when

$$p_1(x) \cdots p_m(x) = q_1(x) \cdots q_n(x).$$

The proof is by induction on $\max\{m, n\} \geq 1$. The base step is obviously true, for the given equation is now $p_1(x) = q_1(x)$. For the inductive step, the given equation shows that $p_m(x) \mid q_1(x) \cdots q_n(x)$. By Theorem 3.68 (Euclid's lemma for polynomials), there is some i with $p_m(x) \mid q_i(x)$. But $q_i(x)$, being monic irreducible, has no monic divisors other than 1 and itself, and so $q_i(x) = p_m(x)$. Reindexing, we may assume that $q_n(x) = p_m(x)$. Canceling this factor, we have $p_1(x) \cdots p_{m-1}(x) = q_1(x) \cdots q_{n-1}(x)$. By the inductive hypothesis, $m - 1 = n - 1$ (hence $m = n$) and, after possible reindexing, $q_i = p_i$ for all i. •

Example 3.85.
The reader may check, in $\mathbb{I}_8[x]$, that

$$x^2 - 1 = (x - 1)(x + 1) = (x - 3)(x + 3).$$

It is true, but not obvious, that each of the linear factors is irreducible (of course, $\mathbb{I}_8$ is not a field). Therefore, unique factorization does not hold in $\mathbb{I}_8[x]$. ◄

Remark.   A domain R is called a UFD, *unique factorization domain*, if every nonzero nonunit $r \in R$ is a product of irreducibles and, moreover, such a factorization of r is unique in the sense of Theorem 3.84. The domains $\mathbb{Z}[\zeta_p]$ in Example 3.10, where $\zeta = e^{2\pi i/p}$ for p an odd prime, are quite interesting in this regard. Positive integers a, b, c for which

$$a^2 + b^2 = c^2,$$

for example, 3, 4, 5 and 5, 12, 13, are called *Pythagorean triples*, and they have been recognized for at least four thousand years (a Babylonian tablet of roughly this age has been found containing a dozen of them), and they were classified by Diophantus about two thousand years ago (see Exercise 1.67 on page 55). Around 1637, Fermat wrote in the margin of his copy of a book by Diophantus what is nowadays called *Fermat's Last Theorem*: for all integers $n \geq 3$, there do not exist positive integers a, b, c for which

$$a^n + b^n = c^n.$$

Fermat claimed that he had a wonderful proof of this result, but that the margin was too small to contain it. Elsewhere, he did prove it for $n = 4$ and, later, others proved it for small values of n. However, the general statement challenged mathematicians for centuries.

Call a positive integer $n \geq 2$ *good* if there are no positive integers a, b, c with $a^n + b^n = c^n$. If n is good, then so is any multiple nk of it. Otherwise, there are positive integers r, s, t with $r^{nk} + s^{nk} = t^{nk}$; this gives the contradiction $a^n + b^n = c^n$, where $a = r^k, b = s^k$, and $c = t^k$. For example, any integer of the form $4k$ is good. Since every positive integer is a product of primes, Fermat's Last Theorem would follow if every odd prime is good.

As in Exercise 3.85 on page 281, a solution $a^p + b^p = c^p$, for p an odd prime, gives a factorization

$$c^p = (a+b)(a + \zeta b)(a + \zeta^2 b) \cdots (a + \zeta^{p-1} b),$$

where $\zeta = \zeta_p = e^{2\pi i / p}$. In the 1840s, E. Kummer considered this factorization in the domain $\mathbb{Z}[\zeta_p]$ (described in Example 3.10 on page 224). He proved that if unique factorization into irreducibles holds in $\mathbb{Z}[\zeta_p]$, then there do not exist positive integers a, b, c (none of which is divisible by p) with $a^p + b^p = c^p$. Kummer realized, however, that even though unique factorization does hold in $\mathbb{Z}[\zeta_p]$ for some primes p, it does not hold in all $\mathbb{Z}[\zeta_p]$. To extend his proof, he invented what he called "ideal numbers," and he proved that there is unique factorization of ideal numbers as products of "prime ideal numbers." These ideal numbers motivated R. Dedekind to define ideals in arbitrary commutative rings (our definition of ideal is that of Dedekind), and he proved that ideals in the special rings $\mathbb{Z}[\zeta_p]$ correspond to Kummer's ideal numbers. Over the years these investigations have been vastly developed, and at last, in 1995, A. Wiles proved Fermat's Last Theorem. ◀

If k is a field, it is often convenient to collect terms and write prime factorizations as

$$f(x) = a p_1(x)^{e_1} \cdots p_m(x)^{e_m},$$

where $a \in k$, $e_i \geq 1$ for all i, and $p_1(x), \ldots, p_m(x)$ are distinct monic irreducible polynomials in $k[x]$. In particular, if $f(x)$ is a product of linear factors in $k[x]$, that is, if all the roots of $f(x)$ lie in k, then

$$f(x) = a(x - r_1)^{e_1} (x - r_2)^{e_2} \cdots (x - r_s)^{e_s},$$

where $e_j \geq 1$ for all j. We call e_j the **multiplicity** of the root r_j. As linear polynomials over a field are always irreducible, unique factorization shows that multiplicities of roots are well-defined.

There are formulas for gcd's and lcm's of two polynomials in $k[x]$. When considering two polynomials, we may allow exponents $e_i = 0$ in their prime factorizations to allow the same set of monic irreducible polynomials to appear.

→ **Proposition 3.86.** *Let k be a field and let $g(x) = a p_1^{e_1} \cdots p_n^{e_n} \in k[x]$ and let $h(x) = b p_1^{f_1} \cdots p_n^{f_n} \in k[x]$, where $a, b \in k$, the p_i are distinct monic irreducible polynomials, and $e_i, f_i \geq 0$ for all i. Define*

$$m_i = \min\{e_i, f_i\} \quad and \quad M_i = \max\{e_i, f_i\}.$$

Then

$$(g, h) = p_1^{m_1} \cdots p_n^{m_n} \quad and \quad [g, h] = p_1^{M_1} \cdots p_n^{M_n}.$$

Proof.   The proof is a just an adaptation of the proof of Proposition 1.55.   •

The next result is an analog of Proposition 1.47: for $b \geq 2$, every positive integer has an expression in base b.

Lemma 3.87.   *Let k be a field, and let $b(x) \in k[x]$ have $\deg(b) \geq 1$. Each nonzero $f(x) \in k[x]$ has an expression*

$$f(x) = d_m(x)b(x)^m + \cdots + d_j(x)b(x)^j + \cdots + d_0(x),$$

where, for every j, either $d_j(x) = 0$ or $\deg(d_j) < \deg(b)$.

Proof.   By the division algorithm, there are $g(x), d_0(x) \in k[x]$ with

$$f(x) = g(x)b(x) + d_0(x),$$

where either $d_0(x) = 0$ or $\deg(d_0) < \deg(b)$. Now $\deg(f) = \deg(gb)$, so that $\deg(b) \geq 1$ gives $\deg(g) < \deg(f)$. By induction, there are $d_j(x) \in k[x]$ with each $d_j(x) = 0$ or $\deg(d_j) < \deg(b)$, such that

$$g(x) = d_m b^{m-1} + \cdots + d_2 b + d_1.$$

Therefore,

$$\begin{aligned}
f &= gb + d_0 \\
&= (d_m b^{m-1} + \cdots + d_2 b + d_1)b + d_0 \\
&= d_m b^m + \cdots + d_2 b^2 + d_1 b + d_0. \quad •
\end{aligned}$$

Just as for integers, it can be proved that the "digits" $d_i(x)$ are unique (see Proposition 1.47).

Definition.   Polynomials $q_1(x), \ldots, q_n(x)$ in $k[x]$, where k is a field, are ***pairwise relatively prime*** if $(q_i, q_j) = 1$ for all $i \neq j$.

It is easy to see that if $q_1(x), \ldots, q_m(x)$ are pairwise relatively prime, then $q_i(x)$ and the product $q_1(x) \cdots \widehat{q_i}(x) \cdots q_m(x)$ are relatively prime for all i.

Lemma 3.88.   *Let k be a field, let $f(x)/g(x) \in k(x)$, and suppose that $g(x) = q_1(x) \cdots q_m(x)$, where $q_1(x), \ldots, q_m(x) \in k[x]$ are pairwise relatively prime. Then there are $a_i(x) \in k[x]$ with*

$$\frac{f(x)}{g(x)} = \sum_{i=1}^{m} \frac{a_i(x)}{q_i(x)}.$$

Proof. The proof is by induction on $m \geq 1$. The base step $m = 1$ is clearly true. Since q_1 and $q_2 \ldots q_m$ are relatively prime, there are polynomials s and t with $1 = sq_1 + tq_2 \cdots q_m$. Therefore,

$$
\begin{aligned}
\frac{f}{g} &= (sq_1 + tq_2 \cdots q_m)\frac{f}{g} \\
&= \frac{sq_1 f}{g} + \frac{tq_2 \cdots q_m f}{g} \\
&= \frac{sq_1 f}{q_1 q_2 \cdots q_m} + \frac{tq_2 \cdots q_m f}{q_1 q_2 \cdots q_m} \\
&= \frac{sf}{q_2 \cdots q_m} + \frac{tf}{q_1}.
\end{aligned}
$$

The polynomials $q_2(x), \ldots, q_m(x)$ are pairwise relatively prime, and the inductive hypothesis now rewrites the first summand.   ●

We now prove the algebraic portion of the method of partial fractions used in calculus to integrate rational functions.

Theorem 3.89 (Partial Fractions).   *Let k be a field, and let the factorization into irreducibles of a monic polynomial $g(x) \in k[x]$ be*

$$ g(x) = p_1(x)^{e_1} \cdots p_m(x)^{e_m}. $$

If $f(x)/g(x) \in k(x)$, then

$$ \frac{f(x)}{g(x)} = h(x) + \sum_{i=1}^{m} \left(\frac{d_{i1}(x)}{p_i(x)} + \frac{d_{i2}(x)}{p_i(x)^2} + \cdots + \frac{d_{ie_i}(x)}{p_i(x)^{e_i}} \right), $$

where $h(x) \in k[x]$ and either $d_{ij}(x) = 0$ or $\deg(d_{ij}) < \deg(p_i)$ for all j.

Proof. Clearly, the polynomials $p_1(x)^{e_1}, p_2(x)^{e_2}, \ldots, p_m(x)^{e_m}$ are pairwise relatively prime. By Lemma 3.88, there are $a_i(x) \in k[x]$ with

$$ \frac{f(x)}{g(x)} = \sum_{i=1}^{m} \frac{a_i(x)}{p_i(x)^{e_i}}. $$

For each i, the division algorithm gives polynomials $Q_i(x)$ and $R_i(x)$ with $a_i(x) = Q_i(x)p_i(x)^{e_i} + R_i(x)$, where $R_i(x) = 0$ or $\deg(R_i) < \deg(p_i(x)^{e_i})$. Hence,

$$ \frac{a_i(x)}{p_i(x)^{e_i}} = Q_i(x) + \frac{R_i(x)}{p_i(x)^{e_i}}. $$

By Lemma 3.87,

$$ R_i(x) = d_{im}(x)p_i(x)^m + d_{i,m-1}(x)p_i(x)^{m-1} + \cdots + d_{i0}(x), $$

where, for all j, either $d_{ij}(x) = 0$ or $\deg(d_{ij}) < \deg(p_i)$; moreover, since $\deg(R_i) < \deg(p_i^{e_i})$, we have $m \leq e_i$. Therefore,

$$\frac{a_i(x)}{p_i(x)^{e_i}} = Q_i(x) + \frac{d_{im}(x)p_i(x)^m + d_{i,m-1}(x)p_i(x)^{m-1} + \cdots + d_{i0}(x)}{p_i(x)^{e_i}}$$

$$= Q_i(x) + \frac{d_{im}(x)p_i(x)^m}{p_i(x)^{e_i}} + \frac{d_{i,m-1}(x)p_i(x)^{m-1}}{p_i(x)^{e_i}} + \cdots + \frac{d_{i0}(x)}{p_i(x)^{e_i}}.$$

After cancellation, each of the summands $d_{ij}(x)p_i(x)^j / p_i(x)^{e_i}$ is either a polynomial or a rational function of the form $d_{ij}(x)/p_i(x)^s$, where $1 \leq s \leq e_i$. If we call $h(x)$ the sum of all those polynomials which are not rational functions, then this is the desired expression.   •

It is known that the only irreducible polynomials in $\mathbb{R}[x]$ are linear or quadratic, so that all the numerators in the partial fraction decomposition in $\mathbb{R}[x]$ are either constant or linear. Theorem 3.89 is used in proving that all rational functions in $\mathbb{R}(x)$ can be integrated in closed form using polynomials, logs, and arctans.

Here is the integer version of partial fractions. If a/b is a positive rational number and if the prime factorization of b is $b = p_1^{e_1} \cdots p_m^{e_m}$, then

$$\frac{a}{b} = h + \sum_{i=1}^{m} \left(\frac{c_{i1}}{p_i} + \frac{c_{i2}}{p_i^2} + \cdots + \frac{c_{ie_i}}{p_i^{e_i}} \right),$$

where $h \in \mathbb{Z}$ and $0 \leq c_{ij} < p_i$ for all j.

EXERCISES

H **3.82** True or false with reasons.

(i)   Every element of $\mathbb{Z}[x]$ is a product of a constant in $\mathbb{Z}$ and monic irreducible polynomials in $\mathbb{Z}[x]$.

(ii)   Every element of $\mathbb{Z}[x]$ is a product of a constant in $\mathbb{Z}$ and monic irreducible polynomials in $\mathbb{Q}[x]$.

(iii)   If k is a field and $f(x) \in k[x]$ can be written both as $ap_1(x) \cdots p_m(x)$ and as $bq_1(x) \cdots q_n(x)$, where a, b are constants in k, $p_1(x), \ldots, p_m(x)$ are monic irreducible polynomials, and $q_1(x), \ldots, q_n(x)$ are monic nonconstant polynomials, then $q_1(x), \ldots, q_n(x)$ are irreducible.

(iv)   If k is a field and $f(x) \in k[x]$ can be written both as $ap_1(x) \cdots p_m(x)$ and as $bq_1(x) \cdots q_n(x)$, where a, b are constants in k, $p_1(x), \ldots, p_m(x)$ are monic irreducible polynomials, and $q_1(x), \ldots, q_n(x)$ are monic nonconstant polynomials, then $m \geq n$.

(v) If k is a subfield of K and $f(x) \in k[x]$ has the factorization $f(x) = ap_1^{e_1} \cdots p_n^{e_n}$, where a is a constant and the $p_i(x)$ are monic irreducible in $k[x]$, then $f(x) = ap_1^{e_1} \cdots p_n^{e_n}$ is also the factorization of $f(x)$ in $K[x]$ as a product of a constant and monic irreducible polynomials.

(vi) If $f(x)$ is a polynomial over a field K whose factorization into a constant and monic irreducible polynomials in $K[x]$ is $f(x) = ap_1^{e_1} \cdots p_n^{e_n}$, and if all the coefficients of $f(x)$ and of the polynomials $p_i(x)$ lie in some subfield $k \subseteq K$, then $f(x) = ap_1^{e_1} \cdots p_n^{e_n}$ is also the factorization of $f(x)$ in $k[x]$ as a product of a constant and monic irreducible polynomials.

3.83 In $k[x]$, where k is a field, let $g = p_1^{e_1} \cdots p_m^{e_m}$ and $h = p_1^{f_1} \cdots p_m^{f_m}$, where the p_i's are distinct monic irreducibles and e_i, $f_i \geq 0$ for all i. Prove that $g \mid h$ if and only if $e_i \leq f_i$ for all i.

*3.84 H **(i)** If $f(x) \in \mathbb{R}[x]$, show that $f(x)$ has no repeated roots in $\mathbb{C}$ if and only if $(f, f') = 1$.

(ii) Prove that if $p(x) \in \mathbb{Q}[x]$ is an irreducible polynomial, then $p(x)$ has no repeated roots.

*3.85 Let $\zeta = e^{2\pi i/n}$.

H **(i)** Prove that

$$x^n - 1 = (x - 1)(x - \zeta)(x - \zeta^2) \cdots (x - \zeta^{n-1})$$

and, if n is odd, that

$$x^n + 1 = (x + 1)(x + \zeta)(x + \zeta^2) \cdots (x + \zeta^{n-1}).$$

H **(ii)** For numbers a and b, prove that

$$a^n - b^n = (a - b)(a - \zeta b)(a - \zeta^2 b) \cdots (a - \zeta^{n-1} b)$$

and, if n is odd, that

$$a^n + b^n = (a + b)(a + \zeta b)(a + \zeta^2 b) \cdots (a + \zeta^{n-1} b).$$

→ 3.7 IRREDUCIBILITY

Although there are some techniques to help decide whether an integer is prime, the general problem of factoring (large) integers is a very difficult one. It is also very difficult to determine whether a polynomial is irreducible, but we now present some useful techniques that frequently work.

We know that if $f(x) \in k[x]$ and r is a root of $f(x)$ in a field k, then there is a factorization $f(x) = (x - r)g(x)$ in $k[x]$, and so $f(x)$ is not irreducible. In Corollary 3.65, we saw that this decides the matter for quadratic and cubic polynomials in $k[x]$: such polynomials are irreducible in $k[x]$ if and only if they have no roots in k. On the other hand, there may be polynomials having no roots which are, nevertheless, not irreducible; for example, $(x^2 + 1)^2 \in \mathbb{R}[x]$.

→ **Theorem 3.90.** *Let $f(x) = a_0 + a_1x + \cdots + a_nx^n \in \mathbb{Z}[x] \subseteq \mathbb{Q}[x]$. Every rational root r of $f(x)$ has the form $r = b/c$, where $b \mid a_0$ and $c \mid a_n$.*

Proof. We may assume that $r = b/c$ is in lowest terms, that is, $(b, c) = 1$. Substituting r into $f(x)$ gives

$$0 = f(b/c) = a_0 + a_1b/c + \cdots + a_nb^n/c^n,$$

and multiplying through by c^n gives

$$0 = a_0c^n + a_1bc^{n-1} + \cdots + a_nb^n.$$

Hence, $a_0c^n = b(-a_1c^{n-1} - \cdots - a_nb^{n-1})$, that is, $b \mid a_0c^n$. Since b and c are relatively prime, it follows that b and c^n are relatively prime, and so Corollary 1.40 to Euclid's lemma in $\mathbb{Z}$ gives $b \mid a_0$. Similarly, $a_nb^n = c(-a_{n-1}b^{n-1} - \cdots - a_0c^{n-1})$, $c \mid a_nb^n$, and $c \mid a_n$. •

Definition. A complex number α is called an ***algebraic integer*** if α is a root of a monic polynomial $f(x) \in \mathbb{Z}[x]$.

We note that it is crucial, in the definition of algebraic integer, that $f(x) \in \mathbb{Z}[x]$ be monic. Every algebraic number β, that is, every complex number β that is a root of some polynomial $g(x) \in \mathbb{Q}[x]$, is necessarily a root of some polynomial $h(x) \in \mathbb{Z}[x]$; just clear the denominators of the coefficients of $g(x)$.

Corollary 3.91. *If an algebraic integer α is rational, then $\alpha \in \mathbb{Z}$. More precisely, if $f(x) \in \mathbb{Z}[x] \subseteq \mathbb{Q}[x]$ is a monic polynomial, then every rational root α of $f(x)$ is an integer that divides the constant term.*

Proof. If $f(x) = a_0 + a_1x + \cdots + a_nx^n$ is monic, then $a_n = 1$, and Theorem 3.90 applies at once. •

For example, consider $f(x) = x^3 + 4x^2 - 2x - 1 \in \mathbb{Q}[x]$. By Corollary 3.65, this cubic is irreducible if and only if it has no rational root. As $f(x)$ is monic, the candidates for rational roots are ± 1, for these are the only divisors of -1 in $\mathbb{Z}$. But $f(1) = 2$ and $f(-1) = 4$, so that neither 1 nor -1 is a root. Thus, $f(x)$ has no roots in $\mathbb{Q}$, and hence $f(x)$ is irreducible in $\mathbb{Q}[x]$.

Corollary 3.91 gives a new solution of Exercise 1.70 on page 58. If m is an integer that is not a perfect square, then the polynomial $x^2 - m$ has no integer roots, and so $\sqrt{m}$ is irrational. Indeed, the reader can now generalize to nth roots: if m is not an nth power of an integer, then $\sqrt[n]{m}$ is irrational, for any rational root of $x^n - m$ must be an integer.

We are now going to find several conditions that imply that a polynomial $f(x) \in \mathbb{Z}[x]$ does not factor in $\mathbb{Z}[x]$ as a product of polynomials of smaller degree. Since $\mathbb{Z}$ is not a field, this does not force $f(x)$ be irreducible in $\mathbb{Z}[x]$; for example, $f(x) = 2x + 2$ does not so factor in $\mathbb{Z}[x]$, yet it is not irreducible there. However, C. F. Gauss proved that if $f(x) \in \mathbb{Z}[x]$ does not factor as a product of polynomials in $\mathbb{Z}[x]$ of smaller degree, then $f(x)$ is irreducible in $\mathbb{Q}[x]$. We prove this result after first proving several lemmas.

→ **Definition.** A polynomial $f(x) = a_0 + a_1 x + a_2 x^2 + \cdots + a_n x^n \in \mathbb{Z}[x]$ is called *primitive* if the gcd of its coefficients is 1.

Of course, every monic polynomial in $\mathbb{Z}[x]$ is primitive. It is easy to see that if d is the gcd of the coefficients of $f(x)$, then $(1/d)f(x)$ is a primitive polynomial in $\mathbb{Z}[x]$.

→ **Lemma 3.92 (Gauss's Lemma).** *If $f(x)$, $g(x) \in \mathbb{Z}[x]$ are both primitive, then their product $f(x)g(x)$ is also primitive.*

Proof. Let $f(x) = \sum a_i x^i$, $g(x) = \sum b_j x^j$, and $f(x)g(x) = \sum c_k x^k$. If $f(x)g(x)$ is not primitive, then there is a prime p that divides every c_k. Since $f(x)$ is primitive, at least one of its coefficients is not divisible by p; let a_i be the first such (i.e., i is the smallest index of such a coefficient). Similarly, let b_j be the first coefficient of $g(x)$ that is not divisible by p. The definition of multiplication of polynomials gives

$$a_i b_j = c_{i+j} - (a_0 b_{i+j} + \cdots + a_{i-1} b_{j+1} + a_{i+1} b_{j-1} + \cdots + a_{i+j} b_0).$$

Each term on the right side is divisible by p, and so p divides $a_i b_j$. As p divides neither a_i nor b_j, however, this contradicts Euclid's lemma in $\mathbb{Z}$. •

Here is a more elegant proof of this last lemma. If a polynomial $h(x) \in \mathbb{Z}[x]$ is not primitive, then there exists a prime p that divides each of its coefficients (if the gcd is $d > 1$, then take for p any prime divisor of d). In this case, all the coefficients of $h(x)$ are 0 in $\mathbb{F}_p$. If $\varphi \colon \mathbb{Z} \to \mathbb{F}_p$ is the natural map $a \mapsto [a]$, then Theorem 3.33 shows that the function $\varphi^* \colon \mathbb{Z}[x] \to \mathbb{F}_p[x]$, which reduces all the coefficients of a polynomial mod p, is a ring homomorphism. If the product $f(x)g(x)$ of primitive polynomials is not primitive, then $0 = \varphi^*(fg) = \varphi^*(f)\varphi^*(g)$ in $\mathbb{F}_p[x]$. On the other hand, neither $\varphi^*(f)$ nor $\varphi^*(g)$ is 0, because they are primitive. We have contradicted the fact that $\mathbb{F}_p[x]$ is a domain.

→ **Lemma 3.93.** *Every nonzero $f(x) \in \mathbb{Q}[x]$ has a unique factorization*

$$f(x) = c(f) f^{\#}(x),$$

where $c(f) \in \mathbb{Q}$ is positive and $f^{\#}(x) \in \mathbb{Z}[x]$ is primitive.

Proof. There are integers a_i and b_i with

$$f(x) = (a_0/b_0) + (a_1/b_1)x + \cdots + (a_n/b_n)x^n \in \mathbb{Q}[x].$$

Define $B = b_0 b_1 \ldots b_n$, so that $g(x) = Bf(x) \in \mathbb{Z}[x]$. Now define $D = \pm d$, where d is the gcd of the coefficients of $g(x)$; the sign is chosen to make the rational number D/B positive. Now $(B/D)f(x) = (1/D)g(x)$ lies in $\mathbb{Z}[x]$, and it is a primitive polynomial. If we define $c(f) = D/B$ and $f^\#(x) = (B/D)f(x)$, then $f(x) = c(f)f^\#(x)$ is a desired factorization.

Suppose that $f(x) = eh(x)$ is a second such factorization, so that e is a positive rational number and $h(x) \in \mathbb{Z}[x]$ is primitive. Now $c(f)f^\#(x) = f(x) = eh(x)$, so that $f^\#(x) = [e/c(f)]h(x)$. Write $e/c(f)$ in lowest terms: $e/c(f) = u/v$, where u and v are relatively prime positive integers. The equation $vf^\#(x) = uh(x)$ holds in $\mathbb{Z}[x]$; equating like coefficients, v is a common divisor of each coefficient of $uh(x)$. Since $(u, v) = 1$, Euclid's lemma in $\mathbb{Z}$ shows that v is a (positive) common divisor of the coefficients of $h(x)$. Since $h(x)$ is primitive, it follows that $v = 1$. A similar argument shows that $u = 1$. Since $e/c(f) = u/v = 1$, we have $e = c(f)$ and hence $h(x) = f^\#(x)$. •

→ **Definition.**   The rational $c(f)$ in Lemma 3.93 is called the ***content*** of $f(x)$.

Corollary 3.94.   *If $f(x) \in \mathbb{Z}[x]$, then $c(f) \in \mathbb{Z}$.*

Proof.   If d is the gcd of the coefficients of $f(x)$, then $(1/d)f(x) \in \mathbb{Z}[x]$ is primitive. Since $d[(1/d)f(x)]$ is a factorization of $f(x)$ as the product of a positive rational d (which is even a positive integer) and a primitive polynomial, the uniqueness in Lemma 3.93 gives $c(f) = d \in \mathbb{Z}$. •

Corollary 3.95.   *If $f(x) \in \mathbb{Q}[x]$ factors as $f(x) = g(x)h(x)$, then*

$$c(f) = c(g)c(h) \qquad and \qquad f^\#(x) = g^\#(x)h^\#(x).$$

Proof.   We have

$$f(x) = g(x)h(x)$$
$$c(f)f^\#(x) = [c(g)g^\#(x)][c(h)h^\#(x)]$$
$$= c(g)c(h)g^\#(x)h^\#(x).$$

Since $g^*(x)h^\#(x)$ is primitive, by Lemma 3.92, and $c(g)c(h)$ is a positive rational, the uniqueness of the factorization in Lemma 3.93 gives $c(f) = c(g)c(h)$ and $f^\#(x) = g^\#(x)h^\#(x)$. •

→ **Theorem 3.96 (Gauss).** *Let $f(x) \in \mathbb{Z}[x]$. If*

$$f(x) = G(x)H(x) \text{ in } \mathbb{Q}[x],$$

then there is a factorization

$$f(x) = g(x)h(x) \text{ in } \mathbb{Z}[x],$$

where $\deg(g) = \deg(G)$ *and* $\deg(h) = \deg(H)$. *Therefore, if* $f(x)$ *does not factor into polynomials of smaller degree in* $\mathbb{Z}[x]$, *then* $f(x)$ *is irreducible in* $\mathbb{Q}[x]$.

Proof. By Corollary 3.95, there is a factorization

$$f(x) = c(G)c(H)G^{\#}(x)H^{\#}(x) \text{ in } \mathbb{Q}[x],$$

where $G^{\#}(x)$, $H^{\#}(x) \in \mathbb{Z}[x]$ are primitive polynomials. But $c(G)c(H) = c(f)$, by Corollary 3.95, and $c(f) \in \mathbb{Z}$, by Corollary 3.94 (which applies because $f(x) \in \mathbb{Z}[x]$). Therefore, $f(x) = g(x)h(x)$ is a factorization in $\mathbb{Z}[x]$, where $g(x) = c(f)G^{\#}(x)$ and $h(x) = H^{\#}(x)$. •

Remark. Gauss used these ideas to prove Theorem 7.21: $k[x_1, \ldots, x_n]$, the ring of all polynomials in n variables with coefficients in a field k, is a unique factorization domain. ◀

The next criterion for irreducibility uses the integers mod p.

→ **Theorem 3.97.** *Let* $f(x) = a_0 + a_1 x + a_2 x^2 + \cdots + x^n \in \mathbb{Z}[x]$ *be monic, and let* p *be a prime. If* $f^*(x) = [a_0] + [a_1]x + [a_2]x^2 + \cdots + x^n$ *is irreducible in* $\mathbb{F}_p[x]$, *then* $f(x)$ *is irreducible in* $\mathbb{Q}[x]$.

Proof. By Theorem 3.33, the natural map $\varphi \colon \mathbb{Z} \to \mathbb{F}_p$ defines a homomorphism $\varphi^* \colon \mathbb{Z}[x] \to \mathbb{F}_p[x]$ by

$$\varphi^*(b_0 + b_1 x + b_2 x^2 + \cdots) = [b_0] + [b_1]x + [b_2]x^2 + \cdots,$$

that is, just reduce all the coefficients mod p. If $g(x) \in \mathbb{Z}[x]$, denote its image $\varphi^*(g(x)) \in \mathbb{F}_p[x]$ by $g^*(x)$. Suppose that $f(x)$ factors in $\mathbb{Z}[x]$; say, $f(x) = g(x)h(x)$, where $\deg(g) < \deg(f)$ and $\deg(h) < \deg(f)$ [of course, $\deg(f) = \deg(g) + \deg(h)$]. Now $f^*(x) = g^*(x)h^*(x)$, because φ^* is a ring homomorphism, so that $\deg(f^*) = \deg(g^*) + \deg(h^*)$. Since $f(x)$ is monic, $f^*(x)$ is also monic, and so $\deg(f^*) = \deg(f)$. [Of course, the hypothesis that $f(x)$ be monic can be relaxed; we may assume, instead, that p does not divide its leading coefficient.] Thus, both $g^*(x)$ and $h^*(x)$ have degrees less than $\deg(f^*)$, contradicting the irreducibility of $f^*(x)$. Therefore, $f(x)$ is not a product of polynomials in $\mathbb{Z}[x]$ of smaller degree, and so Gauss's theorem says that $f(x)$ is irreducible in $\mathbb{Q}[x]$. •

The converse of Theorem 3.97 is false; its criterion does not always work. It is not difficult to find an irreducible $f(x) \in \mathbb{Z}[x] \subseteq \mathbb{Q}[x]$ with $f(x)$ factoring mod p for some prime p. Exercise 3.100 on page 304 shows that $x^4 + 1$ is an irreducible polynomial in $\mathbb{Q}[x]$ which factors in $\mathbb{F}_p[x]$ for every prime p.

Theorem 3.97 says that if one can find a prime p with $f^*(x)$ irreducible in $\mathbb{F}_p[x]$, then $f(x)$ is irreducible in $\mathbb{Q}[x]$. Until now, the finite fields $\mathbb{F}_p$ have been oddities; $\mathbb{F}_p$ has appeared only as a curious artificial construct. Now the finiteness of $\mathbb{F}_p$ is a genuine advantage, for there are only a finite number of polynomials in $\mathbb{F}_p[x]$ of any given degree. In principle, then, one can test whether a polynomial of degree n in $\mathbb{F}_p[x]$ is irreducible by just looking at *all* the possible factorizations of it.

Since it becomes tiresome not to do so, we are now going to write the elements of $\mathbb{F}_p$ without brackets.

Example 3.98.
We determine the irreducible polynomials in $\mathbb{F}_2[x]$ of small degree.

As always, the linear polynomials x and $x + 1$ are irreducible.

There are four quadratics: x^2; $x^2 + x$; $x^2 + 1$; $x^2 + x + 1$ (more generally, there are p^n monic polynomials of degree n in $\mathbb{F}_p[x]$, for there are p choices for each of the n coefficients $a_0, \ldots, a_{n-1}$). Since each of the first three has a root in $\mathbb{F}_2$, there is only one irreducible quadratic.

There are eight cubics, of which four are reducible because their constant term is 0. The remaining polynomials are

$$x^3 + 1; \qquad x^3 + x + 1; \qquad x^3 + x^2 + 1; \qquad x^3 + x^2 + x + 1.$$

Since 1 is a root of the first and fourth, the middle two are the only irreducible cubics.

There are 16 quartics, of which eight are reducible because their constant term is 0. Of the eight with nonzero constant term, those having an even number of nonzero coefficients have 1 as a root. There are now only four surviving polynomials $f(x)$, and each of them has no roots in $\mathbb{F}_2$; that is, they have no linear factors. If $f(x) = g(x)h(x)$, then both $g(x)$ and $h(x)$ must be irreducible quadratics. But there is only one irreducible quadratic, namely, $x^2 + x + 1$, and so $(x^2 + x + 1)^2 = x^4 + x^2 + 1$ is reducible while the other three quartics are irreducible. The following list summarizes these observations.

Irreducible Polynomials of Low Degree over $\mathbb{F}_2$

degree 2: $x^2 + x + 1$.
degree 3: $x^3 + x + 1$; $x^3 + x^2 + 1$.
degree 4: $x^4 + x^3 + 1$; $x^4 + x + 1$; $x^4 + x^3 + x^2 + x + 1$.   ◄

Example 3.99.
Here is a list of the monic irreducible quadratics and cubics in $\mathbb{F}_3[x]$. The reader can verify that the list is correct by first enumerating all such polynomials; there are 6 monic quadratics having nonzero constant term, and there are 18 monic cubics having nonzero constant term. It must then be checked which of these have 1 or -1 as a root (it is more convenient to write -1 instead of 2).

Monic Irreducible Quadratics and Cubics over $\mathbb{F}_3$

degree 2:	$x^2 + 1$;	$x^2 + x - 1$;	$x^2 - x - 1.$
degree 3:	$x^3 - x + 1$;	$x^3 + x^2 - x + 1$;	$x^3 - x^2 + 1$;
	$x^3 - x^2 + x + 1$;	$x^3 + x^2 - 1$;	$x^3 - x^2 - 1$;
	$x^3 + x^2 + x - 1$;	$x^3 - x^2 - x - 1.$ ◀	

Example 3.100.

(i) We show that $f(x) = x^4 - 5x^3 + 2x + 3$ is an irreducible polynomial in $\mathbb{Q}[x]$. By Corollary 3.91, the only candidates for rational roots of $f(x)$ are $1, -1, 3, -3$, and the reader may check that none of these is a root. Since $f(x)$ is a quartic, one cannot yet conclude that $f(x)$ is irreducible, for it might be a product of (irreducible) quadratics.

Let us try the criterion of Theorem 3.97. Since $f^*(x) = x^4 + x^3 + 1$ in $\mathbb{F}_2[x]$ is irreducible, by Example 3.98, it follows that $f(x)$ is irreducible in $\mathbb{Q}[x]$. [It was not necessary to check that $f(x)$ has no rational roots; irreducibility of $f^*(x)$ is enough to conclude irreducibility of $f(x)$.]

(ii) Let $\Phi_5(x) = x^4 + x^3 + x^2 + x + 1 \in \mathbb{Q}[x]$.
In Example 3.98, we saw that $(\Phi_5)^*(x) = x^4 + x^3 + x^2 + x + 1$ is irreducible in $\mathbb{F}_2[x]$, and so $\Phi_5(x)$ is irreducible in $\mathbb{Q}[x]$. ◀

As any linear polynomial over a field, $\Phi_2(x) = x + 1$ is irreducible in $\mathbb{Q}[x]$; $\Phi_3(x) = x^2 + x + 1$ is irreducible in $\mathbb{Q}[x]$ because it has no rational roots; we have just seen that $\Phi_5(x)$ is irreducible in $\mathbb{Q}[x]$. Let us introduce another irreducibility criterion in order to prove that $\Phi_p(x)$ is irreducible in $\mathbb{Q}[x]$ for all primes p.

Lemma 3.101. *Let $g(x) \in \mathbb{Z}[x]$. If there is $c \in \mathbb{Z}$ with $g(x + c)$ irreducible in $\mathbb{Z}[x]$, then $g(x)$ is irreducible in $\mathbb{Q}[x]$.*

Proof. By Theorem 3.33, the function $\varphi: \mathbb{Z}[x] \to \mathbb{Z}[x]$, given by

$$f(x) \mapsto f(x + c),$$

is an isomorphism. If $g(x) = s(x)t(x)$, then $g(x + c) = \varphi(g(x)) = \varphi(st) = \varphi(s)\varphi(t)$ is a forbidden factorization of $g(x + c)$. Therefore, Gauss's theorem says that $g(x)$ is irreducible in $\mathbb{Q}[x]$. •

→ **Theorem 3.102 (Eisenstein Criterion).**

Let $f(x) = a_0 + a_1 x + \cdots + a_n x^n \in \mathbb{Z}[x]$. If there is a prime p dividing a_i for all $i < n$ but with $p \nmid a_n$ and $p^2 \nmid a_0$, then $f(x)$ is irreducible in $\mathbb{Q}[x]$.

Proof.   Assume, on the contrary, that

$$f(x) = (b_0 + b_1 x + \cdots + b_m x^m)(c_0 + c_1 x + \cdots + c_k x^k),$$

where $m < n$ and $k < n$; by Theorem 3.96, we may assume that both factors lie in $\mathbb{Z}[x]$. Now $p \mid a_0 = b_0 c_0$, so that Euclid's lemma in $\mathbb{Z}$ gives $p \mid b_0$ or $p \mid c_0$; since $p^2 \nmid a_0$, only one of them is divisible by p, say, $p \mid c_0$ but $p \nmid b_0$. By hypothesis, the leading coefficient $a_n = b_m c_k$ is not divisible by p, so that p does not divide c_k (or b_m). Let c_r be the first coefficient not divisible by p (so that p does divide $c_0, \ldots, c_{r-1}$). If $r < n$, then $p \mid a_r$, and so $b_0 c_r = a_r - (b_1 c_{r-1} + \cdots + b_r c_0)$ is also divisible by p. This contradicts Euclid's lemma, for $p \mid b_0 c_r$, but p divides neither factor. It follows that $r = n$; hence $n \geq k \geq r = n$, and so $k = n$, contradicting $k < n$. Therefore, $f(x)$ is irreducible in $\mathbb{Q}[x]$.   •

Remark.   R. Singer found the following elegant proof of Eisenstein's criterion.

Let $\varphi^*: \mathbb{Z}[x] \to \mathbb{F}_p[x]$ be the ring homomorphism that reduces coefficients mod p, and let $f^*(x)$ denote $\varphi^*(f(x))$. If $f(x)$ is not irreducible in $\mathbb{Q}[x]$, then Gauss's theorem gives polynomials $g(x), h(x) \in \mathbb{Z}[x]$ with $f(x) = g(x)h(x)$, where $g(x) = b_0 + b_1 x + \cdots + b_m x^m$, $h(x) = c_0 + c_1 x + \cdots + c_k x^k$, and $m, k > 0$. There is thus an equation $f^*(x) = g^*(x)h^*(x)$ in $\mathbb{F}_p[x]$.

Since $p \nmid a_n$, we have $f^*(x) \neq 0$; in fact, $f^*(x) = u x^n$ for some unit $u \in \mathbb{F}_p$, because all its coefficients aside from its leading coefficient are 0. By Theorem 3.84, unique factorization in $\mathbb{F}_p[x]$, we must have $g^*(x) = v x^m$ and $h^*(x) = w x^k$, where v and w are units in $\mathbb{F}_p$, because the only monic divisors of x^n are powers of x. It follows that each of $g^*(x)$ and $h^*(x)$ has constant term 0; that is, $[b_0] = 0 = [c_0]$ in $\mathbb{F}_p$; equivalently, $p \mid b_0$ and $p \mid c_0$. But $a_0 = b_0 c_0$, and so $p^2 \mid a_0$, a contradiction. Therefore, $f(x)$ is irreducible in $\mathbb{Q}[x]$.   ◄

Recall, for $n \geq 1$, that $x^n - 1 = \prod_{d \mid n} \Phi_d(x)$, where $\Phi_d(x)$ is the dth *cyclotomic polynomial* (we proved, in Proposition 3.47, that $\Phi_d(x) \in \mathbb{Z}[x]$ for all $d \geq 1$). In particular, if $n = p$ is prime, then

$$\Phi_p(x) = (x^p - 1)/(x - 1) = x^{p-1} + x^{p-2} + \cdots + x + 1.$$

→ **Corollary 3.103 (Gauss).**   *For every prime p, the pth cyclotomic polynomial $\Phi_p(x)$ is irreducible in $\mathbb{Q}[x]$.*

Remark.   The cyclotomic polynomial $\Phi_d(x)$ is irreducible in $\mathbb{Q}[x]$ for every (not necessarily prime) $d \geq 1$ (see Tignol, *Galois' Theory of Algebraic Equations*, Theorem 12.31).   ◄

Proof. Since $\Phi_p(x) = (x^p - 1)/(x - 1)$, we have

$$\Phi_p(x + 1) = [(x + 1)^p - 1]/x$$

$$= x^{p-1} + \binom{p}{1}x^{p-2} + \binom{p}{2}x^{p-3} + \cdots + p.$$

Since p is prime, Proposition 1.39 shows that Eisenstein's criterion applies; we conclude that $\Phi_p(x + 1)$ is irreducible in $\mathbb{Q}[x]$. By Lemma 3.101, $\Phi_p(x)$ is irreducible in $\mathbb{Q}[x]$. •

We do not say that $x^{n-1} + x^{n-2} + \cdots + x + 1$ is irreducible when n is not prime. For example, when $n = 4$, $x^3 + x^2 + x + 1 = (x + 1)(x^2 + 1)$.

EXERCISES

H **3.86** True or false with reasons.
- **(i)** $\sqrt{3}$ is an algebraic integer.
- **(ii)** $\frac{13}{78}$ is a rational root of $1 + 5x + 6x^2$.
- **(iii)** If $f(x) = 3x^4 + ax^3 + bx^2 + cx + 7$ with $a, b, c \in \mathbb{Z}$, then the roots of $f(x)$ in $\mathbb{Q}$, if any, lie in $\{\pm 1, \pm 7, \pm \frac{1}{3}, \pm \frac{7}{3}\}$.
- **(iv)** If $f(x) = 3x^4 + ax^3 + bx^2 + cx + 7$ with $a, b, c \in \mathbb{Q}$, then the roots of $f(x)$ in $\mathbb{Q}$, if any, lie in $\{\pm 1, \pm 7, \pm \frac{1}{3}, \pm \frac{7}{3}\}$.
- **(v)** $6x^2 + 10x + 15$ is a primitive polynomial.
- **(vi)** Every primitive polynomial in $\mathbb{Z}[x]$ is irreducible.
- **(vii)** Every irreducible polynomial in $\mathbb{Z}[x]$ is primitive.
- **(viii)** Every monic polynomial in $\mathbb{Z}[x]$ is primitive.
- **(ix)** The content of $3x + \frac{1}{5}$ is $\frac{3}{5}$.
- **(x)** The content of $3x + \frac{6}{5}$ is $\frac{3}{5}$.
- **(xi)** If $f(x) = g(x)h(x)$ in $\mathbb{Q}[x]$, and if $f(x)$ has all its coefficients in $\mathbb{Z}$, then all the coefficients of $g(x)$ and $h(x)$ also lie in $\mathbb{Z}$.
- **(xii)** For every integer c, the polynomial $(x+c)^2 - (x+c) - 1$ is irreducible in $\mathbb{Q}[x]$.
- **(xiii)** For all integers n, the polynomial $x^8 + 5x^3 + 5n$ is irreducible in $\mathbb{Q}[x]$.
- **(xiv)** The polynomial $x^7 + 9x^3 + (9n + 6)$ is irreducible in $\mathbb{Q}[x]$ for every integer n.

*H **3.87** Determine whether the following polynomials are irreducible in $\mathbb{Q}[x]$.
- **(i)** $f(x) = 3x^2 - 7x - 5$.
- **(ii)** $f(x) = 350x^3 - 25x^2 + 34x + 1$.
- **(iii)** $f(x) = 2x^3 - x - 6$.
- **(iv)** $f(x) = 8x^3 - 6x - 1$.
- **(v)** $f(x) = x^3 + 6x^2 + 5x + 25$.
- **(vi)** $f(x) = x^5 - 4x + 2$.
- **(vii)** $f(x) = x^4 + x^2 + x + 1$.

(viii) $f(x) = x^4 - 10x^2 + 1$.

(ix) $f(x) = x^6 - 210x - 616$.

(x) $f(x) = 350x^3 + x^2 + 4x + 1$.

3.88 If p is a prime, prove that there are exactly $\frac{1}{3}\left(p^3 - p\right)$ monic irreducible cubic polynomials in $\mathbb{F}_p[x]$.

H **3.89** Prove that there are exactly 6 irreducible quintics in $\mathbb{F}_2[x]$.

3.90   H (i)   If $a \neq \pm 1$ is a squarefree integer, show that $x^n - a$ is irreducible in $\mathbb{Q}[x]$ for every $n \geq 1$. Conclude that there are irreducible polynomials in $\mathbb{Q}[x]$ of every degree $n \geq 1$.

(ii)   If $a \neq \pm 1$ is a squarefree integer, prove that $\sqrt[n]{a}$ is irrational.

H **3.91** Let k be a field, and let $f(x) = a_0 + a_1x + \cdots + a_nx^n \in k[x]$ have degree n. If $f(x)$ is irreducible, then so is $a_n + a_{n-1}x + \cdots + a_0x^n$.

→ 3.8 QUOTIENT RINGS AND FINITE FIELDS

The fundamental theorem of algebra states that every nonconstant polynomial $f(x) \in \mathbb{C}[x]$ has all its roots in $\mathbb{C}$ (more precisely, $f(x)$ is a product of linear polynomials in $\mathbb{C}[x]$). We now return to the study of ideals and homomorphisms in order to prove a "local" analog of the Fundamental Theorem of Algebra for polynomials over an arbitrary field k: given a polynomial $f(x) \in k[x]$, then there is some field K containing k and all the roots of $f(x)$. We call this a *local* analog, for even though all the roots of $f(x)$ are in K, there may be other polynomials in $k[x]$ which do not have all their roots in K. The main idea behind the construction of K involves quotient rings, a straightforward generalization of the construction of $\mathbb{I}_m$ which we now review.

Given $\mathbb{Z}$ and an integer m, the congruence relation on $\mathbb{Z}$ is defined by

$$a \equiv b \bmod m \quad \text{if and only if} \quad m \mid (a - b).$$

This definition can be rewritten: $a \equiv b \bmod m$ if and only if $a - b \in (m)$, where (m) denotes the principal ideal in $\mathbb{Z}$ generated by m. Congruence is an equivalence relation on $\mathbb{Z}$, its equivalence classes $[a]$ are called *congruence classes*, and the set $\mathbb{I}_m$ is the family of all the congruence classes.

We now begin the new construction. Given a commutative ring R and an ideal I, define the relation **congruence** mod I on R:

$$a \equiv b \bmod I \quad \text{if and only if} \quad a - b \in I.$$

→ **Lemma 3.104.** *If R is a commutative ring and I is an ideal in R, then congruence mod I is an equivalence relation on R.*

Proof.

(i) *Reflexivity*: if $a \in R$, then $a - a = 0 \in I$; hence, $a \equiv a \bmod I$.

(ii) *Symmetry*: if $a \equiv b \bmod I$, then $a - b \in I$. Since $-1 \in R$, we have $b - a = (-1)(a - b) \in I$, and so $b \equiv a \bmod I$.

(iii) *Transitivity*: if $a \equiv b \bmod I$ and $b \equiv c \bmod I$, then $a - b \in I$ and $b - c \in I$. Hence, $a - c = (a - b) + (b - c) \in I$, and $a \equiv c \bmod I$. •

→ **Definition.** If R is a commutative ring and I is an ideal in R, then the equivalence class of $a \in R$, namely, $[a] = \{b \in R : b \equiv a \bmod I\}$ is called the ***congruence class*** of $a \bmod I$. The set of all the congruence classes mod I is denoted by R/I:

$$R/I = \big\{[a] : a \in R\big\}.$$

Recall that addition and multiplication were defined on $\mathbb{I}_m$ by the formulas

$$[a] + [b] = [a + b] \qquad \text{and} \qquad [a][b] = [ab].$$

It was not obvious that these functions $\mathbb{I}_m \times \mathbb{I}_m \to \mathbb{I}_m$ are well-defined, and we were obliged to prove that they are (see Propositions 2.103 and 2.105). These formulas also give addition and multiplication operations on R/I.

Lemma 3.105. *The functions*

$$\alpha \colon (R/I) \times (R/I) \to R/I, \text{ given by } ([a], [b]) \mapsto [a + b],$$

and

$$\mu \colon (R/I) \times (R/I) \to R/I, \text{ given by } ([a], [b]) \mapsto [ab],$$

are well-defined operations on R/I.

Proof. Let us prove that α and μ are well-defined. Recall Lemma 2.19: if $\equiv$ is an equivalence relation on a set X, then $[a] = [a']$ if and only if $a \equiv a'$; here, $[a] = [a']$ if and only if $a - a' \in I$. Is addition well-defined? If $[a] = [a']$ and $[b] = [b']$, is $[a + b] = [a' + b']$; that is, if $a - a' \in I$ and $b - b' \in I$, is $(a + b) - (a' + b') \in I$? The answer is "yes," for $(a + b) - (a' + b') = (a - a') + (b - b') \in I$. Hence, $[a + b] = [a' + b']$. Is multiplication well-defined; that is, if $[a] = [a']$ and $[b] = [b']$, is $[ab] = [a'b']$? Now $a - a' \in I$ and $b - b' \in I$, and so

$$ab - a'b' = (ab - ab') + (ab' - a'b') = a(b - b') + (a - a')b' \in I,$$

for ideals are closed under products ri for $r \in R$ and $i \in I$. Therefore, $[ab] = [a'b']$. •

The proof that R/I, equipped with the operations in Lemma 3.105, is a commutative ring is, mutatis mutandis, precisely the proof that $\mathbb{I}_m$ is a commutative ring. In essence, the ring axioms hold in R/I because they are inherited from the ring axioms in R.

→ **Theorem 3.106.**   *If I is an ideal in a commutative ring R, then R/I, equipped with the addition and multiplication defined in* Lemma 3.105, *is a commutative ring.*

Proof.   We check each axiom of the definition of commutative ring on page 219.
(i) Since $a + b = b + a$ in R, we have

$$[a] + [b] = [a + b] = [b + a] = [b] + [a].$$

(ii) Now $[a] + ([b] + [c]) = [a] + [b + c] = [a + (b + c)]$, while $([a] + [b]) + [c] = [a + b] + [c] = [(a + b) + c]$, and the result follows because $a + (b + c) = (a + b) + c$ in R.
(iii) Define $0 = [0]$, where the 0 in brackets is the zero element of R. Now $0 + [a] = [0 + a] = [a]$, because $0 + a = a$ in R.
(iv) Define $[a]' = [-a]$. Now $[-a] + [a] = [-a + a] = [0] = 0$.
(v) Since $ab = ba$ in R, we have $[a][b] = [ab] = [ba] = [b][a]$.
(vi) Now $[a]([b][c]) = [a][bc] = [a(bc)]$, while $([a][b])[c] = [ab][c] = [(ab)c]$, and the result follows because $a(bc) = (ab)c$ in R.
(vii) Define $1 = [1]$, where the 1 in brackets is the *one* in R. Now $1[a] = [1a] = [a]$, because $1a = a$ in R.
(viii) We use the distributivity in R: $[a]([b] + [c]) = [a][b + c] = [a(b + c)]$ and $[a][b] + [a][c] = [ab] + [ac] = [ac + ab] = [a(b + c)]$.   •

→ **Definition.**   The commutative ring R/I just constructed is called the ***quotient ring*** of R modulo I (pronounced R mod I).

Congruence classes $[a] = \{b \in \mathbb{Z} : b = a + km \text{ for } k \in \mathbb{Z}\}$ in $\mathbb{I}_m$ can also be described as *cosets*, $[a] = a + (m)$, and this description generalizes to describe the elements of arbitrary quotient rings.

→ **Definition.**   If R is a commutative ring and I is an ideal, then a ***coset*** is a subset of R of the form

$$a + I = \{b \in R : b = a + i \text{ for some } i \in I\}.$$

We now show that cosets are the same as congruence classes.

→ **Lemma 3.107.**   *If R is a commutative ring and I is an ideal, then for each $a \in R$, the congruence class $[a]$ in R/I is the coset $a + I$.*

Proof.   If $b \in [a]$, then $b - a \in I$. Hence, $b = a + (b - a) \in a + I$, and so $[a] \subseteq a + I$. For the reverse inclusion, if $c \in a + I$, then $c = a + i$ for some $i \in I$, and so $c - a \in I$. Hence, $c \equiv a \mod I$, $c \in [a]$, and $a + I \subseteq [a]$. Therefore, $[a] = a + I$.   •

The coset notation $a + I$ is the notation most commonly used; thus,

$$R/I = \{a + I : a \in R\}.$$

Remark.   If we forget the multiplication in a commutative ring R, then R is an additive abelian group, and every ideal I in R is a subgroup. Since every subgroup of an abelian group is a normal subgroup, the quotient group R/I is defined. We claim that this quotient group coincides with the additive group of the quotient ring R/I. The elements of either group are the same (they are cosets of I), and the addition of cosets in each is the same as well. In particular, the principal ideal (m) in $\mathbb{Z}$ is often denoted by $m\mathbb{Z}$ and we have been denoting the quotient ring $\mathbb{Z}/m\mathbb{Z}$ by $\mathbb{I}_m$.   ◄

The natural map $\pi : \mathbb{Z} \to \mathbb{I}_m$, defined by $\pi(a) = [a] = a + (m)$, is a ring homomorphism, and so is its generalization to quotient rings.

→ **Definition.**   If R is a commutative ring and I is an ideal, then the ***natural map*** $\pi : R \to R/I$ is defined by $\pi(a) = a + I$ for all $a \in R$.

Let us show that $\pi : R \to R/I$ is a homomorphism. We have $\pi(1) = 1 + I$, which is the identity in R/I. The definition of addition in R/I gives

$$\pi(a) + \pi(b) = (a + I) + (b + I) = a + b + I = \pi(a + b),$$

and the definition of multiplication gives

$$\pi(a)\pi(b) = (a + I)(b + I) = ab + I = \pi(ab).$$

We can now prove a converse to Proposition 3.38.

→ **Proposition 3.108.**   *Every ideal I in a commutative ring R is the kernel of some homomorphism. More precisely, the natural map $\pi : R \to R/I$ is a surjective homomorphism whose kernel is I.*

Proof.   The elements of R/I are cosets $a + I$; since $a + I = \pi(a)$, the natural map is surjective. Clearly, $I \subseteq \ker \pi$, for if $a \in I$, then $\pi(a) = a + I = 0 + I$. For the reverse inclusion, if $a \in \ker \pi$, then $\pi(a) = a + I = I$, and so $a \in I$. Therefore, $\ker \pi = I$.

●

Proposition 2.101 describes the congruence classes $[a]$ in $\mathbb{I}_m = \mathbb{Z}/(m)$ in a very simple way; they are the cosets of all possible remainders after dividing by m:

$$\mathbb{I}_m = \big\{ [0], [1], [2], \dots, [m-1] \big\}.$$

In general, there is no such easy description of the elements of R/I. On the other hand, we shall soon see (Theorem 3.114) that there is a version of this description for $k[x]/(f(x))$, where k is a field and $f(x) \in k[x]$.

Remark.   For readers familiar with groups, the proof of Theorem 3.106 can be short-ened a bit. If we forget the multiplication in a commutative ring R, then an ideal I is a subgroup of the additive group R; since R is an abelian group, the subgroup I is neces-sarily normal, and so the quotient group R/I is defined. Thus, axioms (i) through (iv) in the definition of commutative ring hold. One now defines multiplication, shows that it is well-defined, and verifies axioms (v) through (viii). Presumably, the term *quotient ring* arose by analogy with the term *quotient group*. The proof of Proposition 3.108 can also be shortened a bit. The natural map $\pi: R \to R/I$ is a group homomorphism be-tween the additive groups of R and of R/I, and one need check only that $\pi(1) = 1 + I$ and π preserves multiplication, as we have just done.   ◄

$\rightarrow$ **Theorem 3.109 (First Isomorphism Theorem).**   *If $\varphi: R \to S$ is a homomorphism of commutative rings, then $\ker \varphi$ is an ideal in R, $\operatorname{im} \varphi$ is a subring of S, and there is an isomorphism*

$$\widetilde{\varphi}: R/\ker \varphi \to \operatorname{im} \varphi$$

defined by $\widetilde{\varphi}: a + \ker \varphi \mapsto \varphi(a)$.

Proof.   Let $I = \ker \varphi$. We have already seen, in Proposition 3.38, that I is an ideal in R and that $\operatorname{im} \varphi$ is a subring of A.

$\widetilde{\varphi}$ *is well-defined.*

If $a + I = b + I$, then $a - b \in I = \ker \varphi$, so that $\varphi(a - b) = 0$. But $\varphi(a - b) = \varphi(a) - \varphi(b)$; hence, $\widetilde{\varphi}(a + I) = \varphi(a) = \varphi(b) = \widetilde{\varphi}(b + I)$.

$\widetilde{\varphi}$ *is a homomorphism.*

First, $\widetilde{\varphi}(1 + I) = \varphi(1) = 1$.

Second,

$$\begin{aligned}
\widetilde{\varphi}((a + I) + (b + I)) &= \widetilde{\varphi}(a + b + I) \\
&= \varphi(a + b) \\
&= \varphi(a) + \varphi(b) \\
&= \widetilde{\varphi}(a + I) + \widetilde{\varphi}(b + I).
\end{aligned}$$

Third,

$$\begin{aligned}
\widetilde{\varphi}((a + I)(b + I)) &= \widetilde{\varphi}(ab + I) \\
&= \varphi(ab) \\
&= \varphi(a)\varphi(b) \\
&= \widetilde{\varphi}(a + I)\widetilde{\varphi}(b + I).
\end{aligned}$$

$\widetilde{\varphi}$ *is surjective.* If $x \in \operatorname{im} \varphi$, then $x = \varphi(a)$ for some $a \in R$, and so $x = \widetilde{\varphi}(a + I)$.

$\widetilde{\varphi}$ *is injective.* If $a + I \in \ker \widetilde{\varphi}$, then $\widetilde{\varphi}(a + I) = 0$. But $\widetilde{\varphi}(a + I) = \varphi(a)$. Hence, $\varphi(a) = 0$, $a \in \ker \varphi = I$, and $a + I = I = 0 + I$. Therefore, $\ker \widetilde{\varphi} = \{0 + I\}$, and $\widetilde{\varphi}$ is an injection.   ●

Remark. The proof of the first isomorphism theorem can be shortened for readers familiar with group theory. If we forget the multiplication, then the proof of Theorem 2.116 shows that the function $\widetilde{\varphi} \colon R/I \to S$, given by $\widetilde{\varphi}(r + I) = \varphi(r)$, is a group isomorphism of the additive groups. Since $\varphi(1 + I) = \varphi(1) = 1$, it is only necessary to prove that $\widetilde{\varphi}$ preserves multiplication, as we have just done. ◄

The first isomorphism theorem says that if $\varphi \colon R \to S$ is a homomorphism, then there is no significant difference between the quotient ring $R/\ker \varphi$ and the subring $\operatorname{im} \varphi$, for they are isomorphic rings. Another viewpoint is that one can create an isomorphism from a homomorphism once one knows its kernel and image. Hence, given a homomorphism, one should first try to describe its kernel and its image. (There are analogs for commutative rings of the second and third isomorphism theorems for groups, but they are less useful for rings than they are for groups. There is an analog for commutative rings of the correspondence theorem for groups. We do not need it now, but we prove it in Proposition 7.1.)

Recall that the *prime field* of a field k is the intersection of all the subfields of k.

→ **Proposition 3.110.** *If k is a field, then its prime field is isomorphic either to $\mathbb{Q}$ or to $\mathbb{F}_p$ for some prime p.*

Proof. Consider the homomorphism $\chi \colon \mathbb{Z} \to k$ defined by $\chi(n) = n \cdot 1$, where 1 is the identity in R. Since every ideal in $\mathbb{Z}$ is principal, there is an integer $m \geq 0$ with $\ker \chi = (m)$. If $m = 0$, then χ is an injection, and so $\operatorname{im} \chi$ is a subring of k isomorphic to $\mathbb{Z}$. By Exercise 3.51 on page 252, $\mathbb{Q} = \operatorname{Frac}(\mathbb{Z}) \cong \operatorname{Frac}(\operatorname{im} \chi)$. Since $\mathbb{Q}$ is the smallest field containing $\mathbb{Z}$ as a subring, it follows from Exercise 3.26 on page 235 that the prime field of k is isomorphic to $\mathbb{Q}$ in this case. If $m \neq 0$, the first isomorphism theorem gives $\mathbb{F}_m = \mathbb{Z}/(m) \cong \operatorname{im} \chi \subseteq k$. Since k is a field, $\operatorname{im} \chi$ is a domain, and so Proposition 3.12 gives m prime. If we now write p instead of m, then $\operatorname{im} \chi = \{0, 1, 2 \cdot 1, \ldots, (p - 1) \cdot 1\}$ is a subfield of k isomorphic to $\mathbb{F}_p$. Thus, Exercise 3.26 shows that $\operatorname{im} \chi \cong \mathbb{F}_p$ is the prime field of k in this case. •

This last result is the first step in classifying different types of fields.

→ **Definition.** A field k has ***characteristic* 0** if its prime field is isomorphic to $\mathbb{Q}$; if its prime field is isomorphic to $\mathbb{F}_p$ for some prime p, then one says that k has ***characteristic p***.

The fields $\mathbb{Q}$, $\mathbb{R}$, $\mathbb{C}$, $\mathbb{C}(x)$ have characteristic 0, as does any subfield of the latter three. Every finite field has characteristic p for some prime p, as does the function field $\mathbb{F}_p(x)$ of all rational functions over $\mathbb{F}_p$.

Recall that if R is a commutative ring, $r \in R$, and $n \in \mathbb{N}$, then $nr = r + \cdots + r$, where there are n summands. If 1 is the one in R, then $nr = (n1)r$. Thus, if $n1 = 0$ in R, then $nr = 0$ for all $r \in R$. Now, in $\mathbb{F}_p$, we have $p[1] = [p] = [0]$, and so

$p[r] = [0]$ for all $[r] \in \mathbb{F}_p$. More generally, if k is any field of characteristic $p > 0$, then $pa = 0$ for all $a \in k$. In particular, when $p = 2$, we have $0 = 2a = a + a$, and so, in this case, $-a = a$ for all $a \in k$.

Example 3.111.

Consider the evaluation homomorphism $\varphi \colon \mathbb{R}[x] \to \mathbb{C}$, defined by $f(x) \mapsto f(i)$, where $i^2 = -1$; that is, $\varphi \colon \sum_k a_k x^k \mapsto \sum_k a_k i^k$. The first isomorphism theorem teaches us to seek $\operatorname{im} \varphi$ and $\ker \varphi$.

First, φ is surjective: if $a + bi \in \mathbb{C}$, then $a + bi = \varphi(a + bx) \in \operatorname{im} \varphi$. Second,

$$\ker \varphi = \{ f(x) \in \mathbb{R}[x] : f(i) = 0 \},$$

the set of all polynomials having i as a root. Of course, $x^2 + 1 \in \ker \varphi$, and we claim that $\ker \varphi = (x^2 + 1)$. Since $\mathbb{R}[x]$ is a PID, the ideal $\ker \varphi$ is generated by the monic polynomial of least degree in it. If $x^2 + 1$ does not generate $\ker \varphi$, then there would be a linear divisor of $x^2 + 1$ in $\mathbb{R}[x]$; that is, $x^2 + 1$ would have a real root. The first isomorphism theorem now gives $\mathbb{R}[x]/(x^2 + 1) \cong \mathbb{C}$.

Thus, the quotient ring construction builds the complex numbers from the reals; that is, if one did not know the field of complex numbers, it could be defined as $\mathbb{R}[x]/(x^2 + 1)$. One advantage of constructing $\mathbb{C}$ in this way is that it is not necessary to check all the field axioms; $R[x]/(x^2 + 1)$ is automatically a commutative ring, and one need prove only that nonzero elements are units. ◀

→ **Theorem 3.112.** *If k is a field and $I = (p(x))$, where $p(x) \in k[x]$ is nonconstant, then the following statements are equivalent.*

(i) $k[x]/I$ *is a field.*

(ii) $k[x]/I$ *is a domain.*

(iii) $p(x)$ *is irreducible in $k[x]$.*

Proof.
(i) $\Rightarrow$ (ii). Every field is a domain.
(ii) $\Rightarrow$ (iii). If $p(x)$ is not irreducible, then there is a factorization $p(x) = g(x)h(x)$ in $k[x]$ with $\deg(g) < \deg(p)$ and $\deg(h) < \deg(p)$. If $g(x) \in I = (p)$, then $p(x) \mid g(x)$ and $\deg(p) \leq \deg(g)$, a contradiction; thus, $g(x) + I \neq 0 + I$ in $k[x]/I$. Similarly, $h(x) + I \neq 0$ in $k[x]/I$. However, the product

$$(g(x) + I)(h(x) + I) = p(x) + I = 0 + I$$

is zero in the quotient ring, contradicting $k[x]/I$ being a domain. Therefore, $p(x)$ must be an irreducible polynomial.

(iii) $\Rightarrow$ (i). Assume that $p(x)$ is irreducible. Since $p(x)$ is not a unit, the ideal $I = (p(x))$ does not contain 1; that is, $1+I \neq 0$ in $k[x]/I$. If $f(x)+I \in k[x]/I$ is nonzero, then $f(x) \notin I$, that is, $f(x)$ is not a multiple of $p(x)$ or, to say it another way, $p \nmid f$. By Lemma 3.67, p and f are relatively prime, and so there are polynomials s and t with $sf + tp = 1$. Thus, $sf - 1 \in I$, and so $1 + I = sf + I = (s + I)(f + I)$. Therefore, every nonzero element of $k[x]/I$ has an inverse, and so $k[x]/I$ is a field. $\bullet$

Compare this theorem with Proposition 3.19, which can be rephrased as giving the equivalence of the statements: $\mathbb{I}_m$ is a field; $\mathbb{I}_m$ is a domain; m is a prime.

Proposition 3.113.

(i) *If k is a field and $p(x) \in k[x]$ is irreducible, then $k[x]/(p(x))$ is a field containing (an isomorphic copy of) k and a root z of $p(x)$.*

(ii) *If $g(x) \in k[x]$ and z is also a root of $g(x)$, then $p(x) \mid g(x)$.*

Proof.
(i) Denote $(p(x))$ by I. The quotient ring $k[x]/I$ is a field, by Theorem 3.112, because $p(x)$ is irreducible. Define $\varphi \colon k \to k[x]/I$ by $\varphi(a) = a + I$; φ is a homomorphism because it is the restriction to k of the natural map $k[x] \to k[x]/I$. By Corollary 3.45, φ is an injection (because k is a field), and so it is an isomorphism from k to the subfield $k^* = \{a + I : a \in k\} \subseteq k[x]/I$.

Remember that x is a particular element of $k[x]$; we claim that $z = x + I \in k[x]/I$ is a root of $p(x)$. More precisely, let

$$p(x) = a_0 + a_1 x + \cdots + a_n x^n \in k[x],$$

where $a_i \in k$ for all i. We will not identify the subfield $k^* \subseteq k[x]/I$ with its isomorphic replica k. Instead, we define $p^*(x) \in k^*[x]$ by

$$p^*(x) = (a_0 + I) + (a_1 + I)x + \cdots + (a_n + I)x^n \in k^*[x],$$

and we prove that z is a root of $p^*(x)$:

$$
\begin{aligned}
p^*(z) &= (a_0 + I) + (a_1 + I)z + \cdots + (a_n + I)z^n \\
&= (a_0 + I) + (a_1 + I)(x + I) + \cdots + (a_n + I)(x + I)^n \\
&= (a_0 + I) + (a_1 x + I) + \cdots + (a_n x^n + I) \\
&= a_0 + a_1 x + \cdots + a_n x^n + I \\
&= p(x) + I = I,
\end{aligned}
$$

because $p(x) \in I = (p(x))$. But $I = 0 + I$ is the zero element of $k[x]/I$, and so z is a root of $p^*(x)$.

(ii) Since z is a root of $g(x)$, we have $g(x) \in \ker \pi$, where $\pi \colon k[x] \to k[x]/(p(x))$ is the natural map, and so $p(x) \mid g(x)$. $\bullet$

Here is a compact description of $k[x]/(f(x))$ that is similar to Corollary 1.59, the description of $\mathbb{I}_m$ as $\{[0], [1], \ldots, [m-1]\}$. Although the next theorem is true for any, not necessarily irreducible, polynomial $f(x)$, the most important case is when $f(x)$ is irreducible.

→ **Theorem 3.114.** *Let k be a field, let $f(x) \in k[x]$ be a nonzero polynomial of degree $n \geq 1$, and let $I = (f(x))$. Then every element in $k[x]/(f(x))$ has a unique expression of the form*

$$b_0 + b_1 z + \cdots + b_{n-1} z^{n-1},$$

where $z = x + I$ is a root of $f(x)$ and all $b_i \in k$.

Proof. Every element of $k[x]/I$ has the form $h(x) + I$, where $h(x) \in k[x]$. By the division algorithm, there are polynomials $q(x), r(x) \in k[x]$ with $h(x) = q(x)f(x) + r(x)$ and either $r(x) = 0$ or $\deg(r) < n = \deg(f)$. Since $h - r = qf \in I$, it follows that $h(x) + I = r(x) + I$. As in the proof of Proposition 3.113, we may rewrite $r(x) + I$ as $r(z) = b_0 + b_1 z + \cdots + b_{n-1} z^{n-1}$ with all $b_i \in k$.

To prove uniqueness, suppose that

$$r(z) = b_0 + b_1 z + \cdots + b_{n-1} z^{n-1} = c_0 + c_1 z + \cdots + c_{n-1} z^{n-1},$$

where all $c_i \in k$. Define $g(x) \in k[x]$ by $g(x) = \sum_{i=0}^{n-1} (b_i - c_i)x^i$. Since z is a root of $g(x)$, Proposition 3.113(ii) gives $f(x) \mid g(x)$. If $g(x)$ is not the zero polynomial, then $\deg(g) \geq n = \deg(f)$, contradicting $\deg(g) < n$. Therefore, $g(x) = 0$ and $b_i = c_i$ for all i. •

Applying this theorem to Example 3.111, in which $f(x) = x^2 + 1 \in \mathbb{R}[x]$, $n = 2$, and the coset $x + I$ [where $I = (x^2 + 1)$] is denoted by i, we see that every complex number has a unique expression of the form $a + bi$, where $a, b \in \mathbb{R}$, and that $i^2 + 1 = 0$, that is, $i^2 = -1$. The easiest way to multiply in $\mathbb{C}$ is to first treat i as a variable and then to impose the condition $i^2 = -1$. For example, to compute $(a + bi)(c + di)$, first write $ac + (ad + bc)i + bdi^2$, and then observe that $i^2 = -1$. The proper way to multiply $(b_0 + b_1 z + \cdots + b_{n-1} z^{n-1})(c_0 + c_1 z + \cdots + c_{n-1} z^{n-1})$ in the quotient ring $k[x]/(p(x))$ is to first regard the factors as polynomials in z and then to impose the condition that $p(z) = 0$. These remarks follow from the natural map $\pi: f(x) \mapsto f(x) + I$ being a homomorphism. Since $\pi: f(x) \mapsto f(z)$, we see that $\pi(f)\pi(g)$ is the product $f(z)g(z)$. On the other hand, $\pi(fg)$ first multiplies $f(x)g(x)$ and then sets $x = z$.

→ **Corollary 3.115.** *If $g(x) \in \mathbb{F}_p[x]$ is an irreducible polynomial of degree n, then $k = \mathbb{F}_p[x]/(g(x))$ is a finite field with exactly p^n elements.*

Proof. By Theorem 3.112, k is a field, and by Theorem 3.114, there are exactly p^n elements in k. •

We cannot yet assert the existence of a field with exactly p^n elements, for we do not yet know that there exists an irreducible polynomial of degree n in $\mathbb{F}_p[x]$ (see Exercise 3.104 on page 305).

We are now going to generalize Example 3.111.

→ **Definition.**   Let E be a field and let k be a subfield. If $z \in E$, then we define $k(z)$ to be the smallest subfield of E containing k and z; that is, $k(z)$ is the intersection of all the subfields of E containing k and z. We call $k(z)$ the field obtained from k by *adjoining* z.

For example, $\mathbb{C} = \mathbb{R}(i)$; the complex numbers are obtained from $\mathbb{R}$ by adjoining i. In Theorem 3.114, we have $K = k(z)$, where $z = x + I$.

The next result shows that quotient rings occur in nature.

→ **Proposition 3.116.**   *Let k be a subfield of a field K and let $z \in K$.*

(i) *If z is a root of some nonzero polynomial $f(x) \in k[x]$, then z is a root of an irreducible polynomial $p(x) \in k[x]$, and $p(x) \mid f(x)$.*

(ii) *If $p(x)$ is an irreducible polynomial in $k[x]$ having a root z in K, then there is an isomorphism*

$$\varphi \colon k[x]/(p(x)) \to k(z)$$

with $\varphi(x + (p(x))) = z$ and $\varphi(a) = a$ for all $a \in k$.

(iii) *If z and z' are roots of $p(x)$ lying in K, then there is an isomorphism $\theta \colon k(z) \to k(z')$ with $\theta(z) = z'$ and with $\theta(a) = a$ for all $a \in k$.*

(iv) *Each element in $k(z)$ has a unique expression of the form*

$$b_0 + b_1 z + \cdots + b_{n-1} z^{n-1},$$

where $b_i \in k$ and $n = \deg(p)$.

Proof.
(i) By Theorem 3.84, there is a factorization

$$f(x) = a p_1(x) \cdots p_m(x),$$

where $a \in k$ is a nonzero constant and $p_1(x), \ldots, p_m(x)$ are monic irreducible polynomials in $k[x]$. Since evaluation at z is a ring homomorphism $k[x] \to K$, we have

$$0 = f(z) = a p_1(z) \cdots p_m(z).$$

But K is a field, hence a domain, and so $p_i(z) = 0$ for some i; that is, z is a root of $p_i(x)$.

(ii) Since $p(x)$ is irreducible, Theorem 3.112 shows that $k[x]/(p(x))$ is a field and, hence, $\operatorname{im}\varphi$ is a field; that is, $\operatorname{im}\varphi$ is a subfield of K containing k and z, and so $k(z) \subseteq \operatorname{im}\varphi$. On the other hand, every element in $\operatorname{im}\varphi$ has the form $g(x) + I$, where $g(x) \in k[x]$, so that any subfield S of K which contains k and z must also contain $\operatorname{im}\varphi$. Hence, $\operatorname{im}\varphi = k(z)$.

(iii) As in part (ii), there is an isomorphism $\psi : k[x]/(p(x)) \to k(z')$ with $\psi(a) = a$ for all $a \in k$ and $\psi(x + (p(x))) = z$. The composite $\theta = \psi \circ \varphi^{-1}$ is the desired isomorphism.

(iv) By Theorem 3.114, each element in $k[x]/I$, where $I = (p(x))$, has a unique expression of the form $b_0 + b_1(x + I) + \cdots + b_{n-1}(x + I)^{n-1}$, and injectivity of the isomorphism $k[x]/I \to k(z)$ sending $x + I \mapsto z$ preserves this uniqueness. •

Corollary 3.117. *Let k be a field, let $p(x) \in k[x]$ be irreducible, and let $\alpha \in k[x]/(p)$. If α is a root of some nonzero polynomial in $k[x]$, then there exists a unique monic irreducible polynomial $h(x) \in k[x]$ having α as a root.*

Remark.   Proposition 4.32 shows that the hypothesis on elements α is redundant. ◄

Proof.   Proposition 3.116(i) applies to give an irreducible polynomial $h(x) \in k[x]$ having α as a root; clearly, we may assume that $h(x)$ is monic.

To prove uniqueness of $h(x)$, suppose that $g(x) \in k[x]$ is another monic irreducible polynomial having α as a root. Over $k[x]/(p)$, the gcd $(h, g) \neq 1$ (for $x - \alpha$ is a common divisor), and so Corollary 3.75 says that $(h, g) \neq 1$ in $k[x]$. Since $h(x)$ is irreducible, its only monic divisors are 1 and itself, and so $(h, g) = h$. Therefore, $h(x) \mid g(x)$. But since $g(x)$ is monic irreducible, we must have $h(x) = g(x)$. •

We can now prove two important results: the first, due to Kronecker, says that if $f(x) \in k[x]$, where k is any field, then there is some larger field E containing k and all the roots of $f(x)$; the second, due to Galois, constructs finite fields other than $\mathbb{F}_p$.

→ **Definition.**   A polynomial $f(x) \in k[x]$ *splits* over a larger field K if $f(x)$ is a product of linear polynomials in $K[x]$.

→ **Theorem 3.118 (Kronecker).**   *If k is a field and $f(x) \in k[x]$ is nonconstant, then then there is a field K containing k over which $f(x)$ splits.*

Proof.   We prove the theorem by induction on $\deg(f)$, and we modify the statement a bit to enable us to prove the inductive step more easily: if E is any field containing k as a subfield (so that $f(x) \in k[x] \subseteq E[x]$), then there is a field K containing E such that $f(x)$ is a product of linear polynomials in $K[x]$. If $\deg(f) = 1$, then $f(x)$ is linear and we can choose $K = E$. For the inductive step, we consider two cases. If $f(x)$ is not irreducible, then $f(x) = g(x)h(x)$ in $k[x]$, where $\deg(g) < \deg(f)$ and $\deg(h) < \deg(f)$. By induction, there is a field E containing k with $g(x)$ splitting over E; a second use of the inductive hypothesis provides a field K containing E with $h(x)$

splitting over K. Thus, $f(x) = g(x)h(x)$ splits over K. In the second case, $p(x)$ is irreducible in $k[x]$, and Proposition 3.113(i) provides a field E containing k and a root z of $p(x)$. Thus, there is a factorization $p(x) = (x - z)\ell(x)$ in $E[x]$. By induction, there is a field K containing E so that $\ell(x)$, and hence $f(x) = (x - z)\ell(x)$, splits over K. •

For the familiar fields $\mathbb{Q}$, $\mathbb{R}$, and $\mathbb{C}$, Kronecker's theorem offers nothing new. The Fundamental Theorem of Algebra, first proved by Gauss in 1799 (completing earlier attempts of Euler and of Lagrange), says that every nonconstant $f(x) \in \mathbb{C}[x]$ has a root in $\mathbb{C}$. It follows, by induction on the degree of $f(x)$, that all the roots of $f(x)$ lie in $\mathbb{C}$; that is, $f(x)$ splits over $\mathbb{C}$. On the other hand, if $k = \mathbb{F}_p$ or $k = \mathbb{C}(x) = \text{Frac}(\mathbb{C}[x])$, then Kronecker's theorem does apply to tell us, for any given $f(x)$, that there is always some larger field E that contains all the roots of $f(x)$. For example, there is some field containing $\mathbb{C}(x)$ and $\sqrt{x}$. There is a general version of the Fundamental Theorem of Algebra: every field k is a subfield of an ***algebraically closed*** field K, that is, K is a field containing k such that every $f(x) \in k[x]$ is a product of linear polynomials in $K[x]$. In contrast, Kronecker's theorem gives roots of just one polynomial at a time.

We now consider finite fields; that is, fields having only finitely many elements. We first prove that the number of elements in a finite field is a power of a prime; we give a group-theoretic proof of this in Exercise 3.105 on page 305.

→ **Definition.** A ***primitive element*** π of a finite field E is an element $\pi \in E$ with every nonzero $a \in E$ equal to some power of π.

→ **Proposition 3.119.**

(i) *If E is a finite field, then E has a primitive element.*

(ii) *If E is a finite field of characteristic p with prime field k, then $|E| = p^n$ for some $n \geq 1$.*

Proof.
(i) Since E is finite, its multiplicative group $E^\times$ is cyclic, by Theorem 3.55, and a generator π of $E^\times$ is a primitive element of E.
(ii) Since every element of E is equal to some power of π, we have $E = k(\pi)$. Finiteness of E implies that there is a repetition on the list $1, \pi, \pi^2, \pi^3, \ldots$; there are positive integers $r > s$ with $\pi^r = \pi^s$. Hence, $\pi^{r-s} = 1$, and so π is a root of $x^{r-s} - 1$. By Theorem 3.116(i), there is an irreducible polynomial $g(x) \in k[x]$ having π as a root, and by Theorem 3.116(ii), there is an isomorphism $k[x]/(g(x)) \cong k(\pi) = E$. If $\deg(g) = n$, then Corollary 3.115 shows that $|E| = p^n$. •

We are now going to display the finite fields.

→ **Theorem 3.120 (Galois).** *If p is a prime and n is a positive integer, then there exists a field that has exactly p^n elements.*

Proof. Write $q = p^n$, and consider the polynomial

$$g(x) = x^q - x \in \mathbb{F}_p[x].$$

By Kronecker's theorem, there is a field E containing $\mathbb{F}_p$ such that $g(x)$ is a product of linear factors in $E[x]$. Define

$$F = \{\alpha \in E : g(\alpha) = 0\};$$

thus, F is the set of all the roots of $g(x)$. Since the derivative $g'(x) = qx^{q-1} - 1 = p^n x^{q-1} - 1 = -1$, it follows that the gcd $(g, g') = 1$. By Exercise 3.67 on page 274, all the roots of $g(x)$ are distinct; that is, F has exactly $q = p^n$ elements.

We claim that F is a subfield of E, which will complete the proof. If $a, b \in F$, then $a^q = a$ and $b^q = b$. Therefore, $(ab)^q = a^q b^q = ab$, and $ab \in F$. By Exercise 3.97 on page 304, $(a - b)^q = a^q - b^q = a - b$, so that $a - b \in F$. Finally, if $a \neq 0$, then the cancellation law applied to $a^q = a$ gives $a^{q-1} = 1$, and so the inverse of a is a^{q-2} (which lies in F because F is closed under multiplication).   •

In Corollary 5.25, we will see that any two finite fields k with the same number of elements are isomorphic. Here are some special cases of this theorem.

Example 3.121.
In Exercise 3.19 on page 234, we constructed a field k with 4 elements as all the matrices $\begin{bmatrix} a & b \\ b & a+b \end{bmatrix} = a \begin{bmatrix} 1 & 0 \\ 0 & 1 \end{bmatrix} + b \begin{bmatrix} 0 & 1 \\ 1 & 1 \end{bmatrix}$, where $a, b \in \mathbb{I}_2$. (We remark that $\begin{bmatrix} 0 & 1 \\ 1 & 1 \end{bmatrix}$ is a primitive element, as is $\begin{bmatrix} 1 & 1 \\ 1 & 0 \end{bmatrix}$.)

On the other hand, we may construct a field with 4 elements as the quotient $F = \mathbb{F}_2[x]/(x^2 + x + 1)$, for $x^2 + x + 1 \in \mathbb{F}_2[x]$ is irreducible. By Corollary 3.114, F is a field consisting of all $a + bz$, where z is a root of $x^2 + x + 1$ and $a, b \in \mathbb{F}_2$. Since $z^2 + z + 1 = 0$, we have $z^2 = -z - 1 = z + 1$; moreover, $z^3 = zz^2 = z(z + 1) = z^2 + z = 1$. It is now easy to see that there is a ring isomorphism $\varphi : k \to F$ with $\varphi : \begin{bmatrix} a & b \\ b & a+b \end{bmatrix} \mapsto a + bz$.   ◄

Example 3.122.
According to the table in Example 3.99, there are three monic irreducible quadratics in $\mathbb{F}_3[x]$, namely,

$$p(x) = x^2 + 1, \quad q(x) = x^2 + x - 1, \quad \text{and} \quad r(x) = x^2 - x - 1;$$

each gives rise to a field with $9 = 3^2$ elements. Let us look at the first two in more detail. Corollary 3.114 says that $E = \mathbb{F}_3[x]/(q(x))$ is given by

$$E = \{a + b\alpha : \text{ where } \alpha^2 + 1 = 0\}.$$

Similarly, if $F = \mathbb{F}_3[x]/(p(x))$, then

$$F = \{a + b\beta : \text{ where } \beta^2 + \beta - 1 = 0\}.$$

The reader can show that these two fields are isomorphic by checking that the function $\varphi \colon E \to F$, defined by

$$\varphi(a + b\alpha) = a + b(1 - \beta),$$

is an isomorphism.

Now $\mathbb{F}_3[x]/(x^2 - x - 1)$ is also a field with nine elements, and one can show that it is isomorphic to both of the two fields E and F given above.

In Example 3.99, we exhibited eight monic irreducible cubics $p(x) \in \mathbb{F}_3[x]$; each of them gives rise to a field $\mathbb{F}_3[x]/(p(x))$ having $27 = 3^3$ elements. All eight of these fields are isomorphic. ◀

In the section on codes in Chapter 4, we shall see that finite fields are an essential ingredient in sending data from outer space to Earth.

EXERCISES

H **3.92** True or false with reasons.

(i) If I is a proper ideal in a commutative ring R and $\pi \colon R \to R/I$ is the natural map, then $\ker \pi = I$.

(ii) If I is a proper ideal in a commutative ring R and $\pi \colon R \to R/I$ is the natural map, then π is surjective.

(iii) If $f \colon R \to S$ is a homomorphism of commutative rings, then S has a subring isomorphic to $R/(\ker f)$.

(iv) If I is a proper ideal in a commutative ring R, then R has a subring isomorphic to R/I.

(v) If p is a prime number, then every field of characteristic p is finite.

(vi) Every field of characteristic 0 is infinite.

(vii) If $f(x)$ is an irreducible polynomial over a field k, then $k[x]/(f(x))$ is a field.

(viii) If $f(x)$ is a nonconstant polynomial over a field k and if the quotient ring $k[x]/(f(x))$ is a field, then $f(x)$ is irreducible.

(ix) If $f(x)$ is an irreducible polynomial over a field k, then every element $z \in k[x]/(f(x))$ is a root of $f(x)$.

(x) If $k \subseteq K$ are fields and $z \in K$ is a root of some nonzero polynomial $p(x) \in k[x]$, then $p(x)$ is irreducible in $k[x]$.

(xi) There is a field containing $\mathbb{C}(x)$ and $\sqrt{x+i}$.

(xii) For every positive integer n, there exists a field with exactly 11^n elements.

(xiii) For every positive integer n, there exists a field with exactly 10^n elements.

(xiv) For every positive integer n, there exists a field with exactly 9^n elements.

(xv) There exists a field E of characteristic 2 such that $x^4 + x + 1$ is a product of linear factors in $E[x]$.

3.93 For every commutative ring R, prove that $R[x]/(x) \cong R$.

3.94 (***Chinese Remainder Theorem in*** $k[x]$)

H **(i)**   Prove that if k is a field and $f(x)$, $f'(x) \in k[x]$ are relatively prime, then given $b(x), b'(x) \in k[x]$, there exists $c(x) \in k[x]$ with

$$c - b \in (f) \quad \text{and} \quad c - b' \in (f');$$

moreover, if $d(x)$ is another common solution, then $c - d \in (ff')$.

H **(ii)**   Prove that if k is a field and $f(x)$, $g(x) \in k[x]$ are relatively prime, then

$$k[x]/(f(x)g(x)) \cong k[x]/(f(x)) \times k[x]/(g(x)).$$

* H **3.95**   Generalize Exercise 3.84 on page 281 by proving that if k is a field of characteristic 0 and if $p(x) \in k[x]$ is an irreducible polynomial, then $p(x)$ has no repeated roots.

3.96   **(i)**   Prove that a field K cannot have subfields k' and k'' with $k' \cong \mathbb{Q}$ and $k'' \cong \mathbb{F}_p$ for some prime p.

   (ii)   Prove that a field K cannot have subfields k' and k'' with $k' \cong \mathbb{F}_p$ and $k'' \cong \mathbb{F}_q$, where $p \neq q$.

* **3.97**   Let p be a prime and let $q = p^n$ for some $n \geq 1$.

H **(i)**   Show that the function $F : \mathbb{F}_q \to \mathbb{F}_q$, given by $F(a) = a^p$, is an isomorphism (F is called the ***Frobenius map***).

   (ii)   Show that every element $a \in \mathbb{F}_q$ has a pth root, i.e., there is $b \in \mathbb{F}_q$ with $a = b^p$.

   (iii)   Let k be a field of characteristic $p > 0$. For every positive integer n, show that the ring homomorphism $F_n : k \to k$, given by $F_n(a) = a^{p^n}$, is injective.

* H **3.98**   Prove that every element z in a finite field E is a sum of two squares. (If $z = a^2$ is a square, then we may write $z = a^2 + 0^2$.)

* H **3.99**   If p is a prime and $p \equiv 3 \bmod 4$, prove that either $a^2 \equiv 2 \bmod p$ is solvable or $a^2 \equiv -2 \bmod p$ is solvable.

* **3.100**   **(i)**   Prove that $x^4 + 1$ factors in $\mathbb{F}_2[x]$.

H **(ii)**   If $x^4 + 1 = (x^2 + ax + b)(x^2 + cx + d) \in \mathbb{F}_p[x]$, where p is an odd prime, prove that $c = -a$ and

$$d + b - a^2 = 0$$
$$a(d - b) = 0$$
$$bd = 1.$$

H **(iii)**   Prove that $x^4 + 1$ factors in $\mathbb{F}_p[x]$, where p is an odd prime, if any of the following congruences are solvable:

$$b^2 \equiv -1 \bmod p,$$
$$a^2 \equiv \pm 2 \bmod p.$$

H **(iv)**   Prove that $x^4 + 1$ factors in $\mathbb{F}_p[x]$ for all primes p.

* **3.101**   Generalize Proposition 3.116(iii) as follows. Let $\varphi : k \to k'$ be an isomorphism of fields, let E/k and E'/k' be extensions, let $p(x) \in k[x]$ and $p^*(x) \in k'[x]$ be irreducible polynomials (as in Theorem 3.33, if $p(x) = \sum a_i x^i$, then $p^*(x) = \sum \varphi(a_i)x^i$), and let

$z \in E$ and $z' \in E'$ be roots of $p(x)$, $p^*(x)$, respectively. Then there exists an isomorphism $\widetilde{\varphi} \colon k(z) \to k'(z')$ with $\widetilde{\varphi}(z) = z'$ and with $\widetilde{\varphi}$ extending φ.

$$
\begin{array}{ccc}
k(z) & \xrightarrow{\ \widetilde{\varphi}\ } & k'(z') \\
\big| & & \big| \\
k & \xrightarrow[\varphi]{} & k'
\end{array}
$$

*3.102 Let $f(x) = a_0 + a_1 x + \cdots + a_{n-1} x^{n-1} + x^n \in k[x]$, where k is a field, and suppose that $f(x) = (x - r_1)(x - r_2) \ldots (x - r_n) \in E[x]$, where E is some field containing k. Prove that

$$
a_{n-1} = -(r_1 + r_2 + \cdots + r_n) \quad \text{and} \quad a_0 = (-1)^n r_1 r_2 \cdots r_n.
$$

Conclude that the sum and the product of all the roots of $f(x)$ lie in k.

H 3.103 If $E = \mathbb{F}_2[x]/(p(x))$, where $p(x) = x^3 + x + 1$, then E is a field with 8 elements. Show that a root π of $p(x)$ is a primitive element of E by writing every nonzero element of E as a power of π.

*3.104 (i)   Prove, for all $n \geq 1$, that there is an irreducible polynomial of degree n in $\mathbb{Q}[x]$.

    H (ii) Prove, for all $n \geq 1$ and every prime p, that there is an irreducible polynomial of degree n in $\mathbb{F}_p[x]$.

    (iii) Prove, for all $n \geq 1$ and every finite field k, that there is an irreducible polynomial of degree n in $k[x]$.

*H 3.105 If E is a finite field, use Theorem 2.147, Cauchy's theorem, to prove that $|E| = p^n$ for some prime p and some $n \geq 1$.

3.9   A MATHEMATICAL ODYSSEY

Latin Squares

In 1782, L. Euler posed the following problem in an article he was writing about magic squares. Suppose there are 36 officers of 6 ranks and from 6 regiments. If the regiments are numbered 1 through 6 and the ranks are captain, major, lieutenant, ..., then each officer has a double label (e.g., captain 3 or major 4). Euler asked whether there is a 6×6 formation of these officers so that each row and each column contains exactly one officer of each rank and one officer from each regiment. Thus, no row can have two captains in it, nor can any column; no row can have two officers from the same regiment, nor can any column.

The problem is made clearer by the following definitions.

Definition.   An $n \times n$ **Latin square** is an $n \times n$ matrix whose entries are taken from a set X with n elements (e.g., $X = \{0, 1, \ldots, n - 1\}$) so that no element occurs twice in any row or in any column.

It is easy to see that an $n \times n$ matrix A (whose entries lie in a set X with $|X| = n$) is a Latin square if and only if every row and every column of A is a permutation of X.

As is customary, we may denote a matrix A by $A = [a_{ij}]$, where a_{ij} are its entries; the first index i of a_{ij} describes the ith row and the second index j describes the jth column.

$$a_{i1} \; a_{i2} \; \ldots \; a_{in} \qquad \begin{matrix} a_{1j} \\ a_{2j} \\ \vdots \\ a_{nj} \end{matrix}$$

Example 3.123.
There are exactly two 2×2 Latin squares having entries 0 and 1:

$$A = \begin{bmatrix} 0 & 1 \\ 1 & 0 \end{bmatrix} \quad \text{and} \quad B = \begin{bmatrix} 1 & 0 \\ 0 & 1 \end{bmatrix}. \quad \blacktriangleleft$$

Example 3.124.
Here are two 4×4 Latin squares.

$$A = \begin{bmatrix} 0 & 1 & 2 & 3 \\ 1 & 0 & 3 & 2 \\ 2 & 3 & 0 & 1 \\ 3 & 2 & 1 & 0 \end{bmatrix} \quad \text{and} \quad B = \begin{bmatrix} 0 & 1 & 2 & 3 \\ 2 & 3 & 0 & 1 \\ 3 & 2 & 1 & 0 \\ 1 & 0 & 3 & 2 \end{bmatrix} \quad \blacktriangleleft$$

Example 3.125.
The multiplication table of a finite group $G = \{a_1, \ldots, a_n\}$ of order n, namely, $[a_i a_j]$, is a Latin square. Since the cancellation laws hold in groups, $a_i a_j = a_i a_k$ implies $a_j = a_k$, and so the ith row is a permutation of G; since $a_i a_\ell = a_j a_\ell$ implies $a_i = a_j$, the jth column is a permutation of G. $\quad \blacktriangleleft$

We are going to use the following construction, which is usually the first attempt of a neophyte defining matrix multiplication.

Definition. If $A = [a_{ij}]$ and $B = [b_{ij}]$ are $m \times n$ matrices, then their **Hadamard product**, denoted by $A \circ B$, is the $m \times n$ matrix whose ij entry is the ordered pair (a_{ij}, b_{ij}).

If the entries of A and of B lie in a commutative ring R, then one often replaces the ij entry of the Hadamard product, (a_{ij}, b_{ij}), by the product $a_{ij} b_{ij}$ in R.

Suppose that the entries of A lie in a set X with $|X| = n$, and the entries of B lie in a set Y with $|Y| = n$. There are exactly n^2 ordered pairs in $X \times Y$, and we say that A and B are *orthogonal* if every ordered pair occurs as an entry in their Hadamard product $A \circ B$.

Definition. Two $n \times n$ Latin squares $A = [a_{ij}]$ and $B = [b_{ij}]$, with entries in sets X and Y with $|X| = n = |Y|$, respectively, are called **orthogonal** if all the entries in their Hadamard product $A \circ B$, namely, all the ordered pairs (a_{ij}, b_{ij}), are distinct.

For example, there is no orthogonal pair of 2×2 Latin squares: as we saw in Example 3.123, there are only two 2×2 Latin squares with entries in $X = \{0, 1\}$, and their Hadamard product is

$$A \circ B = \begin{bmatrix} 01 & 10 \\ 10 & 01 \end{bmatrix}.$$

There are only two distinct ordered pairs, not four as the definition requires.

Example 3.126.
The two 4×4 Latin squares in Example 3.124 are orthogonal, for all 16 ordered pairs are distinct.

$$A \circ B = \begin{bmatrix} 00 & 11 & 22 & 33 \\ 12 & 03 & 30 & 21 \\ 23 & 32 & 01 & 10 \\ 31 & 20 & 13 & 02 \end{bmatrix} \quad \blacktriangleleft$$

Let A be a matrix whose entries lie in a set X. If $\alpha \colon x \mapsto x'$ is a permutation of X, then applying α to each entry in A yields a new matrix A'. In more detail, if $x = a_{ij}$ is the ij entry of A, then x' is the ij entry of A', and so $A' = [(a_{ij})']$.

Lemma 3.127. *Let $A = [a_{ij}]$ be a Latin square whose entries lie in a set X with n elements. If $x \mapsto x'$ is a permutation of X, then $A' = [(a_{ij})']$ is a Latin square. Moreover, if A and $B = [b_{ij}]$ are orthogonal Latin squares, then A' and B are also orthogonal.*

Proof. As the ith row $(a_{i1}, \ldots, a_{in})$ of A is a permutation of X, so is the ith row $((a_{i1})', \ldots, (a_{in})')$ of A' (the composite of two permutations is again a permutation). A similar argument shows that the columns of A' are permutations of X, and so A' is a Latin square.

If A' and B are not orthogonal, then two entries of $A' \circ B$ are equal: say, $(a'_{ij}, b_{ij}) = (a'_{k\ell}, b_{k\ell})$, so that $a'_{ij} = a'_{k\ell}$ and $b_{ij} = b_{k\ell}$. Since priming is a permutation, $a_{ij} = a_{k\ell}$. Thus, there is a repeated ordered pair in $A \circ B$, contradicting the orthogonality of A and B. Hence, A' and B are orthogonal.   •

Euler's problem asks whether there is a pair of orthogonal 6×6 Latin squares (the first index denotes the rank and the second index denotes the regiment). Euler was more interested in orthogonal Latin squares than he was in officers. To see why he cared about the case $n = 6$, let us first construct some orthogonal pairs.

Proposition 3.128.

(i) *If k is a finite field and $a \in k^\times = k - \{0\}$, then the $|k| \times |k|$ matrix*

$$L_a = [\ell_{xy}] = [ax + y],$$

where $x, y \in k$, is a Latin square.

(ii) *If $a, b \in k^\times$ and $a \neq b$, then L_a and L_b are orthogonal Latin squares.*

Proof.
(i) The xth row of L_a consists of the elements $ax + y$, where x is fixed. These are all distinct, for if $ax + y = ax + y'$, then $y = y'$. Similarly, the yth column of L_a consists of elements $ax + y$, where y is fixed, and these are distinct because $ax + y = ax' + y$ implies $ax = ax'$. Since $a \neq 0$, the cancellation law gives $x = x'$.
(ii) Suppose that two ordered pairs coincide; say,

$$(ax + y, bx + y) = (ax' + y', bx' + y').$$

Thus, $ax + y = ax' + y'$ and $bx + y = bx' + y'$. There result equations

$$a(x - x') = y' - y = b(x - x').$$

Since $a \neq b$, the cancellation law says that $x - x' = 0$, and so $y' - y = 0$, i.e., $x' = x$ and $y' = y$. Therefore, L_a and L_b are orthogonal Latin squares.   •

Corollary 3.129.   *For every prime power $p^e > 2$, there exists a pair of orthogonal $p^e \times p^e$ Latin squares.*

Proof.   By Galois's theorem, there exists a finite field k with $|k| = p^e$. To have an orthogonal pair of Latin squares, we need $|k^\times| \geq 2$; that is, $p^e - 1 \geq 2$, and hence $p^e > 2$.   •

Remark.   Galois invented finite fields around 1830, so that Euler, in 1782, constructed orthogonal $p^e \times p^e$ Latin squares in a different (more complicated) way.   ◄

We now show how to create large orthogonal Latin squares from small ones. Let K and L be sets with $|K| = k$ and $|L| = \ell$. If $B = [b_{ij}]$ is an $\ell \times \ell$ matrix with entries in L, then aB is the $\ell \times \ell$ matrix whose ij entry is ab_{ij} [where ab_{ij} abbreviates the ordered pair (a, b_{ij})]. If $A = [a_{st}]$ is a $k \times k$ matrix whose entries lie in K, then the **Kronecker product** $A \otimes B$ of A and B is the $k\ell \times k\ell$ matrix

$$\begin{bmatrix} a_{11}B & a_{12}B & \ldots & a_{1k}B \\ a_{21}B & a_{22}B & \ldots & a_{2k}B \\ \ldots & \ldots & \ldots & \ldots \\ a_{k1}B & a_{k2}B & \ldots & a_{kk}B \end{bmatrix}.$$

Theorem 3.130 (**Euler**).   *If $n \not\equiv 2$ mod 4, then there exists an orthogonal pair of $n \times n$ Latin squares.*

Proof.   We merely state the main steps of the proof. One shows first that if A and B are Latin squares, then $A \otimes B$ is a Latin square. Second, one proves that if A and A' are orthogonal $k \times k$ Latin squares, and if B and B' are orthogonal $\ell \times \ell$ Latin squares, then $A \otimes B$ and $A' \otimes B'$ are orthogonal $k\ell \times k\ell$ Latin squares. Neither of these steps is challenging. Of course, one can form the Kronecker product of a finite number of matrices.

If n is a positive integer, we claim that $n \equiv 2$ mod 4 if and only if there is an odd integer m with $n = 2m$. If $n \equiv 2$ mod 4, then $n - 2 = 4k$ for some integer k, and $n = 2(2k + 1)$. Conversely, if $n = 2m$ for m odd, then $n = 2m = 2(2d + 1) = 4d + 2$, and so $n - 2 = 4d$. It follows that $n \not\equiv 2$ mod 4 if and only if $n = 2^e p_1^{e_1} \cdots p_t^{e_t}$, where $e \neq 1$ and the p_i are odd primes. By Corollary 3.129, there is an orthogonal pair of $p_i^{e_i} \times p_i^{e_i}$ Latin squares for each i and, if $e \geq 2$, an orthogonal pair of $2^e \times 2^e$ Latin squares. Taking the Kronecker product of these gives a pair of orthogonal $n \times n$ Latin squares.   ●

The smallest n not covered by Euler's theorem is $n = 6$, and this is why Euler posed the question of the 36 officers. Indeed, he conjectured that there is no orthogonal pair of $n \times n$ Latin squares if $n \equiv 2$ mod 4. In 1901, G. Tarry proved that there does not exist an orthogonal pair of 6×6 Latin squares, thereby answering Euler's question posed at the beginning of this section: there is no such formation of 36 officers. However, in 1958, E. T. Parker discovered an orthogonal pair of 10×10 Latin squares, thereby disproving Euler's conjecture. Parker's example is displayed on the front cover of this book. Table 3.1 is a less colorful version of it; note that every number less that 100 appears, in decimal notation, as an entry. Parker, R. C. Bose, and S. S. Shrikhande went on to prove that there exists a pair of orthogonal $n \times n$ Latin squares for all n except 2 and 6.

00	15	23	32	46	51	64	79	87	98
94	77	10	25	52	49	01	83	68	36
71	34	88	17	20	02	43	65	96	59
45	81	54	66	18	27	72	90	39	03
82	40	61	04	99	16	28	37	53	75
26	62	47	91	74	33	19	58	05	80
13	29	92	48	31	84	55	06	70	67
69	93	35	50	07	78	86	44	12	21
57	08	76	89	63	95	30	11	24	42
38	56	09	73	85	60	97	22	41	14

Table 3.1

Magic Squares

We are now going to use orthogonal Latin squares to construct some magic squares.

Definition.   An $n \times n$ **magic square** is an $n \times n$ matrix $A = [a_{ij}]$ whose entries consist of all the numbers $0, 1, \ldots, n^2 - 1$ and whose row sums and columns sums are the same; that is, there is a number σ, called the **magic number**, with

$$\sum_{j=1}^{n} a_{ij} = \sigma \text{ for all } i \quad \text{and} \quad \sum_{i=1}^{n} a_{ij} = \sigma \text{ for all } j.$$

The 1514 engraving *Melencolia I*, by Albrecht Dürer (see Figure 3.2) contains the following square in its upper right corner.

16	3	2	13
5	10	11	8
9	6	7	12
4	15	14	1

Notice the date 1514 in the bottom row.[15] The row and column sums all equal 34; in fact, the sum $\sum_i a_{ii}$ of the diagonal terms is also 34, as is the sum $\sum_i a_{in-i}$ of the terms on the **back diagonal** (going up from the bottom left corner to the top right corner). This is not a magic square, for its entries range from 1 to 16 instead of from 0 to 15, but this is easily remedied: subtract 1 from each entry and get a magic square with magic number 30.

Proposition 3.131.   *If A is an $n \times n$ magic square, then its magic number is*

$$\sigma = \tfrac{1}{2}n(n^2 - 1).$$

Proof.   If ρ_i denotes the sum of the entries in the ith row of A, then $\rho_i = \sigma$ for all i, and so $\sum_{i=1}^{n} \rho_i = n\sigma$. But this last number is the sum of all the entries in A; that is,

$$n\sigma = 1 + 2 + \cdots + (n^2 - 1) = \tfrac{1}{2}(n^2 - 1)n^2.$$

Therefore, $\sigma = \tfrac{1}{2}n(n^2 - 1)$.   •

If $n = 4$, then $\sigma = \tfrac{1}{2}4 \cdot 15 = 30$.

There is a minor disagreement about terminology. In order that a square be magic, some authors also require that the diagonal entries and the back diagonal entries each add up to the magic number, as in the modified Dürer square.

---

[15]Dürer was familiar with *Qabala*, Hebrew mysticism, in which each letter of the alphabet is assigned a number, and each word is assigned the sum of the values of its letters. The values assigned to the letters of the Latin alphabet are $1, 2, \ldots, 26$. Notice that 4 and 1 flank 1514 in the magic square; these are the initials of the artist Dürer, Albrecht.

Figure 3.2

Melencholia I, by Albrecht Dürer
Grunwald Center for the Graphic Arts
UCLA Hammer Museum

Definition.   A *diabolic square* is a magic square whose diagonal and back diagonal sums are each equal to the magic number.

We will construct some diabolic squares below, but let us first return to magic squares. There are many methods of constructing magic squares. For example, in 1693, De la Loubère showed how to construct an $n \times n$ magic square, for any odd n, in which 0 can occur in any ij position (see Stark, *An Introduction to Number Theory*, Chapter 4). We now use orthogonal Latin squares to construct magic squares.

Proposition 3.132.   *If $A = [a_{ij}]$ and $B = [b_{ij}]$ are orthogonal Latin squares with entries in $0, 1, \ldots, n - 1$, then the matrix $M = [a_{ij}n + b_{ij}]$ is an $n \times n$ magic square.*

Proof.   Since A and B are orthogonal, the entries (a_{ij}, b_{ij}) of their Hadamard product $A \circ B$ are all distinct. It follows from Proposition 1.47, which says that the n-adic digits of a nonnegative number are unique, that every number from 0 through $n^2 - 1$ occurs in M (note that $0 \le a_{ij} < n$ and $0 \le b_{ij} < n$). Now A being a Latin square says that each row and column of A is a permutation of $0, 1, \ldots, n - 1$, and so each row sum and column sum equals $s = \sum_{i=0}^{n-1} i = \frac{1}{2}(n - 1)n$; similarly, each row sum and column sum of B equals s. Hence, each row sum of M is equal to $sn + s$, as is each column sum. Therefore, M is a magic square.   •

The magic number of M is $\sigma = s(n + 1)$, where $s = \frac{1}{2}n(n - 1)$. This agrees with the value of the magic number in Proposition 3.131, for $s(n+1) = \frac{1}{2}n(n - 1)(n + 1) = \frac{1}{2}n(n^2 - 1)$.

Parker's 10×10 orthogonal Latin squares have been converted into decimal digits in Table 3.1, which is an example of a 10×10 magic square as just constructed.

Example 3.133.
Proposition 3.132 is not the only way to construct magic squares. For example, here is a 6×6 magic square (whose magic number is, of course, 105); it is even diabolic. This magic square does not arise from an orthogonal pair of 6×6 Latin squares, for Tarry has shown us that there aren't any!

34	0	5	25	18	23
2	31	6	20	22	24
30	8	1	21	26	19
7	27	32	16	9	14
29	4	23	11	13	15
3	35	28	12	17	10

◀

We now construct some diabolic squares from orthogonal Latin squares. It is known that $n \times n$ diabolic squares exist for all $n \ge 3$, but we will construct them for only certain n.

Definition. An $n \times n$ Latin square $A = [a_{ij}]$ with entries in a set X with $|X| = n$ is a *diagonal Latin square* if its diagonal and its back diagonal are permutations of X.

Lemma 3.134. *If n is an odd positive integer that is not a multiple of 3, then there exists an orthogonal pair of $n \times n$ diagonal Latin squares.*

Proof. We begin by constructing a diagonal $n \times n$ Latin square. It will be convenient to label the rows and columns so that $0 \le i, j \le n - 1$. Define A to be the $n \times n$ matrix whose ij entry is the congruence class $[i + 2j] \bmod n$; we simplify notation by omitting the brackets of the entries. Thus,

$$A = \begin{bmatrix} 0 & 2 & 4 & \cdots & 2(n-1) \\ 1 & 3 & 5 & \cdots & 1+2(n-1) \\ 2 & 4 & 6 & \cdots & 2+2(n-1) \\ \vdots & \vdots & \vdots & \vdots & \vdots \\ n-1 & n+1 & n+3 & \cdots & 3(n-1) \end{bmatrix}.$$

We now show that A is a diagonal Latin square; remember that its entries lie in $\mathbb{I}_n$. Each row is a permutation: for fixed i, if $i + 2j = i + 2j'$, then $2(j - j') \equiv 0 \bmod n$. But $(2, n) = 1$, because n is odd, so that $[j] = [j']$. Each column is a permutation: for fixed j, if $i + 2j = i' + 2j$, then $i - i' \equiv 0 \bmod n$; that is, $[i] = [i']$. For the main diagonal, if $i + 2i = i' + 2i'$, then $3i = 3i'$; since $3 \nmid n$, we have $[i] = [i']$. Finally, for the back diagonal, if $i + 2(n - i) = i' + 2(n - i')$, then $-i = -i'$, and so $[i] = [i']$.

It is obvious that the transpose A^T of A is also a diagonal Latin square, and we now show that A and A^T are orthogonal.[16] Note that the ij entry of A^T is $j + 2i$, so that the ij entry of the Hadamard product $A \circ A^T$ is $(i + 2j, j + 2i)$. To check orthogonality, suppose that $(i + 2j, j + 2i) = (i' + 2j', j' + 2i')$. Now $i + 2j = i' + 2j'$ and $j + 2i = j' + 2i'$ (remember that entries lie in $\mathbb{I}_n$), so that $[i - i'] = [j' - j]$ and $[2(j' - j)] = [i - i']$. Multiplying the second equation by $[2]$ gives $[4(j' - j)] = [2(i - i')] = [j' - j]$. Now $[4(j' - j)] = [j' - j]$ and, hence, $3(j' - j) \equiv 0 \bmod n$. But $3 \nmid n$, so that $j' - j \equiv 0 \bmod n$ and $[j] = [j']$. A similar argument gives $[i] = [i']$. $\bullet$

Proposition 3.135. *If n is an odd positive integer which is not a multiple of 3, then there exists an $n \times n$ diabolic square.*

Proof. Let $A = [a_{ij}]$ and $B = [b_{ij}]$ be diagonal orthogonal $n \times n$ Latin squares, which exist, by Lemma 3.134. By Proposition 3.132, the matrix $M = [a_{ij}n + b_{ij}]$ is a magic square with magic number $\sigma = s(n + 1)$, where $s = \sum_{i=0}^{n-1} i$. As the main

---

[16]If A is a Latin square, it is not always true that A and A^T are orthogonal. For example, the 4×4 Latin square A in Example 3.126 has all 0s on its main diagonal, and so does its transpose. Since all the diagonal entries of $A \circ A^T$ equal $(0, 0)$, the Latin squares A and A^T are not orthogonal.

diagonal of A and of B are permutations of $\{0, 1, \ldots, n - 1\}$, the sum of the diagonal terms is $s(n + 1)$, and the same is true of the back diagonal.   •

Design of Experiments

Here is a fertilizer story that will ultimately be seen to be related to Latin squares. To maximize his corn production, a farmer has to choose the best type of seed. But he knows that the amount of fertilizer also affects his crop. How can he design an experiment to show him what is the best combination? We give a simple illustration. Suppose there are three types of seed: A, B, and C. To measure the effect of using different amounts of fertilizer, the farmer can divide a plot into 9 subplots, as follows:

Amount of Fertilizer	Seed Type		
High	A	B	C
Medium	A	B	C
Low	A	B	C

In each position, an observation x_{sf} is made, where x_{sf} is the number of ears harvested according to the seed type s and level f of fertilizer.

The farmer now wants to see the effect of differing dosages of pesticide. He could have 27 observations x_{sfp} (more generally, if he had n different dosages and n different seed types, there would be n^3 observations). On the other hand, suppose he arranges his experiment as follows (again, we illustrate with $n = 3$).

	Amount of Pesticide		
Amount of Fertilizer	High	Medium	Low
High	A	B	C
Medium	C	A	B
Low	B	C	A

The seed types are now arranged in a Latin square. For example, the observation from the northwest subplot is the number of ears from seed type A, with a high level of fertilizer, and high level of pesticide. There are only 9 observations instead of 27 (more generally, there are n^2 observations instead of n^3). Obviously we do not have all possible observations. To infer properties about a large collection from measuring a small sample is what statistics is all about. And it turns out that the Latin square organization of data gives essentially the same statistical information as that given by the complete set of all n^3 observations. A discussion of the analysis of variance for such designs can be found, for example, in Li, *An Introduction to Experimental Statistics*.

The farmer now wants to consider water amounts. Again we illustrate with $n = 3$. In addition to the seed types A, B, C, and the various levels of fertilizer and pesticide, let there be three water levels: $\alpha > \beta > \gamma$.

	Amount of Pesticide		
Amount of Fertilizer	High	Medium	Low
High	$A\alpha$	$B\beta$	$C\gamma$
Medium	$C\beta$	$A\gamma$	$B\alpha$
Low	$B\gamma$	$C\alpha$	$A\beta$

The observation from the northwest subplot, for example, is the number of ears of seed type A, high level of fertilizer, high level of pesticide, and high level of water. The experiment has been designed so that the plots arise from orthogonal Latin squares. Again, the statistical data arising from this small number, namely 9, of observations are essentially the same as what one would get from 81 observations x_{sfpw} (more generally, n^2 observations instead of n^4). Euler called such matrices *Graeco-Latin squares*, because he described them, as above, using Latin and Greek fonts; he coined the term *Latin square* for this same notational reason. One could test more variables if one could find an *orthogonal set* of Latin squares as defined below.

Definition.   A set $A_1, A_2, \ldots, A_t$ of $n \times n$ Latin squares is an **orthogonal set** if each pair of them is orthogonal.

Lemma 3.136.   *If $A_1, A_2, \ldots, A_t$ is an orthogonal set of $n \times n$ Latin squares, then $t \leq n - 1$.*

Proof.   There is no loss in generality in assuming that each A_ν has entries lying in $X = \{0, 1, \ldots, n-1\}$. Permute the entries of A_1 so that its first row is $0, 1, \ldots, n-1$ in this order. By Lemma 3.127, this new matrix A_1' is a Latin square which is orthogonal to each of $A_2, \ldots, A_t$. Now permute the entries of A_2 so that its first row is $0, 1, \ldots, n-1$ in this order. This new matrix A_2' is a Latin square, and it is orthogonal to each of $A_1', A_3, \ldots, A_t$. Continuing in this way, we may assume that the top row of each A_ν is $0, 1, \ldots, n-1$ in this order.

If $\nu \neq \lambda$, then the first row of $A_\nu \circ A_\lambda$, their Hadamard product, is

$$(0, 0), (1, 1), \ldots, (n-1, n-1).$$

We claim that A_ν and A_λ do not have the same 2, 1 entry. Otherwise, there is some k with $a_{21}^\nu = k = a_{21}^\lambda$ (where a_{ij}^ν denotes the ij entry of A_ν) so that

$$(a_{21}^\nu, a_{21}^\lambda) = (k, k).$$

This contradicts the orthogonality of A_ν and A_λ, for the ordered pair (k, k) already occurs in the first row of $A_\nu \circ A_\lambda$ as $(a_{1k}^\nu, a_{1k}^\lambda)$. Therefore, distinct A_ν have distinct entries in the 2, 1 position. In any A_ν, however, there are only $n - 1$ choices for its 2, 1 entry, because 0 already occurs in its 1, 1 position, and so there are at most $n - 1$ distinct A_ν's. •

Definition.   A *complete orthogonal set* of $n \times n$ Latin squares is an orthogonal set of $n - 1$ Latin squares.

Theorem 3.137.   *If $q = p^e$, then there exists a complete orthogonal set of $q - 1$ $q \times q$ Latin squares.*

Proof.   If k is a finite field with q elements, then there are $q - 1$ elements $a \in k^\times$, and so there are $q - 1$ Latin squares L_a, each pair of which is orthogonal, by Theorem 3.128.
•

One Latin square can test two variables (e.g., levels of fertilizer and pesticide) on different varieties (e.g., of seed). A Graeco-Latin square (i.e., a pair of orthogonal Latin squares) allows testing for a third variable (e.g., levels of water). More generally, a set of t orthogonal Latin squares allows one to test levels of $t + 1$ different variables on different varieties.

Projective Planes

And now another stream enters the story. By the early 1800s, mathematicians were studying the problems of perspective arising from artists painting pictures of three-dimensional scenes on two-dimensional canvases. To the eye, parallel lines seem to meet at the horizon, and this suggests adjoining a new construct, a "line at infinity," to the ordinary plane. Every line is parallel to a line ℓ passing through the origin O. For each such line, define a new point, ω_ℓ, and form the new set

$$\mathbb{P}^2(\mathbb{R}) = \mathbb{R}^2 \cup H,$$

where $H = \{\omega_\ell : \ell \text{ is a line through } O\}$. In $\mathbb{P}^2(\mathbb{R})$, we define new *lines*: H is a line (the *line at infinity*, or the horizon); for each (ordinary) line L in $\mathbb{R}^2$, define $L^* = L \cup \{\omega_\ell\}$, where ℓ is the line through the origin which is parallel to L.

Let us show that every pair of (new) lines in $\mathbb{P}^2(\mathbb{R})$ intersect in a point. If $L^* = L \cup \{\omega_\ell\}$, then $L^* \cap H = \{\omega_\ell\}$. Now consider $L^* \cap M^*$, where $M^* = M \cup \{\omega_m\}$. If L and M are parallel, then $L \cap M = \emptyset$ and $\{\omega_\ell\} = \{\omega_m\}$; hence, $L^* \cap M^* = \{\omega_\ell\}$. If L and M are not parallel, then there is a point Q in the plane with $L \cap M = \{Q\}$. Since $\omega_\ell \neq \omega_\ell$, we have $L^* \cap M^* = \{Q\}$.

It is also true that every two distinct points $Q, R \in \mathbb{P}^2(\mathbb{R})$ determine a line. If $Q = \omega_\ell$ and $R = \omega_m$, then Q, R determine H. If $Q = \omega_\ell$ and $R \in \mathbb{R}^2$, then Q, R

determine the ordinary line L through R which is parallel to ℓ; hence, Q, R determine L^*. Finally, if both $Q, R \in \mathbb{R}^2$, then they determine an ordinary line L in the plane, and hence they determine the new line L^*.

Since we are now interested in finite structures, let us replace the plane $\mathbb{R} \times \mathbb{R}$ by a finite "plane" $k \times k$, where k is a finite field with q elements. We regard this finite plane as the direct sum of additive abelian groups. Define a **line ℓ through the origin** $O = (0, 0)$ to be a subset of the form

$$\ell = \{(ax, ay) : a \in k \text{ and } (x, y) \neq O\},$$

and, more generally, define a **line** to be a coset

$$(u, v) + \ell = \big\{(u + ax, v + ay) : a \in k\big\}.$$

Since k is finite, we can do some counting. There are q^2 points in the plane, and there are q points on every line. As usual, two points determine a line. Call two lines **parallel** if they do not intersect, and say that two lines have the same **direction** if they are parallel. How many directions are there? Every line, being a coset of a line ℓ through the origin, has the same direction as ℓ, whereas distinct lines through the origin have different directions, for they intersect. Thus, the number of directions is the same as the number of lines through the origin. There are $q^2 - 1$ points $V \neq O$, each of which determines a line $\ell = OV$ through the origin. Since there are q points on ℓ, there are $q - 1$ points on ℓ other than O, and each of them determines ℓ. There are thus

$$(q^2 - 1)/(q - 1) = q + 1$$

directions. We adjoin $q + 1$ new points ω_ℓ to $k \times k$, one for each direction; that is, one for each line ℓ through the origin. Define H, the **line at infinity**, by

$$H = \{\omega_\ell : \ell \text{ is a line through the origin}\},$$

and define the **projective plane** over k:

$$\mathbb{P}^2(k) = (k \times k) \cup H.$$

Define a (projective) **line** in $\mathbb{P}^2(k)$ to be either H or an old line $(u, v) + \ell$ in $k \times k$ with the point ω_ℓ adjoined, where ℓ is a line through the origin. It follows that $|\mathbb{P}^2(k)| = q^2 + q + 1$, every line has $q + 1$ points, and any two points determine a unique line. In Example 4.26, we shall use linear algebra to give another construction of the projective plane $\mathbb{P}^2(k)$.

Example 3.138.
If $k = \mathbb{F}_2$, then $k \times k$ has 4 points: $O = (0, 0)$, $a = (1, 0)$, $b = (0, 1)$, and $c = (1, 1)$, and 6 lines, each with two points, as in Figure 3.3.

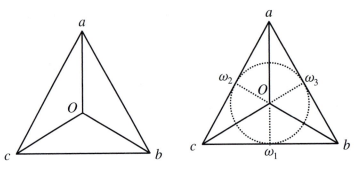

Figure 3.3   Affine Plane   **Figure 3.4**   Projective Plane

There are three sets of parallel lines: Oa and bc, Ob and ac, and Oc and ab. The projective plane $\mathbb{P}(\mathbb{F}_2)$ is obtained by adding new points ω_1, ω_2, ω_3 and forcing parallel lines to meet. There are now 7 lines: the 6 original lines (each lengthened) and the line at infinity $\{\omega_1, \omega_2, \omega_3\}$.   ◄

We now abstract the features we need.

Definition.   A **projective plane of order** n is a set X with $|X| = n^2 + n + 1$, a family of subsets called **lines**, each having $n + 1$ points, such that every two points determine a unique line.

We have seen above that if k is a finite field with q elements, then $\mathbb{P}(k)$ is a projective plane of order q. It is possible to construct projective planes without using finite fields. For example, it is known that there are four projective planes of order 9, only one of which arises from the finite field with 9 elements.

The following theorem is the reason we have introduced projective planes.

Theorem.   *If $n \geq 3$, then there exists a projective plane of order n if and only if there exists a complete orthogonal set of $n \times n$ Latin squares.*

Proof.   See Ryser, *Combinatorial Mathematics*, p. 92.   •

A natural question is to find those n for which there exists a projective plane of order n. Notice that this is harder than Euler's original question; instead of asking whether there is an orthogonal pair of $n \times n$ Latin squares, we are now asking whether there is an orthogonal set of $n - 1$ $n \times n$ Latin squares. If $n = p^e$, then we have constructed a projective plane of order n above. Since Tarry proved that there is no orthogonal pair of 6×6 Latin squares, there is no set of 5 pairwise orthogonal 6×6

Latin squares, and so there is no projective plane of order 6. The following theorem was proved in 1949.

Theorem (Bruck-Ryser).   *If either* $n \equiv 1$ mod 4 *or* $n \equiv 2$ mod 4 *and, further, if* n *is not a sum of two squares, then there does not exist a projective plane of order* n.

Proof.   See Ryser, *Combinatorial Mathematics*, p. 111.   •

The first few $n \equiv 1$ or 2 mod 4 with $n \geq 3$ are

$$5, \ 6, \ 9, \ 10, \ 13, \ 14, \ 17, \ 18, \ 21, \ 22.$$

Some of these are primes or prime powers, and so they must be sums of two squares[17] because projective planes of these orders do exist; and so it is:

$$5 = 1 + 4; \qquad 9 = 0 + 9; \qquad 13 = 4 + 9; \qquad 17 = 1 + 16.$$

Of the remaining numbers, $10 = 1 + 9$ and $18 = 9 + 9$ are sums of two squares (and the theorem does not apply), but the others are not. It follows that there is no projective plane of order 6, 14, 21, or 22 (thus, Tarry's result follows from the Bruck-Ryser theorem).

The smallest n not covered by the Bruck–Ryser theorem is $n = 10$. The question whether there exists a projective plane of order 10 was the subject of much investigation (after Tarry, 10 was also the first open case of Euler's conjecture). This is a question about a set with 111 points, and so one would expect that a computer could solve it quickly. But it is really a question about 11-point subsets of a set with 111 points, the order of magnitude of which is the binomial coefficient $\binom{111}{11}$, a huge number. In spite of this, C. Lam was able to show, in 1988, that there does not exist a projective plane of order 10. He used a massive amount of calculation: 19,200 hours on VAX 11/780 followed by 3000 hours on CRAY-1S. Thus, two and half years of actual computer running time (not counting the years of human thought and ingenuity involved in instructing the machines) solved the problem. As of this writing, it is unknown whether a projective plane of order 12 exists ($12 \equiv 0$ mod 4, and so it is not covered by the Bruck-Ryser theorem).

---

[17]Recall Theorem 3.83, Fermat's two-square theorem: if $p \equiv 1$ mod 4, then p is a sum of two squares. Since there exists a projective plane of order p, the Bruck-Ryser theorem implies the two-square theorem. In fact, the Bruck-Ryser theorem implies that if $p \equiv 1$ mod 4, then p^e is a sum of two squares for all $e \geq 1$.

4

Linear Algebra

→ **4.1 VECTOR SPACES**

Linear algebra is the study of vector spaces and their homomorphisms (called linear transformations, and which are concretely described by matrices), with applications to systems of linear equations. Most readers have had some course involving matrices with real entries or complex entries and, hence, Sections 4.1, 4.3, and 4.4 can safely be skipped (most of the results here, while proved for vector spaces with scalars in arbitrary fields, have essentially the same proofs as their special cases for real vector spaces). Note, however, that the discussion of ruler-compass constructions in Section 4.2 uses scalars in certain unfamiliar subfields of $\mathbb{C}$, and the discussion of codes in Section 4.5 (whose ideas enable us to see photographs sent from outer space) uses scalars in finite fields.

→ **Definition.**   If k is a field, then a ***vector space over k*** is an (additive) abelian group V equipped with a ***scalar multiplication***: there is a function $k \times V \to V$, denoted by $(a, v) \mapsto av$, such that, for all $a, b \in k$ and all $u, v \in V$,

(i) $a(u + v) = au + av$;

(ii) $(a + b)v = av + bv$;

(iii) $(ab)v = a(bv)$;

(iv) $1v = v$, where 1 is the one in k.

Besides the 5 axioms explicitly mentioned [scalar multiplication is defined plus axioms (i) through (iv)], there are several more axioms implicit in the statement that a vector space is an additive abelian group. Addition is a function $V \times V \to V$, denoted by $(u, v) \mapsto u + v$, satisfying the following equations for all $u, v, w \in V$:

(i)$'$ $(u + v) + w = u + (v + w)$;

(ii)$'$ $u + v = v + u$;

(iii)$'$ there is $0 \in V$ with $0 + v = v$;

(iv)$'$ for each $v \in V$, there is $v' \in V$ with $v + v' = 0$.

Thus, the definition of vector space involves ten axioms.

Elements of V are called ***vectors***[1] and elements of k are called ***scalars***.

$\rightarrow$ **Example 4.1.**

(i) Euclidean space $\mathbb{R}^n$ is a vector space over $\mathbb{R}$; its vectors are the n-tuples $v = (a_1, \ldots, a_n)$, where $a_i \in \mathbb{R}$ for all i. Picture a vector v as an arrow from the origin to the point having coordinates $(a_1, \ldots, a_n)$. Addition is given by

$$(a_1, \ldots, a_n) + (b_1, \ldots, b_n) = (a_1 + b_1, \ldots, a_n + b_n);$$

geometrically, the sum of two vectors is described by the ***parallelogram law*** (see Figure 2.5 on page 101).

If $c \in \mathbb{R}$, then scalar multiplication by c is given by

$$cv = c(a_1, \ldots, a_n) = (ca_1, \ldots, ca_n).$$

Scalar multiplication $v \mapsto cv$ "stretches" v by a factor $|c|$, reversing its direction when c is negative (we put quotes around *stretches* because cv is shorter than v when $|c| < 1$).

(ii) The example in part (i) can be generalized. If k is any field, define $V = k^n$, the set of all n-tuples $v = (a_1, \ldots, a_n)$, where $a_i \in k$ for all i. Unless we say otherwise, we will usually view vectors in k^n as $n \times 1$ columns. Addition and scalar multiplication by $c \in k$ are given by the same formulas as in (i):

$$(a_1, \ldots, a_n) + (b_1, \ldots, b_n) = (a_1 + b_1, \ldots, a_n + b_n);$$
$$c(a_1, \ldots, a_n) = (ca_1, \ldots, ca_n).$$

(iii) If R is a commutative ring and k is a subring that is a field, then R is a vector space over k. Regard the elements of R as vectors and the elements of k as scalars; define scalar multiplication cv, where $c \in k$ and $v \in R$, to be the given product of two elements in R. Notice that the axioms in the definition of vector space are just particular cases of some of the axioms holding in the commutative ring R.

---

[1]The word *vector* comes from the Latin word meaning "to carry"; vectors in Euclidean space carry the data of length and direction. The word *scalar* comes from regarding $v \mapsto cv$ as a change of scale. The terms *scale* and *scalar* come from the Latin word meaning "ladder," for the rungs of a ladder are evenly spaced.

For example, if k is a field, then the polynomial ring $R = k[x]$ is a vector space over k. Vectors are polynomials $f(x)$, scalars are elements $c \in k$, and scalar multiplication gives the polynomial $cf(x)$; that is, if

$$f(x) = b_n x^n + \cdots + b_1 x + b_0,$$

then

$$cf(x) = cb_n x^n + \cdots + cb_1 x + cb_0.$$

In particular, if a field k is a subfield of a larger field E, then E is a vector space over k. For example, $\mathbb{C}$ is a vector space over $\mathbb{R}$.

(iv) If k is a field, let $\mathrm{Mat}_{m \times n}(k)$ denote the set of all $m \times n$ matrices having entries in k. Define the sum $A + B$ of two matrices A and B by adding entries in the same position: if $A = [a_{ij}]$ and $B = [b_{ij}]$, then

$$A + B = [a_{ij} + b_{ij}].$$

If $c \in k$, then multiplying each entry of $A = [a_{ij}]$ by c gives

$$cA = [ca_{ij}].$$

It is routine to check that $\mathrm{Mat}_{m \times n}(k)$ is a vector space over k.

If $n = 1$, then $\mathrm{Mat}_{n \times 1}(k) = k^n$. If $m = n$, then we write $\mathrm{Mat}_n(k)$ instead of $\mathrm{Mat}_{n \times n}(k)$. ◄

A *subspace* of a vector space V is a subset of V that is a vector space under the addition and scalar multiplication in V. However, we give a simpler definition that is more convenient to use.

→ **Definition.**   If V is a vector space over a field k, then a ***subspace*** of V is a subset U of V such that

(i) $0 \in U$;

(ii) $u, u' \in U$ imply $u + u' \in U$;

(iii) $u \in U$ and $c \in k$ imply $cu \in U$.

→ **Proposition 4.2.**   *Every subspace U of a vector space V over a field k is itself a vector space over k.*

Proof.   By hypothesis, U is closed under scalar multiplication: if $u \in U$ and $c \in k$, then $cu \in U$. Axioms (i) through (iv) in the definition of vector space hold for all scalars and for all vectors in V; in particular, they hold for all vectors in U. For example, axiom (iii) says that $(ab)v = a(bv)$ holds for all $a, b \in k$ and all $v \in V$; in particular, this equation holds for all $u \in U$.

By hypothesis, U is closed under addition: if $u, u' \in U$, then $u + u' \in U$. Axioms (i)$'$ through (iv)$'$ in the definition of vector space hold for all scalars and for all vectors in V; in particular, they hold for all vectors in U. Finally, axiom (iii)$'$ requires $0 \in U$, and this, too, is part of the hypothesis. •

$\rightarrow$ **Example 4.3.**

(i) The extreme cases $U = V$ and $U = \{0\}$ (where $\{0\}$ denotes the subset consisting of the zero vector alone) are always subspaces of a vector space. A subspace $U \subseteq V$ with $U \neq V$ is called a ***proper subspace*** of V; we may write $U \subsetneq V$ to denote U being a proper subspace of V.

(ii) If $v = (a_1, \ldots, a_n)$ is a nonzero vector in $\mathbb{R}^n$, then

$$\ell = \{av : a \in \mathbb{R}\}$$

is a line through the origin, and ℓ is a subspace of $\mathbb{R}^n$. For example, the diagonal $\{(a, a) : a \in \mathbb{R}\}$ is a subspace of the plane $\mathbb{R}^2$.

Similarly, a plane through the origin consists of all vectors $av_1 + bv_2$, where v_1, v_2 is a fixed pair of noncollinear vectors, and a, b vary over $\mathbb{R}$. It is easy to check that planes through the origin are subspaces of $\mathbb{R}^n$.

By Proposition 4.2, lines and planes through the origin are vector spaces; without this proposition, one would be obliged to check each of the ten axioms in the definition of vector space.

(iii) If $m \leq n$ and $\mathbb{R}^m$ is regarded as the set of all those vectors in $\mathbb{R}^n$ whose last $n - m$ coordinates are 0, then $\mathbb{R}^m$ is a subspace of $\mathbb{R}^n$. For example, we may regard $\mathbb{R}^1 = \mathbb{R}$ as all points $(x, 0)$ in $\mathbb{R}^2$; that is, $\mathbb{R}$ can be viewed as the real axis in the plane.

(iv) If k is a field, then a ***linear system over*** k of m equations in n unknowns is a set of equations

$$a_{11}x_1 + \cdots + a_{1n}x_n = b_1$$
$$a_{21}x_1 + \cdots + a_{2n}x_n = b_2$$
$$\vdots \qquad \qquad \vdots \qquad\qquad (1)$$
$$a_{m1}x_1 + \cdots + a_{mn}x_n = b_m,$$

where $a_{ij}, b_i \in k$. A ***solution*** is a vector $s = (s_1, \ldots, s_n) \in k^n$ with $\sum_j a_{ij}s_j = b_i$ for all i. The set of all solutions is a subset of k^n, called the ***solution set*** of the system; if there are no solutions, one calls the system ***inconsistent***. A linear system is ***homogeneous*** if all the b_i are 0. Since the zero vector is always a solution of a homogeneous linear system, a homogeneous system is always consistent. These definitions can be written more compactly in matrix notation. The ***coefficient matrix*** of system (1) is $A = [a_{ij}]$. If b is the column vector

$b = (b_1, \ldots, b_m)$, then $s = (s_1, \ldots, s_n)$ is a solution if and only if $As = b$. A solution s of a homogeneous system $Ax = 0$ is called **nontrivial** if some $s_j \neq 0$.

The set of all solutions of a homogeneous linear system forms a subspace of k^n, called the **solution space** (or *nullspace*) of the system. To see this, let $Ax = 0$ be a homogeneous system, and let U be its solution space. Now $0 \in U$ because $A0 = 0$. If $u, u' \in U$, then $Au = 0 = Au'$, and so $A(u + u') = Au + Au' = 0+0 = 0$; hence, $u+u' \in U$. If $c \in k$ and $u \in U$, then $A(cu) = c(Au) = c0 = 0$, and so $cu \in U$. Therefore, U is a subspace of k^n.

We can solve systems of linear equations over the field $\mathbb{F}_p$, where p is a prime; that is, we can treat a system of congruences mod p just as we treat an ordinary system of equations over $\mathbb{R}$.

For example, the system of congruences

$$3x - 2y + z \equiv 1 \bmod 7$$
$$x + y - 2z \equiv 0 \bmod 7$$
$$-x + 2y + z \equiv 4 \bmod 7$$

can be regarded as a system of equations over the field $\mathbb{F}_7$. This system can be solved just as in high school, for inverses mod 7 are now known: $[2][4] = [1]$; $[3][5] = [1]$; $[6][6] = [1]$. The solution is

$$(x, y, z) = ([5], [4], [1]).$$

(v) Recall that the **transpose** of an $m \times n$ matrix $A = [a_{ij}]$ is the $n \times m$ matrix A^T whose ij entry is a_{ji}; the ith row of A is the ith column of A^T; the jth column of A is the jth row of A^T. The basic properties of transposing are

$$(A + B)^T = A^T + B^T; \quad (cA)^T = cA^T;$$
$$(AB)^T = B^T A^T; \quad (A^T)^T = A.$$

An $n \times n$ matrix A is **symmetric** if $A^T = A$. If k is a field, we show that the set S of all symmetric $n \times n$ matrices is a subspace of $\mathrm{Mat}_n(k)$. If 0 denotes the matrix all of whose entries are 0, then $0^T = 0$, and so $0 \in S$. If $A, B \in S$, then

$$(A + B)^T = A^T + B^T = A + B,$$

so that $A + B \in S$. Finally, if $c \in k$ and $A \in S$, then

$$(cA)^T = c(A^T) = cA,$$

so that $cA \in S$. Therefore, S is a subspace of $\mathrm{Mat}_n(k)$. By Proposition 4.2, the set of all $n \times n$ symmetric matrices with entries in a field k is a vector space. ◄

The following type of square matrix is important.

→ **Definition.**   An $m \times m$ matrix A is **nonsingular** if there exists an $m \times m$ matrix B such that $AB = I$ and $BA = I$. One calls B the **inverse** of A and denotes it by A^{-1}.

Recall that the *dot product* of two vectors $v = (a, b, c)$, $v' = (a', b', c')$ in $\mathbb{R}^3$ is defined as $v \cdot v' = aa' + bb' + cc' \in \mathbb{R}$. There is a geometric interpretation of this number:

$$v \cdot v' = \|v\| \, \|v'\| \cos \theta,$$

where $\|v\|$ is the length of v and θ is the angle between v and v'. It follows that if $v \cdot v' = 0$, then either $v = 0$, $v' = 0$, or that v and v' are **orthogonal**.[2] We can adapt dot product to more general spaces.

→ **Definition.**   If k is a field and V is a vector space over k, then an **inner product** on V is a function $f \colon V \times V \to k$, usually denoted by $f(v, w) = (v, w)$, such that

(i) $(v, w + w') = (v, w) + (v, w')$ for all $v, w, w' \in V$;

(ii) $(v, aw) = a(v, w)$ for all $v, w \in V$ and $a \in k$;

(iii) $(v, w) = (w, v)$ for all $v, w \in V$.

An inner product is called **nondegenerate** (or **nonsingular**) if, for all $v \in V$, $(v, v) = 0$ implies $v = 0$. (See Theorem 4.104 for an application of this construction.)

→ **Example 4.4.**

(i) Let k be any field, let $V = k^n$, and let $v = (a_1, \ldots, a_n)$, $v' = (a_1', \ldots, a_n') \in V$. Then

$$(v, v') = a_1 a_1' + \cdots + a_n a_n'$$

is an inner product on k^n. If $k = \mathbb{R}$, then this inner product is nondegenerate, for if $\sum_i a_i^2 = 0$, then each $a_i = 0$. However, if $k = \mathbb{C}$, then this inner product is **degenerate** (not nondegenerate). For example, if $n = 2$ and $v = (1, i)$, then $(v, v) = 1 + i^2 = 0$. One usually repairs this for vector spaces $V = \mathbb{C}^n$ by defining $(v, v') = \sum a_j \bar{a}_j'$, where $\bar{a}$ is complex conjugate. This does not give an inner product [because axiom (ii) of the definition may not hold: $(v, aw) = \bar{a}(v, w)$], but it does give $(v, v) = 0$ implies $v = 0$.

The same phenomenon can occur for inner products defined on vector spaces over a finite field k. For example, let $k = \mathbb{F}_2$; if n is even and $v = (1, 1, \ldots, 1) \in k^n$, then $(v, v) = 0$; if n is odd and $v = (0, 1, 1, \ldots, 1)$, then $(v, v) = 0$.

---

[2]In Greek, *ortho* means "right" and *gon* means "angle." Thus, *orthogonal* means "right angled" or "perpendicular".

(ii) Let k be a field, and regard a vector in k^n as an $n \times 1$ column matrix. If A is an $n \times n$ symmetric matrix with entries in k, define an inner product on $V = k^n$ by

$$(v, w) = v^T A w.$$

The reader may prove that this is an inner product, and that it is nondegenerate if and only if A is a nonsingular matrix. ◄

→ **Example 4.5.**
Let V be a vector space with an inner product, and let $W \subseteq V$ be a subspace. Define

$$W^\perp = \{v \in V : (w, v) = 0 \text{ for all } w \in W\}.$$

Let us check that $W^\perp$ (pronounced W perp) is a subspace. Clearly, $0 \in W^\perp$. If $v, v' \in W^\perp$, then $(w, v) = 0$ and $(w, v') = 0$ for all $w \in W$. Hence, $(w, v + v') = (w, v) + (w, v') = 0$ for all $w \in W$, and so $v + v' \in W^\perp$. Finally, if $v \in W^\perp$ and $a \in k$, then $(w, av) = a(w, v) = 0$, so that $av \in W^\perp$. Therefore, $W^\perp$ is a subspace of V; it is called the ***orthogonal complement*** of W, to remind us that $(v, w) = 0$ in Euclidean space does imply that u and v are orthogonal vectors. It is easy to see that $W \cap W^\perp = \{0\}$ if and only if the inner product is nondegenerate. ◄

Dimension is a rather subtle idea. One thinks of a curve in the plane, that is, the image of a continuous function $f \colon \mathbb{R} \to \mathbb{R}^2$, as a one-dimensional subset of a two-dimensional space. Imagine the confusion at the end of the 19th century when a "space-filling curve" was discovered: there exists a continuous $f \colon \mathbb{R} \to \mathbb{R}^2$ with image the whole plane! We will give a way of defining dimension that works for vector spaces over $\mathbb{R}$ (there are topological ways of defining dimension of more general spaces).

The key observation in getting the "right" definition of dimension is to understand why $\mathbb{R}^3$ is three-dimensional. Every vector (x, y, z) is a linear combination of the three vectors $e_1 = (1, 0, 0)$, $e_2 = (0, 1, 0)$, and $e_3 = (0, 0, 1)$; that is,

$$(x, y, z) = xe_1 + ye_2 + ze_3.$$

It is not so important that every vector is a linear combination of these specific vectors; what is important is that there are three of them, for it turns out that 3 is the smallest number of vectors in $\mathbb{R}^3$ with the property that every vector is a linear combination of them.

→ **Definition.** A ***list***[3] in a vector space V is a finite sequence $X = v_1, \ldots, v_n$ of vectors in V, where $n \in \mathbb{N}$. In particular, we allow the ***empty list*** having no terms (it is the list with $n = 0$).

---

[3]A list $X = a_1, \ldots, a_n$ is exactly the same thing as an n-tuple $(a_1, \ldots, a_n)$. We write n-tuples with parentheses to conform to standard notation.

More precisely,we are saying that a list $X = v_1, \ldots, v_n$ in V is a function

$$\varphi \colon \{1, 2, \ldots, n\} \to V,$$

for some $n \in \mathbb{N}$, with $\varphi(i) = v_i$ for all i. Note that X is ordered in the sense that there is a first vector v_1, a second vector v_2, and so forth. A vector may appear several times on a list; that is, φ need not be injective. The empty list φ has im $\varphi = \varnothing$.

→ **Definition.** Let V be a vector space over a field k. A ***linear combination*** of a nonempty list $v_1, \ldots, v_n$ in V is a vector v of the form

$$v = a_1 v_1 + \cdots + a_n v_n,$$

where $n \in \mathbb{N}$ and $a_i \in k$ for all i. A linear combination of the empty list is defined to be 0, the zero vector.

→ **Definition.** If $X = v_1, \ldots, v_m$ is a list in a vector space V, then

$$\langle X \rangle = \langle v_1, \ldots, v_m \rangle$$

is the set of all the linear combinations of $v_1, \ldots, v_m$; it is called the ***subspace spanned by X***. We also say that $v_1, \ldots, v_m$ ***spans*** $\langle v_1, \ldots, v_m \rangle$.

Example 4.6.
If A is an $m \times n$ matrix over a field k, then its ***row space*** $Row(A)$ is the subspace of k^n spanned by the rows of A. The ***column space*** $Col(A)$ of A is the subspace of k^m spanned by the columns of A. Note that $Row(A) = Col(A^T)$ and $Col(A) = Row(A^T)$, for the columns of A are the rows of A^T (and the rows of A are the columns of A^T).

If A is an $m \times n$ matrix, its row space $Row(A)$, its column space $Col(A)$, and the ***solution space*** $Sol(A)$ of $Ax = 0$ are related. If the usual inner product on k^n is nondegenerate, then $Row(A)^{\perp} = Sol(A)$, $Col(A)^{\perp} = Sol(A^T)$, and $Sol(A)^{\perp} = Row(A)$ (see Leon, *Linear Algebra with Applications*, pages 242–244). ◄

→ **Proposition 4.7.** *If $X = v_1, \ldots, v_m$ is a list in a vector space V, then $\langle X \rangle$ is a subspace of V containing the subset $\{v_1, \ldots, v_m\}$.*

Proof. Let us write $L = \langle v_1, \ldots, v_m \rangle$. Now $0 \in L$, for

$$0 = 0 v_1 + \cdots + 0 v_m.$$

If $u = a_1 v_1 + \cdots + a_m v_m$ and $v = b_1 v_1 + \cdots + b_m v_m \in L$, then

$$u + v = a_1 v_1 + \cdots + a_n v_n + b_1 v_1 + \cdots + b_m v_m$$
$$= a_1 v_1 + b_1 v_1 \cdots + a_m v_m + b_m v_m$$
$$= (a_1 + b_1) v_1 + \cdots + (a_m + b_m) v_m \in L.$$

Finally, if $c \in k$, then

$$c(a_1 v_1 + \cdots + a_m v_m) = (ca_1)v_1 + \cdots + (ca_m)v_m \in L.$$

Therefore, L is a subspace.

To see that each $v_i \in L$, choose the linear combination having $a_i = 1$ and all other coefficients 0.   •

If $X = v_1, \ldots, v_n$ is a list in a vector space V, then its **underlying set** is the subset $\{v_1, \ldots, v_n\}$. Note that v_1, v_2, v_3 and v_2, v_1, v_3 are distinct lists having the same underlying set. Moreover, v_1, v_2, v_2 and v_1, v_2 are also distinct lists having the same underlying set. One reason we are being so fussy about lists and underlying sets can be found in our discussion of coordinates on page 332.

Lemma 4.8.   *If $X = v_1, \ldots, v_n$ is a list in a vector space V, then $\langle X \rangle$ depends only on its underlying set $\{v_1, \ldots, v_n\}$.*

Proof.   If $\sigma \in S_n$ is a permutation, then define a list $X^\sigma = v_{\sigma(1)}, \ldots, v_{\sigma(n)}$. Now a linear combination of X is a *vector* $v = a_1 v_1 + \cdots + a_n v_n$. Since addition in V is commutative, v is also a linear combination of the list X^σ. Therefore, $\langle X \rangle = \langle X^\sigma \rangle$, for both subsets are comprised of the same vectors.

If the list X has a repetition; say, $v_i = v_j$ for some $i \neq j$, then

$$a_1 v_1 + \cdots + a_n v_n = a_1 v_1 + \cdots + (a_i + a_j)v_i + \cdots + \widehat{v}_j + \cdots + a_n v_n,$$

where $a_1 v_1 + \cdots + \widehat{v}_j + \cdots + a_n v_n$ is the shorter sum with v_j deleted. It follows that the set of linear combinations of X is the same as the set of all linear combinations of the shorter list obtained from X by deleting v_j.   •

We now extend the definition of $\langle Y \rangle$ to arbitrary, possibly infinite, subsets $Y \subseteq V$.

→ **Definition.**   If Y is a subset of a vector space V, then $\langle Y \rangle$ is the set of all finite linear combinations of lists $v_1, \ldots, v_n$, for $n \in \mathbb{N}$, whose terms all lie in Y.

If Y is finite, then Lemma 4.8 shows that this definition coincides with our earlier definition of $\langle Y \rangle$ on page 327.

→ **Lemma 4.9.**   *Let V be a vector space over a field k.*

(i) *Every intersection of subspaces of V is itself a subspace.*

(ii) *If Y is a subset of V, then $\langle Y \rangle$ is the intersection of all the subspaces of V containing Y.*

(iii) *If Y is a subset of V, then $\langle Y \rangle$ is the **smallest** subspace of V containing Y; that is, if U is any subspace of V containing Y, then $\langle Y \rangle \subseteq U$.*

Proof.

(i) Let $\mathcal{S}$ be a family of subspaces of V, and denote $\bigcap_{S \in \mathcal{S}} S$ by W. Since $0 \in S$ for every $S \in \mathcal{S}$, we have $0 \in W$. If $x, y \in W$, then $x, y \in S$ for every $S \in \mathcal{S}$; as S is a subspace, we have $x + y \in S$ for all $S \in \mathcal{S}$, and so $x + y \in W$. Finally, if $x \in W$, then $x \in S$ for every $S \in \mathcal{S}$; if $c \in k$, then $cx \in S$ for all S, and so $cx \in W$. Therefore, W is a subspace of V.

(ii) Let $\mathcal{S}'$ denote the family of all the subspaces of V containing subset Y. We claim that

$$\langle Y \rangle = \bigcap_{S \in \mathcal{S}'} S.$$

The inclusion $\subseteq$ is clear: if $v_1, \ldots, v_n$ is a list with each $v_i \in Y$ and $\sum_i c_i v_i \in \langle Y \rangle$, then $\sum_i c_i v_i \in S$ for every $S \in \mathcal{S}'$, because a subspace contains the linear combinations of any list of its vectors. [This argument even holds if $Y = \varnothing$, for then $\langle Y \rangle = \{0\}$.] The reverse inclusion follows from a general fact about intersections: for any $S_0 \in \mathcal{S}'$, we have $\bigcap_{S \in \mathcal{S}'} S \subseteq S_0$. In particular, $S_0 = \langle Y \rangle \in \mathcal{S}'$, by Proposition 4.7.

(iii) A subspace U containing Y is one of the subspaces S involved in the intersection $\langle Y \rangle = \bigcap_{S \in \mathcal{S}'} S$.   •

Were all terminology in algebra consistent, we would call $\langle Y \rangle$ the subspace *generated by* Y. The reason for the different terms is that the theories of groups, rings, and vector spaces developed independently of each other.

$\xrightarrow{}$ **Example 4.10.**

(i) Let $V = \mathbb{R}^2$, let $e_1 = (1, 0)$, and let $e_2 = (0, 1)$. Now $V = \langle e_1, e_2 \rangle$, for if $v = (a, b) \in V$, then

$$\begin{aligned} v &= (a, 0) + (0, b) \\ &= a(1, 0) + b(0, 1) \\ &= ae_1 + be_2 \in \langle e_1, e_2 \rangle. \end{aligned}$$

(ii) If k is a field and $V = k^n$, define e_i as the n-tuple having 1 in the ith coordinate and 0's elsewhere. The reader may adapt the argument in part (i) to show that $e_1, \ldots, e_n$ spans k^n.

The list $e_1, \ldots, e_n$ is called the **standard basis** of k^n. Every vector in k^n is a linear combination of the standard basis: $(a_1, \ldots, a_n) = a_1 e_1 + \cdots + a_n e_n$.

(iii) A vector space V need not be spanned by a finite sequence. For example, let $V = k[x]$, and suppose that $X = f_1(x), \ldots, f_m(x)$ is a finite list in V. If d is the largest degree of any of the $f_i(x)$, then every (nonzero) linear combination of $\sum_i a_i f_i(x)$, where $a_i \in k$, has degree at most d. Thus, x^{d+1} is not a linear combination of vectors in X, and so X does not span $k[x]$.   ◀

The following definition makes sense even though we have not yet defined *dimension*.

→ **Definition.**   A vector space V is called *finite dimensional* if it is spanned by a finite list; otherwise, V is called *infinite dimensional*.

Example 4.10(ii) shows that k^n is finite dimensional, while part (iii) of this Example shows that $k[x]$ is infinite dimensional. By Example 4.1(iii), both $\mathbb{R}$ and $\mathbb{C}$ are vector spaces over $\mathbb{Q}$, and each is infinite dimensional.

Given a subspace U of a vector space V, we seek a list X which spans U. Notice that U can have many such lists; for example, if $X = v_1, v_2, \ldots, v_m$ spans U and u is any vector in U, then $v_1, v_2, \ldots, v_m, u$ also spans U. Let us, therefore, seek a *shortest list* that spans U.

→ **Definition.**   A list $X = v_1, \ldots, v_m$ in a vector space V is a *shortest spanning list* (or a *minimal spanning list*) if no proper sublist $v_1, \ldots, \widehat{v_i} \ldots, v_m$ spans $\langle v_1, \ldots, v_m \rangle \subseteq V$.

→ **Proposition 4.11.**   *If V is a vector space, then the following conditions on a list $X = v_1, \ldots, v_m$ spanning V are equivalent:*

(i) *X is not a shortest spanning list; that is, a proper sublist spans $\langle X \rangle$.*

(ii) *Some v_i is in the subspace spanned by the others; that is,*

$$v_i \in \langle v_1, \ldots, \widehat{v_i}, \ldots, v_m \rangle;$$

(iii) *There are scalars $a_1, \ldots, a_m$, not all zero, with*

$$\sum_{j=1}^{m} a_j v_j = 0.$$

Proof.   (i) $\Rightarrow$ (ii). If X is not a shortest spanning list, then one of the vectors in X, say, v_i, can be thrown out, and $v_i \in \langle v_1, \ldots, \widehat{v_i}, \ldots, v_m \rangle$.

(ii) $\Rightarrow$ (iii). If $v_i = \sum_{j \neq i} c_j v_j$, then define $a_i = -1 \neq 0$ and $a_j = c_j$ for all $j \neq i$.

(iii) $\Rightarrow$ (i). The given equation implies that one of the vectors, say, v_i, is a linear combination of the others, say,

$$v_i = \sum_{j \neq i} a_i^{-1} a_j v_j.$$

Deleting v_i gives a shorter list, which still spans: if $v \in V$, then we know that $v = \sum_{j=1}^{m} b_j v_j$, for the list $v_1, \ldots, v_m$ spans V. We rewrite:

$$v = b_i v_i + \sum_{j \neq i} b_j v_j$$

$$= b_i \left(\sum_{j \neq i} a_i^{-1} a_j v_j \right) + \sum_{j \neq i} b_j v_j \in \langle v_1, \ldots, \widehat{v_i}, \ldots, v_m \rangle. \quad \bullet$$

Definition. A list $X = v_1, \ldots, v_m$ in a vector space V is **linearly dependent** if there are scalars $a_1, \ldots, a_m$, not all zero, with $\sum_{j=1}^{m} a_j v_j = 0$; otherwise, X is called **linearly independent**.

The empty set $\varnothing$ is defined to be linearly independent (we interpret $\varnothing$ as a list of length 0).

Example 4.12.

(i) Any list $X = v_1, \ldots, v_m$ containing the zero vector is linearly dependent.

(ii) A list v_1 of length 1 is linearly dependent if and only if $v_1 = 0$; hence, a list v_1 of length 1 is linearly independent if and only if $v_1 \neq 0$.

(iii) A list v_1, v_2 is linearly dependent if and only if one of the vectors is a scalar multiple of the other.

(iv) If there is a repetition on the list $v_1, \ldots, v_m$ (that is, if $v_i = v_j$ for some $i \neq j$), then $v_1, \ldots, v_m$ is linearly dependent: define $c_i = 1$, $c_j = -1$, and all other $c = 0$. Therefore, if $v_1, \ldots, v_m$ is linearly independent, then all the vectors v_i are distinct. ◄

The contrapositive of Proposition 4.11 is worth stating.

Corollary 4.13. *If $X = v_1, \ldots, v_m$ is a list spanning a vector space V, then X is a shortest spanning list if and only if X is linearly independent.*

Linear independence has been defined indirectly as not being linearly dependent. Because of the importance of linear independence, let us define it directly. A list $X = v_1, \ldots, v_m$ is **linearly independent** if, whenever a linear combination $\sum_{j=1}^{m} a_j v_j = 0$, then every $a_j = 0$. Informally, this says that every "sublist" of a linearly independent list is itself linearly independent (this is one reason for decreeing that $\varnothing$ be linearly independent).

We have arrived at the notion we have been seeking.

Definition. A *basis* of a finite dimensional vector space V is a linearly independent list which spans V.

Thus, bases are shortest spanning lists. Of course, all the vectors in a linearly independent list $v_1, \ldots, v_n$ are distinct, by Example 4.12(iv).

Example 4.14.

In Example 4.10(ii), we saw that the standard basis $E = e_1, \ldots, e_n$ spans k^n, where e_i is the n-tuple having 1 in the ith coordinate and 0s elsewhere. To see that E is linearly independent, note that $\sum_{i=1}^{n} a_i e_i = (a_1, \ldots, a_n)$, so that $\sum_{i=1}^{n} a_i e_i = 0$ if and only if each $a_i = 0$. Therefore, E is a basis of k^n. ◄

→ **Proposition 4.15.** *Let $X = v_1, \ldots, v_n$ be a list in a vector space V over a field k. Then X is a basis if and only if each vector in V has a unique expression as a linear combination of vectors in X.*

Proof. If a vector $v = \sum a_i v_i = \sum b_i v_i$, then $\sum (a_i - b_i) v_i = 0$, and so independence gives $a_i = b_i$ for all i; that is, the expression is unique.

Conversely, existence of an expression shows that the list of v_i spans. Moreover, if $0 = \sum c_i v_i$ with not all $c_i = 0$, then the vector 0 does not have a unique expression as a linear combination of the v_i. •

→ **Definition.** If $X = v_1, \ldots, v_n$ is a basis of a vector space V and if $v \in V$, then there are unique scalars $a_1, \ldots, a_n$ with $v = \sum_{i=1}^{n} a_i v_i$. The n-tuple $(a_1, \ldots, a_n)$ is called the ***coordinate list*** of a vector $v \in V$ relative to the basis X.

If $E = e_1, \ldots, e_n$ is the standard basis of $V = k^n$, then each vector $v \in V$ has a unique expression

$$v = a_1 v_1 + a_2 v_2 + \cdots + a_n v_n,$$

where $a_i \in k$ for all i. The coordinate list of $v \in k^n$ coincides with its usual coordinates, for

$$v = (a_1, \ldots, a_n) = a_1 e_1 + \cdots + a_n e_n.$$

Since there is a first vector v_1, a second vector v_2, and so forth, the coefficients in this linear combination determine a unique n-tuple $(a_1, a_2, \ldots, a_n)$. Were a basis merely a subset of V and not a list, then there would be $n!$ coordinate lists for every vector.

We are going to define the dimension of a vector space V to be the number of vectors in a basis. Two questions arise at once.

(i) Does every vector space have a basis?

(ii) Do all bases of a vector space have the same number of elements?

The first question is easy to answer; the second needs some thought.

→ **Theorem 4.16.** *Every finite dimensional vector space V has a basis.*

Proof. A finite spanning list X exists, since V is finite dimensional. If it is linearly independent, it is a basis; if not, X can be shortened to a spanning sublist X', by Proposition 4.11. If X' is linearly independent, it is a basis; if not, X' can be shortened to a spanning sublist X''. Eventually, we arrive at a shortest spanning sublist, which is independent and hence it is a basis. •

Remark.   The definitions of spanning and linear independence can be extended to infinite dimensional vector spaces (when dealing with infinite dimensional vector spaces, one usually speaks of subspaces spanned by subsets rather than by lists). We can prove that these vector spaces also have bases. For example, it turns out that a basis of $k[x]$ is $1, x, x^2, \ldots, x^n, \ldots$.   ◄

We are now going to prove invariance of dimension, one of the most important results about vector spaces.

→ **Lemma 4.17.**   *Let* $u_1, \ldots, u_n$ *span a vector space* V. *If* $v_1, \ldots, v_m \in V$ *and* $m > n$, *then* $v_1, \ldots, v_m$ *is a linearly dependent list.*

Proof.   The proof is by induction on $n \geq 1$.

Base Step. If $n = 1$, then there are at least two vectors v_1, v_2, for $m > n$, and $v_1 = a_1 u_1$ and $v_2 = a_2 u_1$. If $u_1 = 0$, then $v_1 = 0$ and the list of v's is linearly dependent. Suppose $u_1 \neq 0$. We may assume that $v_1 \neq 0$, or we are done; hence, $a_1 \neq 0$. Therefore, $u_1 = a_1^{-1} v_1$, and so v_1, v_2 is linearly dependent (for $v_2 - a_2 a_1^{-1} v_1 = 0$), and hence the larger list $v_1, \ldots, v_m$ is linearly dependent.

Inductive Step. There are equations, for $i = 1, \ldots, m$,

$$v_i = a_{i1} u_1 + \cdots + a_{in} u_n.$$

We may assume that some $a_{i1} \neq 0$, otherwise $v_1, \ldots, v_m \in \langle u_2, \ldots, u_n \rangle$, and the inductive hypothesis applies. Changing notation if necessary (that is, by reordering the v's), we may assume that $a_{11} \neq 0$. For each $i \geq 2$, define

$$v_i' = v_i - a_{i1} a_{11}^{-1} v_1 \in \langle u_2, \ldots, u_n \rangle,$$

for the coefficient of u_1 in v_i' is $a_{i1} - (a_{i1} a_{11}^{-1}) a_{11} = 0$. Since $m - 1 > n - 1$, the inductive hypothesis gives scalars $b_2, \ldots, b_m$, not all 0, with

$$b_2 v_2' + \cdots + b_m v_m' = 0.$$

Rewrite this equation using the definition of v_i':

$$\left(-\sum_{i \geq 2} b_i a_{i1} a_{11}^{-1} \right) v_1 + b_2 v_2 + \cdots + b_m v_m = 0.$$

Not all the coefficients are 0, and so $v_1, \ldots, v_m$ is linearly dependent.   •

→ **Theorem 4.18 (Invariance of Dimension).**   *If $X = x_1, \ldots, x_n$ and $Y = y_1, \ldots, y_m$ are bases of a vector space V, then $m = n$.*

Proof. If $m \neq n$, then either $n < m$ or $m < n$. In the first case, $y_1, \ldots, y_m \in \langle x_1, \ldots, x_n \rangle$, because X spans V, and Lemma 4.17 gives Y linearly dependent, a contradiction. A similar contradiction arises if $m < n$, and so we must have $m = n$.
•

It is now permissible to make the following definition.

→ **Definition.**   If V is a finite dimensional vector space over a field k, then its ***dimension***, denoted by $\dim(V)$, is the number of elements in a basis of V.

→ **Example 4.19.**

   (i) Example 4.14 shows that k^n has dimension n, for the standard basis has n elements, and this agrees with our intuition when $k = \mathbb{R}$. Thus, the plane $\mathbb{R} \times \mathbb{R}$ is two-dimensional!

   (ii) If $V = \{0\}$, then $\dim(V) = 0$, for there are no elements in its basis $\varnothing$. (This is the reason for defining $\varnothing$ to be linearly independent.)

   (iii) Let $X = \{x_1, \ldots, x_n\}$ be a finite set. Define

$$k^X = \{\text{functions } f : X \to k\}.$$

   Now k^X is a vector space if we define addition $f + f'$ to be

$$f + f' : x \mapsto f(x) + f'(x)$$

   and scalar multiplication af, for $a \in k$ and $f : X \to k$, by

$$af : x \mapsto af(x).$$

   It is easy to check that the set of n functions of the form f_x, where $x \in X$, defined by

$$f_x(y) = \begin{cases} 1 & \text{if } y = x; \\ 0 & \text{if } y \neq x, \end{cases}$$

   forms a basis, and so $\dim(k^X) = n = |X|$.
       This is not a new example: after all, an n-tuple $(a_1, \ldots, a_n)$ is really a function $f : \{1, \ldots, n\} \to k$ with $f(i) = a_i$ for all i. Thus, the functions f_x comprise the standard basis. ◀

The following proof illustrates the intimate relation between linear algebra and systems of linear equations.

Corollary 4.20. *A homogeneous system of linear equations over a field k with more unknowns than equations has a nontrivial solution.*

Proof. An n-tuple $(s_1, \ldots, s_n)$ is a solution of a system

$$a_{11}x_1 + \cdots + a_{1n}x_n = 0$$

$$\vdots \quad \vdots \quad \vdots$$

$$a_{m1}x_1 + \cdots + a_{mn}x_n = 0$$

if $a_{i1}s_1 + \cdots + a_{in}s_n = 0$ for all i. In other words, if $C_1, \ldots, C_n$ are the columns of the $m \times n$ coefficient matrix $A = [a_{ij}]$, then the definition of matrix multiplication gives

$$s_1 C_1 + \cdots + s_n C_n = 0.$$

Note that $C_i \in k^m$. Now k^m can be spanned by m vectors (the standard basis, for example). Since $n > m$, by hypothesis, Lemma 4.17 shows that the list $C_1, \ldots, C_n$ is linearly dependent; there are scalars $p_1, \ldots, p_n$, not all zero, with $p_1 C_1 + \cdots + p_n C_n = 0$. Therefore, $(p_1, \ldots, p_n)$ is a nontrivial solution of the system. •

→ **Definition.** A list $u_1, \ldots, u_m$ in a vector space V is a ***longest linearly independent list*** (or a ***maximal linearly independent list***) if there is no vector $v \in V$ such that $u_1, \ldots, u_m, v$ is linearly independent.

→ **Lemma 4.21.** *Let V be a finite dimensional vector space.*

(i) *Let $v_1, \ldots, v_m$ be a linearly independent list in V, and let $v \in V$. If $v \notin \langle v_1, \ldots, v_m \rangle$, then $v_1, \ldots, v_m, v$ is linearly independent.*

(ii) *If a longest linearly independent list $X = v_1, \ldots, v_n$ exists, then it is a basis of V. Conversely, if a basis exists, then it is a longest linearly independent list.*

Proof.
(i) Let $av + \sum_i a_i v_i = 0$. If $a \neq 0$, then $v = -a^{-1} \sum_i a_i v_i \in \langle v_1, \ldots, v_m \rangle$, a contradiction. Therefore, $a = 0$ and $\sum_i a_i v_i = 0$. But linear independence of $v_1, \ldots, v_m$ implies each $a_i = 0$, and so the longer list $v_1, \ldots, v_m, v$ is linearly independent.
(ii) If X is not a basis, then it does not span: there is $w \in V$ with $w \notin \langle v_1, \ldots, v_n \rangle$. But the longer list X, w is linearly independent, by part (i), contradicting X being a longest independent list. We let the reader prove the converse. •

It is not obvious that longest linearly independent lists always exist; that they do exist follows from the next result, which is quite useful in its own right.

→ **Proposition 4.22.** *Let V be an n-dimensional vector space. If* $Z = u_1, \ldots, u_m$ *is a linearly independent list in V, then Z can be extended to a basis; that is, there are vectors* $v_1, \ldots, v_{n-m}$ *so that* $u_1, \ldots, u_m, v_1, \ldots, v_{n-m}$ *is a basis of V.*

Proof. If $m > n$, then Lemma 4.17 implies that Z is linearly dependent, a contradiction; therefore, $m \leq n$. If the linearly independent list Z does not span V, there is $v_1 \in V$ with $v_1 \notin \langle Z \rangle$, and the longer list $Z, v_1 = u_1, \ldots, u_m, v_1$ is linearly independent, by Lemma 4.21. If Z, v_1 does not span V, there is $v_2 \in V$ with $v_2 \notin \langle Z, v_1 \rangle$. This process eventually stops, for the length of these lists can never exceed $n = \dim(V)$. •

→ **Corollary 4.23.** *If* $\dim(V) = n$, *then any list of* $n + 1$ *or more vectors is linearly dependent.*

Proof. Otherwise, such a list could be extended to a basis having too many elements. •

In Exercise 4.11 on page 342, it is shown that if $\text{Mat}_{m \times n}(k)$ is the vector space of all $m \times n$ matrices over k. then $\dim(\text{Mat}_{m \times n}(k)) = mn$. It now follows that if B is an $n \times n$ matrix over k, then the list $I, B, B^2, \ldots, B^{n^2}$ is linearly dependent. Hence, there are scalars $a_0, a_1, \ldots, a_{n^2}$, not all 0, with

$$a_0 I + a_1 B + \cdots + a_{n^2} B^{n^2} = 0.$$

Therefore, there is a polynomial $f(x) \in k[x]$ pf degree $\leq n^2$, with $f(B) = 0$. This is a "poor man's version" of the Cayley-Hamilton theorem, which states that there is a polynomial (the *characteristic polynomial*) $h_B(x) \in k[x]$ of degree n with $h_B(B) = 0$.

Corollary 4.24. *Let V be a vector space with* $\dim(V) = n$.

(i) *A list X of n vectors which spans V must be linearly independent.*

(ii) *Any linearly independent list Y of n vectors must span V.*

Proof.
(i) If the list X is linearly dependent, then it could be shortened to give a basis of V which is too small.
(ii) If the list Y does not span V, then it could be lengthened to give a basis of V which is too large. •

→ **Corollary 4.25.** *Let U be a subspace of a vector space V of dimension n.*

(i) *U is finite dimensional.*

(ii) $\dim(U) \leq \dim(V)$.

(iii) *If* $\dim(U) = \dim(V)$, *then* $U = V$.

Proof.
(i) Take $u_1 \in U$. If $U = \langle u_1 \rangle$, then U is finite dimensional. Otherwise, there is $u_2 \notin \langle u_1 \rangle$. By Lemma 4.21, u_1, u_2 is linearly independent. If $U = \langle u_1, u_2 \rangle$, we are done. This process cannot be repeated $n + 1$ times, for then $u_1, \ldots, u_{n+1}$ would be a linearly independent list in $U \subseteq V$, contradicting Corollary 4.23.
(ii) A basis of U is linearly independent, and so it can be extended to a basis of V. Therefore, $\dim(U) \leq \dim(V)$.
(iii) If $\dim(U) = \dim(V)$, then a basis of U is already a basis of V (otherwise it could be extended to a basis of V that would be too large). •

Example 4.26.
A *projective plane of order n* was defined, in Chapter 3, as a set X with $|X| = n^2+n+1$, and a family of subsets of X, called *lines*, each having $n + 1$ points, such that every two points determine a unique line. If q is a prime power, we constructed a projective plane of order q by adjoining a line at infinity to $\mathbb{F}_q^2$.

We now give a second construction of a projective plane. Let k be a field and let $W = k^3$. A line L in k^3 through the origin consists of all the scalar multiples of any one of its nonzero vectors: if $v = (a, b, c) \in L$ and $v \neq (0, 0, 0)$, then

$$L = \{rv = (ra, rb, rc) : r \in k\}.$$

Of course, if v' is another nonzero vector in L, then $L = \{rv' : r \in k\}$. Thus, both v and v' span L if and only if both are nonzero and $v' = tv$ for some nonzero $t \in k$. Define a relation on the set of all nonzero vectors in k^3 by

$$v = (a, b, c) \sim v' = (a', b', c') \quad \text{if there exists } t \in k \text{ with } v' = tv.$$

Note that $t \neq 0$ lest $tv = (0, 0, 0)$. It is easy to check that $\sim$ is an equivalence relation on $W - \{(0, 0, 0)\}$, and we denote the equivalence class of $v = (a, b, c)$ by

$$[v] = [a, b, c].$$

An equivalence class $[v]$ is called a ***projective point***, and the set of all projective points is called the ***projective plane*** over k, denoted by $\mathbb{P}^2(k)$. If π is a plane in k^3 through the origin (that is, if π is a 2-dimensional subspace of k^3), then we define the ***projective line*** $[\pi]$ to consist of all the projective points $[v]$ for which $v \in \pi$.

Corollary 4.27. *Let k be a field.*

(i) *Every two distinct projective points $[v]$ and $[v']$ in $\mathbb{P}^2(k)$ lie in a unique projective line.*

(ii) *Two distinct projective lines $[\pi]$ and $[\pi']$ in $\mathbb{P}^2(k)$ intersect in a unique projective point.*

Proof.
(i) That $[v]$ and $[v']$ are projective points says that v and v' are nonzero vectors in k^3; that $[v] \neq [v']$ says that $v \not\sim v'$; that is, there is no scalar $t \neq 0$ with $v' = tv$, so that v, v' is a linearly independent list. Therefore, there is a unique plane $\pi = \langle v, v' \rangle$ through the origin containing v and v', and so $[\pi]$ is a projective line containing $[v]$ and $[v']$. This projective line is unique, for if $[v], [v'] \in [\pi']$, then $v, v' \in \pi'$, and so $\pi \subseteq \pi'$. Corollary 4.25(iii) gives $\pi = \pi'$, and so $[\pi] = [\pi']$.
(ii) Consider π and π' in k^3. By Exercise 4.19 on page 343,

$$\dim(\pi + \pi') + \dim(\pi \cap \pi') = \dim(\pi) + \dim(\pi').$$

Since $\pi \neq \pi'$, we have $\pi \subsetneq \pi + \pi'$; hence, $2 = \dim(\pi) < \dim(\pi + \pi') \leq 3 = \dim(k^3)$, and so $\dim(\pi + \pi') = 3$. Hence, $\dim(\pi \cap \pi') = 2 + 2 - 3 = 1$, so that $[\pi \cap \pi'] = [\pi] \cap [\pi']$ is a projective point. The point of intersection is unique, lest we contradict part (i).   •

Proposition 4.28.   *If $q = p^n$ for some prime p, then there exists a projective plane of order q.*

Proof.   Let $X = \mathbb{P}^2(k)$, where $k = \mathbb{F}_q$. Now $|k^3| = q^3$, and so there are $q^3 - 1$ nonzero vectors in k^3. If $v \in k^3$ is nonzero, then $|[v]| = q - 1$, for there are exactly $q - 1$ nonzero scalars in k. Thus, $|X| = (q^3 - 1)/(q - 1) = q^2 + q + 1$. Finally, a plane π through the origin in k^3 has $q^2 - 1$ nonzero vectors, and so $|[\pi]| = (q^2 - 1)/(q - 1) = q + 1$. By Corollary 4.27, X is a projective plane of order q.   •

Here is the usual definition of a projective plane.

Definition.   Let X be a set and let $\mathcal{L}$ be a family of subsets of X, called **lines**. Then $(X, \mathcal{L})$ is a **projective plane** if

  (i) Every two lines intersect in a unique point.

  (ii) Every two points determine a unique line.

  (iii) There exist 4 points in X no three of which are collinear.

  (iv) There are 4 lines in $\mathcal{L}$ no three of which contain the same point.

In the special case when X is finite, this definition is equivalent to the definition given in Chapter 3.

Define the **dual** of a statement S about $(X, \mathcal{L})$ to be the statement obtained from S in which the terms *point* and *line* are interchanged as are the terms *containing* and *contained in*. The dual of each axiom in the definition of projective plane is another axiom. We conclude that any theorem about projective planes yields a **dual theorem** whose proof is obtained from the original proof by dualizing each of the statements in its proof.

One can also see duality by comparing the construction of $\mathbb{P}^2(k)$ with the construction of $k^2 \cup \omega$ in Chapter 3, where

$$\omega = \{\omega_\ell : \ell \text{ is a line through the origin}\}$$

is the line at infinity. In more detail, let $\ell = \{r(a, b) : r \in k\}$ be a line in k^2 through the origin, where $(a, b) \neq (0, 0)$. We may denote ℓ by $[a, b]$ and

$$\omega_\ell = \omega_{[a,b]}.$$

Note that this notation is consistent with $[a, b, c] \in \mathbb{P}^2(k)$; that is, if $\ell = \{r(a', b') : r \in k\}$, then there is a nonzero $t \in k$ with $(a', b') = t(a, b)$. Define a function $\varphi: \mathbb{P}^2(k) \to k^2 \cup \omega$ by

$$\varphi([a, b, c]) = \begin{cases} (ac^{-1}, bc^{-1}) & \text{if } c \neq 0; \\ \omega_{[a,b]} & \text{if } c = 0. \end{cases}$$

It is straightforward to check that φ is a (well-defined) bijection.

A proof of the following lemma is not difficult.

Lemma.   *A subset $\pi \subseteq k^3$ is a plane through the origin if and only if there are $p, q, r \in k$, not all zero, with $\pi = \{(a, b, c) \in k^3 : pa + qb + rc = 0\}$. Moreover, if $\pi' = \{(a, b, c) \in k^3 : p'a + q'b + r'c = 0\}$, then $\pi = \pi'$ if and only if there is a nonzero $t \in k$ with $(p', q', r') = t(p, q, r)$.*

Projective points almost have coordinates: if $v = (a, b, c)$, then we call $[a, b, c]$ the ***homogeneous coordinates*** of the projective *point* $[v]$ (these are defined only up to nonzero scalar multiple). In light of the preceding Lemma, projective lines, too, almost have coordinates: if $\pi = \{(a, b, c) \in k^3 : pa + qb + rc = 0\}$, then we call $[p, q, r]$ the ***homogeneous coordinates*** of the projective *line* $[\pi]$ (these are defined only up to nonzero scalar multiple). The bijection $\varphi: \mathbb{P}^2(k) \to k^2 \cup \omega$ preserves lines, and the duality in projective planes can be viewed as replacing a projective point with homogeneous coordinates $[a, b, c]$ with the projective line having these same homogeneous coordinates.   ◄

We are now going to apply linear algebra to fields.

→ **Proposition 4.29 (= Proposition 3.119).**   *If E is a finite field, then $|E| = p^n$ for some prime p and some $n \geq 1$.*

Proof.   By Proposition 3.110, the prime field of E is isomorphic to $\mathbb{F}_p$ for some prime p. Since E is finite, it is finite dimensional; say, $\dim(E) = n$. If $v_1, \ldots, v_n$ is a basis, then there are exactly p^n vectors $a_1 v_1 + \cdots + a_n v_n \in E$, where $a_i \in \mathbb{F}_p$ for all i.   •

→ **Definition.**   If k is a subfield of a field K, then we usually say that K is an **extension of** k . We abbreviate this by writing "K/k is an extension." [4]

If K/k is an extension, then K may be regarded as a vector space over k, as in Example 4.1(iii). We say that K is a **finite extension** of k if K is a finite dimensional vector space over k; the dimension of K, denoted by $[K : k]$, is called the **degree** of K/k.

Here is the reason $[K : k]$ is called the degree.

→ **Proposition 4.30.**   *Let E/k be an extension, let $z \in E$ be a root of an irreducible polynomial $p(x) \in k[x]$, and let $k(z)$ be the smallest subfield of E containing k and z. Then*

$$[k(z) : k] = \dim_k(k(z)) = \deg(p).$$

Proof.   Proposition 3.116(iv) says that each element in $k(z)$ has a unique expression of the form $b_0 + b_1 z + \cdots + b_{n-1} z^{n-1}$, where $b_i \in k$ and $n = \deg(p)$. By Proposition 4.15, the list $1, z, z^2, \ldots, z^{n-1}$ is a basis of $k(z)$.   •

The following formula is quite useful, especially when one is proving a theorem by induction on degrees.

→ **Theorem 4.31.**   *Let $k \subseteq K \subseteq E$ be fields, with K a finite extension of k and E a finite extension of K. Then E is a finite extension of k, and*

$$[E : k] = [E : K][K : k].$$

Proof.   If $A = a_1, \ldots, a_n$ is a basis of K over k and if $B = b_1, \ldots, b_m$ is a basis of E over K, then it suffices to prove that a list X of all $a_i b_j$ is a basis of E over k.

To see that X spans E, take $e \in E$. Since B is a basis of E over K, there are scalars $\lambda_j \in K$ with $e = \sum_j \lambda_j b_j$. Since A is a basis of K over k, there are scalars $\mu_{ji} \in k$ with $\mu_j = \sum_i \mu_{ji} a_i$. Therefore, $e = \sum_{ij} \mu_{ji} a_i b_j$, and X spans E over k.

To prove that X is linearly independent over k, assume that there are scalars $\lambda_{ji} \in k$ with $\sum_{ij} \lambda_{ji} a_i b_j = 0$. If we define $\lambda_j = \sum_i \mu_{ji} a_i$, then $\lambda_j \in K$ and $\sum_j \lambda_j b_j = 0$. Since B is linearly independent over K, it follows that

$$0 = \lambda_j = \sum_i \mu_{ji} a_i$$

for all j. Since A is linearly independent over k, it follows that $\mu_{ji} = 0$ for all j and i, as desired.   •

---

[4] We pronounce K/k as "K over k"; there should be no confusing this notation with that of a quotient ring, for K is a field and hence it has no proper nonzero ideals.

→ **Definition.** Assume that K/k is an extension and that $z \in K$. We call z **algebraic** over k if there is some nonzero polynomial $f(x) \in k[x]$ having z as a root; otherwise, z is called **transcendental** over k.

When one says that a real number is transcendental, one usually means that it is transcendental over $\mathbb{Q}$. For example, F. Lindemann (1852–1939) proved, in 1882, that π is transcendental, so that $[\mathbb{Q}(\pi) : \mathbb{Q}]$ is infinite (see A. Baker, *Transcendental Number Theory*, p. 5). Using this fact, we can see that $\mathbb{R}$, viewed as a vector space over $\mathbb{Q}$, is infinite dimensional. (For a proof of the irrationality of π, a more modest result, we refer the reader to Niven and Zuckerman, *An Introduction to the Theory of Numbers*.)

→ **Proposition 4.32.** *If K/k is a finite extension, then every $z \in K$ is algebraic over k.*

Proof. If $[K : k] = n$, then the list $1, z, z^2, \ldots, z^n$ has length $n + 1$, by Corollary 4.23. Hence, there are $a_i \in k$, not all zero, with $\sum_{i=0}^{n} a_i z^i = 0$. If we define $f(x) = \sum_{i=0}^{n} a_i x^i$, then $f(x)$ is not the zero polynomial and $f(z) = 0$. Therefore, z is algebraic over k. •

EXERCISES

H **4.1** True or false with reasons.

(i) If k is a field, then the subset E of all all polynomials of odd degree is a subspace of $k[x]$.

(ii) If A and B are $n \times n$ matrices over a field k, and if the homogeneous system $Ax = 0$ has a nontrivial solution, then the homogeneous system $(BA)x = 0$ has a nontrivial solution.

(iii) If A and B are $n \times n$ matrices over a field k, and if the homogeneous system $Ax = 0$ has a nontrivial solution, then the homogeneous system $(AB)x = 0$ has a nontrivial solution.

(iv) If v_1, v_2, v_3, v_4 spans a vector space V, then $\dim(V) = 4$.

(v) If k is a field, then the list $1, x, x^2, \ldots, x^{100}$ is linearly independent in $k[x]$.

(vi) There is a linearly independent list of 4 matrices in $\mathrm{Mat}_2(\mathbb{R})$.

(vii) There is a linearly independent list of 5 matrices in $\mathrm{Mat}_2(\mathbb{R})$.

(viii) $[\mathbb{Q}(E^{2\pi i/5}) : \mathbb{Q}] = 5$.

(ix) There is an inner product on $\mathbb{R}^2$ with $(v, v) = 0$ for some nonzero $v \in \mathbb{R}^2$.

(x) The set of all $f : \mathbb{R} \to \mathbb{R}$ with $f(1) = 0$ is a subspace of $\mathcal{F}(\mathbb{R})$.

*4.2 (i) If $f : k \to k$ is a function, where k is a field, and if $\alpha \in k$, define a new function $\alpha f : k \to k$ by $a \mapsto \alpha f(a)$. Prove that with this definition of scalar multiplication, the ring $\mathcal{F}(k)$ of all functions on k is a vector space over k.

(ii) If $\mathcal{PF}(k) \subseteq \mathcal{F}(k)$ denotes the family of all polynomial functions $a \mapsto \alpha_n a^n + \cdots + \alpha_1 a + \alpha_0$, prove that $\mathcal{PF}(k)$ is a subspace of $\mathcal{F}(k)$.

4.3 Prove that $\dim(V) \leq 1$ if and only if the only subspaces of a vector space V are $\{0\}$ and V itself.

H **4.4** Prove, in the presence of all the other axioms in the definition of vector space, that the commutative law for vector addition is redundant; that is, if V satisfies all the other axioms, then $u + v = v + u$ for all $u, v \in V$.

4.5 Is L a subspace of $\mathrm{Mat}_n(k)$ if L is the subset consisting of all the $n \times n$ Latin squares?

4.6 (i) If V is a vector space over $\mathbb{F}_2$ and if $v_1 \neq v_2$ are nonzero vectors in V, prove that v_1, v_2 is linearly independent. Is this true for vector spaces over any other field?

(ii) Let k be a field, and let $P_2(k)$ be the projective plane consisting of all points $[x]$ for $x \in k^3$ (as in Example 4.26). Prove that $[x] \neq [y]$ in $P_2(k)$ if and only if x, y is a linearly independent list in k^3.

4.7 Prove that the columns of an $m \times n$ matrix A over a field k are linearly dependent in k^m if and only if the homogeneous system $Ax = 0$ has a nontrivial solution.

*__4.8__ H (i) Prove that the list of polynomials $1, x, x^2, x^3, \ldots, x^{100}$ is a linearly independent list in $k[x]$, where k is a field.

(ii) Define $V_n = \langle 1, x, x^2, \ldots, x^n \rangle$. Prove that $1, x, x^2, \ldots, x^n$ is a basis of V_n, and conclude that $\dim(V_n) = n + 1$.

H **4.9** It is shown in analytic geometry that if ℓ_1 and ℓ_2 are nonvertical lines with slopes m_1 and m_2, respectively, then ℓ_1 and ℓ_2 are perpendicular if and only if $m_1 m_2 = -1$. If

$$\ell_i = \{\alpha v_i + u_i : \alpha \in \mathbb{R}\},$$

for $i = 1, 2$, prove that $m_1 m_2 = -1$ if and only if the dot product $v_1 \cdot v_2 = 0$.

4.10 (i) A *line in space passing through a point u* is defined as

$$\{u + \alpha w : \alpha \in \mathbb{R}\} \subseteq \mathbb{R}^3,$$

where w is a fixed nonzero vector. Show that every line through u is a coset of a one-dimensional subspace of $\mathbb{R}^3$.

H (ii) A *plane in space passing through a point u* is defined as the subset

$$\{v \in \mathbb{R}^3 : (v - u) \cdot n = 0\} \subseteq \mathbb{R}^3,$$

where $n \neq 0$ is a fixed *normal vector* and $(v - u) \cdot n$ is a dot product. Prove that a plane through u is a coset of a two-dimensional subspace of $\mathbb{R}^3$.

*__4.11__ (i) Prove that $\dim(\mathrm{Mat}_{m \times n}(k)) = mn$.

(ii) Determine $\dim(S)$, where S is the subspace of $\mathrm{Mat}_n(k)$ consisting of all the symmetric matrices.

4.12 Let $A \in \mathrm{Mat}_n(k)$. If the characteristic of k is not 2, then A is called **skew symmetric** if $A^T = -A$, where A^T is the transpose of A. In case k has characteristic 2, then A is **skew symmetric** if it is symmetric and if all its diagonal entries are 0.

(i) Prove that the subset K of $\mathrm{Mat}_n(k)$, consisting of all the skew symmetric matrices, is a subspace of $\mathrm{Mat}_n(k)$.

H (ii) Determine $\dim(K)$.

H **4.13** If p is a prime with $p \equiv 1 \bmod 4$, prove that there is a nonzero vector $v \in \mathbb{F}_p^2$ with $(v, v) = 0$, where (v, v) is the usual inner product of v with itself [see Example 4.4(i)].

*H **4.14** Let k be a field, and let k^n have the usual inner product. Prove that if $v = a_1e_1 + \cdots + a_ne_n$, then $a_i = (v, e_i)$ for all i.

*H **4.15** If $f(x) = c_0 + c_1x + \cdots + c_mx^m \in k[x]$ and if $A \in \mathrm{Mat}_n(k)$, define

$$f(A) = c_0I + c_1A + \cdots + c_mA^m \in \mathrm{Mat}_n(k).$$

Prove that there is some nonzero $f(x) \in k[x]$ with $f(A) = 0$.

* **4.16** If U is a subspace of a vector space V over a field k, then U is a subgroup of V (viewed as an additive abelian group). Define a scalar multiplication on the cosets in the quotient group V/U by

$$\alpha(v + U) = \alpha v + U,$$

where $\alpha \in k$ and $v \in V$. Prove that this is a well-defined function that makes V/U into a vector space over k (V/U is called a ***quotient space***).

*H **4.17** If V is a finite dimensional vector space and U is a subspace, prove that

$$\dim(U) + \dim(V/U) = \dim(V).$$

Conclude that $\dim(V/U) = \dim(V) - \dim(U)$.

* **4.18** Let $Ax = b$ be a linear system of equations, and let s be a solution. If U is the solution space of the homogeneous linear system $Ax = 0$, prove that every solution of $Ax = b$ has a unique expression of the form $s + u$ for $u \in U$. Conclude that the solution set of $Ax = b$ is the coset $s + U$.

* **4.19** If U and W are subspaces of a vector space V, define

$$U + W = \{u + w : u \in U \text{ and } w \in W\}.$$

(i) Prove that $U + W$ is a subspace of V.

H **(ii)** If U and U' are subspaces of a finite dimensional vector space V, prove that

$$\dim(U) + \dim(U') = \dim(U \cap U') + \dim(U + U').$$

(iii) A subspace $U \subseteq V$ has a ***complement*** S if $S \subseteq V$ is a subspace such that $U + S = V$ and $U \cap S = \{0\}$; one says that U is a ***direct summand*** of V if U has a complement. If V is finite dimensional, prove that every subspace U of V is a direct summand. (This is true for infinite dimensional vector spaces as well, but a proof requires Zorn's lemma.)

* **4.20** If U and W are vector spaces over a field k, define their ***direct sum*** to be the set of all ordered pairs,

$$U \oplus W = \{(u, w) : u \in U \text{ and } w \in W\},$$

with addition $(u, w) + (u', w') = (u + u', w + w')$ and scalar multiplication $\alpha(u, w) = (\alpha u, \alpha w)$.

(i) Prove that $U \oplus W$ is a vector space.

(ii) If U and W are finite dimensional vector spaces over a field k, prove that

$$\dim(U \oplus W) = \dim(U) + \dim(W).$$

* **4.21** Assume that V is an n-dimensional vector space over a field k and that V has a nondegenerate inner product. If W is an r-dimensional subspace of V, prove that $V = W \oplus W^{\perp}$. (See Example 4.5.) Conclude that $\dim(W^{\perp}) = n - r$.

4.22 Here is a theorem of Pappus holding in k^2, where k is a field. Let ℓ and m be distinct lines, let A_1, A_2, A_3 be distinct points on ℓ, and let B_1, B_2, B_3 be distinct points on m. Define C_1 to be $A_2 B_3 \cap A_3 B_2$, C_2 to be $A_1 B_3 \cap A_3 B_1$, and C_3 to be $A_1 B_2 \cap A_2 B_1$. Then C_1, C_2, C_3 are collinear.

State the dual of the theorem of Pappus (*dual* is defined on page 338).

→ Gaussian Elimination

The following homogeneous system of equations over a field k can be solved at once:

$$
\begin{aligned}
x_1 + \quad &+ u_{1,m+1} x_{m+1} + \cdots + u_{1n} x_n = 0 \\
x_2 + \quad &+ u_{2,m+1} x_{m+1} + \cdots + u_{2n} x_n = 0 \\
&\vdots \qquad\qquad \vdots \qquad\qquad \vdots \\
x_m + u_{m,m+1} x_{m+1} &+ \cdots + u_{mn} x_n = 0.
\end{aligned}
$$

Replacing $x_{m+1}, \ldots, x_n$ by constants $c_{m+1}, \ldots, c_n \in k$, we have

$$
x_i = -(u_{i\,m+1} c_{m+1} + \cdots + u_{in} c_n) \qquad \text{for all } i \leq m,
$$

so that an arbitrary solution has the form

$$
\left(-\sum_{j=m+1}^{n} u_{1j} c_j, \ldots, -\sum_{j=m+1}^{n} u_{mj} c_j, c_{m+1}, \ldots, c_n \right).
$$

The coefficient matrix U of this system,

$$
U = \begin{bmatrix}
1 & 0 & \ldots & 0 & u_{1,m+1} & \ldots & u_{1n} \\
0 & 1 & \ldots & 0 & u_{2,m+1} & \ldots & u_{2n} \\
 & \vdots & & \vdots & & \vdots & \\
0 & 0 & \ldots & 1 & u_{m,m+1} & \ldots & u_{mn}
\end{bmatrix},
$$

is an example of a matrix in *echelon form*.

→ Definition.

An $m \times n$ matrix U is in *row reduced echelon*[5] *form* if

(i) each row of all zeros, if any, lies below every nonzero row;

(ii) the *leading entry* of each nonzero row (its first nonzero entry) is 1;

(iii) every other entry in a *leading column* (a column containing a leading entry) is 0;

---

[5]The word *echelon* means "wing," for the staggering of the leading entries suggests the shape of a bird's wing.

(iv) the leading columns are $\text{COL}(t_1), \ldots, \text{COL}(t_r)$, where $t_1 < t_2 < \cdots < t_r$ and $r \leq m$.

We say that U is in *echelon form* if the leading columns are $\text{COL}(1), \ldots, \text{COL}(r)$; that is, $t_i = i$ for all $i \leq r$.

Consider the matrices

$$A = \begin{bmatrix} 1 & 2 & 3 & 0 & 0 \\ 0 & 0 & 0 & 1 & 5 \\ 0 & 0 & 0 & 0 & 0 \end{bmatrix} \quad \text{and} \quad B = \begin{bmatrix} 1 & 0 & 2 & 3 & 0 \\ 0 & 1 & 0 & 0 & 5 \\ 0 & 0 & 0 & 0 & 0 \end{bmatrix}.$$

Both A and B are in row reduced echelon form, but only B is in echelon form.

→ **Definition.** There are three *elementary row operations* $A \xrightarrow{\ o\ } A'$ changing a matrix A into a matrix A':

Type I: o adds a scalar multiple of one row of A to another row; that is, o replaces $\text{ROW}(i)$ by $\text{ROW}(i) + c\text{ROW}(j)$, where $c \in k$ is nonzero and $j \neq i$;

Type II: o multiplies one row of A by a nonzero $c \in k$; that is, o replaces $\text{ROW}(i)$ by $c\text{ROW}(i)$, where $c \in k$ and $c \neq 0$.

Type III: o interchanges two rows of A.

There are analogous *elementary column operations* on a matrix.

An interchange (Type III) can be accomplished by operations of types I and II (in spite of this redundancy, interchanges are still regarded as elementary operations because they arise frequently). We indicate this schematically.

$$\begin{bmatrix} a & b \\ c & d \end{bmatrix} \rightarrow \begin{bmatrix} a-c & b-d \\ c & d \end{bmatrix} \rightarrow \begin{bmatrix} a-c & b-d \\ a & b \end{bmatrix} \rightarrow \begin{bmatrix} -c & -d \\ a & b \end{bmatrix} \rightarrow \begin{bmatrix} c & d \\ a & b \end{bmatrix}$$

Recall that $Row(A)$, the *row space* of an $m \times n$ matrix A over a field k, is the subspace of k^n spanned by the rows of A, and the *column space*, $Col(A)$, is the subspace of K^m spanned by the columns of A.

→ **Proposition 4.33.** *If $A \rightarrow A'$ is an elementary row operation, then A and A' have the same row space*: $Row(A) = Row(A')$.

Proof. Suppose that $A \rightarrow A'$ is an elementary operation of Type I. The row space of A is $Row(A) = \langle \alpha_1, \ldots, \alpha_m \rangle$, where α_i is the ith row of A; the row space $Row(A')$ is spanned by $\alpha_i + c\alpha_j$ and $\alpha_1, \ldots, \widehat{\alpha_i}, \ldots, \alpha_m$, where $c \in k$ and $j \neq i$. It is obvious that $Row(A') \subseteq Row(A)$. For the reverse inclusion, observe that $\alpha_i = (\alpha_i + c\alpha_j) - c\alpha_j \in Row(A')$.

If $A \to A'$ is an elementary operation of Type II, then $Row(A')$ is spanned by $c\alpha_i$ and $\alpha_1, \ldots, \widehat{\alpha_i}, \ldots, \alpha_m$, where $c \neq 0$. It is obvious that $Row(A') \subseteq Row(A)$. For the reverse inclusion, observe that $\alpha_i = c^{-1}(c\alpha_i) \in Row(A')$.

There is no need to consider elementary operations of Type III, for we have seen that any such can be obtained as a sequence of elementary operations of the other two types.   •

→ **Definition.** If A is an $m \times n$ matrix over a field k with row space $Row(A)$, then

$$\text{rank}(A) = \dim(Row(A)).$$

→ **Corollary 4.34.** *If $A \to A'$ is an elementary row operation, then*

$$\text{rank}(A) = \text{rank}(A').$$

Proof. Even more is true; the row spaces of A and of A' are equal, and so they certainly have the same dimension.   •

We remark that if $A \to A'$ is an elementary row operation, then A and A' may not have the same column space. For example, consider $\begin{bmatrix} 1 & 0 \\ 1 & 0 \end{bmatrix} \to \begin{bmatrix} 1 & 0 \\ 0 & 0 \end{bmatrix}$. However, it is true that both the row space and the column space of a matrix have the same dimension [see Corollary 4.84(iii)].

We are going to show that if $A \to A'$ is an elementary row operation, then the homogeneous systems $Ax = 0$ and $A'x = 0$ have the same solution space. To prove this, we introduce *elementary matrices*.

→ **Definition.** Let o be an elementary row operation, so that A' can be denoted by $A' = o(A)$. An ***elementary matrix*** is an $m \times m$ matrix E of the form $E = o(I)$, where I is the $m \times m$ identity matrix. If o is of Type I, II, or III, then we say that $o(I)$ is an ***elementary matrix of* Type I, II, or III**.

Here are the 2×2 elementary matrices, where c is a nonzero scalar.

$$\begin{bmatrix} 1 & 0 \\ c & 1 \end{bmatrix}, \quad \begin{bmatrix} 1 & c \\ 0 & 1 \end{bmatrix}, \quad \begin{bmatrix} c & 0 \\ 0 & 1 \end{bmatrix}, \quad \begin{bmatrix} 1 & 0 \\ 0 & c \end{bmatrix}, \quad \begin{bmatrix} 0 & 1 \\ 1 & 0 \end{bmatrix}$$

Applying elementary column operations to the identity matrix yields the same family of elementary matrices.

The next lemma shows that the effect of an elementary row operation on a matrix A is the same as multiplying A on the *left* by an elementary matrix, while the effect of an elementary column operation on A is the same as multiplying A on the *right* by an elementary matrix.

→ **Lemma 4.35.** *If A is an m ×n matrix and if A $\xrightarrow{o}$ A' is an elementary row operation, then o(A) = o(I)A; if A $\xrightarrow{o}$ A' is an elementary column operation, then o(A) = Ao(I).*

Proof.   We will merely illustrate the result, leaving the proof to the reader:

$$\text{Type I} \quad \begin{bmatrix} 1 & 0 & 0 \\ 0 & 1 & 0 \\ u & 0 & 1 \end{bmatrix} \begin{bmatrix} a & b & c \\ d & e & f \\ g & h & i \end{bmatrix} = \begin{bmatrix} a & b & c \\ d & e & f \\ ua+g & ub+h & uc+i \end{bmatrix};$$

$$\text{Type II} \quad \begin{bmatrix} 1 & 0 & 0 \\ 0 & u & 0 \\ 0 & 0 & 1 \end{bmatrix} \begin{bmatrix} a & b & c \\ d & e & f \\ g & h & i \end{bmatrix} = \begin{bmatrix} a & b & c \\ ud & ue & uf \\ g & h & i \end{bmatrix}.$$

As before, the result is true for elementary row operations of Type III.

We illustrate an elementary column operation:

$$\text{Type I} \quad \begin{bmatrix} a & b & c \\ d & e & f \\ g & h & i \end{bmatrix} \begin{bmatrix} 1 & 0 & 0 \\ 0 & 1 & 0 \\ u & 0 & 1 \end{bmatrix} = \begin{bmatrix} a+cu & b & c \\ d+df & e & f \\ g+iu & h & i \end{bmatrix}. \quad \bullet$$

Recall that an $n \times n$ matrix A is *nonsingular* if there exists an $m \times m$ matrix B such that $AB = I$ and $BA = I$; one calls B the *inverse* of A and denotes it by A^{-1}.

→ **Proposition 4.36.**   *Every elementary matrix E is a nonsingular matrix. In fact, E^{-1} is an elementary matrix of the same type as E.*

Proof.   If o is an elementary row operation of Type I, then o replaces ROW(i) by ROW(i) + cROW(j). Define o' to be the elementary row operation which replaces ROW(i) by ROW(i) − cROW(j). We claim that the inverse of the elementary matrix $E = o(I)$ is $o'(E)$. By Lemma 4.35, which says that $o(I)A = o(A)$ for every matrix A, we have

$$E'E = o'(I)E = o'(E) = o'(o(I) = I,$$

and, similarly, $EE' = I$. Note that E' is an elementary matrix of Type I.

If o is an elementary row operation of Type II, then o replaces ROW(i) by cROW(i). Define o' to be the elementary row operation which replaces ROW(i) by c^{-1}ROW(i) (this is why we insist that $c \neq 0$). As in the preceding paragraph, if $E' = o'(I)$, then $EE' = I = E'E$. Note that E' is an elementary matrix of Type II.

An elementary matrix E of Type III is equal to its own inverse: $EE = I$. $\bullet$

The next lemma is the key to Gaussian elimination.

Lemma 4.37. *If $A = A_0 \to A_1 \to \cdots \to A_p = B$ is a sequence of elementary row operations, then the linear systems $Ax = 0$ and $Bx = 0$ have the same solution space.*

Proof. The proof is by induction on $p \geq 1$. Let S and S_1 be the solution spaces of $Ax = 0$ and $A_1x = 0$, respectively. If $A_1 = o(A)$, then $A_1 = EA$, by Lemma 4.35, where E is the elementary matrix $o(I)$. If $v \in S$, then $Av = 0$; hence, $0 = EAv = A_1v$, and so $v \in S_1$. The reverse inclusion $S_1 \subseteq S$ follows from the equation $A = E^{-1}A_1$, for E^{-1} is also an elementary matrix. The proof of the inductive step is easy. •

→ **Corollary 4.38.** *If A and B are $m \times n$ matrices over a field k, and if there is a sequence of elementary row operations*

$$A = A_0 \to A_1 \to \cdots \to A_q = B,$$

then there is a nonsingular matrix P with $B = PA$. If there is a sequence of elementary column operations

$$B = B_0 \to B_1 \to \cdots \to B_r = C,$$

then there is a nonsingular matrix Q with $C = BQ$.

Proof. There are elementary matrices E_i with $A_i = E_i A_{i-1}$ for all $i \geq 1$. Therefore, $B = A_q = E_q \cdots E_2 E_1 A$. Define $P = E_q \cdots E_2 E_1$, so that $B = PA$. Now P is nonsingular, for the product of nonsingular matrices is nonsingular $[(E_q \cdots E_2 E_1)^{-1} = E_1^{-1} E_2^{-1} \cdots E_q^{-1}]$. The second statement is proved similarly. •

Recall that if $\sigma \in S_n$ is a permutation, then an $n \times n$ matrix Q_σ is called a *permutation matrix* if it arises from the $n \times n$ identity matrix I by permuting its columns by σ.

If $\tau \in S_n$ is a transposition, then Q_τ interchanges two columns, and so it is an elementary matrix of Type III (remember that applying an elementary column operation to I yields an elementary matrix). Since every permutation σ is a product of transpositions (Proposition 2.35), Q_σ is a product of elementary matrices.

If $Ax = 0$ is a homogeneous system, then the columns of A correspond to the labels on the variables: COL(i) corresponds to x_i. Thus, AQ_σ, which is the matrix whose columns have been permuted by σ, corresponds to the "same" homogeneous system $(AQ_\sigma)y = 0$ whose variables $y_i = x_{\sigma(i)}$ are merely the original variables relabeled.

→ **Definition.**   A matrix A is *Gaussian equivalent* to a matrix B if there is a sequence of elementary row operations

$$A = A_0 \to A_1 \to \cdots \to A_p = B.$$

It is easy to show that Gaussian equivalence is an equivalence relation on the set of all $m \times n$ matrices. If A and B are Gaussian equivalent matrices, then Corollary 4.34 implies that A and B have the same rank, and Lemma 4.37 shows that A and B have the same solution space.

→ **Theorem 4.39 (Gaussian Elimination).**

(i) *Every $m \times n$ matrix A over a field k is Gaussian equivalent to a matrix U in row reduced echelon form.*

(ii) *The matrix U in part (i) is uniquely determined by A.*

(iii) *There is a nonsingular matrix P and a permutation matrix Q_σ with PAQ_σ in echelon form.*

Proof.
(i) The proof is by induction on n, the number of columns of A. Let $n = 1$. If $A = 0$, we are done. If $A \neq 0$, then $a_{j1} \neq 0$ for some j. Multiply ROW(j) by a_{j1}^{-1}, and then interchange ROW(j) with ROW(1), so that the new matrix $A' = [a'_{p1}]$ has $a'_{11} = 1$. For each $p > 1$, replace a'_{p1} by $a'_{p1} - a'_{p1}a'_{11} = 0$. We have arrived at an $m \times 1$ row reduced echelon matrix, for the entry in its first row is 1, while all the other rows are 0.

For the inductive step, let A be an $m \times (n + 1)$ matrix. If the first column of A is 0, then put the matrix comprised of the last n columns into row reduced echelon form, by induction. The resulting matrix is itself in row reduced echelon form. If the first column of A is not 0, put its first column in row reduced echelon form (as in the base step), so that the new matrix $A' = \left[\begin{smallmatrix} 1 & Y \\ 0 & M \end{smallmatrix}\right]$, where M is an $(m - 1) \times n$ matrix. Your first guess is to apply the inductive hypothesis to the matrix comprised of the last n columns, as in the first case. This may not be convenient, for one of the elementary row operations may have added a multiple of ROW(1) to another row, thereby changing the first column. Instead, we use the inductive hypothesis to replace M by D, where D is a row reduced echelon matrix Gaussian equivalent to M. Thus, A' is Gaussian equivalent to $N = \left[\begin{smallmatrix} 1 & Y \\ 0 & D \end{smallmatrix}\right]$. Let the leading columns of N be COL(t_2), ..., COL(t_r), where $2 \leq t_2 < \cdots < t_r$ (the first column of D is the second column of N). It is possible that the entry $y_{i,t_2} \neq 0$. If so, replace ROW(1) of N by ROW(1)$-y_{i,t_1}$ROW(2) (the first row of D is the second row of N). Since the leading entry of COL(t_2) has only 0 entries to its left, this operation does not change any columns of N to the left of COL(t_2). Thus, the leading entry of COL(t_2) is now the only nonzero entry in its column, while COL(1) has not been changed. Next, make $y_{1,t_3} = 0$ in the same way (so that COL(1) and COL(t_2) are unchanged), and continue until all $y_{1,t_i} = 0$. We have arrived at a row reduced echelon matrix with leading columns COL(1), COL(t_2), ..., COL(t_r).

(ii) Suppose that U is a row reduced echelon matrix Gaussian equivalent to A. Let the nonzero rows of U be $\beta_1, \ldots, \beta_r$, let the leading columns of U be COL(t_1), ...,

COL(t_r), and let $\beta_i = e_{t_i} + v_i$, where $v_i \in \langle e_v : v > t_i \rangle$ (as usual, $e_1, \ldots, e_n$ is the standard basis of k^n). We claim that COL(t_1), ..., COL(t_r) are precisely those columns of U in which the leading entry of a nonzero vector in $\langle \beta_1, \ldots, \beta_r \rangle = Row(U)$ can occur. It will then follow that the leading columns are determined by $Row(U)$. Clearly, COL(t_i) contains a leading entry (namely, that of β_i). On the other hand, if γ is a nonzero vector in $\langle \beta_1, \ldots, \beta_r \rangle$, then $\gamma = c_1\beta_1 + \cdots + c_r\beta_r$. If we picture each β_i as the ith row of U, then multiply ROW(i) by c_i and add: γ is the sum, and its jth coordinate is just the sum of the entries in the jth column. Thus, for each i, the t_i coordinate of γ is c_i, for there is no other nonzero entry in COL(t_i). Since $\gamma \neq 0$, some $c_i \neq 0$; we claim that the first such, $c_{t_{i_0}}$, is its leading coefficient. Now all $c_i = 0$ for $i < i_0$, and so $\gamma = c_{i_0} e_{t_{i_0}} + \omega$, where $\omega = \sum_{i > i_0} c_i v_i \in \langle e_v : v > t_{i_0} \rangle$. Hence, the leading coefficient of γ lies in COL(t_{i_0}).

If U' is another row reduced echelon matrix Gaussian equivalent to A, then Proposition 4.33 says that $Row(U') = Row(U)$. Since we have just proved that the row space determines the leading columns, it follows that the leading columns of U' and of U are the same. Let the nonzero rows of U' be $\beta'_1, \ldots, \beta'_r$. Now $\beta'_i \in Row(U) = Row(U')$, so there are $c_v \in k$ with $\beta'_i = \sum_v c_v \beta_v$. But we saw in the preceding paragraph that for each v, the t_vth coordinate of β'_i is c_v. Hence, $c_i = 1$ and all other $c_v = 0$, so that $\beta'_i = \beta_i$ for all i. Therefore, $U' = U$.

(iii) Choose σ to be any permutation with $\sigma(t_i) = i$ for $i = 1, \ldots, r$. $\bullet$

Corollary 4.40.   *Let A be an $m \times n$ matrix over a field k. If* rank(A) $= m$, *then the (inhomogeneous) linear system $Ax = b$ is consistent for every $b \in k^n$.*

Proof.   If U is an $m \times n$ echelon matrix of rank m, then it has no zero rows. For any $c \in k^n$, the linear system $Ux = c$ is

$$
\begin{aligned}
x_1 + \quad\quad\quad +u_{1,m+1}x_{m+1} + \cdots + u_{1n}x_n &= c_1 \\
x_2 + \quad\quad\quad +u_{2,m+1}x_{m+1} + \cdots + u_{2n}x_n &= c_2 \\
\vdots \quad\quad\quad \vdots \quad\quad\quad \vdots \quad\quad\quad \\
x_m + u_{m,m+1}x_{m+1} + \cdots + u_{mn}x_n &= c_m.
\end{aligned}
$$

A solution is given by $x_i = c_i - (u_{i,m+1}d_{m+1} + \cdots + u_{in}d_n)$ for $d_{m+1}, \ldots, d_n \in k$.

By Gaussian elimination, there is a nonsingular matrix P and a permutation matrix Q with $PAQ = U$ in echelon form. We have just seen that the system $Ux = PbQ$ has a solution, and so $P^{-1}UQ^{-1}x = P^{-1}(PbQ)Q^{-1}$, that is, $Ax = b$, has a solution. $\bullet$

→ **Corollary 4.41.**   *Let A be an $m \times n$ matrix over a field k. If U is the row reduced echelon form of A, then a basis of $Row(A)$ consists of the nonzero rows of U.*

Proof.   Now $Row(A) = Row(U)$, by Proposition 4.33, and so it is spanned by the rows of U. But it is obvious that the nonzero rows of U are linearly independent, and so they form a basis of $Row(U) = Row(A)$.   •

We have just seen how to find rank(A): put A into echelon form U, and rank(A) is the number of nonzero rows in U.

→ **Definition.**   If U is an $m \times n$ matrix in row reduced echelon form with leading columns $COL(t_1), \ldots, COL(t_r)$, then $x_{t_1}, \ldots, x_{t_r}$ are called *fixed variables* (or *lead variables*) and the other variables are called *free variables*.

Recall that Gaussian elimination solves problems in linear algebra by replacing a matrix A by the echelon matrix U which is Gaussian equivalent to it. For example, the solution space of a homogeneous system $Ax = 0$ is the same as the solution space of $Ux = 0$. Now $Ux = 0$ is easily solved, as on page 344.

The next theorem shows that an arbitrary homogeneous system of linear equations is no more difficult to solve that the simplest system on page 344.

→ **Theorem 4.42.**   *Let $Ax = 0$ be a system of linear equations, where A is an $m \times n$ matrix over a field k, and let $U = [u_{ij}]$ be the (unique) echelon form Gaussian equivalent to A. Relabeling the variables if necessary, the solution space $Sol(A)$ of $Ax = 0$ consists of all vectors of the form*

$$\left(-\sum_{\ell=r+1}^{n} u_{1\ell}c_\ell, \ldots, -\sum_{\ell=r+1}^{n} u_{r\ell}c_\ell, c_{r+1}, \ldots, c_n\right).$$

Proof.   By Lemma 4.37, an n-tuple $s = (c_1, \ldots, c_n)$ is a solution of $Ax = 0$ if and only if it is a solution of $Ux = 0$. By Theorem 4.39, there is a permutation σ such that UQ_σ is in echelon form. But permuting the columns merely relabels the variables, and so it is no loss in generality to assume that U is in echelon form; that is, the first r variables are the fixed variables and the last $n - r$ variables are the free variables. Now U has the form

$$B = \begin{bmatrix} 1 & 0 & \cdots & 0 & u_{1,r+1} & \cdots & u_{1n} \\ 0 & 1 & \cdots & 0 & u_{2,r+1} & \cdots & u_{2n} \\ & \vdots & & \vdots & \vdots & & \\ 0 & 0 & \cdots & 1 & u_{r,r+1} & \cdots & u_{rn} \\ 0 & 0 & \cdots & 0 & 0 & \cdots & 0 \\ & \vdots & & \vdots & \vdots & & \vdots \\ 0 & 0 & \cdots & 0 & 0 & \cdots & 0 \end{bmatrix},$$

and the result follows as on page 349.   •

→ **Theorem 4.43 (Rank-Nullity Theorem).** *Let A be an m × n matrix over a field k. If Sol(A) is the solution space of the homogeneous linear system Ax = 0, then*

$$\dim(Sol(A)) = n - r,$$

where $r = \text{rank}(A)$.

Proof. We may assume that the variables have been relabeled so that the fixed variables precede all the free variables. For each ℓ with $1 \leq \ell \leq n - r$, define s_ℓ to be the solution $(c_1, \ldots, c_n)$ with $c_{p_\ell} = 1$ and $c_{p_\nu} = 0$ for all $\nu \neq \ell$. Thus,

$$s_1 = (-u_{1,r+1}, -u_{2,r+1}, \ldots, -u_{r,r+1}, 1, 0, \ldots, 0)$$
$$s_2 = (-u_{1,r+2}, -u_{2,r+2}, \ldots, -u_{r,r+2}, 0, 1, \ldots, 0)$$
$$\vdots$$
$$s_{n-r} = (-u_{1,n}, -u_{2,n}, \ldots, -u_{r,n}, 0, 0, \ldots, 1).$$

These $n - r$ vectors are linearly independent (look at their last $n - r$ coordinates), while Theorem 4.42 shows that they span $Sol(A)$:

$$\left(-\sum_{\ell=r+1}^{n} u_{1\ell}c_\ell, \ldots, -\sum_{\ell=r+1}^{n} u_{r\ell}c_\ell, c_{r+1}, \ldots, c_n\right) = \sum_{\ell=r+1}^{n} c_\ell s_\ell. \quad \bullet$$

The dimension $n - r$ of the system $Ax = 0$ is often called the number of ***degrees of freedom*** in the general solution.

Example 4.44.
Consider the matrix

$$A = \begin{bmatrix} 0 & 0 & 1 & 1 & 0 \\ -2 & -4 & 1 & 0 & -3 \\ 3 & 6 & -1 & 1 & 5 \end{bmatrix}.$$

Find rank(A), find a basis of its row space, and find a basis of the solution space of the homogeneous system $Ax = 0$.

The matrix A is Gaussian equivalent to

$$B = \begin{bmatrix} 1 & 2 & 0 & 0 & 1 \\ 0 & 0 & 1 & 0 & -1 \\ 0 & 0 & 0 & 1 & 1 \end{bmatrix}.$$

Thus, rank$(A) = 3$ and a basis of the row space is $(1, 2, 0, 0, 1)$, $(0, 0, 1, 0, -1)$, $(0, 0, 0, 1, 1)$. Note that the rows of A are also linearly independent, for they span a 3-dimensional space (see Corollary 4.24). The fixed variables are x_1, x_3, and x_4, while the free variables are x_2 and x_5. The solution space has dimension $5 - 3 = 2$. The system $Bx = 0$ is

$$x_1 + 2x_2 + x_5 = 0$$
$$x_3 - x_5 = 0$$
$$x_4 + x_5 = 0$$

The general solution is $(-2c - d, c, d, -d, d)$. ◀

EXERCISES

H **4.23** True or false with reasons.

   (i) There is a solution to an $n \times n$ inhomogeneous system $Ax = b$ if A is a triangular matrix.

   (ii) Gaussian equivalent matrices have the same row space.

   (iii) Gaussian equivalent matrices have the same column space.

   (iv) The matrix $A = \begin{bmatrix} 1 & 0 & 0 \\ 0 & 1 & -1 \\ 0 & 0 & 1 \end{bmatrix}$ is nonsingular.

   (v) Every nonsingular matrix over a field is a product of elementary matrices.

   (vi) If A is an $m \times n$ matrix, then $Row(A^T) = Col(A)$.

4.24 **(i)** Prove that a list $v_1, \ldots, v_m$ in a vector space V is linearly independent if and only if it spans an m-dimensional subspace of V.

   H **(ii)** Determine whether the list

$$v_1 = (1, 1, -1, 2), \quad v_2 = (2, 2, -3, 1), \quad v_3 = (-1, -1, 0, -5)$$

   in k^4 is linearly independent.

H **4.25** Do the vectors $v_1 = (1, 4, 3)$, $v_2 = (-1, -2, 0)$, $v_3 = (2, 2, 3)$ span k^3?

*****4.26** **(i)** An $n \times n$ matrix is *triangular* if either all its entries below the diagonal are 0 or all its entries above the diagonal are 0. Prove that every $n \times n$ row reduced echelon matrix is triangular.

   (ii) Use Theorem 4.39 to prove that every $n \times n$ matrix A is Gaussian equivalent to a triangular matrix.

H **4.27** Let k be a field, and let A be an $n \times m$ matrix over k. An (inhomogeneous) linear system $Ax = \beta$, where $\beta \in k^m$, is called *consistent* if there is $v \in k^n$ with $Av = \beta$. Prove that $Ax = \beta$ is consistent if and only if β lies in the column space of A. (Recall Exercise 4.18 on page 343: the solution set of a consistent inhomogeneous system $Ax = b$ is a coset of the solution space of $Ax = 0$.)

4.28 If A is an $n \times n$ nonsingular matrix, prove that any system $Ax = b$ has a unique solution, namely, $x = A^{-1}b$.

4.29 Let $\alpha_1, \ldots, \alpha_n$ be the columns of an $m \times n$ matrix A over a field k, and let $\beta \in k^m$.

   (i) Prove that $\beta \in \langle \alpha_1, \ldots, \alpha_n \rangle$ if and only if the inhomogeneous system $Ax = \beta$ has a solution.

H **(ii)** Define the **augmented matrix** $[A|\beta]$ to be the $m \times (n + 1)$ matrix whose first n columns are A and whose last column is β. Prove that β lies in the column space of A if and only if $\mathrm{rank}([A|\beta]) = \mathrm{rank}(A)$.

(iii) Does $\beta = (0, -3, 5)$ lie in the subspace spanned by $\alpha_1 = (0, -2, 3)$, $\alpha_2 = (0, -4, 6)$, $\alpha_3 = (1, 1, -1)$?

4.30 **(i)** Prove that an $n \times n$ matrix A over a field k is nonsingular if and only if it is Gaussian equivalent to the identity I.

H **(ii)** Find the inverse of
$$A = \begin{bmatrix} 2 & 3 & 1 \\ -1 & 1 & 0 \\ 1 & 0 & 1 \end{bmatrix}.$$

*4.31 **(i)** Let $Ax = b$ be an $m \times n$ linear system over a field k. Prove that there exists a solution $x = (x_1, \ldots, x_n)$ with $x_{j_1} = 0 = x_{j_2} = \cdots = x_{j_s}$, where $s \leq n$, if and only if there is a solution to the $m \times (n - s)$ system $A^* x^* = b$, where A^* is obtained from A be deleting columns $j_1, j_2, \ldots, j_s$ from A.

H **(ii)** Prove that if the matrix A^* in part (i) has rank m, then there exists a solution to $Ax = b$ with $x_{j_1} = 0 = x_{j_2} = \cdots = x_{j_s}$.

4.2 EUCLIDEAN CONSTRUCTIONS

There are myths in several ancient civilizations in which the gods demand precise solutions of mathematical problems in return for granting relief from catastrophes. We quote from van der Waerden, *Geometry and Algebra in Ancient Civilizations.*

> In the dialogue 'Platonikos' of Eratosthenes, a story was told about the problem of doubling the cube. According to this story, as Theon of Smyrna recounts it in his book 'Exposition of mathematical things useful for the reading of Plato', the Delians asked for an oracle in order to be liberated from a plague. The god (Apollo) answered through the oracle that they had to construct an altar twice as large as the existing one without changing its shape. The Delians sent a delegation to Plato, who referred them to the mathematicians Eudoxos and Helikon of Kyzikos.

The altar was cubical in shape, and so the problem involves constructing $\sqrt[3]{2}$. The gods were cruel, for although there is a geometric construction of $\sqrt{2}$ (it is the length of the diagonal of a square with sides of length 1), we are going to prove that it is impossible to construct $\sqrt[3]{2}$ by the methods of Euclidean geometry–that is, by using only straightedge and compass. Actually, the gods were not so cruel, for the Greeks did use other methods. Thus, Menaechmus constructed $\sqrt[3]{2}$ with the intersection of the parabolas $y^2 = 2x$ and $x^2 = y$; this is elementary for us, but it was an ingenious feat when there was no analytic geometry and no algebra.

There are several other geometric problems handed down from the Greeks. Can one trisect every angle? Can one construct a regular n-gon? Can one "square the circle"; that is, can one construct a square whose area is equal to the area of a given circle?

Notation. Let P and Q be points in the plane; we denote the line segment with endpoints P and Q by PQ, and we denote the length of this segment by $|PQ|$. Let $L[P, Q]$ denote the line determined by P and Q, and let $C[P; PQ]$ denote the circle with center P and radius $|PQ|$.

Without a precise definition of constructibility, some of the classical problems appear ridiculously easy. For example, a $60°$ angle can be trisected using a protractor: just find $20°$ and draw the angle. Thus, it is essential to state the problems carefully and to agree on certain ground rules. The Greek problems specify that only two tools are allowed, and each must be used in only one way. Given distinct points P and Q in the plane, a ***straightedge*** is a tool that can draw the line $L[P, Q]$; a ***compass*** draws the circles $C[P; PQ]$ and $C[Q; QP]$ with radius $|PQ| = |QP|$ and centers P and Q, respectively.

What we are calling a *straightedge*, others call a ***ruler***; we use the first term to avoid possible confusion, for a ruler has extra properties: one can mark two points on it, say, U and V, and the marked point U is allowed to slide along a curve. These added functions of a ruler makes it a more powerful instrument. About 425 B.C., Hippias of Elis was able to square the circle by drawing a certain curve as well as lines and circles. Nicomedes solved the Delian problem of doubling the cube using a ruler and compass, and both Nicomedes and Archimedes were able to trisect arbitrary angles with these tools (we present Archimedes' proof later in this section).

Informally, one constructs a new point T from (not necessarily distinct) old points P, Q, R, and S by using the first pair P, Q to draw a line or circle, the second pair R, S to draw a line or circle, and then obtaining T as one of the points of intersection of the two drawn lines, the drawn line and the drawn circle, or the two drawn circles. More generally, a point is called constructible if it is obtained from $(1, 0)$ and $(-1, 0)$ by a finite number of such steps. Given a pair of constructible points, we do *not* assert that every point on the drawn line or the drawn circles they determine is constructible.

When we say it is *impossible* to solve the classical problems (using only straightedge and compass), we mean what we say; we do not mean that it is merely very difficult. The reader should ponder how one might prove that something is impossible.

Here is the formal discussion.

Given the plane, we establish a coordinate system by first choosing two distinct points, A and $\bar{A}$; call the line they determine the ***x-axis***. Use a compass to draw the two circles $C[A; A\bar{A}]$ and $C[\bar{A}; \bar{A}A]$ of radius $|A\bar{A}|$ with centers A and $\bar{A}$, respectively. These two circles intersect in two points; the line they determine is called the ***y-axis***; it is the perpendicular bisector of $A\bar{A}$, and it intersects the x-axis in a point O, called the *origin*. We define the distance $|OA|$ to be 1. We have introduced coordinates in the plane; in particular, $A = (1, 0)$ and $\bar{A} = (-1, 0)$.

Definition.   Let $E \neq F$ and $G \neq H$ be points in the plane. A point Z is ***built from*** E, F, G, and H if one of the following conditions hold:

  (i) $Z \in L[E, F] \cap L[G, H]$, where $L[E, F] \neq L[G, H]$;

  (ii) $Z \in L[E, F] \cap C[G; GH]$ or $Z \in L[G, H] \cap C[E; EF]$;

  (iii) $Z \in C[E; EF] \cap C[G; GH]$, where $C[E; EF] \neq C[G; GH]$.

A point Z is ***constructible*** if $Z = A$ or $Z = \bar{A}$ or if there are points $P_1, \ldots, P_n$ with $Z = P_n$ so that, for all $j \geq 1$, the point P_{j+1} is built from points in $\{A, \bar{A}, P_1, \ldots, P_j\}$.

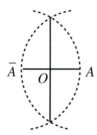

Figure 4.1   The First Constructible Points

Example 4.45.
Let us show that $Z = (0, 1)$ is constructible. Figure 4.1 illustrates that the points $P_1 = (0, \sqrt{3})$ and $P_2 = (0, -\sqrt{3})$ are constructible, for both lie in the intersection $C[A; A\bar{A}] \cap C[\bar{A}; \bar{A}A]$, and so the y-axis $L[P_1, P_2]$ can be drawn. Finally,

$$Z = (0, 1) \in L[P_1, P_2] \cap C[O; OA]. \quad \blacktriangleleft$$

In our discussion, we shall freely use any standard result of Euclidean geometry. For example, angles can be bisected with straightedge and compass; that is, if $(\cos \theta, \sin \theta)$ is constructible, then so is $(\cos \tfrac{1}{2}\theta, \sin \tfrac{1}{2}\theta)$.

Definition.   A complex number $z = x + iy$ is ***constructible*** if the point (x, y) is a constructible point.

Example 4.45 shows that the numbers $1, -1, 0, i\sqrt{3}, -i\sqrt{3}, i$, and $-i$ are constructible complex numbers.

Lemma 4.46. *A complex number $z = x + iy$ is constructible if and only if its real part x and its imaginary part y are constructible.*

Proof. If z is constructible, then a standard Euclidean construction draws the vertical line L through (x, y) which is parallel to the y-axis. It follows that x is constructible, for the point $(x, 0)$ is constructible, being the intersection of L and the x-axis. Similarly, the point $(0, y)$ is the intersection of the y-axis and a line L' through (x, y) which is parallel to the x-axis. It follows that $P = (y, 0)$ is constructible, for it is in the intersection of $C[O; OP]$ with the x-axis. Hence, y is a constructible number.

Conversely, assume that x and y are constructible numbers; that is, $Q = (x, 0)$ and $P = (y, 0)$ are constructible points. The point $(0, y)$ is constructible, being the intersection of the y-axis and $C[O; OP]$. One can draw the vertical line through $(x, 0)$ as well as the horizontal line through $(0, y)$, and (x, y) is the intersection of these lines. Therefore, (x, y) is a constructible point, and so $z = x + iy$ is a constructible number.

•

Notation. The subset of $\mathbb{C}$ of all the ***constructible numbers*** will be denoted by K.

Theorem 4.47. *The set $K \cap \mathbb{R}$ of all constructible real numbers is a subfield of $\mathbb{R}$ that is closed under square roots of its positive elements.*

Proof. Let a and b be constructible reals.

(i) $-a$ is constructible. If $P = (a, 0)$ is a constructible point, then $(-a, 0)$ is the other intersection of the x-axis and $C[O; OP]$.

(ii) $a + b$ and $-a + b$ are constructible.

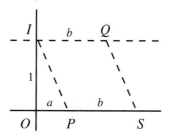

Figure 4.2 $a + b$

Assume that a and b are positive. Let $I = (0, 1)$, $P = (a, 0)$ and $Q = (b, 1)$. Now Q is constructible: it is the intersection of the horizontal line through I and the vertical line through $(b, 0)$ [the latter point is constructible, by hypothesis]. The line through Q parallel to IP intersects the x-axis in $S = (a + b, 0)$, as desired.

The same procedure constructs $-a + b$ if we replace $P = (a, 0)$ by $P' = (-a, 0)$. The point $(-a + b, 0)$ is the intersection of the x-axis and the line through $Q = (b, 0)$ parallel to I.

(iii) *ab* is constructible.

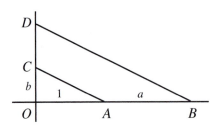

Figure 4.3 *ab*

By part (i), we may assume that both *a* and *b* are positive. In Figure 4.3, $A = (1, 0)$, $B = (1 + a, 0)$, and $C = (0, b)$. Define D to be the intersection of the *y*-axis and the line through B parallel to AC. Since the triangles $\triangle OAC$ and $\triangle OBD$ are similar,

$$|OB|/|OA| = |OD|/|OC|;$$

hence $(a + 1)/1 = (b + |CD|)/b$, and $|CD| = ab$. Therefore, $b + ab$ is constructible. Since $-b$ is constructible, by part (i), we have $ab = (b + ab) - b$ constructible, by part (ii).

(iv) If $a \neq 0$, then a^{-1} is constructible. Let $A = (1, 0)$, $S = (0, a)$, and $T = (0, 1 + a)$. Define B as the intersection of the *x*-axis and the line through T parallel to AS; thus,

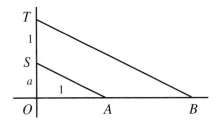

Figure 4.4 a^{-1}

$B = (1 + u, 0)$ for some u. Similarity of the triangles $\triangle OSA$ and $\triangle OTB$ gives

$$|OT|/|OS| = |OB|/|OA|.$$

Hence, $(1 + a)/a = (1 + u)/1$, and so $u = a^{-1}$. Therefore, $1 + a^{-1}$ is constructible, and so $(1 + a^{-1}) - 1 = a^{-1}$ is constructible.

(v) If $a \geq 0$, then $\sqrt{a}$ is constructible. Let $A = (1, 0)$ and $P = (1 + a, 0)$; construct Q, the midpoint of OP. Define R as the intersection of the circle $C[Q; QO]$ with the vertical line through A. The (right) triangles $\triangle AOR$ and $\triangle ARP$ are similar, so that

$$|OA|/|AR| = |AR|/|AP|,$$

and, hence, $|AR| = \sqrt{a}$. •

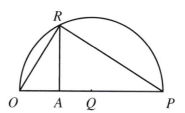

Figure 4.5 $\sqrt{a}$

Corollary 4.48. *The set K of all constructible numbers is a subfield of $\mathbb{C}$ that is closed under square roots.*

Proof. If $z = a+ib$ and $w = c+id$ are constructible, then a, b, c, d are constructible, by Theorem 4.47, and so $a, b, c, d \in K \cap \mathbb{R}$. Hence, $a \pm c, b \pm d \in K \cap \mathbb{R}$, because $K \cap \mathbb{R}$ is a subfield of $\mathbb{R}$, and so $(a + c) \pm i(b + d) \in K$, by Lemma 4.46. Similarly, $zw = (ac - bd) + i(ad + bc) \in K$. If $z \neq 0$, then $z^{-1} = (a/z\bar{z}) - i(b/z\bar{z})$. But $z = a + ib \in K$ implies $\bar{z} = a - ib \in K$. Therefore, $z^{-1} \in K$, and so K is a subfield of $\mathbb{C}$.

If $z = a+ib \in K$, then $a, b \in K \cap \mathbb{R}$, by Lemma 4.46, and so $r^2 = a^2+b^2 \in K \cap \mathbb{R}$. Since r^2 is nonnegative, we have $\sqrt{r} \in K \cap \mathbb{R}$. Now $z = re^{i\theta}$, so that $e^{i\theta} = r^{-1}z \in K$, because K is a subfield of $\mathbb{C}$. That every angle can be bisected gives $e^{i\theta/2} \in K$, and so $\sqrt{z} = \sqrt{r}e^{i\theta/2} \in K$, as desired. $\bullet$

Corollary 4.49. *If a, b, c are constructible, then the roots of the quadratic ax^2+bx+c are also constructible.*

Proof. This follows from the quadratic formula and Corollary 4.48. $\bullet$

We are now going to give an algebraic characterization of the geometric idea of constructibility. Recall that if E/k is an extension of fields (that is, k is a subfield of a field E), then E may be regarded as a vector space over k. The dimension of E, denoted by $[E : k]$, is called the *degree* of E/k. In particular, if E/k be an extension and $z \in E$ is a root of an irreducible polynomial $p(x) \in k[x]$, then Proposition 4.30 gives $[k(z) : k] = \dim_k(k(z)) = \deg(p)$.

Definition. A *2-tower* is an ascending sequence of subfields of $\mathbb{C}$,

$$\mathbb{Q}(i) = F_0 \subseteq F_1 \subseteq \cdots \subseteq F_n,$$

with $[F_j : F_{j-1}] \leq 2$ for all $j \geq 1$. A complex number z is *polyquadratic* if there is a 2-tower $\mathbb{Q}(i) = F_0 \subseteq F_1 \subseteq \cdots \subseteq F_n$ with $z \in F_n$. Denote the set of all polyquadratic complex numbers by $\mathcal{P}$.

We now begin a series of lemmas which culminates in Theorem 4.54, which says that a complex number is constructible if and only if it is polyquadratic.

Lemma 4.50. *If F/k is a field extension, then $[F : k] \leq 2$ if and only $F = k(u)$, where $u \in F$ is a root of some quadratic polynomial $f(x) \in k[x]$.*

Proof. If $[F : k] = 1$, then $F = k$, so that $F = k(u)$ for any $u \in k$; define $f(x) = (x - u)^2$. If $[F : k] = 2$, then $F \neq k$ and there is some $u \in F$ with $u \notin k$. By Proposition 4.32, there is some irreducible polynomial $f(x) \in k[x]$ having u as a root. Now $2 = [F : k] = [F : k(u)][k(u) : k]$, by Theorem 4.31; but $[k(u) : k] = 2$, by Proposition 4.30, and so $[F : k(u)] = 1$. Therefore, $F = k(u)$.

Conversely, let $F = k(u)$, where u is a root of a quadratic polynomial $f(x) \in k[x]$. If $f(x)$ factors in $k[x]$, then $u \in k$, $F = k(u) = k$, and $[F : k] = 1$. If $f(x)$ is irreducible, then $[F : k] = [k(u) : u] = 2$, by Proposition 4.30.   •

Lemma 4.51.

(i) *The set $\mathcal{P}$ of all polyquadratic numbers is a subfield of $\mathbb{C}$ which is closed under square roots.*

(ii) *A complex number $z = a + ib$, where $a, b \in \mathbb{R}$, is polyquadratic if and only if both a and b are polyquadratic.*

Proof.
(i) If $z, z' \in \mathcal{P}$, then there are 2-towers $\mathbb{Q}(i) = F_0 \subseteq F_1 \subseteq \cdots \subseteq F_n$ and $\mathbb{Q}(i) = F_0' \subseteq F_1' \subseteq \cdots \subseteq F_m'$ with $z \in F_n$ and $z' \in F_m'$. Now $[F_j : F_{j-1}] \leq 2$ implies $F_j = F_{j-1}(u_j)$, where $u_j \in F_j$ is a root of some quadratic $f_j(x) \in F_{j-1}[x]$. For all j with $1 \leq j \leq n$, define $F_j'' = F_m'(u_1, \ldots, u_j)$. Since $F_j'' = F_{j-1}''(u_j)$, we have $F_{j-1} = F_0(u_1, \ldots, u_{j-1}) \subseteq F_m'(u_1, \ldots, u_j) = F_{j-1}''$, so that $f_j(x) \in F_{j-1}''[x]$ and $[F_j'' : F_{j-1}''] \leq 2$. Hence,

$$\mathbb{Q}(i) = F_0' \subseteq F_1' \subseteq \cdots \subseteq F_m' \subseteq F_1'' \subseteq \cdots \subseteq F_n''$$

is a 2-tower. Of course, every element of F_n'' is polyquadratic; since F_n'' contains both z and z', it contains their inverses and their sum and product. Therefore, $\mathcal{P}$ is a subfield.

Let $z \in \mathcal{P}$. If $\mathbb{Q}(i) = F_0 \subseteq F_1 \subseteq \cdots \subseteq F_n$ is a 2-tower with $z \in F_n$, then $\mathbb{Q}(i) = F_0 \subseteq F_1 \subseteq \cdots \subseteq F_n \subseteq F_n(\sqrt{z})$ is also a 2-tower.

(ii) If both $a, b \in \mathcal{P}$, then $z = a + ib \in \mathcal{P}$, for $\mathcal{P}$ is a subfield containing i. Conversely, let $\mathbb{Q}(i) = F_0 \subseteq F_1 \subseteq \cdots \subseteq F_n$ be a 2-tower with $z \in F_n$. Since complex conjugation is an automorphism of $\mathbb{C}$, $\mathbb{Q}(i) = \overline{F_0} \subseteq \overline{F_1} \subseteq \cdots \subseteq \overline{F_n}$ is a 2-tower with $\overline{z} \in \overline{F_n}$; hence, $\overline{z}$ is polyquadratic. Therefore, $a = \frac{1}{2}(z + \overline{z}) \in \mathcal{P}$ and $b = \frac{1}{2i}(z - \overline{z}) \in \mathcal{P}$.   •

Lemma 4.52. *Let $P = a + ib$ and $Q = c + id$ be polyquadratic.*

(i) *The line $L[P, Q]$ has equation $x = a$ if it is vertical ($c = a$) or $y = mx + q$ if it is not vertical ($c \neq a$), where m, q are polyquadratic.*

(ii) *The circle $C[P; PQ]$ has equation $(x - a)^2 + (y - b)^2 = r^2$, where a, b, r are polyquadratic.*

Proof. Lemma 4.51 gives $a, b, c, d \in \mathcal{P}$.
(i) If $L[P, Q]$ is not vertical, then its equation is $y = mx + q$, where $m = (d - b)/(c - a)$ and $q = -ma + b$. Hence $m, q \in \mathcal{P}$.
(ii) The circle $C[P; PQ]$ has equation $(x - a)^2 + (y - b)^2 = r^2$, where r is the distance from P to Q. Now $a, b \in \mathcal{P}$, by Lemma 4.51(ii), and since $\mathcal{P}$ is closed under square roots, $r = \sqrt{(c - a)^2 + (d - b)^2} \in \mathcal{P}$. •

Proposition 4.53. *Every polyquadratic number z is constructible.*

Proof. If $z \in \mathcal{P}$, then there is a 2-tower $\mathbb{Q}(i) = F_0 \subseteq F_1 \subseteq \cdots \subseteq F_n$ with $z \in F_n$; we prove that $z \in K$ by induction on $n \geq 0$. The base step is true, for $F_0 = \mathbb{Q}(i) \subseteq K$, by Corollary 4.48. Now $F_n = F_{n-1}(u)$, where u is a root of a quadratic $f(x) = x^2 + bx + c \in F_{n-1}[x]$. The quadratic formula gives $u \in F_{n-1}\left(\sqrt{b^2 - 4c}\right)$; but K is closed under square roots, by Corollary 4.48, so that $\sqrt{b^2 - 4c} \in K$. The inductive hypothesis $F_{n-1} \subseteq K$ now gives $z \in F_{n-1}\left(\sqrt{b^2 - 4c}\right) \subseteq K\left(\sqrt{b^2 - 4c}\right) \subseteq K$. •

Here is the result we have been seeking.

Theorem 4.54. *A number $z \in \mathbb{C}$ is constructible if and only if z is polyquadratic.*

Proof. In light of Proposition 4.53, $\mathcal{P} \subseteq K$, and so it suffices to prove that $K \subseteq \mathcal{P}$; that is, every constructible z is polyquadratic. There are complex numbers $1, w_0 = -1, w_1, \ldots, w_m = z$ with w_j built from $w_0, w_1, \ldots, w_{j-1}$ for all $j \geq 0$. We prove, by induction on $m \geq 0$, that w_m is polyquadratic. Since $w_0 = -1$ is polyquadratic, we may pass to the inductive step. By the inductive hypothesis, we may assume that $w_0, \ldots, w_{m-1}$ are polyquadratic, and so it suffices to prove that if z is built from P, Q, R, S, where P, Q, R, S are polyquadratic, then z is polyquadratic.
Case 1: $z \in L[P, Q] \cap L[R, S]$.

If $L[P, Q]$ is vertical, then it has equation $x = a$; if $L[P, Q]$ is not vertical, then Lemma 4.31 says that $L[P, Q]$ has equation $y = mx + q$, where $m, q \in \mathcal{P}$. Similarly, $L[R, S]$ has equation $x = c$ or $y = m'x + p$, where $m', p \in \mathcal{P}$. Since these lines are not parallel, one can solve the linear system

$$y = mx + q$$
$$y = m'x + p$$

for $z = x_0 + iy_0 \in L[P, Q] \cap L[R, S]$. Therefore, $z = x_0 + iy_0 \in \mathcal{P}$.

Case 2: $z \in L[P, Q] \cap C[R; RS]$.

The circle $C[R; RS]$ has equation $(x - u)^2 + (y - v)^2 = \rho^2$, where $R = (u, v)$ and $S = (s, t)$ and $\rho^2 = (u - s)^2 + (v - t)^2$; moreover, all coefficients lie in $\mathcal{P}$, by Lemma 4.52. If the line $L[P, Q]$ is vertical, its equation is $x = a$. If $z = x_0 + iy_0 \in L[P, Q] \cap C[R; RS]$, then $(x_0 - u)^2 + (y_0 - v)^2 = \rho^2$, so that y_0 is a root of a quadratic in $\mathcal{P}[x]$, and $z = a + iy_0 \in \mathcal{P}$. If the line $L[P, Q]$ is not vertical, its equation is $y = mx + q$, where $m, q \in \mathcal{P}$. If $z = x_0 + iy_0 \in L[P, Q] \cap C[R; RS]$, then $(x_0 - u)^2 + (mx_0 + q - v)^2 = \rho^2$, and so $x_0 \in \mathcal{P}$, for it is a root of a quadratic in $\mathcal{P}[x]$. Hence, $y_0 = mx_0 + q \in \mathcal{P}$ and $z = x_0 + iy_0 \in \mathcal{P}$.

Case 3: $z \in C[P; PQ] \cap C[R; RS]$.

If $R = (u, v)$ and $S = (s, t)$, the circle $C[R; RS]$ has equation $(x - u)^2 + (y - v)^2 = \rho^2$, where $\rho^2 = (u - s)^2 + (v - t)^2$; similarly, if $P = (a, b)$ and $Q = (c, d)$, the circle $C[P; PQ]$ has equation $(x - a)^2 + (y - b)^2 = r^2$, where $r^2 = (u - s)^2 + (v - t)^2$. By Lemma 4.52, all the coefficients lie in $\mathcal{P}$. If $z = x_0 + iy_0 \in C[P; PQ] \cap C[R; RS]$, then expanding the equations of the circles gives an equation of the form

$$x_0^2 + y_0^2 + \alpha x_0 + \beta y_0 + \gamma = 0 = x_0^2 + y_0^2 + \alpha' x_0 + \beta' y_0 + \gamma'.$$

Canceling $x_0^2 + y_0^2$ gives a linear equation $\lambda x + \mu y + \nu = 0$ with $\lambda, \mu, \nu \in \mathcal{P}$; indeed, $\lambda x + \mu y + \nu = 0$ is the equation of a line $L[P', Q']$ with $P', Q' \in \mathcal{P}$ [for example, take $P' = (0, -\nu/\mu)$ and $Q' = (-\nu/\lambda, 0)$]. Thus, the points $z \in C[P; PQ] \cap C[R; RS]$ are the points of intersection of the line $L[P', Q']$ and either circle. The argument in Case 2 now shows that $z \in \mathcal{P}$.   •

Corollary 4.55.   *If a complex number z is constructible, then $[\mathbb{Q}(z) : \mathbb{Q}]$ is a power of* 2.

Remark.   The converse of this corollary is false; it can be shown that there are non-constructible numbers z with $[\mathbb{Q}(z) : \mathbb{Q}] = 4$.   ◀

Proof.   This follows from Theorems 4.54 and 4.31.   •

Remark.   It was proved, by G. Mohr in 1672 and by L. Mascheroni in 1797, that every geometric construction carried out by straightedge and compass can be done without the straightedge. There is a short proof of this theorem given by N. Hungerbühler in *The American Mathematical Monthly*, 101 (1994), pp. 784–787.   ◀

Two of the classical Greek problems were solved by P. L. Wantzel in 1837.

Theorem 4.56 (Wantzel). *It is impossible to duplicate the cube using only straight-edge and compass.*

Proof. [6] The question is whether $z = \sqrt[3]{2}$ is constructible. Since $x^3 - 2$ is irreducible, $[\mathbb{Q}(z) : \mathbb{Q}] = 3$, by Corollary 4.55; but 3 is not a power of 2.   •

Consider how ingenious this proof is. At the beginning of this section, we asked the reader to ponder how one might prove impossibility. The idea here is to translate the geometric problem of constructibility into a statement of algebra, and then to show that the existence of a geometric construction produces a contradiction in algebra.

A student in one of my classes, imbued with the idea of continual progress through technology, asked me, "Will it ever be possible to duplicate the cube with straightedge and compass?" *Impossible* here is used in its literal sense.

Theorem 4.57 (Wantzel). *It is impossible to trisect a 60° angle using only straight-edge and compass.*

Proof. We may assume that one side of the angle is on the x-axis, and so the question is whether $z = \cos 20° + i \sin 20°$ is constructible. If z is constructible, then Lemma 4.46 would show that $\cos 20°$ is constructible. Corollary 1.26, the triple angle formula, gives $\cos 3\alpha = 4 \cos^3 \alpha - 3 \cos \alpha$. Setting $\alpha = 20°$, we have $\cos 3\alpha = \frac{1}{2}$, so that $\cos 20°$ is a root of $4x^3 - 3x - \frac{1}{2}$; equivalently, $\cos 20°$ is a root of $f(x) = 8x^3 - 6x - 1 \in \mathbb{Z}[x]$. Now $f(x) \in \mathbb{Z}[x]$ is irreducible in $\mathbb{Q}[x]$ because $f(x)$ is irreducible mod 7 (Theorem 3.97). Therefore, $3 = [\mathbb{Q}(z) : \mathbb{Q}]$, by Theorem 3.116(iv), and so $\cos 20°$ is not constructible, because 3 is not a power of 2.   •

If the rules of constructibility are relaxed, then an angle can be trisected.

Theorem 4.58 (Archimedes). *Every angle can be trisected using ruler and compass, where a ruler is a straightedge on which points U and V can be marked, and the point U is allowed to slide along a circle.*

Proof. Since it is easy to construct 30°, 60°, and 90°, it suffices to trisect an acute angle α, for if $3\beta = \alpha$, then $3(\beta + 30°) = \alpha + 90°$, $3(\beta + 60°) = \alpha + 180°$, and $3(\beta + 90°) = \alpha + 270°$.

Draw the given angle $\alpha = \angle AOE$, where the origin O is the center of the unit circle. Take a ruler on which the distance 1 has been marked; that is, there are points U and V on the ruler with $|UV| = 1$. There is a chord through A parallel to EF; place the ruler so that this chord is AU. Since α is acute, U lies in the first quadrant. Keeping A on the sliding ruler, move the point U down the circle; the ruler intersects

---

[6]The notion of dimension of a vector space was not known in the early 19th century; in place of Corollary 4.55, Wantzel proved that if a number is constructible, then it is a root of an irreducible polynomial in $\mathbb{Q}[x]$ of degree some power of 2.

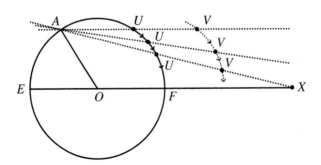

Figure 4.6 A Ruler Sliding

the extended diameter EF in some point X with $|UX| > 1$. Continue moving U down the circle, as in Figure 4.6, keeping A on the sliding ruler, until the ruler intersects EF in the point $X = C$ (with V becoming X).

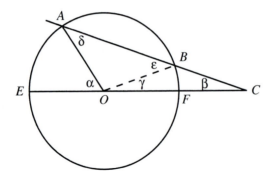

Figure 4.7 Trisecting α

Relabel the points as in Figure 4.7, so that $U = B$ and $|BC| = 1$. We claim that $\beta = \angle BCO = \frac{1}{3}\alpha$. Now

$$\alpha = \delta + \beta,$$

because α is an exterior angle of $\triangle AOC$, and hence it is the sum of the two opposite internal angles. Since $\triangle OAB$ is isosceles (OA and OB are radii), $\delta = \varepsilon$, and so

$$\alpha = \varepsilon + \beta.$$

But $\varepsilon = \gamma + \beta = 2\beta$, for it is an exterior angle of the isosceles triangle $\triangle BCO$; therefore,

$$\alpha = 2\beta + \beta = 3\beta. \quad \bullet$$

Theorem 4.59 (Lindemann). *It is impossible to square the circle with straightedge and compass.*

Proof. The problem is whether one can construct a square whose area is the same as the area of the unit circle. If a side of the square has length z, then one is asking whether $z = \sqrt{\pi}$ is constructible. Now $\mathbb{Q}(\pi)$ is a subspace of $\mathbb{Q}(\sqrt{\pi})$. We have already mentioned that Lindemann proved that π is transcendental (over $\mathbb{Q}$), so that $[\mathbb{Q}(\pi) : \mathbb{Q}]$ is infinite. It follows from Corollary 4.25(iii) that $[\mathbb{Q}(\sqrt{\pi}) : \mathbb{Q}]$ is also infinite. Thus, $[\mathbb{Q}(\sqrt{\pi}) : \mathbb{Q}]$ is surely not a power of 2, and so $\sqrt{\pi}$ is not constructible.
$\bullet$

Sufficiency of the following result was discovered, around 1796, by Gauss, when he was still in his teens (he wrote that this result led to his decision to become a mathematician). He claimed necessity as well, but none of his published papers contains a complete proof of it. The first published proof of necessity is due to P. L. Wantzel, in 1837.

Theorem 4.60 (Gauss-Wantzel). *Let p be an odd prime. A regular p-gon is constructible if and only if $p = 2^{2^t} + 1$ for some $t \geq 0$.*

Proof. We prove necessity only; for sufficiency, see Theorem 5.41. The problem is whether $z = e^{2\pi i/p}$ is constructible. Now z is a root of the cyclotomic polynomial $\Phi_p(x)$, which is an irreducible polynomial in $\mathbb{Q}[x]$ of degree $p-1$, by Corollary 3.103.

Since z is constructible, $p - 1 = 2^s$ for some s (by Corollary 4.55), so that

$$p = 2^s + 1.$$

We claim that s itself is a power of 2. Otherwise, there is an odd number $k > 1$ with $s = km$. Now k odd implies that -1 is a root of $x^k + 1$; in fact, there is a factorization in $\mathbb{Z}[x]$:

$$x^k + 1 = (x + 1)(x^{k-1} - x^{k-2} + x^{k-3} - \cdots + 1).$$

Thus, setting $x = 2^m$ gives a forbidden factorization of p in $\mathbb{Z}$:

$$p = 2^s + 1 = (2^m)^k + 1$$
$$= [2^m + 1][(2^m)^{k-1} - (2^m)^{k-2} + (2^m)^{k-3} - \cdots + 1]. \quad \bullet$$

Gauss constructed a regular 17-gon explicitly, a feat the Greeks would have envied. On the other hand, it follows, for example, that it is impossible to construct regular 7-gons, 11-gons, or 13-gons.

Numbers F_t of the form $F_t = 2^{2^t} + 1$ are called ***Fermat primes*** if they are prime. For $0 \leq t \leq 4$, one can check that F_t is, indeed, prime; they are

$$3, \ 5, \ 17, \ 257, \ \text{and} \ 65{,}537.$$

It is known that the next few values of t give composite numbers, and it is unknown whether any other Fermat primes exist.

The following result is known.

Theorem. *A regular n-gon is constructible if and only if n is a product of a power of 2 and distinct Fermat primes.*

Proof. See Hadlock, *Field Theory and Its Classical Problems*, p. 106.   •

→ **4.3   LINEAR TRANSFORMATIONS**

Homomorphisms between vector spaces are called *linear transformations*.

→ **Definition.**   A function $T: V \to W$, where V and W are vector spaces over a field k, is a ***linear transformation*** if, for all vectors $u, v \in V$ and all scalars $a \in k$,

(i) $T(u + v) = T(u) + T(v)$;

(ii) $T(av) = aT(v)$.

We say that a linear transformation T is ***nonsingular*** (or is an ***isomorphism***) if T is a bijection. Two vector spaces V and W over k are ***isomorphic***, denoted by $V \cong W$, if there is a nonsingular linear transformation $T: V \to W$.

We will soon see how linear transformations determine matrices; Corollary 4.73 will show that nonsingular linear transformations correspond to nonsingular matrices.

It is easy to see that a linear transformation T preserves all linear combinations:

$$T(a_1v_1 + \cdots + a_mv_m) = a_1T(v_1) + \cdots + a_mT(v_m).$$

→ **Example 4.61.**

(i)   The identity function $1_V: V \to V$ on any vector space V is a nonsingular linear transformation.

(ii) If T is nonsingular, then its inverse function $T^{-1}: V \to U$ is a linear transformation, and it is also nonsingular. If $T: U \to V$ and $S: V \to W$ are linear transformations, then their composite $S \circ T: U \to W$ is also a linear transformation. If both S and T are nonsingular, then so is their composite, and $(S \circ T)^{-1} = T^{-1} \circ S^{-1}$.

(iii) If V and W are vector spaces over a field k, write

$$\text{Hom}_k(V, W) = \{\text{all linear transformations } V \to W\}.$$

Define $S + T$ by $S + T: v \mapsto S(v) + T(v)$ for all $v \in V$, and define cT, where $c \in k$, by $cT: v \mapsto cT(v)$ for all $v \in V$. It is routine to check that both $S + T$ and cT are linear transformations and that $\text{Hom}_k(V, W)$ is a vector space over k.

(iv) Let A be an $m \times n$ matrix over a field k. The function $T_A \colon k^n \to k^m$, defined by $T_A(x) = Ax$, where x is an $n \times 1$ column vector and Ax is matrix multiplication, is easily seen to be a linear transformation. We shall see, in Proposition 4.64, that every linear transformation $k^n \to k^m$ is equal to T_A for some $m \times n$ matrix A.

◀

We now show how to construct linear transformations $T \colon V \to W$, where V and W are vector spaces over a field k. The next theorem says that there is a linear transformation that does anything to a basis; the reader should compare this theorem with Theorem 3.33.

→ **Theorem 4.62.**   *Let $v_1, \ldots, v_n$ be a basis of a vector space V over a field k. If W is a vector space over k and $w_1, \ldots, w_n$ is a list in W, then there exists a unique linear transformation $T \colon V \to W$ with $T(v_i) = w_i$ for all i.*

Proof.   By Theorem 4.15, each $v \in V$ has a unique expression of the form $v = \sum_i a_i v_i$, and so $T \colon V \to W$, given by $T(v) = \sum a_i w_i$, is a (well-defined!) function. It is now a routine verification to check that T is a linear transformation.

To prove uniqueness of T, assume that $S \colon V \to W$ is a linear transformation with $S(v_i) = w_i = T(v_i)$ for all i. If $v \in V$, then $v = \sum a_i v_i$ and

$$S(v) = S\left(\sum a_i v_i\right) = \sum S(a_i v_i)$$
$$= \sum a_i S(v_i) = \sum a_i T(v_i) = T(v).$$

Since v is arbitrary, $S = T$.   •

→ **Corollary 4.63.**   *If linear transformations $S, T \colon V \to W$ agree on a basis, then $S = T$.*

Proof.   If $v_1, \ldots, v_n$ is a basis of V and if $S(v_i) = T(v_i)$ for all i, then the uniqueness statement in Theorem 4.62 gives $S = T$.   •

We have already used this corollary in the proof of Theorem 2.65, which shows that the symmetry group of a regular n-gon is the dihedral group D_{2n}; every isometry of the plane which fixes the origin is a linear transformation (Proposition 2.59), and so it is determined by its values on a linearly independent list of two vectors.

Linear transformations $k^n \to k^m$ are easy to describe; every one arises from matrix multiplication, as in Example 4.61(iv).

→ **Proposition 4.64.**   *If $T \colon k^n \to k^m$ is a linear transformation, then there exists a unique $m \times n$ matrix A such that*

$$T(y) = Ay$$

for all $y \in k^n$ (here, y is an $n \times 1$ column matrix and Ay is matrix multiplication).

Proof. If $e_1, \ldots, e_n$ is the standard basis of k^n and $e'_1, \ldots, e'_m$ is the standard basis of k^m, define $A = [a_{ij}]$ to be the matrix whose jth column is the coordinate list of $T(e_j)$. If $S: k^n \to k^m$ is defined by $S(y) = Ay$, then $S = T$ because they agree on a basis: $T(e_j) = \sum_i a_{ij} e'_i = Ae_j = S(e_j)$, the jth column of A.

Uniqueness of A follows from Corollary 4.63, for the jth column of A is the coordinate list of $T(e_j)$. •

Let $T: V \to W$ be a linear transformation, and let $X = v_1, \ldots, v_n$ and $Y = w_1, \ldots, w_m$ be bases of V and W, respectively. The matrix for T is set up from the equation

$$T(v_j) = a_{1j}w_1 + a_{2j}w_2 + \cdots + a_{mj}w_m = \sum_i a_{ij}w_i.$$

This is the reason we write $T(v_j) = \sum_i a_{ij}w_i$ instead of $T(v_j) = \sum_i a_{ji}w_i$, which appears to be more natural.

Example 4.65.

We show that $R_\psi : \mathbb{R}^2 \to \mathbb{R}^2$, counterclockwise rotation about the origin by ψ radians, is a linear transformation (we gave a geometric proof that R_ψ is a linear transformation in Proposition 2.59). If we identify $\mathbb{R}^2$ with $\mathbb{C}$, then every point can be written (in polar form) as $(r\cos\theta, r\sin\theta)$, and we have the formula:

$$R_\psi(r\cos\theta, r\sin\theta) = (r\cos(\theta + \psi), r\sin(\theta + \psi)).$$

Denote the standard basis of $\mathbb{R}^2$ by e_1, e_2, where

$$e_1 = (1, 0) = (\cos 0, \sin 0) \quad \text{and} \quad e_2 = (0, 1) = (\cos\tfrac{\pi}{2}, \sin\tfrac{\pi}{2}).$$

Thus,

$$R_\psi(e_1) = R_\psi(\cos 0, \sin 0) = (\cos\psi, \sin\psi),$$

and

$$
\begin{aligned}
R_\psi(e_2) &= R_\psi(\cos\tfrac{\pi}{2}, \sin\tfrac{\pi}{2}) \\
&= (\cos(\tfrac{\pi}{2} + \psi), \sin(\tfrac{\pi}{2} + \psi)) \\
&= (-\sin\psi, \cos\psi).
\end{aligned}
$$

On the other hand, if T is the linear transformation with

$$T(e_1) = (\cos\psi, \sin\psi) \quad \text{and} \quad T(e_2) = (-\sin\psi, \cos\psi),$$

then the addition formulas for cosine and sine give

$$
\begin{aligned}
T(r\cos\theta, r\sin\theta) &= r\cos\theta T(e_1) + r\sin\theta T(e_2) \\
&= r\cos\theta(\cos\psi, \sin\psi) + r\sin\theta(-\sin\psi, \cos\psi) \\
&= (r[\cos\theta\cos\psi - \sin\theta\sin\psi], r[\cos\theta\sin\psi + \sin\theta\cos\psi]) \\
&= (r\cos(\theta + \psi), r\sin(\theta + \psi)) \\
&= R_\psi(r\cos\theta, r\sin\theta).
\end{aligned}
$$

Therefore, $R_\psi = T$, and so R_ψ is a linear transformation. ◄

Here is the connection between linear transformations and matrices.

→ **Definition.**   Let $X = v_1, \ldots, v_n$ be a basis of V and let $Y = w_1, \ldots, w_m$ be a basis of W. If $T: V \to W$ is a linear transformation, then the ***matrix of T*** with respect to X and Y is the $m \times n$ matrix $A = [a_{ij}]$ whose jth column $a_{1j}, a_{2j}, \ldots, a_{mj}$ is the coordinate list of $T(v_j)$ relative to Y: $T(v_j) = a_{1j}w_1 + \cdots + a_{nj}w_n$. The matrix A does depend on the choice of bases X and Y, and we denote it by

$$A = {}_Y[T]_X.$$

In case $V = W$, we usually let the bases $X = v_1, \ldots, v_n$ and $w_1, \ldots, w_m$ coincide. If $1_V: V \to V$, given by $v \mapsto v$, is the identity linear transformation, then ${}_X[1_V]_X$ is the $n \times n$ ***identity matrix*** I_n (usually, the subscript n is omitted), defined by $I = [\delta_{ij}]$, where δ_{ij} is the ***Kronecker delta***:

$$\delta_{ij} = \begin{cases} 1 & \text{if } i = j; \\ 0 & \text{if } i \neq j. \end{cases}$$

Thus, I has 1s on the diagonal and 0s elsewhere. On the other hand, if X and Y are different bases, then ${}_Y[1_V]_X$ is not the identity matrix; its columns are the coordinate lists of the x's with respect to the basis Y (this matrix is often called the ***transition matrix*** from X to Y).

Example 4.66.
Let V be a vector space with basis $X = v_1, \ldots, v_n$, and let $\sigma \in S_n$ be a permutation. By Theorem 4.62, there is a linear transformation $T: V \to V$ with $T(v_i) = v_{\sigma(i)}$ for all i. The reader should check that $P_\sigma = {}_X[T]_X$ is the permutation matrix obtained from the $n \times n$ identity matrix I by permuting its columns by σ. ◄

→ **Example 4.67.**
Let k be a field and let k^n be equipped with the usual inner product: if $v = (a_1, \ldots, a_n)$ and $u = (b_1, \ldots, b_n)$, then $(v, u) = a_1b_1 + \cdots + a_nb_n$. Define the ***adjoint***[7] of a linear transformation $T: k^n \to k^n$ to be a linear transformation $T^*: k^n \to k^n$ such that

$$(Tu, v) = (u, T^*v)$$

for all $u, v \in k^n$.

We begin by showing that T^* exists. Let $E = e_1, \ldots, e_n$ be the standard basis. If T^* does exist, then it would have to satisfy

$$(Te_j, e_i) = (e_j, T^*e_i)$$

---

[7]There is another notion of *adjoint*, unrelated to this notion, on page 386.

for all i, j. But if $Te_j = a_{j1}e_1 + \cdots + a_{jn}e_n$, then $(Te_j, e_i) = a_{ji}$, by Exercise 4.14 on page 343. With this in mind, let us *define* $T^*e_i = a_{1i}e_1 + \cdots + a_{ni}e_n$ for each i. By Theorem 4.62, we have defined a linear transformation T^*.

If $A = [a_{ij}] = {}_E[T]_E$, then the defining equation for T^* shows that ${}_E[T^*]_E = A^T$; that is, the matrix of the adjoint of A is the transpose of A.

The definition of *adjoint* can be generalized. If $T\colon V \to W$ is a linear transformation, where V and W are vector spaces equipped with inner products, then its **adjoint** is a linear transformation $T^*\colon W \to V$ satisfying $(Tv, w) = (v, T^*w)$ for all $v \in V$ and $w \in W$. ◄

Example 4.68.

(i) In Example 4.65, we considered $R_\psi\colon \mathbb{R}^2 \to \mathbb{R}^2$, counterclockwise rotation about the origin by ψ radians. The matrix of R_ψ with respect to the standard basis $E = e_1, e_2$ is

$$_E[R_\psi]_E = \begin{bmatrix} \cos\psi & -\sin\psi \\ \sin\psi & \cos\psi \end{bmatrix}.$$

(ii) This example shows that matrices assigned to a given linear transformation can actually be different. Let $T\colon \mathbb{R}^2 \to \mathbb{R}^2$ be counterclockwise rotation about the origin by $\frac{\pi}{2}$ radians. As in part (i), the matrix of T relative to the standard basis $X = e_1, e_2$ is

$$_X[T]_X = \begin{bmatrix} 0 & -1 \\ 1 & 0 \end{bmatrix}.$$

The list $Y = v_1, v_2$, where $v_1 = (4, 1)^T$ and $v_2 = (-2, 1)^T$ are column vectors, is a basis. We compute $_Y[T]_Y$ by writing $T(v_1)$ and $T(v_2)$ as linear combinations of v_1, v_2. Now

$$T(v_1) = \begin{bmatrix} 0 & -1 \\ 1 & 0 \end{bmatrix}\begin{bmatrix} 4 \\ 1 \end{bmatrix} = \begin{bmatrix} -1 \\ 4 \end{bmatrix}$$

$$T(v_2) = \begin{bmatrix} 0 & -1 \\ 1 & 0 \end{bmatrix}\begin{bmatrix} -2 \\ 1 \end{bmatrix} = \begin{bmatrix} -1 \\ -2 \end{bmatrix}.$$

We must find numbers a, b, c, d with

$$T(v_1) = \begin{bmatrix} -1 \\ 4 \end{bmatrix} = av_1 + bv_2$$

$$T(v_2) = \begin{bmatrix} -1 \\ -2 \end{bmatrix} = cv_1 + dv_2.$$

Each of these vector equations gives a system of linear equations:

$$4a - 2b = -1$$
$$a + b = 4$$

and

$$4c - 2d = -1$$
$$c + d = -2.$$

These are easily solved:

$$a = \tfrac{7}{6}, \quad b = \tfrac{17}{6}, \quad c = -\tfrac{5}{6}, \quad d = -\tfrac{7}{6}.$$

It follows that

$$_Y[T]_Y = \tfrac{1}{6} \begin{bmatrix} 7 & -5 \\ 17 & -7 \end{bmatrix}.$$

These computations will be revisited in Example 4.75. ◀

Example 4.69.
We have illustrated, given a linear transformation $T : V \rightarrow V$ and a basis X of V, how to set up the matrix $A = {}_X[T]_X$. We now reverse the procedure and show how to construct a linear transformation from an $n \times n$ matrix over k.

Consider the matrix

$$C = \begin{bmatrix} 0 & 0 & 8 \\ 1 & 0 & -6 \\ 0 & 1 & 12 \end{bmatrix}.$$

To define a linear transformation $T : k^3 \rightarrow k^3$, it suffices to specify $T(e_i)$ for each vector in the standard basis $E = e_1, e_2, e_3$. Using the columns of C, we define

$$T(e_1) = e_2, \quad T(e_2) = e_3, \quad T(e_3) = 8e_1 - 6e_2 + 12e_3.$$

Of course, $C = {}_E[T]_E$.

We now find the matrix of T with respect to a new basis. Define $X = x_0, x_1, x_2$ by

$$x_0 = e_1, \quad x_1 = (C - 2I)e_1, \quad x_2 = (C - 2I)^2 e_1.$$

We prove that X spans k^3 by showing that $\langle X \rangle = k^3$. Clearly, $e_1 = x_0 \in \langle X \rangle$, while $x_1 = Ce_1 - 2e_1 = e_2 - 2x_0$; hence,

$$e_2 = 2x_0 + x_1 \in \langle X \rangle.$$

Now $x_2 = C^2 e_1 - 4Ce_1 + 4e_1 = e_3 - 4e_2 + 4e_1$, so that

$$e_3 = x_2 + 4e_2 - 4e_1$$
$$= x_2 + 4(2x_0 + x_1) - 4x_0$$
$$= 4x_0 + 4x_1 + x_2 \in \langle X \rangle.$$

But a spanning list of 3 vectors in a 3-dimensional space must be a basis, and so X is a basis of k^3.

What is the matrix $J = {}_X[T]_X$? Using the preceding equations, the reader may verify that

$$
\begin{aligned}
T(x_0) &= 2x_0 + x_1 \\
T(x_1) &= 2x_1 + x_2 \\
T(x_2) &= 2x_2.
\end{aligned}
$$

It follows that the matrix of T with respect to the basis X is

$$
J = \begin{bmatrix} 2 & 0 & 0 \\ 1 & 2 & 0 \\ 0 & 1 & 2 \end{bmatrix}. \quad \blacktriangleleft
$$

The following proposition is a paraphrase of Theorem 4.62.

→ **Proposition 4.70.** *Let V and W be vector spaces over a field k, and let X and Y be bases of V and W, respectively. The function*

$$
\mu_{X,Y} \colon \operatorname{Hom}_k(V, W) \to \operatorname{Mat}_{m \times n}(k),
$$

given by

$$
T \mapsto {}_Y[T]_X,
$$

is an isomorphism of vector spaces.

Proof. First, let us show that $\mu_{X,Y}$ is surjective. Given a matrix A, its columns define vectors in W. In more detail, if $X = v_1, \ldots, v_n$ and $Y = w_1, \ldots, w_m$, then the jth column of A is $(a_{1j}, \ldots, a_{mj})^T$; define $z_j = \sum_{i=1}^m a_{ij} w_i$. By Theorem 4.62, there exists a linear transformation $T \colon V \to W$ with $T(v_j) = z_j$, and ${}_Y[T]_X = A$. To see that $\mu_{X,Y}$ is injective, suppose that ${}_Y[T]_X = A = {}_Y[S]_X$. Since the columns of A determine $T(v_j)$ and $S(v_j)$ for all j, Corollary 4.63 gives $S = T$.

Finally, we show that $\mu_{X,Y}$ is a linear transformation. Since, for all j, the jth column of $S + T$ is $(S+T)(v_j) = S(v_j) + T(v_j)$, we have $\mu_{X,Y}(S+T) = \mu_{X,Y}(S) + \mu_{X,Y}(T)$. A similar argument shows that $\mu_{X,Y}(cT) = c\mu_{X,Y}(T)$. •

The next theorem shows where the definition of matrix multiplication comes from: the product of two matrices is the matrix of a composite.

→ **Theorem 4.71.** *Let $T \colon V \to W$ and $S \colon W \to U$ be linear transformations. Choose bases $X = x_1, \ldots, x_n$ of V, $Y = y_1, \ldots, y_m$ of W, and $Z = z_1, \ldots, z_\ell$ of U. Then*

$$
{}_Z[S \circ T]_X = \big({}_Z[S]_Y\big)\big({}_Y[T]_X\big).
$$

Proof. Let $_Y[T]_X = [a_{ij}]$, so that $T(x_j) = \sum_p a_{pj} y_p$, and let $_Z[S]_Y = [b_{qp}]$, so that $S(y_p) = \sum_q b_{qp} z_q$. Then

$$(S \circ T)(x_j) = S(T(x_j)) = S\left(\sum_p a_{pj} y_p\right)$$

$$= \sum_p a_{pj} S(y_p) = \sum_p \sum_q a_{pj} b_{qp} z_q = \sum_q c_{qj} z_q,$$

where $c_{qj} = \sum_p b_{qp} a_{pj}$. Therefore,

$$_Z[S \circ T]_X = [c_{qj}] = \left(_Z[S]_Y\right)\left(_Y[T]_X\right). \quad \bullet$$

Corollary 4.72. *Matrix multiplication is associative:* $A(BC) = (AB)C$.

Proof. Let A be an $m \times n$ matrix, let B be an $n \times p$ matrix, and let C be a $p \times q$ matrix. By Theorem 4.62, there are linear transformations

$$k^q \xrightarrow{T} k^p \xrightarrow{S} k^n \xrightarrow{R} k^m$$

with $C = [T]$, $B = [S]$, and $A = [R]$ (in order that the proof not be cluttered, we abbreviate notation by omitting bases: we write $[T]$ instead of $_Y[T]_X$).
    Then

$$[R \circ (S \circ T)] = [R][S \circ T] = [R]([S][T]) = A(BC).$$

On the other hand,

$$[(R \circ S) \circ T] = [R \circ S][T] = ([R][S])[T] = (AB)C.$$

Since composition of functions is associative,

$$R \circ (S \circ T) = (R \circ S) \circ T,$$

and so

$$A(BC) = [R \circ (S \circ T)] = [(R \circ S) \circ T] = (AB)C. \quad \bullet$$

We can prove Corollary 4.72 directly, manipulating summations, but the connection with composition of linear transformations is the real reason why matrix multiplication is associative.

Corollary 4.73. *Let* $T : V \to W$ *be a linear transformation of vector spaces* V *and* W *over a field* k, *and let* X *and* Y *be bases of* V *and* W, *respectively. If* T *is nonsingular and* $A = {}_Y[T]_X$, *then* A *is a nonsingular matrix and the matrix of* T^{-1} *is*

$$_X[T^{-1}]_Y = A^{-1} = ({}_Y[T]_X)^{-1}.$$

Conversely, if $A = {}_Y[T]_X$ *is a nonsingular matrix for some bases* X *of* V *and* Y *of* W, *then* T *is a nonsingular linear transformation.*

Proof.

$$I = {}_Y[1_W]_Y = \big({}_Y[T]_X\big)\big({}_X[T^{-1}]_Y\big)$$

and

$$I = {}_X[1_V]_X = \big({}_X[T^{-1}]_Y\big)\big({}_Y[T]_X\big).$$

Therefore, ${}_X[T^{-1}]_Y = ({}_Y[T]_X)^{-1}$.

Let $B = [b_{ij}]$ be a matrix with $BA = I = AB$. As in Theorem 4.62, there is a unique linear transformation $S \colon W \to V$ with $S(y_i) = \sum_p b_{pi} x_p$; by definition of the matrix of S with respect to Y and X, we have $B = {}_X[S]_Y$. Hence,

$$I = BA = {}_X[S]_Y \, {}_Y[T]_X = {}_X[S \circ T]_X.$$

Thus, $(S \circ T)(x_i) = Ix_i = x_i$ for all i, and so $S \circ T = 1_V$. A similar argument, using $I = AB$, shows that $T \circ S = 1_W$. We conclude that T is a bijection and, hence, that it is a nonsingular linear transformation.   •

The next corollary determines all the matrices arising from the same linear transformation.

→ **Corollary 4.74.** *Let $T \colon V \to V$ be a linear transformation on a vector space V over a field k. If X and Y are bases of V, then there is a nonsingular matrix P with entries in k, namely, $P = {}_Y[1_V]_X$, so that*

$$ {}_Y[T]_Y = P\big({}_X[T]_X\big)P^{-1}.$$

Conversely, if $B = PAP^{-1}$, where B, A, and P are $n \times n$ matrices with entries in k and P is nonsingular, then there is a linear transformation $T \colon k^n \to k^n$ and bases X and Y of k^n such that $B = {}_Y[T]_Y$ and $A = {}_X[T]_X$.

Proof.   The first statement follows from Theorem 4.71 and associativity:

$$ {}_Y[T]_Y = {}_Y[1_V T 1_V]_Y = ({}_Y[1_V]_X)({}_X[T]_X)({}_X[1_V]_Y).$$

Set $P = {}_Y[1_V]_X$, and note that Corollary 4.73 gives $P^{-1} = {}_X[1_V]_Y$.

For the converse, let $E = e_1, \ldots, e_n$ be the standard basis of k^n, and define $T \colon k^n \to k^n$ by $T(e_j) = Ae_j$ (remember that vectors in k^n are column vectors, so that Ae_j is matrix multiplication). Since Ae_j is the jth column of A, we have $A = {}_E[T]_E$. Define a list $Y = y_1, \ldots, y_n$ by $y_j = P^{-1}e_j$; that is, the vectors in Y are the columns of P^{-1}.

Let us show that $Y = P^{-1}e_1, \ldots, P^{-1}e_n$ is a basis of k^n. If $\sum_j a_j P^{-1}e_j = 0$, then $P^{-1}(\sum_j a_j e_j) = 0$; multiplying on the left by P gives $\sum_j a_j e_j = 0$, and linear independence of the standard basis gives all $a_j = 0$. Thus, Y is linearly independent.

To see that Y spans k^n, take $w \in k^n$. Now $Pw = \sum_j b_j e_j$, and so $w = P^{-1}Pw = \sum_j b_j P^{-1}e_j \in \langle Y \rangle$. Therefore, Y is a basis.

It remains to prove that $B = {}_Y[T]_Y$; that is, $T(y_j) = \sum_i b_{ij}y_i$, where $B = [b_{ij}]$.

$$T(y_j) = Ay_j = AP^{-1}e_j = P^{-1}Be_j$$
$$= P^{-1}\sum_i b_{ij}e_i = \sum_i b_{ij}P^{-1}e_i = \sum_i b_{ij}y_i \quad \bullet$$

→ **Definition.** Two $n \times n$ matrices B and A over a field k are *similar* if there is a nonsingular matrix P over k with $B = PAP^{-1}$.

Corollary 4.74 says that two matrices are similar if and only if they arise from the same linear transformation on a vector space V (from different choices of basis). For example, the matrices C and J in Example 4.69 are similar. The first matrix C arises from a linear transformation $T: k^3 \to k^3$ relative to the standard basis E; that is, $C = {}_E[T]_E$. The second matrix J arises from the basis X in that example; that is, $J = {}_X[T]_X$.

Example 4.75.
We can now simplify the calculations in Example 4.68(ii). Recall that we have two bases of $\mathbb{R}^2$: the standard basis $E = e_1, e_2$ and $F = v_1, v_2$, where $v_1 = \begin{bmatrix} 4 \\ 1 \end{bmatrix}$ and $v_2 = \begin{bmatrix} -2 \\ 1 \end{bmatrix}$, and the linear transformation $T: \mathbb{R}^2 \to \mathbb{R}^2$, rotation by 90°, with matrix

$$_E[T]_E = \begin{bmatrix} 0 & -1 \\ 1 & 0 \end{bmatrix}.$$

Now the transition matrices are

$$P^{-1} = {}_E[1]_F = \begin{bmatrix} 4 & -2 \\ 1 & 1 \end{bmatrix}$$

and

$$P = {}_F[1]_E = {}_E[1]_F^{-1} = \frac{1}{6}\begin{bmatrix} 1 & 2 \\ -1 & 4 \end{bmatrix}.$$

Therefore,

$$_F[T]_F = P\,_E[T]_E\,P^{-1} = \frac{1}{6}\begin{bmatrix} 7 & -5 \\ 17 & -7 \end{bmatrix},$$

which agrees with our earlier result on page 370. ◀

Just as for group homomorphisms and ring homomorphisms, we can define the kernel and image of linear transformations.

→ **Definition.**   If $T: V \to W$ is a linear transformation, then the **kernel** (or the **null space**) of T is

$$\ker T = \{v \in V : T(v) = 0\},$$

and the **image** of T is

$$\operatorname{im} T = \{w \in W : w = T(v) \text{ for some } v \in V\}.$$

As in Example 4.61(iv), an $m \times n$ matrix A with entries in a field k determines a linear transformation $T_A: k^n \to k^m$, namely, $T_A(y) = Ay$, where y is an $n \times 1$ column vector. The kernel of T_A is the solution space $Sol(A)$ [see Example 4.3(iv)], and the image of T_A is the column space $Col(A)$.

The proof of the next proposition is straightforward.

Proposition 4.76.   *Let $T: V \to W$ be a linear transformation.*

(i) *$\ker T$ is a subspace of V and $\operatorname{im} T$ is a subspace of W.*

(ii) *T is injective if and only if $\ker T = \{0\}$.*

We can now give a new proof of Corollary 4.20 that a homogeneous system over a field k with r equations in n unknowns has a nontrivial solution if $r < n$. If A is the $r \times n$ coefficient matrix of the system, then $T: x \mapsto Ax$ is a linear transformation $T: k^n \to k^r$. If there is only the trivial solution, then $\ker T = \{0\}$, so that k^n is isomorphic to the subspace $\operatorname{im} T \subseteq k^r$, and this contradicts Corollary 4.25(ii).

Lemma 4.77.   *The following statements are equivalent for a linear transformation $T: V \to W$.*

(i) *T is nonsingular (that is, T is an isomorphism).*

(ii) *For every basis X of V, $T(X)$ is a basis of W.*

(iii) *For some basis X of V, $T(X)$ is a basis of W.*

Proof.
(i) $\Rightarrow$ (ii). Let $X = v_1, v_2, \ldots, v_n$ be a basis of V. If $\sum c_i T(v_i) = 0$, then $T(\sum c_i v_i) = 0$, and so $\sum c_i v_i \in \ker T = \{0\}$. Hence each $c_i = 0$, because X is linearly independent. If $w \in W$, then the surjectivity of T provides $v \in V$ with $w = T(v)$. But $v = \sum a_i v_i$, and so $w = T(v) = T(\sum a_i v_i) = \sum a_i T(v_i)$. Therefore, $T(X)$ is a basis of W.
(ii) $\Rightarrow$ (iii) Obvious.
(iii) $\Rightarrow$ (i) If $w \in W$, then $w = \sum c_i T(v_i) = T(\sum c_i v_i)$, since $T(v_1), \ldots, T(v_n)$ is a basis of W, and so T is surjective. If $\sum c_i v_i \in \ker T$, then $\sum c_i T(v_i) = 0$, and so linear independence gives all $c_i = 0$; hence, $\sum c_i v_i = 0$ and $\ker T = \{0\}$. Therefore, T is nonsingular.   •

→ **Theorem 4.78.** *If V is an n-dimensional vector space over a field k, then V is isomorphic to k^n.*

Proof. Choose a basis $v_1, \ldots, v_n$ of V. If $e_1, \ldots, e_n$ is the standard basis of k^n, then Theorem 4.62 says that there is a linear transformation $T : V \to k^n$ with $T(v_i) = e_i$ for all i; by Lemma 4.77, T is nonsingular. •

Theorem 4.78 not only says that every finite dimensional vector space is essentially the familiar vector space of all n-tuples; it also says that a choice of basis in V is tantamount to a choice of coordinate list for each vector in V. We want the freedom to change coordinates because the usual coordinates may not be the most convenient ones for a given problem, as the reader has probably seen (in a calculus course) when rotating axes to simplify the equation of a conic section.

→ **Corollary 4.79.** *Two finite dimensional vector spaces V and W over a field k are isomorphic if and only if $\dim(V) = \dim(W)$.*

Proof. Assume that there is a nonsingular $T : V \to W$. If $X = v_1, \ldots, v_n$ is a basis of V, then Lemma 4.77 says that $T(v_1), \ldots, T(v_n)$ is a basis of W. Therefore, $\dim(W) = |X| = \dim(V)$.

If $n = \dim(V) = \dim(W)$, then there are isomorphisms $T : V \to k^n$ and $S : W \to k^n$, by Theorem 4.78. Hence, the composite $S^{-1} \circ T : V \to W$ is an isomorphism. •

→ **Proposition 4.80.** *Let V be a finite dimensional vector space over a field k with $\dim(V) = n$, and let $T : V \to V$ be a linear transformation. The following statements are equivalent:*

(i) *T is an isomorphism;*

(ii) *T is surjective;*

(iii) *T is injective.*

Remark. Compare this proposition with the pigeonhole principle, Exercise 2.13 on page 105. ◄

Proof.
(i) ⇒ (ii) This implication is obvious, for isomorphisms are bijections.
(ii) ⇒ (iii) Assume that T is surjective. If $X = v_1, \ldots, v_n$ is a basis of V, we claim that $T(X) = T(v_1), \ldots, T(v_n)$ spans V. If $w \in V$, then surjectivity of T gives $v \in V$ with $w = T(v)$. Now $v = \sum_i a_i v_i$ for scalars $a_i \in k$, and so $w = T(v) = \sum_i a_i T(v_i)$. Since $\dim(V) = n$, it follows from Corollary 4.24 that $T(X)$ is a basis of V. Lemma 4.77 now says that T is an isomorphism, and so T is injective.

(iii) $\Rightarrow$ (i) Assume that T is injective. If $X = v_1, \ldots, v_n$ is a basis of V, then we claim that $T(X) = T(v_1), \ldots, T(v_n)$ is linearly independent. If $\sum c_i T(v_i) = 0$, then $T(\sum_i c_i T v_i) = 0$, so that $\sum_i c_i v_i \in \ker T = \{0\}$. Hence, $\sum_i c_i v_i = 0$, and linear independence of X gives all $c_i = 0$. Therefore, $T(X)$ is linearly independent. Since $\dim(V) = n$, it follows from Corollary 4.24 that $T(X)$ is a basis of V. Lemma 4.77 now says that T is an isomorphism.   •

Call a linear transformation $T : V \to V$ **singular** if T is not an isomorphism; that is, T is not nonsingular.

Corollary 4.81.   *Let V be a finite dimensional vector space, and let $T : V \to V$ be a linear transformation on V. Then T is singular if and only if there exists a nonzero vector $v \in V$ with $T(v) = 0$.*

Proof.   If T is singular, then $\ker T \neq \{0\}$, by Proposition 4.80. Conversely, if there is a nonzero vector v with $T(v) = 0$, then $\ker T \neq \{0\}$ and T is not an isomorphism.   •

This corollary says that a homogeneous $n \times n$ linear system $Ax = 0$ with a singular coefficient matrix A always has a nontrivial solution.

Recall that an $n \times n$ matrix A with entries in a field k is nonsingular if there is a matrix B (its inverse) with entries in k with $AB = I = BA$. The next corollary shows that "one-sided inverses" are enough.

Corollary 4.82.   *Let A and B be $n \times n$ matrices with entries in a field k. If $AB = I$, then $BA = I$. Therefore, A is nonsingular and $B = A^{-1}$.*

Proof.   There are linear transformations $T, S : k^n \to k^n$ with $_X[T]_X = A$ and $_X[S]_X = B$, where X is the standard basis. Let us abbreviate $_X[T]_X$ to $[T]$ in this proof. Since $AB = I$, Proposition 4.70 gives

$$[T \circ S] = [T][S] = I = [1_{k^n}].$$

Since $T \mapsto [T]$ is a bijection, by Proposition 4.70, it follows that $T \circ S = 1_{k^n}$. By Proposition 2.9, T is a surjection and S is an injection. But Proposition 4.80 says that both T and S are isomorphisms, so that $S = T^{-1}$ and $T \circ S = 1_{k^n} = S \circ T$. Therefore, $I = [S \circ T] = [S][T] = BA$, as desired.   •

→ **Proposition 4.83.**   *Let $T : V \to W$ be a linear transformation, where V and W are vector spaces over a field k of dimension n and m, respectively. Then*

$$\dim(\ker T) + \dim(\operatorname{im} T) = n.$$

Proof.   Choose a basis $u_1, \ldots, u_p$ of $\ker T$, and extend it to a basis of V by adjoining vectors $w_1, \ldots, w_q$. Since V is spanned by the list $u_1, \ldots, u_p, w_1, \ldots, w_q$, the subspace $\operatorname{im} T$ is spanned by the list $T(u_1), \ldots, T(u_p), T(w_1), \ldots, T(w_q)$; but $T(u_i) = 0$ for all i, so that $\operatorname{im} T$ is spanned by the shorter list $T(w_1), \ldots, T(w_q)$. Since $\dim(\ker T) = p$ and $p + q = n$, it suffices to prove that $T(w_1), \ldots, T(w_q)$ is a linearly independent list.

If $c_1 T(w_1) + \cdots + c_q T(w_q) = 0$, then $T(c_1 w_1 + \cdots + c_q w_q) = 0$ and $c_1 w_1 + \cdots + c_q w_q \in \ker T$. Hence, there are $a_1, \ldots, a_p \in k$ with

$$c_1 w_1 + \cdots + c_q w_q = a_1 u_1 + \cdots + a_p u_p.$$

Since $u_1, \ldots, u_p, w_1, \ldots, w_q$ is a basis of V, it is a linearly independent list, so that $0 = c_1 = \cdots = c_q$ (and also $0 = a_1 = \cdots = a_p$). Therefore, $T(w_1), \ldots, T(w_q)$ is a basis of $\operatorname{im} T$, and $\dim(\operatorname{im} T) = q$. •

→ **Corollary 4.84.**   *Let A be an $m \times n$ matrix over a field k.*

(i) $\operatorname{rank}(A) = \dim(\operatorname{im} T_A)$, *where $T_A \colon k^n \to k^m$ is defined by $T_A(x) = Ax$.*

(ii) $\operatorname{rank}(A) = \dim(Col(A))$.

(iii) $\operatorname{rank}(A) = \operatorname{rank}(A^T)$; *i.e., $Row(A)$ and $Col(A)$ have the same dimension.*

Proof.
(i) By Theorem 4.43, the rank-nullity theorem, $\dim(Sol(A)) = n - \operatorname{rank}(A)$; that is, $\operatorname{rank}(A) = n - \dim(Sol(A))$. But $\ker T_A = Sol(A)$, and so Proposition 4.83 gives $\dim(\operatorname{im} T_A) = n - \dim(Sol(A)) = \operatorname{rank}(A)$.
(ii) $\operatorname{im} T_A = \langle T_A(e_1), \ldots, T_A(e_n) \rangle = \langle A e_1, \ldots, A e_n \rangle = Col(A)$.
(iii) By definition, $\operatorname{rank}(A) = \dim(Row(A))$, while $\operatorname{rank}(A) = \dim(Col(A))$ was proved in part (ii). •

→ **Definition.**   If V is a vector space over a field k, then the ***general linear group***, denoted by $GL(V)$, is the set of all nonsingular linear transformations $V \to V$.

A composite $S \circ T$ of linear transformations S and T is again a linear transformation, and $S \circ T$ is nonsingular if both S and T are; moreover, the inverse of a nonsingular linear transformation is again nonsingular. It follows that $GL(V)$ is a group with composition as operation, for composition of functions is always associative.

→ **Definition.**   The set of all nonsingular $n \times n$ matrices with entries in a field k is denoted by $GL(n, k)$.

It is easy to prove that $GL(n, k)$ is a group under matrix multiplication.

A choice of basis gives an isomorphism between the general linear group and the group of nonsingular matrices.

→ **Proposition 4.85.** *Let V be an n-dimensional vector space over a field k, and let $X = v_1, \ldots, v_n$ be a basis of V. Then $\mu\colon \mathrm{GL}(V) \to \mathrm{GL}(n, k)$, defined by $T \mapsto {}_X[T]_X$, is an isomorphism of groups.*

Proof. By Proposition 4.70, the function $\mu_{X,X}\colon T \mapsto [T] = {}_X[T]_X$ is an isomorphism of vector spaces

$$\mathrm{Hom}_k(V, V) \to \mathrm{Mat}_n(k).$$

Moreover, Theorem 4.71 says that $\mu_{X,X}(T \circ S) = \mu_{X,X}(T)\mu_{X,X}(S)$ for all $T, S \in \mathrm{Hom}_k(V, V)$.

If $T \in \mathrm{GL}(V)$, then $\mu_{X,X}(T) = {}_X[T]_X$ is a nonsingular matrix, by Corollary 4.73; thus, if μ is the restriction of $\mu_{X,X}$, then $\mu\colon \mathrm{GL}(V) \to \mathrm{GL}(n, k)$ is an injective homomorphism.

It remains to prove that μ is surjective. Since $\mu_{X,X}$ is surjective, if $A \in \mathrm{GL}(n, k)$, then $A = {}_X[T]_X$ for some $T\colon V \to V$. It suffices to show that T is an isomorphism, for then $T \in \mathrm{GL}(V)$. Since A is a nonsingular matrix, there is a matrix B with $AB = I$. Now $B = {}_X[S]_X$ for some $S\colon V \to V$, and

$$\mu_{X,X}(T \circ S) = \mu_{X,X}(T)\mu_{X,X}(S) = AB = I = \mu_{X,X}(1_V).$$

Therefore, $T \circ S = 1_V$, since $\mu_{X,X}$ is an injection, and so $T \in \mathrm{GL}(V)$, by Corollary 4.82.   •

The center of the general linear group is easily identified; we now generalize Exercise 2.84 on page 170.

Definition. A linear transformation $T\colon V \to V$ is a *scalar transformation* if there is $c \in k$ with $T(v) = cv$ for all $v \in V$; that is, $T = c1_V$. A *scalar matrix* is a matrix of the form cI, where $c \in k$ and I is the identity matrix.

A scalar transformation $T = c1_V$ is nonsingular if and only if $c \neq 0$ (its inverse is $c^{-1}1_V$).

Corollary 4.86. *The center of the group $\mathrm{GL}(V)$ consists of all the nonsingular scalar transformations. The center of the group $\mathrm{GL}(n, k)$ consists of all the nonsingular scalar matrices.*

Proof. If $T \in \mathrm{GL}(V)$ is not scalar, then there is some vector $v \in V$ with $T(v)$ not a scalar multiple of v. Of course, this forces $v \neq 0$. We claim that the list $X = v, T(v)$ is linearly independent. We know that $T(v)$ is not a scalar multiple of v. If $v = dT(v)$, for some $d \in k$, then $d \neq 0$ (lest $v = 0$), and so $T(v) = d^{-1}v$, a contradiction. Hence, $v, T(v)$ is a linearly independent list, by Example 4.12(iii). By Proposition 4.22, this list can be extended to a basis $v, T(v), u_3, \ldots, u_n$ of V. It is easy to see that $v, v + T(v), u_3, \ldots, u_n$ is also a basis of V, and so there is a nonsingular linear transformation S with $S(v) = v$, $S(T(v)) = v + T(v)$, and $S(u_i) = u_i$ for all i. Now S and T do not commute, for $ST(v) = v + T(v)$ while $TS(v) = T(v)$. Therefore, T is not in the center of $\mathrm{GL}(V)$.

If $f : G \to H$ is a group isomorphism between groups G and H, then $f(Z(G)) = Z(H)$. In particular, if $T = c1_V$ is a nonsingular scalar transformation, then $_X[T]_X$ is in the center of $\mathrm{GL}(n, k)$ for any basis $X = v_1, \ldots, v_n$ of V. But $T(v_i) = cv_i$ for all i, so that $_X[T]_X = cI$ is a scalar matrix.   •

EXERCISES

H **4.32** True or false with reasons.

(i) Every linear transformation $T : V \to V$, where V is a finite dimensional vector space over $\mathbb{R}$, is represented by infinitely many matrices.

(ii) Every matrix over $\mathbb{R}$ is similar to infinitely many different matrices.

(iii) If S and T are linear transformations on the plane $\mathbb{R}^2$ that agree on two nonzero points, then $S = T$.

(iv) If A and B are $n \times n$ nonsingular matrices, then $A + B$ is nonsingular.

(v) If A and B are $n \times n$ nonsingular matrices, then AB is nonsingular.

(vi) If k is a field, then

$$\{A \in \mathrm{Mat}_n(k) : AB = BA \text{ for all } B \in \mathrm{Mat}_n(k)\}$$

is a 1-dimensional subspace of $\mathrm{Mat}_n(k)$.

(vii) The vector space of all 3×3 symmetric matrices over $\mathbb{R}$ is isomorphic to the vector space consisting of 0 and all $f(x) \in \mathbb{R}[x]$ with $\deg(f) \le 5$.

(viii) If X and Y are bases of a finite dimensional vector space over a field k, then $_Y[1_V]_X$ is the identity matrix.

(ix) Transposition $\mathrm{Mat}_{m \times n}(\mathbb{C}) \to \mathrm{Mat}_{n \times m}(\mathbb{C})$, given by $A \mapsto A^T$, is a nonsingular linear transformation.

(x) If V is the vector space of all continuous $f : [0, 1] \to \mathbb{R}$, then integration $f \mapsto \int_0^1 f(x) \, dx$ is a linear transformation $V \to \mathbb{R}$.

4.33 Let k be a field, let $V = k[x]$, the polynomial ring viewed as a vector space over k, and let $V_n = \langle 1, x, x^2, \ldots, x^n \rangle$. By Exercise 4.8 on page 342, we know that $X_n = 1, x, x^2, \ldots, x^n$ is a basis of V_n.

(i) Prove that differentiation $T : V_3 \to V_3$, defined by $T(f(x)) = f'(x)$, is a linear transformation, and find the matrix $A = {}_{X_3}[T]_{X_3}$ of differentiation.

(ii) Prove that integration $S : V_3 \to V_4$, defined by $S(f) = \int_0^x f(t) \, dt$, is a linear transformation, and find the matrix $A = {}_{X_4}[S]_{X_3}$ of integration.

4.34 If $\sigma \in S_n$ and $P = P_\sigma$ is the corresponding permutation matrix (see Example 4.66), prove that $P^{-1} = P^T$.

*4.35 Let V and W be vector spaces over a field k, and let $S, T : V \to W$ be linear transformations.

  (i) If V and W are finite dimensional, prove that

$$\dim(\operatorname{Hom}_k(V, W)) = \dim(V)\dim(W).$$

  (ii) The **dual space** V^* of a vector space V over k is defined by

$$V^* = \operatorname{Hom}_k(V, k).$$

  If $X = v_1, \ldots, v_n$ is a basis of V, define $\delta_1, \ldots, \delta_n \in V^*$ by

$$\delta_i(v_j) = \begin{cases} 0 & \text{if } j \neq i \\ 1 & \text{if } j = i. \end{cases}$$

  Prove that $\delta_1, \ldots, \delta_n$ is a basis of V^* (it is called the **dual basis** arising from $v_1, \ldots, v_n$).

  (iii) If $\dim(V) = n$, prove that $\dim(V^*) = n$, and hence that $V^* \cong V$.

*4.36 **(i)** If $S : V \to W$ is a linear transformation and $f \in W^*$, then the composite $V \xrightarrow{S} W \xrightarrow{f} k$ lies in V^*. Prove that $S^* : W^* \to V^*$, defined by $S^* : f \mapsto f \circ S$, is a linear transformation.

  (ii) If $X = v_1, \ldots, v_n$ and $Y = w_1, \ldots, w_m$ are bases of V and W, respectively, denote the dual bases by X^* and Y^* (see Exercise 4.35). If $S : V \to W$ is a linear transformation, prove that the matrix of S^* is a transpose:

$$_{X^*}[S^*]_{Y^*} = \big(_Y[S]_X\big)^T.$$

  Remark. Here is a convincing reason why targets are necessary in a function's definition. We have just seen that every linear transformation $S : V \to W$ defines a linear transformation $S^* : W^* \to V^*$ whose domain is W^*. Thus, changing the target of S changes the domain of S^*, and so S^* is changed in an essential way. We conclude that the target of a function should be an essential part of its definition. ◄

4.37 **(i)** If $A = \begin{bmatrix} a & b \\ c & d \end{bmatrix}$, define $\det(A) = ad - bc$. Given a system of linear equations $Ax = 0$ with coefficients in a field,

$$ax + by = p$$
$$cx + dy = q,$$

  prove that there exists a unique solution if and only if $\det(A) \neq 0$.

  (ii) If V is a vector space with basis $X = v_1, v_2$, define $T : V \to V$ by $T(v_1) = av_1 + bv_2$ and $T(v_2) = cv_1 + dv_2$. Prove that T is a nonsingular linear transformation if and only if $\det(_X[T]_X) \neq 0$.

*4.38 Let U be a subspace of a vector space V.

  (i) Prove that the **natural map** $\pi : V \to V/U$, given by $v \mapsto v + U$, is a linear transformation with kernel U. (Quotient spaces were defined in Exercise 4.16 on page 343.)

  H **(ii)** State and prove the **first isomorphism theorem** for vector spaces.

4.39 Let k be a field and let $k^\times$ be its multiplicative group of nonzero elements. Prove that det: $\mathrm{GL}(2, k) \to k^\times$ is a surjective group homomorphism whose kernel is $\mathrm{SL}(2, k)$. Conclude that $\mathrm{SL}(2, k) \lhd \mathrm{GL}(2, k)$ and $\mathrm{GL}(2, k)/\mathrm{SL}(2, k) \cong k^\times$.

H **4.40** Let V be a finite dimensional vector space over a field k, and let $\mathcal{B}$ denote the family of all the bases of V. Prove that $\mathcal{B}$ is a transitive $\mathrm{GL}(V)$-set.

4.41 Recall that if U and W are subspaces of a vector space V such that $U \cap W = \{0\}$ and $U + W = V$, then U is called a *direct summand* of V and W is called a *complement* of U. In Exercise 4.20 on page 343, we saw that every subspace of a finite dimensional vector space is a direct summand.

   (i)   Let $U = \{(a, a) : a \in \mathbb{R}\}$. Find all the complements of U in $\mathbb{R}^2$.

   (ii)   If U is a subspace of a finite dimensional vector space V, prove that any two complements of U are isomorphic.

*4.42** If A is an $m \times n$ matrix and B is an $p \times m$ matrix, prove that

$$\mathrm{rank}(BA) \leq \mathrm{rank}(A).$$

4.43 Let $\mathbb{R}^n$ be equipped with the usual inner product: if $v = (a_1, \dots, a_n)$ and $u = (b_1, \dots, b_n)$, then $(v, u) = a_1 b_1 + \cdots + a_n b_n$.

   (i)   A linear transformation $U : \mathbb{R}^n \to \mathbb{R}^n$ is called *orthogonal* if $(Uv, Uw) = (v, w)$ for all $v, w \in \mathbb{R}^n$.

   Prove that every orthogonal transformation is nonsingular.

   (ii)   An *orthonormal basis* of k^n is a basis $v_1, \dots, v_n$ such that

$$(v_i, v_j) = \delta_{ij},$$

   where (v_i, v_j) is the inner product and δ_{ij} is the Kronecker delta. For example, the standard basis is an orthonormal basis for the usual inner product.

   Prove that a linear transformation $U : \mathbb{R}^n \to \mathbb{R}^n$ is orthogonal if and only if $U(v_1), \dots, U(v_n)$ is an orthonormal basis whenever $v_1, \dots, v_n$ is an orthonormal basis.

   (iii)   If $w \in \mathbb{R}^n$ and $v_1, \dots, v_n$ is an orthonormal basis, then $w = \sum_{i=1}^{n} c_i v_i$. Prove that $c_i = (w, v_i)$.

4.44 Let $U : \mathbb{R}^n \to \mathbb{R}^n$ be an orthogonal transformation, and let $X = v_1, \dots, v_n$ be an orthonormal basis. If $O = {}_X[U]_X$, prove that $O^{-1} = O^T$. (The matrix O is called an *orthogonal matrix*.)

4.45 Let A be an $n \times n$ real symmetric matrix.

   (i)   Give an example of a nonsingular matrix P for which PAP^{-1} is not symmetric.

   (ii)   Prove that OAO^{-1} is symmetric for every $n \times n$ real orthogonal matrix O.

→ **4.4   EIGENVALUES**

We introduce determinants of square matrices, and we will use them to investigate invertibility. Several important results in this section will be stated without proof.

The usual, though inelegant, definition of the determinant of an $n \times n$ real matrix $A = [a_{ij}]$ is

$$\det(A) = \sum_{\sigma \in S_n} \operatorname{sgn}(\sigma) a_{\sigma(1),1} a_{\sigma(2),2} \cdots a_{\sigma(n),n}.$$

Recall that $\operatorname{sgn}(\sigma) = \pm 1$: it is $+1$ if σ is an even permutation, and it is -1 if σ is odd. The term $a_{\sigma(1),1} a_{\sigma(2),2} \cdots a_{\sigma(n),n}$ has exactly one factor from each column of A because all the second subscripts are distinct, and it has exactly one factor from each row because all the first subscripts are distinct. We often call this formula the *complete expansion* of the determinant. From this definition, we see that the formula for $\det(A)$ makes sense for $n \times n$ matrices A with entries in any commutative ring R.

Another way to view the determinant is to consider it as a function

$$D = D_n \colon \operatorname{Mat}_n(R) \to R.$$

After axiomatizing some desirable properties of D, one proves that these properties characterize D, and one then proves that such a function D exists. Regard an $n \times n$ matrix $A = [a_{ij}]$, not as n^2 entries, but rather as the list $\alpha_1, \ldots, \alpha_n$ of its rows, where $\alpha_i = (a_{i1}, \ldots, a_{in}) \in R^n$. Thus, $D \colon R^n \times \cdots \times R^n \to R$, where there are n factors R^n. Given any function of n variables, we can construct n functions of one variable by fixing each of the other $n - 1$ variables. In more detail, given $\alpha_1, \ldots, \alpha_n$, there are functions $d_i \colon R^n \to R$, one for each i, defined by

$$d_i(\beta) = D(\alpha_1, \ldots, \alpha_{i-1}, \beta, \alpha_{i+1}, \ldots, \alpha_n).$$

Of course, the notation d_i is too abbreviated; it depends on D and on the list of the other rows $\alpha_1, \ldots, \widehat{\alpha_i}, \ldots, \alpha_n$.

→ **Definition.** Let R be a commutative ring, and view an $n \times n$ matrix A with entries in R as the list of its rows: $A = (\alpha_1, \ldots, \alpha_n)$. An $n \times n$ *determinant function* is a function $D \colon \operatorname{Mat}_n(R) \to R$ with the following properties:

(i) D is *alternating*: $D(A) = 0$ if two rows of A are equal; that is, if $\alpha_i = \alpha_j$ for $i \neq j$, then $D(\alpha_1, \ldots, \alpha_n) = 0$;

(ii) D is *multilinear*: for each list $\alpha_1, \ldots, \alpha_n$, the functions $d_i \colon R^n \to R$, given by $d_i(\beta) = D(\alpha_1, \ldots, \alpha_{i-1}, \beta, \alpha_{i+1}, \ldots, \alpha_n)$, satisfy

$$d_i(\beta + \gamma) = d_i(\beta) + d_i(\gamma) \quad \text{and} \quad d_i(c\beta) = c d_i(\beta)$$

for all $c \in R$ and all $\beta, \gamma \in R^n$;

(iii) $D(e_1, \ldots, e_n) = 1$, where $e_1, \ldots, e_n$ is the standard basis; that is, if I is the identity matrix, then $D(I) = 1$.

One can prove, for any determinant function D, that

$$D(AB) = D(A)D(B) \tag{1}$$

and

$$D(A^T) = D(A), \tag{2}$$

where A^T is the transpose of A. Moreover, it can be shown that $D(A)$ must equal the complete expansion, and so D is unique, if it exists. More precisely, for each $n \geq 1$, there is at most one determinant function $D\colon \mathrm{Mat}_n(R) \to R$.

Trying to prove that the function $D\colon \mathrm{Mat}_n(R) \to R$, defined by the complete expansion, is a determinant function looks hopeless. Instead, one proves existence of a determinant function by induction on $n \geq 1$. If $A = [a_{11}]$ is a 1×1 matrix, define $\det(A) = a_{11}$. For the inductive step, assume that there exists a (necessarily unique) determinant function det defined on all $(n-1) \times (n-1)$ matrices over R. Define, for any fixed i,

$$D_i^n(A) = \sum_j (-1)^{i+j} a_{ij} \det(A_{ij}), \tag{3}$$

where A_{ij} denotes the $(n-1) \times (n-1)$ matrix obtained from A by deleting its ith row and jth column.

→ **Definition.**   Formula (3) is called the ***Laplace expansion*** of $\det(A)$ across the ith row.

The proof that D_i^n is a determinant function for each i is rather long (see Curtis, *Linear Algebra*, for example, which considers the special case when R is a field. For a completely different proof for arbitrary commutative rings, which uses *exterior algebra*, see my book, *Advanced Modern Algebra*). Since, for each i, Laplace expansion across the ith row is a determinant function, uniqueness implies that the determinant can be computed using Laplace expansion across any row. Moreover, Eq. (2), which we can now write as $\det(A^T) = \det(A)$, implies that $\det(A)$ can be computed by Laplace expansion down any column (for transposing interchanges rows and columns). One of the key virtues of Laplace expansion is that it is amenable to inductive proofs. For example, induction is the easiest way to solve Exercise 4.51 on page 397: if A is a triangular matrix, then $\det(A)$ is the product of its diagonal entries.

When k is a field, there is an efficient way of computing $\det(A)$, using elementary row operations $A \to A'$ that change a matrix A into a matrix A':

Type I add a scalar multiple of one row of A to another row;

Type II multiply one row of A by a nonzero $c \in k$;

Type III interchange two rows of A.

If $A \to A'$ is an elementary row operation, then $\det(A') = r \det(A)$ for some $r \in k$: if the operation is of type II, then the multilinearity in the definition of a determinant function shows that $r = c$; Exercise 4.47 shows that $r = 1$ for an elementary operation of type I; Exercise 4.49 shows that $r = -1$ for an elementary operation of type III. When k is a field, Exercise 4.26 on page 353 says that one can put A into triangular form by Gaussian elimination: there is a sequence of elementary row operations

$$A \to A_1 \to A_2 \to \cdots \to A_q = \Delta,$$

where Δ is triangular. Therefore, we can compute $\det(A)$ in terms of $\det(\Delta)$ and this sequence of operations, for Exercise 4.51 on page 397 shows that $\det(\Delta)$ is the product of its diagonal entries.

Let us return to matrices with entries in an arbitrary commutative ring R. We now modify the definition of an elementary operation of Type II so that it multiplies one row of A by a *unit* $c \in R$.

$\to$ **Definition.**   An $n \times n$ matrix A over a commutative ring R is **invertible** if there exists a matrix B (with entries in R) with $AB = I = BA$.

When R is a field, invertible is called *nonsingular*; Corollary 4.88 shows that a matrix over a commutative ring R is invertible if and only if its determinant is a unit in R.

$\to$ **Definition.**   Let $A = [a_{ij}]$ be an $n \times n$ matrix with entries in a commutative ring R. Then the **adjoint**[8] of A is the matrix

$$\mathrm{adj}(A) = [c_{ij}],$$

where

$$c_{ji} = (-1)^{i+j} \det(A_{ij})$$

and A_{ij} denotes the $(n-1) \times (n-1)$ matrix obtained from A by deleting its ith row and jth column.

The reversing of indices is deliberate. In words, $\mathrm{adj}(A)$ is the transpose of the matrix whose ij entry is $(-1)^{i+j} \det(A_{ij})$. We often call c_{ij} the ij-**cofactor** of A.

For example, if $A = \begin{bmatrix} a & b \\ c & d \end{bmatrix}$, then

$$\mathrm{adj}(A) = \begin{bmatrix} d & -c \\ -b & a \end{bmatrix}^T = \begin{bmatrix} d & -b \\ -c & a \end{bmatrix}.$$

---

[8]There is another notion of *adjoint*, unrelated to this notion, on page 369.

→ **Proposition 4.87.** *If A is an n × n matrix with entries in a commutative ring R, then*

$$A \operatorname{adj}(A) = \det(A)I = \operatorname{adj}(A)A.$$

Proof. If $A = [a_{ij}]$, denote a_{ij} by $(A)_{ij}$ in this proof. Thus, if $C = [c_{ij}]$, then $(AC)_{ij} = \sum_k a_{ik}c_{kj}$. If we now define $C = [c_{ij}]$ by $c_{ij} = (-1)^{i+j} \det(A_{ji})$ [so that $C = \operatorname{adj}(A)$], then Laplace expansion across the ith row of A gives

$$(AC)_{ii} = \sum_k (-1)^{i+k} a_{ik} \det(A_{ik}) = \det(A).$$

We pause a moment before computing $(AC)_{ij}$ for $j \neq i$. Define $M = [m_{pq}]$ to be the matrix obtained from A by replacing its jth row $(a_{j1}, \dots, a_{jn})$ by its ith row $(a_{i1}, \dots, a_{in})$; thus, $m_{jk} = a_{ik}$ for all k. Note that $M_{jk} = A_{jk}$ for all k (because M and A differ only in the jth row, which is deleted to obtain the smaller matrices M_{jk} and A_{jk}).

When $j \neq i$,

$$(AC)_{ij} = \sum_k a_{ik}(-1)^{i+k} \det(A_{jk})$$

$$= \sum_k (-1)^{i+k} m_{jk} \det(M_{jk})$$

$$= \det(M),$$

because $a_{ik} = m_{jk}$ and $A_{jk} = M_{jk}$. But $\det(M) = 0$, because two of its rows are equal. Therefore, $A \operatorname{adj}(A) = AC$ is the scalar matrix have diagonal entries all equal to $\det(A)$. ●

→ **Corollary 4.88.** *If A is an n × n matrix with entries in a commutative ring R, then A is invertible if and only if* $\det(A)$ *is a unit in R. Moreover,*

$$\det(A^{-1}) = \det(A)^{-1}.$$

Proof. If A is invertible, then there is a matrix B with $AB = I$. By Eq. (1), $1 = \det(I) = \det(AB) = \det(A) \det(B)$, so that $\det(A)$ is a unit in R. Conversely, assume that $\det(A)$ is a unit in R. If $B = \det(A)^{-1}\operatorname{adj}(A)$, then Proposition 4.87 shows that $AB = I = BA$.

Now $I = AA^{-1}$ gives $1 = \det(I) = \det(AA^{-1}) = \det(A) \det(A^{-1})$. Therefore, $\det(A^{-1}) = \det(A)^{-1}$. ●

In the special case when R is a field, we see that A is invertible (i.e., A is nonsingular) if and only if $\det(A) \neq 0$ (a familiar result from a standard linear algebra course). On the other hand, if A is an $n \times n$ matrix with entries in $\mathbb{Z}$, then A is invertible if and only if $\det(A) = \pm 1$; that is, A^{-1} exists and has only integer entries if and only if $\det(A) = \pm 1$. If $R = k[x]$, where k is a field, then A is invertible if and only if its determinant is a nonzero constant.

Corollary 4.89. *Let P and M be n × n matrices with entries in a commutative ring R. If P is invertible, then*

$$\det(PMP^{-1}) = \det(M).$$

Proof. We have $\det(P^{-1}) = \det(P)^{-1}$, by Corollary 4.88. Since determinants lie in the commutative ring R,

$$\det(PMP^{-1}) = \det(P)\det(M)\det(P^{-1})$$
$$= \det(M)\det(P)\det(P^{-1}) = \det(M). \quad \bullet$$

Corollary 4.90. *Let $T: V \to V$ be a linear transformation on a vector space V over a field k, and let X and Y be bases of V. If $A = {}_X[T]_X$ and $B = {}_Y[T]_Y$, then $\det(A) = \det(B)$.*

Proof. By Corollary 4.74, A and B are similar; that is, there is a nonsingular (hence, invertible) matrix P with $B = PAP^{-1}$. $\quad \bullet$

It follows from this corollary that every matrix associated to a linear transformation T has the same determinant, and so we can now define the determinant of a linear transformation.

→ **Definition.** If $T: V \to V$ is a linear transformation on a finite dimensional vector space V, then

$$\det(T) = \det(A),$$

where $A = {}_X[T]_X$ for some basis X of V.

As we have just remarked, this definition does not depend on the choice of X.

Perhaps the simplest linear transformations $T: V \to V$ are the scalar transformations $T = c1_V$; that is, there is a scalar $c \in k$ such that $T(v) = cv$ for all $v \in V$. We now ask, for an arbitrary linear transformation $T: V \to V$, whether there exists $c \in k$ with $T(v) = cv$ for some $v \in V$ (of course, this can only be interesting if $v \neq 0$).

→ **Definition.** Let $T: V \to V$ be a linear transformation, where V is a finite dimensional vector space over a field k. A scalar $c \in k$ is called an ***eigenvalue***[9] of T if there exists a nonzero vector v, called an ***eigenvector***, with

$$T(v) = cv.$$

---

[9]The word *eigenvalue* is a partial translation of the original German word *Eigenwert* (*Wert* means *value*). A translation of *eigen* is *characteristic* or *proper*, and one often sees *characteristic value* used instead of *eigenvalue*. This partial translation applies to other such words as well (e.g., the German *Eigenvektor*) giving our *eigenvector* and *characteristic vector*.

Proposition 4.91. *Let $T: V \to V$ be a linear transformation, where V is a vector space over a field k, and let $c_1, \ldots, c_r$ be distinct eigenvalues of T lying in k. If v_i is an eigenvector of T for c_i, then the list $X = v_1, \ldots, v_r$ is linearly independent.*

Proof. We use induction on $r \geq 1$. The base step $r = 1$ is true, for any nonzero vector is a linearly independent list of length 1, and eigenvectors are, by definition, nonzero. For the inductive step, assume that

$$a_1 v_1 + \cdots + a_{r+1} v_{r+1} = 0.$$

Applying T to this equation gives

$$a_1 c_1 v_1 + \cdots + a_{r+1} c_{r+1} v_{r+1} = 0.$$

Multiply the first equation by c_{r+1}, and then subtract from the second to obtain

$$a_1 (c_1 - c_{r+1}) v_1 + \cdots + a_r (c_r - c_{r+1}) v_r = 0.$$

By the inductive hypothesis, $a_i (c_i - c_{r+1}) = 0$ for all $i \leq r$. Since all the eigenvalues are distinct, $c_i - c_{r+1} \neq 0$, and so $a_i = 0$ for all $i \leq r$. The original equation now reads $a_{r+1} v_{r+1} = 0$, and so $a_{r+1} = 0$, by the base step. Thus, all the coefficients a_i are zero, and $v_1, \ldots, v_{r+1}$ is linearly independent. $\quad \bullet$

Lemma 4.92. *Let $T: V \to V$ be a linear transformation on a vector space V over a field k. Then $c \in k$ is an eigenvalue of T if and only if $c1_V - T$ is singular.*

Proof. If c is an eigenvalue of T, there existswith a nonzero vector $v \in V$ with $T(v) = cv$; hence, $(c1_V - T)(v) = 0$, and $c1_V - T$ is singular. Conversely, if $c1_V - T$ is singular, then Corollary 4.81 provides a nonzero vector v with $(c1_V - T)(v) = 0$. Hence, $T(v) = cv$, and c is an eigenvalue of T. $\quad \bullet$

We have been led to linear transformations of the form $cI - T$ for scalars $c \in k$, and this leads us to consider matrices $xI - A$, where A is a matrix representing T. Since we have been treating matrices with entries in commutative rings, it is legitimate for us to compute the determinant of $xI - A$, a matrix whose entries lie in $k[x]$.

→ **Definition.** If A is an $n \times n$ matrix with entries in a field k, then its **characteristic polynomial**[10] is

$$h_A(x) = \det(xI - A).$$

If $T: V \to V$ is a linear transformation on an n-dimensional vector space V, then the **characteristic polynomial** $h_T(x)$ is defined to be $h_A(x)$, where $A = {}_X[T]_X$ is any matrix representing T.

---

[10]No one calls the *characteristic polynomial* the *eigenpolynomial*.

If R is a commutative ring and A is an $n \times n$ matrix with entries in R, then $\det(A) \in R$. In particular, the entries of $xI - A$ lie in $R = k[x]$, where k is a field, and so $h_A(x) = \det(xI - A) \in k[x]$; that is, the characteristic polynomial really is a polynomial.

The next proposition shows that $h_T(x) = \det(xI - A)$ is well-defined by proving that $\det(xI - A)$ is independent of the choice of matrix A representing T.

Proposition 4.93. *If A and B are similar $n \times n$ matrices with entries in a field k, then they have the same characteristic polynomial*:

$$h_A(x) = \det(xI - A) = \det(xI - B) = h_B(x).$$

Proof. If $B = PAP^{-1}$, then

$$P(xI - A)P^{-1} = PxIP^{-1} - PAP^{-1} = xI - B.$$

Therefore, $\det(P(xI - A)P^{-1}) = \det(xI - B)$. But

$$\det(P(xI - A)P^{-1}) = \det(P)\det(xI - A)\det(P^{-1}) = \det(xI - A). \quad \bullet$$

Corollary 4.94.

(i) *Let A be an $n \times n$ matrix A over a field k. If its characteristic polynomial $h_A(x)$ splits over k, then a scalar $c \in k$ is an eigenvalue of A if and only if c is a root of $h_A(x)$.*

(ii) *Similar matrices have the same eigenvalues with multiplicities.*

Proof.
(i) By Corollary 4.88, $cI - A$ is singular if and only if $h_A(c) = \det(cI - A) = 0$, and so c is an eigenvalue if and only if c is a root of the characteristic polynomial, by Lemma 4.92.
(ii) By Proposition 4.93, both A and B have the same characteristic polynomial. By part (i), A and B have the same eigenvalues occurring with the same multiplicities. $\quad \bullet$

Every eigenvalue of a matrix A is a root of $h_A(x)$, but there may be roots of the characteristic polynomial that do not lie in k. For example, regard $A = \begin{bmatrix} 0 & 1 \\ -1 & 0 \end{bmatrix}$ as a matrix over $\mathbb{R}$; its characteristic polynomial $h_A(x) = x^2 + 1$, and its eigenvalues are $\pm i$. Since these eigenvalues do not lie in $\mathbb{R}$, there is no eigenvector in $\mathbb{R}^2$ for either of them; there do not exist real numbers a and b with $\begin{bmatrix} 0 & 1 \\ -1 & 0 \end{bmatrix} \begin{bmatrix} a \\ b \end{bmatrix} = \begin{bmatrix} ia \\ ib \end{bmatrix}$. However, if we regard A as a complex matrix (whose entries happen to be real), then we can find eigenvectors; for example, $(1, i)^T$ is an eigenvector. Almost everyone extends the definition of *eigenvalue* to include such roots [of course, nothing new occurs if all the roots of $h_A(x)$ lie in k].

→ **Remark.** The following "trick" allows us to dispense with the hypothesis in Corollary 4.94 that the characteristic polynomial splits over k. Let A be an $n \times n$ matrix over a field k, and let $T : k^n \to k^n$ be the linear transformation $T(x) = Ax$, where $x \in k^n$ is a column vector. By Kronecker's theorem (Theorem 3.118), there is an extension field K/k containing all the roots of $h_A(x)$; that is, K contains all the eigenvalues of A. Now $\widetilde{T} : K^n \to K^n$, defined by $\widetilde{T}(\widetilde{x}) = A\widetilde{x}$, is a linear transformation, where $\widetilde{x} \in K^n$ is a column vector. Our original discussion of eigenvalues shows that if $c \in K$ is an eigenvalue of A, then there is an eigenvector $\widetilde{v} \in K^n$ with $\widetilde{T}(\widetilde{v}) = c\widetilde{v}$. ◀

→ **Definition.** If $A = [a_{ij}]$ is an $n \times n$ matrix, then its **trace** is the sum of its diagonal entries:

$$\mathrm{tr}(A) = \sum_{i=1}^{n} a_{ii}.$$

Proposition 4.95. *Let k be a field, and let A be an $n \times n$ matrix with entries in k. Then $h_A(x)$ is a monic polynomial of degree n. Moreover, the coefficient of x^{n-1} in $h_A(x)$ is $-\mathrm{tr}(A)$ and the constant term is $(-1)^n \det(A)$.*

Proof. Let $A = [a_{ij}]$ and let $B = xI - A$; thus, $B = [b_{ij}] = [x\delta_{ij} - a_{ij}]$, where δ_{ij} is the Kronecker delta. The complete expansion is

$$\det(B) = \sum_{\sigma \in S_n} \mathrm{sgn}(\sigma) b_{\sigma(1),1} b_{\sigma(2),2} \cdots b_{\sigma(n),n}.$$

If σ is the identity $(1) \in S_n$, then the corresponding term in the complete expansion of $\det(B) = \det(xI - A)$ is

$$b_{11} \cdots b_{nn} = (x - a_{11})(x - a_{22}) \cdots (x - a_{nn}) = \prod_i (x - a_{ii}),$$

a monic polynomial in $k[x]$ of degree n. If $\sigma \neq (1)$, then the σth term in the complete expansion cannot have exactly $n - 1$ factors from the diagonal of $xI - A$, for if σ fixes $n - 1$ indices, then $\sigma = (1)$. Therefore, the sum of the terms over all $\sigma \neq (1)$ is either 0 or a polynomial in $k[x]$ of degree at most $n - 2$; hence, $\deg(h_A) = n$. By Exercise 3.102 on page 305, the coefficient of x^{n-1} is $-\sum_i a_{ii} = -\mathrm{tr}(A)$, and the constant term of $h_A(x)$ is $h_A(0) = \det(-A) = (-1)^n \det(A)$. •

Corollary 4.96. *If A and B are similar $n \times n$ matrices with entries in a field k, then A and B have the same trace and the same determinant.*

Proof. Now A and B have the same characteristic polynomial, by Proposition 4.93, and so Proposition 4.95 applies to give $\mathrm{tr}(A) = \mathrm{tr}(B)$ and $\det(A) = \det(B)$. •

Let A and B be similar matrices. Another proof that $\mathrm{tr}(A) = \mathrm{tr}(B)$ is described in Exercise 4.56 on page 398, while Corollary 4.89 shows that $\det(A) = \det(B)$.

Corollary 4.97. *If* $T: V \to V$ *is a linear transformation, where* $\dim(V) = n$, *then* T *has at most* n *eigenvalues.*

Proof. If $\dim(V) = n$, then $\deg(h_T) = n$, by Proposition 4.95, and so the result follows from Theorem 3.50. •

The next corollary interprets the trace as the sum (with multiplicities) of the eigenvalues and the determinant as the product (with multiplicities) of the eigenvalues.

Corollary 4.98. *Let* $A = [a_{ij}]$ *be an* $n \times n$ *matrix with entries in a field* k, *and let* $h_A(x) = \prod_{i=1}^{n}(x - \alpha_i)$ *be its characteristic polynomial. Then*

$$\mathrm{tr}(A) = \sum_i \alpha_i \quad and \quad \det(A) = \prod_i \alpha_i.$$

Proof. By Exercise 3.102 on page 305, if $f(x) = \sum_j c_j x^j \in k[x]$ is a monic polynomial with $f(x) = \prod_{i=1}^{n}(x - \alpha_i)$, then $c_{n-1} = -\sum_j \alpha_j$ and $c_0 = (-1)^n \prod_j \alpha_j$. This is true, in particular, for $f(x) = h_A(x)$. The result now follows from Proposition 4.95, which identifies c_{n-1} with $-\mathrm{tr}(A)$ and c_0 with $(-1)^n \det(A)$ (in each case, the sign cancels). •

Example 4.99.
Consider $A = \left[\begin{smallmatrix} 1 & 2 \\ 3 & 4 \end{smallmatrix}\right]$ as a matrix in $\mathrm{Mat}_2(\mathbb{Q})$. Its characteristic polynomial is

$$h_A(x) = \det\left(\begin{bmatrix} x-1 & -2 \\ -3 & x-4 \end{bmatrix}\right) = x^2 - 5x - 2.$$

The eigenvalues $\frac{1}{2}(5 \pm \sqrt{33})$ of A can be found by the quadratic formula. Note that

$$-\mathrm{tr}(A) = -\tfrac{1}{2}(5 + \sqrt{33}) + \tfrac{1}{2}(5 - \sqrt{33}) = 5;$$
$$\det(A) = \tfrac{1}{2}(5 + \sqrt{33})\tfrac{1}{2}(5 - \sqrt{33}) = -2. \quad \blacktriangleleft$$

If scalar matrices are the simplest matrices, then *diagonal matrices* are the next simplest, where an $n \times n$ matrix $D = [d_{ij}]$ is **diagonal** if all its off-diagonal entries $d_{ij} = 0$ for $i \neq j$.

→ **Definition.**   An $n \times n$ matrix A is **diagonalizable** if A is similar to a diagonal matrix.

Of course, every diagonal matrix is diagonalizable.

Proposition 4.100.

  (i) *An $n \times n$ matrix A over a field k is diagonalizable if and only if there is a basis of k^n comprised of eigenvectors of A.*

 (ii) *If A is similar to a diagonal matrix D, then the diagonal entries of D are the eigenvalues of A (with the same multiplicities).*

Proof.
(i) As usual, define a linear transformation $T : k^n \to k^n$ by $T(v) = Av$. If A is similar to a diagonal matrix $D = [d_{ij}]$, then there is a basis $X = v_1, \ldots, v_n$ of k^n with $D = {}_X[T]_X$; that is, $T(v_j) = d_{1j}v_1 + \cdots + d_{nj}v_n$. Since D is diagonal, however, we have $T(v_j) = d_{jj}v_j$, and so the basis X is comprised of eigenvectors. (All the v_i are nonzero, for 0 is never a part of a basis.)

Conversely, let $X = v_1, \ldots, v_n$ be a basis of k^n be comprised of eigenvectors; say, $T(v_j) = c_j v_j$ for all j. The jth column of $B = {}_X[T]_X$ is $[0, \ldots, 0, c_j, 0, \ldots, 0]^T$, for all j, and so B is a diagonal matrix with diagonal entries $c_1, \ldots, c_n$. Finally, A and B are similar, for both represent the linear transformation T relative to different bases.

(ii) If D is a diagonal matrix with diagonal entries d_{ii}, then $\det(xI - D) = \prod_i (x - d_{ii})$, and so the eigenvalues of D are its diagonal entries. Since A and D are similar, Proposition 4.93 shows that they have the same eigenvalues (with the same multiplicities). •

Example 4.101.
Here is an example of an 2×2 matrix which is not diagonalizable. If $A = \begin{bmatrix} 1 & 1 \\ 0 & 1 \end{bmatrix}$, then it has only one eigenvalue, namely, 1 (with multiplicity 2). By Proposition 4.100(ii), if A were similar to a diagonal matrix D, then $D = \begin{bmatrix} 1 & 0 \\ 0 & 1 \end{bmatrix} = I$. This gives $A = PIP^{-1} = I$, a contradiction. ◄

→ **Corollary 4.102.** *Let A be an $n \times n$ matrix over a field k which contains all the eigenvalues of A. If the characteristic polynomial of A has no repeated roots, then A is diagonalizable.*

Remark. Recall Exercise 3.67 on page 274: a polynomial $f(x) \in k[x]$ has no repeated roots if and only if the gcd $(f, f') = 1$, where $f'(x)$ is the derivative of $f(x)$. ◄

Proof. Since $\deg(h_A) = n$, there are n distinct eigenvalues $c_1, \ldots, c_n$. As these eigenvalues all lie in k, there are corresponding eigenvectors $v_1, \ldots, v_n$ in k^n; that is, $Av_i = c_i v_i$. By Proposition 4.91, the list $v_1, \ldots, v_n$ is linearly independent, and hence it is a basis of k^n. The result now follows from Proposition 4.100. •

The converse of Corollary 4.102 is false. For example, the 2×2 identity matrix I is obviously diagonalizable (it is actually diagonal), yet its characteristic polynomial $x^2 - 2x + 1$ has repeated roots.

Example 4.103.

Let $A = \begin{bmatrix} a & b \\ b & c \end{bmatrix}$ be a real (symmetric) matrix. We claim that the eigenvalues of A are real. Now $h_A(x) = x^2 - (a+c)x + (ac - b^2)$. By the quadratic formula, the eigenvalues are

$$x = \tfrac{1}{2}\left[(a + c) \pm \sqrt{(a + c)^2 - 4(ac - b^2)}\right].$$

But $(a + c)^2 - 4(ac - b^2) = (a - c)^2 + 4b^2 \geq 0$, and so its square root is real. Thus, the eigenvalues x are real. ◀

It is not obvious how to generalize the argument in Example 4.103 for $n \geq 3$, but the result is true: the eigenvalues of a symmetric real matrix are real. The result is called the *principal axis theorem* because of an application of it to find normal forms for (higher-dimensional) conic sections.

Recall Example 4.67: if $T : \mathbb{R}^n \to \mathbb{R}^n$ is a linear transformation, then its *adjoint* is the linear transformation $T^* : \mathbb{R}^n \to \mathbb{R}^n$ such that

$$(Tu, v) = (u, T^*v)$$

for all $u, v \in \mathbb{R}^n$, where (u, v) is the usual inner product. This example also shows that if $A = {}_E[T]_E$, then ${}_E[T^*]_E = A^T$. Hence, if A is a symmetric matrix, then $T = T^*$.

Since eigenvalues of a real matrix may be complex, we begin by extending the inner product on $\mathbb{R}^n$ to an inner product on $\mathbb{C}^n$. The usual formula $(u, v) = a_1 c_1 + \cdots + a_n c_n$, where $u = (a_1, \ldots, a_n)$ and $v = (c_1, \ldots, c_n) \in \mathbb{C}^n$, is an inner product, but it is degenerate. For example, the nonzero vector $(1, i)$ in $\mathbb{C}^2$ satisfies $(u, u) = 1^2 + i^2 = 0$.

→ **Definition.**   The *hermitian form* on $\mathbb{C}^n$ is defined by

$$(u, v) = a_1 \bar{c}_1 + \cdots + a_n \bar{c}_n,$$

where $u = (a_1, \ldots, a_n)$ and $v = (c_1, \ldots, c_n) \in \mathbb{C}^n$.

The hermitian form is nondegenerate, for $(u, u) = a_1 \bar{a}_1 + \cdots + a_n \bar{a}_n$ is a sum of squares of real numbers. It follows that $(u, u) = 0$ if and only if $u = 0$. It is easy to check that $(u + u', v) = (u, v) + (u', v)$ and that $(qu, v) = q(u, v)$, where $u, u', v \in \mathbb{C}^n$ and $q \in \mathbb{C}$. However, this is *not* an inner product because $(v, u) = (u, v)$ may not always hold. In fact, $(v, u) = c_1 \bar{a}_1 + \cdots + c_n \bar{a}_n = \overline{(v, u)}$. Hence, $(u, qv) = \overline{(qv, u)} = \overline{q(v, u)} = \bar{q}\overline{(v, u)} = \bar{q}(u, v)$. Therefore, for all $q \in \mathbb{C}$,

$$(u, qv) = \bar{q}(u, v).$$

→ **Theorem 4.104 (Principal Axis Theorem).**   *If A is an $n \times n$ real symmetric matrix, then all its eigenvalues are real, and A is diagonalizable.*

Proof. Let (,) be the hermitian form on $\mathbb{C}^n$. Exercise 4.54 on page 398 shows that if $T: \mathbb{C}^n \to \mathbb{C}^n$ is a linear transformation, then there exists a linear transformation $T^\#: \mathbb{C}^n \to \mathbb{C}^n$ with $(Tu, v) = (u, T^\# v)$, In particular, matrix multiplication $v \mapsto Av$ gives a linear transformation $T: \mathbb{C}^n \to \mathbb{C}^n$. Exercise 4.54 also shows that if A is a real symmetric matrix, then $T^\# = T$; that is, for all $u, v \in \mathbb{C}^n$,

$$(Tu, v) = (u, Tv).$$

We can now generalize Example 4.103 by proving that all the eigenvalues of T (and of A) are real. If $c \in \mathbb{C}$ is an eigenvalue of T (which exists because $\mathbb{C}$ is algebraically closed), then there is a nonzero $v \in \mathbb{C}^n$ with $T(v) = cv$. We evaluate (Tv, v) in two ways. On the one hand, $(Tv, v) = (cv, v) = c(v, v)$. On the other hand, since $T = T^\#$, we have $(Tv, v) = (v, Tv) = (v, cv) = \overline{c}(v, v)$. Now $(v, v) \neq 0$ because $v \neq 0$, so that $c = \overline{c}$; that is, c is real.

We now prove, by induction on $n \geq 1$, that A is diagonalizable. Since the base step $n = 1$ is obvious, we proceed to the inductive step. Choose an eigenvalue c of A; since c is real, there is an eigenvector $v \in \mathbb{R}^n$ with $Av = cv$. Note that $\mathbb{R}^n = \langle v \rangle \oplus \langle v \rangle^\perp$, by Exercise 4.21 on page 343, so that $\dim(\langle v \rangle^\perp) = n - 1$. We claim that $T(\langle v \rangle^\perp) \subseteq \langle v \rangle^\perp$. If $w \in \langle v \rangle^\perp$, then $(w, v) = 0$; we must show that $(Tw, v) = 0$. Now $(Tw, v) = (w, T^\# v) = (w, Tv)$, since A is symmetric. But $(w, Tv) = (w, cv) = \overline{c}(w, v) = 0$, so that $T(w) \in \langle v \rangle^\perp$, as desired. If T' is the restriction of T to $\langle v \rangle^\perp$, then $T': \langle v \rangle^\perp \to \langle v \rangle^\perp$. Since $(Tu, w) = (u, Tw)$ for all $u, w \in \mathbb{R}^n$, we have the equation $(T'u, w) = (u, T'w)$, in particular, for all $u, w \in \langle v \rangle^\perp$. Therefore, the inductive hypothesis applies, and T' is diagonalizable. Proposition 4.100 says that there is a basis $v_2, \ldots, v_n$ of $\langle v \rangle^\perp$ consisting of eigenvectors of T', hence of T. But $v, v_2, \ldots, v_n$ is a basis of $\mathbb{R}^n = \langle v \rangle \oplus \langle v \rangle^\perp$, and so A is diagonalizable (using Proposition 4.100 again). •

To understand why some matrices are not diagonalizable, it is best to consider the more general question when two arbitrary $n \times n$ matrices are similar. Given a matrix A over a field k, the basic idea to find a "simplest" matrix C similar to A; such a matrix C is called a *canonical form* for A. One's first candidate for a canonical form is a diagonal matrix, but Example 4.101 says this is not adequate. It turns out that every matrix has two useful canonical forms: its **rational canonical form** and its **Jordan canonical form**. Each of these canonical forms is tailored for particular types of application. For example, the entries of the rational canonical form always lie in the field k (whereas, all the eigenvalues of A occur as entries of the Jordan canonical form), and this is needed in proving the following result.

→ **Theorem.** *Let A and B be $n \times n$ matrices over a field k and let K/k be a field extension. If A and B are similar over K, then they are similar over k. That is, if there is a nonsingular matrix P over K with $PAP^{-1} = B$, then there is a nonsingular matrix Q over k with $QAQ^{-1} = B$.*

For example, two real matrices that are similar over $\mathbb{C}$ must be similar over the reals. Both forms are used to prove the theorem that every $n \times n$ matrix is similar to its transpose.

It turns out that powers of rational canonical forms are complicated, while powers of Jordan forms are easily computed. Obviously, this property of Jordan forms is useful in finding the order of a nonsingular matrix in the general linear group. It is also used to prove the Cayley-Hamilton theorem.

→ **Theorem (Cayley-Hamilton).**   *Let A be an $n \times n$ matrix with characteristic polynomial $h_A(x) = c_0 + c_1 x + c_2 x^2 + \cdots + x^n$. Then*

$$c_0 I + c_1 A + c_2 A^2 + \cdots + A^n = 0.$$

There are proofs of the Cayley-Hamilton theorem without using canonical forms (see, for example, Birkhoff-Mac Lane).

EXERCISES

H **4.46** True or false with reasons.

(i)   If a matrix A is similar to a symmetric matrix, then A is symmetric.

(ii) $\begin{bmatrix} 2 & 1 \\ 0 & 1 \end{bmatrix}$ is invertible over $\mathbb{Q}$.

(iii) $\begin{bmatrix} 2 & 1 \\ 0 & 1 \end{bmatrix}$ is invertible over $\mathbb{Z}$.

(iv)   If A is a 2×2 matrix over $\mathbb{R}$ all of whose entries are positive, then $\det(A)$ is positive.

(v)   If A is a 2×2 matrix over $\mathbb{R}$ all of whose entries are positive, then $\det(A) \geq 0$.

(vi)   If A and B are $n \times n$ matrices, then $\mathrm{tr}(A + B) = \mathrm{tr}(A) + \mathrm{tr}(B)$.

(vii)   If two $n \times n$ matrices over a field k have the same characteristic polynomial, then they are similar.

(viii)   If $A = \begin{bmatrix} 1 & 2 \\ 3 & 4 \end{bmatrix}$, then $A^2 - 5A - 2I = 0$.

(ix)   Every $n \times n$ matrix over $\mathbb{R}$ has a real eigenvalue.

(x) $\begin{bmatrix} 2 & 1 & 7 \\ 0 & 1 & 8 \\ 0 & 0 & 0 \end{bmatrix}$ is diagonalizable.

*4.47 Let R be a commutative ring, let $D: \mathrm{Mat}_n(R) \to R$ be a determinant function, and let A be an $n \times n$ matrix with rows $\alpha_1, \ldots, \alpha_n$. Define $d_i : R^n \to R$ by $d_i(\beta) = D(\alpha_i, \ldots, \alpha_{i-1}, \beta, \alpha_{i+1}, \ldots, \alpha_n)$.

(i)   If $i \neq j$ and $r \in R$, prove that

$$d_i(r\alpha_j) = 0.$$

(ii)   If $i \neq j$ and $r \in R$, prove that $d_i(\alpha_i + r\alpha_j) = D(A)$.

(iii)   If $r_j \in R$, prove that

$$d_i\left(\alpha_i + \sum_{j \neq i} r_j \alpha_j\right) = D(A).$$

4.48 If O is an orthogonal matrix, prove that $\det(O) = \pm 1$.

*H **4.49** If A' is obtained from an $n \times n$ matrix by interchanging two of its rows, prove that $\det(A') = -\det(A)$.

4.50 If A is an $n \times n$ matrix over a commutative ring R and if $r \in R$, prove that $\det(rA) = r^n \det(A)$. In particular, $\det(-A) = (-1)^n \det(A)$.

***4.51** If $A = [a_{ij}]$ is an $n \times n$ triangular matrix, prove that

$$\det(A) = a_{11}a_{22}\cdots a_{nn}.$$

***4.52** If $u_1, \ldots, u_n$ is a list in a field k, then the corresponding **Vandermonde matrix** is the $n \times n$ matrix

$$V = \text{Van}(u_1, \ldots, u_n) = \begin{bmatrix} 1 & u_1 & u_1^2 & u_1^3 & \cdots & u_1^{n-1} \\ 1 & u_2 & u_2^2 & u_2^3 & \cdots & u_2^{n-1} \\ \vdots & \vdots & \cdots & \cdots & \vdots & \vdots \\ 1 & u_n & u_n^2 & u_n^3 & \cdots & u_n^{n-1} \end{bmatrix}.$$

(i) Prove that

$$\det(V) = \prod_{i<j}(u_j - u_i).$$

Conclude that V is nonsingular if all the u_i are distinct.

H **(ii)** If ω is a primitive nth root of unity ($\omega^n = 1$ and $\omega^i \neq 1$ for $i < n$), prove that $\text{Van}(1, \omega, \omega^2, \ldots, \omega^{n-1})$ is nonsingular and that

$$\text{Van}(1, \omega, \omega^2, \ldots, \omega^{n-1})^{-1} = \tfrac{1}{n}\text{Van}(1, \omega^{-1}, \omega^{-2}, \ldots, \omega^{-n+1}).$$

(iii) Let $f(x) = a_0 + a_1 x + a_2 x^2 + \cdots + a_n x^n \in k[x]$, and let $y_i = f(u_i)$. Prove that the coefficient vector $a = (a_0, \ldots, a_n)$ is a solution of the linear system

$$Vx = y, \qquad (4)$$

where $y = (y_0, \ldots, y_n)$. Conclude that if all the u_i are distinct, then $f(x)$ is determined by Eq. (4).

4.53 Define a **tridiagonal matrix** to be an $n \times n$ matrix of the form

$$T[x_1, \ldots, x_n] = \begin{bmatrix} x_1 & 1 & 0 & 0 & \cdots & 0 & 0 & 0 & 0 \\ -1 & x_2 & 1 & 0 & \cdots & 0 & 0 & 0 & 0 \\ 0 & -1 & x_3 & 1 & \cdots & 0 & 0 & 0 & 0 \\ 0 & 0 & -1 & x_4 & \cdots & 0 & 0 & 0 & 0 \\ & & & \vdots & \ddots & & \vdots & & \\ 0 & 0 & 0 & 0 & \cdots & x_{n-3} & 1 & 0 & 0 \\ 0 & 0 & 0 & 0 & \cdots & -1 & x_{n-2} & 1 & 0 \\ 0 & 0 & 0 & 0 & \cdots & 0 & -1 & x_{n-1} & 1 \\ 0 & 0 & 0 & 0 & \cdots & 0 & 0 & -1 & x_n \end{bmatrix}.$$

(i) If $D_n = \det(T[x_1, \ldots, x_n])$, prove that $D_1 = x_1$, $D_2 = x_1 x_2 + 1$, and, for all $n > 2$,

$$D_n = x_n D_{n-1} + D_{n-2}.$$

(ii)   Prove that if all $x_i = 1$, then $D_n = F_{n+1}$, the nth Fibonacci number. (Recall that $F_0 = 0$, $F_1 = 1$, and $F_n = F_{n-1} + F_{n-2}$ for all $n \geq 2$.)

*4.54 Let $T: \mathbb{C}^n \to \mathbb{C}^n$ be a linear transformation, and let $A = [a_{ij}]$ be its matrix relative to the standard basis $e_1, \ldots, e_n$.

H **(i)**   If $(\ ,\ )$ is the hermitian form on $\mathbb{C}^n$, prove that there exists a linear transformation $T^\#: \mathbb{C}^n \to \mathbb{C}^n$ with
$$(Tu, v) = (u, T^\# v)$$
for all $u, v \in \mathbb{C}^n$.

(ii)   Prove that the matrix of $T^\#$ relative to the standard basis is $A^\# = [\bar{a}_{ji}]$. (Thus, $A^\#$ is obtained from A^T by conjugating every entry. An $n \times n$ complex matrix A is called *hermitian* if $A^\# = A$. Of course, every real symmetric matrix is hermitian.)

(iii)   A real $n \times n$ matrix A defines a linear transformation $T: \mathbb{C}^n \to \mathbb{C}^n$ by matrix multiplication: $T(v) = Av$. Prove that if A is symmetric, then $T^\# = T$.

H **(iv)**   Prove that every hermitian matrix A over $\mathbb{C}$ is diagonalizable.

*4.55 If A is an $m \times n$ matrix over a field k, prove that rank$(A) \geq d$ if and only if A has a nonsingular $d \times d$ submatrix. Conclude that rank(A) is the maximum such d.

*4.56   **(i)**   If A and B are $n \times n$ matrices with entries in a commutative ring R, prove that tr$(AB) = $ tr(BA).

(ii)   Using part (i) of this exercise, give another proof of Corollary 4.96: if A and B are similar matrices with entries in a field k, then tr$(A) = $ tr(B).

4.57 If A is an $n \times n$ matrix over a field k, where $n \geq 2$, prove that det$(\text{adj}(A)) = \det(A)^{n-1}$.

*4.58 If $g(x) = x + c_0$, then its *companion matrix* $C(g)$ is the 1×1 matrix $[-c_0]$; if $s \geq 2$ and $g(x) = x^s + c_{s-1}x^{s-1} + \cdots + c_1 x + c_0$, then its *companion matrix* $C(g)$ is the $s \times s$ matrix

$$\begin{bmatrix} 0 & 0 & 0 & \cdots & 0 & -c_0 \\ 1 & 0 & 0 & \cdots & 0 & -c_1 \\ 0 & 1 & 0 & \cdots & 0 & -c_2 \\ 0 & 0 & 1 & \cdots & 0 & -c_3 \\ \vdots & \vdots & \vdots & \vdots & \vdots & \vdots \\ 0 & 0 & 0 & \cdots & 1 & -c_{s-1} \end{bmatrix}.$$

If $C = C(g)$ is the companion matrix of $g(x) \in k[x]$, prove that the characteristic polynomial $h_C(x) = \det(xI - C) = g(x)$.

*4.59 Let R be a commutative ring. If A $n \times n$ matrix over R and B is an $m \times m$ matrix over R, then their *direct sum* is defined to be the $(m + n) \times (m + n)$ matrix

$$A \oplus B = \begin{bmatrix} A & 0 \\ 0 & B \end{bmatrix}.$$

If $A_1, \ldots, A_t$ are square matrices over R, prove that

$$\det(A_1 \oplus \cdots \oplus A_t) = \det(A_1) \cdots \det(A_t).$$

*H 4.60 If $A_1, \ldots, A_t$ and $B_1, \ldots, B_t$ are square matrices with A_i similar to B_i for all i, prove that $A_1 \oplus \cdots \oplus A_t$ is similar to $B_1 \oplus \cdots \oplus B_t$. (The direct sum of matrices is defined in Exercise 4.59.)

4.61 Prove that an $n \times n$ matrix A with entries in a field k is singular if and only if 0 is an eigenvalue of A.

H **4.62** Let A be an $n \times n$ matrix over a field k. If c is an eigenvalue of A, prove, for all $m \geq 1$, that c^m is an eigenvalue of A^m.

4.63 Find all possible eigenvalues of $n \times n$ matrices A over $\mathbb{R}$ for which A and A^2 are similar.

4.64 An $n \times n$ matrix N is called **nilpotent** if $N^m = 0$ for some $m \geq 1$. Prove that all the eigenvalues of a nilpotent matrix are 0. Use the Cayley-Hamilton theorem to prove the converse: if all the eigenvalues of a matrix A are 0, then A is nilpotent.

H **4.65** If N is a nilpotent matrix, prove that $I + N$ is nonsingular.

4.5 CODES

When we discussed codes earlier, in Example 1.79, our emphasis was on security: how can we prevent an unauthorized person from reading our messages? We now leave the world of spies in order to consider the accuracy of a received message. Suppose that Pat asks Mike for Ella's phone number, but a dog barks as Mike answers. Because of this noise, Pat isn't sure whether he has heard the number correctly, and he asks Mike to repeat it. Most likely, one or two repetitions will ensure that Pat will have Ella's number. But simply repeating a message several times may not be practical. A better paradigm involves scientists on Earth wanting to see photographs sent from Mars or Saturn. In 2004, robotic cameras sent to these planets encoded each photograph as a bitstring in the following way. A photograph is divided into 1024×1024 pixels p_{ij} (these are the numbers actually used); thus, there are $2^{10} \times 2^{10} = 2^{20} = 1,048,576$ pixels. Each pixel p_{ij} is equipped with a 12-digit bitstring c_{ij} describing its color, intensity, and so on. The $2^{10} \times 2^{10}$ matrix $[c_{ij}]$ is then written as a single bitstring: row 1, row 2, ..., row 1024. In this way, one photograph is converted into a message having roughly 12,000,000 bits. This binary number must be transmitted across space, and space is "noisy" because cosmic rays interfere with electronic signals. Clearly, it is not economical to send such a long message across space several times and, even if it were sent repeatedly, it is most likely that no two of the received messages would be identical. Still, as we saw with Ella's phone number, it is natural to repeat, and redundancy is the key to accurate reception. We seek some practical ways of encoding messages efficiently so that mistakes in a received message can be detected and, even better, corrected. This is what is done to enable us to see reasonably faithful photographs sent from the planets.

Block Codes

The mathematical study of codes began, in the 1940s, with the work of C. E. Shannon, R. W. Hamming, and M. J. E. Golay. Sending a message over a noisy channel involves three steps: *encoding* the message (incorporating some redundancy); *transmitting* the message; *decoding* the received message.

Notation.   We will usually denote the finite field $\mathbb{F}_2$ by $\mathbb{B}$ in this section.

→ **Definition.**   Call a finite set $\mathcal{A}$ an **alphabet**, and call its elements **letters**. If m and n are positive integers, then an **encoding function** is an injective function $E\colon \mathcal{A}^m \to \mathcal{A}^n$. Elements $w \in \mathcal{A}^m$ or in $\mathcal{A}^n$ are called **words**, the set $C = \operatorname{im} E \subseteq \mathcal{A}^n$ is called an $[n, m]$ **block code**[11] **over** $\mathcal{A}$, and the elements of C are called **codewords**. If $\mathcal{A} = \mathbb{B}$, then an $[n, m]$ block code is called a **binary code**.

Since encoding involves redundancy, it is usually the case that $m < n$. The choice of m is not restrictive, for any long message can be subdivided into shorter subwords of lengths $\leq m$. A transmitted message may be a photograph sent from outer space to Earth and, of course, we want to read it. Were there no noise in the transmission, then any codeword $c = E(w)$ could be decoded as $w = E^{-1}(c)$, for encoding functions are injective. Since errors may be introduced, however, the task is to equip a code with sufficient redundancy so that one can efficiently recapture a codeword from its transmitted version.

→ **Example 4.105.**

(i) **Parity Check** $[m + 1, m]$ Code

Define an encoding function $E\colon \mathbb{B}^m \to \mathbb{B}^{m+1}$ by

$$w = (a_1, \ldots, a_m) \in \mathbb{B}^m \mapsto E(w) = (a_1, \ldots, a_m, b),$$

where $b = \sum_{i=1}^m a_i$. It is clear that E is injective, and it is easy to check that the code $C = \operatorname{im} E \subseteq \mathbb{B}^{m+1}$ is given by

$$C = \big\{ (b_1, \ldots, b_{m+1}) \in \mathbb{B}^{m+1} : b_1 + \cdots + b_{m+1} = 0 \big\}.$$

If $w \in \mathbb{B}^m$, then the **parity** of $E(w)$ is *even* in the sense that the sum of its coordinates in $\mathbb{B}$ is 0. If one receives a message of *odd* parity (the coordinate sum is 1), then one knows that there has been a mistake in transmission. Thus, this code can *detect* the presence of single errors; however, a double error in the received message cannot be detected because the parity is unchanged if, for example, two 0s are replaced by two 1s.

(ii) **Triple Repetition** $[12, 4]$ Code

Consider the encoding function $E\colon \mathbb{B}^4 \to \mathbb{B}^{12}$ defined by $E(w) = (w, w, w)$; that is, C consists of all words of the form

$$E(a_1, a_2, a_3, a_4) = (a_1, a_2, a_3, a_4, a_1, a_2, a_3, a_4, a_1, a_2, a_3, a_4).$$

---

[11]This is called a *block code* because all codewords have the same length, namely, n; it is not difficult to modify block codes to allow words of varying length.

Now transmit (a_1, a_2, a_3, a_4), and let the received message be $y = (r_1, \ldots, r_{12})$. Because of interference, it is possible that

$$(r_1, \ldots, r_{12}) \neq E(a_1, a_2, a_3, a_4).$$

Decode the received word $y = (r_1, \ldots, r_{12})$ as follows. If there were no errors in transmission, then $r_1 = r_5 = r_9$. In any event, since there are only two possible values for r_1, r_5, r_9, at least two of these values must be the same; define b_1 to be this popular value. Similarly, define b_2, b_3, and b_4, so that y is decoded as (b_1, b_2, b_3, b_4). (Since messages must be decoded, we do not consider a double repetition [8, 4] code, for there is no natural candidate for decoding $(r_1, \ldots, r_8)$ if, say, $r_1 \neq r_5$.) Notice that this coding scheme can detect errors: if r_i, r_{i+4}, r_{i+8} are not all equal, then there has been an error (alas, a really bad error may not be detected). Indeed, this code can *correct* an error in the sense that a received word y with one mistake can be replaced by a codeword all but one of whose letters coincide with the letters of y.

(iii) *Two-Dimensional Parity* [9, 4] *Code*

Let us write a word $y = (r_1, \ldots, r_9) \in \mathbb{B}^9$ as a 3×3 matrix

$$\begin{bmatrix} r_1 & r_2 & r_3 \\ r_4 & r_5 & r_6 \\ r_7 & r_8 & r_9 \end{bmatrix}.$$

Consider the encoding function $E : \mathbb{B}^4 \to \mathbb{B}^9$ defined by

$$E(a, b, c, d) = \begin{bmatrix} a & b & r_3 \\ c & d & r_6 \\ r_7 & r_8 & r_9 \end{bmatrix} = \begin{bmatrix} a & b & a+b \\ c & d & c+d \\ a+c & b+d & a+b+c+d \end{bmatrix}.$$

Thus, r_3 and r_6 are parity checks for the first two rows, r_7 and r_8 are parity checks for the first two columns, and r_9 is the parity check for r_7 and r_8 as well as for r_3 and r_6. We have constructed a [9, 4] code $C = \operatorname{im} E$ consisting of all 3×3 matrices over $\mathbb{B}$ whose rows and columns have even parity. Suppose that a received matrix is

$$y = \begin{bmatrix} 1 & 0 & 1 \\ 1 & 1 & 1 \\ 1 & 1 & 0 \end{bmatrix}.$$

We see that the second row has an error, as does the first column, for their entries do not sum to 0 in $\mathbb{B}$. Hence, an error in the 2,1 entry has been detected, and it can be corrected. We now show that 2 errors can be detected. For example, suppose that a received matrix is

$$y' = \begin{bmatrix} 1 & 0 & 1 \\ 1 & 0 & 1 \\ 1 & 1 & 0 \end{bmatrix}.$$

The parity checks on the rows are correct, but the parity checks detect errors in the first two columns.

Comparing the triple repetition $[12, 4]$ code with this one, we see that the present code is more efficient in that a word of length 4 is encoded into a word of length 9 instead of a longer word of length 12. ◀

We are going to measure the *distance* between words in $\mathcal{A}^n$.

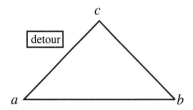

Figure 4.8   The Triangle Inequality

Definition.   If X is a set, then a *metric* on X is a function $\delta : X \times X \to \mathbb{R}$ such that

(i) $\delta(a, b) \geq 0$ for all $a, b \in X$, and $\delta(a, b) = 0$ if and only if $a = b$;

(ii) $\delta(a, b) = \delta(b, a)$ for all $a, b \in X$;

(iii) *Triangle inequality*:
$\delta(a, b) \leq \delta(a, c) + \delta(c, b)$ for all $a, b, c \in X$.

A metric has the essential properties that any self-respecting notion of distance should have. Distances are nonnegative (two points should not be -5 units apart), and there should be a positive distance between distinct points. The distance from Urbana to Chicago should be the same as the distance from Chicago to Urbana. The triangle inequality says that a "straight line" is the shortest path between two points.

Example 4.106.

(i) If $X = \mathbb{R}$, then $\delta(x, y) = |x - y|$ is a metric.

This is the reason why absolute value is introduced in calculus. In this case, a point z is between x and y if and only if $\delta(x, y) = \delta(x, z) + \delta(z, y)$.

(ii) *Euclidean metric*

If $X = \mathbb{R}^n$ and $x = (x_1, \ldots, x_n)$, then $\delta(x, y) = \sqrt{\sum_{i=1}^{n}(x_i - y_i)^2}$ is a metric. Note that $\sqrt{x^2} = |x|$, so that this definition agrees with the metric in part (i) when $n = 1$.

(iii) L^2 *metric*

Let $L^2[a, b]$ be the set of all *square-integrable functions* on $[a, b]$; that is, $L^2[a, b] = \{f : [a, b] \to \mathbb{R} : \int_a^b f^2(x)\, dx < \infty\}$. Then

$$\delta(f, g) = \sqrt{\int_a^b \left(f(x) - g(x)\right)^2 dx}$$

is a metric on $L^2[a, b]$.

(iv) *p-adic metric*

If p is a prime and $n \in \mathbb{Z}$ is nonzero, then $n = p^k u$, where $k \geq 0$ and $p \nmid u$; that is, p^k is the largest power of p dividing n. Write $k = k(n)$. If we define $\delta(n, n) = 0$ and $\delta(n, m) = p^{-k(n-m)}$ if $m \neq n$, then δ is a metric on $\mathbb{Z}$. ◄

→ **Definition.** Let $\mathcal{A}$ be an alphabet, and let $w = (a_1, \ldots, a_n)$, $w' = (a_1', \ldots, a_n') \in \mathcal{A}^n$. The function $\delta : \mathcal{A}^n \times \mathcal{A}^n \to \mathbb{R}$, defined by

$$\delta(w, w') = \text{the number of } i \text{ with } a_i \neq a_i',$$

is called the ***Hamming distance***.[12]

→ **Proposition 4.107.** *If $\mathcal{A}$ is an alphabet and $n \geq 1$, then the Hamming distance is a metric on $\mathcal{A}^n$.*

Proof. Let $w = (a_1, \ldots, a_n)$, $w' = (a_1', \ldots, a_n') \in \mathcal{A}^n$. Clearly, $\delta(w, w') \geq 0$ and $\delta(w, w) = 0$. On the other hand, if $\delta(w, w') = 0$, then $a_i = a_i'$ for all i, and $w = w'$. It is obvious that $\delta(w, w') = \delta(w', w)$, and it remains only to prove the triangle inequality.

If we define $\delta_i(w, w') = 1$ if $a_i \neq a_i'$ and $\delta_i(w, w') = 0$ if $a_i = a_i'$, then

$$\delta(w, w') = \sum_{i=1}^n \delta_i(w, w').$$

It suffices to prove $\delta_i(w, w') \leq \delta_i(w, z) + \delta_i(z, w')$ for each i, where $z = (b_1, \ldots, b_n)$. If $\delta_i(w, w') = 0$, then this inequality holds; otherwise, $\delta_i(w, w') = 1$. Now $\delta_i(w, z) + \delta_i(z, w')$ is either 0, 1, or 2, so that it suffices to prove $\delta_i(w, z) + \delta_i(z, w') \neq 0$. If this sum is 0, then $\delta_i(w, z) = 0 = \delta_i(z, w')$; that is, $a_i = b_i$ and $b_i = a_i'$. But if $\delta_i(w, w') = 1$, then $a_i \neq a_i'$, a contradiction. •

---

[12]After R. W. Hamming.

→ **Definition.**   If $\mathcal{A}$ is an alphabet and $C \subseteq \mathcal{A}^n$ is a code, then its ***minimum distance*** is

$$d = d(C) = \min_{w, w' \in C, \; w \neq w'} \delta(w, w'),$$

where δ is the Hamming distance.

The minimum distance is important enough that it is usually incorporated in the parameters used to describe codes.

Notation.   An (n, M, d)-***code over*** $\mathcal{A}$ is a code $C \subseteq \mathcal{A}^n$, where $\mathcal{A}$ is an alphabet, $M = |C|$, and d is its minimum distance.

Note that if $\mathcal{A} = q$, then $M = q^m$, for the encoding function $E : \mathcal{A}^m \to \mathcal{A}^n$ is an injection. Thus, an (n, M, d)-code is an $[n, \log_q M]$ code in our earlier notation.

We can now give precise definitions of *error detecting* and *error correcting*.

→ **Definition.**   Let $\mathcal{A}$ be an alphabet, and let $C \subseteq \mathcal{A}^n$ be a code. The code C can ***detect up to*** $s > 0$ ***errors*** if changing a codeword $c \in C$ in at most s places does not give a codeword.

For example, the parity check $[m + 1, m]$ code can detect up to 1 error because changing one coordinate of a codeword converts an even word into an odd word.

A code C detects up to s errors if, for each $c \in C$ and $y \in \mathcal{A}^n$, we have $0 < \delta(y, c) \leq s$ implies $y \notin C$.

→ **Definition.**   Let $\mathcal{A}$ be an alphabet, and let $C \subseteq \mathcal{A}^n$ be a code. If $y \in \mathcal{A}^n$, then $c \in C$ is a ***nearest codeword*** to y if $\delta(y, c) \leq \delta(y, c')$ for all $c' \in C$.

Given y, there may not exist a unique nearest codeword to y (Exercise 4.73 on page 431 asks for an example of distinct codewords equidistant from a word). When a received word y does have a unique nearest codeword c, then we shall decode y as $E^{-1}(c)$. We are not saying that $c = E(w)$, but it is the best (and most natural) candidate for the truth.

→ **Definition.**   Let $\mathcal{A}$ be an alphabet, and let $C \subseteq \mathcal{A}^n$ be a code. The code C can ***correct up to*** t ***errors*** if changing a codeword c in at most t places gives a word $y \in \mathcal{A}^n$ whose unique nearest codeword is c.

Thus, a code C corrects up to t errors if, given a codeword c and a word y with $\delta(y, c) \leq t$, then $\delta(y, c) < \delta(y, c')$ for every codeword $c' \neq c$.

→ **Proposition 4.108.** *Let A be an alphabet, and let $C \subseteq A^n$ be an (n, M, d)-code.*

(i) *If $d \geq 2t + 1$ and if $y \in A^n$, then a codeword $c \in C$ with $\delta(y, c) \leq t$, should one exist, is unique.*

(ii) *If $d \geq s + 1$, then C can detect up to s errors.*

(iii) *If $d \geq 2t + 1$, then C can correct up to t errors.*

Proof.

(i) Suppose that c, c' are codewords with $\delta(y, c) = \delta(y, c') \leq t$. The triangle inequality gives $\delta(c, c') \leq \delta(c, y) + \delta(y, c') \leq 2t$. But the minimum distance between distinct codewords is $d(C) \geq 2t + 1$; therefore, $c = c'$.

(ii) If $w \neq c$ differs from c in at most s places, then $0 < \delta(c, w) \leq s$. But if $w \in C$, then

$$s \geq \delta(c, w) \geq d > s,$$

a contradiction.

(iii) If w is obtained from c by changing at most t places, then $\delta(c, w) \leq t$. If there is a codeword c' with $\delta(c', w) < \delta(c, w)$, then the triangle inequality gives

$$\begin{aligned}
2t + 1 &\leq d \\
&\leq \delta(c, c') \\
&\leq \delta(c, w) + \delta(w, c') \\
&\leq 2\delta(c, w) \leq 2t.
\end{aligned}$$

This contradiction shows that C corrects up to t errors. •

→ **Example 4.109.**

(i) The parity check $[m + 1, m]$ code of Example 4.105(i), consisting of all words in $\mathbb{B}^{m+1}$ having an even number of 1s, is an $(m + 1, 2^m, 2)$-code. The minimum distance is at least 2, for changing one place in a word with an even number of 1s yields a word with an odd number of 1s. By Proposition 4.108(ii), this code detects 1 error. However, we see no error correcting.

(ii) The triple repetition $[3m, m]$ code of Example 4.105(ii), consisting of all words in $\mathbb{B}^{3m}$ of the form (w, w, w), where $w \in \mathbb{B}^m$, is a $(3m, 2^m, 3)$-code, for one must change a codeword in at least 3 places to obtain another codeword. By Proposition 4.108(iii), this code can detect 2 errors; this code can also correct 1 error.

(iii) The two-dimensional parity $[9, 4]$ code C in Example 4.105(iii) consists of all 3×3 matrices over $\mathbb{B}$ such that the sum of the entries in each row is 0 and the sum of the entries in each column are 0. Exercise 4.67 on page 430 asks you to check that at least 3 changes are needed to pass from one codeword (here, a matrix) to another codeword. Thus, $d \geq 4$, and so C detects 2 errors; C can also correct 1 error. ◄

A code with large minimum distance d can correct many errors. For example, a 101-times binary repetition code C is a $[101m, m]$ code with encoding function $E \colon \mathbb{B}^m \to \mathbb{B}^{101m}$ which repeats an m-letter word 101 times. Here, $d = 101$, and so Proposition 4.108(iii) shows that C can correct up to 50 errors. Obviously, C is a rather impractical code. We can measure this impracticality.

Definition.   The **rate of information** of an $[n, m]$ code is m/n. If $|\mathcal{A}| = q$, then we have seen that $M = q^m$, and so the **rate of information** of an (n, M, d)-code is $(\log_q M)/n$.

This notion of rate is a natural one: it says that n letters are used to send an m-letter message. The multiple repetition $[101m, m]$ code just described is inefficient, for it has rate $\frac{1}{101}$: sending a short message requires a very large number of letters. On the other hand, the irredundant $[m, m]$ code with $E = 1_{\mathcal{A}^m} \colon \mathcal{A}^m \to \mathcal{A}^m$, which merely repeats a message verbatim, has rate 1. Thus, small rates may correct many errors, but they are inefficient; large rates (i.e., rates near 1) may not even detect errors. An (n, M, d)-code over $\mathcal{A}$ can correct t errors if $d \geq 2t + 1$; hence, it is more accurate when d is large. Exercise 4.68 on page 430 gives the *Singleton bound*: if $|\mathcal{A}| = q$, then $M = q^m \leq q^{n-d+1}$. Therefore, $m + d \leq n + 1$ and $\frac{m}{n} + \frac{d}{n} \leq 1 + \frac{1}{n}$. If m is large, then the rate m/n is close to 1, but now d is small. On the other hand, if m is small, then the rate is small, but now d may be large; that is, the code may correct many errors. Thus, we seek a compromise. Given d, we seek (n, q^m, d)-codes in which m is "large"–such codes are relatively efficient; given m, we seek (n, q^m, d)-codes in which d is "large"–such codes are relatively accurate.

Linear Codes

If $\mathcal{A}$ is an arbitrary set, then defining a function $E \colon \mathcal{A}^m \to \mathcal{A}^n$ can be quite complicated. On the other hand, if $\mathcal{A}$ is a field, then $\mathcal{A}^m$ and $\mathcal{A}^n$ are vector spaces equipped with the standard bases. Moreover, if E is a linear transformation, then it can be efficiently described by a formula: there is an $m \times n$ matrix G with $E(w) = wG$, where w is an $1 \times m$ row vector.

→ **Definition.**   An $[n, m]$ **linear code** C over a finite field k is an m-dimensional subspace of k^n. An **encoding function** $E \colon k^m \to k^n$ for C is an injective linear transformation $E \colon k^m \to k^n$ with $E(k^m) = C$.

If $k = \mathbb{F}_q$ is the finite field with q elements, then an $[n, m]$ linear code is an (n, q^m, d)-code, where d is its minimum distance. In a linear code, there is another way of finding its minimum distance.

→ **Definition.** If $w = (a_1, \ldots, a_n) \in k^n$, where k is a field, then the ***support*** of w is defined by

$$\text{Supp}(w) = \{\text{indices } i : a_i \neq 0\}.$$

If C is a linear (n, M, d)-code, then the ***Hamming weight*** of w is

$$\text{wt}(w) = |\text{Supp}(w)|;$$

that is, $\text{wt}(w)$ is the number of nonzero coordinates in w. The ***zero set*** of w is the complement of $\text{Supp}(w)$:

$$\mathcal{Z}(w) = \{\text{indices } i : a_i = 0\}.$$

Note that $\text{wt}(w) = \delta(w, 0)$, where δ is the Hamming distance and $0 = (0, \ldots, 0)$ (which lies in C because C is a subspace of k^n).

→ **Proposition 4.110.** *If C is a linear (n, M, d)-code over a finite field $\mathbb{F}_q$, then*

$$d = \min_{c \in C,\, c \neq 0} \{\text{wt}(c) : c \in C\}.$$

Thus, d is the smallest weight among nonzero codewords.

Proof. Since C is a subspace, $w, w' \in C$ implies $w - w' \in C$. Thus,

$$
\begin{aligned}
d &= \min_{w, w' \in C,\, w \neq w'} \delta(w, w') \\
&= \min_{w, w' \in C,\, w \neq w'} \{\text{wt}(w - w') : w, w' \in C\} \\
&= \min_{c \in C,\, c \neq 0} \{\text{wt}(c) : c \in C\}. \quad \bullet
\end{aligned}
$$

We are going to introduce a matrix which describes a given linear code, but we first introduce a ***partition notation*** for a matrix U.

Notation. If A is an $m \times r$ matrix and B is an $m \times s$ matrix, then

$$U = [A | B]$$

is the $m \times (r + s)$ matrix whose first r columns give the matrix A and whose last s columns give the matrix B.

If N is a $q \times m$ matrix and A is an $m \times r$ matrix, then the definition of matrix multiplication says that the ij entry of NA is the dot product $\text{ROW}_N(i) \cdot \text{COL}_A(j)$. Hence, the jth column of NA does not involve any column of A other than $\text{COL}_A(j)$; it follows that if $U = [A | B]$, then

$$N[A | B] = [NA | NB]. \tag{1}$$

Since it is customary in coding theory to consider vectors in k^m as rows instead of as columns (in contrast to earlier sections), we will now view elements $w \in k^n$ as $1 \times n$ row vectors, and we will denote $n \times 1$ column vectors by w^T.

Let $\sigma \in S_n$ be a permutation, and let Q_σ be the $n \times n$ permutation matrix obtained from the identity matrix I by permuting its columns by σ. If k is a field and $c = (c_1, \ldots, c_n) \in k^n$ is a $1 \times n$ row vector, then

$$cQ_\sigma = (c_1, \ldots, c_n)Q_\sigma = (c_{\sigma(1)}, \ldots, c_{\sigma(n)});$$

the jth coordinate of cQ_σ is the dot product of c with the jth column of Q_σ. But if $e_{\sigma(j)}$ is the vector having 1 in the $\sigma(j)$th coordinate and 0s elsewhere, then $c \cdot e_{\sigma(j)} = (c_1, \ldots, c_n) \cdot (0, \ldots, 1, \ldots, 0)$. Thus, $c \cdot e_{\sigma(j)} = c_{\sigma(j)}$. If Q_σ is an $n \times n$ permutation matrix, define $\sigma_* : k^n \to k^n$ by

$$\sigma_*(c) = cQ_\sigma.$$

Definition. Two $[n, m]$ linear codes C, C' over a field k are **permutation equivalent** if there is $\sigma \in S_n$ with $\sigma_*(C) = C'$; that is, $c = (a_1, \ldots, a_n) \in C$ if and only if $(a_{\sigma(1)}, \ldots, a_{\sigma(n)}) \in C'$.

It is easy to see that permutation equivalence is an equivalence relation on the family of all linear codes in k^n. Permutation equivalent codes are essentially the same: for example, if all words in a code C are reversed, then the new, reversed, code has the same parameters as the original code.

$\to$ **Proposition 4.111.** *If C is a linear $[n, m]$ code over a field k, then there are a linear code C' permutation equivalent to C and an $m \times n$ matrix G of the form $G = [I|B]$, where I is the $m \times m$ identity matrix, such that*

$$C' = \{w'G : w' \in k^m\}.$$

Thus, C' is the row space of the matrix G.

Proof. If $e_1, \ldots, e_m$ is the standard basis of k^m and $\gamma_1, \ldots, \gamma_m$ is some basis of C, define a linear transformation $E : k^m \to k^n$ by $E(e_i) = \gamma_i$. Now E is an injection, by Lemma 4.77(iii); in fact, E is an isomorphism $k^m \to C$. By Proposition 4.64, $E(w) = Aw^T$, where $w \in k^m$ is an $1 \times m$ row vector, and A is the $n \times m$ matrix whose columns are the vectors $E(e_i)^T$. Since we now consider elements in k^m as rows instead of as columns, we write $E(w) = wN$, where $N = A^T$ is an $m \times n$ matrix.

By Proposition 4.39, Gaussian elimination converts the matrix N to a matrix G in echelon form. There is a nonsingular $m \times m$ matrix P and an $n \times n$ permutation matrix Q_σ with $G = PNQ_\sigma = [U|B]$, where U is an $m \times m$ matrix in echelon form and B is an $m \times (n - m)$ matrix. Since E is injective, the echelon matrix U has no zero rows, and so it is the identity; thus, $G = [I|B]$. Define

$$C' = \{w'G : w' \in k^m\}.$$

Now $e_1 G, \ldots, e_m G$ is a linearly independent list, and so $m \leq \dim(C')$. We claim that C and C' are permutation equivalent; that is, $C' = \sigma_*(C)$. Let $c' = w'G \in C'$, define $w = w'P$, and define $c = wN$. Note that $c \in C$ because $wN = E(w) \in C$. Then

$$
\begin{aligned}
c' &= w'G \\
&= w'PNQ_\sigma \\
&= wNQ_\sigma \\
&= cQ_\sigma \\
&= \sigma_*(c).
\end{aligned}
$$

Hence, $C' \subseteq \sigma_*(C)$. Therefore, $m \leq \dim(C') \leq \dim(\sigma_*(C)) = \dim(C) = m$. By Corollary 4.25(iii), we have $C' = \sigma_*(C)$, and so C and C' are permutation equivalent codes. •

→ **Definition.** If C is an $[n, m]$ linear code over a field k, then an $m \times n$ matrix G with $C = Row(G) = \{wG : w \in k^n\}$ is called a ***generating matrix*** of C. An ***echelon generating matrix*** of C is a generating matrix of the form $G = [I|B]$, where I is the $m \times m$ identity matrix.

Every linear code C has a generating matrix: for example, any $m \times n$ matrix whose rows form a basis of C is a generating matrix for C. In light of Proposition 4.111, we may assume that every linear code has an echelon generating matrix.

The examples of codes we have given so far are, in fact, binary linear codes; that is, they are linear codes over $k = \mathbb{B} = \mathbb{F}_2$.

→ **Example 4.112.**

(i) Consider the parity check $[m+1, m]$ code C in Example 4.105(i). The codewords are all those $c = (b_1, \ldots, b_{m+1}) \in \mathbb{B}^{m+1}$ for which $\sum b_i = 0$; the codewords form a subspace, and so C is a linear code. Recall that the encoding function $E: \mathbb{B}^m \to \mathbb{B}^{m+1}$ is defined by

$$
E: (a_1, \ldots, a_m) \mapsto (a_1, \ldots, a_m, b),
$$

where $b = \sum_{i=1}^m a_i$. It is easy to see that E is a linear transformation. Moreover, an echelon generating matrix is the $m \times (m+1)$ matrix

$$
G = \begin{bmatrix}
1 & 0 & 0 & \cdots & 0 & 1 \\
0 & 1 & 0 & \cdots & 0 & 1 \\
& \vdots & \vdots & \vdots & \vdots & \\
0 & 0 & 0 & \cdots & 1 & 1
\end{bmatrix}.
$$

In partition notation, $G = [I|B]$, where I is the $m \times m$ identity matrix and B is the column of all 1s.

(ii) Consider the two-dimensional parity [9, 4] code in Example 4.105(iii). By definition,

$$E: (a, b, c, d) \mapsto \begin{bmatrix} a & b & a+b \\ c & d & c+d \\ a+c & b+d & a+b+c+d \end{bmatrix};$$

it is easy to check that E is a linear transformation. We find a generating matrix G by evaluating E on the standard basis of $\mathbb{B}^4$, for the ith row of G is $e_i G$.

$$G = \begin{bmatrix} 1 & 0 & 0 & 0 & 1 & 0 & 1 & 0 & 1 \\ 0 & 1 & 0 & 0 & 1 & 0 & 0 & 1 & 1 \\ 0 & 0 & 1 & 0 & 0 & 1 & 1 & 0 & 1 \\ 0 & 0 & 0 & 1 & 0 & 1 & 0 & 1 & 1 \end{bmatrix}$$

Thus, G is an echelon generating matrix.

(iii) Consider the triple repetition $[3m, m]$ code in Example 4.105(ii). An echelon generating matrix for C is $G = [I|I|I]$, where I is the $m \times m$ identity matrix.

(iv) The examples above are too simple. Given a linear $[n, m]$ code C with encoding function $E: k^m \to k^n$, the matrix G having rows $E(e_1), \dots, E(e_m)$ (where $e_1, \dots, e_m$ is the standard basis of k^m) is a generating matrix for C. In general, Gaussian elimination is needed to convert G into an echelon generating matrix for C (or a linear code C' permutation equivalent to C). ◄

If $G = [I|B]$ is an echelon generating matrix of a linear $[n, m]$ code C, then Eq. (1) on page 407 gives

$$wG = w[I|B] = [w|wB]$$

for all $w \in k^m$. If there were no errors in transmitting C, then it is obvious how to decode a codeword wG: just take its first m coordinates. The last $n - m$ columns of an echelon generating matrix $G = [I|B]$ should be viewed as generalizing the last column of the echelon generating matrix of the $[m + 1, m]$ parity check code in Example 4.112(i); thus, the last columns B in $G = [I|B]$ are a generalized parity check providing redundancy to help decode a message sent over a noisy channel.

In Example 4.112, we started with some linear codes and found an echelon generating matrix for each of them. Now we start with an echelon generating matrix G and use it to construct a code $C = \{wG : w \in k^m\}$.

→ **Example 4.113.**
Consider the 4×7 matrix

$$G = \begin{bmatrix} 1 & 0 & 0 & 0 & 0 & 1 & 1 \\ 0 & 1 & 0 & 0 & 1 & 0 & 1 \\ 0 & 0 & 1 & 0 & 1 & 1 & 0 \\ 0 & 0 & 0 & 1 & 1 & 1 & 1 \end{bmatrix},$$

and define the **Hamming** $[7, 4]$ **code** to be $C = \{wG : w \in \mathbb{B}^4\}$.

Obviously, the rate of information is $r = \frac{4}{7}$. Let us compute the minimum distance d of C. Denote the rows of G by $\gamma_1, \gamma_2, \gamma_3, \gamma_4$. Since $k = \mathbb{B}$ here, linear combinations $\sum_i a_i \gamma_i$ are just sums (for a_i is either 0 or 1). By Proposition 4.110, we can compute d by computing weights of codewords. Now

$$\mathrm{wt}(\gamma_1) = 3; \quad \mathrm{wt}(\gamma_2) = 3; \quad \mathrm{wt}(\gamma_3) = 3; \quad \mathrm{wt}(\gamma_4) = 4.$$

There are $\binom{4}{2} = 6$ sums of two rows, and a short calculation shows that the minimum weight of these is 3; there are $\binom{4}{3} = 4$ sums of three rows, and the minimum weight is 3; the sum of all four rows has weight 7. We conclude that the minimum distance d of C is 3, and so C can detect 2 errors; the code C can also correct 1 error, by Proposition 4.108(iii). See Example 4.117 for another way to compute $d(C)$.

This construction can be generalized. If $\ell \geq 3$, there are $2^\ell - 1$ nonzero words in $\mathbb{B}^\ell$. Define a $(2^\ell - 1 - \ell) \times (2^\ell - 1)$ matrix $G = [I|B]$, where the rows of B are all $w \in \mathbb{B}^{2^\ell - 1 - \ell}$ with $\mathrm{wt}(w) \geq 2$. Define the **Hamming** $[2^\ell - 1, 2^\ell - 1 - \ell]$ **code** to be $C = \{wG : w \in \mathbb{B}^{2^\ell - 1 - \ell}\}$. Of course, G is an echelon generating matrix for C. The rate of information of a Hamming code is

$$\frac{2^\ell - 1 - \ell}{2^\ell - 1} = 1 - \frac{\ell}{2^\ell - 1};$$

as ℓ gets large, the rate of information of the ℓth Hamming code gets very close to 1. It can be shown that the minimum distance of every Hamming code is 3, as in the $[7, 4]$ code. ◀

The next proposition gives a criterion for testing whether a word is a codeword.

→ **Proposition 4.114.** *If $G = [I|B]$ is an $m \times (m + p)$ echelon generating matrix of a linear $[m + p, m]$ code C over a field k (so that B is a $m \times p$ matrix), then $w \in k^{m+p}$ lies in C if and only if $w[-B^T|J]^T = 0$, where J is the $p \times p$ identity matrix.*

Proof. Denote the $(m + p) \times p$ matrix $[-B^T|J]$ by H. Since C is the row space of G, every codeword c is a linear combination of the rows of G. But the ith row of G is $e_i G$, where $e_1, \ldots, e_m$ is the standard basis of k^m (in this section, elements in k^m are regarded as $1 \times m$ row vectors), so that $c = \sum_i a_i(e_i G)$, where $a_i \in k$. Hence, it suffices to show that $GH^T = 0$, for then $cH^T = \sum_i a_i e_i GH^T = 0$.

The dot product of $v, w \in k^{m+p}$ is equal to a matrix product: $v \cdot w = vw^T$. In particular, the ij entry of GH^T is $\mathrm{ROW}_G(i) \cdot \mathrm{COL}_{H^T}(j) = \mathrm{ROW}_G(i)\mathrm{COL}_{H^T}(j)^T$. The ith row of G is

$$\mathrm{ROW}_G(i) = e_i G = e_i[I|B] = [e_i|e_i B] = (e_i, b_{i1}, b_{i2}, \ldots, b_{ip}).$$

The jth column of H^T is $H^T(e'_j)^T$, where $e'_1, \ldots, e'_p$ is the standard basis of k^p and $(e'_j)^T$ is a column vector. Now

$$H^T(e'_j)^T = [-B^T | J]^T (e'_j)^T = \left(e'_j [-B^T | J] \right)^T = [-e'_j B^T | e'_j]^T.$$

But $e'_j B^T$ is the jth row of B^T, which is the jth column of B. Hence,

$$\text{COL}_{H^T}(j) = (-b_{1j}, -b_{2j}, \ldots, -b_{pj}, e'_j)^T,$$

and

$$\text{COL}_{H^T}(j)^T = (-b_{1j}, -b_{2j}, \ldots, -b_{pj}, e'_j).$$

The ij entry of GH^T is thus

$$\begin{aligned}
\text{ROW}_G(i) \cdot \text{COL}_{H^T}(j) &= \text{ROW}_G(i) \left(\text{COL}_{H^T}(j) \right)^T \\
&= (e_i, b_{i1}, b_{i2}, \ldots, b_{ip}) \cdot (-b_{1j}, -b_{2j}, \ldots, -b_{pj}, e'_j) \\
&= b_{ij} - b_{ij} \\
&= 0.
\end{aligned}$$

Therefore, $GH^T = 0$ and $cH^T = c[-B^T | J]^T = 0$, as desired.

Conversely, consider the homogeneous system $[-B^T | J]^T x^T = 0$ and its solution space $S = \{v^T \in k^{m+p} : [-B^T | J]^T v^T = 0\}$. Now $v \in S$ if and only if $v[-B^T | J] = 0$, so that the first part of the proof shows that $C \subseteq S$. But $\dim(C) = m$, while $\dim(S) = m + p - r$, where $r = \text{rank}([-B^T | J]^T) = p$. By Theorem 4.43, we have $\dim(S) = m + p - p = m$, and so $C = S$, by Corollary 4.25. Therefore, if $w[-B^T | J]^T = 0$, then $w \in S$ and, hence, $w \in C$. •

→ **Definition.** Let C be a linear $[(m+p), m]$ linear code over a field k. A $p \times (m+p)$ matrix H is a ***parity check matrix*** for C if, for all $w \in k^{m+p}$, we have $wH^T = 0$ if and only if $w \in C$.

Proposition 4.114 says that if $G = [I | B]$ is an $m \times (m+p)$ echelon generating matrix of a linear $[m+p, p]$ code C, then $H = [-B^T | J]$ is a parity check matrix for C. A parity check matrix for C need not be unique, as we will see in Proposition 4.132.

Remark. If $C \subseteq k^n$ is a code, define its ***dual code*** to be the orthogonal complement

$$C^\perp = \{y \in k^n : (y, c) = 0 \text{ for all } c \in C\},$$

where $(y, c) = y_1 c_1 + \cdots + y_n c_n$ is the usual dot product of $y = (y_1, \ldots, y_n)$ and $c = (c_1, \ldots, c_n)$. Proposition 4.114 is involved in showing that if $G = [I | B]$ is a generating matrix for C, then the parity check matrix $H = [-B^T | J]$ is a generating matrix for the dual code $C^\perp$. ◄

The next corollary computes the smallest weight $d(C)$ of a linear code C in terms of a curious number associated to a parity check matrix.

Definition. If A is an $m \times n$ matrix over a field, where $m < n$, then the list of its columns is linearly dependent. Define

$$\mu(A) = \text{the minimal number of linearly dependent columns of } A.$$

If A is an $m \times n$ matrix with $m < n$ and $\text{rank}(A) = r$, then any $r + 1$ columns are linearly dependent. Therefore,

$$\mu(A) \leq r + 1. \tag{2}$$

Example 4.115.
Consider the matrices

$$A = \begin{bmatrix} 1 & 0 & 1 & 0 \\ 0 & 1 & 0 & 0 \end{bmatrix}, \quad B = \begin{bmatrix} 1 & 0 & 1 & 1 \\ 0 & 1 & 0 & 0 \end{bmatrix}, \quad C = \begin{bmatrix} 1 & 0 & 1 & -1 \\ 0 & 1 & 1 & 1 \end{bmatrix}.$$

All three matrices have rank 2, but $\mu(A) = 1$, $\mu(B) = 2$, and $\mu(C) = 3$. ◀

Corollary 4.116. *Let C be a linear $[n, m]$ code over a field k, and let H be a parity check matrix for C.*

(i) $d(C) = \mu(H)$.

(ii) $d(C) \leq m + 1$.

Proof.
(i) Let $\beta_1, \ldots, \beta_n$ be the columns of H. Since H is a parity check matrix for C, a word $y = (y_1, \ldots, y_n) \in C$ if and only if $yH^T = 0$. But $yH^T = 0$ if and only if $Hy^T = 0$; that is,

$$y_1 \beta_1 + \cdots + y_n \beta_n = 0.$$

Let $d(C) = d$, choose $y \in C$ of weight d, and let $y_{i_1}, \ldots, y_{i_d}$ be the nonzero coordinates of y. Since y is a nonzero codeword,

$$0 = Hy^T = y_1 \beta_1 + \cdots + y_n \beta_n = y_{i_1} \beta_{i_1} + \cdots + y_{i_d} \beta_{i_d},$$

and the list $\beta_{i_1}, \ldots, \beta_{i_d}$ is linearly dependent. Thus, $\mu(H) \leq d$. Suppose there is a linearly dependent list $\beta_{j_1}, \ldots, \beta_{j_p}$ for some $p < d$. Then there are scalars $z_{j_1}, \ldots, z_{j_p}$, not all 0, with $z_{j_1} \beta_{j_1}, \ldots, z_{j_p} \beta_{j_p} = 0$. Define $\bar{z} = (\bar{z}_1, \ldots, \bar{z}_n) \in k^n$ by $\bar{z}_{j_\nu} = z_{j_\nu}$ for $\nu = 1, \ldots, p$, and $\bar{z}_j = 0$ otherwise; then $H\bar{z}^T = 0$. But then $\bar{z} \in C$, since H is a parity check matrix for C, and this is a contradiction because $\text{wt}(\bar{z}) < d$.
(ii) By Eq. (2), $\mu(H) \leq r + 1$, where $\text{rank}(H) = r$; here, $r = m$. ●

→ **Example 4.117.**
In Example 4.113, we computed the minimum distance of the Hammming [7, 4] code C to be 3. Using the echelon generating matrix for C displayed in that example, we see that a parity check matrix for C is

$$H = \begin{bmatrix} 0 & 1 & 1 & 1 & 1 & 0 & 0 \\ 1 & 0 & 1 & 1 & 0 & 1 & 0 \\ 1 & 1 & 0 & 1 & 0 & 0 & 1 \end{bmatrix}$$

(since the entries lie in $\mathbb{B} = \mathbb{F}_2$, we have $-1 = 1$). It is easy to see that any two columns of H are linearly independent. Since columns 1, 4, and 5 are linearly dependent, $\mu(H) = 3$, and so $d(C) = 3$.   ◀

We now wish to construct relatively efficient (linear) codes which can correct many errors.

Recall Theorem 3.114: if k is a field, if $f(x) \in k[x]$, and if $I = (f(x))$ is the principal ideal generated by $f(x)$, then the quotient ring $k[x]/I$ is a vector space over k with basis the list $1, z, z^2, \ldots, z^{n-1}$, where $z = x + I$. Thus, $k[x]/I$ is n-dimensional, and there is a (vector space) isomorphism $k^n \to k[x]/I$. If we denote words in k^n by $(a_0, a_1, \ldots, a_{n-1})$ instead of by $(a_1, a_2, \ldots, a_n)$, then $(a_0, a_1, \ldots, a_{n-1}) \mapsto a_0 + a_1 z + \cdots + a_{n-1}z^{n-1}$ is an isomorphism $k^n \to k[x]/I$.

→ **Definition.**   A *cyclic code of length* n over a field is a linear code C such that

$$(a_0, a_1, \ldots, a_{n-1}) \in C \text{ implies } (a_{n-1}, a_0, \ldots, a_{n-2}) \in C.$$

That $k[x]/I$ is a commutative ring in addition to being a vector space will now be exploited. Compare the next proof with that of Lemma 3.26.

→ **Proposition 4.118.**   *Let k be a finite field, let $I = (x^n - 1)$ be the principal ideal in $k[x]$ generated by $x^n - 1$, and let $z = x + I$. Then $C \subseteq k[x]/I$ is a cyclic code if and only if C is an ideal in the commutative ring $k[x]/I$. Moreover, $C = (g(z))$ for some monic divisor $g(x)$ of $x^n - 1$ in $k[x]$.*

Proof.   Let C be an ideal in $k[x]/I$, and let $c = a_0 + a_1 z + \cdots + a_{n-1}z^{n-1} \in C$. Since C is an ideal, C contains $zc = a_0 z + a_1 z^2 + \cdots + a_{n-2}z^{n-1} + a_{n-1}z^n$. But $z^n = 1$ (because z is a root of $x^n - 1$). Hence, $a_{n-1} + a_0 z + \cdots + a_{n-2}z^{n-1} \in C$, and C is cyclic.

Conversely, assume that C is a cyclic code. Since C is a linear code, C is closed under addition and scalar multiplication by elements in k. As we have just seen, multiplication by z corresponds to shifting coefficients one step to the right (and making a_{n-1} the constant term). The reader may prove, by induction on i, that C is closed under multiplication by all elements $b_0 + b_1z + \cdots + b_{i-1}z^{i-1} \in k[x]/I$. Therefore, C is an ideal.

Let $\beta \colon k[x] \to k[x]/I$ be the natural map, and consider the inverse image $J = \beta^{-1}(C) = \{f(x) \in k[x] : f(z) \in C\}$. By Exercise 3.47 on page 251, J is an ideal in $k[x]$ containing $x^n - 1$. But every ideal in $k[x]$ is a principal ideal, by Theorem 3.59, so that there is a monic $g(x) \in k[x]$ with $J = (g(x))$. Since $x^n - 1 \in J$, we have $x^n - 1 = h(x)g(x)$ for some polynomial $h(x)$; that is, $g(x) \mid (x^n - 1)$. Finally, since J is generated by $g(x)$, its image $C = \beta(J)$ is generated by $\beta(g(x)) = g(z)$. •

→ **Definition.** Let $C \subseteq k[x]/I$ be a cyclic code, where $I = (x^n - 1)$. A monic polynomial $g(x) \in k[x]$ is called a ***generating polynomial for*** C if $C = (g(z))$, where $z = x + I$.

As in Proposition 4.118, a generating polynomial $g(x)$ of a cyclic code of length n can be chosen to be a divisor of $x^n - 1$, and so all its roots are nth roots of unity.

→ **Corollary 4.119.** *If C is a cyclic code of length n over a field k with generating polynomial $g(x)$, then* $\dim(C) = n - \deg(g)$.

Proof. Since $g(x) \mid (x^n - 1)$, there is an inclusion of ideals $I = (x^n - 1) \subseteq (g(x)) = J$. Regarding $k[x]$ and its quotients merely as vector spaces over k, we see that "enlargement of coset" $\gamma \colon k[x]/I \to k[x]/J$, given by $h(x) + I \mapsto h(x) + J$, is a surjective linear transformation. To compute $\ker \gamma$, consider the diagram

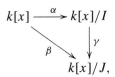

where α and β are the natural maps. Now $\beta = \gamma \circ \alpha$ and α, β, and γ are surjective. Thus, the hypothesis of Exercise 2.102 on page 190 holds, and so

$$\ker \gamma = \alpha(\ker \beta) = \alpha((g(x)) = (g(z)) = C,$$

where $z = x + I$. As vector spaces, $(k[x]/I)/C \cong k[x]/J$ (this is just the first isomorphism theorem). Hence,

$$\dim(k[x]/I) - \dim(C) = \dim(k[x]/J).$$

But $\dim(k[x]/I) = \deg(x^n - 1) = n$ and $\dim(k[x]/J) = \dim(k[x]/(g(x))) = \deg(g)$, so that $\dim(C) = n - \deg(g)$. •

Corollary 4.120.   *If C is a cyclic code of length n with generating polynomial $g(x) = g_0 + g_1 x + \cdots + g_s x^s$ with $\deg(g) = s$, then a generating matrix for C is the $(n - s) \times n$ matrix*

$$G = \begin{bmatrix} g_0 & g_1 & g_2 & \cdots & g_s & 0 & 0 & \cdots & 0 \\ 0 & g_0 & g_1 & g_2 & \cdots & g_s & 0 & \cdots & 0 \\ 0 & 0 & g_0 & g_1 & g_2 & \cdots & g_s & \cdots & 0 \\ & & \vdots & & \vdots & & \vdots & & \\ 0 & 0 & \cdots & 0 & g_0 & g_1 & g_2 & \cdots & g_s \end{bmatrix}.$$

Proof. Since C is an ideal, $g(x), xg(x), x^2 g(x), \ldots, x^{n-s} g(x)$ are codewords, and these codewords correspond to the rows of G. Write $G = [X|T]$, where T is the $s \times s$ submatrix consisting of the last s columns of G. As T is a lower triangular matrix with all diagonal entries g_s, we have $\det(T) = g_s^s$, by Exercise 4.51 on page 397. But $g_s = 1$, since generating polynomials are monic polynomials. Hence, $\det(T) = 1 \neq 0$, and the list of $n - s$ rows of G is linearly independent. Since $\dim(C) = n - s$, by Corollary 4.119, the list of rows of G is a basis of C, and so G is a generating matrix for C.   •

$\rightarrow$ **Example 4.121.**

Let C be the cyclic code of length $n = 6$ over $\mathbb{F}_7$ with generating polynomial

$$g(x) = (x - 3)(x - 3^2)(x - 3^3)(x - 3^4) = x^4 + 6x^3 + 3x^2 + 2x + 4.$$

By Corollary 4.120, a generating matrix for C is

$$N = \begin{bmatrix} 4 & 2 & 3 & 6 & 1 & 0 \\ 0 & 4 & 2 & 3 & 6 & 1 \end{bmatrix}.$$

By Gaussian elimination, an echelon generating matrix for C is

$$G = \begin{bmatrix} 1 & 0 & 4 & 2 & 3 & 6 \\ 0 & 1 & 4 & 6 & 5 & 2 \end{bmatrix}.$$

A parity check matrix is

$$H = \begin{bmatrix} 3 & 3 & 1 & 0 & 0 & 0 \\ 5 & 1 & 0 & 1 & 0 & 0 \\ 4 & 2 & 0 & 0 & 1 & 0 \\ 1 & 5 & 0 & 0 & 0 & 1 \end{bmatrix}.$$

The reader may check that $GH^T = 0$.   ◀

The roots of $x^n - 1$ are *nth roots of unity*. Recall that an element z in a field k is a ***primitive nth root of unity*** if $z^n = 1$ but $z^i \neq 1$ for all i with $0 < i < n$.

Lemma 4.122. *Let $\mathbb{F}_q$ denote the finite field of q elements. If n is a positive integer, then there exists a primitive nth root of unity in some extension field of $\mathbb{F}_q$ if and only if $\gcd(n, q) = 1$.*

Proof. Assume that $(n, q) = 1$, and let $E/\mathbb{F}_q$ be a splitting field of $f(x) = x^n - 1$ over $\mathbb{F}_q$. Now the derivative $f'(x) = nx^{n-1} - 1 \neq 0$, so that $\gcd(f, f') = 1$ (they have no common root); hence, $f(x)$ has no repeated roots, by Exercise 3.67 on page 274. Thus, if K is the set of all the roots of $f(x) = x^n - 1$, then K is a multiplicative group of order n. But K is cyclic, by Theorem 3.55, and a generator of K must be a primitive nth root of unity.

Conversely, assume that there exists a primitive nth root of unity. Now $q = p^s$ for some prime p. If $(n, q) \neq 1$, then $p \mid n$; that is, $n = pu$ for some integer u. Hence, $x^n - 1 = x^{pu} - 1 = (x^u - 1)^p$, and so the multiplicative group of all nth roots of unity has fewer than n elements. Therefore, there is no primitive nth root of unity. $\bullet$

→ **Corollary 4.123.** *Let $C \subseteq \mathbb{F}_q^n$ be a cyclic code with generating polynomial $g(x)$, where $(n, q) = 1$. Then $a = (a_0, \dots, a_{n-1}) \in \mathbb{F}_q^n$ lies in C if and only if $a(\eta) = 0$ for every root η of $g(x)$, where $a(x) = a_0 + a_1 x + \cdots + a_{n-1} x^{n-1}$.*

Proof. By Proposition 4.118, $C = (g)$, the principal ideal in $k[x]/I$ generated by $g(x) + I$ [where $I = (x^n - 1)$]. If $a \in C$, then $a \in (g)$, so there is $f(x) \in k[x]$ with $a(x) + I = f(x)g(x) + I$. Hence, $a(x) - f(x)g(x) \in I$: there is some $h(x) \in k[x]$ with

$$a(x) = f(x)g(x) + h(x)(x^n - 1). \tag{3}$$

Since $g(x) \mid (x^n - 1)$, every root η of $g(x)$ satisfies $\eta^n = 1$. Therefore, $a(\eta) = 0$ because the right side of Eq. (3) is zero.

Since $(n, q) = 1$, Lemma 4.122 says that $x^n - 1$ has no repeated roots; since $g(x) \mid (x^n - 1)$, the generating polynomial $g(x)$ has no repeated roots. Therefore, the polynomials $x - \eta$, as η ranges over all the roots of $g(x)$, are pairwise relatively prime. If $a(\eta) = 0$ for every root η of $g(x)$, then $(x - \eta) \mid a(x)$ for all η; by Exercise 3.59 on page 273, $a(x)$ is divisibly by $\prod_\eta (x - \eta) = g(x)$; that is, $a \in (g) = C$. $\bullet$

The following theorem will enable us to construct relatively efficient codes which can correct many errors.

→ **Theorem 4.124 (Bose-Chaudhuri-Hocquenghem).** *Let C be a cyclic code over $\mathbb{F}_q$ of length n with generating polynomial $g(x)$; let $(q, n) = 1$ and let ζ be a primitive nth root of unity. If consecutive powers $\zeta^u, \zeta^{u+1}, \dots, \zeta^{u+\ell}$ are roots of $g(x)$, where $0 \leq u$ and $u + \ell < n$, then $d = d(C) \geq \ell + 2$.*

Proof. A codeword $c = (c_0, c_1, \dots, c_{n-1}) \in \mathbb{F}_q^n$ is identified with the polynomial $c(x) = c_0 + c_1 x + \dots + c_{n-1} x^{n-1} \in \mathbb{F}_q[x]$, and its weight $\text{wt}(c)$ is the number of its nonzero coefficients. By Proposition 4.110, it suffices to prove that every nonzero codeword has at least $\ell + 2$ nonzero coefficients. Suppose, on the contrary, that there exists a nonzero codeword c with wt$(c) < \ell + 2$; thus, $c(x) = c_{i_1} x^{i_1} + \cdots + c_{i_{\ell+1}} x^{i_{\ell+1}}$, where $i_1 < \cdots < i_{\ell+1}$. If $\beta \in \mathbb{F}_q$, then $c(\beta)$ is the dot product

$$c(\beta) = (c_{i_1}, c_{i_2}, \ldots, c_{i_{\ell+1}}) \cdot (\beta^{i_1}, \beta^{i_2}, \ldots, \beta^{i_{\ell+1}}).$$

Now form the $(\ell + 1) \times (\ell + 1)$ matrix W whose jth row, for $0 \le j \le \ell$, arises from ζ^{u+j}:

$$W = \begin{bmatrix} \zeta^{u i_1} & \zeta^{u i_2} & \cdots & \zeta^{u i_{\ell+1}} \\ \zeta^{(u+1)i_1} & \zeta^{(u+1)i_2} & \cdots & \zeta^{(u+1)i_{\ell+1}} \\ \vdots & \vdots & & \\ \zeta^{(u+\ell)i_1} & \zeta^{(u+\ell)i_2} & \cdots & \zeta^{(u+\ell)i_{\ell+1}} \end{bmatrix}.$$

Thus, if $c_* = (c_{i_1}, c_{i_2}, \ldots, c_{i_{\ell+1}})$, then

$$W c_*^T = \left(c(\zeta^u), c(\zeta^{u+1}), \ldots, c(\zeta^{u+\ell}) \right)^T = 0.$$

Factoring out $\zeta^{u i_j}$ from the jth column of W, for all j, gives the (transpose of the) $(\ell + 1) \times (\ell + 1)$ Vandermonde matrix

$$V = \text{Van}(\zeta^{i_1}, \ldots, \zeta^{i_{\ell+1}}) = \begin{bmatrix} 1 & 1 & \cdots & 1 \\ \zeta^{i_1} & \zeta^{i_2} & \cdots & \zeta^{i_{\ell+1}} \\ \zeta^{2i_1} & \zeta^{2i_2} & \cdots & \zeta^{2i_{\ell+1}} \\ \vdots & \vdots & & \\ \zeta^{\ell i_1} & \zeta^{\ell i_2} & \cdots & \zeta^{\ell i_{\ell+1}} \end{bmatrix}.$$

By Exercise 4.52 on page 397, we have

$$\det(W) = \zeta^{u i_1} \cdots \zeta^{u i_\ell} \det(V) = \zeta^{u i_1} \cdots \zeta^{u i_\ell} \prod_{j<k} (\zeta^{i_k} - \zeta^{i_j}).$$

We claim that all the ζ^{i_j} are distinct. If $j < k$, then $0 \le i_k - i_j < n$ (because $i_j < i_k < n$), and so $\zeta^{i_k - i_j} = 1$. Hence, if $\zeta^{i_k} = \zeta^{i_j}$, then we contradict ζ being a primitive nth root of unity. We conclude that $\det(W) \ne 0$. But $c_*^T \ne 0$ and $W c_*^T = 0$, contradicting the nonsingularity of W. Therefore, no codeword of weight $< \ell + 2$ exists, and $d(C) \ge \ell + 2$. •

To use Theorem 4.124 wisely, we seek efficient codes assuming that d and n are given. That is, we want to maximize the rate of information, and this is accomplished by minimizing deg(g) [for $m = n - \deg(g)$, by Corollary 4.119].

→ **Definition.** A linear code over a field k is a **BCH *code*** of length n if it is a cyclic code having a generating polynomial $g(x)$ of smallest possible degree which has consecutive powers $\zeta^u, \zeta^{u+1}, \ldots, \zeta^{u+\ell}$ among its roots, where ζ is a primitive nth root of unity and $0 \leq u \leq u + \ell < n$.

→ **Corollary 4.125.** *Let C be a BCH code of length n with generating polynomial $g(x)$. If consecutive powers $\zeta^u, \zeta^{u+1}, \ldots, \zeta^{u+2t}$ occur among the roots of $g(x)$, where $0 \leq u$ and $u + 2t < n$, then C corrects up to t errors. Moreover, if y is a word, then any codeword c with $\delta(y, c) \leq t$, should one exist, is unique.*

Proof. In the notation of Theorem 4.124, we have $\ell = 2t$, and so the theorem gives $d(C) \geq \ell + 2 \geq 2t + 1$. Proposition 4.108(iii) shows that C corrects up to t errors, and Proposition 4.108(i) shows the uniqueness of a codeword c with $\delta(y, c) \leq t$, should one exist. •

→ **Corollary 4.126.** *For any prime p and any positive integer t, there exists a BCH code C over $\mathbb{F}_p$ which corrects up to t errors.*

Proof. Let $k = \mathbb{F}_q$, where q is a power of p and $2t + 1 < q - 1$. By Theorem 3.55, the multiplicative group $k^\times$ is a cyclic group of order $q - 1$, and a generator ζ is a primitive $(q - 1)$th root of unity; hence, $\zeta, \zeta^2, \ldots, \zeta^{2t+1}$ are distinct. Each ζ^j is a root of some polynomial in $\mathbb{F}_p[x]$, by Proposition 4.32, so that Corollary 3.117 gives a unique monic irreducible polynomial $h_j(x) \in \mathbb{F}_p[x]$ having ζ^j as a root. Finally, define

$$g(x) = \text{lcm}\{h_1(x), \ldots, h_{2t+1}(x)\},$$

and define C to be the BCH code with generating polynomial $g(x)$. The result now follows from Corollary 4.125. •

No simple closed formula is known which gives the degrees of generating polynomials of BCH codes, but there are extensive tables of them.

To see whether a cyclic code with generating polynomial $g(x) \in \mathbb{F}_q[x]$ is a BCH code, we must determine the roots of $g(x)$, and this will force us to investigate finite fields in more detail.

→ **Example 4.127.**
Let us describe $\mathbb{F}_8$. We know that its multiplicative group of nonzero elements is cyclic of order 7, and a generator ζ is a primitive 7th root of unity. There is an irreducible polynomial $m(x) \in \mathbb{F}_2[x]$ having ζ as a root. As in Proposition 3.116, $\deg(m) = 3 = \dim_{\mathbb{F}_2}(\mathbb{F}_8)$. In Example 3.98, we saw that there are only two irreducible cubics in $\mathbb{F}_2[x]$, namely, $x^3 + x + 1$ and $x^3 + x^2 + 1$. To be explicit, we choose ζ to be a root of the first, so that

$$\zeta^3 = \zeta + 1.$$

+	1	ζ	ζ^2	ζ^3	ζ^4	ζ^5	ζ^6
1	0	ζ^3	ζ^6	ζ	ζ^5	ζ^4	ζ^2
ζ		0	ζ^4	1	ζ^3	ζ^6	ζ^5
ζ^2			0	ζ^5	ζ	ζ^3	1
ζ^3				0	ζ^6	ζ^2	ζ^4
ζ^4					0	1	ζ^3
ζ^5						0	ζ
ζ^6							0

Table 4.1: Addition Table for $\mathbb{F}_8$

Table 4.1 is an addition table for $\mathbb{F}_8$. Since $\mathbb{F}_8$ has characteristic 2, the diagonal entries have the form $\zeta^i + \zeta^i = 2\zeta^i = 0$; since addition is commutative, the addition table is a symmetric matrix, and so it is only necessary to compute its upper triangular half. Let us compute the first row of Table 4.1 consisting of $\zeta^j + 1$ for $1 \le j \le 6$. We have $\zeta + 1 = \zeta^3$, and $\zeta^3 + 1 = (\zeta + 1) + 1 = \zeta$. Next,

$$\zeta^2 + 1 = (\zeta + 1)^2 = (\zeta^3)^2 = \zeta^6;$$
$$\zeta^4 + 1 = (\zeta^2 + 1)^2 = (\zeta^6)^2 = \zeta^{12} = \zeta^5;$$
$$\zeta^6 + 1 = (\zeta^3 + 1)^2 = \zeta^2$$

It follows that $\zeta^5 + 1 = \zeta^4$, for all the other powers of ζ have occurred. Now use the first row to compute the second row. For example, $\zeta^j + \zeta = \zeta(\zeta^{j-1} + 1)$. The reader is invited to verify the remaining entries. ◀

→ **Example 4.128.**

(i) The factorization into irreducibles of $x^7 - 1$ in $\mathbb{F}_2[x]$ is

$$x^7 - 1 = x^7 + 1 = (x + 1)(x^3 + x + 1)(x^3 + x^2 + 1).$$

Let C be the cyclic binary code of length 7 having generating polynomial $g(x) = x^3 + x + 1$. Now one root of $g(x)$ is a primitive 7th root of unity, say, ζ. To find the other roots, evaluate $f(\zeta^i)$ for each i, using Table 4.1. We see that ζ, ζ^2 are consecutive roots of $g(x)$, and so C is a BCH code with $\ell = 1$. In fact, C is a $[7, 4]$ code [for $7 - \deg(g) = 4$] with $d(C) \ge \ell + 2 = 3$. By Corollary 4.120, a generating matrix for C is

$$G = \begin{bmatrix} 1 & 0 & 1 & 1 & 0 & 0 & 0 \\ 0 & 1 & 0 & 1 & 1 & 0 & 0 \\ 0 & 0 & 1 & 0 & 1 & 1 & 0 \\ 0 & 0 & 0 & 1 & 0 & 1 & 1 \end{bmatrix}.$$

The echelon form of G is

$$G' = \begin{bmatrix} 1 & 0 & 0 & 0 & 0 & 1 & 1 \\ 0 & 1 & 0 & 0 & 1 & 0 & 1 \\ 0 & 0 & 1 & 0 & 1 & 1 & 0 \\ 0 & 0 & 0 & 1 & 1 & 1 & 1 \end{bmatrix}.$$

Thus, C is the Hamming $[7, 4]$ code in Example 4.113, and so this code is a BCH code. (It can be proved that all the Hamming codes are BCH codes.)

(ii) Consider the cyclic binary code C of length 7 having generator polynomial

$$g(x) = (x + 1)(x^3 + x + 1) = x^4 + x^3 + x^2 + 1.$$

Now $1 = \zeta^0, \zeta, \zeta^2$ are roots of $g(x)$, and so C is a BCH code; in fact, C is a $[7, 3]$ code with $d(C) \geq 4$. By Corollary 4.120, a generating matrix for C is

$$G = \begin{bmatrix} 1 & 1 & 1 & 0 & 1 & 0 & 0 \\ 0 & 1 & 1 & 1 & 0 & 1 & 0 \\ 0 & 0 & 1 & 1 & 1 & 0 & 1 \end{bmatrix}. \quad \blacktriangleleft$$

By using finite fields $\mathbb{F}_q$ other than the prime fields $\mathbb{F}_p$, we may choose generating polynomials that are simpler than those for general BCH codes.

→ **Corollary 4.129 (Reed-Solomon).**   *Let q be a prime power, and let t be a positive integer with $2t \leq q - 1$. There is a BCH code over $\mathbb{F}_q$ of length $q - 1$ which corrects up to t errors and whose rate of information is $1 - \frac{2t}{q-1}$.*

Proof.   Let $\zeta \in \mathbb{F}_q$ be a primitive $(q - 1)$st root of unity, let

$$g(x) = (x - \zeta)(x - \zeta^2) \cdots (x - \zeta^{2t}) \in \mathbb{F}_q[x],$$

where $2t < q - 1$, and let $C \subseteq \mathbb{F}_q^{q-1}$ be the code with generating polynomial $g(x)$. In the notation of Theorem 4.124, $1 + \ell = 2t$, so that this theorem gives $d(C) \geq \ell + 2 = 2t + 1$. By Proposition 4.108(iii), the code C corrects up to t errors. Since $\deg(g) = 2t$, Corollary 4.119 gives $\dim(C) = q - 1 - 2t$. Hence, the rate of information of C is $\frac{q-1-2t}{q-1} = 1 - \frac{2t}{q-1}$.   •

→ **Definition.**   A *t-error correcting Reed-Solomon code* over $\mathbb{F}_q$ is a BCH code of length $q - 1$ with generating polynomial

$$g(x) = (x - \zeta)(x - \zeta^2) \cdots (x - \zeta^{2t}),$$

where ζ is a primitive $(q - 1)$th root of unity in $\mathbb{F}_q$ and $2t < q - 1$.

We have mentioned that there is no simple formula for the degree of the generating polynomial of a general BCH code. In contrast, the degree of the generating polynomial in Corollary 4.129 is $2t$. If C is a t-error correcting Reed-Solomon code over $\mathbb{F}_q$, then $\dim(C) = q - 1 - 2t$, by Corollary 4.119.

→ **Example 4.130.**

A primitive 6th root of unity in $\mathbb{F}_7$ is [3]. The BCH [6, 2] code C over $\mathbb{F}_7$ with generating polynomial

$$g(x) = (x - 3)(x - 3^2)(x - 3^3)(x - 3^4) = 4 + 2x + 3x^2 + 6x^3 + x^4$$

is a 2-error correcting Reed-Solomon code over $\mathbb{F}_7$. By Example 4.121, a generating matrix for C is

$$G = \begin{bmatrix} 4 & 2 & 3 & 6 & 1 & 0 \\ 0 & 4 & 2 & 3 & 6 & 1 \end{bmatrix}.$$

Thus, C is the row space $Row(G)$:

$$C = \big\{ (4a,\ 2a + 4b,\ 3a + 2b,\ 6a + 3b,\ a + 6b,\ b) : a, b \in \mathbb{F}_7 \big\}. \quad \blacktriangleleft$$

→ **Example 4.131.**

Let $\zeta \in \mathbb{F}_8$ be a primitive 7th root of unity. The BCH [7, 3] code C over $\mathbb{F}_8$ with generating polynomial

$$g(x) = (x + \zeta)(x + \zeta^2)(x + \zeta^3)(x + \zeta^4)$$

is a 2-error correcting Reed-Solomon code over $\mathbb{F}_8$ (since $\mathbb{F}_8$ has characteristic 2, we have $-1 = 1$). Using Table 4.1, we can see that $g(x) = \zeta^3 + \zeta x + x^2 + \zeta^3 x^3 + x^4$. By Corollary 4.120, a generating matrix for C is

$$G = \begin{bmatrix} \zeta^3 & \zeta & 1 & \zeta^3 & 1 & 0 & 0 \\ 0 & \zeta^3 & \zeta & 1 & \zeta^3 & 1 & 0 \\ 0 & 0 & \zeta^3 & \zeta & 1 & \zeta^3 & 1 \end{bmatrix}. \quad \blacktriangleleft$$

A long message sent from space may be hit by a cosmic ray, causing a *burst*, a scrambling of a sequence of consecutive letters. To cope with bursts, one first encodes a message using a Reed-Solomon code over $\mathbb{F}_q$, where q is a power of 2. For example, let C be the 5-error correcting Reed-Solomon code over $\mathbb{F}_{256}$ with generating polynomial $g(x) = \prod_{i=1}^{10}(x - \zeta^i)$, where ζ is a primitive 255th root of unity. Now every element in $\mathbb{F}_{256}$ can be written as a bitstring of length eight, since $256 = 2^8$. Form a binary code C' of length $8 \cdot 255 = 2040$ by replacing the letters (lying in $\mathbb{F}_{256}$) of codewords in C by bitstrings of length eight. Transmit a message in the binary code C', and decode it back into the Reed-Solomon code C. Since C corrects up to five errors, this corresponds to correcting a binary interval of length up to 33 binary symbols in the received binary message (an interval of length 34 could involve six letters of the Reed-Solomon code). In this way, Reed-Solomon codes use finite fields to correct binary error bursts.

Decoding

Let $\mathcal{A}$ be a finite alphabet, and let $C \subseteq \mathcal{A}^n$ be an $[n, m]$ block code. A word $y \in \mathcal{A}^n$ can be decoded in the following inefficient way. Enumerate all the words $c \in C$, say, $c_1, \ldots, c_r$, where $r = |\mathcal{A}|^m$, and compute $\delta(y, c_i)$ for all i. Decode y as c_i, where c_i is a nearest codeword to y (if there are several nearest codewords, choose the first such in the enumeration).

If $C \subseteq \mathbb{F}_q^n$ is a linear code, the naive decoding just described can be better organized, but it is still inefficient. We aim to decode a received word y by replacing it by a nearest codeword c, and this suggests looking at words of the form $y - c$ for $c \in C$, for $\delta(y, c) = \mathrm{wt}(y - c)$.

$\rightarrow$ **Definition.** Let $C \subseteq \mathbb{F}_q^n$ be a linear $[n, m]$ code. If $y \in \mathbb{F}_q^n$ and $c \in C$, then the **error vector** is

$$e = e(y, c) = y - c.$$

Now the vector space $\mathbb{F}_q^n$ is an additive abelian group, and C is a subgroup. Given y, the totality of all error vectors $y - c$ for $c \in C$ is the coset $y + C$. To say that $c \in C$ is a nearest codeword to y is to say that the error vector $e(y, c) = y - c$ has minimum weight in $y + C$.

Definition. If $C \subseteq \mathbb{F}_q^n$ is a linear $[n, m]$ code, and if $y + C$ is a coset of C in $\mathbb{F}_q^n$, then a **coset leader** is a vector $e \in y + C$ of minimal weight.

The vectors in $\mathbb{F}_q^n$ can be organized to make the naive decoding of a block code slightly more efficient. Enumerate all the words $c \in C$, say, $c_1, \ldots, c_{q^m}$, and enumerate a set of coset representatives, say, $w_1, \ldots, w_{q^{n-m}}$. Make a table whose jth column consists of the enumeration of the jth coset $w_j + C = \{w_j + c_i : i = 1, \ldots, q^m\}$. Given a vector y, locate the (unique) column $w_j + C$ in which it lives (the group $\mathbb{F}_q^n$ is the disjoint union of cosets). If e_j is a coset leader, then $e_j = y - c$ for some $c \in C$, and c is a nearest codeword to y (should there be more than one nearest codeword, choose the first one in the enumeration of the coset $w_j + C$). We can make this procedure a bit more efficient. If H is a parity check matrix for C, then $y \in C$ if and only if $y H^T = 0$, by Proposition 4.114. Now two vectors $y, w \in \mathbb{F}_q^n$ lie in the same coset of C if and only if $y - w = c \in C$. But $y H^T - w H^T = (y - w) H^T = c H^T = 0$, so that $y, w \in \mathbb{F}_q^n$ lie in the same coset of C if and only if $y H^T = w H^T$.

$\rightarrow$ **Definition.** Let H be a parity check matrix for a linear $[n, m]$ code $C \subseteq \mathbb{F}_q^n$. If $y \in \mathbb{F}_q^n$, then its **syndrome** is $S(y) = y H^T$.

A parity check matrix H of a linear code C need not be unique, and a syndrome $y H^T$ does depend on H. To decode a received word y, first compute its syndrome $S(y) = y H^T$, and then compute the syndromes $S(e_j)$ of the coset leaders e_j until one is found for which $S(e_j) = S(y)$. There can be only one such coset $e_j + C$, for if $S(e_j) = S(e_k)$, then $S(e_j - e_k) = 0$, $e_j - e_k \in C$, and $e_j + C = e_k + C$. Thus, $c = y - e_j$ is a codeword nearest to y, and y is decoded as c. Although this method is better than the naive decoding of a general block code, it is still not very practical. After all, if C is an $[n, m]$ linear code over $\mathbb{F}_q$, then there are q^{n-m} cosets of C.

There are efficient decoding procedures for all BCH codes, but we will focus on Reed-Solomon codes. More precisely, if C is a t-error correcting Reed-Solomon code over $\mathbb{F}_q$, and if y is a received word with $\delta(y, c) \leq t$ for some codeword c, then we will show how to find c (without more information, it would be foolish to try to decode words y which are not close to any codeword).

As usual, we may view a vector $y = (y_0, \ldots, y_{q-2}) \in \mathbb{F}_q^{q-1}$ as a polynomial $y(x) = y_0 + y_1 x + \cdots + y_{q-2} x^{q-2} \in \mathbb{F}_q[x]$.

→ **Proposition 4.132.** *Let ζ be a primitive element of $\mathbb{F}_q$, and let $C \subseteq \mathbb{F}_q^{q-1}$ be a t-error correcting Reed-Solomon code over $\mathbb{F}_q$ with generating polynomial $g(x) = (x - \zeta)(x - \zeta^2) \cdots (x - \zeta^{2t})$. Define the $2t \times (q - 1)$ matrix*

$$U = \begin{bmatrix} 1 & \zeta & \zeta^2 & \cdots & \zeta^{q-2} \\ 1 & \zeta^2 & \zeta^4 & \cdots & \zeta^{2(q-2)} \\ \vdots & \vdots & \vdots & \vdots & \vdots \\ 1 & \zeta^{2t} & \zeta^{4t} & \cdots & \zeta^{2t(q-2)} \end{bmatrix}$$

(i) *If $y = (y_0, \ldots, y_{q-2}) \in \mathbb{F}_q^{q-1}$, then*

$$yU^T = \left(y(\zeta), y(\zeta^2), \ldots, y(\zeta^{2t}) \right),$$

where $y(\zeta^i) = y_0 + y_1 \zeta^i + \cdots + y_{q-2} \zeta^{i(q-2)}$.

(ii) *Let $f(x) = f_0 + f_1 x + \cdots + f_r x^r \in \mathbb{F}_q[x]$, where $r \leq t$. If we write $f = (f_0, \ldots, f_r, 0, \ldots, 0) \in \mathbb{F}_q^{2t}$, then*

$$fU = \left(f(1), \zeta f(\zeta), \zeta^2 f(\zeta^2), \ldots, \zeta^{q-2} f(\zeta^{q-2}) \right).$$

(iii) *U is a parity check matrix for C.*

(iv) *If $e = y - c$ is an error vector, then e and y have the same syndrome:*

$$S(e) = eU^T = yU^T = S(y),$$

and so $e(\zeta^j) = y(\zeta^j)$ for all $j \leq 2t$.

Proof.
(i) Note that U is a matrix over $\mathbb{F}_q$ because $\zeta \in \mathbb{F}_q$ (this would not necessarily be true if C were only a BCH code). The ij entry of yU^T is the dot product

$$y \cdot \text{ROW}_U(i) = y_0 + y_1\zeta^i + \cdots + y_{q-2}\zeta^{i(q-2)} = y(\zeta^i).$$

Therefore,

$$yU^T = \big(y(\zeta), y(\zeta^2), \ldots, y(\zeta^{2t})\big).$$

(ii) The ij entry of fU is the dot product

$$(f_0, \ldots, f_r, 0, \ldots, 0) \cdot \text{COL}_U(j) = (f_0, \ldots, f_r, 0, \ldots, 0) \cdot (\zeta^j, \zeta^{2j}, \ldots, \zeta^{2jt})$$
$$= f_0\zeta^j + f_1\zeta^{2j} + \cdots + f_r\zeta^{2jr}$$
$$= \zeta^j\big(f_0 + f_1\zeta^j + \cdots + f_r\zeta^{jr}\big)$$
$$= \zeta^j f(\zeta^j).$$

(iii) By Corollary 4.123, if $C \subseteq \mathbb{F}_q^{q-1}$ is a cyclic code with generating polynomial $g(x)$, then $y = (y_0, \ldots, y_{q-2}) \in \mathbb{F}_q^{q-1}$ lies in C if and only if $y(\eta) = 0$ for every root η of $g(x)$. Therefore, $y \in C$ if and only if $yU^T = 0$.
(iv)
$$eU^T = (y - c)U^T = yU^T - cU^T = yU^T.$$

The last statement follows because $yU^T = \big(y(\zeta), y(\zeta^2), \ldots, y(\zeta^{2t})\big)$. •

Let C be a t-error correcting Reed-Solomon code over $\mathbb{F}_q$ and let U be the parity check matrix in Proposition 4.132. A received word y is decoded by finding its error vector $e = y - c$. The most difficult step in the decoding process is to locate the nonzero coordinates of e. After all, if C were a binary code, then the nonzero coordinates of e actually determine e (of course, a Reed-Solomon code $C \subseteq \mathbb{F}_q^{q-1}$ is never a binary code lest $q = 2$ and $q - 1 = 1$).

Recall the following definition from Chapter 3: if $A = [a_{ij}]$ and $B = [b_{ij}]$ are $m \times n$ matrices over a field k, then their **Hadamard product** is $A \circ B = [a_{ij}b_{ij}]$. In particular, the Hadamard product of $1 \times n$ vectors is defined:

$$[a_1, \ldots, a_n] \circ [b_1, \ldots, b_n] = [a_1b_1, \ldots, a_nb_n].$$

→ **Definition.** If e is an error vector, then nonzero vectors u with $e \circ u = 0$ are called *error locator vectors*.

The support of e is always contained in the zero set of an error locator vector u.

Lemma 4.133. *If $e = (e_0, \ldots, e_{q-2})$ and $u = (u_0, \ldots, u_{q-2})$ lie in $\mathbb{F}_q^{q-1}$, then $e \circ u = (0, \ldots, 0)$ implies*

$$\{j : e_j \neq 0\} = \operatorname{Supp}(e) \subseteq \mathcal{Z}(u) = \{j : u_j = 0\}.$$

Proof. The hypothesis $e \circ u = (0, \ldots, 0)$ says that $e_j u_j = 0$ for all j. If $j \in \operatorname{Supp}(e)$, then $e_j \neq 0$. Since $e_j u_j = 0$, we must have $u_j = 0$, and so $j \in \mathcal{Z}(u)$.   •

We create a polynomial whose roots occur at the powers ζ^{j_v}, where $j_v \in \operatorname{Supp}(e)$.

→ **Definition.** Let C be a t-error correcting Reed-Solomon code over $\mathbb{F}_q$, let y be a received word, and let $e = y - c = (e_0, e_1, \ldots, e_{q-2})$ be an error vector. If $\operatorname{wt}(e) \leq t$ and $\operatorname{Supp}(e) = \{j_1, \ldots, j_r\}$, where $r \leq t$, then the ***error locator polynomial*** is

$$f(x) = (x - \zeta^{j_1})(x - \zeta^{j_2}) \cdots (x - \zeta^{j_r}) = f_0 + f_1 x + \cdots + f_{r-1}x^{r-1} + f_r x^r,$$

where $f_r = 1$. Let us write $f = (f_0, f_1, \ldots, f_{r-1}, 1, 0, \ldots, 0) \in \mathbb{F}_q^{2t}$ (so that the matrix product fU is defined, where U is the parity check matrix in Proposition 4.132).

We now improve Lemma 4.133.

→ **Lemma 4.134.** *Let C be a t-error correcting Reed-Solomon code over $\mathbb{F}_q$, let U be the parity check matrix in* Proposition 4.132, *let y be a received word, and let $e = y - c = (e_0, e_1, \ldots, e_{q-2})$ be an error vector with $\operatorname{wt}(e) \leq t$. If $u = fU = \big(f(1), \zeta f(\zeta), \ldots, \zeta^{q-2} f(\zeta^{q-2})\big)$, where $f = (f_0, f_1, \ldots, f_{r-1}, 1, 0, \ldots, 0) \in \mathbb{F}_q^{2t}$ is the error locator polynomial, then u is an error locator vector and*

$$\operatorname{Supp}(e) = \mathcal{Z}(u).$$

Proof. Let $\operatorname{Supp}(e) = \{j_1, \ldots, j_r\}$, where $r \leq t$, and let $f(x) = \sum_{i=1}^r f_i x^i$ be the error locating polynomial. We claim that $e \circ u = 0$. If $j_v \in \operatorname{Supp}(e)$, then $u_{j_v} = f(\zeta^{j_v}) = 0$, by Proposition 4.132(ii), and so $e_{j_v} u_{j_v} = 0$. If $j \notin \operatorname{Supp}(e)$, then $e_j = 0$, and so $e_j u_j = 0$. Therefore, $e \circ u = [0\ 0\ \ldots\ 0]$; that is, u is an error locator vector. It follows from Lemma 4.133 that $\operatorname{Supp}(e) \subseteq \mathcal{Z}(u)$. This inclusion cannot be proper: if $u_j = 0$ for some $j \notin \{j_1, \ldots, j_r\}$, then $\zeta^j f(\zeta^j) = 0$, hence $f(\zeta^j) = 0$, giving the polynomial $f(x)$ of degree $r \leq t$ too many roots. Therefore, $\operatorname{Supp}(e) = \mathcal{Z}(u)$.   •

We are now going to show that the error locator polynomial, and hence an error locator vector, can be found by solving a system of linear equations.

→ **Definition.** Let C be a t-error correcting Reed-Solomon code over $\mathbb{F}_q$, let ζ be a primitive element of $\mathbb{F}_q$, let $y = (y_0, \ldots, y_{q-2}) \in \mathbb{F}_q^{q-1}$, and let $e = y - c$ be an error vector with $\text{wt}(e) = r \leq t$. The ***syndrome matrix*** is the $r \times r$ matrix

$$\Sigma(y) = \begin{bmatrix} y(\zeta) & y(\zeta^2) & \cdots & y(\zeta^r) \\ y(\zeta^2) & y(\zeta^3) & \cdots & y(\zeta^{r+1}) \\ \vdots & \vdots & \cdots & \vdots \\ y(\zeta^r) & y(\zeta^{r+1}) & \cdots & y(\zeta^{2r-1}) \end{bmatrix}.$$

→ **Proposition 4.135.**   *Let C be a t-error correcting Reed-Solomon code over $\mathbb{F}_q$, let $y = (y_0, \ldots, y_{q-2}) \in \mathbb{F}_q^{q-1}$, and let $e = y - c$ be an error vector with $\text{wt}(e) = r \leq t$.*

(i) *Both e and y have the same syndrome matrix: $\Sigma(e) = \Sigma(y)$.*

(ii) *If $f(x) = f_0 + f_1 x + \cdots + f_{r-1} x^{r-1} + x^r$ is the error polynomial, then*

$$\dot{f} = (f_0, f_1, \ldots, f_{r-1}) \in \mathbb{F}_q^r$$

is a solution to the linear system

$$\Sigma(y)\dot{f}^T = h^T,$$

where $h = [-y(\zeta^{r+1}), -y(\zeta^{r+2}), \ldots, -y(\zeta^{2r})]$.

(iii) *The syndrome matrix $\Sigma(y)$ is nonsingular.*

Proof.
(i) Both e and y have the same syndrome, by Proposition 4.132(iv), and we have $e(\zeta^j) = y(\zeta^j)$ for all $j \leq 2t$, by Proposition 4.132(ii). Thus, $\Sigma(e) = \Sigma(y)$.
(ii) Each ζ^{j_v}, for $v = 1, \ldots, r$, is a root of the error locator polynomial $f(x)$, so that

$$f_0 + f_1 \zeta^{j_v} + f_2 \zeta^{2j_v} + \cdots + f_{r-1}\zeta^{(r-1)j_v} + \zeta^{rj_v} = 0.$$

For each $i = 1, \ldots, r$, multiply this equation by ζ^{ij_v} to obtain

$$f_0 \zeta^{ij_v} + f_1 \zeta^{(i+1)j_v} + \cdots + f_{r-1}\zeta^{(i+r-1)j_v} + \zeta^{(i+r)j_v} = 0.$$

Recall that $e(x) = e_{j_1} x^{j_1} + e_{j_2} x^{j_2} + \cdots + e_{j_r} x^{j_r}$, and sum these equations over $v = 1, \ldots, r$ to obtain

$$f_0 e(\zeta^i) + f_1 e(\zeta^{i+1}) + \cdots + f_{r-1} e(\zeta^{i+r-1}) + \zeta^{i+r} = 0.$$

Hence, these r equations, for $i = 1, \ldots, r$, form an inhomogeneous $r \times r$ linear system

$$\Sigma(e)\dot{f}^T = h^T,$$

where $h = [-y(\zeta^{r+1}), -y(\zeta^{r+2}), \ldots, -y(\zeta^{2r})]^T$. But $\Sigma(e) = \Sigma(y)$, by part (i).

(iii) We claim that there is a factorization $\Sigma(y) = VDV^T$, where V is the $r \times r$ Vandermonde matrix

$$V = \begin{bmatrix} 1 & 1 & \cdots & 1 \\ \zeta^{j_1} & \zeta^{j_2} & \cdots & \zeta^{j_r} \\ \zeta^{2j_1} & \zeta^{2j_2} & \cdots & \zeta^{2j_r} \\ \vdots & \vdots & \cdots & \vdots \\ \zeta^{(r-1)j_1} & \zeta^{(r-1)j_2} & \cdots & \zeta^{(r-1)j_r} \end{bmatrix}$$

and $D = \text{diag}\{e_{j_1}\zeta^{j_1}, e_{j_2}\zeta^{j_2}, \ldots, e_{j_r}\zeta^{j_r}\}$. Now the matrix VD is obtained from V by multiplying its νth column, for all $\nu = 1, \ldots, r$, by $e_{j_\nu}\zeta^{j_\nu}$. Consider the product

$$(VD)V^T = \begin{bmatrix} e_{j_1}\zeta^{j_1} & e_{j_2}\zeta^{j_2} & \cdots & e_{j_r}\zeta^{j_r} \\ e_{j_1}\zeta^{2j_1} & e_{j_2}\zeta^{2j_2} & \cdots & e_{j_r}\zeta^{2j_r} \\ \vdots & \vdots & \cdots & \vdots \\ e_{j_1}\zeta^{rj_1} & e_{j_2}\zeta^{rj_2} & \cdots & e_{j_r}\zeta^{rj_r} \end{bmatrix} \begin{bmatrix} 1 & \zeta^{j_1} & \cdots & \zeta^{rj_1} \\ 1 & \zeta^{j_2} & \cdots & \zeta^{rj_2} \\ \vdots & \vdots & \cdots & \vdots \\ 1 & \zeta^{j_r} & \cdots & \zeta^{rj_r} \end{bmatrix}.$$

The 1,1 entry of $(VD)V^T$ is

$$e_{j_1}\zeta^{j_1} + e_{j_2}\zeta^{j_2} + \cdots + e_{j_r}\zeta^{j_r} = e(\zeta).$$

In fact, a similar calculation, using Proposition 4.132(iv), shows that all the entries of VDV^T match those of $\Sigma(y)$.

As all the ζ^{j_ν} are distinct, Exercise 4.52 on page 397 shows that the Vandermonde matrix V is nonsingular. The diagonal matrix D is nonsingular because its diagonal entries are nonzero. Therefore, $\Sigma(y) = VDV^T$ is nonsingular. •

Proposition 4.135 finds the error positions $j_1, \ldots, j_r$ in the error vector e. The error locating polynomial $f(x)$ is found by solving the linear system $\Sigma(y)\dot{f}^T = h^T$ [the uniqueness of $\dot{f}$ follows from the nonsingularity of $\Sigma(y)$]. Since $\dot{f} = (f_0, \ldots, f_{r-1})$, we have $f(x) = f_0 + f_1 x + \cdots + f_{r-1}x^{r-1} + x^r$. The monic polynomial $f(x)$ determines the error locator vector $u = fU$, where U is the parity check matrix in Proposition 4.132, and Lemma 4.134 gives $\{j_1, \ldots, j_r\} = \text{Supp}(e) = \mathcal{Z}(u)$. Erase the entries $y_{j_1}, \ldots, y_{j_r}$; to decode y, it remains to replace these erroneous entries by the true values.

The next theorem decodes Reed-Solomon codes. It is known that all BCH codes can be decoded using a generalization of this method.

→ **Theorem 4.136.** *If C is a t-error correcting Reed-Solomon code over $\mathbb{F}_q$ and if y is a word having an error vector e with $\text{wt}(e) \leq t$, then y can be efficiently decoded.*

Proof. Proposition 4.135 finds the error positions $j_1, \ldots, j_r$ in the error vector e, for $\text{Supp}(e) = \mathcal{Z}(fU)$, where $f(x)$ is the error locator polynomial and U is the parity check matrix for C in Proposition 4.132.

Assume first that $r = t$. Let U^* be the matrix obtained from U by deleting all columns except $j_1, j_2, \ldots, j_t$. By Exercise 4.31 on page 354, one can find the nonzero coordinates of e by solving $U^*e^* = Uy^T$ [our notation sets e^* to be the $1 \times t$ vector $(e_{j_1}, \ldots, e_{j_t})$]. Now Exercise 4.31 goes on to say that this smaller inhomogenous system can be solved (by Gaussian elimination) if $\text{rank}(U^*) = t$. But this is so; any t columns of U form a nonsingular $t \times t$ Vandermonde matrix, for it has distinct rows; thus, any t columns form a linearly independent list. Having found e^* and hence e, we have the codeword $c = y - e$. By Corollary 4.125, c is the unique nearest codeword to y. Decode y as $E^{-1}(c)$, where E is the encoding function.

If $r < t$, then this procedure will fail [the syndrome matrix $\Sigma(y)$ will be singular, and the error locating polynomial will not be determined]. Find the largest $r \le t$ for which the $r \times r$ matrix $\Sigma(y)$ is nonsingular, and then complete the decoding. •

→ **Example 4.137.**
Let C be the 2-error correcting Reed-Solomon code over $\mathbb{F}_7$ given in Example 4.130. Note that 3 is a primitive 6th root of unity in $\mathbb{F}_7$:

$$3^2 = 9 \equiv 2, \quad 3^3 = 27 \equiv 6, \quad 3^4 = 81 \equiv 4, \quad 3^5 = 243 \equiv 5, \quad 3^6 \equiv 1.$$

Here, $t = 2$ and $q = 7$, so that the parity check matrix U in Proposition 4.132 is the 4×6 matrix:

$$U = \begin{bmatrix} 1 & 3 & 2 & 6 & 4 & 5 \\ 1 & 2 & 4 & 1 & 2 & 4 \\ 1 & 6 & 1 & 6 & 1 & 6 \\ 1 & 4 & 2 & 1 & 4 & 2 \end{bmatrix}.$$

We are now going to decode the word $y = (4, 0, 5, 1, 0, 1)$ *assuming that there is an error vector e with* $\text{wt}(e) \le 2$.

(i) The syndrome is $yU^T = S(y) = (4, 1, 0, 3)$.

(ii) The syndrome matrix is $\Sigma(y) = \begin{bmatrix} 4 & 1 \\ 1 & 0 \end{bmatrix}$.

(iii) Solve $\Sigma(y)\dot{f}^T = h^T$, which here is $\begin{bmatrix} 4 & 1 \\ 1 & 0 \end{bmatrix}\begin{bmatrix} f_0 \\ f_1 \end{bmatrix} = \begin{bmatrix} 4 \\ 5 \end{bmatrix}$, and obtain $\dot{f} = (4, 5)$.

Thus, the error locator polynomial is $f(x) = 4 + 5x + x^2$.

(iv) The error locator vector is $u = fU = (3, 0, 1, 0, 6, 4)$. Thus, $\mathcal{Z}(u) = \{1, 3\}$, and so $\text{Supp}(e) = \{1, 3\}$.

(v) Solve the system $U^*e^* = Uy^T$, which here is

$$\begin{bmatrix} 3 & 6 \\ 2 & 1 \\ 6 & 6 \\ 4 & 1 \end{bmatrix}\begin{bmatrix} e_1 \\ e_3 \end{bmatrix} = \begin{bmatrix} 4 \\ 1 \\ 0 \\ 3 \end{bmatrix}.$$

The solution is $e^* = (1, 6)$, and the error vector is $e = (0, 1, 0, 6, 0, 0)$.

We now can decode y.

$$
\begin{aligned}
c &= y - e \\
&= (4, 0, 5, 1, 0, 1) - (0, 1, 0, 6, 0, 0) \\
&= (4, -6, 5, 2, 0, 1) \\
&= (4, 1, 5, 2, 0, 1)
\end{aligned}
$$

Thus, y is decoded as $E^{-1}(c)$, where E is the encoding function.   ◄

EXERCISES

4.66 Let $\mathcal{A}$ be an alphabet with $|\mathcal{A}| = q \geq 2$, let $T: \mathcal{A}^n \to \mathcal{A}^n$ be a transmission function, and let the probability of error in each letter of a transmitted word be p, where $0 < p < 1$.

   (i)   Prove that the probability P of the occurrence of exactly ℓ erroneous letters in a transmitted word of length n is $P = \left(\frac{p}{q-1}\right)^\ell (1 - p)^{n-\ell}$.

   (ii)   Prove that $P = \binom{n}{\ell} p^\ell (1 - p)^{n-\ell}$, and conclude that the probability P is independent of q.

*4.67 Prove that $d \geq 3$, where d is the minimum distance of the two-dimensional parity code in Example 4.105(iii).

*4.68 Let $\mathcal{A}$ be an alphabet with $|\mathcal{A}| = q$, and let $C \subseteq \mathcal{A}^n$ be an (n, M, d)-code.

   (i)   Prove that $\pi: C \to \mathcal{A}^{n-d+1}$, defined by $\pi(c_1, \ldots, c_n) = (c_d, \ldots, c_n)$, is an injection.

   (ii)   (*Singleton bound.*)   Prove that

$$
M \leq q^{n-d+1}.
$$

*4.69 If $\mathcal{A}$ is an alphabet with $|\mathcal{A}| = q$, and if $u \in \mathcal{A}^n$, define the (closed) **ball** of radius r with center u by

$$
B_r(u) = \{w \in \mathcal{A}^n : \delta(w, u) \leq r\},
$$

where δ is the Hamming distance.

   (i)   Prove that

$$
\left|\{w \in \mathcal{A}^n : \delta(u, w) = i\}\right| = \binom{n}{i}(q - 1)^i.
$$

   (ii)   Prove that

$$
|B_r(u)| = \sum_{i=0}^{r} \binom{n}{i}(q - 1)^i.
$$

H **4.70** (***Hamming bound***) If $C \subseteq \mathcal{A}^n$ is an (n, M, d)-code, where $|\mathcal{A}| = q$ and $d = 2t + 1$, prove that

$$M \leq \frac{q^n}{\sum_{i=0}^{t} \binom{n}{i}(q-1)^i}.$$

4.71 An (n, M, d)-code over an alphabet $\mathcal{A}$ with $|\mathcal{A}| = q$ is called a ***perfect code*** if it attains the Hamming bound:

$$M = \frac{q^n}{\sum_{i=0}^{t} \binom{n}{i}(q-1)^i}.$$

Prove that the Hamming $[2^\ell - 1, 2^\ell - 1 - \ell]$ codes in Example 4.113 are perfect codes.

4.72 Suppose that a code C detects up to s errors and that it corrects up to t errors. Prove that $t \leq s$.

*H **4.73** If $C \subseteq \mathbb{F}^n$ is a linear code and $w \in \mathbb{F}^n$, define $r = \min_{c \in C} \delta(w, c)$. Give an example of a linear code $C \subseteq \mathbb{F}^n$, which corrects up to t errors, and a word $w \in \mathbb{F}^n$ with $w \notin C$, such that there are distinct codewords $c, c' \in C$ with $\delta(w, c) = r = \delta(x, c')$. Conclude that correcting a transmitted word by choosing the codeword nearest to it may not be well-defined.

4.74 Let C be an $[n, m]$ linear code over a finite field $\mathbb{F}$, and let G be a generating matrix of C. Prove that an $m \times n$ matrix A is also a generating matrix of C if and only if $A = GH$ for some matrix $H \in GL(n, \mathbb{F})$.

4.75 Prove that the BCH code of length $m + 1$ over $\mathbb{F}_2$ having generating polynomial $x - 1$ is the $[m + 1, m]$ parity check code.

4.76 H **(i)** Write $x^{15} - 1$ as a product of irreducible polynomials in $\mathbb{F}_2[x]$.

    H **(ii)** Find an irreducible quartic polynomial $g(x) \in \mathbb{F}_2[x]$, and use it to define a primitive 15th root of unity $\zeta \in \mathbb{F}_{16}$.

    H **(iii)** Find a BCH code C over $\mathbb{F}_2$ of length 15 having minimum distance $d(C) \geq 3$.

H **4.77** Let C be the 2-error correcting Reed-Solomon $[7, 3]$ code over $\mathbb{F}_8$ in Example 4.131. Decode the word $y = (\zeta^3, \zeta, 1, \zeta^3 + \zeta, 0, \zeta^3, \zeta^3)$ assuming that it has an error vector of weight 2.

5

Fields

The study of the roots of polynomials is intimately related to the study of fields. If $f(x) \in k[x]$, where k is a field, then it is natural to consider the relation between k and the larger field E, where E is obtained from k by adjoining all the roots of $f(x)$. For example, if $E = k$, then $f(x)$ is a product of linear factors in $k[x]$. We shall see that the pair E and k has a *Galois group*, $\mathrm{Gal}(E/k)$, and that this group determines whether there exists a formula for the roots of $f(x)$ which generalizes the quadratic formula.

→ 5.1 CLASSICAL FORMULAS

Revolutionary events were changing the western world in the early sixteenth century: the printing press had been invented around 1450; trade with Asia and Africa was flourishing; Columbus had just discovered the New World; and Martin Luther was challenging papal authority. The Reformation and the Renaissance were beginning.

The Italian peninsula was not one country but a collection of city states with many wealthy and cosmopolitan traders. Public mathematics contests, sponsored by the dukes of the cities, were an old tradition; there are records from 1225 of Leonardo of Pisa (c. 1180–c. 1245), also called Fibonacci, approximating roots of $x^3 + 2x^2 + 10x - 20$ with good accuracy. One of the problems frequently set involved finding roots of a given cubic equation

$$X^3 + bX^2 + cX + d = 0,$$

where a, b, and c were real numbers, usually integers.[1]

---

[1] Around 1074, Omar Khayyam (1048–1123), a Persian mathematician now more famous for his poetry, used intersections of conic sections to give geometric constructions of roots of cubics.

Modern notation did not exist in the early 1500s, and so the feat of finding the roots of a cubic involved not only mathematical ingenuity but also an ability to surmount linguistic obstacles. Designating variables by letters was invented in 1591 by F. Viète (1540–1603) who used consonants to denote constants and vowels to denote variables (the modern notation of using letters $a, b, c, \ldots$ at the beginning of the alphabet to denote constants and letters x, y, z at the end of the alphabet to denote variables was introduced in 1637 by R. Descartes in his book *La Géométrie*). The exponential notation $A^2, A^3, A^4, \ldots$ was essentially introduced by J. Hume in 1636 (he used $A^{ii}, A^{iii}, A^{iv}, \ldots$). The symbols $+$, $-$, and $\sqrt{}$, as well as the symbol / for division, as in a/b, were introduced by J. Widman in 1486. The symbol $\times$ for multiplication was introduced by W. Oughtred in 1631, and the symbol $\div$ for division by J. H. Rahn in 1659. The symbol $=$ was introduced by the Oxford don Robert Recorde in 1557, in his *Whetstone of Wit*:

> And to avoide the tediouse repetition of these woordes: is equal to: I will lette as I doe often in woorke use, a paire of paralleles, or gemowe[2] lines of one lengthe, thus: $=$, because noe 2 thynges, can be moare equalle.

These symbols were not adopted at once, and often there were competing notations. Most of this notation did not become universal in Europe until the next century, with the publication of Descartes's *La Géométrie*.

Let us return to cubic equations. The lack of good notation was a great handicap. For example, the cubic equation $X^3 + 2X^2 + 4X - 1 = 0$ would be given, roughly, as follows:

> Take the cube of a thing, add to it twice the square of the thing, to this add 4 times the thing, and this must all be set equal to 1.

Complicating matters even more, negative numbers were not accepted; an equation of the form $X^3 - 2X^2 - 4X + 1 = 0$ would only be given in the form $X^3 + 1 = 2X^2 + 4X$. Thus, there were many forms of cubic equations, depending (in our notation) on whether coefficients were positive, negative, or zero.

The following history is from the excellent account in J.-P. Tignol, *Galois' Theory of Equations*.

> The algebraic solution of $X^3 + mX = n$ was first obtained around 1515 by Scipione del Ferro, professor of mathematics in Bologna. Not much is known about him nor about his solution as, for some reason, he decided not to publicize his result. After his death in 1526, his method passed to some of his pupils.

---

[2]*Gemowe* is an obsolete word meaning *twin* or, in this case, *parallel*.

The second discovery of the solution is much better known, through the accounts of its author himself, Niccolò Fontana (c. 1500–1557), from Brescia, nicknamed "Tartaglia" ("Stammerer"). In 1535, Tartaglia, who had dealt with some very particular cases of cubic equations, was challenged to a public problem-solving contest by Antonio Maria Fior, a former pupil of Scipione del Ferro. When he heard that Fior had received the solution of cubic equations from his master, Tartaglia threw all his energy and his skill into the struggle. He succeeded in finding the solution just in time to inflict upon Fior a humiliating defeat.

The news that Tartaglia had found the solution of cubic equation reached Giralamo Cardano (1501–1576), a very versatile scientist, who wrote a number of books on a wide variety of subjects, including medicine, astrology, astronomy, philosophy, and mathematics. Cardano then asked Tartaglia to give him his solution, so that he could include it in a treatise on arithmetic, but Tartaglia flatly refused, since he was himself planning to write a book on this topic. It turns out that Tartaglia later changed his mind, at least partially, since in 1539 he handed on to Cardano the solution of $X^3 + mX = n$, $xX^3 = mX + n$, and a very brief indication of $X^3 + n = mX$ in verses. ...

Having received Tartaglia's poem, Cardano set to work; he not only found justifications for the formulas, but he also solved all the other types of cubics. He then published his results, giving due credit to Tartaglia and to del Ferro, in the epoch-making book *Ars Magna, sive de regulis algebraicis* (The Great Art, or the Rules of Algebra).

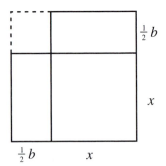

Figure 5.1   Completing the Square

Let us now derive the formulas for the roots of polynomials of low degree. The usual way to derive the quadratic formula is by "completing the square," which can be taken literally. Consider the quadratic equation $x^2 + bx + c = 0$ with $b \geq 0$, and view $x^2 + bx$ as the area pictured in Figure 5.1. One completes the square by adding on the corner square having area $\frac{1}{4}b^2$. The area of the large square is $(x + \frac{1}{2}b)^2$; if $c + \frac{1}{4}b^2 \geq 0$, then we have constructed a square with sides of length $x + \frac{1}{2}b$ and area $c + \frac{1}{4}b^2$. This geometric construction can be done algebraically without assuming that certain quantities are non-negative. Let $f(x) = x^2 + bx + c$.

$$x^2 + bx + c = x^2 + bx + \tfrac{1}{4}b^2 + c - \tfrac{1}{4}b^2$$
$$= (x + \tfrac{1}{2}b)^2 + \tfrac{1}{4}(4c - b^2).$$

Therefore, if z is a root of $f(x)$, then

$$z + \tfrac{1}{2}b = \pm\tfrac{1}{2}\sqrt{b^2 - 4c}.$$

We now present a different derivation of the quadratic formula which begins by replacing a given polynomial by a simpler one.

Definition. A polynomial $f(x) \in \mathbb{R}[x]$ of degree n is **reduced**[3] if it has no x^{n-1} term; that is, $f(x) = a_n x^n + a_{n-2}x^{n-2} + \cdots + a_0$.

Lemma 5.1. *The substitution $X = x - \frac{1}{n}a_{n-1}$ changes*

$$f(X) = X^n + a_{n-1}X^{n-1} + h(X),$$

where $h(X) = 0$ or $\deg(h) \leq n - 2$, into a reduced polynomial

$$f^*(x) = f(x - \tfrac{1}{n}a_{n-1}).$$

Moreover, if u is a root of $f^*(x)$, then $u - \frac{1}{n}a_{n-1}$ is a root of $f(X)$.

Proof. The substitution $X = x - \frac{1}{n}a_{n-1}$ gives

$$f^*(x) = f(x - \tfrac{1}{n}a_{n-1})$$
$$= (x - \tfrac{1}{n}a_{n-1})^n + a_{n-1}(x - \tfrac{1}{n}a_{n-1})^{n-1} + h(x - \tfrac{1}{n}a_{n-1})$$
$$= \left(x^n - a_{n-1}x^{n-1} + g_1(x)\right) + a_{n-1}\left(x^{n-1} + g_2(x)\right) + h(x - \tfrac{1}{n}a_{n-1})$$
$$= x^n + g_1(x) + a_{n-1}g_2(x) + h(x - \tfrac{1}{n}a_{n-1}),$$

where each of $g_1(x)$, $g_2(x)$, $h(x - \frac{1}{n}a_{n-1})$, and $g_1(x) + a_{n-2}g_2(x) + h(x - \frac{1}{n}a_{n-1})$ is either 0 or a polynomial of degree $\leq n - 2$. It follows that the polynomial $f^*(x) = f(x - \frac{1}{n}a_{n-1})$ is reduced.

Finally, if u is a root of $f^*(x)$, then $0 = f^*(u) = f(u - \frac{1}{n}a_{n-1})$; that is, $u - \frac{1}{n}a_{n-1}$ is a root of $f(X)$. ●

---

[3]If $f(x) = x^n + c_{n-1}x^{n-1} + \cdots + c_1 x + c_0 = (x - r_1)\cdots(x - r_n)$, then $c_{n-1} = -(r_1 + \cdots + r_n)$. Thus, $f(x)$ is reduced if and only if the sum of its roots is 0.

Here is another proof of the quadratic formula.

→ **Corollary 5.2 (Quadratic Formula).**   *If $f(X) = X^2 + bX + c$, then its roots are*

$$\tfrac{1}{2}\left(-b \pm \sqrt{b^2 - 4c}\right).$$

Proof.   Define x by $X = x - \tfrac{1}{2}b$. Now

$$f^*(x) = (x - \tfrac{1}{2}b)^2 + b(x - \tfrac{1}{2}b) + c.$$

The linear terms cancel, the reduced polynomial is

$$f^*(x) = x^2 - \tfrac{1}{4}(b^2 - 4c),$$

and the roots of $f^*(x)$ are $u = \pm\tfrac{1}{2}\sqrt{b^2 - 4c}$. But Lemma 5.1 says that the roots of $f(X)$ are $u - \tfrac{1}{2}b$; that is, the roots of $f(X)$ are $\tfrac{1}{2}\left(-b \pm \sqrt{b^2 - 4c}\right)$.   •

The following consequence of the quadratic formula will be used in deriving the cubic formula.

Corollary 5.3.   *Given numbers c and d, there exist numbers α and β with $\alpha + \beta = c$ and $\alpha\beta = d$.*

Proof.   If $d = 0$, choose $\alpha = 0$ and $\beta = c$. If $d \neq 0$, then $\alpha \neq 0$ and we may set $\beta = d/\alpha$. Substituting, $c = \alpha + \beta = \alpha + d/\alpha$, so that

$$\alpha^2 - c\alpha + d = 0.$$

The quadratic formula now shows that such an α exists, as does $\beta = d/\alpha$ (of course, α and β might be complex numbers).   •

Lemma 5.1 simplifies the original polynomial and, at the same time, keeps control of its roots. In particular, if $n = 3$, then $f^*(x)$ has the form $x^3 + qx + r$.

Cardano's "trick" in solving a reduced cubic is to write a root u of $x^3 + qx + r$ as

$$u = \alpha + \beta,$$

and then to find α and β. Now

$$0 = u^3 + qu + r$$
$$= (\alpha + \beta)^3 + q(\alpha + \beta) + r.$$

Note that

$$(\alpha + \beta)^3 = \alpha^3 + 3\alpha^2\beta + 3\alpha\beta^2 + \beta^3$$
$$= \alpha^3 + \beta^3 + 3\alpha\beta(\alpha + \beta)$$
$$= \alpha^3 + \beta^3 + 3\alpha\beta u.$$

Therefore, $0 = \alpha^3 + \beta^3 + 3\alpha\beta u + qu + r$, and so

$$0 = \alpha^3 + \beta^3 + u(3\alpha\beta + q) + r. \tag{1}$$

We have already set $\alpha + \beta = u$; by Corollary 5.3, we may impose a second condition,

$$\alpha\beta = -\tfrac{1}{3}q, \tag{2}$$

which makes the u term in Eq. (1) go away, leaving

$$\alpha^3 + \beta^3 = -r. \tag{3}$$

Cubing each side of Eq. (2) gives

$$\alpha^3\beta^3 = -\tfrac{1}{27}q^3. \tag{4}$$

Equations (3) and (4) in the two unknowns α^3 and β^3 can be solved, as in Corollary 5.3. Substituting $\beta^3 = -q^3/(27\alpha^3)$ into Eq. (3) gives

$$\alpha^3 - \frac{q^3}{27\alpha^3} = -r,$$

which may be rewritten as

$$\alpha^6 + r\alpha^3 - \tfrac{1}{27}q^3 = 0, \tag{5}$$

a quadratic $y^2 + ry - \tfrac{1}{27}q^3$ in α^3. The quadratic formula gives

$$\alpha^3 = \tfrac{1}{2}\left(-r + \sqrt{D}\right), \tag{6}$$

where $D = r^2 + \tfrac{4}{27}q^3$. Note that β^3 is also a root of the quadratic in Eq. (5), so that

$$\beta^3 = \tfrac{1}{2}\left(-r - \sqrt{D}\right), \tag{7}$$

Now take a cube root[4] to obtain α. By Eq. (2), $\beta = -q/(3\alpha)$, and so $u = \alpha + \beta$ has been found.

What are the other two roots? Theorem 3.49 says that if u is a root of a polynomial $f(x)$, then $f(x) = (x - u)g(x)$ for some polynomial $g(x)$. After finding one root $u = \alpha + \beta$, divide $x^3 + qx + r$ by $x - u$, and use the quadratic formula on the quadratic quotient $g(x)$ to find the other two roots [any root of $g(x)$ is a root of $f(x)$].

Here is an explicit formula for the other two roots of $f(x)$ (instead of the method just given for finding them). There are three cube roots of unity, namely, 1, $\omega = -\tfrac{1}{2} + i\tfrac{\sqrt{3}}{2}$, and $\omega^2 = -\tfrac{1}{2} - i\tfrac{\sqrt{3}}{2}$. It follows that the other cube roots of α^3 are $\omega\alpha$ and

---

[4]The number $z = \tfrac{1}{2}(-r + \sqrt{D})$ might be complex. The easiest way to find a cube root of z is to write it in polar form $se^{i\theta}$, where $s \geq 0$; a cube root is then $\sqrt[3]{s}\, e^{i\theta/3}$.

$\omega^2\alpha$. If β is the "mate" of α, that is, if $\beta = -q/(3\alpha)$, as in Eq. (2), then the mate of $\omega\alpha$ is

$$-q/(3\omega\alpha) = \beta/\omega = \omega^2\beta,$$

and the mate of $\omega^2\alpha$ is

$$-q/(3\omega^2\alpha) = \beta/\omega^2 = \omega\beta.$$

Therefore, explicit formulas for the roots of $f(x)$ are $\alpha + \beta$, $\omega\alpha + \omega^2\beta$, and $\omega^2\alpha + \omega\beta$. We have proved the *cubic formula* (also called ***Cardano's formula***).

→ **Theorem 5.4 (Cubic Formula).**   *The roots of $x^3 + qx + r$ are*

$$\alpha + \beta, \qquad \omega\alpha + \omega^2\beta, \qquad and \qquad \omega^2\alpha + \omega\beta,$$

where $\alpha^3 = \frac{1}{2}(-r + \sqrt{D})$, $\beta = -\frac{1}{3}\dfrac{q}{\alpha}$, $D = r^2 + \frac{4}{27}q^3$, and $\omega = -\frac{1}{2} + i\frac{\sqrt{3}}{2}$ is a cube root of unity.

Proof.   We have just given the proof when $\alpha \neq 0$. By Eq. (2), we have $\alpha\beta = -q/3$, and so $\alpha = 0$ forces $q = 0$; that is, the reduced cubic is $x^3 + r$. In this case, $\beta^3 = -r$, the roots are β, $\omega\beta$, and $\omega^2\beta$, and the cubic formula holds in this case as well.   •

Recall that Eq. (7) gives $\beta^3 = \frac{1}{2}(-r - \sqrt{D})$.

The quadratic and cubic formulas are not valid over arbitrary coefficient fields. For example, since $2 = 0$ in fields k of characteristic 2, the quadratic formula does not make sense for quadratics in $k[x]$ because $\frac{1}{2}$ is not defined. Similarly, the cubic formula (and the quartic formula below) does not apply to polynomials with coefficients in fields of characteristic 2 or characteristic 3 because the formulas involve $\frac{1}{2}$ and $\frac{1}{3}$, one of which is not defined in these fields.

→ **Example 5.5 (Good Example).**
We find the roots of $x^3 - 15x - 126$. The polynomial is already reduced, for there is no x^2 term, and so it is in the form to which the cubic formula applies (were it not reduced, one would first reduce it, as in Lemma 5.1). Here, $q = -15$, $r = -126$, $D = (-126)^2 + \frac{4}{27}(-15)^3 = 15{,}376$, and $\sqrt{D} = 124$. The cubic formula gives $\alpha^3 = \frac{1}{2}[-(-126) + 124] = 125$ and $\alpha = 5$, while $\beta = -q/3\alpha = 15/(3 \cdot 5) = 1$. The roots are thus $\alpha + \beta = 6$, $\omega\alpha + \omega^2\beta = -3 + 2i\sqrt{3}$, and $\omega^2\alpha + \omega\beta = -3 - 2i\sqrt{3}$.

Alternatively, having found $u = 6$ to be a root, the division algorithm gives

$$x^3 - 15x - 126 = (x - 6)(x^2 + 6x + 21),$$

and the quadratic formula gives $-3 \pm 2i\sqrt{3}$ as the roots of the quadratic factor.   ◄

$\rightarrow$ **Example 5.6 (Bad Example).**

In Example 5.5, the cubic formula gave the roots of $x^3 - 15x - 126$ in a routine way. Let us now try the cubic formula on the polynomial

$$x^3 - 7x + 6 = (x - 1)(x - 2)(x + 3)$$

whose roots are, obviously, 1, 2, and -3. There is no x^2 term, $q = -7$, $r = 6$, and $D = r^2 + \frac{4}{27}q^3 = -\frac{400}{27} < 0$. The cubic formula gives a messy answer: the roots are

$$\alpha + \beta, \qquad \omega\alpha + \omega^2\beta, \qquad \omega^2\alpha + \omega\beta,$$

where $\alpha^3 = \frac{1}{2}(-6 + \sqrt{-\frac{400}{27}})$ and $\beta^3 = \frac{1}{2}(-6 - \sqrt{-\frac{400}{27}})$. Something strange has happened. There are three curious equations saying that each of 1, 2, and -3 is equal to one of the messy expressions displayed above; thus,

$$\omega\sqrt[3]{\frac{1}{2}\left(-6 + \sqrt{-\frac{400}{27}}\right)} + \omega^2\sqrt[3]{\frac{1}{2}\left(-6 - \sqrt{-\frac{400}{27}}\right)}$$

is equal to 1, 2, or -3. Aside from the complex cube roots of unity, this expression involves square roots of the negative number $-\frac{400}{27}$.

This example shows why the cubic formula is rarely used; although it does give the roots of a cubic, it may give them in unrecognizable form. ◀

Until the Middle Ages, mathematicians had no difficulty in ignoring square roots of negative numbers when dealing with quadratic equations. For example, consider the problem of finding the sides x and y of a rectangle having area A and perimeter p. The equations

$$xy = A \qquad \text{and} \qquad 2x + 2y = p$$

lead to the quadratic equation $2x^2 - px + 2A = 0$, and, as in Corollary 5.3, the quadratic formula gives the roots

$$x = \tfrac{1}{4}\left(p \pm \sqrt{p^2 - 16A}\right).$$

If $p^2 - 16A \geq 0$, one has found x (and y); if $p^2 - 16A < 0$, one merely says that there is no rectangle whose perimeter and area are in this relation. But the cubic formula does not allow one to discard "imaginary" roots, for we have just seen that an "honest" real and positive root, even a positive integer, can appear in terms of complex numbers.[5] The Pythagoreans in ancient Greece considered *number* to mean positive integer. By the Middle Ages, *number* came to mean positive real number (although there was little understanding of what real numbers are). The importance of the cubic formula in the

---

[5] We saw a similar phenomenon in Theorem 1.15: the integer terms of the Fibonacci sequence are given in terms of $\sqrt{5}$.

history of mathematics is that it forced mathematicians to take both complex numbers and negative numbers seriously.

The physicist R. P. Feynman (1918–1988), one of the first winners of the annual Putnam national mathematics competition (and also a Nobel laureate in physics), suggested another possible value of the cubic formula. As mentioned earlier, the cubic formula was found in 1515, a time of great change. One of the factors contributing to the Dark Ages was an almost slavish worship of the classical Greek and Roman civilizations. It was believed that that earlier era had been the high point of man's accomplishments; contemporary man was inferior to his forebears (a world view opposite to the modern one of continual progress!). The cubic formula was essentially the first instance of a mathematical formula unknown to the ancients, and so it may well have been a powerful example showing that 16th-century man was the equal of his ancestors.

The quartic formula, discovered by Lodovici Ferrari (1522–1565) in the early 1540s, also appeared in Cardano's book, but it was given much less attention there than the cubic formula. The reason given by Cardano is that cubic polynomials have an interpretation as volumes, whereas quartic polynomials have no such obvious justification. Cardano wrote,

> As the first power refers to a line, the square to a surface, and the cube to a solid body, it would be very foolish for us to go beyond this point. Nature does not permit it. Thus, ..., all those matters up to and including the cubic are fully demonstrated, but for the others which we will add, we do not go beyond barely setting out.

The *discriminant* of a cubic polynomial $f(x) \in \mathbb{R}[x]$ is a number that detects interesting properties; are all the roots of $f(x)$ real; does $f(x)$ have repeated roots.

Definition. If $f(x) = x^3 + qx + r = (x-u)(x-v)(x-w)$, then define the numbers $\Delta = (u-v)(u-w)(v-w)$ and

$$\Delta^2 = [(u-v)(u-w)(v-w)]^2;$$

we call Δ^2 the *discriminant*[6] of $f(x)$.

It is natural to consider Δ^2 instead of Δ, for Δ is a number depending not only on the roots but also on the order in which they are listed. Had we listed the roots as u, w, v, for example, then $(u-w)(u-v)(w-v) = -\Delta$, because the factor $w-v = -(v-w)$ has changed sign. Squaring eliminates this difference.

---

[6]More generally, if $f(x) = (x-u_1)(x-u_2)\ldots(x-u_n)$ is a polynomial of degree n, then the discriminant of $f(x)$ is defined to be Δ^2, where $\Delta = \prod_{i<j}(u_i - u_j)$ (one takes $i < j$ so that each difference $u_i - u_j$ occurs just once in the product). In particular, the quadratic formula shows that the discriminant of $x^2 + bx + c$ is $b^2 - 4c$.

Note that if $\Delta^2 = 0$, then $\Delta = 0$ and the cubic has a repeated root. Can we detect this without first computing the roots? The cubic formula allows us to compute Δ^2 in terms of q and r.

Lemma 5.7. *The discriminant Δ^2 of $f(x) = x^3 + qx + r$ is*

$$\Delta^2 = -27r^2 - 4q^3 = -27D.$$

Proof. If the roots of $f(x)$ are u, v, and w, then the cubic formula gives

$$u = \alpha + \beta; \qquad v = \omega\alpha + \omega^2\beta; \qquad w = \omega^2\alpha + \omega\beta,$$

where $\omega = -\frac{1}{2} - i\frac{\sqrt{3}}{2}$, $D = r^2 + \frac{4}{27}q^3$, $\alpha = \sqrt[3]{\frac{1}{2}(-r + \sqrt{D})}$, and $\beta = -\frac{1}{3}\frac{q}{\alpha}$. We check easily that

$$u - v = \alpha + \beta - \omega\alpha - \omega^2\beta = (1 - \omega)(\alpha - \omega^2\beta);$$
$$u - w = \alpha + \beta - \omega^2\alpha - \omega\beta = -\omega^2(1 - \omega)(\alpha - \omega\beta);$$
$$v - w = \omega\alpha + \omega^2\beta - \omega^2\alpha - \omega\beta = \omega(1 - \omega)(\alpha - \beta).$$

Therefore,

$$\Delta = -\omega^3(1 - \omega)^3(\alpha - \beta)(\alpha - \omega\beta)(\alpha - \omega^2\beta).$$

Of course, $-\omega^3 = -1$, while

$$(1 - \omega)^3 = 1 - 3\omega + 3\omega^2 - \omega^3 = -3(\omega - \omega^2).$$

But $\omega = -\frac{1}{2} + i\frac{\sqrt{3}}{2}$ and $\omega^2 = \overline{\omega} = -\frac{1}{2} - i\frac{\sqrt{3}}{2}$, so that $\omega - \omega^2 = i\sqrt{3}$; hence, $(1 - \omega)^3 = -3(\omega - \omega^2) = -3i\sqrt{3}$, and

$$-\omega^3(1 - \omega)^3 = 3i\sqrt{3}.$$

Finally, Exercise 3.85 on page 281 gives

$$(\alpha - \beta)(\alpha - \omega\beta)(\alpha - \omega^2\beta) = \alpha^3 - \beta^3 = \sqrt{D}.$$

Therefore, $\Delta = 3i\sqrt{3}\sqrt{D}$, and

$$\Delta^2 = -27D = -27r^2 - 4q^3. \quad \bullet$$

It follows, for example, that the cubic formula is not needed to see that the cubic $f(x) = x^3 - 3x + 2$ has a repeated root, for $-27r^2 - 4q^3 = 0$. It also follows that if $f(x) \in k[x]$, then its discriminant lies in k as well.

We are now going to use the discriminant to detect whether the roots of a cubic are all real.

Lemma 5.8. *Every $f(x) \in \mathbb{R}[x]$ of odd degree has a real root.*

Remark.   The proof we give assumes that $f(x)$ has a complex root (which follows from the Fundamental Theorem of Algebra).   ◀

Proof.   The proof is by induction on $n \geq 0$, where $\deg(f) = 2n + 1$. The base step $n = 0$ is obviously true. Let $n \geq 1$ and let u be a complex root of $f(x)$. If u is real, we are done. Otherwise $u = a + ib$, and Exercise 5.6 on page 448 shows that the complex conjugate $\bar{u} = a - ib$ is also a root; moreover, $u \neq \bar{u}$ because u is not real. Both $x - u$ and $x - \bar{u}$ are divisors of $f(x)$; as these divisors are relatively prime, their product is also a divisor; there is a factorization in $\mathbb{C}[x]$

$$f(x) = (x - u)(x - \bar{u})g(x).$$

Now $(x - u)(x - \bar{u}) = x^2 - 2ax + a^2 + b^2 \in \mathbb{R}[x]$, and so the division algorithm gives $g(x) = f(x)/(x - u)(x - \bar{u}) \in \mathbb{R}[x]$. Since $\deg(g) = (2n + 1) - 2 = 2(n - 1) + 1$, the inductive hypothesis says that $g(x)$, and hence $f(x)$, has a real root.   •

Proposition 5.9.   *All the roots u, v, w of $x^3 + qx + r \in \mathbb{R}[x]$ are real numbers if and only if the discriminant $\Delta^2 \geq 0$; that is, $27r^2 + 4q^3 \leq 0$.*

Proof.   If u, v, and w are real numbers, then $\Delta = (u - v)(u - w)(v - w)$ is a real number. Therefore, $-27r^2 - 4q^3 = \Delta^2 \geq 0$, and $27r^2 + 4q^3 \leq 0$.

Conversely, assume that $w = s + ti$ is not real (i.e., $t \neq 0$); by Exercise 5.6 on page 448, the complex conjugate of a root is also a root, say, $v = s - ti$; by Lemma 5.8, the other root u is real. Now

$$\begin{aligned} \Delta &= (u - s + ti)(u - s - ti)(s - ti - [s + ti]) \\ &= (-2ti)[(u - s)^2 + t^2]. \end{aligned}$$

Since u, s, and t are real numbers,

$$\begin{aligned} \Delta^2 &= (-2ti)^2[(u - s)^2 + t^2]^2 \\ &= 4t^2 i^2 [(u - s)^2 + t^2]^2 \\ &= -4t^2[(u - s)^2 + t^2]^2 < 0, \end{aligned}$$

and so $0 > \Delta^2 = -27r^2 - 4q^3$. We have shown that if there is a nonreal root, then $27r^2 + 4q^3 > 0$; equivalently, if all the roots are real, then $27r^2 + 4q^3 \leq 0$.   •

We present the derivation of the quartic formula given by Descartes.

→ **Theorem 5.10 (Quartic Formula).**   *There is a method to compute the four roots of a quartic*

$$X^4 + bX^3 + cX^2 + dX + e.$$

Proof. As with the cubic, the quartic can be simplified, by setting $X = x - \frac{1}{4}b$, to

$$x^4 + qx^2 + rx + s; \tag{8}$$

moreover, if a number u is a root of the second polynomial, then $u - \frac{1}{4}b$ is a root of the first.

Factor the quartic in Eq. (8) into quadratics:

$$x^4 + qx^2 + rx + s = (x^2 + jx + \ell)(x^2 - jx + m) \tag{9}$$

(the coefficient of x in the second factor is $-j$ because the quartic has no x^3 term). If j, ℓ, and m can be found, then the quadratic formula can be used to find the roots of the quartic in Eq. (8).

Expanding the right-hand side of Eq. (9) and equating coefficients of like terms gives the equations

$$\begin{cases} m + \ell - j^2 & = q; \\ j(m - \ell) & = r; \\ \ell m & = s. \end{cases} \tag{10}$$

Adding and subtracting the top two equations in Eqs. (10) yield

$$\begin{cases} 2m & = j^2 + q + r/j; \\ 2\ell & = j^2 + q - r/j. \end{cases} \tag{11}$$

Now substitute these into the bottom equation of Eqs. (10):

$$\begin{aligned} 4s = 4\ell m &= (j^2 + q + r/j)(j^2 + q - r/j) \\ &= (j^2 + q)^2 - r^2/j^2 \\ &= j^4 + 2j^2 q + q^2 - r^2/j^2. \end{aligned}$$

Clearing denominators and transposing gives

$$j^6 + 2qj^4 + (q^2 - 4s)j^2 - r^2 = 0, \tag{12}$$

a cubic equation in j^2. The cubic formula allows one to solve for j^2, and one then finds ℓ and m using Eqs. (11).   •

→ **Example 5.11.**
Consider

$$x^4 - 2x^2 + 8x - 3 = 0,$$

so that $q = -2$, $r = 8$, and $s = -3$. If we factor this quartic into

$$(x^2 + jx + \ell)(x^2 - jx + m),$$

then Eq. (12) gives

$$j^6 - 4j^4 + 16j^2 - 64 = 0.$$

One could use the cubic formula to find j^2, but this would be very tedious, for one must first get rid of the j^4 term before doing the rest of the calculations. It is simpler, in this case, to observe that $j = 2$ is a root, for the equation can be rewritten

$$j^6 - 4j^4 + 16j^2 - 64 = j^6 - 2^2 j^4 + 2^4 j^2 - 2^6 = 0$$

(many elementary texts are fond of saying, in such circumstances, that $j = 2$ is found "by inspection"). We now find ℓ and m using Eqs. (11).

$$2\ell = 4 - 2 + (8/2) = 6$$
$$2m = 4 - 2 - (8/2) = -2$$

Thus, the original quartic factors into

$$(x^2 - 2x + 3)(x^2 + 2x - 1).$$

The quadratic formula now gives the roots of the quartic:

$$-1 + i\sqrt{2}, \qquad -1 - i\sqrt{2}, \qquad 1 + i\sqrt{2}, \qquad 1 - i\sqrt{2}. \quad \blacktriangleleft$$

Do not be misled by this example. Just as with the cubic formula, the quartic formula may give the roots in recognizable form. The quartic formula gives complicated versions of the roots of $x^4 - 25x^2 + 60x - 36 = (x - 1)(x - 2)(x - 3)(x + 6)$, as the reader may check.

It is now very tempting, as it was for our ancestors, to seek the roots of a quintic $g(X) = X^5 + bX^4 + cX^3 + dX^2 + eX + f$. Begin by reducing the polynomial with the substitution $X = x - \frac{1}{5}b$. It is natural to expect that some further ingenious substitution together with the formulas for the roots of lower degree polynomials will yield the roots of $g(X)$. But quintics resisted all such attempts for almost 300 years. We shall continue this story later in this chapter.

Viète's Cubic Formula

A formula involving extraction of roots is not necessarily the simplest way to find the roots of a cubic. We shall now give another formula for the roots of $x^3 + qx + r$, due to Viète, which replaces the operations of extraction of roots (which are, after all, "infinatary" in the sense that their evaluation requires limits in $\mathbb{R}$, in contrast to the "finitary" field operations) by evaluation of cosines. By Corollary 1.26, we have

$$\cos(3\theta) = 4\cos^3\theta - 3\cos\theta.$$

It follows that one root of the cubic

$$y^3 - \tfrac{3}{4}y - \tfrac{1}{4}\cos(3\theta) \tag{13}$$

is $u = \cos\theta$. By Exercise 5.8 on page 449, the other two roots of this particular cubic are $u = \cos(\theta + 120°)$ and $u = \cos(\theta + 240°)$.

Now let $f(x) = x^3 + qx + r$ be a cubic all of whose roots are real (Proposition 5.9 gives a way of checking when this is the case). We want to force $f(x)$ to look like Eq. (13). If v is a root of $f(x)$, set

$$v = tu,$$

where t and u will be chosen[7] in a moment. Substituting,

$$0 = f(tu) = t^3u^3 + qtu + r,$$

and so

$$u^3 + (q/t^2)u + r/t^3 = 0;$$

that is, u is a root of $g(y) = y^3 + (q/t^2)y + r/t^3$. If we can choose t so that

$$q/t^2 = -\tfrac{3}{4} \tag{14}$$

and

$$r/t^3 = -\tfrac{1}{4}\cos(3\theta) \tag{15}$$

for some θ, then $g(y) = y^3 - \tfrac{3}{4}y - \tfrac{1}{4}\cos(3\theta)$ and its roots are

$$u = \cos\theta, \quad u = \cos(\theta + 120°), \quad u = \cos(\theta + 240°).$$

But if $u^3 + (q/t^2)u + r/t^3 = 0$, then $t^3u^3 + qtu + r = 0$; that is, the roots $v = tu$ of $f(x) = x^3 + qx + r = 0$ are

$$v = tu = t\cos\theta, \quad v = t\cos(\theta + 120°), \quad v = t\cos(\theta + 240°).$$

We now find t and u. Eq. (14) gives $t^2 = -4q/3$, and so

$$t = \sqrt{-4q/3}. \tag{16}$$

One immediate consequence of Proposition 5.9, $27r^2 + 4q^3 \le 0$, is

$$4q^3 \le -27r^2;$$

---

[7] Scipione's trick writes a root as a sum $\alpha + \beta$, while Viète's trick writes it as a product.

as the right side is negative, q must also be negative. Therefore, $-4q/3$ is positive, and $t = \sqrt{-4q/3}$ is a real number. Equation (15) gives

$$\cos(3\theta) = -4r/t^3,$$

and this determines θ if $|-4r/t^3| \leq 1$. Since $27r^2 \leq -4q^3$, we have $9r^2/q^2 \leq -4q/3$; taking square roots,

$$\left|\frac{3r}{q}\right| \leq \sqrt{\frac{-4q}{3}} = t,$$

because Eq. (16) gives $t = \sqrt{-4q/3}$. Now $t^2 = -4q/3$, and so

$$\left|\frac{-4r}{t^3}\right| = \left|\frac{-4r}{(-4q/3)t}\right| = \left|\frac{3r}{q} \cdot \frac{1}{t}\right| \leq \frac{t}{t} = 1,$$

as desired. We have proved the following theorem.

Theorem 5.12 (Viète).   *Let* $f(x) = x^3 + qx + r$ *be a cubic polynomial for which* $27r^2 + 4q^3 \leq 0$. *If* $t = \sqrt{-4q/3}$ *and* $\cos 3\theta = -4r/t^3$, *then the roots of* $f(x)$ *are*

$$t \cos \theta, \quad t \cos(\theta + 120°), \quad and \quad t \cos(\theta + 240°).$$

Example 5.13.
Consider once again the cubic $x^3 - 7x + 6 = (x-1)(x-2)(x+3)$ that was discussed in Example 5.6; of course, its roots are 1, 2, and -3. The cubic formula gave rather complicated expressions for these roots in terms of cube roots of complex numbers involving $\sqrt{-\frac{400}{27}}$. Let us now find the roots using Theorem 5.12 (which applies because $27r^2 + 4q^3 = -400 \leq 0$). We first compute t and θ:

$$t = \sqrt{-4q/3} = \sqrt{-4(-7)/3} = \sqrt{28/3} \approx 3.055$$

and

$$\cos(3\theta) = -4r/t^3 \approx -24/(3.055)^3 \approx -.842;$$

since $\cos(3\theta) \approx -.842$, a trigonometric table or a calculator gives $3\theta \approx 148°$ and

$$\theta \approx 49°.$$

The roots of the cubic equation are, approximately,

$$3.055 \cos 49°, \quad 3.055 \cos 169°, \quad and \quad 3.055 \cos 289°.$$

These are good approximations to the true answers. Using a trigonometric table or a calculator once again, we find that

$$\cos 49° \approx .656 \quad and \quad 3.055 \cos 49° \approx 2.004 \approx 2.00;$$
$$\cos 169° \approx -.982 \quad and \quad 3.055 \cos 169° \approx -3.00;$$
$$\cos 289° \approx .326 \quad and \quad 3.055 \cos 289° \approx .996 \approx 1.00. \quad \blacktriangleleft$$

Remark. By Lemma 5.8, every cubic $f(x) \in \mathbb{R}[x]$ has a real root; a variation of the proof of Viète's theorem shows how to find it in case $f(x)$ has complex roots; that is, when the discriminant condition is

$$-4q^3 < 27r^2.$$

Recall the hyperbolic functions

$$\cosh \theta = \tfrac{1}{2}(e^\theta + e^{-\theta})$$

and

$$\sinh \theta = \tfrac{1}{2}(e^\theta - e^{-\theta}).$$

We saw, on page 9, that $\cosh \theta \geq 1$ for all θ; we can prove that $\sinh \theta$ can take on any real number as a value. These functions satisfy cubic equations (see Exercise 5.9 on page 449):

$$\cosh(3\theta) = 4\cosh^3(\theta) - 3\cosh(\theta)$$

and

$$\sinh(3\theta) = 4\sinh^3(\theta) + 3\sinh(\theta).$$

From the first of these cubic equations, we see that $h(y) = y^3 - \tfrac{3}{4}y - \tfrac{1}{4}\cosh(3\theta)$ has $u = \cosh(\theta)$ as a root. To force $f(x) = x^3 + qx + r$ to look like $h(y)$, we write the real root v of $f(x)$ as $v = tu$. As in the proof of Viète's theorem, we want $t^2 = -4q/3$ and $\cosh(3\theta) = -4r/t^3$.

If $-4q/3 \geq 0$, then t is real. Using the discriminant condition $-4q^3 < 27r^2$, we can show that $-4r/t^3 \geq 1$, and so there is a number φ with $\cosh(\varphi) = -4r/t^3$. It follows that the real root of $f(x)$ is given by

$$v = t \cosh(\varphi/3),$$

where $t = \sqrt{-4q/3}$. [Of course, the other two (complex) roots of $f(x)$ are the roots of the quadratic $f(x)/(x-v)$.]

If $-4q/3 < 0$, then we use the hyperbolic sine. We know that $\sinh(\theta)$ is a root of $k(y) = y^3 + \tfrac{3}{4}y - \tfrac{1}{4}\sinh(3\theta)$. To force $f(x)$ to look like $k(y)$, we write the real root v of $f(x)$ as $v = tu$, where $t = \sqrt{4q/3}$ (our present hypothesis gives $4q/3 > 0$) and $\sinh(3\theta) = -4r/t^3$. As we remarked earlier, there exists a number γ with $\sinh(\gamma) = -4r/t^3$, and so the real root of $f(x)$ in this case is

$$v = t \sinh(\gamma/3). \quad \blacktriangleleft$$

EXERCISES

5.1 **(i)** Find the roots of $f(x) = x^3 - 3x + 1$.

　　H **(ii)** Find the roots of $f(x) = x^3 - 9x + 28$. Answer: $-4, 2 \pm i\sqrt{3}$.

　　(iii) Find the roots of $f(x) = x^3 - 24x^2 - 24x - 25$. Answer: $17, -\frac{1}{2} \pm i\frac{\sqrt{3}}{2}$.

5.2 **(i)** Find the roots of $f(x) = x^3 - 15x - 4$ using the cubic formula. Answer: $g = \sqrt[3]{2 + \sqrt{-121}}$ and $h = \sqrt[3]{2 - \sqrt{-121}}$.

　　(ii) Find the roots of $f(x)$ using the trigonometric formula. Answer: $4, -2 \pm \sqrt{3}$.

5.3 Find the roots of $f(x) = x^3 - 6x + 4$. Answer: $2, -1 \pm \sqrt{3}$.

5.4 Find the roots of $x^4 - 15x^2 - 20x - 6$. Answer: $-3, -1, 2 \pm \sqrt{6}$.

*5.5 The following *castle problem* appears in an old Chinese text; it was solved by the mathematician Qín Jiǔsháo in 1247.

> There is a circular castle whose diameter is unknown; it is provided with four gates, and two lengths out of the north gate there is a large tree, which is visible from a point six lengths east of the south gate. What is the length of the diameter?

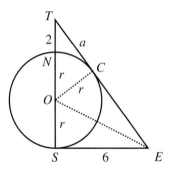

Figure 5.2   The Castle Problem

　　(i) Prove that the radius r of the castle is a root of the cubic $X^3 + X^2 - 36$.

　　(ii) Show that one root of $f(X) = X^3 + X^2 - 36$ is an integer and find the other two roots. Compare your method with Cardano's formula and with Viète's trigonometric solution.

*H 5.6 Show that if u is a root of a polynomial $f(x) \in \mathbb{R}[x]$, then the complex conjugate $\bar{u}$ is also a root of $f(x)$.

*5.7 Assume that $0 \le 3\alpha < 360°$.

　　(i) If $\cos 3\alpha$ is positive, show that there is an acute angle β with $3\alpha = 3\beta$ or $3\alpha = 3(\beta + 90°)$, and that the sets of numbers

$$\cos\beta, \quad \cos(\beta + 120°), \quad \cos(\beta + 240°)$$

and

$$\cos(\beta + 90°), \quad \cos(\beta + 210°), \quad \cos(\beta + 330°)$$

coincide.

(ii) If $\cos 3\alpha$ is negative, show that there is an acute angle β with $3\alpha = 3(\beta + 30°)$ or $3\alpha = 3(\beta + 60°)$, and that the sets of numbers

$$\cos(\beta + 30°), \quad \cos(\beta + 150°), \quad \cos(\beta + 270°)$$

and

$$\cos(\beta + 60°), \quad \cos(\beta + 180°), \quad \cos(\beta + 270°)$$

coincide.

*H **5.8** Show that if $\cos 3\theta = r$, then the roots of $4x^3 - 3x - r$ are

$$\cos\theta, \quad \cos(\theta + 120°), \quad \text{and} \quad \cos(\theta + 240°).$$

***5.9** H **(i)** Prove that $\cosh(3\theta) = 4\cosh^3(\theta) - 3\cosh(\theta)$.

      H **(ii)** Prove that $\sinh(3\theta) = 4\sinh^3(\theta) + 3\sinh(\theta)$.

H **5.10** Find the roots of $x^3 - 9x + 28$.

H **5.11** Find the roots of $x^3 - 24x^2 - 24x - 25$.

5.12 H **(i)** Find the roots of $x^3 - 15x - 4$ using the cubic formula.

      H **(ii)** Find the roots using the trigonometric formula.

H **5.13** Find the roots of $x^3 - 6x + 4$.

H **5.14** Find the roots of $x^4 - 15x^2 - 20x - 6$.

$\rightarrow$ **5.2 INSOLVABILITY OF THE GENERAL QUINTIC**

For almost 300 years, from the early 1500s to the early 1800s, mathematicians sought some generalization of the quadratic, cubic, and quartic formulas that would give the roots of any polynomial. Finally, P. Ruffini (1765–1822), in 1799, and N. H. Abel (1802–1829), in 1824, proved that no such formula exists for the general quintic polynomial (both proofs had some gaps, but Abel's proof was accepted by his contemporaries and Ruffini's proof was not). Just before his untimely death, E. Galois (1811–1832) was able to determine precisely those polynomials whose roots can be found by a formula involving square, cube, fourth, . . . roots of numbers as well as the usual field operations of adding, subtracting, multiplying, and dividing. In so doing, he also founded the Theory of Groups.

If $f(x) \in k[x]$ is a monic polynomial, where k is a field containing all the roots $z_1, z_2, \ldots, z_n$ of $f(x)$ (with possible repetitions), then

$$f(x) = x^n + a_{n-1}x^{n-1} + \cdots + a_1 x + a_0 = (x - z_1) \ldots (x - z_n).$$

By induction on $n \geq 1$, one can easily generalize Exercise 3.102 on page 305:

$$a_{n-1} = -\sum_i z_i$$

$$a_{n-2} = \sum_{i<j} z_i z_j$$

$$a_{n-3} = -\sum_{i<j<k} z_i z_j z_k \tag{1}$$

$$\vdots$$

$$a_0 = (-1)^n z_1 z_2 \cdots z_n.$$

Notice that $-a_{n-1}$ is the sum of the roots and that $\pm a_0$ is the product of the roots. Given the coefficients of $f(x)$, can one find its roots; that is, given the a's, can one solve the system (1) of n equations in n unknowns? If $n = 2$ and k does not have characteristic 2, the answer is "yes": the quadratic formula works (this is precisely Corollary 5.3). If $n = 3$ or 4 and k does not have characteristic 2 or 3, the answer is still "yes," for the cubic and quartic formulas work. But if $n \geq 5$, we shall see that no *analogous* solution exists.

We did not say that no solution of system (1) exists if $n \geq 5$; we said that no solution analogous to the solutions of the classical formulas exists. We have already seen that the classical Greek problems are impossible to solve if we limit ourselves to using particular tools in a particular way; but these problems can be solved if we relax the restrictions (for example, we have seen how Archimedes trisected angles). Similarly, it is quite possible that there is some way of finding the roots of a polynomial if one does not limit oneself to field operations and extraction of roots only. For example, we have seen Viète's trigonometric solution of the cubic. Indeed, one can find the real roots of any $f(x) \in \mathbb{R}[x]$ by *Newton's method*: if r is a real root of a polynomial $f(x)$ and if h_0 is a "good" approximation to r, then $r = \lim_{n\to\infty} h_n$, where one defines $h_{n+1} = h_n - f(h_n)/f'(h_n)$. There is a method of Hermite finding roots of quintics using elliptic modular functions, and there are methods for finding the roots of many polynomials of higher degree using hypergeometric functions. We are going to show, once we give precise definitions, that if $n \geq 5$, then finding roots "by radicals" is not always possible.

Let us recall several definitions and propositions from earlier chapters. If k is a subfield of a field K, then we also say that K is an **extension** of k, and we abbreviate this by writing "K/k is an extension." If K/k is an extension, then K may be regarded as a vector space over k, as in Example 4.1(iii). We say that K is a **finite extension** of k if K is a finite-dimensional vector space over k. The dimension of K, denoted by $[K : k]$, is called the **degree** of K/k.

→ **Example 5.14.**

Let $p(x) \in k[x]$ be an irreducible polynomial of degree n, where k is a field, and let $k(z)/k$ be an extension obtained by adjoining a root z of $p(x)$. Proposition 3.116(iv) says that each element in $k(z)$ has a unique expression of the form $b_0 + b_1 z + \cdots + b_{n-1} z^{n-1}$, where $b_i \in k$. Thus, the list $1, z, z^2, \ldots, z^{n-1}$ is a basis of $k(z)/k$, and so $\dim(k(z)) = n = \deg(p)$. ◄

For the reader's convenience, we now display several results from Chapter 4 that we will be using.

Theorem 4.31. *Let $k \subseteq K \subseteq E$ be fields, with K a finite extension of k and E a finite extension of K. Then E is a finite extension of k, and*

$$[E : k] = [E : K][K : k].$$

Definition. Assume that K/k is an extension and that $z \in K$. We call z **algebraic** over k if there is some nonzero polynomial $f(x) \in k[x]$ having z as a root; otherwise, z is called **transcendental** over k.

In Chapter 3, we considered adjoining one element to a field, examining $k(z)$ in some detail. Let us now generalize this construction of adjoining one element to a field to adjoining a set of elements to a field. This will be especially interesting when we adjoin the set of all roots of a given polynomial.

→ **Definition.** Let k be a subfield of a field K and let $\{z_1, \ldots, z_n\}$ be a subset of K. The subfield of K obtained by **adjoining** $z_1, \ldots, z_n$ **to** k, denoted by $k(z_1, \ldots, z_n)$, is the intersection of all the subfields of K containing k and $z_1, \ldots, z_n$.

Of course, $k(z_1, \ldots, z_n)$ is the *smallest* subfield of K containing k and all the z_i in the sense that if S is any other subfield of K containing k and the z_i, then $k(z_1, \ldots, z_n) \subseteq S$.

Proposition 4.32. *If K/k is a finite extension, then every $z \in K$ is algebraic over k. Conversely, if $K = k(z_1, \ldots, z_n)$, and if each z_i is algebraic over k, then K/k is a finite extension.*

By Kronecker's theorem, given $f(x) \in k[x]$, where k is a field, there is an extension K/k containing all the roots of $f(x)$; that is, the polynomial $f(x)$ is a product of linear factors in $K[x]$.

→ **Definition.** Let k be a subfield of a field K, and let $f(x) \in k[x]$. We say that $f(x)$ **splits over K** if

$$f(x) = a(x - z_1) \ldots (x - z_n),$$

where $z_1, \ldots, z_n$ are in K and $a \in k$.

An extension E/k is called a **splitting field** of $f(x)$ **over** k if $f(x)$ splits over E, but $f(x)$ does not split over any proper subfield of E.

$\rightarrow$ **Example 5.15.**

Let $m \geq 1$, let k be a field, and let $f(x) = x^m - 1 \in k[x]$. By Kronecker's theorem, there is an extension K/k over which $f(x)$ splits. The roots of $f(x)$ are, of course, the mth roots of unity. Recall Theorem 3.55, which says that K contains a primitive mth root of unity; that is, there is some mth root of unity, say, $z \in K$, with every mth root of unity being a power of z. In other words, the mth roots of unity form a multiplicative cyclic group, and a primitive mth root of unity is a generator.

Let p be a prime, and consider $g(x) = x^p - 1$. If k has characteristic $\neq p$, then $g(x)$ has no repeated roots [by Exercise 3.67 on page 274, $g(x)$ has no repeated roots if and only if $(g, g') = 1$, where $g'(x)$ is the derivative of $g(x)$]. On the other hand, if k has characteristic p, then $x^p - 1 = (x - 1)^p$, and so there is only one pth root of unity, namely, 1.

Now consider $h(x) = x^p - a \in k[x]$, and let $k(u)$ be the extension obtained from k by adjoining u, where $u^p = a$. If k has characteristic $\neq p$ and if k contains the pth roots of unity, then we claim that $k(u)$ is a splitting field of $h(x)$ over k. If z is a primitive root of unity, then the roots of $h(x)$ are $u, zu, z^2u, \ldots, z^{p-1}u$. Therefore, $k(u)$ is a splitting field of $h(x)$ over k. On the other hand, if k has characteristic p, then $h(x) = x^p - a = x^p - u^p = (x - u)^p$. Thus, there is only one root of $h(x)$, and so $k(u)$ is a splitting field of $h(x)$ over k in this case as well. $\blacktriangleleft$

$\rightarrow$ **Proposition 5.16.** *If $f(x) \in k[x]$, where k is a field, then a splitting field E/k of $f(x)$ exists.*

Proof. By Kronecker's theorem, Theorem 3.118, there exists an extension K/k with $f(x) = a(x - z_1) \cdots (x - z_n)$ in $K[x]$. If we define $E = k(z_1, \ldots, z_n)$, where $z_1, \ldots, z_n$ are the roots of $f(x)$, then $f(x)$ splits over E. If $B \subsetneq E$ is a proper subfield of E, then some $z_i \notin B$, and so $f(x)$ does not split over B. Therefore, E is a splitting field of $f(x)$. $\bullet$

A splitting field of $f(x) \in k[x]$ is the smallest subfield E of K containing k and all the roots of $f(x)$. For example, consider $f(x) = x^2 + 1 \in \mathbb{Q}[x]$. The roots of $f(x)$ are $\pm i$, and so $f(x)$ splits over $\mathbb{C}$; that is, $f(x) = (x - i)(x + i)$ is a product of linear polynomials in $\mathbb{C}[x]$. However, $\mathbb{C}$ is not a splitting field because $\mathbb{C}$ is not the *smallest* field containing $\mathbb{Q}$ and all the roots of $f(x)$; here, $\mathbb{Q}(i)$ is a splitting field.

The reason we say "a" splitting field instead of "the" splitting field is that the definition involves not only $f(x)$ and k, but the larger field K as well. If $f(x)$ splits in $K[x]$, where K/k is a field extension, then the proof of Proposition 5.16 shows that there is a unique splitting field E of $f(x)$ contained in K, namely, $E = k(z_1, \ldots, z_n)$. However, if no such extension K is given, then splitting fields may be distinct. In Theorem 5.23, we shall see that any two splitting fields of $f(x)$ over k are, in fact, isomorphic. Analysis of this technical point will enable us to prove that any two finite fields with the same number of elements are isomorphic.

→ **Example 5.17.**

Let $E = F(y_1, \ldots, y_n)$ be the field of all rational functions in n variables $y_1, \ldots, y_n$ with coefficients in a field F; that is, $E = \text{Frac}(F[y_1, \ldots, y_n])$, the fraction field of the polynomial ring in n variables. The coefficients of $f(x) = (x-y_1)(x-y_2)\ldots(x-y_n)$, which we denote by a_i, are given explicitly in terms of the y's by Eqs. (1) on page 450. Define $k = F(a_0, \ldots, a_{n-1})$. Notice that E is a splitting field of $f(x)$ over k, for it arises from k by adjoining to it all the roots of $f(x)$, namely, all the y's. ◄

→ **Definition.** Let E be a field containing a subfield k. An ***automorphism***[8] of E is an isomorphism $\sigma : E \to E$; we say that σ ***fixes*** k if $\sigma(a) = a$ for every $a \in k$.

→ **Remark.** If E/k is a field extension, then Example 4.1(iii) shows that E is a vector space over k. If $\sigma : E \to E$ is an automorphism fixing k, then σ is a linear transformation. Clearly, $\sigma(z + z') = \sigma(z) + \sigma(z')$ for all $z, z' \in E$, but σ also preserves scalar multiplication: if $a \in k$, then

$$\sigma(az) = \sigma(a)\sigma(z) = a\sigma(z),$$

because σ fixes k. ◄

We have seen that a splitting field of $x^2 + 1 \in \mathbb{Q}[x]$ is $E = \mathbb{Q}(i)$. Complex conjugation $\sigma : a \mapsto \overline{a}$ is an example of an automorphism of E fixing $\mathbb{Q}$.

→ **Proposition 5.18.** *Let k be a subfield of a field K, let*

$$f(x) = x^n + a_{n-1}x^{n-1} + \cdots + a_1 x + a_0 \in k[x],$$

and let $E = k(z_1, \ldots, z_n)$ be a splitting field. If $\sigma : E \to E$ is an automorphism fixing k, then σ permutes the roots $z_1, \ldots, z_n$ of $f(x)$.

Proof. If z is a root of $f(x)$, then

$$0 = f(z) = z^n + a_{n-1}z^{n-1} + \cdots + a_1 z + a_0.$$

Applying σ to this equation gives

$$0 = \sigma(z)^n + \sigma(a_{n-1})\sigma(z)^{n-1} + \cdots + \sigma(a_1)\sigma(z) + \sigma(a_0)$$
$$= \sigma(z)^n + a_{n-1}\sigma(z)^{n-1} + \cdots + a_1\sigma(z) + a_0,$$

because σ fixes k. Therefore, $\sigma(z)$ is a root of $f(x)$; if Z is the set of all the roots, then $\sigma' : Z \to Z$, where σ' is the restriction $\sigma|Z$. But σ' is injective (because σ is), so that Exercise 2.13 on page 105 says that σ' is a permutation. •

---

[8]The word *automorphism* is made up of two Greek roots: *auto* meaning "self" and *morph* meaning "shape" or "form." Just as an isomorphism carries one group onto an identical replica, an automorphism carries a group onto itself.

→ **Corollary 5.19.** *Let $k \subseteq B \subseteq F$ be a tower of fields, where B is the splitting field of some polynomial $f(x) \in k[x]$. If $\sigma: F \to F$ is an automorphism fixing k, then $\sigma(B) = B$.*

Proof. Note that $\sigma(B) \subseteq B$, by Proposition 5.18, for σ permutes the roots $z_1, \ldots, z_n$ of $f(x)$. As vector spaces over k, we have $B \cong \sigma(B)$, for σ is an injective linear transformation. Since $[B : k] < \infty$, by Exercise 5.24 on page 470, both B and $\sigma(B)$ are finite-dimensional and $\dim(B) = \dim(\sigma(B))$. Corollary 4.25(iii) now gives $B = \sigma(B)$.   •

The following proposition will be useful.

→ **Proposition 5.20.** *Let $E = k(z_1, \ldots, z_n)$. If $\sigma: E \to E$ is an automorphism fixing k and if $\sigma(z_i) = z_i$ for all i, then σ is the identity.*

Proof. We prove the proposition by induction on $n \geq 1$. If $n = 1$, then each $u \in E$ has the form $f(z_1)/g(z_1)$, where $f(x), g(x) \in k[x]$ and $g(z_1) \neq 0$. But σ fixes z_1 as well as the coefficients of $f(x)$ and of $g(x)$, so that σ fixes all $u \in E$. For the inductive step, write $K = k(z_1, \ldots, z_{n-1})$, and note that $E = K(z_n)$ [for $K(z_n)$ is the smallest subfield containing k and $z_1, \ldots, z_{n-1}, z_n$]. The inductive step is just a repetition of the base step with k replaced by K.   •

→ **Definition.** Let k be a subfield of a field E. The **Galois group** of E over k, denoted by $\mathrm{Gal}(E/k)$, is the set of all those automorphisms of E that fix k. If $f(x) \in k[x]$, and if $E = k(z_1, \ldots, z_n)$ is a splitting field, then the **Galois group** of $f(x)$ over k is defined to be $\mathrm{Gal}(E/k)$.

It is easy to check that $\mathrm{Gal}(E/k)$ is a group with operation composition of functions. This definition is due to E. Artin (1898–1962), in keeping with his and E. Noether's emphasis on "abstract" algebra. Galois's original version (a group isomorphic to this one) was phrased, not in terms of automorphisms, but in terms of certain permutations of the roots of a polynomial (see Tignol, *Galois' Theory of Algebraic Equations*, pp. 235–254).

For example, if $f(x) = x^2 + 1 \in \mathbb{Q}[x]$, then complex conjugation σ is an automorphism of its splitting field $\mathbb{Q}(i)$, which fixes $\mathbb{Q}$ (it interchanges the roots i and $-i$). Since $\mathrm{Gal}(\mathbb{Q}(i)/\mathbb{Q})$ is a subgroup of the symmetric group S_2, which has order 2, it follows that $\mathrm{Gal}(\mathbb{Q}(i)/\mathbb{Q}) = \langle \sigma \rangle \cong \mathbb{I}_2$. One should regard the elements of $\mathrm{Gal}(E/k)$ as generalizations of complex conjugation.

→ **Theorem 5.21.** *If $f(x) \in k[x]$ has degree n, then its Galois group $\mathrm{Gal}(E/k)$ is isomorphic to a subgroup of S_n.*

Proof. Let E/k be a splitting field of $f(x)$ over k, and let $X = \{z_1, \ldots, z_n\}$ be the set of the distinct roots of $f(x)$ in E. If $\sigma \in \mathrm{Gal}(E/k)$, then Proposition 5.18 shows that its restriction $\sigma|X$ is a permutation of X. Define $\varphi \colon \mathrm{Gal}(E/k) \to S_X$ by $\varphi \colon \sigma \mapsto \sigma|X$. To see that φ is a homomorphism, note that both $\varphi(\sigma\tau)$ and $\varphi(\sigma)\varphi(\tau)$ are functions $X \to X$, and hence they are equal if they agree on each $z_i \in X$. But $\varphi(\sigma\tau) \colon z_i \mapsto (\sigma\tau)(z_i)$, while $\varphi(\sigma)\varphi(\tau) \colon z_i \mapsto \sigma(\tau(z_i))$, and these are the same.

The image of φ is a subgroup of $S_X \cong S_m$, where $m = |X| \le n$ [if $f(x)$ has repeated roots, then $m < n$]. The kernel of φ is the set of all $\sigma \in \mathrm{Gal}(E/k)$ such that σ is the identity permutation on X; that is, σ fixes each of the roots z_i. As σ also fixes k, by definition of the Galois group, Proposition 5.20 gives $\ker \varphi = \{1\}$. Therefore, φ is injective; that is, $\mathrm{Gal}(E/k)$ is isomorphic to a subgroup of S_m. If $m = n$, the proof is done. If $m < n$, that is, if $f(x)$ has repeated roots, use the fact that S_m is isomorphic to a subgroup of S_n. For example, S_m is isomorphic to the subgroup of all permutations in S_n that fix each of $m+1, \ldots, n$. Thus, the theorem is true even if $f(x)$ has repeated roots. ●

We are now going to compare different splitting fields of a polynomial over a given field k. The definition of a splitting field E of $f(x) \in k[x]$ was given in terms of some field extension K/k over which $f(x)$ is a product of linear factors. But what if K is not given at the outset? For example, suppose that $k = \mathbb{C}(x)$ and $f(y) = y^2 - x$, or that $k = \mathbb{F}_3$ and $f(x) = x^9 - x \in \mathbb{F}_3[x]$. Now Kronecker's theorem, Theorem 3.118, gives a field extension of $\mathbb{C}(x)$ containing $\sqrt{x}$, and it gives a field extension of $\mathbb{F}_3$ containing all the roots of $f(x) = x^9 - x$. Neither of these field extensions is unique; for example, several splitting fields of $f(x)$ over $\mathbb{F}_3$ are given in Example 3.121. Nevertheless, we are going to show that, to isomorphism, splitting fields do not depend on the choice of extension field K.

The next result constructs automorphisms in $\mathrm{Gal}(E/k)$, and it also counts the number of them when k has characteristic 0.

Recall Theorem 3.33: if R and S are commutative rings and $\varphi \colon R \to S$ is a homomorphism, then $\varphi^* \colon R[x] \to S[x]$, defined by

$$\varphi^* \colon f(x) = r_0 + r_1 x + r_2 x^2 + \cdots$$
$$\mapsto \varphi(r_0) + \varphi(r_1)x + \varphi(r_2)x^2 + \cdots = f^*(x),$$

is a homomorphism; if φ is an isomorphism, so is φ^*.

→ **Proposition 5.22.** *Let $f(x) \in k[x]$, and let E be a splitting field of $f(x)$ over k. Let $\varphi \colon k \to k'$ be an isomorphism of fields, let $\varphi^* \colon k[x] \to k'[x]$ be the isomorphism $g(x) \mapsto g^*(x)$ given by Theorem 3.33, and let E' be a splitting field of $f^*(x)$ over k'.*

(i) *There exists an isomorphism* $\Phi\colon E \to E'$ *extending* φ.

$$
\begin{array}{ccc}
E & \overset{\Phi}{\dashrightarrow} & E' \\
\big| & & \big| \\
k & \underset{\varphi}{\longrightarrow} & k'
\end{array}
$$

(ii) *If k has characteristic* 0, *then there are exactly* $[E; k]$ *isomorphisms* $\Phi\colon E \to E'$ *extending* φ.

Proof.
(i) The proof is by induction on $[E : k]$. If $[E : k] = 1$, then $f(x)$ is a product of linear polynomials in $k[x]$, and it follows easily that $f^*(x)$ is also a product of linear polynomials in $k'[x]$. Thus, we may set $\Phi = \varphi$.

For the inductive step, choose a root z of $f(x)$ in E that is not in k, and let $p(x)$ be the irreducible polynomial in $k[x]$ of which z is a root [Proposition 3.116(i)]. Since $z \notin k$, $\deg(p) > 1$; moreover, $[k(z) : k] = \deg(p)$, by Example 5.14. Let $p^*(x)$ be the corresponding polynomial in $k'[x]$, and let z' be a root of $p^*(x)$ in E'. Note that $p^*(x)$ is irreducible, because the isomorphism $\varphi^*\colon k[x] \to k'[x]$ takes irreducible polynomials to irreducible polynomials.

By Exercise 3.101 on page 304, there is an isomorphism $\widetilde{\varphi}\colon k(z) \to k'(z')$ extending φ with $\widetilde{\varphi}(z) = z'$. We now regard $f(x)$ as a polynomial over $k(z)$ (for $k \subseteq k(z)$ implies $k[x] \subseteq k(z)[x]$). We claim that E is a splitting field of $f(x)$ over $k(z)$; that is,

$$
E = k(z)(z_1, \ldots, z_n),
$$

where $z_1, \ldots, z_n$ are the roots of $f(x)$. Clearly,

$$
E = k(z_1, \ldots, z_n) \subseteq k(z)(z_1, \ldots, z_n).
$$

For the reverse inclusion, since $z \in E$, we have

$$
k(z)(z_1, \ldots, z_n) \subseteq k(z_1, \ldots, z_n) = E.
$$

But $[E : k(z)] < [E : k]$, by Theorem 4.31, so that the inductive hypothesis gives an isomorphism $\Phi\colon E \to E'$ that extends $\widetilde{\varphi}$ and, hence, φ.

(ii) The proof in this part is again an induction on $[E : k]$. If $[E : k] = 1$, then $E = k$ and there is only one extension, namely, $\Phi = \varphi$. If $[E : k] > 1$, let $f(x) = p(x)g(x)$ in $k[x]$, where $p(x)$ is an irreducible factor of largest degree, say, d. We may assume that $d > 1$, otherwise $f(x)$ splits over k and $[E : k] = 1$. Choose a root $z \in E$ of $p(x)$ [this is possible because E/k is a splitting field of $f(x) = p(x)g(x)$]. As in part (i), the polynomial $p^*(x) \in k'[x]$ is irreducible, and there is some root z' of $p^*(x)$ in E'. Since k has characteristic 0, Exercise 3.95 on page 304 shows that $p(x)$ and $p^*(x)$

have no repeated roots; that is, each has d distinct roots. By Proposition 3.116(iii), there exist d isomorphisms $\widetilde{\varphi} \colon k(z) \to k'(z')$ extending φ, one for each of the roots z'; there are no other isomorphisms extending φ, for any such extension must send z into some z', in which case Proposition 5.20 shows it is already one of the $\widetilde{\varphi}$. As in part (i), E is a splitting field of $f(x)$ over $k(z)$, and E' can be viewed as a splitting field of $f^*(x)$ over $k'(z')$. But $[E : k] = [E : k(z)][k(z) : k] = [E : k(z)]d$, so that $[E : k(z)] < [E : k]$. By induction, each $\widetilde{\varphi}$ has exactly $[E : k(z)]$ extensions $\Phi \colon E \to E'$. Thus, we have exhibited $[E : k(z)][k(z) : k] = [E : k]$ such extensions Φ. If $\tau \colon E \to E'$ is another extension of φ, then $\tau(z) = z'$ for some root z' of $p^*(x)$, and so τ is an extension of that $\widetilde{\varphi}$ with $\widetilde{\varphi}(z) = z'$. But all such extensions $E \to E'$ have already been counted.   •

The hypothesis that k have characteristic 0 is used in proving Proposition 5.22(ii) by guaranteeing that irreducible polynomials in $k[x]$ have no repeated roots. This weaker statement, called *separability*, gives a better theorem. For example, it is true that every finite field k satisfies this hypothesis [see Exercise 5.31(iii)].

→ **Theorem 5.23.**   *If k is a field and $f(x) \in k[x]$, then any two splitting fields of $f(x)$ over k are isomorphic.*

Proof.   Let E and E' be splitting fields of $f(x)$ over k. If φ is the identity, then Proposition 5.22(i) applies at once.   •

Corollary 5.24.   *The Galois group $\mathrm{Gal}(E/k)$ of a polynomial $f(x) \in k[x]$ with splitting field E depends only on $f(x)$ and k, but not upon the choice of E.*

Proof.   If $\varphi \colon E \to E'$ is an isomorphism fixing k, then there is an isomorphism $\mathrm{Gal}(E/k) \to \mathrm{Gal}(E'/k)$ given by $\sigma \mapsto \varphi\sigma\varphi^{-1}$.   •

It is remarkable that the next theorem was not proved until the 1890s, 60 years after Galois discovered finite fields.

→ **Corollary 5.25** (**E. H. Moore**).   *Any two finite fields having exactly p^n elements are isomorphic.*

Proof.   If E is a field with $q = p^n$ elements, then Lagrange's theorem applied to the multiplicative group $E^\times$ shows that $a^{q-1} = 1$ for every $a \in E^\times$. It follows that every element of E, including $a = 0$, is a root of $f(x) = x^q - x = x(x^{q-1} - 1) \in \mathbb{F}_p[x]$, and so E is a splitting field of $f(x)$ over $\mathbb{F}_p$.   •

It follows that if $g(x), h(x) \in \mathbb{F}_p[x]$ are irreducible polynomials of degree n, then $\mathbb{F}_p[x]/(g(x)) \cong \mathbb{F}_p[x]/(h(x))$, for both are fields with exactly p^n elements.

E. H. Moore (1862–1932) began his mathematical career as an algebraist, but he did important work in many other parts of mathematics as well; for example, the analytical notion of Moore-Smith convergence is named in part after him.

We can now compute the order of the Galois group $\text{Gal}(E/k)$ when k has characteristic 0.

→ **Theorem 5.26.**   *If E/k is the splitting field of some polynomial in $k[x]$, where k is a field of characteristic 0, then $|\text{Gal}(E/k)| = [E : k]$.*

Proof.   This is the special case of Proposition 5.22(ii) when $k = k'$, $E = E'$, and $\varphi = 1_k$.   •

→ **Remark.**   Theorem 5.26 may not be true if k is a field of characteristic $p > 0$. It is true if k is a finite field, but it is false for $k = \mathbb{F}_p(x)$, all rational functions over $\mathbb{F}_p$: a counterexample is described in Exercise 5.32 on page 471. As we mentioned after the proof of Proposition 5.22, the key to investigating this question involves the notion of *separability*.   ◄

→ **Corollary 5.27.**   *Let $f(x) \in k[x]$ be an irreducible polynomial of degree n, where k is a field of characteristic 0. If E/k is a splitting field of $f(x)$ over k, then n is a divisor of $|\text{Gal}(E/k)|$.*

Proof.   If $z \in E$ is a root of $f(x)$, then $[k(z) : k] = n$, as in Example 5.14. But $[E : k] = [E : k(z)][k(z) : k]$, so that $n \mid [E : k]$. Since k has characteristic 0, Theorem 5.26 gives $|\text{Gal}(E/k)| = [E : k]$.   •

If k is a field, then the factorization into irreducibles of a polynomial in $k[x]$ can change as one enlarges the ground field k.

Lemma 5.28.   *Let B/k be a splitting field of some polynomial $g(x) \in k[x]$. If $p(x) \in k[x]$ is irreducible, and if*

$$p(x) = q_1(x) \cdots q_t(x)$$

is the factorization of $p(x)$ into irreducibles in $B[x]$, then all the $q_i(x)$ have the same degree.

Proof.   Regard $p(x)$ as a polynomial in $B[x]$ (for $k \subseteq B$ implies $k[x] \subseteq B[x]$), and let $E = B(z_1, \ldots, z_n)$ be a splitting field of $p(x)$, where $z_1, \ldots, z_n$ are the roots of $p(x)$. If $p(x)$ does not factor in $B[x]$, we are done. Otherwise, choose z_1 to be a root of $q_1(x)$ and, for each $j \neq 1$, choose z_j to be a root of $q_j(x)$. Since both z_1 and z_j are roots of the irreducible $p(x)$, Proposition 3.116(iii) gives an isomorphism $\varphi_j \colon k(z_1) \to k(z_j)$ with $\varphi_j(z_1) = z_j$ which fixes k pointwise. Now Proposition 5.22(i) says that φ_j extends to an automorphism Φ_j of E, and Corollary 5.19 gives $\Phi_j(B) = B$. Hence, Φ_j induces an isomorphism $\Phi_j^* \colon B[x] \to B[x]$ (by letting Φ_j act on the coefficients of a polynomial). It follows that

$$p^*(x) = q_1^*(x) \cdots q_t^*(x),$$

where $p^*(x) = \Phi_j^*(p)$ and $q_i^*(x) = \Phi_j^*(q_i)$ for all i. Note that all the $q_i^*(x)$ are irreducible, because isomorphisms take irreducible polynomials into irreducible polynomials. Now $p^*(x) = p(x)$, because Φ_j fixes k pointwise, and so unique factorization in $B[x]$ gives $q_1^*(x) = q_\ell(x)$ for some ℓ. But $z_j = \Phi_j(z_1)$ is a root of $q_1^*(x)$, so that $q_1^*(x) = q_j(x)$. Therefore, $\deg(q_1) = \deg(q_1^*) = \deg(q_j)$, and all the q_j have the same degree. $\bullet$

This lemma allows us to characterize those field extensions which are splitting fields.

$\rightarrow$ **Theorem 5.29.** *Let E/k be a finite field extension. Then E/k is a splitting field of some polynomial in $k[x]$ if and only if every irreducible polynomial in $k[x]$ having a root in E splits in $E[x]$.*

Proof. Suppose that E/k is a splitting field of some polynomial in $k[x]$. Let $p(x) \in k[x]$ be irreducible, and let $p(x) = q_1(x) \cdots q_t(x)$ be its factorization into irreducibles in $E[x]$. If $p(x)$ has a root in E, then it has a linear factor in $E[x]$; by Lemma 5.28, all the $q_i(x)$ are linear, and so $p(x)$ splits in $E[x]$.

Conversely, assume that every irreducible polynomial in $k[x]$ having a root in E splits in $E[x]$. Choose $\beta_1 \in E$ with $\beta_1 \notin k$. Since E/k is finite, Proposition 3.116(i) gives an irreducible polynomial $p_1(x) \in k[x]$ having β_1 as a root. By hypothesis, $p_1(x)$ splits in $E[x]$; let $B_1 \subseteq E$ be a splitting field of $p_1(x)$. If $B_1 = E$, we are done. Otherwise, choose $\beta_2 \in E$ with $\beta_2 \notin B_1$. As above, there is an irreducible $p_2(x) \in k[x]$ having β_2 as a root. Define $B_2 \subseteq E$ to be the splitting field of $p_1(x)p_2(x)$, so that $k \subseteq B_1 \subseteq B_2 \subseteq E$. Since E/k is finite, this process eventually ends with $E = B_r$ for some $r \geq 1$. $\bullet$

$\rightarrow$ **Definition.** A field extension E/k is a ***normal extension*** if every irreducible polynomial $p(x) \in k[x]$ having a root in E splits in $E[x]$.

Here is the basic strategy for showing that there are polynomials of degree 5 for which there is no formula, analogous to the classical formulas, giving their roots. First, we will translate the classical formulas (giving the roots of $f(x) \in k[x]$) in terms of subfields of a splitting field E over k. Second, this translation into the language of fields will itself be translated into the language of groups: if there is a formula for the roots of $f(x)$, then $\mathrm{Gal}(E/k)$ must be a *solvable* group (which we will soon define). Finally, polynomials of degree at least 5 can have Galois groups which are not solvable.

$\rightarrow$ ## Formulas and Solvability by Radicals

Without further ado, here is the translation of the existence of a formula for the roots of a polynomial in terms of subfields of a splitting field.

→ **Definition.**   A *pure extension* of *type* m is an extension $k(u)/k$, where $u^m \in k$ for some $m \geq 1$. An extension K/k is a *radical extension* if there is a tower of fields

$$k = K_0 \subseteq K_1 \subseteq \cdots \subseteq K_t = K \tag{2}$$

in which each K_{i+1}/K_i is a pure extension of type m_i. We call Eq. (2) a *radical tower*.

It is easy to see that any field extension K/k with $[K : k] \leq 2$ is a pure extension. By Theorem 4.54, a complex number z is constructible if and only if it is *polyquadratic*; that is, there is a tower of fields $\mathbb{Q}(i) = F_0 \subseteq F_1 \subseteq \cdots \subseteq F_n$ with $z \in F_n$ and with $[F_i : F_{i-1}] \leq 2$ for all i. Exercise 5.17 on page 470 asks you to prove that $\mathbb{Q}(i, z)/\mathbb{Q}$ is a radical extension.

When we say that there is a *formula* for the roots of a polynomial $f(x)$ analogous to the quadratic, cubic, and quartic formulas, we mean that there is some expression giving the roots of $f(x)$ in terms of the coefficients of $f(x)$. As in the classical formulas, the expression may involve the field operations, constants, and extraction of roots, but it should not involve any other operations such as cosines, definite integrals, or limits, for example. We maintain that a formula as informally described above exists precisely when $f(x)$ is *solvable by radicals* in the following sense.

→ **Definition.**   Let $f(x) \in k[x]$ have a splitting field E. We say that $f(x)$ is *solvable by radicals* if there is a radical extension

$$k = K_0 \subseteq K_1 \subseteq \cdots \subseteq K_t$$

with $E \subseteq K_t$.

→ **Example 5.30.**
For every field k and every $m \geq 1$, we show that the polynomial $f(x) = x^m - 1 \in k[x]$ is solvable by radicals. Recall that the set Γ_m of all mth roots of unity in a splitting field E/k of $f(x)$ is a cyclic group, by Theorem 3.55; a generator ζ is called a *primitive root of unity*. Note that $|\Gamma_m| = m$ unless k has characteristic $p > 0$ and $p \mid m$, in which case $|\Gamma_m| = m'$, where $m = p^e m'$ and $p \nmid m'$ [because $x^m - 1 = x^{p^e m'} - 1 = (x^{m'} - 1)^{p^e}$]. Now $E = k(\zeta)$, so that E is a pure extension of k, and hence E/k is a radical extension. Therefore, $f(x) = x^m - 1$ is solvable by radicals. ◄

Let us illustrate solvability by radicals by considering the classical formulas for polynomials of small degree.

Quadratics

Let $f(x) = x^2 + bx + c \in \mathbb{Q}[x]$. Define $K_1 = \mathbb{Q}(u)$, where $u = \sqrt{b^2 - 4c}$. Then K_1 is a radical extension of $\mathbb{Q}$, for $u^2 \in \mathbb{Q}$. Moreover, the quadratic formula implies that K_1 is the splitting field of $f(x)$, and so $f(x)$ is solvable by radicals.

Cubics

Let $f(X) = X^3 + bX^2 + cX + d \in \mathbb{Q}[x]$. The change of variable $X = x - \frac{1}{3}b$ yields a new polynomial $f^*(x) = x^3 + qx + r \in \mathbb{Q}[x]$ having the same splitting field E [for if u is a root of $f^*(x)$, then $u - \frac{1}{3}b$ is a root of $f(x)$]. Define $K_1 = \mathbb{Q}(\sqrt{D})$, where $D = r^2 + \frac{4}{27}q^3$, and define $K_2 = K_1(\alpha)$, where $\alpha^3 = \frac{1}{2}(-r + \sqrt{D})$. The cubic formula shows that K_2 contains the root $\alpha + \beta$ of $f^*(x)$, where $\beta = -q/3\alpha$. Finally, define $K_3 = K_2(\omega)$, where $\omega^3 = 1$. The other roots of $f^*(x)$ are $\omega\alpha + \omega^2\beta$ and $\omega^2\alpha + \omega\beta$, both of which lie in K_3, and so $E \subseteq K_3$.

An interesting aspect of the cubic formula is the so-called *casus irreducibilis*; the formula for the roots of an irreducible cubic in $\mathbb{Q}[x]$ having all roots real (as in Example 5.6) requires the presence of complex numbers (see Rotman, *Galois Theory*, 2d ed.).

Casus Irreducibilis. If $f(x) = x^3 + qx + r \in \mathbb{Q}[x]$ is an irreducible polynomial having real roots, then any radical extension $K_t/\mathbb{Q}$ containing the splitting field of $f(x)$ is not real; that is, $K_t \subsetneq \mathbb{R}$.

It follows that one cannot modify the definition of $f(x)$ being solvable by radicals so that a splitting field E is equal to the last term K_t in a tower of pure extensions (instead of $E \subseteq K_t$).

Quartics

Let $f(x) = X^4 + bX^3 + cX^2 + dX + e \in \mathbb{Q}[x]$. The change of variable $X = x - \frac{1}{4}b$ yields a new polynomial $f^*(x) = x^4 + qx^2 + rx + s \in \mathbb{Q}[x]$; moreover, the splitting field E of $f(x)$ is equal to the splitting field of $f^*(x)$, for if u is a root of $f^*(x)$, then $u - \frac{1}{4}b$ is a root of $f(x)$. Recall

$$f^*(x) = x^4 + qx^2 + rx + s = (x^2 + jx + \ell)(x^2 - jx + m),$$

and Eq. (12) on page 443 shows that j^2 is a root of the cubic

$$(j^2)^3 + 2q(j^2)^2 + (q^2 - 4s)j^2 - r^2.$$

Define pure extensions

$$\mathbb{Q} = K_0 \subseteq K_1 \subseteq K_2 \subseteq K_3,$$

as in the cubic case, so that $j^2 \in K_3$. Define $K_4 = K_3(j)$, and note that Eqs. (11) on page 443 give $\ell, m \in K_4$; define $K_5 = K_4(\sqrt{j^2 - 4\ell})$ and $K_6 = K_5(\sqrt{j^2 - 4m})$. The quartic formula gives $E \subseteq K_6$ (this tower can be shortened).

We have seen that quadratics, cubics, and quartics are solvable by radicals. Conversely, if $f(x) \in \mathbb{Q}[x]$ is a polynomial that is solvable by radicals, then there is a formula of the desired kind that expresses its roots in terms of its coefficients. For suppose that

$$\mathbb{Q} = K_0 \subseteq K_1 \subseteq \cdots \subseteq K_t$$

is a radical extension with splitting field $E \subseteq K_t$. Let z be a root of $f(x)$. Now $K_t = K_{t-1}(u)$, where u is an mth root of some element $\alpha \in K_{t-1}$; hence, z can be expressed in terms of u and K_{t-1}; that is, z can be expressed in terms of $\sqrt[m]{\alpha}$ and K_{t-1}. But $K_{t-1} = K_{t-2}(v)$, where some power of v lies in K_{t-2}. Hence, z can be expressed in terms of u, v, and K_{t-2}. Ultimately, z is expressed by a formula analogous to those occurring in the classical formulas.

→ Translation into Group Theory

The second stage of the strategy involves investigating the effect of $f(x)$ being solvable by radicals on its Galois group.

Suppose that $k(u)/k$ is a pure extension of type 6; that is, $u^6 \in k$. Now $k(u^3)/k$ is a pure extension of type 2, for $(u^3)^2 = u^6 \in k$, and $k(u)/k(u^3)$ is obviously a pure extension of type 3. Thus, $k(u)/k$ can be replaced by a tower of pure extensions $k \subseteq k(u^3) \subseteq k(u)$ of types 2 and 3. More generally, one may assume, given a tower of pure extensions, that each field is of prime type over its predecessor: if $k \subseteq k(u)$ is of type m, then factor $m = p_t \ldots p_q$, where the p's are (not necessarily distinct) primes, and replace $k \subseteq k(u)$ by

$$k \subseteq k(u^{m/p_1}) \subseteq k(u^{m/p_1 p_2}) \subseteq \cdots \subseteq k(u).$$

The next theorem, a key result allowing us to translate solvability by radicals into the language of Galois groups, shows why normal extensions are so called. The reader should recognize that extensions of fields seem to be playing the same role as do subgroups of groups.

→ **Theorem 5.31.**   *Let $k \subseteq K \subseteq E$ be a tower of fields, where both K/k and E/k are normal extensions. Then $\mathrm{Gal}(E/K)$ is a normal subgroup of $\mathrm{Gal}(E/k)$, and*

$$\mathrm{Gal}(E/k)/\mathrm{Gal}(E/K) \cong \mathrm{Gal}(K/k).$$

Proof.   Since K/k is a normal extension, it is a splitting field of some polynomial in $k[x]$, by Theorem 5.29. Hence, if $\sigma \in \mathrm{Gal}(E/k)$, then $\sigma(K) = K$, by Corollary 5.19. Define $\rho \colon \mathrm{Gal}(E/k) \to \mathrm{Gal}(K/k)$ by $\sigma \mapsto \sigma|K$. It is easy to see, as in the proof of Theorem 5.21, that ρ is a homomorphism and that $\ker \rho = \mathrm{Gal}(E/K)$. It follows that $\mathrm{Gal}(E/K)$ is a normal subgroup of $\mathrm{Gal}(E/k)$. Now ρ is surjective: if $\tau \in \mathrm{Gal}(K/k)$, then Proposition 5.22(i) applies to show that there is $\sigma \in \mathrm{Gal}(E/k)$ extending τ; that is, $\rho(\sigma) = \sigma|K = \tau$. The first isomorphism theorem completes the proof.   •

The next (technical) result will be needed when we apply Theorem 5.31.

Lemma 5.32. *Let B be a finite extension of a field k.*

(i) *There is a finite extension F/B with F/k a normal extension.*

(ii) *If B is a radical extension of k, then there is a tower of fields $k \subseteq B \subseteq F$ with F/k both a normal extension and a radical extension. Moreover, the set of types of the pure extensions occurring in a radical tower of F/k is the same as the set of types in the radical tower of B/k.*

Proof.
(i) Since B is a finite extension, $B = k(z_1, \ldots, z_\ell)$ for elements $z_1, \ldots, z_\ell$. For each i, Theorem 3.116 gives an irreducible polynomial $p_i(x) \in k[x]$ with $p_i(z_i) = 0$. Define $f(x) = p_1(x) \cdots p_\ell(x) \in k[x] \subseteq B[x]$, and define F to be a splitting field of $f(x)$ over B. Since $f(x) \in k[x]$, we have F/k a splitting field of $f(x)$ over k as well, and so F/k is a normal extension.

(ii) Now
$$F = k(z_1, z_1', z_1'', \ldots; z_2, z_2', z_2'', \ldots; \ldots; z_\ell, z_\ell', z_\ell'', \ldots),$$

where $z_i, z_i', z_i'', \ldots$ are the roots of $p_i(x)$. We claim that

$$F = k\big(\{\sigma(z_1), \ldots, \sigma(z_\ell) : \sigma \in \mathrm{Gal}(F/k)\}\big).$$

Clearly, the right-hand side is contained in F, and so it suffices to prove the reverse inclusion. In fact, it suffices to prove that $z_i' = \sigma(z_i)$ [where z_i' now denotes any root of $p_i(x)$ for some i]. By Proposition 3.116(iii), there is an isomorphism $\gamma : k(z_i) \to k(z_i')$ fixing k and taking $z_i \mapsto z_i'$, and, by Proposition 5.22(i), each such γ extends to an isomorphism $\sigma \in \mathrm{Gal}(F/k)$. Therefore, $z_i' = \sigma(z_i)$, as desired.

Since B is a radical extension of k, there are $u_1, \ldots, u_t \in B$ and a radical tower,

$$k \subseteq k(u_1) \subseteq k(u_1, u_2) \cdots \subseteq k(u_1, \ldots, u_t) = B, \tag{3}$$

with each $k(u_1, \ldots, u_{i+1})$ a pure extension of $k(u_1, \ldots, u_i)$. We now show that F is a radical extension of k. Let $\mathrm{Gal}(F/k) = \{\sigma_1 = 1, \sigma_2, \ldots, \sigma_n\}$. Define

$$B_1 = k(u_1, \sigma_2(u_1), \sigma_3(u_1), \ldots, \sigma_n(u_1)).$$

There is a radical tower

$$k \subseteq k(u_1) \subseteq k(u_1, \sigma_2(u_1)) \subseteq k(u_1, \sigma_2(u_1), \sigma_3(u_1)) \subseteq \cdots \subseteq B_1$$

displaying B_1 as a radical extension of k. In more detail, if u_1^p lies in k, then $\sigma_j(u_1^p) = \sigma_j(u_1)^p \in \sigma_j(k) = k \subseteq k(u_1, \sigma_2(u_1), \ldots, \sigma_{j-1}(u_1))$. Note that these pure extensions all have the same type, namely, p, which is a type in the original radical tower (3). Define

$$B_2 = k(u_2, \sigma_2(u_2), \sigma_3(u_2), \ldots, \sigma_n(u_2));$$

there is a radical tower:

$$B_1 \subseteq B_1(u_2) \subseteq B_1(u_2, \sigma_2(u_2)) \subseteq B_1(u_2, \sigma_2(u_2), \sigma_3(u_2)) \subseteq \cdots \subseteq B_2.$$

Now B_2 is a radical extension of B_1: if $u_2^q \in k(u_1) \subseteq B_1$, then $\sigma_j(u_2^q) = \sigma_j(u_2)^q \in \sigma_j(B_1) \subseteq B_1 \subseteq B_1(u_2, \sigma_2(u_2), \ldots, \sigma_{j-1}(u_2))$. Again, these pure extensions have the same type, namely, q, which is a type in the radical tower (3). Since B_1 is a radical extension of k, the radical tower from k to B_1 followed by the radical tower from B_1 to B_2 displays B_2 as a radical extension of k. For each $i \geq 2$, define B_{i+1} to be B_i with $u_i, \sigma_2(u_i), \sigma_3(u_i), \ldots$ adjoined. The argument above shows that B_{i+1} is a radical extension of k. Finally, since $F = B_t$, we have shown that F is a radical extension of k, and that the statement about the types of its pure extensions is correct. $\bullet$

$\rightarrow$ **Lemma 5.33.** *Let $k(u)/k$ be a pure extension of prime type p distinct from the characteristic of k. If k contains the pth roots of unity and if $u \notin k$, then $\mathrm{Gal}(k(u)/k) \cong \mathbb{I}_p$.*

Proof. Denote $\mathrm{Gal}(k(u)/k)$ by G. Let $a = u^p \in k$. If ω is a primitive pth root of unity, then the roots $1, \omega, \ldots, \omega^{p-1}$ are distinct [because $p \neq \mathrm{char}(k)$], and the roots of $f(x) = x^p - a$ are $u, \omega u, \omega^2 u, \ldots, \omega^{p-1} u$; since $\omega \in k$, it follows that $k(u)$ is the splitting field of $f(x)$ over k. If $\sigma \in G$, then $\sigma(u) = \omega^i u$ for some i, by Theorem 5.18(i). Define $\varphi \colon G \to \mathbb{I}_p$ by $\varphi(\sigma) = [i]$, the congruence class of $i \bmod p$. To see that φ is a homomorphism, suppose that $\tau \in G$ and $\varphi(\tau) = [j]$. Then $\sigma\tau(u) = \sigma(\omega^j u) = \omega^{i+j} u$, so that $\varphi(\sigma\tau) = [i+j] = [i] + [j] = \varphi(\sigma) + \varphi(\tau)$. Now $\ker \varphi = \{1\}$, for if $\varphi(\sigma) = [0]$, then $\sigma(u) = u$; since σ fixes k, by the definition of $G = \mathrm{Gal}(k(u)/k)$, Proposition 5.20 gives $\sigma = 1$. Finally, we show that φ is a surjection. Since $u \notin k$, the automorphism taking $u \mapsto \omega u$ is not the identity, so that $\mathrm{im}\, \varphi \neq \{[0]\}$. But $\mathbb{I}_p$, having prime order p, has no subgroups aside from $\{[0]\}$ and $\mathbb{I}_p$, so that $\mathrm{im}\, \varphi = \mathbb{I}_p$. Therefore, φ is an isomorphism. $\bullet$

Here is the heart of the translation we have been seeking.

$\rightarrow$ **Theorem 5.34.** *Let $k = K_0 \subseteq K_1 \subseteq K_2 \subseteq \cdots \subseteq K_t$ be a radical extension of a field k. Assume, for each i, that each K_i is a pure extension of prime type p_i over K_{i-1}, where $p_i \neq \mathrm{char}(k)$, and that k contains all the p_ith roots of unity.*

(i) *If K_t is a splitting field over k, then there is a sequence of subgroups*

$$\mathrm{Gal}(K_t/k) = G_0 \geq G_1 \geq G_2 \geq \cdots \geq G_t = \{1\},$$

with each G_{i+1} a normal subgroup of G_i and with G_i/G_{i+1} cyclic of prime order.

(ii) *If $f(x)$ is solvable by radicals, then its Galois group $\mathrm{Gal}(E/k)$ is a quotient of a solvable group.*

Proof.

(i) Defining $G_i = \mathrm{Gal}(K_t/K_i)$ gives a sequence of subgroups of $\mathrm{Gal}(K_t/k)$. Since $K_1 = k(u)$, where $u^{p_1} \in k$, the assumption that k contains a primitive pth root of unity shows that K_1 is a splitting field of $x^{p_1} - u^{p_1}$ (see Example 5.15). We may thus apply Theorem 5.31 to see that $G_1 = \mathrm{Gal}(K_t/K_1)$ is a normal subgroup of $G_0 = \mathrm{Gal}(K_t/k)$ and that $G_0/G_1 \cong \mathrm{Gal}(K_1/k) = \mathrm{Gal}(K_1/K_0)$. By Lemma 5.33, $G_0/G_1 \cong \mathbb{I}_{p_1}$. This argument can be repeated for each i.

(ii) There is a radical tower,

$$k = K_0 \subseteq K_1 \subseteq K_2 \subseteq \cdots \subseteq K_t,$$

with each K_i/K_{i+1} a pure extension of prime type, and with $E \subseteq K_t$. By Lemma 5.32, this radical tower can be lengthened; there is a radical tower

$$k = K_0 \subseteq K_1 \subseteq \cdots \subseteq K_t \subseteq \cdots \subseteq F,$$

where F/k is a normal extension. Moreover, the (prime) types of the pure extensions in this longer radical extension are the same as those occurring in the original radical tower. Therefore, k contains all those roots of unity required in the hypothesis of part (i) to show that $\mathrm{Gal}(F/k)$ is a solvable group.

Since E is a splitting field, if $\sigma \in \mathrm{Gal}(F/k)$, then $\sigma|E \in \mathrm{Gal}(E/k)$, and so $\rho \colon \sigma \mapsto \sigma|E$ is a homomorphism $\mathrm{Gal}(F/k) \to \mathrm{Gal}(E/k)$. Finally, Proposition 5.22(i) shows that ρ is surjective; since F is a splitting field over k, every $\sigma \in \mathrm{Gal}(E/k)$ extends to some $\widetilde{\sigma} \in \mathrm{Gal}(F/k)$.   •

We shall see that not every group satisfies the conclusion of Theorem 5.34(i); those groups that do enjoy that property have a name.

→ **Definition.**   A *normal series* of a group G is a sequence of subgroups

$$G = G_0 \geq G_1 \geq G_2 \geq \cdots \geq G_t = \{1\}$$

with each G_{i+1} a normal subgroup of G_i; the ***factor groups*** of this series are the quotient groups

$$G_0/G_1, \quad G_1/G_2, \quad \ldots, \quad G_{t-1}/G_t.$$

A finite group G is called ***solvable*** if $G = \{1\}$ or if G has a normal series each of whose factor groups has prime order.

In this language, Theorem 5.34 says that $\mathrm{Gal}(K_t/k)$ is a solvable group if K_t is a radical extension of k and k contains enough roots of unity.

$\rightarrow$ **Example 5.35.**

    (i) S_4 *is a solvable group.*

        Consider the chain of subgroups

$$S_4 \geq A_4 \geq \mathbf{V} \geq W \geq \{1\},$$

where $\mathbf{V}$ is the four-group and W is any subgroup of $\mathbf{V}$ of order 2. This is a normal series: first, it begins with S_4 and ends with $\{1\}$; second, each term is a normal subgroup of its predecessor: $A_4 \lhd S_4$; $\mathbf{V} \lhd A_4$ (in fact, $\mathbf{V} \lhd S_4$, a stronger statement); $W \lhd \mathbf{V}$ because $\mathbf{V}$ is abelian. Now $|S_4/A_4| = |S_4|/|A_4| = 24/12 = 2$, $|A_4/\mathbf{V}| = |A_4|/|\mathbf{V}| = 12/4 = 3$, $|\mathbf{V}/W| = |\mathbf{V}|/|W| = 4/2 = 2$, and $|W/\{1\}| = |W| = 2$. Thus, each factor group has prime order, and so S_4 is solvable.

   (ii) *Every finite abelian group G is solvable.*

        We prove this by induction on $|G|$; the base step $|G| = 1$ is trivially true. For the inductive step, recall Proposition 2.124: if G is a finite abelian group, then G has a subgroup of order d for every divisor d of $|G|$. Since $|G| > 1$, there is a factorization $|G| = pd$ for some prime p, and so there is a subgroup H of G of order d. Now $H \lhd G$, because G is abelian, and $|G/H| = |G|/|H| = pd/d = p$. By induction, there is a normal series from H to $\{1\}$ with factor groups of prime orders, from which it follows that G is a solvable group.

  (iii) *Every simple nonabelian group G is not solvable.*

        Since the only normal subgroups of G are $\{1\}$ and G, every normal series of G has the form

$$G = G_0 = G_1 = \cdots = G_m > G_{m-1} = \cdots = G_t = \{1\}.$$

Therefore, all the factor groups are $\{1\}$ except $G_m/G_{m-1} \cong G$. As G is not cyclic (it is not even abelian), G is not solvable.

  (iv) S_5 *is not a solvable group (in fact, S_n is not solvable for $n \geq 5$).*

        In Exercise 2.135 on page 208, we saw, for all $n \geq 5$, that A_n is the only proper nontrivial normal subgroup of S_n (the key fact in the proof is the simplicity of A_n for $n \geq 5$). It follows that S_n has only one normal series, namely,

$$S_n > A_n > \{1\}$$

(this is not quite true; another normal series is $S_n > A_n \geq A_n > \{1\}$, which repeats a term; of course, this repetition only contributes the new factor group $A_n/A_n = \{1\}$). But the factor groups of this normal series are $S_n/A_n \cong \mathbb{I}_2$ and $A_n/\{1\} \cong A_n$, and the latter group is not of prime order. Therefore, S_n is not a solvable group for $n \geq 5$. ◀

→ **Proposition 5.36.** *Every quotient G/N of a solvable group G is itself a solvable group.*

Remark. One can also prove that every subgroup of a solvable group is itself a solvable group (see my book, *Advanced Modern Algebra*, Proposition 4.22). Since A_n is simple for $n \geq 5$, it is not solvable, by Example 5.35(iii). This gives a second proof that S_n is not solvable for $n \geq 5$. ◀

Proof. By the first isomorphism theorem for groups, quotient groups are isomorphic to homomorphic images, and so it suffices to prove that if $f : G \to H$ is a surjection (for some group H), then H is a solvable group.

Let $G = G_0 \geq G_1 \geq G_2 \geq \cdots \geq G_t = \{1\}$ be a sequence of subgroups as in the definition of solvable group. Then

$$H = f(G_0) \geq f(G_1) \geq f(G_2) \geq \cdots \geq f(G_t) = \{1\}$$

is a sequence of subgroups of H. If $f(x_{i+1}) \in f(G_{i+1})$ and $u_i \in f(G_i)$, then $u_i = f(x_i)$ and $u_i f(x_{i+1}) u_i^{-1} = f(x_i) f(x_{i+1}) f(x_i)^{-1} = f(x_i x_{i+1} x_i^{-1}) \in f(G_i)$, because $G_{i+1} \triangleleft G_i$; that is, $f(G_{i+1})$ is a normal subgroup of $f(G_i)$. The map $\varphi \colon G_i \to f(G_i)/f(G_{i+1})$, defined by $x_i \mapsto f(x_i) f(G_{i+1})$, is a surjection, for it is the composite of the surjection $G_i \to f(G_i)$ and the natural map $f(G_i) \to f(G_i)/f(G_{i+1})$. Since $G_{i+1} \leq \ker \varphi$, this map induces a surjection $G_i/G_{i+1} \to f(G_i)/f(G_{i+1})$, namely, $x_i G_{i+1} \mapsto f(x_i) f(G_{i+1})$. Now G_i/G_{i+1} is cyclic of prime order, so that its quotient $f(G_i)/f(G_{i+1})$ is a cyclic group of order 1 or order a prime. Thus, deleting any repetitions if necessary, $H = f(G)$ has a series in which all the quotient groups are cyclic of prime order; therefore, H is a solvable group. •

Here is the main criterion.

→ **Theorem 5.37 (Galois).** *Let k be a field and let $f(x) \in k[x]$. If $f(x)$ is solvable by radicals, then its Galois group $\mathrm{Gal}(E/k)$ is a solvable group if k has "enough" roots of unity.*

Remark. Let $/k$ be a splitting field of $f(x)$. Since $f(x)$ is solvable by radicals, there is a radical extension $k = K_0 \subseteq K_1 \subseteq \cdots \subseteq K_t$ with every $[K_{i+1} : K_i]$ prime and with $E \subseteq K_t$. By "enough" roots of unity, we mean that k contains all pth roots of unity with p equal to some $[K_{i+1} : K_i]$. Exercise 5.28 on page 470 shows how to eliminate this hypothesis. ◀

Proof. By Lemma 5.34(ii), $\mathrm{Gal}(E/k)$ is a quotient of a solvable group and, by Proposition 5.36, any quotient of a solvable group is itself solvable. •

If k has characteristic 0, then the converse of Theorem 5.37 is true; it was also proved by Galois (see my book *Advanced Modern Algebra*, p. 235). However, the converse is false in characteristic p. If $f(x) = x^p - x - t \in k[x]$, where $k = \mathbb{F}_p(t)$, then the Galois group of $f(x)$ over k is cyclic of order p, but $f(x)$ is not solvable by radicals (see Proposition 4.56 in *Advanced Modern Algebra*).

In 1827, Abel proved a theorem saying, in group-theoretic language not known to him, that if the Galois group of a polynomial $f(x)$ is commutative, then $f(x)$ is solvable by radicals. This is why abelian groups are so called. Since every finite abelian group is solvable [Example 5.35(ii)], Abel's theorem is a special case of Galois's theorem.

It is not difficult to prove that every subgroup of S_2, S_3, and S_4 is a solvable group. Thus, Theorem 5.21 shows that the Galois group of every quadratic, cubic, and quartic polynomial is a solvable group. Therefore, the converse of Galois's theorem shows that if k has characteristic 0, then every polynomial $f(x) \in k[x]$ with $\deg(f) \leq 4$ is solvable by radicals (of course, we already know this because we have proved the classical formulas).

We now complete our discussion by showing, for $n \geq 5$, that the general polynomial of degree n is not solvable by radicals.

→ **Theorem 5.38 (Abel-Ruffini).**   *For all $n \geq 5$, the general polynomial of degree n,*

$$f(x) = (x - y_1)(x - y_2) \cdots (x - y_n),$$

is not solvable by radicals.

Proof.   Let F be a field, let $E = F(y_1, \ldots, y_n)$ be the field of all rational functions in n variables $y_1, \ldots, y_n$ over F, and let $k = F(a_0, \ldots, a_n)$, where the a_i are the coefficients of $f(x)$. In Example 5.17, we saw that E is the splitting field of $f(x)$ over k. In particular, if we choose $F = \mathbb{C}$, then k is an extension field of $\mathbb{C}$, and hence k contains all the roots of unity.

We claim that S_n is isomorphic to a subgroup of $\mathrm{Gal}(E/k)$. Exercise 3.51(ii) on page 252 says that if A and R are domains and $\varphi \colon A \to R$ is an isomorphism, then $a/b \mapsto \varphi(a)/\varphi(b)$ is an isomorphism $\mathrm{Frac}(A) \to \mathrm{Frac}(R)$. If $\sigma \in S_n$, then $f(y_1, \ldots, y_n) \mapsto f(y_{\sigma(1)}, \ldots, y_{\sigma(n)})$ is an automorphism $\widetilde{\sigma}$ of $\mathbb{C}[y_1, \ldots, y_n]$ fixing $\mathbb{C}$, by Theorem 3.33; of course, $\widetilde{\sigma}$ permutes the variables of a polynomial in several variables. By Exercise 3.51 on page 252, automorphisms of a domain R extend to automorphisms of $\mathrm{Frac}(R)$. In particular, $\widetilde{\sigma}$ extends to an automorphism σ^* of $E = \mathrm{Frac}(\mathbb{C}[y_1, \ldots, y_n])$. Now Eqs. (1) on page 450 show that σ^* fixes k, and so $\sigma^* \in \mathrm{Gal}(E/k)$. Using Proposition 5.20, it is easy to see that $\sigma \mapsto \sigma^*$ is an injection $S_n \to \mathrm{Gal}(E/k)$, so that $n! \leq |\mathrm{Gal}(E/k)|$. But the reverse inequality holds as well, by Theorem 5.21, so that $n! = |\mathrm{Gal}(E/k)|$ and $\mathrm{Gal}(E/k) \cong S_n$. Therefore, $\mathrm{Gal}(E/k)$ is not a solvable group for $n \geq 5$, by Example 5.35(iv), and Theorem 5.37 shows that $f(x)$ is not solvable by radicals.   •

We have proved that there is no generalization of the classical formulas to polynomials of degree $n \geq 5$.

→ **Example 5.39.**

Here is an explicit example of a quintic polynomial which is not solvable by radicals. Let $f(x) = x^5 - 4x + 2 \in \mathbb{Q}[x]$. By Eisenstein's criterion (Theorem 3.102), $f(x)$ is irreducible over $\mathbb{Q}$ Let $E/\mathbb{Q}$ be the splitting field of $f(x)$ contained in $\mathbb{C}$, and let $G = \text{Gal}(E/\mathbb{Q})$.

We now use some calculus. There are exactly two real roots of the derivative $f'(x) = 5x^4 - 4$, namely, $\pm\sqrt[4]{4/5} \sim \pm.946$, and so $f(x)$ has two critical points. Now $f(\sqrt[4]{4/5}) < 0$ and $f(-\sqrt[4]{4/5}) > 0$, so that $f(x)$ has one relative maximum and one relative minimum. It follows easily that $f(x)$ has exactly three real roots (although we will not need to know their values, they are, approximately, -1.5185, 0.5085, and 1.2435; the complex roots are $-.1168 \pm 1.4385i$). The restriction of complex conjugation to E, call it τ, is a transposition, for τ interchanges the two complex roots while it fixes the three real roots.

The Galois group G is isomorphic to a subgroup of $S_X \cong S_5$, where X is the set of 5 roots of $f(x)$. Now Corollary 5.27 gives $|G| = [E : \mathbb{Q}]$ divisible by 5, so that G has an element σ of order 5, by Cauchy's theorem (Theorem 2.147); σ must be a 5-cycle, for these are the only elements of order 5 in S_5. Exercise 2.126 on page 207 says that S_5 is generated by any transposition and any 5-cycle, so that $G = \text{Gal}(E/\mathbb{Q}) \cong S_5$. Therefore, $\text{Gal}(E/\mathbb{Q})$ is not a solvable group, by Example 5.35(iv), and so Theorem 5.37 (with the unnecessary hypothesis on roots of unity removed) says that $f(x)$ is not solvable by radicals. ◄

An (impractical) algorithm computing Galois groups is given in van der Waerden, *Modern Algebra*, vol. I, pp. 189-192. However, more advanced expositions of Galois theory show how to compute explicitly Galois groups of $f(x) \in \mathbb{Q}[x]$ when $\deg(f) \leq 4$.

EXERCISES

H **5.15** True or false with reasons.

(i)   Every algebraically closed field contains n distinct nth roots of unity, where $n \geq 1$.

(ii) There are no 5th roots of unity in a field of characteristic 5.

(iii) $\mathbb{R}$ is a splitting field of $x^2 - 5$ over $\mathbb{Q}$.

(iv) $\mathbb{Q}(\sqrt{5})$ is a normal extension of $\mathbb{Q}$.

(v)   No polynomial of degree ≥ 5 in $\mathbb{Q}[x]$ is solvable by radicals.

(vi) $\mathbb{F}_2(x) = \text{Frac}(\mathbb{F}_2[x])$ is an infinite field of characteristic 2.

(vii) A polynomial $f(x) \in \mathbb{Q}[x]$ can have two splitting fields inside of $\mathbb{C}$.

(viii) The alternating group A_4 is a solvable group.

(ix) The alternating group A_5 is a solvable group.

*5.16 Let $\varphi\colon A \to H$ be a group homomorphism. If $B \lhd A$ and $B \leq \ker\varphi$, prove that the *induced map* $\varphi^*\colon A/B \to H$, given by $aB \mapsto \varphi(a)$, is a well-defined homomorphism with $\operatorname{im}\varphi^* = \operatorname{im}\varphi$.

*5.17 If $z \in \mathbb{C}$ is a constructible number, prove that $\mathbb{Q}(i, z)/\mathbb{Q}$ is a radical extension.

5.18 Let k be a field and let $f(x) \in k[x]$. Prove that if E and E' are splitting fields of $f(x)$ over k, then $\operatorname{Gal}(E/k) \cong \operatorname{Gal}(E'/k)$.

5.19 Prove that $\mathbb{F}_3[x]/(x^3 - x^2 - 1) \cong \mathbb{F}_3[x]/(x^3 - x^2 + x - 1)$.

H 5.20 Is $\mathbb{F}_4$ a subfield of $\mathbb{F}_8$?

5.21 Let k be a field of characteristic $p > 0$, and define the *Frobenius map* $F\colon k \to k$ by $F\colon a \mapsto a^p$.

    (i)   Prove that $F\colon k \to k$ is an injection.

    H (ii)   When k is finite, prove that F is an automorphism fixing the prime field $\mathbb{F}_p$. Conclude that $F \in \operatorname{Gal}(k/\mathbb{F}_p)$.

    H (iii)   Prove that if k is finite, then every $a \in k$ has a pth root; that is, there is $b \in k$ with $b^p = a$.

5.22 Let $q = p^n$ for some prime p and some $n \geq 1$.

    (i)   If α is a generator of $\mathbb{F}_q^\times$, prove that $\mathbb{F}_q = \mathbb{F}_p(\alpha)$.

    H (ii)   Prove that the irreducible polynomial $p(x) \in \mathbb{F}_p[x]$ of α has degree n.

    H (iii)   Prove that if $G = \operatorname{Gal}(\mathbb{F}_q/\mathbb{F}_p)$, then $|G| \leq n$.

    H (iv)   Prove that $\operatorname{Gal}(\mathbb{F}_q/\mathbb{F}_p)$ is cyclic of order n with generator the Frobenius F.

5.23 Given $f(x) = ax^2 + bx + c \in \mathbb{Q}[x]$, prove that the following statements are equivalent.

    (i) $f(x)$ is irreducible.

    (ii) $\sqrt{b^2 - 4ac}$ is not rational.

    (iii) $\operatorname{Gal}(\mathbb{Q}(\sqrt{b^2 - 4ac})/\mathbb{Q})$ has order 2.

*5.24 Let E/k be a splitting field of a polynomial $f(x) \in k[x]$. If $\deg(f) = n$, prove that $[E : k] \leq n!$. Conclude that E/k is a finite extension.

H 5.25 What is the degree of the splitting field of $x^{30} - 1$ over $\mathbb{F}_5$?

5.26 Prove that if $f(x) \in \mathbb{Q}[x]$ has a rational root a, then its Galois group is the same as the Galois group of $f(x)/(x - a)$.

*5.27   **(i)**   Let H be a normal subgroup of a finite group G. If both H and G/H are solvable groups, prove that G is a solvable group.

    (ii)   If H and K are solvable groups, prove that $H \times K$ is solvable.

*5.28 We are going to improve Theorem 5.37 by eliminating the hypothesis involving roots of unity: if k is a field and $f(x) \in k[x]$ is solvable by radicals, then its Galois group $\operatorname{Gal}(E/k)$ is a solvable group.

Since $f(x)$ is solvable by radicals, there is a radical tower $k = K_0 \subseteq \cdots \subseteq F$ with $E \subseteq F$; moreover, we were able to assume that F/k a splitting field of some polynomial. Finally, if k contains a certain set Ω of mth roots of unity, then $\operatorname{Gal}(E/k)$ is solvable.

    (i)   Define E^*/E to be a splitting field of $x^m - 1$, and define $k^* = k(\Omega)$. Prove that E^* is a splitting field of $f(x)$ over k^*, and conclude that $\operatorname{Gal}(E^*/k^*)$ is solvable.

    (ii)   Prove that $\operatorname{Gal}(E^*/k^*) \lhd \operatorname{Gal}(E^*/k)$ and $\operatorname{Gal}(E^*/k)/\operatorname{Gal}(E^*/k^*) \cong \operatorname{Gal}(k^*/k)$.

    (iii)   Use Exercise 5.27 to prove that $\operatorname{Gal}(E^*/k)$ is solvable.

   (iv) Prove that $\mathrm{Gal}(E^*/E) \lhd \mathrm{Gal}(E^*/k)$ and $\mathrm{Gal}(E^*/k)/\mathrm{Gal}(E^*/E) \cong \mathrm{Gal}(E/k)$. Conclude that $\mathrm{Gal}(E/k)$ is solvable.

*5.29 Let $f(x) \in \mathbb{Q}[x]$ be an irreducible cubic with Galois group G.

   H **(i)** Prove that if $f(x)$ has exactly one real root, then $G \cong S_3$.

   H **(ii)** Find the Galois group of $f(x) = x^3 - 2 \in \mathbb{Q}[x]$.

   H **(iii)** Find a cubic polynomial $g(x) \in \mathbb{Q}[x]$ whose Galois group has order 3.

*5.30 **(i)** If k is a field and $f(x) \in k[x]$ has derivative $f'(x)$, prove that either $f'(x) = 0$ or $\deg(f') < \deg(f)$.

   H **(ii)** If k is a field of characteristic 0, prove that an irreducible polynomial $p(x) \in k[x]$ has no repeated roots; that is, if E is the splitting field of $p(x)$, then there is no $a \in E$ with $(x - a)^2 \mid p(x)$ in $E[x]$.

*5.31 Let k be a field of characteristic p.

   (i) Prove that if $f(x) = \sum_i a_i x^i \in k[x]$, then $f'(x) = 0$ if and only if the only nonzero coefficients are those a_i with $p \mid i$.

   (ii) If k is finite and $f(x) = \sum_i a_i x^i \in k[x]$, prove that $f'(x) = 0$ if and only if there is $g(x) \in k[x]$ with $f(x) = g(x)^p$.

   (iii) Prove that if k is a finite field, then every irreducible polynomial $p(x) \in k[x]$ has no repeated roots.

*5.32 H **(i)** If $k = \mathbb{F}_p(t)$, the field of rational functions over $\mathbb{F}_p$, prove that $x^p - t \in k[t]$ has repeated roots. (It can be shown that $x^p - t$ is an irreducible polynomial.)

   (ii) Prove that $E = k(\alpha)$ is a splitting field of $x^p - t$ over k.

   (iii) Prove that $\mathrm{Gal}(E/k) = \{1\}$.

5.3 EPILOG

Further investigation of these ideas is the subject of Galois theory, which studies the relationship between extension fields and their Galois groups. Aside from its intrinsic beauty, Galois theory is used extensively in algebraic number theory.

The following technical notion turns out to be important.

Definition. A polynomial $f(x) \in k[x]$ is **separable** if its irreducible factors have no repeated roots (thus, an *irreducible* polynomial is separable if it has no repeated roots). A finite field extension E/k is **separable** if every $\alpha \in E$ is a root of an irreducible polynomial in $k[x]$ having no repeated roots.

We have seen that E/k is separable (over any subfield k) if E has characteristic 0 [Exercise 5.30(ii) on page 471] or if E is finite [Exercise 5.31(iii) on page 471]. On the other hand, there are extensions E of the function field $\mathbb{F}_p(x)$ which are not separable, as we have seen in Exercise 5.32 on page 471. The following generalization of Theorem 5.26 shows why separable polynomials are interesting (there is a proof in my book *Advanced Modern Algebra*, Theorem 4.7).

Theorem.   *Let k be a field and let $f(x) \in k[x]$ be a separable polynomial. If E/k is a splitting field of $f(x)$, then $|\operatorname{Gal}(E/k)| = [E : k]$.*

Proof.   The hypothesis in Theorem 5.26 that k have characteristic 0 is only used to guarantee separability.   •

Definition.   Let E/k be a field extension with Galois group $G = \operatorname{Gal}(E/k)$. If $H \leq G$, then the *fixed field* E^H is defined by

$$E^H = \{u \in E : \sigma(u) = u \text{ for all } \sigma \in H\}.$$

The following theorems can be proved (for example, see Section 4.2 of my book *Advanced Modern Algebra*). Theorem 5.29, which characterizes splitting fields, can be modified in the presence of separability.

Theorem.   *Let E/k be a field extension with Galois group $G = \operatorname{Gal}(E/k)$. Then the following statements are equivalent.*

(i) *E is a splitting field of some separable polynomial $f(x) \in k[x]$.*

(ii) *Every irreducible $p(x) \in k[x]$ having one root in E is separable and it splits in $E[x]$.*

(iii) *$k = E^G$; that is, if $a \in E$ and $\sigma(a) = a$ for all $\sigma \in G$, then $a \in k$.*

Definition.   A finite field extension E/k is a *Galois extension* if it satisfies any of the equivalent conditions in this theorem.

The following theorem shows that there is an intimate connection between the intermediate fields B in a Galois extension E/k (that is, subfields B with $k \subseteq B \subseteq E$) and the subgroups of the Galois group.

Theorem (Fundamental Theorem of Galois Theory).   *Let E/k be a finite Galois extension with Galois group $G = \operatorname{Gal}(E/k)$.*

(i) *The function $H \mapsto E^H$ is a bijection, from the set of all subgroups of $\operatorname{Gal}(E/k)$ to the set of all intermediate fields, which reverses inclusions:*

$$H \leq L \text{ if and only if } E^L \subseteq E^H.$$

For every intermediate field B and every $H \leq G$,

$$E^{\operatorname{Gal}(E/B)} = B \quad and \quad \operatorname{Gal}(E/E^H) = H.$$

(ii) *For every intermediate field B and every subgroup H of G,*

$$[B : k] = [G : \operatorname{Gal}(E/B)] \quad and \quad [G : H] = [E^H : k].$$

(iii) *An intermediate field B is a Galois extension of k if and only if* $\mathrm{Gal}(E/B)$ *is a normal subgroup of G.*

Here are some consequences.

Theorem (Theorem of the Primitive Element). *If E/k is a finite separable extension, then there is* **primitive element** $\alpha \in E$; *that is, $E = k(\alpha)$.*

In particular, every finite extension of $\mathbb{Q}$ has a primitive element. This follows from a theorem of E. Steinitz which says, given a finite extension E/k, that there exists $\alpha \in E$ with $E = k(\alpha)$ if and only if there are only finitely many intermediate fields $k \subseteq B \subseteq E$. But the Fundamental Theorem says that there is a bijection between the family of all the intermediate fields and the family of all subgroups of the finite group $\mathrm{Gal}(E/k)$.

Theorem. *The finite field $\mathbb{F}_q$, where $q = p^n$, has exactly one subfield of order p^d for every divisor d of n, and no others.*

This follows from $\mathrm{Gal}(\mathbb{F}_q/\mathbb{F}_p)$ being cyclic of order n and Proposition 2.75: if G is a cyclic group of order n, then G has a unique subgroup of order d for each divisor d of n.

Theorem. *If E/k is a Galois extension whose Galois group is abelian, then every intermediate field is a Galois extension.*

This follows from the Fundamental Theorem because every subgroup of an abelian group is normal.

There are many proofs of the Fundamental Theorem of Algebra, and there is one using Galois theory (see my book *Advanced Modern Algebra*, p. 233).

Theorem (Fundamental Theorem of Algebra). *If $f(x) \in \mathbb{C}[x]$ is not a constant, then $f(x)$ has a root in $\mathbb{C}$.*

We now use the Fundamental Theorem of Galois Theory to complete the discussion of constructibility in Chapter 4.

Recall that a prime p is a *Fermat prime* if p has the form $p = 2^m + 1$ (in which case $m = 2^t$; see the proof of Corollary 3.103). We end with a proof of Gauss's theorem that if p is a Fermat prime, then a regular p-gon can be constructed with straightedge and compass.

Lemma 5.40. *Let E/k be a Galois extension with Galois group $G = \mathrm{Gal}(E/k)$. Given subgroups $G \geq H \geq L$, then*

$$[E^L : E^H] = [H : L].$$

Proof. Since $H \mapsto E^H$ is order reversing, there is a tower of fields

$$k = E^G \subseteq E^H \subseteq E^L \subseteq E$$

(we have $k = E^G$ because E/k is a Galois extension). Theorem 4.31 gives $[E^L : k] = [E^L : E^H][E^H : k]$, and so the Fundamental Theorem of Galois Theory gives

$$[E^L : E^H] = \frac{[E^L : k]}{[E^H : k]} = \frac{[G : L]}{[G : H]} = \frac{|G|/|L|}{|G|/|H|} = \frac{|H|}{|L|} = [H : L]. \quad \bullet$$

Theorem 5.41 (Gauss).   *Let p be an odd prime. A regular p-gon is constructible if and only if $p = 2^m + 1$ for some $m \geq 0$.*

Proof.   Necessity was proved in Theorem 4.60, where it was shown that m must be a power of 2 when $m > 0$.

If p is a prime, then $x^p - 1 = (x - 1)\Phi_p(x)$, where $\Phi_p(x)$ is the pth cyclotomic polynomial. A primitive pth root of unity ζ is a root of $\Phi_p(x)$, and $\mathbb{Q}(\zeta)$ is a splitting field of $\Phi_p(x)$ over $\mathbb{Q}$. Since $\Phi_p(x)$ is an irreducible polynomial of degree $p - 1$ (Corollary 3.103), we have $[\mathbb{Q}(\zeta) : \mathbb{Q}] = p - 1 = 2^m$. By Theorem 5.26, we have $|\operatorname{Gal}(\mathbb{Q}(\zeta)/\mathbb{Q})| = 2^m$. As any 2-group, $\operatorname{Gal}(\mathbb{Q}(\zeta)/\mathbb{Q})$ has a normal series

$$\operatorname{Gal}(\mathbb{Q}(\zeta)/\mathbb{Q}) = G_0 \geq G_1 \geq \cdots \geq G_t = \{1\}$$

with every factor group of order 2; that is, $[G_{i-1} : G_i] = 2$ for all $i \geq 1$. By the Fundamental Theorem of Galois Theory, there is a tower of subfields

$$\mathbb{Q} = K_0 \subseteq K_1 \subseteq \cdots \subseteq K_t = \mathbb{Q}(\zeta).$$

Moreover, Lemma 5.40 gives $[K_i : K_{i-1}] = [G_{i-1} : G_i] = 2$ for all $i \geq 1$. This says that ζ is polyquadratic, and hence ζ is constructible, by Theorem 4.54. $\bullet$

6

Groups II

6.1 FINITE ABELIAN GROUPS

We continue our study of groups by considering finite abelian groups; as is customary, these groups are written additively. We are going to prove that every finite abelian group is a direct sum of cyclic groups, and so we begin by considering direct sums.

Definition. The *external direct sum* of two abelian groups S and T is the abelian group $S \times T$ whose underlying set is the cartesian product of S and T and whose operation is given by $(s, t) + (s', t') = (s + s', t + t')$.

It is routine to check that the external direct sum is an (abelian) group; the identity is $(0, 0)$ and the inverse of (s, t) is $(-s, -t)$. For example, the plane $\mathbb{R}^2$ is a group under vector addition, and $\mathbb{R}^2 = \mathbb{R} \times \mathbb{R}$.

Definition. If S and T are subgroups of an abelian group G, then G is the *internal direct sum*, denoted by $G = S \oplus T$, if each element $g \in G$ has a unique expression of the form $g = s + t$, where $s \in S$ and $t \in T$.

If S and T are subgroups of an abelian group G, define

$$S + T = \{s + t : s \in S \text{ and } t \in T\}.$$

Now $S + T$ is always a subgroup of G, for it is $\langle S \cup T \rangle$, the subgroup generated by S and T (see Exercise 6.5 on page 488). Saying that $G = S + T$ means that each $g \in G$ has an expression of the form $g = s + t$, where $s \in S$ and $t \in T$; saying that $G = S \oplus T$ means that such expressions are unique. Here is the additive version of Proposition 2.127. We need not say that S and T are normal subgroups, for every subgroup of an abelian group is normal.

Lemma 6.1. *If S and T are subgroups of an abelian group G, then $G = S \oplus T$ if and only if $S + T = G$ and $S \cap T = \{0\}$.*

Proof. Assume that $G = S \oplus T$. Every $g \in G$ has a unique expression of the form $g = s + t$, where $s \in S$ and $t \in T$; hence, $G = S + T$. If $x \in S \cap T$, then x has two expressions as $s + t$, namely, $x = x + 0$ and $x = 0 + x$. Since expressions are unique, we must have $x = 0$, and so $S \cap T = \{0\}$.

Conversely, $G = S + T$ implies that each $g \in G$ has an expression of the form $g = s + t$, where $s \in S$ and $t \in T$. To see that this expression is unique, suppose also that $g = s' + t'$, where $s' \in S$ and $t' \in T$. Then $s + t = s' + t'$ gives $s - s' = t' - t \in S \cap T = \{0\}$. Therefore, $s = s'$ and $t = t'$, as desired.   •

Definition. A subgroup S of an abelian group G is called a ***direct summand*** if there exists a subgroup T of G with $G = S \oplus T$; that is, $S + T = G$ and $S \cap T = \{0\}$. Such a subgroup T is called a ***complement*** of S.

Note that $S \times T$ cannot equal $S \oplus T$, for neither S nor T is a subgroup of $S \times T$; indeed, they are not even subsets of the cartesian product. This is easily remedied. Given abelian groups S and T, define subgroups S^* and T^* of the external direct sum $S \times T$ by

$$S^* = \{(s, 0) : s \in S\} \quad \text{and} \quad T^* = \{(0, t) : t \in T\}.$$

Of course, $S \cong S^*$ via $s \mapsto (s, 0)$ and $T \cong T^*$ via $t \mapsto (0, t)$. It is easy to see that $S \times T = S^* \oplus T^*$, for $S^* + T^* = S \times T$, because $(s, t) = (s, 0) + (0, t)$, and $S^* \cap T^* = \{(0, 0)\}$. Thus, the external direct sum can be viewed as an internal direct sum (of subgroups isomorphic to S and to T). The next result shows, conversely, that an internal direct sum is isomorphic to an external one.

Proposition 6.2. *Let S and T be subgroups of an abelian group G with $G = S + T$. If $G = S \oplus T$ (that is, $S \cap T = \{0\}$), then there is an isomorphism $\varphi \colon S \oplus T \to S \times T$ with $\varphi(S) = S^*$ and $\varphi(T) = T^*$.*

Proof. If $g \in S \oplus T$, then Lemma 6.1 says that g has a unique expression of the form $g = s + t$. Define $\varphi \colon S \oplus T \to S \times T$ by $\varphi(g) = \varphi(s + t) = (s, t)$. Uniqueness of the expression $g = s + t$ implies that φ is a well-defined function. It is obvious that $\varphi(S) = S^*$ and $\varphi(T) = T^*$. Let us check that φ is a homomorphism. If $g' = (s', t')$, then $(s, t) + (s', t') = (s + s', t + t')$; hence,

$$\begin{aligned}
\varphi(g + g') &= \varphi(s + s' + t + t') \\
&= (s + s', t + t') \\
&= (s, t) + (s', t') \\
&= \varphi(g) + \varphi(g').
\end{aligned}$$

If $\varphi(g) = (s, t) = (0, 0)$, then $s = 0, t = 0$, and $g = s + t = 0$; hence, φ is injective. Finally, φ is surjective, for if $(s, t) \in S \times T$, then $\varphi(s + t) = (s, t)$. $\quad\bullet$

We now extend this discussion to more than two summands.

Definition. The *external direct sum* of abelian groups $S_1, S_2, \ldots, S_n$ is the abelian group $S_1 \times S_2 \times \cdots \times S_n$ whose underlying set is the cartesian product of $S_1, S_2, \ldots, S_n$, and whose operation is given by

$$(s_1, s_2, \ldots, s_n) + (s_1', s_2', \ldots, s_n') = (s_1 + s_1', s_2 + s_2', \ldots, s_n + s_n').$$

For example, Euclidean n-space $\mathbb{R}^n$ is the external direct sum of $\mathbb{R}$ with itself n times: $\mathbb{R}^n = \mathbb{R} \times \cdots \times \mathbb{R}$.

Definition. If $S_1, \ldots, S_n$ are subgroups of an abelian group G, then G is the *internal direct sum*, denoted by

$$G = S_1 \oplus \cdots \oplus S_n,$$

if, for each $g \in G$, there are unique $s_i \in S_i$ with $g = s_1 + \cdots + s_n$.

Example 6.3.
Let k be a field and let $G = k^n$ be the external direct sum of k with itself n times. As usual, let $e_1, \ldots, e_n$ be the standard basis; that is, $e_i = (0, \ldots, 0, 1, 0, \ldots, 0)$, the n-tuple having ith coordinate 1 and all other coordinates 0. If V_i is the one-dimensional subspace spanned by e_i, that is, $V_i = \{ae_i : a \in k\}$, then k^n is the internal direct sum $k^n = V_1 \oplus \cdots \oplus V_n$, for every vector has a unique expression as a linear combination of a basis. $\quad\blacktriangleleft$

We now show that every external direct sum can be viewed as an internal direct sum. If $S_1, \ldots, S_n$ are abelian groups, define, for each i,

$$S_i^* = \big\{(0, \ldots, 0, s_i, 0, \ldots, 0) : s_i \in S_i\big\} \subseteq S_1 \times \cdots \times S_n;$$

that is, S_i^* consists of all those n-tuples in the cartesian product whose only nonzero coordinates occur in the ith position. Of course, S_i and S_i^* are isomorphic, for all i, via $s_i \mapsto (0, \ldots, 0, s_i, 0, \ldots, 0)$. Let us check that G is the internal direct sum

$$G = S_1^* \oplus \cdots \oplus S_n^*.$$

If $g = (s_1, \ldots, s_n) \in S_1 \times \cdots \times S_n$, then

$$g = (s_1, 0 \ldots, 0) + (0, s_2, 0, \ldots, 0) + \cdots + (0, \ldots, 0, s_n).$$

Such an expression is unique, for if $(s_1, \ldots, s_n) = (t_1, \ldots, t_n)$, then the definition of equality of n-tuples gives $s_i = t_i$ for all i.

How do we generalize Lemma 6.1 to several summands? If an abelian group G is generated by subgroups $S_1, S_2, \ldots, S_n$, one's first guess is that $G = S_1 \oplus \cdots \oplus S_n$ if $S_i \cap S_j = \{0\}$ for all $i \neq j$. We now show that this is not adequate.

Let V be a 2-dimensional vector space over a field k, and let x, y be a basis; hence, $V = \langle x \rangle \oplus \langle y \rangle$. It is easy to check that the intersection of any two of the subspaces $\langle x \rangle$, $\langle y \rangle$, and $\langle x + y \rangle$ is $\{0\}$. On the other hand, we do not have $V = \langle x \rangle \oplus \langle y \rangle \oplus \langle x + y \rangle$ because 0 has two expressions in $\langle x \rangle + \langle y \rangle + \langle x + y \rangle$, namely, $0 = 0 + 0 + 0$ and $0 = -x - y + (x + y)$.

We are now going to show that every internal direct sum is isomorphic to an external one. Here is the generalization of Lemma 6.1 and Proposition 6.2.

Proposition 6.4. *Let $G = S_1 + S_2 + \cdots + S_n$, where the S_i are subgroups; that is, each $g \in G$ has an expression of the form*

$$g = s_1 + s_2 + \cdots + s_n,$$

where $s_i \in S_i$ for all i. Then the following conditions are equivalent.

(i) $G = S_1 \oplus S_2 \oplus \cdots \oplus S_n$; *that is, for every element $g \in G$, the expression $g = s_1 + \cdots + s_n$, where $s_i \in S_i$ for all i, is unique.*

(ii) *There is an isomorphism $\varphi \colon G \to S_1 \times S_2 \times \cdots \times S_n$ with $\varphi(S_i) = S_i^*$ for all i.*

(iii) *If we define $G_i = S_1 + \cdots + \widehat{S_i} + \cdots + S_n$, where $\widehat{S_i}$ means that the term S_i is omitted from the sum, then $S_i \cap G_i = \{0\}$ for each i.*

Proof.
(i) $\Rightarrow$ (ii) If $g \in G$ and $g = s_1 + \cdots + s_n$, then define $\varphi \colon G \to S_1 \times \cdots \times S_n$ by $\varphi(g) = \varphi(s_1 + \cdots + s_n) = (s_1, \ldots, s_n)$. Uniqueness of the expression for g shows that φ is well-defined. It is straightforward to prove that φ is an isomorphism with $\varphi(S_i) = S_i^*$ for all i.
(ii) $\Rightarrow$ (iii) If $g \in S_i \cap G_i$, then $\varphi(g) \in S_i^* \cap (S_1^* + \cdots + \widehat{S_i^*} + \cdots + S_n^*)$. But if $\varphi(g) \in S_1^* + \cdots + \widehat{S_i^*} + \cdots + S_n^*$, then its ith coordinate is 0; if $\varphi(g) \in S_i^*$, then its jth coordinates are 0 for all $j \neq i$. Therefore, $\varphi(g) = 0$. Since φ is an isomorphism, it follows that $g = 0$.
(iii) $\Rightarrow$ (i) Let $g \in G$, and suppose that

$$g = s_1 + \cdots + s_n = t_1 + \cdots + t_n,$$

where $s, t_i \in S_i$ for all i. Now $s_i - t_i = \sum_{j \neq i}(t_j - s_j) \in S_i \cap (S_1 + \cdots + \widehat{S_i} + \cdots + S_n) = \{0\}$ for each i. Therefore, $s_i = t_i$ for all i, and the expression $g = \sum_i s_i$ is unique. •

Notation.   From now on, we shall use the notation $S_1 \oplus \cdots \oplus S_n$ to denote either version of the direct sum, external or internal, because our point of view is almost always internal. We will also write[1]

$$\bigoplus_{i=1}^{n} S_i = S_1 \oplus \cdots \oplus S_n.$$

The notation $G = \sum_{i=1}^{n} S_i$ abbreviates $G = S_1 + \cdots + S_n = \langle S_1 \cup \cdots \cup S_n \rangle$. Thus, $G = \sum_i S_i$ if every $g \in G$ has an expression of the form $g = \sum_i s_i$ for $s_i \in S_i$, while $G = \bigoplus_i S_i$ if $G = \sum_i S_i$ and expressions $g = \sum_i s_i$ are unique.

It will be convenient to analyze groups "one prime at a time."

Definition.   If p is a prime, then an abelian group G is ***p-primary***[2] if, for each $a \in G$, there is $n \geq 1$ with $p^n a = 0$.

Definition.   If G is any abelian group, then its ***p-primary component*** is

$$G_p = \{a \in G : p^n a = 0 \text{ for some } n \geq 1\}.$$

If we do not want to specify the prime p, we may call an abelian group *primary* (instead of *p-primary*). It is clear that primary components are subgroups. This is not true for nonabelian groups. For example, if $G = S_3$, then $G_2 = \{(1), (1\ 2), (1\ 3), (2\ 3)\}$, which is not a subgroup of S_3 because $(1\ 2)(1\ 3) = (1\ 3\ 2) \notin G_2$.

Theorem 6.5 (Primary Decomposition).

(i) *Every finite abelian group G is the direct sum of its p-primary components*:

$$G = \bigoplus_p G_p.$$

(ii) *Two finite abelian groups G and G' are isomorphic if and only if $G_p \cong G'_p$ for every prime p.*

---

[1]In *Advanced Modern Algebra*, the sequel to the previous edition of this book, I denote the direct sum by $\sum_i S_i$. I now think that it is clearer to denote the direct sum by $\bigoplus_i S_i$ (which is one of several notations commonly used today) and to denote the *sum*, the subgroup generated by $\bigcup_i S_i$. by $\sum_i S_i$. If I have a chance to redo the sequel, then I will use this notation there as well.

[2]In Chapter 2, we called a finite group G a *p-group* if $|G|$ is a power of p, and we proved, in Exercise 2.117 on page 207, that a finite group G is a *p*-group if and only if each $g \in G$ has order some power of p. Thus, a *p*-primary abelian group is just an abelian *p*-group. If one is working wholly in the context of abelian groups, as we are now, then the term *p*-primary is used; if one is working with general groups, then the usage of the term *p*-group is preferred.

Proof.
(i) Let $x \in G$ be nonzero, and let its order be d. By the fundamental theorem of arithmetic, there are distinct primes $p_1, \ldots, p_n$ and positive exponents $e_1, \ldots, e_n$ with

$$d = p_1^{e_1} \cdots p_n^{e_n}.$$

Define $r_i = d/p_i^{e_i}$, so that $p_i^{e_i} r_i = d$. It follows that $r_i x \in G_{p_i}$ for each i. But the gcd d of $r_1, \ldots, r_n$ is 1 (the only possible prime divisors of d are $p_1, \ldots, p_n$, but no p_i is a common divisor because $p_i \nmid r_i$); hence, there are integers $s_1, \ldots, s_n$ with $1 = \sum_i s_i r_i$. Therefore,

$$x = \sum_i s_i r_i x \in G_{p_1} + \cdots + G_{p_n}.$$

Write $H_i = G_{p_1} + \cdots + \widehat{G}_{p_i} + \cdots + G_{p_n}$. By Proposition 6.4, it suffices to prove that if

$$x \in G_{p_i} \cap H_i,$$

then $x = 0$. Since $x \in G_{p_i}$, we have $p_i^\ell x = 0$ for some $\ell \geq 0$; since $x \in H_i$, we have $ux = 0$, where $u = \prod_{j \neq i} p_j^{g_j}$ for $g_j \geq 0$. But p_i^ℓ and u are relatively prime, so there exist integers s and t with $1 = sp_i^\ell + tu$. Therefore,

$$x = (sp_i^\ell + tu)x = sp_i^\ell x + tux = 0.$$

(ii) If $f : G \to G'$ is a homomorphism, then $f(G_p) \subseteq G'_p$ for every prime p, for if $p^\ell a = 0$, then $0 = f(p^\ell a) = p^\ell f(a)$. If f is an isomorphism, then $f^{-1} : G' \to G$ is also an isomorphism (so that $f^{-1}(G'_p) \subseteq G_p$ for all p). It follows that each restriction $f | G_p : G_p \to G'_p$ is an isomorphism, with inverse $f^{-1} | G'_p$.

Conversely, if there are isomorphisms $f_p : G_p \to G'_p$ for all p, then there is an isomorphism $\varphi : \bigoplus_p G_p \to \bigoplus_p G'_p$ given by $\sum_p a_p \mapsto \sum_p f_p(a_p)$. •

Notation.   If G is an abelian group and m is an integer, then

$$mG = \{ma : a \in G\}.$$

It is easy to see that mG is a subgroup of G.
The next type of subgroup will play an important role.

Definition.   Let p be a prime and let G be a p-primary[3] abelian group. A subgroup $S \subseteq G$ is a ***pure***[4] ***subgroup*** if, for all $n \geq 0$,

$$S \cap p^n G = p^n S.$$

---

[3] If G is not a primary group, then a pure subgroup $S \subseteq G$ is defined to be a subgroup which satisfies $S \cap mG = mS$ for all $m \in \mathbb{Z}$ (see Exercises 6.16 and 6.18 on page 489).

[4] A polynomial equation is called ***pure*** if it has the form $x^n = a$; pure subgroups are defined in terms of such equations, and they are probably so called because of this resemblance.

The inclusion $S \cap p^n G \supseteq p^n S$ is true for every subgroup $S \subseteq G$, and so it is only the reverse inclusion $S \cap p^n G \subseteq p^n S$ that is significant. It says that if $s \in S$ satisfies an equation $s = p^n g$ for some $g \in G$, then there exists $s' \in S$ with $s = p^n s'$; that is, if an equation $s = p^n x$ is solvable for $x \in G$, then it is solvable for $x \in S$.

Example 6.6.

(i) Every direct summand S of G is a pure subgroup. Let $G = S \oplus T$, and suppose that $s = p^n g$, where $s \in S$ and $g \in G$. Now $g = u + v$, where $u \in S$ and $v \in T$, and so $s = p^n u + p^n v$. Hence, $p^n v = s - p^n u \in S \cap T = \{0\}$, so that $p^n v = 0$. Therefore, $s = p^n u$, and S is pure in G.

(ii) If $G = \langle g \rangle$ is a cyclic group of order p^2, where p is a prime, then we claim that $S = \langle pg \rangle$ is not a pure subgroup of G. Every element $s' \in S$ has the form $s' = mpg$ for some $m \in \mathbb{Z}$. Now $s = pg \in S$; if there is an element $s' \in S$ with $s = ps'$, then $s = ps' = p(mpg) = mp^2 g = 0$, a contradiction.   ◀

In Exercise 6.12 on page 488, we shall see that the converse of Example 6.6(i) is true: if G is a finite abelian group and S is a subgroup of G, then S is a pure subgroup if and only if S is a direct summand. This is the reason we have introduced pure subgroups, for it is easier to prove that S is a direct summand by verifying whether certain equations are solvable than to construct a subgroup T with $S + T = G$ and $S \cap T = \{0\}$.

Lemma 6.7.   *If p is a prime and $G \neq \{0\}$ is a finite p-primary abelian group, then G has a nonzero pure cyclic subgroup.*

Proof.   Let $G = \langle x_1, \ldots, x_q \rangle$. The order of x_i is p^{n_i} for all i, because G is p-primary. If $x \in G$, then $x = \sum_i a_i x_i$, where $a_i \in \mathbb{Z}$, so that if ℓ is the largest of the n_i, then $p^\ell x = 0$. Now choose any $y \in G$ of largest order p^ℓ (for example, y could be one of the x_i). We claim that $S = \langle y \rangle$ is a pure subgroup of G.

Suppose that $s \in S$, so that $s = mp^t y$, where $t \geq 0$ and $p \nmid m$, and let

$$s = p^n a$$

for some $a \in G$. If $t \geq n$, define $s' = mp^{t-n} y \in S$, and note that

$$p^n s' = p^n m p^{t-n} y = mp^t y = s.$$

If $t < n$, then

$$p^\ell a = p^{\ell-n} p^n a = p^{\ell-n} s = p^{\ell-n} mp^t y = mp^{\ell-n+t} y.$$

But $p \nmid m$ and $\ell - n + t < \ell$, because $-n + t < 0$, and so $p^\ell a \neq 0$. This contradicts y having largest order, and so this case cannot occur. Therefore, S is a pure subgroup of G.   •

Proposition 6.8.   *If G is an abelian group and p is a prime, then G/pG is a vector space over $\mathbb{F}_p$ which is finite dimensional when G is finite.*

Proof.   If $[r] \in \mathbb{F}_p$ and $a \in G$, define scalar multiplication

$$[r](a + pG) = ra + pG.$$

This formula is well-defined, for if $k \equiv r \bmod p$, then $k = r + pm$ for some integer m, and so

$$ka + pG = ra + pma + pG = ra + pG,$$

because $pma \in pG$. It is now routine to check that the axioms for a vector space do hold. If G is finite, then so is G/pG, and it is clear that G/pG has a finite basis.   •

Definition.   If p is a prime and G is a finite p-primary abelian group, then

$$d(G) = \dim(G/pG).$$

Observe that d is additive over direct sums,

$$d(G \oplus H) = d(G) + d(H),$$

for Proposition 2.126 gives

$$\frac{G \oplus H}{p(G \oplus H)} = \frac{G \oplus H}{pG \oplus pH} \cong \frac{G}{pG} \oplus \frac{H}{pH}.$$

The dimension of the left side is $d(G \oplus H)$ and the dimension of the right side is $d(G) + d(H)$, for the union of bases of G/pG and of H/pH, respectively, is a basis of $(G/pG) \oplus (H/pH)$.

The abelian groups G with $d(G) = 1$ are easily characterized.

Lemma 6.9.   *A p-primary abelian group G is cyclic if and only if $d(G) = 1$.*

Proof.   If G is cyclic, then so is any quotient of G; in particular, G/pG is cyclic, so that $\dim(G/pG) = 1$ and $d(G) = 1$.

Conversely, assume that $d(G) = 1$; that is, $G/pG \cong \mathbb{I}_p$. Since $\mathbb{I}_p$ is a simple group, the correspondence theorem says that pG is a maximal subgroup of G. We claim that pG is the only maximal subgroup of G. If $L \subseteq G$ is any maximal subgroup, then $G/L \cong \mathbb{I}_p$, for G/L is a simple p-primary abelian group, and hence it has order p, by Proposition 2.153 on page 203. Thus, if $a \in G$, then $p(a + L) = 0$ in G/L, so that $pa \in L$; hence $pG \subseteq L$. But pG is maximal, and so $pG = L$. It follows that every proper subgroup of G is contained in pG (for every proper subgroup is contained in some maximal subgroup). Now $G/pG \cong \mathbb{I}_p$ is cyclic; let $G/pG = \langle z + pG \rangle$ for some $z \in G$ If $\langle z \rangle$ is a proper subgroup of G, then $\langle z \rangle \subseteq pG$ (the unique maximal subgroup of G), contradicting $z + pG$ being a generator of G/pG. Therefore, $G = \langle z \rangle$, and so G is cyclic.   •

If $G = (\mathbb{I}_p)^n$, then $pG = \{0\}$, $G/pG \cong G$, and $d(G) = \dim(G)$. More generally, if G is a direct sum of p-primary cyclic groups, say, $G = \bigoplus_i C_i$, then $pG = \bigoplus_i pC_i$, and Proposition 2.126 gives

$$G/pG = \left(\bigoplus_i C_i\right)/\left(\bigoplus_i pC_i\right) \cong \bigoplus_i (C_i/pC_i).$$

We have just seen that $d(C_i) = 1$ for all i, and so additivity of d over direct sums shows that $d(G)$ counts the number of cyclic summands in this decomposition of G.

Lemma 6.10. *Let G be a finite p-primary abelian group.*

(i) *If $S \subseteq G$, then $d(G/S) \leq d(G)$.*

(ii) *If S is a pure subgroup of G, then*

$$d(G) = d(S) + d(G/S).$$

Proof.
(i) By the correspondence theorem, $p(G/S) = (pG + S)/S$, so that

$$(G/S)/p(G/S) = (G/S)/[(pG + S)/S] \cong G/(pG + S),$$

by the third isomorphism theorem. Since $pG \subseteq pG + S$, there is a surjective homomorphism (of vector spaces over $\mathbb{F}_p$),

$$G/pG \to G/(pG + S),$$

namely, $g + pG \mapsto g + (pG + S)$. Hence, $\dim(G/pG) \geq \dim(G/(pG + S))$; that is, $d(G) \geq d(G/S)$.

(ii) We now analyze $(pG + S)/pG$, the kernel of $G/pG \to G/(pG + S)$. By the second isomorphism theorem,

$$(pG + S)/pG \cong S/(S \cap pG).$$

Since S is a pure subgroup, $S \cap pG = pS$. Therefore,

$$(pG + S)/pG \cong S/pS,$$

and so $\dim[(pG + S)/pG] = d(S)$. But if W is a subspace of a finite dimensional vector space V, then $\dim(V) = \dim(W) + \dim(V/W)$, by Exercise 4.17 on page 343. Hence, if $V = G/pG$ and $W = (pG + S)/pG$, we have

$$d(G) = d(S) + d(G/S). \quad \bullet$$

Theorem 6.11 (Basis Theorem).   *Every finite abelian group G is a direct sum of primary cyclic groups.*

Proof.   By the primary decomposition, Theorem 6.5, we may assume that G is p-primary for some prime p (for if every primary component is a direct sum of cyclic groups, so is G). We prove that G is a direct sum of cyclic groups by induction on $d(G) \geq 1$. The base step is easy, for Lemma 6.9 shows that G must be cyclic in this case.

To prove the inductive step, we begin by using Lemma 6.7 to find a nonzero pure cyclic subgroup $S \subseteq G$. By Lemma 6.10, we have

$$d(G/S) = d(G) - d(S) = d(G) - 1 < d(G).$$

By induction, G/S is a direct sum of cyclic groups, say,

$$G/S = \bigoplus_{i=1}^{q} \langle \overline{x}_i \rangle ,$$

where $\overline{x}_i = x_i + S$.

Let $x \in G$ and let $\overline{x}$ have order p^ℓ, where $\overline{x} = x + S$. We claim that there is $z \in G$ with $z + S = \overline{x} = x + S$ such that order $z =$ order $(\overline{x})$. Now x has order p^n, where $n \geq \ell$. But $p^\ell(x + S) = p^\ell \overline{x} = 0$ in G/S, so there is some $s \in S$ with $p^\ell x = s$. By purity, there is $s' \in S$ with $p^\ell x_i = p^\ell s'$. If we define $z = x - s'$, then $z + S = x + S$ and $p^\ell z = 0$. Hence, if $m\overline{x} = 0$ in G/S, then $p^\ell \mid m$, and so $mz = 0$ in G.

For each i, choose $z_i \in G$ with $z_i + S = \overline{x}_i = x_i + S$ and with order $z_i =$ order $\overline{x}_i$; let $T = \langle z_1, \ldots, z_q \rangle$. Now $S + T = G$, because G is generated by S and the z_i's. To see that $G = S \oplus T$, it now suffices to prove that $S \cap T = \{0\}$. If $y \in S \cap T$, then $y = \sum_i m_i z_i$, where $m_i \in \mathbb{Z}$. Now $y \in S$, and so $\sum_i m_i \overline{x}_i = 0$ in G/S. Since this is a direct sum, each $m_i \overline{x}_i = 0$; after all, for each i,

$$-m_i \overline{x}_i = \sum_{j \neq i} m_j \overline{x}_j \in \langle \overline{x}_i \rangle \cap \left(\langle \overline{x}_1 \rangle + \cdots + \widehat{\langle \overline{x}_i \rangle} + \cdots + \langle \overline{x}_q \rangle \right) = \{0\}.$$

Therefore, $m_i z_i = 0$ for all i, and hence $y = 0$.

Finally, $G = S \oplus T$ implies $d(G) = d(S) + d(T) = 1 + d(T)$, so that $d(T) < d(G)$. By induction, T is a direct sum of cyclic groups, and this completes the proof.   •

When are two finite abelian groups G and G' isomorphic? By the basis theorem, such groups are direct sums of cyclic groups, and so one's first guess is that $G \cong G'$ if they have the same number of cyclic summands of each type. But this hope is dashed by Theorem 2.128, which says that if m and n are relatively prime, then $\mathbb{I}_{mn} \cong \mathbb{I}_m \times \mathbb{I}_n$; for example, $\mathbb{I}_6 \cong \mathbb{I}_2 \times \mathbb{I}_3$. Thus, we retreat and try to count *primary* cyclic summands. But how can we do this? As in the Fundamental Theorem of Arithmetic, we must ask whether there is some kind of unique factorization theorem here.

Before stating the next lemma, recall that we have defined

$$d(G) = \dim(G/pG).$$

In particular, $d(pG) = \dim(pG/p^2G)$ and, more generally,

$$d(p^nG) = \dim(p^nG/p^{n+1}G).$$

Lemma 6.12.   *Let G be a finite p-primary abelian group, where p is a prime, and let $G = \bigoplus_j C_j$, where each C_j is cyclic. If b_n is the number of summands C_j having order p^n, then there is some $t \geq 1$ with*

$$d(p^nG) = b_{n+1} + b_{n+2} + \cdots + b_t.$$

Proof.   Let B_n be the direct sum of all C_j, if any, with order p^n. Thus,

$$G = B_1 \oplus B_2 \oplus \cdots \oplus B_t$$

for some t. Now

$$p^nG = p^nB_{n+1} \oplus \cdots \oplus p^nB_t,$$

because $p^nB_j = \{0\}$ for all $j \leq n$. Similarly,

$$p^{n+1}G = p^{n+1}B_{n+2} \oplus \cdots \oplus p^{n+1}B_t.$$

Now Proposition 2.126 shows that $p^nG/p^{n+1}G$ is isomorphic to

$$\left[p^nB_{n+1}/p^{n+1}B_{n+1}\right] \oplus \left[p^nB_{n+2}/p^{n+1}B_{n+2}\right] \oplus \cdots \oplus \left[p^nB_t/p^{n+1}B_t\right].$$

Since d is additive over direct sums, we have

$$d(p^nG) = b_{n+1} + b_{n+2} + \cdots + b_t. \quad \bullet$$

The numbers b_n can now be described in terms of G.

Definition.   If G is a finite p-primary abelian group, where p is a prime, then

$$U_p(n, G) = d(p^nG) - d(p^{n+1}G).$$

Lemma 6.12 shows that

$$d(p^nG) = b_{n+1} + \cdots + b_t$$

and

$$d(p^{n+1}G) = b_{n+2} + \cdots + b_t,$$

so that $U_p(n, G) = b_{n+1}$.

Theorem 6.13.   *If p is a prime, then any two decompositions of a finite p-primary abelian group G into direct sums of cyclic groups have the same number of cyclic summands of each type. More precisely, for each $n \geq 0$, the number of cyclic summands having order p^{n+1} is $U_p(n, G)$.*

Proof.   By the basis theorem, there exist cyclic subgroups C_i with $G = \bigoplus_i C_i$. The lemma shows, for each $n \geq 0$, that the number of C_i having order p^{n+1} is $U_p(n, G)$, a number that is defined without any mention of the given decomposition of G into a direct sum of cyclics. Thus, if $G = \bigoplus_j D_j$ is another decomposition of G, where each D_j is cyclic, then the number of D_j having order p^{n+1} is also $U_p(n, G)$, as desired.

$\bullet$

Corollary 6.14.   *If G and G' are finite p-primary abelian groups, then $G \cong G'$ if and only if $U_p(n, G) = U_p(n, G')$ for all $n \geq 0$.*

Proof.   If $\varphi \colon G \to G'$ is an isomorphism, then $\varphi(p^n G) = p^n G'$ for all $n \geq 0$, and hence it induces isomorphisms of the $\mathbb{F}_p$-vector spaces $p^n G / p^{n+1} G \cong p^n G' / p^{n+1} G'$ for all $n \geq 0$. Hence, their dimensions are the same; that is, $U_p(n, G) = U_p(n, G')$.

Conversely, assume that $U_p(n, G) = U_p(n, G')$ for all $n \geq 0$. If $G = \bigoplus_i C_i$ and $G' = \bigoplus_j C'_j$, where the C_i and C'_j are cyclic, then Lemma 6.12 shows that there are the same number of summands of each type, and so it is a simple matter to construct an isomorphism $G \to G'$. $\bullet$

Definition.   If G is a p-primary abelian group, then the **_elementary divisors_** of G are the numbers p^{n+1}, each repeated with multiplicity $U_p(n, G)$.

If G is a finite abelian group, then its **_elementary divisors_** are the elementary divisors of all its primary components.

For example, the elementary divisors of the abelian group $\mathbb{I}_2 \oplus \mathbb{I}_2 \oplus \mathbb{I}_2$ are $(2, 2, 2)$, and the elementary divisors of $\mathbb{I}_6$ are $(2, 3)$. The elementary divisors of $\mathbb{I}_2 \oplus \mathbb{I}_2 \oplus \mathbb{I}_4 \oplus \mathbb{I}_8$ are $(2, 2, 4, 8)$.

Theorem 6.15 (Fundamental Theorem of Finite Abelian Groups).   *Finite abelian groups G and G' are isomorphic if and only if they have the same elementary divisors; that is, any two decompositions of G and G' as direct sums of primary cyclic groups have the same number of summands of each order.*

Proof.   By the primary decomposition, Theorem 6.5(ii), $G \cong G'$ if and only if, for each prime p, their primary components are isomorphic: $G_p \cong G'_p$. The result now follows from Theorem 6.13. $\bullet$

The results of this section can be generalized from finite abelian groups to finitely generated abelian groups, where an abelian group G is **_finitely generated_** if there are finitely many elements $a_1, \ldots, a_n \in G$ so that every $x \in G$ is a linear combination of them: $x = \sum_i m_i a_i$, where $m_i \in \mathbb{Z}$ for all i. The basis theorem generalizes: every finitely generated abelian group G is a direct sum of cyclic groups, each of which is a finite primary group or an infinite cyclic group. A direct sum of infinite cyclic groups is called a ***free abelian group***. Thus, every finitely generated abelian group is a direct sum of a free abelian group and a finite group. Theorem 6.15 also generalizes: given two decompositions of G into direct sums of infinite and primary cyclic groups, the number of cyclic summands of each type is the same in both decompositions. The basis theorem is no longer true for abelian groups that are not finitely generated; for example, the additive group $\mathbb{Q}$ of rational numbers is not a direct sum of cyclic groups.

The proofs in this section can be extended to prove ***Ulm's theorem***, which classifies all *countable* abelian groups having no elements of infinite order.

EXERCISES

H **6.1** True or false with reasons.

   (i)   If G is a finite abelian group, then $\text{Aut}(G)$ is abelian.

   (ii) If $G = C_1 \oplus \cdots \oplus C_n = C_1' \oplus \cdots \oplus C_m'$, where the C_i and C_j' are cyclic p-primary groups for some prime p, then $m = n$ and, after re-indexing, $C_i = C_i'$ for all i.

   (iii) If G is an abelian group of squarefree order, $G = C_1 \oplus \cdots \oplus C_n$, and $G = C_1' \oplus \cdots \oplus C_m'$, where the C_i and C_j' are cyclic primary groups, then $m = n$ and, after re-indexing, $C_i = C_i'$ for all i.

   (iv) If G and H are finite abelian groups of the same order, and if $pG = \{0\}$ and $pH = \{0\}$, then $G \cong H$.

   (v)   The four-group **V** is a vector space over $\mathbb{F}_2$.

   (vi) Every subgroup of $\mathbb{Z}$ is pure.

   (vii) Every subgroup of $\mathbb{Q}$ is pure.

   (viii) There are five nonisomorphic abelian groups of order 8.

   (ix) If p and q are distinct primes, there are p homomorphisms $\mathbb{I}_p \to \mathbb{I}_q$.

   (x)   Every abelian group of order p^5, where p is a prime, can be generated by 5 or fewer elements.

6.2 Prove that a primary cyclic group G is ***indecomposable***; that is, there do not exist nonzero subgroups S and T with $G = S \oplus T$.

6.3 Let $S \subseteq H \subseteq G$ be abelian groups.

   (i)   If S is a pure subgroup of G, prove that S is a pure subgroup of H.

   (ii) Prove that purity is transitive: if S is a pure subgroup of H and if H is a pure subgroup of G, then S is a pure subgroup of G.

6.4   **(i)**   Give an example of an abelian group $G = S \oplus T$ having a subgroup A such that $A \neq (S \cap A) \oplus (T \cap A)$.

   (ii) Suppose that G is an abelian group and that $G = S \oplus T$. If H is a subgroup with $S \subseteq H \subseteq G$, prove that $H = S \oplus (T \cap H)$.

*6.5   **(i)**   If G is an (additive) abelian group and X is a nonempty subset of G, prove that $\langle X \rangle$, the subgroup generated by X, is the set of all linear combinations of elements in X with coefficients in $\mathbb{Z}$:

$$\langle X \rangle = \left\{ \sum_i m_i x_i : x_i \in X \text{ and } m_i \in \mathbb{Z} \right\}.$$

Compare this exercise with Proposition 2.79.

(ii)   If S and T are subgroups of G, prove that $S + T = \langle S \cup T \rangle$.

6.6   **(i)**   If G and H are finite abelian groups, prove, for all primes p and all $n \geq 0$, that

$$U_p(n, G \oplus H) = U_p(n, G) + U_p(n, H),$$

H **(ii)**   If A, B, and C are finite abelian groups, prove that $A \oplus B \cong A \oplus C$ implies $B \cong C$.

H **(iii)**   If A and B are finite abelian groups, prove that $A \oplus A \cong B \oplus B$ implies $A \cong B$.

6.7   If n is a positive integer, then a **_partition of_** n is a sequence of positive integers $i_1 \leq i_2 \leq \cdots \leq i_r$ with $i_1 + i_2 + \cdots + i_r = n$. If p is a prime, prove that the number of abelian groups of order p^n, to isomorphism, is equal to the number of partitions of n.

H **6.8**   To isomorphism, how many abelian groups are there of order 288?

6.9   Prove the Fundamental Theorem of Arithmetic by applying the Fundamental Theorem of Finite Abelian Groups to $G = \mathbb{I}_n$.

*H **6.10**   If G is a finite abelian group, define

$$\nu_k(G) = \text{ the number of elements in } G \text{ of order } k.$$

Prove that two finite abelian groups G and G' are isomorphic if and only if $\nu_k(G) = \nu_k(G')$ for all integers k. (This result is not true for nonabelian groups; see Proposition 6.29.)

6.11   Regard $\mathbb{Q}$ as an additive abelian group.

(i)   Prove that every finitely generated subgroup of $\mathbb{Q}$ is cyclic.

(ii)   Prove that $\mathbb{Q}$ is not finitely generated.

H **(iii)**   Prove that $\mathbb{Q} \not\cong A \oplus B$, where A and B are nonzero subgroups.

*6.12   H **(i)**   Let S be a subgroup of a p-primary abelian group G, and let $\pi : G \to G/S$ be the natural map $g \mapsto g + S$. Prove that S is a pure subgroup of G if and only if each $g + S \in G/S$ has a preimage $g' \in G$ (that is, $\pi(g') = g + S$) with g' and $g + S$ having the same order.

(ii)   Prove that a subgroup S of a finite p-primary abelian group G is pure if and only if it is a direct summand. (This is not true for infinite abelian groups.)

H **6.13**   Let F and F' be free abelian groups. If F is a direct sum of m infinite cyclic groups and F' is a direct sum of n infinite cyclic groups, prove that $F \cong F'$ if and only if $m = n$.

6.14   **(i)**   If $F = \langle x_1 \rangle \oplus \cdots \oplus \langle x_n \rangle$ is a free abelian group, prove that every $z \in F$ has a unique expression of the form $z = m_1 x_1 + \cdots + m_n x_n$, where $m_i \in \mathbb{Z}$ for all i. One calls $x_1, \ldots, x_n$ a **_basis_** of F.

(ii)   Let $X = x_1, \ldots, x_n$ be a basis of a free abelian group F. Prove that if A is any abelian group and if $a_1, \ldots, a_n$ is any list of elements in A, then there exists a unique homomorphism $f : F \to A$ with $f(x_i) = a_i$ for all i.

6.15 Let p be a prime. Prove that if G is a finite p-primary abelian group, then every subgroup of G is a pure subgroup if and only if $pG = \{0\}$.

*6.16 Let G be an abelian group, not necessarily primary. Define a subgroup $S \subseteq G$ to be a ***pure subgroup*** if $S \cap mG = mS$ for all $m \in \mathbb{Z}$. Prove that if G is a p-primary abelian group, where p is a prime, then a subgroup $S \subseteq G$ is pure as just defined if and only if $S \cap p^n G = p^n S$ for all $n \geq 0$ (the definition in the text).

6.17 Let p be a prime, and let G be a finite p-primary abelian group.

    (i) Prove that pG is the intersection of all the maximal subgroups of G.

    (ii) **(Frattini)**. Prove that every $g \in pG$ is a ***nongenerator***: if $G = \langle X, g \rangle$, that is, G is generated by $X \cup \{g\}$ for some subset $X \subseteq G$, then $G = \langle X \rangle$.

    (iii) **(Burnside)** Prove that $d(G)$ is the number of elements in a *minimal generating set* X of G; that is, X generates G but no proper subset of X generates G.

*6.18 If G is a possibly infinite abelian[5] group, define the ***torsion*[6] *subgroup* tG of G as

$$tG = \{a \in G : a \text{ has finite order}\}.$$

    (i) Prove that tG is a pure subgroup of G. (There exist abelian groups G whose torsion subgroup tG is not a direct summand; hence, a pure subgroup need not be a direct summand.)

    (ii) Prove that G/tG is an abelian group in which every nonzero element has infinite order.

6.19 Let S^1 be the ***circle group***; that is, the multiplicative group of all complex numbers of modulus 1. Prove that the torsion subgroup $G = tS^1$ is an infinite group in which every finite subgroup is cyclic.

6.2 THE SYLOW THEOREMS

We return to nonabelian groups, and so we revert to the multiplicative notation. The Sylow theorems give an analog for finite nonabelian groups of the primary decomposition for finite abelian groups.

Recall that a group G is *simple* if $G \neq \{1\}$ and G has no normal subgroups other than $\{1\}$ and G itself. We saw, in Proposition 2.78, that the abelian simple groups are precisely the cyclic groups $\mathbb{I}_p$ of prime order p, and we saw, in Theorem 2.83, that A_n is a nonabelian simple group for all $n \geq 5$. In fact, A_5 is the nonabelian simple group of smallest order. How can one prove that a nonabelian group G of order less than $60 = |A_5|$ is not simple? Exercise 2.105 states that if G is a group of order $|G| = mp$, where p is prime and $1 < m < p$, then G is not simple. This exercise shows that many

---

[5]If G is not abelian, then tG may not be a subgroup; there is an example on page 136 of a group of matrices containing two elements having finite order whose product has infinite order.

[6]This terminology comes from algebraic topology. To each space X, one assigns a sequence of abelian groups, called *homology groups*, and if X is "twisted," then there are elements of finite order in some of these groups.

of the numbers less than 60 are not orders of simple groups. After throwing out all prime powers (by Exercise 2.118 on page 207, groups of prime power order are never nonabelian simple), the only remaining possibilities are

$$12, 18, 24, 30, 36, 40, 45, 48, 50, 54, 56.$$

The solution to the exercise uses Cauchy's theorem, which says that G has an element of order p, hence a subgroup of order p. We shall see that if G has a subgroup of order p^e instead of p, where p^e is the highest power of p dividing $|G|$, then Exercise 2.105 can be generalized, and the list of candidates can be shortened to 30, 40, and 56.

The first book on group theory was *Traités des Substitutions et des Équations Algébriques*, by C. Jordan; it was published in 1870 (more than half of it is devoted to Galois theory, then called the Theory of Equations). At about the same time, but too late for publication in Jordan's book, three fundamental theorems were discovered. In 1868, E. Schering proved the Basis Theorem: every finite abelian group is a direct product of primary cyclic groups; in 1870, L. Kronecker, unaware of Schering's proof, also proved this result. In 1878, G. Frobenius and L. Stickelberger proved the Fundamental Theorem of Finite Abelian Groups. In 1872, L. Sylow showed, for every finite group G and every prime p, that if p^e is the largest power of p dividing $|G|$, then G has a subgroup of order p^e.

Recall that a *p-group* is a finite group G in which every element has order some power of a prime p; equivalently, G has order p^k for some $k \geq 0$. (When working wholly in the context of abelian groups, as in the last section, one calls G a p-primary group.)

Definition. Let p be a prime. A *Sylow p-subgroup* of a finite group G is a maximal p-subgroup P.

Maximality means that if Q is a p-subgroup of G and $P \leq Q$, then $P = Q$. Sylow p-subgroups always exist: indeed, we now show that if S is any p-subgroup of G (perhaps $S = \{1\}$), then there exists a Sylow p-subgroup P containing S. If there is no p-subgroup strictly containing S, then S itself is a maximal p-subgroup; that is, S is a Sylow p-subgroup. Otherwise, there is a p-subgroup P_1 with $S < P_1$. If P_1 is maximal, it is Sylow, and we are done. Otherwise, there is some p-subgroup P_2 with $P_1 < P_2$; hence, $|P_1| < |P_2|$. This procedure of producing larger and larger p-subgroups P_i must end after a finite number of steps, for $|G|$ is finite; the largest P_i must, therefore, be a Sylow p-subgroup.

Example 6.16.
Let G be a finite group of order $|G| = p^e m$, where p is a prime and $p \nmid m$. We show that if there exists a subgroup P of order p^e, then P is a Sylow p-subgroup of G. If Q is a p-subgroup with $P \leq Q \leq G$, then $|P| = p^e \mid |Q|$. But if $|Q| = p^k$, then $p^k \mid p^e m$ and $k \leq e$; that is, $|Q| = p^e$ and $Q = P$. ◀

Definition. If H is a subgroup of a group G, then a ***conjugate*** of H is a subgroup of G of the form $gHg^{-1} = \{ghg^{-1} : h \in H\}$ for some $g \in G$.

Conjugate subgroups are isomorphic: if $H \leq G$, then $h \mapsto ghg^{-1}$ is an injective homomorphism $H \to G$ with image gHg^{-1}. The converse is false: the four-group **V** contains several subgroups of order 2 which are, of course, isomorphic; they cannot be conjugate because **V** is abelian. On the other hand, all subgroups of order 2 in S_3 are conjugate; for example, $\langle (1\ 3) \rangle = g\langle (1\ 2) \rangle g^{-1}$, where $g = (2\ 3)$.

The ideas of group actions are going to be used, and so we now recall the notions of *orbit* and *stabilizer* which we discussed in Chapter 2.

Definition. If X is a set and G is a group, then G ***acts*** on X if, for each $g \in G$, there is a function $\alpha_g : X \to X$, such that

(i) $\alpha_g \circ \alpha_h = \alpha_{gh}$ for all $g, h \in G$;

(ii) $\alpha_1 = 1_X$, the identity function.

Definition. If G acts on X and $x \in X$, then the ***orbit*** of x, denoted by $\mathcal{O}(x)$, is the subset of X

$$\mathcal{O}(x) = \{\alpha_g(x) : g \in G\} \subseteq X;$$

the ***stabilizer*** of x, denoted by G_x, is the subgroup of G

$$G_x = \{g \in G : \alpha_g(x) = x\} \leq G.$$

A group G acts on $X = \mathbf{Sub}(G)$, the set of all its subgroups, by conjugation: if $g \in G$, then g acts by $\alpha_g(H) = gHg^{-1}$, where $H \leq G$. The orbit of a subgroup H consists of all its conjugates; the stabilizer of H is $\{g \in G : gHg^{-1} = H\}$. This last subgroup has a name.

Definition. If H is a subgroup of a group G, then the ***normalizer*** of H in G is the subgroup

$$N_G(H) = \{g \in G : gHg^{-1} = H\}.$$

The reader should verify that $H \lhd N_G(H)$, for then the quotient group $N_G(H)/H$ is defined.

Proposition 6.17. *If H is a subgroup of a finite group G, then the number of conjugates of H in G is $[G : N_G(H)]$.*

Proof. This is a special case of Theorem 2.143: the size of the orbit of a point is the index of its stabilizer. ●

Lemma 6.18.   *Let P be a Sylow p-subgroup of a finite group G.*

(i) *Every conjugate of P is also a Sylow p-subgroup of G.*

(ii) $p \nmid |N_G(P)/P|$.

(iii) *If $g \in G$ has order some power of p and if $g P g^{-1} = P$, then $g \in P$.*

Proof.
(i) If $g \in G$, then $g P g^{-1}$ is a p-subgroup of G; if it is not a maximal such, then there is a p-subgroup Q with $g P g^{-1} < Q$. Hence, $P < g^{-1} Q g$, contradicting the maximality of P.
(ii) If p divides $|N_G(P)/P|$, then Cauchy's theorem shows that $N_G(P)/P$ contains an element $g P$ of order p, and hence $N_G(P)/P$ contains a (cyclic) subgroup $S^* = \langle g P \rangle$ of order p. By the correspondence theorem (Theorem 2.123), there is a subgroup S with $P \leq S \leq N_G(P)$ such that $S/P \cong S^*$. But S is a p-subgroup of $N_G(P) \leq G$ (by Exercise 2.99 on page 190) strictly larger than P, and this contradicts the maximality of P. We conclude that p does not divide $|N_G(P)/P|$.
(iii) The element g lies in $N_G(P)$, by the definition of normalizer. If $g \notin P$, then the coset $g P$ is a nontrivial element of $N_G(P)/P$ having order some power of p; in light of part (ii), this contradicts Lagrange's theorem.   •

Since every conjugate of a Sylow p-subgroup is also a Sylow p-subgroup, it is natural to let G act by conjugation on a set of Sylow p-subgroups.

Theorem 6.19 (Sylow).   *Let G be a finite group of order $p^e m$, where p is a prime and $p \nmid m$, and let P be a Sylow p-subgroup of G.*

(i) *Every Sylow p-subgroup is conjugate to P.*

(ii) *If r is the number of Sylow p-subgroups, then r is a divisor of $|G|/p^e$ and*

$$r \equiv 1 \bmod p.$$

Proof.   Let $X = \{P_1, \ldots, P_r\}$ be the set of all the conjugates of P, where we have denoted P by P_1. If Q is any Sylow p-subgroup of G, then Q acts on X by conjugation: if $a \in Q$, then it sends

$$\alpha_a(P_i) = \alpha_a(g_i P g_i^{-1}) = a(g_i P g_i^{-1})a^{-1} = (a g_i) P (a g_i)^{-1} \in X.$$

By Corollary 2.144, the number of elements in any orbit is a divisor of $|Q|$; that is, every orbit has size some power of p (because Q is a p-group). If there is an orbit of size 1, then there is some P_i with $a P_i a^{-1} = P_i$ for all $a \in Q$. By Lemma 6.18, we have $a \in P_i$ for all $a \in Q$; that is, $Q \leq P_i$. But Q, being a Sylow p-subgroup, is a maximal p-subgroup of G, and so $Q = P_i$. Applying this argument when $Q = P_1$, we see that every orbit has size an honest power of p except one, the orbit consisting of P_1 alone. Hence, $|X| = r \equiv 1 \bmod p$.

Suppose now that there is some Sylow p-subgroup Q that is not a conjugate of P; thus, $Q \neq P_i$ for any i. Again, we let Q act on X, and again we ask if there is an orbit of size 1, say, $\{P_j\}$. As in the previous paragraph, this implies $Q = P_j$, contrary to our present assumption that $Q \notin X$. Hence, there are no orbits of size 1, which says that each orbit has size an honest power of p. It follows that $|X| = r$ is a multiple of p; that is, $r \equiv 0 \bmod p$, which contradicts the congruence $r \equiv 1 \bmod p$. Therefore, no such Q can exist, and so all Sylow p-subgroups are conjugate to P. Finally, since all Sylow p-subgroups are conjugate, we have $r = [G : N_G(P)]$, and so r is a divisor of $|G| = p^e m$. But $(r, p) = 1$, because $r \equiv 1 \bmod p$, so that $r \mid p^e m$ implies $r \mid m$; that is, $r \mid |G|/p^e$. •

Corollary 6.20. *A finite group G has a unique Sylow p-subgroup if and only if it has a normal Sylow p-subgroup.*

Proof. Assume that P, a Sylow p-subgroup of G, is unique. For each $a \in G$, the conjugate aPa^{-1} is also a Sylow p-subgroup; by uniqueness, $aPa^{-1} = P$ for all $a \in G$, and so $P \lhd G$.

Conversely, assume that P is a normal Sylow p-subgroup of G. If Q is any Sylow p-subgroup, then $Q = aPa^{-1}$ for some $a \in G$; but $aPa^{-1} = P$, by normality, and so $Q = P$. •

The following result gives the order of a Sylow subgroup.

Theorem 6.21 (Sylow). *If G is a finite group of order $p^e m$, where p is a prime and $p \nmid m$, then every Sylow p-subgroup P of G has order p^e.*

Proof. We first show that $p \nmid [G : P]$. Now

$$[G : P] = [G : N_G(P)][N_G(P) : P].$$

The first factor, $[G : N_G(P)] = r$, is the number of conjugates of P in G, and we know that $r \equiv 1 \bmod p$; hence, p does not divide $[G : N_G(P)]$. The second factor is $[N_G(P) : P] = |N_G(P)/P|$; this, too, is not divisible by p, by Lemma 6.18(ii). Therefore, p does not divide $[G : P]$, by Euclid's lemma.

Now $|P| = p^k$ for some $k \leq e$, and so

$$[G : P] = |G|/|P| = p^e m/p^k = p^{e-k} m.$$

Since p does not divide $[G : P]$, we must have $k = e$; that is, $|P| = p^e$. •

Example 6.22.

(i) If G is a finite abelian group, then a Sylow p-subgroup is just its p-primary component. Since G is abelian, every subgroup is normal, and so G has a unique Sylow p-subgroup for every prime p.

(ii) Let $G = S_4$. Now $|S_4| = 24 = 2^3 3$. Thus, a Sylow 2-subgroup of S_4 has order 8. We have seen, in Exercise 2.119 on page 207, that S_4 contains a copy of the dihedral group D_8 consisting of the symmetries of a square. The Sylow theorem says that all subgroups of order 8 are conjugate, hence isomorphic, to D_8. Moreover, the number r of Sylow 2-subgroups is a divisor of 24/8 congruent to 1 mod 2; that is, r is an odd divisor of 3. Since $r \neq 1$ (see Exercise 6.21 on page 499), there are exactly 3 Sylow 2-subgroups; S_4 has exactly 3 subgroups of order 8. ◄

Here is a second proof of the last Sylow theorem, due to Wielandt.

Theorem 6.23 (= Theorem 6.21). *If G is a finite group of order $p^e m$, where p is a prime and $p \nmid m$, then G has a subgroup of order p^e.*

Proof. If X is the family of all those subsets of G having exactly p^e elements, then $|X| = \binom{n}{p^e}$; by Exercise 1.72 on page 58, $p \nmid |X|$. Now G acts on X: define $\alpha_g(B) = gB$, for $g \in G$ and $B \in X$, where $gB = \{gb : b \in B\}$. If p divides $|\mathcal{O}(B)|$ for every $B \in X$, then p is a divisor of $|X|$, for X is the disjoint union of orbits, by Proposition 2.142. As $p \nmid |X|$, there exists a subset B with $|B| = p^e$ and with $|\mathcal{O}(B)|$ not divisible by p. If G_B is the stabilizer of this subset B, then Theorem 2.143 gives $[G : G_B] = |\mathcal{O}(B)|$, and so $|G| = |G_B| \cdot |\mathcal{O}(B)|$. Since $p^e \mid |G|$ and $p \nmid |\mathcal{O}(B)|$, repeated application of Euclid's lemma gives $p^e \mid |G_B|$. Therefore, $p^e \leq |G_B|$.

To prove the reverse inequality, choose an element $b \in B$ and define a function $\tau \colon G_B \to B$ by $g \mapsto gb$. Note that $\tau(g) = gb \in gB = B$, for $g \in G_B$, the stabilizer of B. If $g, h \in G_B$ and $h \neq g$, then $\tau(h) = hb \neq gb = \tau(g)$; that is, τ is an injection. We conclude that $|G_B| \leq |B| = p^e$, and so G_B is a subgroup of G of order p^e. •

If p is a prime not dividing the order of a finite group G, then a Sylow p-subgroup of G has order $p^0 = 1$. Thus, when speaking of Sylow p-subgroups of G, one usually avoids this trivial case and assumes that p is a divisor of $|G|$.

We can now generalize Exercise 2.134 on page 208 and its solution.

Lemma 6.24. *There is no nonabelian simple group G of order $|G| = p^e m$, where p is prime and $m > 1$, $p \nmid m$, and $p^e \nmid (m-1)!$.*

Proof. Suppose that such a simple group G exists. By Sylow's theorem, G contains a subgroup P of order p^e, hence of index m. By Theorem 2.67, the representation of G on the cosets of P, there exists a homomorphism $\varphi \colon G \to S_m$ with $\ker \varphi \leq P$. Since G is simple, however, it has no proper normal subgroups; hence $\ker \varphi = \{1\}$ and φ is an injection; that is, $G \cong \varphi(G) \leq S_m$. By Lagrange's theorem, $p^e m \mid m!$, and so $p^e \mid (m-1)!$, contrary to the hypothesis. •

Proposition 6.25.   *There are no nonabelian simple groups of order less than* 60.

Proof.   If p is a prime, then Exercise 2.118 on page 207 says that every p-group G with $|G| > p$ is not simple.

The reader may check that the only integers n between 2 and 59, neither a prime power nor having a factorization of the form $n = p^e m$ as in the statement of the lemma, are $n = 30, 40$, and 56. By the lemma, these three numbers are the only candidates for orders of nonabelian simple groups of order < 60.

Assume that there is a simple group G of order 30. Let P be a Sylow 5-subgroup of G, so that $|P| = 5$. The number r_5 of conjugates of P is a divisor of $30/5 = 6$ and $r_5 \equiv 1 \bmod 5$. Now $r_5 \neq 1$, lest $P \lhd G$, so that $r_5 = 6$. By Lagrange's theorem, the intersection of any two of these is trivial (intersections of Sylow subgroups can be more complicated; see Exercise 6.22 on page 499). There are 4 nonidentity elements in each of these subgroups, and so there are $6 \times 4 = 24$ nonidentity elements in their union. Similarly, the number r_3 of Sylow 3-subgroups of G is 10 (for $r_3 \neq 1$, r_3 is a divisor of $30/3$, and $r_3 \equiv 1 \bmod 3$). There are 2 nonidentity elements in each of these subgroups, and so the union of these subgroups has 20 nonidentity elements. We have exceeded the number of elements in G, and so G cannot be simple.

Let G be a group of order 40, and let P be a Sylow 5-subgroup of G. If r_5 is the number of conjugates of P, then $r_5 \mid 40/5$ and $r_5 \equiv 1 \bmod 5$. These conditions force $r_5 = 1$, so that $P \lhd G$. Therefore, no simple group of order 40 can exist.

Finally, assume that there is a simple group G of order 56. If P is a Sylow 7-subgroup of G, then P must have $r_7 = 8$ conjugates (for $r_7 \mid 56/7$ and $r_7 \equiv 1 \bmod 7$). Since these groups are cyclic of prime order, the intersection of any pair of them is $\{1\}$, and so there are 48 nonidentity elements in their union. Thus, adding the identity, we have accounted for 49 elements of G. Now a Sylow 2-subgroup Q has order 8, and so it contributes 7 more nonidentity elements, giving 56 elements. But there is a second Sylow 2-subgroup, lest $Q \lhd G$, and we have exceeded our quota. Therefore, there is no simple group of order 56.   •

The "converse" of Lagrange's theorem is false: if G is a finite group of order n, and if $d \mid n$, then G may not have a subgroup of order d. For example, we proved, in Proposition 2.99, that the alternating group A_4 is a group of order 12 having no subgroup of order 6.

Proposition 6.26.   *Let G be a finite group. If p is a prime and if p^k divides $|G|$, then G has a subgroup of order p^k.*

Proof.   If $|G| = p^e m$, where $p \nmid m$, then a Sylow p-subgroup P of G has order p^e. Hence, if p^k divides $|G|$, then p^k divides $|P|$. By Proposition 2.152, P has a subgroup of order p^k; a fortiori, G has a subgroup of order p^k.   •

What examples of p-groups have we seen? Of course, cyclic groups of order p^n are p-groups, as is any direct product of copies of these. By the basis theorem, this describes all finite abelian p-groups. The only nonabelian examples we have seen so far are the dihedral groups D_{2n} (which are 2-groups when n is a power of 2), the quaternions $\mathbf{Q}$ of order 8 (of course, for every 2-group A, the direct products $D_8 \times A$ and $\mathbf{Q} \times A$ are also nonabelian 2-groups), and the groups UT(3, p) in Example 2.150 consisting of all upper triangular 3×3 matrices over $\mathbb{F}_p$ of the form $\begin{bmatrix} 1 & a & b \\ 0 & 1 & c \\ 0 & 0 & 1 \end{bmatrix}$. The obvious generalization of UT(3, p) gives an interesing family of nonabelian p-groups.

Definition.   If k is a field, then an $n \times n$ ***unitriangular*** matrix over k is an upper triangular matrix each of whose diagonal terms is 1. Define UT(n, k) to be the set of all $n \times n$ unitriangular matrices over k.

Proposition 6.27.   *UT(n, k) is a subgroup of GL(n, k) for every field k.*

Proof.   If $A \in$ UT(n, k), then $A = I + N$, where N is *strictly* upper triangular; that is, N is an upper triangular matrix having only 0's on its diagonal. Note that the sum and product of strictly upper triangular matrices is again strictly upper triangular.

Let $e_1, \ldots, e_n$ be the standard basis of k^n. If N is strictly upper triangular, define $T : k^n \to k^n$ by $T(e_i) = Ne_i$, where e_i is regarded as a column matrix. Now T satisfies the equations, for all i,

$$T(e_1) = 0 \quad \text{and} \quad T(e_{i+1}) \in \langle e_1, \ldots, e_i \rangle .$$

It is easy to see, by induction on i, that $T^i(e_j) = 0$ for all $j \leq i$. It follows that $T^n = 0$ and, hence, that $N^n = 0$. Thus, if $A \in$ UT(n, k), then $A = I + N$, where $N^n = 0$.

We can now show that UT(n, k) is a subgroup of GL(n, k). First, if A is unitriangular, then it is nonsingular, for Exercise 4.51 on page 397 shows that $\det(A) = 1$. Here is an alternative argument avoiding determinants. In analogy to the power series expansion $1/(1 + x) = 1 - x + x^2 - x^3 + \cdots$, we try $B = I - N + N^2 - N^3 + \cdots$ as the inverse of $A = I + N$ (we note that the matrix power series stops after $n - 1$ terms because $N^n = 0$), The reader may now check that $BA = I$; therefore, A is nonsingular. Moreover, since N is strictly upper triangular, so is $-N + N^2 - N^3 + \cdots$, so that $A^{-1} \in$ UT(n, k). Finally, $(I + N)(I + M) = I + (N + M + NM)$ is unitriangular, and so UT(n, k) is a subgroup of GL(n, k).   ●

Proposition 6.28.   *Let $q = p^e$, where p is a prime. For each $n \geq 2$, UT$(n, \mathbb{F}_q)$ is a p-group of order $q^{n(n-1)/2}$.*

Proof.   The number of entries in an $n \times n$ unitriangular matrix lying strictly above the diagonal is $\frac{1}{2}(n^2 - n) = n(n-1)/2$. Since each of these entries can be any element of $\mathbb{F}_q$, there are exactly $q^{n(n-1)/2}$ $n \times n$ unitriangular matrices over $\mathbb{F}_q$, and so this is the order of UT$(n, \mathbb{F}_q)$.   ●

In Exercise 2.123 on page 207, we showed that $UT(3, \mathbb{F}_2) \cong D_8$.

Recall Exercise 2.44 on page 147: if G is a group and $x^2 = 1$ for all $x \in G$, then G is abelian. We now ask whether a group G satisfying $x^p = 1$ for all $x \in G$, where p is an odd prime, must also be abelian.

Proposition 6.29. *If p is an odd prime, then there exists a nonabelian group G of order p^3 with $x^p = 1$ for all $x \in G$.*

Proof. If $G = UT(3, \mathbb{F}_p)$, then $|G| = p^3$. If $A \in G$, then $A = I + N$, where $N^3 = 0$; hence $N^p = 0$ because $p \geq 3$. Since $IN = N = NI$, the binomial theorem gives $A^p = (I + N)^p = I^p + N^p = I$. •

In Exercise 6.10 on page 488, we defined $\nu_k(G)$ to be the number of elements of order k in a finite group G, and we proved that if G and H are abelian groups with $\nu_k(G) = \nu_k(H)$ for all k, then G and H are isomorphic. This is false in general, for if p is an odd prime, then each of $UT(3, \mathbb{F}_p)$ and $\mathbb{I}_p \times \mathbb{I}_p \times \mathbb{I}_p$ consist of the identity and $p^3 - 1$ elements of order p.

Theorem 6.30. *Let $\mathbb{F}_q$ denote the finite field with q elements. Then*

$$|GL(n, \mathbb{F}_q)| = (q^n - 1)(q^n - q)(q^n - q^2) \cdots (q^n - q^{n-1}).$$

Proof. Let V be an n-dimensional vector space over $\mathbb{F}_q$. We show first that there is a bijection $\Phi \colon GL(n, \mathbb{F}_q) \to \mathcal{B}$, where $\mathcal{B}$ is the set of all bases of V. Choose, once for all, a basis $e_1, \ldots, e_n$ of V. If $T \in GL(n, \mathbb{F}_q)$, define $\Phi(T) = Te_1, \ldots, Te_n$. By Lemma 4.77, $\Phi(T) \in \mathcal{B}$ because T, being nonsingular, carries a basis into a basis. But Φ is a bijection, for given a basis $v_1, \ldots, v_n$, there is a unique linear transformation S, necessarily nonsingular (by Lemma 4.77), with $Se_i = v_i$ for all i (by Theorem 4.62).

Our problem now is to count the number of bases $v_1, \ldots, v_n$ of V. There are q^n vectors in V, and so there are $q^n - 1$ candidates for v_1 (the zero vector is not a candidate). Having chosen v_1, we see that the candidates for v_2 are those vectors not in $\langle v_1 \rangle$, the subspace spanned by v_1; there are thus $q^n - q$ candidates for v_2. More generally, having chosen a linearly independent list $v_1, \ldots, v_i$, then v_{i+1} can be any vector not in $\langle v_1, \ldots, v_i \rangle$. Thus, there are $q^n - q^i$ candidates for v_{i+1}. The result follows by induction on n. •

Corollary 6.31. $|GL(n, \mathbb{F}_q)| = q^{n(n-1)/2}(q^n - 1)(q^{n-1} - 1) \cdots (q^2 - 1)(q - 1).$

Proof. The number of powers of q in the formula

$$|GL(n, \mathbb{F}_q)| = (q^n - 1)(q^n - q)(q^n - q^2) \cdots (q^n - q^{n-1})$$

is $q^{1+2+\cdots+(n-1)}$, and $1 + 2 + \cdots + (n - 1) = \frac{1}{2}n(n - 1)$. •

Theorem 6.32. *If p is a prime and $q = p^m$, then the unitriangular group* $\mathrm{UT}(n, \mathbb{F}_q)$
is a Sylow p-subgroup of $\mathrm{GL}(n, \mathbb{F}_q)$.

Proof. Now $|\mathrm{UT}(n, \mathbb{F}_q)| = q^{n(n-1)/2}(q^n - 1)(q^{n-1} - 1) \cdots (q^2 - 1)(q - 1)$, by
Corollary 6.31, so that the highest power of p dividing $|\mathrm{GL}(n, \mathbb{F}_q)|$ is $q^{n(n-1)/2}$. But
$|\mathrm{UT}(n, \mathbb{F}_q)| = q^{n(n-1)/2}$, by Corollary 6.28. and so $\mathrm{UT}(n, \mathbb{F}_q)$ must be a Sylow p-
subgroup. •

Corollary 6.33. *If p is a prime, then every finite p-group G is isomorphic to a
subgroup of the unitriangular group* $\mathrm{UT}(m, \mathbb{F}_p)$, *where* $m = |G|$.

Proof. We show first, for every $m \geq 1$, that the symmetric group S_m can be imbedded
in $\mathrm{GL}(m, k)$, where k is a field. Let V be an m-dimensional vector space over k,
and let $v_1, \ldots, v_m$ be a basis of V. Define $\varphi \colon S_m \to \mathrm{GL}(V)$ by $\sigma \mapsto T_\sigma$, where
$T_\sigma \colon v_i \mapsto v_{\sigma(i)}$ for all i. It is easy to see that φ is an injective homomorphism.

By Cayley's theorem, G can be imbedded in S_G; hence, G can be imbedded in
$\mathrm{GL}(m, \mathbb{F}_p)$, where $m = |G|$. Now G is contained in some Sylow p-subgroup P of
$\mathrm{GL}(m, \mathbb{F}_p)$, for every p-subgroup lies in some Sylow p-subgroup. Since all Sylow
p-subgroups are conjugate, $P = a \left(\mathrm{UT}(m, \mathbb{F}_p)\right) a^{-1}$ for some $a \in \mathrm{GL}(m, \mathbb{F}_p)$. Therefore,

$$G \cong aGa^{-1} \leq a^{-1}Pa \leq \mathrm{UT}(m, \mathbb{F}_p). \quad •$$

A natural question is to find the Sylow subgroups of symmetric groups. This can
be done, and the answer is in terms of a construction called *wreath product*.

In an amazing collective effort at the end of the twentieth century, all finite simple
groups were classified. We quote from *The Classification of the Finite Simple Groups*,
by D. Gorenstein, R. Lyons, and R. Solomon.

> The existing proof of the classification of the finite simple groups runs
> to somewhere between 10,000 and 15,000 journal pages, spread across
> some 500 separate articles by more than 100 mathematicians, almost all
> written between 1950 and the early 1980's. Moreover, it was not until the
> 1970's that a global strategy was developed for attacking the complete clas-
> sification problem. In addition, new simple groups were being discovered
> throughout the entire period, ..., so that it was not even possible to state the
> full theorem in precise form ... until the early 1980's. ... Considering the
> significance of the classification theorem, we believe that the present state
> of affairs provides compelling reasons for seeking a simpler proof, more
> coherent and accessible, and with clear foundations. ... The arguments we
> give ... will cover between 3,000 and 4,000 pages.

There is now a list of every finite simple group, and many important properties of
each of them is known. Many questions about arbitrary finite groups can be reduced

to problems about simple groups. Thus, the classification theorem can be used by checking, one by one, whether each simple group on the list satisfies the desired result.

Another important part of group theory is representation theory–the systematic study of homomorphisms of a group into groups of nonsingular matrices. One of the first applications of this theory is a theorem of Burnside: every group of order $p^m q^n$, where p and q are primes, must be solvable.

EXERCISES

H **6.20** True or false with reasons.

(i) If G is a finite group and p is a prime, then G has exactly one Sylow p-subgroup.

(ii) If G is a finite abelian group and p is a prime, then G has exactly one Sylow p-subgroup.

(iii) If G is a finite group and p is a prime, then G has at least one Sylow p-subgroup.

(iv) If a group G acts on a set X, and if $x, y \in X$ belong to the same orbit, then G_x and G_y are conjugate subgroups of G.

(v) If $H \leq G$, then $N_G(H) \triangleleft G$.

(vi) If $H \leq G$, then $H \triangleleft N_G(H)$.

(vii) A Sylow p-subgroup of a group G contains all the other p-subgroups of G.

(viii) If G and H are finite groups of the same order, then, for every prime p, their Sylow p-subgroups have the same order.

(ix) There is a group G of order 400 having exactly 8 Sylow 5-subgroups.

(x) There is a 10×10 unitriangular matrix $A \neq I$ over $\mathbb{F}_7$ such that $AB = BA$ for every 10×10 unitriangular matrix B over $\mathbb{F}_7$.

*6.21 Prove that S_4 has more than one Sylow 2-subgroup.

*H **6.22** Give an example of a finite group G having 3 Sylow p-subgroups (for some prime p) P, Q and R such that $P \cap Q = \{1\}$ and $P \cap R \neq \{1\}$.

6.23 Prove that every finite p-group is solvable.

*H **6.24** (**Frattini argument**) Let K be a normal subgroup of a finite group G. If P is a Sylow p-subgroup of K for some prime p, prove that

$$G = K N_G(P),$$

where $K N_G(P) = \{ab : a \in K \text{ and } b \in N_G(P)\}$.

H **6.25** If F is a field with four elements, prove that the stochastic group $\Sigma(2, F) \cong A_4$.

H **6.26** Show that a Sylow 2-subgroup of S_6 is isomorphic to $D_8 \times \mathbb{I}_2$.

H **6.27** Let Q be a normal p-subgroup of a finite group G. Prove that $Q \leq P$ for every Sylow p-subgroup P of G.

H **6.28** For each prime divisor p of the order of a finite group G, assume that a Sylow p-subgroup Q_p has been chosen. Prove that $G = \left\langle \bigcup_p Q_p \right\rangle$.

6.29 H **(i)**   Let G be a finite group and let P be a Sylow p-subgroup of G. If $H \lhd G$, prove that HP/H is a Sylow p-subgroup of G/H and $H \cap P$ is a Sylow p-subgroup of H.

   H **(ii)**   Let P be a Sylow p-subgroup of a finite group G. Give an example of a subgroup H of G with $H \cap P$ not a Sylow p-subgroup of H.

6.30 Prove that a Sylow 2-subgroup of A_5 has exactly 5 conjugates.

H **6.31** Prove that there are no simple groups of order 300, 312, 616, or 1000.

H **6.32** Prove that if every Sylow subgroup of a finite group G is normal, then G is the direct product of its Sylow subgroups.

6.33 For any group G, prove that if $H \lhd G$, then $Z(H) \lhd G$.

*H **6.34** If p is a prime, prove that every group G of order $2p$ is either cyclic or isomorphic to D_{2p}.

6.35 If $0 \le r \le n$, define the *q-binomial coefficient* $\begin{bmatrix} n \\ r \end{bmatrix}_q$ to be the number of linearly independent r-lists in $(\mathbb{F}_q)^n$.

   H **(i)**   Prove that
$$\begin{bmatrix} n \\ r \end{bmatrix}_q \begin{bmatrix} n \\ n-r \end{bmatrix}_q = \begin{bmatrix} n \\ n \end{bmatrix}_q.$$
   (These coefficients arise in the study of hypergeometric series.)

   H **(ii)**   Prove that there are exactly $\begin{bmatrix} n \\ n-r \end{bmatrix}_q$ r-dimensional subspaces of $(\mathbb{F}_q)^n$.

   (iii)   Prove that
$$\begin{bmatrix} n \\ r \end{bmatrix}_q = \frac{(q^n - 1)(q^{n-1} - 1) \cdots (q - 1)}{(q^r - 1)(q^{r-1} - 1) \cdots (q - 1)(q^{n-r-1})(q^{n-r} - 1) \cdots (q - 1)}.$$

   (iv)   Prove the analog of Lemma 1.17:
$$\begin{bmatrix} n \\ r \end{bmatrix}_q = \begin{bmatrix} n-1 \\ r-1 \end{bmatrix}_q + q^r \begin{bmatrix} n-1 \\ r \end{bmatrix}_q.$$

   (v)   Prove the analog of Exercise 1.34 on page 35:
$$\begin{bmatrix} n \\ r \end{bmatrix}_q = \frac{q^n - 1}{q^r - 1} \begin{bmatrix} n-1 \\ r-1 \end{bmatrix}_q.$$

6.36 Find $Z(\mathrm{UT}(3, \mathbb{F}_q))$ and $Z(\mathrm{UT}(4, \mathbb{F}_q))$.

6.37   **(i)**   Prove that $\mathrm{UT}(n, \mathbb{F}_q)$ has a normal series
$$\mathrm{UT}(n, \mathbb{F}_q) = G_0 \ge G_1 \ge \cdots \ge G_n = \{1\}$$
in which $G_i \cong \mathrm{UT}(i, \mathbb{F}_q)$ consists of all unitriangular $n \times n$ matrixes having all entries on the i superdiagonals 0. For example, if $n = 5$, then G_1 consists of all matrices of the form $\begin{bmatrix} 1&0&*&*&* \\ 0&1&0&*&* \\ 0&0&1&0&* \\ 0&0&0&1&0 \\ 0&0&0&0&1 \end{bmatrix}$ and G_2 consists of all matrices of the form $\begin{bmatrix} 1&0&0&*&* \\ 0&1&0&0&* \\ 0&0&1&0&0 \\ 0&0&0&1&0 \\ 0&0&0&0&1 \end{bmatrix}$.

   (ii)   Prove that the factor groups G_{i-1}/G_i are abelian for all $i \ge 1$.

6.38 H **(i)**   Prove that $|\mathrm{GL}(n, \mathbb{F}_q)| = (q - 1)|\mathrm{SL}(n, \mathbb{F}_q)|$.

   H **(ii)**   Prove that $|\mathrm{SL}(2, \mathbb{F}_5)| = 120$.

   H **(iii)**   Find a Sylow 2-subgroup of $\mathrm{SL}(2, \mathbb{F}_5)$.

6.3   ORNAMENTAL SYMMETRY

In Section 2.3, we defined an isometry of the plane to be a distance-preserving function $\varphi \colon \mathbb{R}^2 \to \mathbb{R}^2$, and we saw, in Proposition 2.61, that $\mathbf{Isom}(\mathbb{R}^2)$, the set of all the isometries of the plane, is a group under composition. For any subset Ω of the plane, its *symmetry group* is defined by

$$\Sigma(\Omega) = \{\varphi \in \mathbf{Isom}(\mathbb{R}^2) : \varphi(\Omega) = \Omega\}.$$

For example, we saw, in Theorem 2.65, that the dihedral group D_{2n} is isomorphic to the symmetry group of a regular n-gon Ω. In this section, we will study symmetry groups of certain designs, called friezes. Our discussion follows that in Burn, *Groups: A Path to Geometry*.

We defined three types of isometry in Example 2.57: *rotations, reflections,* and *translations* (but see Theorem 6.42; there is a fourth type). Identify the plane $\mathbb{R}^2$ with the complex numbers $\mathbb{C}$ via $(a, b) \mapsto a + ib$. Thus, every point $(x, 0)$ is identified with the real number x (in particular, the origin is identified with 0), and the x-axis is identified with $\mathbb{R}$. We will use the notation $e^{i\theta}$ for numbers on the unit circle instead of the normalized notation $e^{2\pi i\theta}$. This identification of $\mathbb{R}^2$ with $\mathbb{C}$ enables us to give simple algebraic formulas for isometries. Compare the following with the geometric descriptions in Example 2.57 on page 139.

Example 6.34.

(i) ***Rotation*** about the origin by θ is the function R_θ sending the point with polar coordinates (r, α) to the point with polar coordinates $(r, \theta + \alpha)$. This isometry can now be written as $R_\theta \colon z \mapsto e^{i\theta}z$, for if $z = re^{i\alpha}$, then

$$R_\theta(z) = e^{i\theta}z = e^{i\theta}re^{i\alpha} = re^{i(\theta+\alpha)}.$$

(ii) An isometry ρ_ℓ is a ***reflection*** if there is a line ℓ, called its ***axis***, each of whose points is fixed by ρ_ℓ, which is the perpendicular bisector of all segments having endpoints $z, \rho_\ell(z)$. In particular, reflection about the x-axis sends each point (a, b) to $(a, -b)$; this is complex conjugation $\sigma \colon z = a + ib \mapsto a - ib = \bar{z}$.

(iii) ***Translation*** by a vector c is $\tau_c \colon z \mapsto z + c$. Remember that the identity $z \mapsto z$ is a translation; it is the only translation having a fixed point. ◀

Recall, if φ is an isometry, that $\varphi(\ell)$ is a line whenever ℓ is a line, and that $\varphi(C)$ is a circle whenever C is a circle. In more detail, if $\ell = L[P, Q]$ is the line determined by distinct points P and Q, then $\varphi(L[P, Q]) = L[\varphi(P), \varphi(Q)]$, by Lemma 2.58; if $C = C[P; PQ]$ is the circle with center P and radius PQ, then $\varphi(C[P; PQ]) = C[\varphi(P); \varphi(P)\varphi(Q)]$.

Here is a geometric lemma.

Lemma 6.35.   *Let A, P, Q be distinct points in the plane, let $C = C[P; PA]$ be the circle with center P and radius PA, and let $C' = C[Q; QA]$ be the circle with center Q and radius QA. Then $C \cap C' = \{A\}$ if and only if A, P, Q are collinear.*

Proof.   We use analytic geometry. Draw P and Q as the points $(0, 0)$ and $(1, 0)$ on the x-axis, and let $A = (a, b)$. The equation of C is $x^2 + y^2 = |PA|^2 = a^2 + b^2$, and the equation of C' is $(x - 1)^2 + y^2 = |QA|^2 = (a - 1)^2 + b^2$. If $B = (p, q) \in C \cap C'$, then there are equations

$$p^2 + q^2 = a^2 + b^2 \qquad \text{and} \qquad (p - 1)^2 + q^2 = (a - 1)^2 + b^2.$$

Hence,

$$(p - 1)^2 + (a^2 + b^2 - p^2) = (a - 1)^2 + b^2.$$

Simplifying, we get $p = a$ and $q = \pm b$. If $b \neq 0$, then there are two points in $C \cap C'$. Hence, if there is only one point in $C \cap C'$, then $b = 0$. But this one point must be A, so that $A = (a, 0)$. Thus, A lies on the x-axis, and A, 0, and 1 are collinear. Conversely, if $C \cap C'$ has more than one point, then $C \cap C' = \{A, B\} \neq \{A\}$. Thus, $B = (a, -b) \neq (a, b) = A$, so that $b \neq 0$ and, hence, A, P, and Q are not collinear. •

Proposition 6.36.   *Let $\varphi \colon \mathbb{C} \to \mathbb{C}$ be an isometry fixing 0.*

(i) *There is some θ with $\varphi(1) = e^{i\theta}$. If $\varphi(1) = 1$, then φ fixes the x-axis pointwise, and φ is either the identity or complex conjugation.*

(ii) *If $\varphi(1) \neq 1$, then φ is either a rotation or a reflection. In more detail, $\varphi \colon z \mapsto e^{i\theta}z$ when φ is a rotation, and $\varphi \colon z \mapsto e^{i\theta}\overline{z}$ when φ is a reflection. In the latter case, the axis of φ is $\ell = \{re^{i\theta/2} : r \in \mathbb{R}\}$.*

In both cases, φ is a linear transformation lying in the orthogonal group $O_2(\mathbb{R})$.

Proof.
(i) Let $z \in \mathbb{R}$ be distinct from 0, and let $C_{|z|}$ be the circle with center 0 and radius $|z| = |0z|$. Since φ is an isometry fixing 0, we have $\varphi(C_{|z|}) = C_{|z|}$, for isometries take circles to circles of the same radius: $\varphi(C[0; 0z]) = C[\varphi(0); \varphi(0)\varphi(z)] = C[0; 0\varphi(z)]$. In particular, $1 \in C_1$ implies that $\varphi(1) \in C_1$, and so $\varphi(1) = e^{i\theta}$ for some θ.

Assume that φ also fixes 1, and let $z \in \mathbb{R}$ be distinct from 0, 1. If $C = C[0, z]$ and $C' = C[1, z]$, then $\varphi(C) = \varphi(C[0, z]) = C[0, \varphi(z)] = C$, because $|0z| = |\varphi(0)\varphi(z)| = |0\varphi(z)|$; similarly, $\varphi(C') = C'$. Since 0, 1, z are collinear, Lemma 6.35 gives $\{z\} = C \cap C'$. Hence,

$$\{\varphi(z)\} = \varphi(C \cap C') = \varphi(C) \cap \varphi(C') = C \cap C' = \{z\}.$$

Therefore, φ fixes $\mathbb{R}$ pointwise.

If $z \notin \mathbb{R}$, let C be the circle with center 0 and radius $0z$ and let C' be the circle with center 1 and radius $1z$. Now $\varphi(C \cap C') = \varphi(C) \cap \varphi(C') = C \cap C'$. But Lemma 6.35 says that $C \cap C' = \{z, \overline{z}\}$, so that either $\varphi(z) = z$ or $\varphi(z) = \overline{z}$. If $\varphi(z) = z$ for some $z \notin \mathbb{R}$, then φ fixes the basis $1, z$ of the vector space $\mathbb{R}^2$ and, hence, φ is the identity (because φ is a linear transformation, by Proposition 2.59). Therefore, if φ is not the identity, then $\varphi(z) = \overline{z}$ for all z.

(ii) If ψ is rotation about 0 by θ, then $\psi^{-1}\varphi$ is an isometry fixing both 0 and 1. By part (i), $\psi^{-1}\varphi$ is either the identity or complex conjugation; that is, $\varphi(z) = e^{i\theta}z$ or $\varphi(z) = e^{i\theta}\overline{z}$.

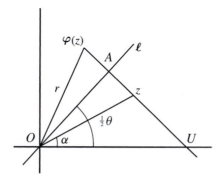

Figure 6.1 $z \mapsto e^{i\theta}\overline{z}$ is a reflection

If $\varphi(z) = e^{i\theta}z$, then Example 6.34(i) shows that φ is a rotation.
If $\varphi \colon z \mapsto e^{i\theta}\overline{z}$, then

$$\varphi(re^{i\theta/2}) = e^{i\theta}\overline{\varphi(re^{i\theta/2})} = re^{i\theta}e^{-i\theta/2} = re^{i\theta/2},$$

so that every point on ℓ is fixed by φ. If $z = re^{i\alpha} \notin \ell$, then $\varphi(z) = re^{i(\theta-\alpha)}$. In Figure 6.1, the intersection of the line $L = L[z, \varphi(z)]$ with ℓ is denoted by A, and the intersection of L with the x-axis is denoted by U. Let us see that ℓ bisects $\angle z0\varphi(z)$. Now $\angle U0\varphi(z) = \theta - \alpha$, so that $\angle z0\varphi(z) = \theta - 2\alpha = 2(\theta - \frac{1}{2}\alpha)$; hence, $\angle \varphi(z)0A = \frac{1}{2}\theta - \alpha = \angle z0A$. Thus, $\triangle z0A$ is congruent to $\triangle \varphi(z)0A$, because $|0\varphi(z)| = r = |0z|$, and so $|\varphi(z)A| = |Az|$. Finally, ℓ is perpendicular to $L = L[\varphi(z), z]$, for $\angle 0A\varphi(z) = \angle 0Az$ and their sum is $180°$. Therefore, φ is a reflection with axis ℓ. •

Having classified all isometries that fix 0, we now investigate arbitrary isometries.

Corollary 6.37. *If φ is an isometry with $\varphi(0) = c$, then there is some θ so that*

$$\varphi(z) = e^{i\theta}z + c \quad or \quad \varphi(z) = e^{i\theta}\overline{z} + c.$$

Proof. If φ is a translation, say, $\varphi \colon z \mapsto z + c$, then φ has the formula $\varphi(z) = e^{i\theta}z + c$ with $\theta = 0$. In general, given $\varphi \colon z \mapsto e^{i\theta}z + c$, define τ to be translation by $c = \varphi(0)$.

Now $\tau^{-1}\varphi$ is an isometry fixing 0, and so it is either a rotation or a reflection, by Proposition 6.36.   •

The isometry $z \mapsto e^{i\theta} + c$ is easily seen to be rotation about c by θ. Your first guess is that isometries of the form $z \mapsto e^{i\theta}\overline{z} + c$ are reflections, but the next proposition shows that this is not always true.

The **direction** of a nonzero complex number $z = re^{i\theta}$ is defined to be θ. Every line ℓ has an equation of the form $z = re^{i\theta} + v_0$, where $r \in \mathbb{R}$ and $v_0 \in \mathbb{C}$, and we say that ℓ has **direction** θ.

Proposition 6.38. *The following statements are equivalent for an isometry with equation* $\varphi \colon z \mapsto e^{i\theta}\overline{z} + c$.

(i) $\varphi^2 = identity$.

(ii) $e^{i\theta}\overline{c} + c = 0$.

(iii) φ *has a fixed point.*

(iv) φ *has a line ℓ comprised of fixed points, and ℓ has direction $\theta/2$.*

(v) φ *is a reflection.*

Proof.
(i) $\Rightarrow$ (ii)

$$\begin{aligned}
\varphi^2(z) &= \varphi(e^{i\theta}\overline{z} + c) \\
&= e^{i\theta}\overline{(e^{i\theta}\overline{z} + c)} + c \\
&= e^{i\theta}(e^{-i\theta}z + \overline{c}) + c \\
&= z + e^{i\theta}\overline{c} + c
\end{aligned}$$

Hence, φ^2 is the identity if and only if $e^{i\theta}\overline{c} + c = 0$.
(ii) $\Rightarrow$ (iii) Since φ is a reflection, the midpoint $\frac{1}{2}(z + \varphi(z))$ of the segment with endpoints z and $\varphi(z)$ lies on the axis of φ and, hence, it is fixed by φ. In particular, the point $\frac{1}{2}c$ is fixed [being the midpoint of 0 and $\varphi(0) = c$]. Indeed, $\varphi(\frac{1}{2}c) = e^{i\theta}\frac{1}{2}\overline{c} + c = \frac{1}{2}(e^{i\theta}\overline{c} + c) + \frac{1}{2}c = \frac{1}{2}c$, because $e^{i\theta}\overline{c} + c = 0$.
(iii) $\Rightarrow$ (iv) Suppose that $\varphi(u) = u$. Let ℓ be the line $\ell = \{u + re^{i\theta/2} : r \in \mathbb{R}\}$; it is clear that ℓ has direction $\theta/2$. If $z \in \ell$, then

$$\begin{aligned}
\varphi(z) &= \varphi(u + re^{i\theta/2}) \\
&= e^{i\theta}\overline{\left(u + re^{i\theta/2}\right)} + c \\
&= e^{i\theta}\overline{u} + re^{i\theta}e^{-i\theta/2} + c
\end{aligned}$$

$$= (e^{i\theta}\overline{u} + c) + re^{i\theta/2}$$
$$= \varphi(u) + re^{i\theta/2}$$
$$= u + re^{i\theta/2}$$
$$= z.$$

(iv) $\Rightarrow$ (v) It suffices to show that φ is a reflection with axis ℓ; since φ fixes every point on ℓ, it suffices to show, for each $z \notin \ell$, that ℓ is the perpendicular-bisector of the segment with endpoints z, $\varphi(z)$. If $\psi \colon z \mapsto e^{i\theta}\overline{z}$, then we saw, in Proposition 6.36, that ψ is the reflection with axis $\ell' = \{re^{i\theta/2} : r \in \mathbb{R}\}$. Hence, ℓ' is the perpendicular-bisector of each segment with endpoints $z - \frac{1}{2}c$, $\psi(z - \frac{1}{2}c)$. If we define τ to be translation by $\frac{1}{2}c$, then $\ell = \tau(\ell')$ is the perpendicular-bisector of the segment with endpoints $\tau(z - \frac{1}{2}c)$, $\tau(\psi(z - \frac{1}{2}c))$. But $\tau(z - \frac{1}{2}c) = z$ and

$$\tau(\psi(z - \tfrac{1}{2}c)) = e^{i\theta}\overline{(z - \tfrac{1}{2}c)} + \tfrac{1}{2}c$$
$$= e^{i\theta}\overline{z} - \tfrac{1}{2}e^{i\theta}\overline{c} + \tfrac{1}{2}c$$
$$= [e^{i\theta}\overline{z} + c] - c - \tfrac{1}{2}e^{i\theta}\overline{c} + \tfrac{1}{2}c$$
$$= \varphi(z) - \tfrac{1}{2}(e^{i\theta}\overline{c} + c)$$
$$= \varphi(z).$$

(v) $\Rightarrow$ (i) The square of a reflection is the identity. $\bullet$

Example 6.39.
We observe that reflections and translations need not commute. Let $\sigma \colon z \mapsto \overline{z}$ be complex conjugation, and let $\tau \colon z \mapsto z + i$ be translation by i. Now $\sigma\tau(z) = \overline{z + i} = \overline{z} - i$, while $\tau\sigma(z) = \overline{z} + i$.   ◄

We now analyze those isometries $\varphi \colon z \mapsto e^{i\theta}\overline{z} + c$ which are not reflections.

Proposition 6.40.   *If $\varphi \colon z \mapsto e^{i\theta}\overline{z} + c$ is not a reflection, then $\varphi = \tau\rho$, where ρ is a reflection, say, with axis ℓ, and τ is a translation $z \mapsto z + \frac{1}{2}w$, where w has direction that of ℓ.*

Proof.   As in the proof of (i) $\Rightarrow$ (ii) in Proposition 6.38, we have $\varphi^2(z) = z + e^{i\theta}\overline{c} + c$. We define $w = e^{i\theta}\overline{c} + c$, so that

$$\varphi^2 \colon z \mapsto z + w. \tag{1}$$

Now define

$$\tau \colon z \mapsto z + \tfrac{1}{2}w,$$

so that $\tau^2 = \varphi^2$.

Note first that

$$e^{i\theta}\overline{w} = e^{i\theta}(e^{-i\theta}c + \overline{c}) = w. \tag{2}$$

It follows that w has direction $\frac{1}{2}\theta$: if $w = re^{i\alpha}$, then substituting $w = e^{i\theta}\overline{w}$ in Eq. (2) gives $re^{i\alpha} = re^{i\theta}e^{-i\alpha}$. Hence, $e^{2i\alpha} = e^{i\theta}$, so that $\alpha = \frac{1}{2}\theta$.

We claim that τ commutes with φ.

$$\begin{aligned}
\varphi(\tau(z)) &= \varphi(z + \tfrac{1}{2}w) \\
&= e^{i\theta}\overline{(z + \tfrac{1}{2}w)} + c \\
&= e^{i\theta}\overline{z} + c + \tfrac{1}{2}e^{i\theta}\overline{w} \\
&= \varphi(z) + \tfrac{1}{2}w \\
&= \tau(\varphi(z)).
\end{aligned}$$

It follows that φ commutes with τ^{-1}:

$$\varphi\tau^{-1} = \tau^{-1}(\tau\varphi)\tau^{-1} = \tau^{-1}(\varphi\tau)\tau^{-1} = \tau^{-1}\varphi.$$

But $\tau^2 = \varphi^2$, so that

$$(\tau^{-1}\varphi)^2 = (\tau^{-1})^2\varphi^2 = \text{identity}.$$

Hence, if we define $\rho = \tau^{-1}\varphi$, then $\rho^2 = \text{identity}$ and

$$\rho(z) = \tau\varphi(z) = e^{i\theta}\overline{z} + (c + \tfrac{1}{2}w).$$

Proposition 6.38 now says that ρ is a reflection whose axis has direction $\frac{1}{2}\theta$, which we have already observed is the direction of w.   ●

Definition.   An isometry φ is a **glide reflection** if $\varphi = \tau_v\rho$, where ρ is a reflection with axis ℓ and τ_v is a translation with v having the same direction as ℓ. Thus,

$$\varphi(z) = e^{i\theta}\overline{z} + v = e^{i\theta}\overline{z} + re^{i\theta/2}$$

for some nonzero $r \in \mathbb{R}$.

Glide reflections are precisely the isometries described in Proposition 6.40. We note that glide reflections φ are not reflections, for φ^2 is not the identity.

Example 6.41.
The isometry $\varphi \colon z \mapsto \overline{z} + 1$ is a glide reflection taking the x-axis to itself: $\varphi(\mathbb{R}) = \mathbb{R}$. If Δ is the triangle with vertices $(0, 0)$, $(\frac{1}{2}, 0)$, $(1, 1)$, then $\varphi(\Delta)$ is the triangle with vertices $(1, 0)$, $(\frac{3}{2}, 0)$, $(2, -1)$, $\varphi^2(\Delta)$ has vertices $(2, 0)$, $(\frac{5}{2}, 0)$, $(3, 1)$, and $\varphi^n(\Delta)$ has vertices $(n, 0)$, $(\frac{2n+1}{2}, 0)$, $(n + 1, (-1)^n)$. The design in Figure 6.2, which goes on infinitely to the left and to the right, is invariant under φ.   ◀

The following statement summarizes our work so far.

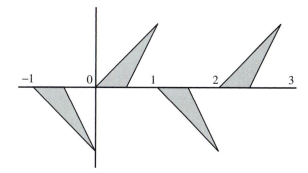

Figure 6.2 Glide Reflection

Theorem 6.42. *Every isometry is either a translation, a rotation, a reflection, or a glide reflection.*

Proof. The theorem follows from Propositions 6.36(ii), 6.38, and 6.40, and Corollary 6.37. •

Corollary 6.43. *Let $\varphi \in \mathbf{Isom}(\mathbb{R}^2)$.*

(i) *If φ has no fixed points, then φ is either a translation or a glide reflection.*

(ii) *If φ has only one fixed point, then φ is a rotation.*

(iii) *If φ has more than one fixed point, then φ is either a reflection or the identity.*

Proof. There are only four types of isometry, by Theorem 6.42: translations, which have no fixed points; rotations, which have exactly one fixed point; reflections, which have infinitely many fixed points, namely, every point on their axes; glide reflections. It suffices to show that a glide reflection φ has no fixed points. If $\varphi(z) = z$, then $\varphi^2(z) = z$; but $\varphi^2 = \tau$, where $\tau \neq$ identity is a translation, by Eq. (1), contradicting the fact that translations have no fixed points. •

Example 6.44.
We use Theorem 6.42 to determine the elements of finite order in $\mathbf{Isom}(\mathbb{R}^2)$. Translations (other than the identity) have infinite order, as do glide reflections (for the square of a glide reflection is a translation). All reflections have order 2. Finally, suppose that $\varphi \colon z \mapsto e^{i\theta}z + c$ is a rotation (about c). By induction, we see that

$$\varphi^n(z) = e^{ni\theta}z + c(1 + e^{i\theta} + e^{2i\theta} + \cdots + e^{(n-1)i\theta}).$$

If $\varphi^n =$ identity, then $\theta = 2\pi/n$ and $\varphi^n(z) = z + c(1 + e^{i\theta} + e^{2i\theta} + \cdots + e^{(n-1)i\theta})$. But now $e^{i\theta}$ is an nth root of unity, and so $1 + e^{i\theta} + e^{2i\theta} + \cdots + e^{(n-1)i\theta} = 0$. Therefore,

508 GROUPS II CH. 6

if φ^n is the identity, then we must have $c = 0$; that is, $\varphi(z) = e^{2\pi i/n}z$. Conversely, if $\theta = 2\pi/n$, then $z \mapsto e^{i\theta}z$ has finite order.

Are there any elements of order 2 besides reflections? Such an isometry φ must have the form $z \mapsto e^{\pi i}z + c$; that is, $\varphi(z) = -z + c$; it is called a **half-turn**. Note that half-turns are not reflections, for a reflection has infinitely many fixed points while a half-turn, being a rotation, has only one fixed point. A half-turn reverses the direction of a line. For example, $\varphi \colon z \mapsto -z + 2$ takes

$$\cdots] - - > --] - - > --] - - > --] - - > --] \cdots$$

to

$$\cdots < - - [-- < - - [-- < - - [-- < - - [-- < \cdots$$

The reader should check that a half-turn φ turns a figure upside down. For example, $\varphi(\vee) = \wedge$ and $\varphi(\wedge) = \vee$. ◄

Recall that if $z_1, \ldots, z_n$ are distinct points in $\mathbb{C}$, then their **center of gravity** is u, where

$$u = \tfrac{1}{n}(z_1 + \cdots + z_n).$$

Lemma 6.45. *Let $\varphi \in \mathbf{Isom}(\mathbb{R}^2)$, and let $z_1, \ldots, z_n$ be distinct points in $\mathbb{R}^2$. Then $\varphi(u) = u'$, where u is the center of gravity of $z_1, \ldots, z_n$ and u' is the center of gravity of $\varphi(z_1), \ldots, \varphi(z_n)$.*

Proof. By Theorem 6.42, φ is either a translation, a rotation, a reflection, or a glide reflection. A rotation about c is a composite $\tau\rho$, where τ is the translation $z \mapsto z + c$ and ρ is a rotation around 0. Proposition 6.40 shows that a glide reflection is also a composite of a translation and a reflection, while every reflection is a composite of a translation and a reflection whose axis passes through 0. We conclude that it suffices to show that $\varphi(u) = u'$ for φ a translation, a rotation about the origin, or a reflection with axis passing through the origin (so that 0 is fixed in either case).

Suppose that φ is a translation: $\varphi(z) = z + a$. Then

$$
\begin{aligned}
\varphi(u) &= u + a \\
&= \tfrac{1}{n}(z_1 + \cdots + z_z) + a \\
&= \tfrac{1}{n}z_1 + \cdots + \tfrac{1}{n}z_n + \tfrac{1}{n}a + \cdots + \tfrac{1}{n}a \\
&= \tfrac{1}{n}(z_1 + a) + \cdots + \tfrac{1}{n}(z_n + a) \\
&= \tfrac{1}{n}\varphi(z_1) + \cdots + \tfrac{1}{n}\varphi(z_n) \\
&= u'.
\end{aligned}
$$

If φ is either a rotation about the origin or a reflection with axis through 0, then Proposition 6.36 shows that φ is a linear transformation. Therefore,

$$\varphi(u) = \varphi(\tfrac{1}{n}[z_1 + \cdots + z_n]) = \tfrac{1}{n}[\varphi(z_1) + \cdots + \varphi(z_n)] = u'. \quad \bullet$$

Lemma 6.46. *If $G \leq$ **Isom**$(\mathbb{R}^2)$ is a finite subgroup, then there is $u \in \mathbb{C}$ with $\varphi(u) = u$ for all $\varphi \in G$.*

Proof. Choose $z \in \mathbb{C}$, and let $\mathcal{O}$ be its orbit:

$$\mathcal{O} = \{\varphi(z) : \varphi \in G\}.$$

Since G is finite, $\mathcal{O}$ is finite: $\mathcal{O} = \{z_1, \ldots, z_n\}$, where $z_1 = z$. Now G acts on $\mathcal{O}$, for if $\psi \in G$, then $\psi(z_j) = \psi\varphi(z_1) \in \mathcal{O}$, because $\psi\varphi \in G$. Thus, each $\varphi \in G$ permutes $\mathcal{O}$, for φ is injective and $\varphi \colon \mathcal{O} \to \mathcal{O}$. Since φ permutes $\mathcal{O}$, the center of gravity u of $\mathcal{O}$ is equal to the center of gravity of $\varphi(\mathcal{O}) = \mathcal{O}$. Therefore, Lemma 6.45 says that $\varphi(u) = u$ for all $\varphi \in G$. •

In his book *Symmetry*, H. Weyl attributes the following theorem to Leonardo da Vinci (1452-1519).

Theorem 6.47 (Leonardo). *If $G \leq$ **Isom**$(\mathbb{R}^2)$ is a finite subgroup, then either $G \cong \mathbb{I}_m$ for some m or $G \cong D_{2n}$ for some n.*

Proof. By Lemma 6.46, there is $c \in \mathbb{C}$ with $\varphi(c) = c$ for all $\varphi \in G$. If $\tau \colon z \mapsto z - c$, then $\tau\varphi\tau^{-1}(0) = \tau\varphi(c) = \tau(c) = 0$. Since $\tau G \tau^{-1} \cong G$, we may assume that every $\varphi \in G$ fixes 0. Thus, Proposition 6.36 applies: we may assume that $G \leq O_2(\mathbb{R})$, and so every $\varphi \in G$ is a linear transformation. Better, we may assume that every $\varphi \in G$ is either a rotation or a reflection.

Suppose that G contains no reflections. Thus, the elements of G are rotations $R_{\theta_1}, \ldots, R_{\theta_m}$, where $\theta_j = 2\pi k_j / n_j$, by Example 6.44. If $n = \max_j \{n_j\}$, then $G \leq \langle R_{2\pi/n} \rangle$. Therefore, G, being a subgroup of a cyclic group, is itself cyclic.

Suppose that G contains a reflection ρ. By Exercise 6.48 on page 517, we may replace G by an isomorphic copy which contains σ, complex conjugation. The subset H, consisting of all the rotations in G, is a subgroup; it is a finite subgroup of **Isom**$(\mathbb{R}^2)$ containing no reflections, and so it is cyclic, say, $H = \langle h \rangle$, where $h(z) = e^{i\theta} z$ has order n, say. Now $\sigma h \sigma^{-1} = h^{-1}$, for

$$\sigma h \sigma^{-1} \colon z \mapsto \overline{z} \mapsto e^{i\theta}\overline{z} \mapsto \overline{e^{i\theta}\overline{z}} = e^{-i\theta} z = h^{-1}(z).$$

Hence, $\langle h, \sigma \rangle = H \cup H\sigma$ is a subgroup isomorphic to D_{2n}. We claim that $\langle h, \sigma \rangle = G$. If $r \in G$ is a reflection, then $r(z) = e^{i\alpha}\overline{z} = R_\alpha \sigma(z)$. But $R_\alpha = r\sigma^{-1} \in H$, for it is a rotation in G, and so $r = R_\alpha \sigma \in \langle h, \sigma \rangle$. •

Leonardo's theorem has found all the finite subgroups of $O_2(\mathbb{R})$ that stabilize a point. We are now going to find all those subgroups of **Isom**$(\mathbb{R}^2)$, called *frieze groups*, that stabilize a line but not a point. To isomorphism, there are only four such subgroups, but when we take geometry into account, we shall see seven types of them.

According to the *Oxford English Dictionary*, a *frieze* is "that member in the entablature of an order which comes between the architrave and the cornice." Fortunately, it goes on to say that a frieze is "any horizontal broad band which is occupied by a sculpture." Now sculptures are three-dimensional, but we use the word *frieze* to mean any (two-dimensional) broad band which repeats some pattern infinitely to the left and to the right. In more precise language, we say that a subset F of the plane is a *band* if F there is some isometry φ (not the identity) in the symmetry group $\Sigma(F)$ which stabilizes a line ℓ; that is, $\varphi(\ell) = \ell$ (we do not insist that φ fix ℓ pointwise). To say that a band F is a frieze means that there is some "design" $D \subseteq F$ so that $F = \bigcup_{n \in \mathbb{Z}} \tau^n(D)$ for some translation $\tau \in \Sigma(F)$. We aim to classify all those subgroups of $\textbf{Isom}(\mathbb{R}^2)$ of the form $\Sigma(F)$ for some frieze F.

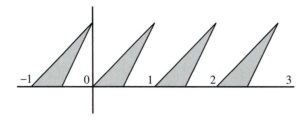

Figure 6.3   The Frieze F

The band F in Figure 6.3 is a frieze: it is stabilized by the translation $\tau : z \mapsto z+1$, and its repeating pattern is the triangle D having base the closed interval $[0, \frac{1}{2}]$.

Consider the band F' in Figure 6.2. It is easy to see that its symmetry group, $\Sigma(F')$, contains the glide reflection $\varphi : z \mapsto \bar{z} + 1$. Note that $\varphi(\mathbb{R}) = \mathbb{R}$ and that $F' = \bigcup_{n \in \mathbb{Z}} \varphi^n(D)$, where D is the triangle with base $[0, \frac{1}{2}]$. This does not show that F' is a frieze because φ is not a translation. However, F' is, indeed, a frieze, for the translation $\tau : z \mapsto z + 2$ lies in $\Sigma(F')$ and $F' = \bigcup_{n \in \mathbb{Z}} \tau^n(D')$, where D' is the union of the triangle with base $[0, \frac{1}{2}]$ and the triangle with base $[1, \frac{3}{2}]$.

Now consider a frieze F'' obtained from F in Figure 6.3 by replacing the triangle D with base $[0, \frac{1}{2}]$ by another figure. For example, let F'' be the frieze in Figure 6.4 (from the palace of Darius in ancient Susa) in which each triangle D in Figure 6.3 has been replaced by a Persian bowman. It is clear that $\Sigma(F'') = \Sigma(F)$. Plainly, there are too many friezes to classify them geometrically; for example, what restrictions, if any, must be imposed on D? In spite of this, we are still able to classify friezes if we do not distinguish triangles and bowmen.

Notation.   The subgroup of all the translations in $\textbf{Isom}(\mathbb{R}^2)$ is denoted by $\textbf{Trans}(\mathbb{R}^2)$.

Informally, a *frieze group* is the symmetry group of a frieze. We will soon replace the following definition by a normalized version.

Definition 1. A *frieze group* is a subgroup G of $\textbf{Isom}(\mathbb{R}^2)$ which stabilizes a line ℓ, that is, $\varphi(\ell) = \ell$ for all $\varphi \in G$, and such that $G \cap \textbf{Trans}(\mathbb{R}^2)$ is infinite cyclic.

Figure 6.4 Persian Bowmen

Saying that each $\varphi \in G$ stabilizes a line ℓ reflects the fact that a frieze is a band; saying that $G \cap \textbf{Trans}(\mathbb{R}^2) = \langle \tau \rangle$ is infinite cyclic reflects the fact that a frieze F has some repeating design $D \subseteq F$ whose $\langle \tau \rangle$-orbit is all of F.

Lemma 6.48. *If $\varphi \in G$, where G is a frieze group, then there is some real number c such that one of the following holds*:

(i) *If φ is a translation, then $\varphi(z) = z + c$.*

(ii) *If φ is a rotation, then φ is a half-turn: $\varphi(z) = -z + c$.*

(iii) *If φ is a reflection, then $\varphi(z) = \overline{z}$ or $\varphi(z) = -\overline{z} + c$.*

(iv) *If φ is a glide reflection, then $\varphi \colon z \mapsto \overline{z} + c$, where $c \neq 0$.*

Proof. We know that $\varphi \colon z \mapsto e^{i\theta} z + c$ or $\varphi \colon z \mapsto e^{i\theta} \overline{z} + c$. Since $\varphi(\mathbb{R}) = \mathbb{R}$, we have $c = \varphi(0) \in \mathbb{R}$ and $\varphi(1) = e^{i\theta} + c \in \mathbb{R}$. Therefore, $e^{i\theta} \in \mathbb{R}$; that is, $e^{i\theta} = \pm 1$. Thus, either $\varphi(z) = \pm z + c$ or $\varphi(z) = \pm \overline{z} + c$.

The remainder of the proof determines the type of isometry corresponding to each of these formulas. Rotations by θ have the form $e^{i\theta} z + c$; since $e^{i\theta} = \pm 1$, we must have $\theta = \pi$, and so rotations here are half-turns. The isometry $\varphi \colon z \mapsto e^{i\theta} \overline{z} + c$ is a reflection if and only if $e^{i\theta} \overline{c} + c = 0$. Here, c is real, so that $\overline{c} = c$, and so φ is a reflection if either $c = 0$ or if $e^{i\theta} = -1$. Thus, either $\varphi(z) = \overline{z}$ or $\varphi(z) = -\overline{z} + c$ for any $c \in \mathbb{R}$. Finally, if $c \neq 0$ and $\varphi(z) = \overline{z} + c$, then $e^{i\theta} \overline{c} + c = 2c \neq 0$ and φ is a glide reflection.   •

We are going to normalize the classification problem for frieze groups, in two ways. First, there is no loss in generality in assuming that the stabilized line ℓ is the real axis $\mathbb{R}$, for we may change the location of the coordinate axes without disturbing symmetries. Second, we will ignore changes in scale. For example, the frieze F in Figure 6.3 has an infinite cyclic symmetry group, namely, $\Sigma(F) = \langle \tau \rangle$, where τ is the translation $\tau: z \mapsto z + 1$. On the other hand, if each vector in $\mathbb{R}^2$ is doubled in size, then F becomes a new frieze Φ with $\Sigma(\Phi) = \langle \tau' \rangle$, where $\tau': z \mapsto z + 2$. Thus, F and Φ are essentially the same frieze, differing only in scale, but their symmetry groups are different because $\tau \notin \Sigma(\Phi)$. Define $\omega: \mathbb{R}^2 \to \mathbb{R}^2$ by $\omega(z) = 2z$. Now ω defines the isomorphism $\Sigma(F) \to \Sigma(\Phi)$ given by $\varphi \mapsto \omega\varphi\omega^{-1}$. Note that

$$\omega\tau\omega^{-1}: z \mapsto \tfrac{1}{2}z \mapsto \tfrac{1}{2}z + 1 \mapsto 2(\tfrac{1}{2}z + 1) = z + 2.$$

Our second normalization assumes that the generator τ of $G \cap \mathbf{Trans}(\mathbb{R}^2)$ is the translation $\tau: z \mapsto z + 1$. In light of the discussion so far, it suffices to classify *normalized* frieze groups.

Definition 2. A normalized *frieze group* is a subgroup $G \leq \mathbf{Isom}(\mathbb{R}^2)$ which stabilizes $\mathbb{R}$ and such that $G \cap \mathbf{Trans}(\mathbb{R}^2) = \langle \tau \rangle$, where $\tau: z \mapsto z + 1$.

Lemma 6.48 simplifies when we assume frieze groups G are normalized. If $\gamma: z \mapsto \overline{z} + c$ is a glide reflection in G, then γ^2 is a translation; in fact, $\gamma^2: z \mapsto z + 2c$. But all translations in G lie in $\langle \tau \rangle$, so that $\gamma^2 = \tau^n: z \mapsto z + n$ for some $n \in \mathbb{Z}$. Hence, $2c = n$, so that $c = m$ or $c = m + \tfrac{1}{2}$ for some $m \in \mathbb{Z}$. Thus, G contains $\tau^{-m}\gamma = \sigma$ if $c = m$, that is, $\sigma(z) = \overline{z}$, or $\tau^{-m}\gamma: z \mapsto \overline{z} + \tfrac{1}{2}$ if $c = m + \tfrac{1}{2}$. In order to distinguish $\gamma \in G$ from $\sigma \in G$, we choose $\gamma: z \mapsto \overline{z} + \tfrac{1}{2}$, so that $\gamma^2 = \tau$. We may also normalize the half-turn R and the reflection ρ so that $R: z \mapsto -z + 1$ and $\rho: z \mapsto -\overline{z} + 1$.

If $\varphi \in \mathbf{Isom}(\mathbb{R}^2)$, let us write $\varphi(z) = e^{i\theta}z^\epsilon + c$, where $\epsilon = \pm 1$, $z^1 = z$, and $z^{-1} = \overline{z}$. If $\psi = e^{i\alpha}z^\eta + d$, then it is easy to see that

$$\varphi\psi(z) = e^{i(\theta+\alpha)}z^{\epsilon\eta} + e^{i\theta}d + c.$$

It follows that the function $\pi: \mathbf{Isom}(\mathbb{R}^2) \to O_2(\mathbb{R})$, defined by

$$\pi: \varphi \mapsto \tau_{\varphi(0)}^{-1}\varphi,$$

is a homomorphism [of course, $\tau_{\varphi(0)}^{-1}\varphi: z \mapsto e^{i\theta}z^\epsilon$], and $\ker \pi = \mathbf{Trans}(\mathbb{R}^2)$, so that $\mathbf{Trans}(\mathbb{R}^2) \lhd \mathbf{Isom}(\mathbb{R}^2)$.

Definition.   Let $\pi: \mathbf{Isom}(\mathbb{R}^2) \to O_2(\mathbb{R})$ be the map just defined (which erases the constant of translation). If G is a frieze group, then its *point group* is $\pi(G)$.

It follows from the second isomorphism theorem that if $T = G \cap \mathbf{Trans}(\mathbb{R}^2)$, then $T \lhd G$ and $G/T \cong \pi(G)$.

Corollary 6.49. *The point group $\pi(G)$ of a frieze group G is a subgroup of* $\text{im}\,\pi = \{1, f, g, h\} \leq O_2(\mathbb{R})$ *(which is isomorphic to the four-group* **V***), where* $f(z) = -z$, $g(z) = -\bar{z}$, *and* $h(z) = \bar{z}$.

Proof. By Lemma 6.48, we have $\text{im}\,\pi = \{1, f, g, h\}$. •

We are now going to classify the (normalized) frieze groups. Since $\text{im}\,\pi = \langle f, g, h \rangle$ is isomorphic to the four-group, there are exactly 5 subgroups of it: $\{1\}$, $\langle f \rangle$, $\langle g \rangle$, $\langle h \rangle$, $\langle f, g, h \rangle = \text{im}\,\pi$. Thus, there are 5 point groups. We will use Exercise 2.101 on page 190: Let $\pi\colon G \to H$ be a surjective homomorphism with $\ker \pi = T$. If $H = \langle X \rangle$, and, for each $x \in X$, a lifting $g_x \in G$ is chosen with $\pi(g_x) = x$, then G is generated by $T \cup \{g_x : x \in X\}$. Here, $T = G \cap \textbf{Isom}(\mathbb{R}^2) = \langle \tau \rangle$, where $\tau\colon z \mapsto z+1$.

Isometry	Formula	Type	Lifts	Order
τ	$z+1$	translation	1	∞
R	$-z+1$	half-turn	f	2
ρ	$-\bar{z}+1$	reflection	g	2
σ	$\bar{z}$	reflection	h	2
γ	$\bar{z}+\frac{1}{2}$	glide reflection	h	∞

Figure 6.5 Normalized Liftings

The following group will appear in the classification of the frieze groups. Recall that the dihedral group D_{2n} is defined as a group of order $2n$ generated by two elements a and b such that $b^2 = 1$, $a^n = 1$, and $bab = a^{-1}$.

Definition. The ***infinite dihedral group*** D_∞ is an infinite group generated by two elements a and b such that $b^2 = 1$ and $bab = a^{-1}$.

By Exercise 6.51 on page 517, any two infinite dihedral groups are isomorphic.

Theorem 6.50. *There are at most 7 types of frieze groups G.*

Proof. We use the notation in Figure 6.5.

Case 1. $\pi(G) = \{1\}$. In this case, $G = G_1 = \langle \tau \rangle$. Of course, $G_1 \cong \mathbb{Z}$.

Case 2. $\pi(G) = \langle f \rangle$. In this case, $G = G_2 = \langle \tau, R \rangle$. Now $R^2 = 1$ and $R\tau R\colon z \mapsto z - 1$; that is, $R\tau R = \tau^{-1}$. Since G_2 is infinite (because τ has infinite order), G_2 is infinite dihedral; that is, $G_2 \cong D_\infty$.

Case 3. $\pi(G) = \langle g \rangle$. In this case, $G_3 = \langle \tau, \rho \rangle$. Now $\rho^2 = 1$ and $\rho\tau\rho\colon z \mapsto z - 1$; that is, $\rho\tau\rho = \tau^{-1}$. Therefore, $G_3 \cong D_\infty$.

The group G_2 is also infinite dihedral, so that $G_3 \cong G_2$, by Exercise 6.51 on page 517; thus, G_2 and G_3 are, algebraically, the same. However, these groups are geometrically distinct, for while G_2 contains only translations and half-turns, the group G_3 contains a reflection.

Cases 4 and 5. $\pi(G) = \langle h \rangle$. There are two possible cases because there are two possible liftings of h, namely, σ and γ.

Cases 4. $G_4 = \langle \tau, \sigma \rangle$. Now τ and σ commute, for each of $\sigma\tau$ and $\tau\sigma$ sends $z \mapsto \bar{z}+1$, so that G_4 is abelian. Moreover, $\sigma^2 = 1$. It follows easily from Proposition 2.127 that

$$G_4 = \langle \sigma \rangle \times \langle \tau \rangle \cong \mathbb{I}_2 \times \mathbb{Z}.$$

Case 5. $G_5 = \langle \tau, \gamma \rangle$. Note that γ and τ commute, for each of $\gamma\tau$ and $\tau\gamma$ sends $z \mapsto \bar{z} + \frac{3}{2}$, so that G_5 is abelian. Since $\gamma^2 = \tau$, we have $G_5 = \langle \tau, \gamma \rangle = \langle \gamma \rangle$ cyclic with generator γ; that is, $G_5 \cong \mathbb{Z}$.

Algebraically, G_5 and G_1 are the same, for both are infinite cyclic. But these groups are different geometrically, for G_5 contains a glide reflection while G_1 has only translations.

Cases 6 and 7. $\pi(G) = \langle f, g, h \rangle$. Again, there are two possible cases because of the two possible liftings of h. Note, in the four-group, that the product of any two nonidentity elements is the third such, and so both $\langle \tau, R, \sigma \rangle$ and $\langle \tau, R, \gamma \rangle$ have point group $\langle f, g, h \rangle$.

Case 6. $G_6 = \langle \tau, R, \sigma \rangle$. Now $\sigma\tau = \tau\sigma$, as in Case 4, while $\sigma R = R\sigma : z \mapsto -\bar{z}+1$. It follows that both $\langle \sigma \rangle \lhd G_6$ and $\langle \tau, R \rangle \lhd G_6$. Since $\langle \sigma \rangle \cap \langle \tau, R \rangle = \{1\}$ and G_6 is generated by these two subgroups, Proposition 2.127 shows that $G_6 = \langle \sigma \rangle \times \langle \tau, R \rangle$. By Case 2, $\langle \tau, R \rangle \cong D_\infty$, and so $G_6 \cong \mathbb{I}_2 \times D_\infty$.

Case 7. $G_7 = \langle \tau, R, \gamma \rangle$. Since $\gamma^2 = \tau$, we have $G_7 = \langle R, \gamma \rangle$. Now $R^2 = 1$ and $R\gamma R : z \mapsto \bar{z} - \frac{1}{2}$, so that $R\gamma R = \gamma^{-1}$. Therefore, $G_7 \cong D_\infty$.

Algebraically, G_7, G_2, and G_4 are the same, for each is isomorphic to D_∞. But these groups are different geometrically, for neither G_2 nor G_3 contains a glide reflection (lest their point group be too big). •

Theorem 6.51.   *Each of the 7 possible frieze groups occurs.*

Proof.   Each of the friezes illustrated in Figure 6.6 has the indicated group of symmetries. One should view each frieze as being bisected by the x-axis, so that half of each letter is above the axis and half below. For example, F_4 is stabilized by σ but not by γ. To prove this theorem, we consider each of the (normalized) isometries τ, R, ρ, σ, and γ, and show that a given frieze is stabilized by some of these and not stabilized by the others.

We remind the reader of the geometric view of the basic isometries. The translation τ is a shift one unit to the right, while σ is the reflection in the x-axis and ρ is the

F_1 :	F	F	F	F	F	F	$G_1 = \langle \tau \rangle$
F_2 :	Z	Z	Z	Z	Z	Z	$G_2 = \langle \tau, R \rangle$
F_3 :	A	A	A	A	A	A	$G_3 = \langle \tau, \rho \rangle$
F_4 :	D	D	D	D	D	D	$G_4 = \langle \tau, \sigma \rangle$
F_5 :	DM	DM	DM	DM	DM	DM	$G_5 = \langle \tau, \gamma \rangle$
F_6 :	I	I	I	I	I	I	$G_6 = \langle \tau, R, \sigma \rangle$
F_7 :	MM	MM	MM	MM	MM	MM	$G_7 = \langle R, \gamma \rangle$

Figure 6.6 The Seven Friezes

reflection in the y-axis. The glide reflection γ reflects in the x-axis and then shifts half a unit to the right, while the half-turn R turns a frieze upside down.

(i) We have $\Sigma(F_1) = \langle \tau \rangle$, because $\tau(F_1) = F_1$, but none of the other isometries stabilize it. Therefore, G_1 is a frieze group.

(ii) We have $\Sigma(F_2) = \langle \tau, R \rangle$, because τ, R stabilize F_2, but ρ, σ, and γ do not stabilize it. Therefore, G_2 is a frieze group.

(iii) We have $\Sigma(F_3) = \langle \tau, \rho \rangle$, because τ, ρ stabilize F_2, but R, σ, and γ do not stabilize it. Therefore, G_3 is a frieze group.

(iv) We have $\Sigma(F_4) = \langle \tau, \sigma \rangle$, because τ, σ stabilize F_2, but R, ρ, and γ do not stabilize it. Therefore, G_4 is a frieze group.

(v) We have $\Sigma(F_5) = \langle \tau, \gamma \rangle$, because τ, γ stabilize F_2, but R, ρ, and σ do not stabilize it. Therefore, G_5 is a frieze group.

(vi) We have $\Sigma(F_6) = \langle \tau, R, \sigma \rangle$, because all the isometries stabilize it. Therefore, G_6 is a frieze group.

(vii) We have $\Sigma(F_7) = \langle \tau, R, \gamma \rangle = \langle R, \gamma \rangle$, because all the isometries except σ stabilize F_7. Therefore, G_7 is a frieze group. •

Corollary 6.52. *To isomorphism, there are 4 frieze groups, namely, $\mathbb{Z}$, D_∞, $\mathbb{I}_2 \times \mathbb{Z}$, and $\mathbb{I}_2 \times D_\infty$.*

Proof. As stated in the proof of Theorem 6.50, $\Sigma(F_1)$ and $\Sigma(F_5)$ are isomorphic to $\mathbb{Z}$, $\Sigma(F_2)$, $\Sigma(F_3)$ and $\Sigma(F_7)$ are isomorphic to D_∞, $\Sigma(F_4)$ is isomorphic to $\mathbb{I}_2 \times \mathbb{Z}$, and $\Sigma(F_6) \cong \mathbb{I}_2 \times D_\infty$. •

Friezes are planar figures involving one axis. The next question is the classification of *wallpaper groups*, which are symmetry groups of planar figures involving two axes. Let $B_r(u) = \{v \in \mathbb{R}^2 : |v - u| < r\}$ be the open ball with center u and radius r. Of course, a subgroup $G \le \mathbf{Isom}(\mathbb{R}^2)$ acts on $\mathbb{R}^2$, and so the orbit $\mathcal{O}(u)$ of any point $u \in \mathbb{R}^2$ makes sense: $\mathcal{O}(u) = \{\varphi(u) : \varphi \in G\}$. A subgroup $G \le \mathbf{Isom}(\mathbb{R}^2)$ is *discrete* if, for each $u \in \mathbb{R}^2$, there is $r > 0$ so that $B_r(u) \cap \mathcal{O}(u) = \{u\}$. One can prove that frieze groups are those discrete subgroups of **Isom**$(\mathbb{R}^2)$ which stabilize a line but not a point (the point groups stabilize a point). *Wallpaper groups* are those discrete subgroups G of **Isom**$(\mathbb{R}^2)$ which do not stabilize a line or a point. If G is a wallpaper group, then the homomorphism $\pi: G \to O_2(\mathbb{R})$ has kernel **Trans**$(\mathbb{R}^2) \cap G$, which is now a free abelian group $\mathbb{Z} \times \mathbb{Z}$. The image of π, still called a *point group*, must be one of $\mathbb{I}_n, D_{2n}$, where $n \in \{1, 2, 3, 4, 6\}$ (this is the so-called *crystallographic restriction*). We refer the interested reader to the final chapter of Burn, *Groups: A Path to Geometry*, where it is shown that there are exactly 17 wallpaper groups.

Similar problems exist in 3-dimensional space. One can classify the five Platonic solids and give their isometry groups: the tetrahedron has symmetry group A_4, the cube and the octahedron each has symmetry group S_4, and the dodecahedron and icosahedron each has symmetry group A_5. *Crystallographic groups* are defined to be the discrete subgroups $G \leq$ **Isom**$(\mathbb{R}^3)$ which do not stabilize a point, a line, or a plane. There is a homomorphism $G \to O_3(\mathbb{R})$, all orthogonal linear transformations on $\mathbb{R}^3$, which generalizes the homomorphism π; its kernel, **Trans**$(\mathbb{R}^3) \cap G$, is a free abelian group $\mathbb{Z} \oplus \mathbb{Z} \oplus \mathbb{Z}$, and its image, a point group, is a finite subgroup of $O_3(\mathbb{R})$. There are 230 crystallographic groups.

EXERCISES

H **6.39** True or false with reasons.
- **(i)** There is a frieze in the plane having a finite symmetry group.
- **(ii)** There is an isometry of the plane having exactly two fixed points.
- **(iii)** An isometry of the plane can be both a translation and a glide reflection.
- **(iv)** An isometry of the plane can be both a reflection and a glide reflection.
- **(v)** There is a subgroup of **Isom**$(\mathbb{R}^2)$ isomorphic to S_3.
- **(vi)** There is a subgroup of **Isom**$(\mathbb{R}^2)$ isomorphic to S_4.
- **(vii)** Two (normalized) frieze groups having the same point group are isomorphic.
- **(viii)** An infinite dihedral group has a subgroup of finite index.

6.40
- **(i)** If $\varphi \in$ **Isom**$(\mathbb{R}^2)$, then $\varphi(z) = e^{i\theta}z + c$ or $\varphi(z) = e^{i\theta}\overline{z} + c$. Prove that θ and c are uniquely determined by φ.
- **(ii)** Prove that the function $f:$ **Isom**$(\mathbb{R}^2) \to O_2(\mathbb{R})$, defined by $\varphi \mapsto \varphi\tau_{\varphi(0)}^{-1}$, is a homomorphism, where $\tau_{\varphi(0)}$ is the translation $z \mapsto z + \varphi(0)$. Prove that the homomorphism f is surjective and that its kernel is the subgroup T of all the translations. Conclude that $T \lhd$ **Isom**$(\mathbb{R}^2)$.

6.41 Prove that $\varphi: (x, y) \mapsto (x + 2, -y)$ is an isometry. What type of isometry is it?

6.42 Verify the following formulas.
- **(i)** If $\tau: z \mapsto z + c$, then $\tau^{-1}: z \mapsto z - c$.
- **(ii)** If $R: z \mapsto e^{i\theta}z + c$, then $R^{-1}: z \mapsto e^{-i\theta}(z - c)$.
- **(iii)** If $\varphi: z \mapsto e^{i\theta}\overline{z} + c$, then $\varphi^{-1}: z \mapsto e^{i\theta}(\overline{z} - c)$.

      (iv) Give an example of isometries α and β such that α and $\beta\alpha\beta^{-1}$ are not isometries of the same type.

6.43   **(i)** Prove that conjugate elements in **Isom**$(\mathbb{R}^2)$ have the same number of fixed points.

      (ii) Prove that if φ is a rotation and ψ is a reflection, then φ and ψ are not conjugate in **Isom**$(\mathbb{R}^2)$.

6.44 If φ and ψ are rotations in **Isom**$(\mathbb{R}^2)$ having different fixed points, prove that the subgroup $\langle\varphi,\psi\rangle$ they generate is infinite.

6.45 If $\varphi\in$ **Isom**$(\mathbb{R}^2)$ fixes three noncollinear points, prove that φ is the identity.

6.46   **(i)** Prove that the composite of two reflections in **Isom**$(\mathbb{R}^2)$ is either a rotation or a translation.

      (ii) Prove that every rotation is a composite of two reflections. Prove that every translation is a composite of two reflections.

      (iii) Prove that every isometry $\mathbb{R}^2\to\mathbb{R}^2$ is a composite of at most three reflections.

6.47 If H denotes the subgroup of **Isom**$(\mathbb{R}^2)$ consisting of all isometries which stabilize $\mathbb{R}$, prove that complex conjugation lies in the center, $Z(H)$.

*6.48   H **(i)** If ρ is a reflection in $O_2(\mathbb{R})$, prove that there is a rotation $R\in O_2(\mathbb{R})$ with $R\rho R^{-1}=\sigma$, where $\sigma(z)=\overline{z}$.

      (ii) If G is a subgroup of $O_2(\mathbb{R})$ containing a reflection ρ, prove that there is a rotation $R\in$ **Isom**$(\mathbb{R}^2)$ with RGR^{-1} containing complex conjugation.

*6.49 Prove that the composite of two reflections is either the identity or a rotation.

*6.50 Prove that if a frieze group G contains two of the following types of isometry–half-turn; glide reflection; reflection with vertical axis–then G contains an isometry of the third type.

*H 6.51 Prove that any two infinite dihedral groups are isomorphic. In more detail, let $G=\langle a,b\rangle$ and $H=\langle c,d\rangle$ be infinite groups in which $a^2=1$, $aba=b^{-1}$, $c^2=1$, and $cdc=d^{-1}$. Prove that $G\cong H$.

6.52 Find the isometry group of the following friezes.

      (i)   SANTACLAUSSANTACLAUSSANTA

      (ii)   HOHOHOHOHOHOHOHOHOHOHO

      (iii) ↗↖ ↗↖ ↗↖ ↗↖ ↗↖ ↗↖

      (iv) ↗   ↗   ↗   ↗   ↗   ↗   ↗
               ↘   ↘   ↘   ↘   ↘   ↘   ↘

      (v) ↗↖ ↗↖ ↗↖ ↗↖ ↗↖ ↗↖
           ↘↙ ↘↙ ↘↙ ↘↙ ↘↙ ↘↙

      (vi) ↗↖   ↗↖   ↗↖   ↗↖
             ↘↙   ↘↙   ↘↙

      (vii) ↗↗↗↗↗↗↗↗↗↗↗
           ↘↘↘↘↘↘↘↘↘↘↘↘

7

Commutative Rings II

7.1 PRIME IDEALS AND MAXIMAL IDEALS

Our main interest in this chapter is the study of polynomials in several variables over a field k. One sees in analytic geometry that polynomials correspond to geometric figures; for example, $f(x, y) = x^2/a^2 + y^2/b^2 - 1$ is intimately related to an ellipse in the plane $\mathbb{R}^2$. But there is a very strong connection between the rings $k[x_1, \ldots, x_n]$ and the geometry of subsets of k^n going far beyond this. Given a set of polynomials $f_1, \ldots, f_t$ of n variables, call the subset $V \subseteq k^n$ consisting of their common zeros an *algebraic set*. Of course, one can study algebraic sets because solutions of systems of polynomial equations (an obvious generalization of systems of linear equations) are intrinsically interesting, but they do arise quite naturally. For example, a problem may lead to a parametrization of its solutions by an algebraic set, and so understanding the algebraic set and its properties (e.g., irreducibility, dimension, genus, singularities, and so forth) leads to an understanding of the original problem. The interplay between $k[x_1, \ldots, x_n]$ and algebraic sets has evolved into what is nowadays called *Algebraic Geometry*, and this chapter may be regarded as an introduction to this subject.

As usual, it is simpler to begin by looking at a more general setting–in this case, commutative rings–before getting involved with polynomial rings. A great deal of the number theory we have presented involves divisibility: given two integers a and b, when does $a \mid b$; that is, when is b a multiple of a? This question translates into a question about principal ideals, for $a \mid b$ if and only if $(b) \subseteq (a)$. We begin investigating ideals by proving the analog of Theorem 2.123, the correspondence theorem for groups. Recall that if $f : X \to Y$ is a function and $B \subseteq Y$ is a subset, then its inverse image is

$$f^{-1}(B) = \{x \in X : f(x) \in B\}.$$

Proposition 7.1 (Correspondence Theorem for Rings). *If I is a proper ideal in a commutative ring R, then the natural map $\pi \colon R \to R/I$ induces an inclusion-preserving bijection π' from the set of all intermediate ideals J (that is, $I \subseteq J \subseteq R$), to the set of all the ideals in R/I, given by*

$$\pi' \colon J \mapsto J/I = \{a + I : a \in J\}.$$

Thus, every ideal in the quotient ring R/I has the form J/I for some unique intermediate ideal J.

Proof. If one forgets its multiplication, the commutative ring R is an additive abelian group and its ideals I are (normal) subgroups. The correspondence theorem for groups, Theorem 2.123, now applies, and it gives an inclusion-preserving bijection

$$\pi_* \colon \{\text{all subgroups of } R \text{ containing } I\} \to \{\text{all subgroups of } R/I\},$$

where $\pi_*(J) = J/I$.

If J is an ideal, then $\pi_*(J)$ is also an ideal, for if $r \in R$ and $a \in J$, then $ra \in J$, and so

$$(r + I)(a + I) = ra + I \in J/I.$$

Let π' be the restriction of π_* to the set of intermediate ideals J. Now π' is an injection because π_* is a bijection. To see that π' is surjective, let J^* be an ideal in R/I. Then $\pi^{-1}(J^*)$ is an intermediate ideal in R, by Exercise 3.47 on page 251 [it contains $I = \pi^{-1}(\{0\})$], and $\pi'(\pi^{-1}(J^*)) = \pi^{-1}(J^*)/I = \pi(\pi^{-1}(J^*)) = J^*$, by Lemma 2.14. Thus, if $J = \pi^{-1}(J^*)$, then $J^* = J/I$. •

Example 7.2.
Let $I = (m)$ be a nonzero ideal in $\mathbb{Z}$. Every ideal J in $\mathbb{Z}$ is principal, say, $J = (a)$, and $(m) \subseteq (a)$ if and only if $a \mid m$. By the correspondence theorem, every ideal in $\mathbb{I}_m$ has the form $([a])$ for some divisor a of m. ◀

We now introduce two especially interesting types of ideal: *prime ideals*, which are related to Euclid's lemma, and *maximal ideals*.

Definition. An ideal I in a commutative ring R is called a ***prime ideal*** if it is a proper ideal, that is, $I \neq R$, and $ab \in I$ implies $a \in I$ or $b \in I$.

Example 7.3.

(i) Recall that a nonzero commutative ring R is a domain if and only if $ab = 0$ in R implies $a = 0$ or $b = 0$. Thus, the ideal $\{0\}$ in R is a prime ideal if and only if R is a domain.

(ii) We claim that the prime ideals in $\mathbb{Z}$ are precisely the ideals (p), where either $p = 0$ or p is a prime. Since m and $-m$ generate the same principal ideal, we may restrict our attention to nonnegative generators. If $p = 0$, then the result follows from part (i), for $\mathbb{Z}$ is a domain. If p is a prime, we show first that (p) is a proper ideal; otherwise, $1 \in (p)$, and there would be an integer a with $ap = 1$, a contradiction. Next, if $ab \in (p)$, then $p \mid ab$. By Euclid's lemma, either $p \mid a$ or $p \mid b$; that is, either $a \in (p)$ or $b \in (p)$. Therefore, (p) is a prime ideal.

Conversely, if $m > 1$ is not a prime, then it has a factorization $m = ab$ with $0 < a < m$ and $0 < b < m$. Thus, neither a nor b is a multiple of m, hence neither lies in (m), and so (m) is not a prime ideal. ◀

The proof in the example works in more generality.

Proposition 7.4. *If k is a field, then $(p(x))$ is a prime ideal if and only if either $p(x) = 0$ or $p(x)$ is irreducible in $k[x]$.*

Proof. If $p(x)$ is a nonzero polynomial which is not irreducible, there is a factorization

$$p(x) = a(x)b(x)$$

with $\deg(a) < \deg(p)$ and $\deg(b) < \deg(p)$. Since every nonzero polynomial $g(x) \in (p)$ has the form $g(x) = d(x)p(x)$ for some $d(x) \in k[x]$, we have $\deg(g) \geq \deg(p)$. It follows that neither $a(x)$ nor $b(x)$ lies in (p), and so (p) is not a prime ideal.

Conversely, if $p(x) = 0$, then $(p(x)) = \{0\}$, which is a prime ideal (because $k[x]$ is a domain). Suppose that $p(x)$ is irreducible. First, (p) is a proper ideal; otherwise, $R = (p)$ and hence $1 \in (p)$, so there is a polynomial $f(x)$ with $1 = p(x)f(x)$. But $p(x)$ has degree at least 1, whereas

$$0 = \deg(1) = \deg(pf) = \deg(p) + \deg(f) \geq \deg(p) \geq 1.$$

This contradiction shows that (p) is a proper ideal.

Second, if $ab \in (p)$, then $p \mid ab$, and so Euclid's lemma in $k[x]$ gives $p \mid a$ or $p \mid b$. Thus, $a \in (p)$ or $b \in (p)$. It follows that (p) is a prime ideal. •

Proposition 7.5. *A proper ideal I in a commutative ring R is a prime ideal if and only if R/I is a domain.*

Proof. Let I be a prime ideal. Since I is a proper ideal, we have $1 \notin I$ and so $1 + I \neq 0 + I$ in R/I. If $0 = (a + I)(b + I) = ab + I$, then $ab \in I$. Since I is a prime ideal, either $a \in I$ or $b \in I$; that is, either $a + I = 0$ or $b + I = 0$. Hence, R/I is a domain. The converse is just as easy. •

Here is a second interesting type of ideal.

Definition. An ideal I in a commutative ring R is a ***maximal ideal*** if I is a proper ideal and there is no ideal J with $I \subsetneqq J \subsetneqq R$.

Thus, if I is a maximal ideal in a commutative ring R and if J is a proper ideal with $I \subseteq J$, then $I = J$.

The prime ideals in the polynomial ring $k[x_1, \ldots, x_n]$ can be quite complicated, but when k is algebraically closed, Hilbert's Nullstellensatz (Theorem 7.45) says that every maximal ideal has the form $(x_1 - a_1, \ldots, x_n - a_n)$ for some point $(a_1, \ldots, a_n) \in k^n$.

We may restate Proposition 3.43 in the present language.

Lemma 7.6. *The ideal $\{0\}$ is a maximal ideal in a commutative ring R if and only if R is a field.*

Proof. It is shown in Proposition 3.43 that every nonzero ideal I in R is equal to R itself if and only if every nonzero element in R is a unit. That is, $\{0\}$ is a maximal ideal if and only if R is a field. •

Proposition 7.7. *A proper ideal I in a commutative ring R is a maximal ideal if and only if R/I is a field.*

Proof. The correspondence theorem for rings shows that I is a maximal ideal if and only if R/I has no ideals other than $\{0\}$ and R/I itself; Lemma 7.6 shows that this property holds if and only if R/I is a field. •

Corollary 7.8. *Every maximal ideal I in a commutative ring R is a prime ideal.*

Proof. If I is a maximal ideal, then R/I is a field. Since every field is a domain, R/I is a domain, and so I is a prime ideal. •

Example 7.9.
The converse of the last corollary is false. For example, consider the principal ideal (x) in $\mathbb{Z}[x]$. By Exercise 3.93 on page 303, we have

$$\mathbb{Z}[x]/(x) \cong \mathbb{Z};$$

since $\mathbb{Z}$ is a domain, (x) is a prime ideal; since $\mathbb{Z}$ is not a field, (x) is not a maximal ideal.

It is not difficult to exhibit a proper ideal J strictly containing (x); let

$$J = \{f(x) \in \mathbb{Z}[x] \colon f(x) \text{ has even constant term}\}.$$

Since $\mathbb{Z}[x]/J \cong \mathbb{F}_2$, which is a field, it follows that J is a maximal ideal containing (x). ◄

Corollary 7.10.   *If k is a field, then $(x_1 - a_1, \ldots, x_n - a_n)$ is a maximal ideal in $k[x_1, \ldots, x_n]$, where $a_i \in k$ for $i = 1, \ldots, n$.*

Proof.   By Theorem 3.33, there is a unique homomorphism

$$\varphi : k[x_1, \ldots, x_n] \to k[x_1, \ldots, x_n]$$

with $\varphi(c) = c$ for all $c \in k$ and with $\varphi(x_i) = x_i - a_i$ for all i. It is easy to see that φ is an isomorphism, for its inverse carries $x_i \mapsto x_i + a_i$ for all i. It follows that I is a maximal ideal in $k[x_1, \ldots, x_n]$ if and only if $\varphi(I)$ is a maximal ideal. But $(x_1, \ldots, x_n)$ is a maximal ideal, for $k[x_1, \ldots, x_n]/(x_1, \ldots, x_n) \cong k$ is a field. Therefore, $(x_1 - a_n, \ldots, x_n - a_n)$ is a maximal ideal.   •

The converse of Corollary 7.8 is true when R is a PID.

Theorem 7.11.   *If R is a PID, then every nonzero prime ideal I is a maximal ideal.*

Proof.   Assume there is a proper ideal J with $I \subseteq J$. Since R is a PID, $I = (a)$ and $J = (b)$ for some $a, b \in R$. Now $a \in J$ implies that $a = rb$ for some $r \in R$, and so $rb \in I$; but I is a prime ideal, so that $r \in I$ or $b \in I$. If $r \in I$, then $r = sa$ for some $s \in R$, and so $a = rb = sab$. Since R is a domain, $1 = sb$, and Exercise 3.24 on page 235 gives $J = (b) = R$, contradicting J being a proper ideal. If $b \in I$, then $J \subseteq I$, and so $J = I$. Therefore, I is a maximal ideal.   •

We can now give a second proof of Proposition 3.112.

Corollary 7.12.   *If k is a field and $p(x) \in k[x]$ is irreducible, then the quotient ring $k[x]/(p(x))$ is a field.*

Proof.   Since $p(x)$ is irreducible, Proposition 7.4 says that the principal ideal $I = (p(x))$ is a nonzero prime ideal; since $k[x]$ is a PID, I is a maximal ideal, and so $k[x]/I$ is a field.   •

Does every commutative ring R contain a maximal ideal? The (positive) answer to this question involves *Zorn's lemma*, a theorem equivalent to the Axiom of Choice, which is usually discussed in a sequel course (but see Corollary 7.27).

EXERCISES

H **7.1** True or false with reasons.

    (i)   If R is a commutative ring and I is an ideal in R, forget their multiplication and apply the correspondence theorem for (additive) groups to R and R/I. The subgroups of R containing I which are ideals in R correspond to the subgroups of R/I which are ideals in R/I.

(ii) If R is a commutative ring and I is an ideal in R, forget their multiplication and apply the correspondence theorem for (additive) groups to R and R/I. The subgroups of R containing I which are not ideals in R correspond to the subgroups of R/I which are not ideals in R/I.

(iii) If I is an ideal in a commutative ring R and if $ab \in I$, where $a, b \in R$, then either $a \in I$ or $b \in I$.

(iv) If I is an ideal in a commutative ring R and if $a^2 \in I$, where $a \in R$, then $a \in I$.

(v) If k is a field and $p(x) \in k[x]$ is irreducible, then $(p(x))$ is a prime ideal in $k[x]$.

(vi) Every prime ideal in $\mathbb{Z}$ is a maximal ideal.

(vii) If R is a commutative ring, k is a field, and $\varphi: R \to k$ is a homomorphism, then $\ker \varphi$ is a maximal ideal in R.

(viii) If R is a commutative ring, k is a field, and $\varphi: R \to k$ is a homomorphism, then $\ker \varphi$ is a prime ideal in R.

(ix) If R is a commutative ring, k is a field, and $\varphi: R \to k$ is a surjective homomorphism, then $\ker \varphi$ is a maximal ideal in R.

(x) $(x, y - 1, z + 1)$ is a maximal ideal in $\mathbb{F}_3[x, y, z]$.

7.2 (i) Find all the maximal ideals in $\mathbb{Z}$.

(ii) Find all the maximal ideals in $k[x]$, where k is a field.

(iii) Find all the maximal ideals in $k[[x]]$, where k is a field.

H **7.3** Recall that a Boolean ring is a commutative ring R for which $a^2 = a$ for all $a \in R$. Prove that every prime ideal in a Boolean ring is a maximal ideal.

*__7.4__ (i) Give an example of a commutative ring containing two prime ideals P and Q for which $P \cap Q$ is not a prime ideal.

(ii) If $P_1 \supseteq P_2 \supseteq \cdots P_n \supseteq P_{n+1} \supseteq \cdots$ is a decreasing sequence of prime ideals in a commutative ring R, prove that $\bigcap_{n \geq 1} P_n$ is a prime ideal.

7.5 Let $f: R \to S$ be a ring homomorphism.

(i) If Q is a prime ideal in S, prove that $f^{-1}(Q)$ is a prime ideal in R. Conclude, in the correspondence theorem, that if J/I is a prime ideal in R/I, where $I \subseteq J \subseteq R$, then J is a prime ideal in R.

H (ii) Give an example to show that if P is a prime ideal in R, then $f(P)$ need not be a prime ideal in S.

7.6 Let k be a field, and let $a = (a_1, \dots, a_n) \in k^n$. Define the *evaluation map* $e_a: k[x_1, \dots, x_n] \to k$ by $e_a: f(x_1, \dots, x_n) \mapsto f(a) = f(a_1, \dots, a_n)$.

(i) Prove that $\ker e_a$ is a maximal ideal in $k[x_1, \dots, x_n]$ by showing that e_a is surjective.

(ii) Prove that $(x_1 - a_1, \dots, x_n - a_n)$ is a maximal ideal in $k[x_1, \dots, x_n]$ by showing that $\ker e_a = (x_1 - a_1, \dots, x_n - a_n)$. (This is a second proof of Corollary 7.10.)

7.7 (i) Find all the maximal ideals in $k[x]$, where k is an algebraically closed field.

(ii) Find all the maximal ideals in $\mathbb{R}[x]$.

(iii) If k is an algebraically closed field, prove that the function

$$k \to \{\text{maximal ideals in } k[x]\},$$

given by $a \mapsto (x - a)$, the principal ideal in $k[x]$ generated by $x - a$, is a bijection.

7.8   **(i)**   Prove that if $x_i - b \in (x_1 - a_1, \ldots, x_n - a_n)$ for some i, where k is a field and $b \in k$, then $b = a_i$.

   (ii)   Prove that $\mu : k^n \to \{\text{maximal ideals in } k[x_1, \ldots, x_n]\}$, given by

$$\mu : (a_1, \ldots, a_n) \mapsto (x_1 - a_1, \ldots, x_n - a_n),$$

is an injection, and give an example of a field k for which μ is not a surjection.

7.9   Prove that if P is a prime ideal in a commutative ring R and if $r^n \in P$ for some $r \in R$ and $n \geq 1$, then $r \in P$.

7.10   Prove that the ideal $(x^2 - 2, y^2 + 1, z)$ in $\mathbb{Q}[x, y, z]$ is a proper ideal.

7.11   Call a nonempty subset S of a commutative ring R *multiplicatively closed* if $0 \notin S$ and, if $s, s' \in S$, then $ss' \in S$. Prove that an ideal I which is maximal with the property that $I \cap S = \varnothing$ is a prime ideal. (The existence of such an ideal I can be proved using Zorn's lemma.)

*7.12   **(i)**   If I and J are ideals in a commutative ring R, define

$$IJ = \Big\{ \sum_\ell a_\ell b_\ell : a_\ell \in I \text{ and } b_\ell \in J \Big\}.$$

Prove that IJ is an ideal in R and that $IJ \subseteq I \cap J$.

   (ii)   Let $R = k[x, y]$, where k is a field and let $I = (x, y) = J$. Prove that $I^2 = IJ \subsetneq I \cap J = I$.

7.13   Let P be a prime ideal in a commutative ring R. If there are ideals I and J in R with $IJ \subseteq P$, prove that $I \subseteq P$ or $J \subseteq P$.

7.14   If I and J are ideals in a commutative ring R, define the *colon ideal*

$$(I : J) = \{r \in R : rJ \subseteq I\}.$$

   (i)   Prove that $(I : J)$ is an ideal containing I.

   (ii)   Let R be a domain and let $a, b \in R$, where $b \neq 0$. If $I = (ab)$ and $J = (b)$, prove that $(I : J) = (a)$.

7.15   Let I and J be ideals in a commutative ring R.

   (i)   Prove that there is an injection $R/(I \cap J) \to (R/I) \times (R/J)$ given by $\varphi : r \mapsto (r + I, r + J)$.

   H **(ii)**   Call I and J *coprime* if $I + J = R$. Prove that the ring homomorphism $\varphi : R/(I \cap J) \to (R/I) \times (R/J)$ is an isomorphism if I and J are coprime.

   (iii)   Let R be a commutative ring and let $I_1, \ldots, I_n$ be pairwise coprime ideals; that is, I_i and I_j are coprime for all $i \neq j$. Prove that

$$R/(I_1 \cap \cdots \cap I_n) \cong (R/I_1) \times \cdots \times (R/I_n).$$

   (iv)   Generalize the *Chinese Remainder Theorem*. If R is a commutative ring and $I_1, \ldots, I_n$ are pairwise coprime ideals, prove that if $a_1, \ldots, a_n \in R$, then there exists $r \in R$ with $r + I_i = a_i + I_i$ for all i.

7.16   A commutative ring R is called a *local ring* if it has a unique maximal ideal.

   (i)   If p is a prime, prove that

$$\{a/b \in \mathbb{Q} : p \nmid b\}$$

is a local ring.

(ii)   If k is a field, prove that $k[[x]]$ is a local ring.

H **(iii)**   If R is a local ring with unique maximal ideal M, prove that $a \in R$ is a unit if and only if $a \notin M$.

7.2   UNIQUE FACTORIZATION

We have proved unique factorization theorems in $\mathbb{Z}$ and in $k[x]$, where k is a field. In fact, we have proved a common generalization of these two results: every Euclidean ring has unique factorization. Our aim now is to generalize this result, first to general PID's, and then to $R[x]$, where R is a ring having unique factorization. It will then follow that there is unique factorization in the ring $k[x_1, \dots, x_n]$ of all polynomials in several variables over a field k. One immediate consequence is that any two polynomials in several variables have a gcd.

We begin by generalizing some earlier definitions. Recall that elements a and b in a commutative ring R are ***associates*** if there exists a unit $u \in R$ with $b = ua$. For example, in $\mathbb{Z}$, the units are ± 1, and so the only associates of an integer m are $\pm m$; in $k[x]$, where k is a field, the units are the nonzero constants, and so the only associates of a polynomial $f(x) \in k[x]$ are the polynomials $uf(x)$, where $u \in k$ and $u \neq 0$. The only units in $\mathbb{Z}[x]$ are ± 1, and so the only associates of a polynomial $f(x) \in \mathbb{Z}[x]$ are $\pm f(x)$.

Consider two principal ideals (a) and (b) in a commutative ring R. It is easy to see that the following are equivalent: $b \mid a$; $a \in (b)$; $a = rb$ for some $r \in R$; $a \in (b)$; $(a) \subseteq (b)$. We can say more when R is a domain.

Proposition 7.13.   *Let R be a domain and let $a, b \in R$.*

(i)   *$a \mid b$ and $b \mid a$ if and only if a and b are associates.*

(ii)   *The principal ideals (a) and (b) are equal if and only if a and b are associates.*

(iii)   *$(a) \subseteq (b)$ if and only if $b \mid a$; that is, $a = cb$ for some $c \in R$. The inclusion is proper, $(a) \subsetneq (b)$, if and only if b is a **proper divisor** of a; that is, neither c nor b is a unit.*

Proof.
(i) This is Proposition 3.15.
(ii) If $(a) = (b)$, then $(a) \subseteq (b)$ and $(b) \subseteq (a)$; hence, $a \in (b)$ and $b \in (a)$. Thus, $a \mid b$ and $b \mid a$; by part (i), a and b are associates. The converse is easy, and one does not need to assume that R is a domain to prove it.
(iii) If $b \mid a$, then $a = cb$ for some $c \in R$; if $x \in (a)$, then $x = ra = rcb \in (b)$ for some $r \in R$, and so $(a) \subseteq (b)$. Conversely, if $(a) \subseteq (b)$, then $a \in (b)$ and $a = cb$; hence, $b \mid a$.

Assume that $(a) \subsetneq (b)$, so that $a = cb$. If c is a unit, then a and b are associates; hence, $(a) = (b)$, by part (i), and this is a contradiction. If b is a unit, then a is also a unit; but any two units are associates, so that a and b are associates, another contradiction. Thus, b is a proper divisor of a.

Conversely, assume that b is a proper divisor of a. That $b \mid a$ gives an inclusion $(a) \subseteq (b)$; if this inclusion is not proper, then $(a) = (b)$ and a and b are associates. But a proper divisor is never an associate, as the reader may easily show, and so $(a) \subsetneq (b)$. •

The notions of prime number in $\mathbb{Z}$ or irreducible polynomial in $k[x]$, where k is a field, have a common generalization.

Definition.   A element p in a commutative ring R is **irreducible** if it is neither 0 nor a unit and if its only factors are associates of p or units.

Thus, if $p \in R$ is neither 0 nor a unit, then p is irreducible if and only if it has no proper divisors. For example, the irreducibles in $\mathbb{Z}$ are the numbers $\pm p$, where p is a prime, and the irreducibles in $k[x]$, where k is a field, are the irreducible polynomials $p(x)$; that is, $\deg(p) \geq 1$ and $p(x)$ has no factorization $p(x) = f(x)g(x)$ where $\deg(f) < \deg(p)$ and $\deg(g) < \deg(g)$. This characterization of irreducible polynomial does not persist in rings $R[x]$ when R is not a field. For example, in $\mathbb{Z}[x]$, the polynomial $f(x) = 2x + 2$ cannot be factored into two polynomials, each having degree smaller than $\deg(f) = 1$, yet $f(x)$ is not irreducible, for in the factorization $2x + 2 = 2(x + 1)$, neither 2 nor $x + 1$ is a unit.

Definition.   If R is a commutative ring, then an element $r \in R$ is a **product of irreducibles** if r is neither 0 nor a unit and there exist irreducible $p_1, \ldots, p_n$, where $n \geq 1$, with $r = p_1 \cdots p_n$.

Note the instance of the definition when $n = 1$; every irreducible element in R is a product of irreducibles (it is a product with one factor!).

Here is the definition we have been seeking.

Definition.   A domain R is a **unique factorization domain** (**UFD**) if

(i) every $r \in R$, neither 0 nor a unit, is a product of irreducibles;

(ii) if $p_1 \cdots p_m = q_1 \cdots q_n$, where p_i and q_j are irreducible, then $m = n$ and there is a permutation $\sigma \in S_n$ with p_i and $q_{\sigma(i)}$ associates for all i.

When we proved that $\mathbb{Z}$ and $k[x]$, for k a field, have unique factorization into irreducibles, we did not mention associates because, in each case, irreducible elements were always replaced by favorite choices of associates: in $\mathbb{Z}$, *positive* irreducibles (i.e., primes) are chosen; in $k[x]$, *monic* irreducible polynomials are chosen. The reader should see, for example, that the statement: "$\mathbb{Z}$ is a UFD" is just a restatement of the Fundamental Theorem of Arithmetic.

The proof that every PID is a UFD uses a new idea: chains of ideals.

Lemma 7.14.   *Let R be a commutative ring.*

(i) *If*
$$I_1 \subseteq I_2 \subseteq \cdots \subseteq I_n \subseteq I_{n+1} \subseteq \cdots$$
is an ascending chain of ideals in R, then $\bigcup_{n=1}^{\infty} I_n$ is an ideal.

(ii) *If R is a PID, then there is no infinite strictly ascending chain of ideals*
$$I_1 \subsetneq I_2 \subsetneq \cdots \subsetneq I_n \subsetneq I_{n+1} \subsetneq \cdots .$$

(iii) *If R is a PID and $r \in R$ is neither 0 nor a unit, then r is a product of irreducibles.*

Proof.
(i) Define $J = \bigcup_{n=1}^{\infty} I_n$. If $a \in J$, then $a \in I_n$ for some n; if $r \in R$, then $ra \in I_n$, because I_n is an ideal; hence, $ra \in J$. If $a, b \in J$, then there are ideals I_n and I_m with $a \in I_n$ and $b \in I_m$; since the chain is ascending, we may assume that $I_n \subseteq I_m$, and so $a, b \in I_m$. As I_m is an ideal, $a - b \in I_m$ and, hence, $a - b \in J$. Therefore, J is an ideal.

(ii) If, on the contrary, an infinite strictly ascending chain exists, then define $J = \bigcup_{n=1}^{\infty} I_n$; by part (i), J is an ideal. Since R is a PID, we have $J = (d)$ for some $d \in J$. Now d got into J by being in I_n for some n. Hence
$$J = (d) \subseteq I_n \subsetneq I_{n+1} \subseteq J,$$
and this is a contradiction.

(iii) Call a nonzero nonunit $a \in R$ *good* if it is a product of irreducibles; otherwise, call a *bad*. We must show that there are no bad elements. If a is bad, it is not irreducible, and so $a = rs$, where both r and s are proper divisors. But the product of good elements is good, and so at least one of the factors, say r, is bad. Proposition 7.13(iii) shows that $(a) \subsetneq (r)$. It follows, by induction, that there exists a sequence $a = a_1, r = a_2, \ldots, a_n, \ldots$ of bad elements with each a_{n+1} a proper divisor of a_n, and this sequence yields a strictly ascending chain
$$(a_1) \subsetneq (a_2) \subsetneq \cdots \subsetneq (a_n) \subsetneq (a_{n+1}) \subsetneq \cdots ,$$
contradicting part (ii) of this lemma.   •

Proposition 7.15.   *Let R be a domain in which every $r \in R$, neither 0 nor a unit, is a product of irreducibles. Then R is a UFD if and only if, for every irreducible element $p \in R$, the principal ideal (p) is a prime ideal in R.*

Proof.   Assume that R is a UFD. If $a, b \in R$ and $ab \in (p)$, then there is $r \in R$ with

$$ab = rp.$$

Factor each of a, b, and r into irreducibles; by unique factorization, the left side of the equation must involve an associate of p. This associate arose as a factor of a or b, and hence $a \in (p)$ or $b \in (p)$.

The proof of the converse is merely an adaptation of the proof of the Fundamental Theorem of Arithmetic. Assume that

$$p_1 \cdots p_m = q_1 \cdots q_n, \tag{1}$$

where the p's and q's are irreducible elements. Let us prove, by induction on $\max\{m, n\} \geq 1$, that $n = m$ and the q's can be reindexed so that q_i and p_i are associates for all i. The base step $\max\{m, n\} = 1$ has $p_1 = q_1$, and the result is obviously true. For the inductive step, the given equation shows that $p_1 \mid q_1 \cdots q_n$. By hypothesis, (p_1) is a prime ideal (which is the analog of Euclid's lemma), and so there is some q_j with $p_1 \mid q_j$. But q_j, being irreducible, has no divisors other than units and associates, so that q_j and p_1 are associates: $q_j = up_1$ for some unit u. Canceling p_1 from both sides of Eq. (1), we have $p_2 \cdots p_m = uq_1 \cdots \widehat{q_j} \cdots q_n$. By the inductive hypothesis, $m - 1 = n - 1$, hence $m = n$, and, after possible reindexing, q_i and p_i are associates for all i.   •

Theorem 7.16.   *If R is a PID, then R is a UFD. In particular, every Euclidean ring is a UFD.*

Proof.   In view of the last two results, it suffices to prove that (p) is a prime ideal whenever p is irreducible. Suppose there is a proper ideal I with $(p) \subsetneq I$. Since R is a PID, $I = (b)$ for some $b \in R$, and b is not a unit. But Proposition 7.13(iii) shows that b is a proper divisor of p, contradicting p being irreducible. Therefore, (p) is a maximal ideal, and hence it is a prime ideal.   •

Recall that the notion of gcd can be defined in any commutative ring.

Definition.   Let R be a commutative ring and let $a_1, \ldots, a_n \in R$. A *common divisor* of $a_1, \ldots, a_n$ is an element $c \in R$ with $c \mid a_i$ for all i. A *greatest common divisor* or *gcd* of $a_1, \ldots, a_n$ is a common divisor d with $c \mid d$ for every common divisor c.

Even in the familiar examples of $\mathbb{Z}$ and $k[x]$, gcd's are not unique unless an extra condition is imposed. For example, if d is a gcd of a pair of integers in $\mathbb{Z}$, as defined above, then $-d$ is also a gcd. To force gcd's to be unique, one defines nonzero gcd's in $\mathbb{Z}$ to be positive; similarly, in $k[x]$, where k is a field, one imposes the condition that nonzero gcd's are monic polynomials. In a general PID, however, elements may not have favorite associates.

If R is a domain and d and d' are gcd's of elements $a_1, \ldots, a_n$, then each is a common divisor, and so $d \mid d'$ and $d' \mid d$. It follows from Proposition 7.13 that d and d' are associates and, hence, that $(d) = (d')$. Thus gcd's, if they exist, are not unique, but they all generate the same principal ideal.

In Exercise 3.81 on page 275, we saw that there exist domains R containing a pair of elements having no gcd. We now show that gcd's always exist in UFDs.

Proposition 7.17. *If R is a UFD, then the gcd of any finite set of elements $a_1, \ldots, a_n$ in R exists.*

Proof.   It suffices to prove that the gcd of two elements a and b exists, for an easy inductive proof then shows that a gcd of any finite number of elements exists. We adapt the proof of Proposition 1.55.

There are units u and v and distinct irreducibles $p_1, \ldots, p_t$ with

$$a = up_1^{e_1} p_2^{e_2} \cdots p_t^{e_t}$$

and

$$b = vp_1^{f_1} p_2^{f_2} \cdots p_t^{f_t},$$

where $e_i \geq 0$ and $f_i \geq 0$ for all i (as usual, allowing some exponents to be 0 permits us to use the same irreducible factors in both factorizations). It is easy to see that if $c \mid a$, then the factorization of c into irreducibles is $c = wp_1^{g_1} p_2^{g_2} \cdots p_t^{g_t}$, where w is a unit and $g_i \leq e_i$ for all i. Thus, c is a common divisor of a and b if and only if $g_i \leq m_i$ for all i, where

$$m_i = \min\{e_i, f_i\}.$$

It is now clear that $p_1^{m_1} p_2^{m_2} \cdots p_t^{m_t}$ is a gcd of a and b.   •

We have not proved that a gcd of elements $a_1, \ldots, a_n$ in a UFD is a linear combination of them; indeed, this is false in $k[x, y]$ (see Exercise 7.23 on page 534).

Definition.   Elements $a_1, \ldots, a_n$ in a UFD are called ***relatively prime*** if every common divisor of $a_1, \ldots, a_n$ is a unit.

We are now going to prove that if R is a UFD, then so is $R[x]$. This theorem was essentially found by Gauss, and the proof uses ideas in the proof of Gauss's theorem (Theorem 3.96). It will follow that $k[x_1, \ldots, x_n]$ is a UFD whenever k is a field.

Definition.   A polynomial $f(x) = a_n x^n + \cdots + a_1 x + a_0 \in R[x]$, where R is a UFD, is called ***primitive*** if its coefficients are relatively prime; that is, the only common divisors of $a_n, \ldots, a_1, a_0$ are units.

If $f(x)$ is not primitive, then there exists an irreducible $q \in R$ that divides each of its coefficients, for if the gcd is a nonunit d, then any irreducible factor q of d will serve.

Example 7.18.
We now show that if R is a UFD, then every irreducible $p(x) \in R[x]$ of positive degree is primitive. If not, then there is an irreducible $q \in R$ with $p(x) = qg(x)$. Since $p(x)$ is irreducible in $R[x]$, its only factors are units in $R[x]$ and associates in $R[x]$. Now q is irreducible in R; can it be a unit in $R[x]$? If there is $f \in R[x]$ with $qf = 1$, then $0 = \deg(1) = \deg(qf) = \deg(q) + \deg(f)$; hence, $\deg(f) = 0$ and $f \in R$. It follows that q is a unit in R, contradicting q being irreducible in R. Therefore, q is not a unit in $R[x]$, and so q is an associate of $p(x)$. But associates in $R[x]$ have the same degree (because units have degree 0). Therefore, $\deg(p) = 0$, contradicting $p(x)$ having positive degree. We conclude that $p(x)$ is primitive. ◄

Here is a generalization of Gauss's lemma (Theorem 3.92).

Lemma 7.19.   *If R is a UFD and $f(x)$, $g(x) \in R[x]$ are both primitive, then their product $f(x)g(x)$ is also primitive.*

Proof.   If, on the contrary, the product $f(x)g(x)$ is not primitive, then there is some irreducible p in R dividing all its coefficients. Define $\pi: R \to R/(p)$ to be the natural map $\pi: a \mapsto a + (p)$; Theorem 3.33 shows that the function $\widetilde{\pi}: R[x] \to (R/(p))[x]$, which replaces each coefficient c of a polynomial by $\pi(c)$, is a ring homomorphism. That $f(x)g(x)$ is not primitive says that $0 = \widetilde{\pi}(fg) = \widetilde{\pi}(f)\widetilde{\pi}(g)$ in $(R/(p))[x]$. Since (p) is a prime ideal, $R/(p)$ is a domain, and hence $(R/(p))[x]$ is also a domain. But, neither $\widetilde{\pi}(f)$ nor $\widetilde{\pi}(g)$ is 0 in $(R/(p))[x]$, because f and g are primitive, and this contradicts $(R/(p))[x]$ being a domain. •

Definition.   If R is a UFD and $f(x) = a_n x^n + \cdots + a_1 x + a_0 \in R[x]$, define $c(f) \in R$ to be a gcd of $a_n, \ldots, a_1, a_0$; one calls $c(f)$ a **content** of $f(x)$.

Note that a content of a polynomial $f(x)$ is not unique, but that any two contents of $f(x)$ are associates. Whenever a polynomial f over a UFD is given, we will let $c(f)$ denote some arbitrary choice of content for f.

Lemma 7.20.   *Let R be a UFD.*

(i) *Every nonzero $f(x) \in R[x]$ has a factorization*

$$f(x) = c(f)f^*(x),$$

where $c(f) \in R$ is a content of f and $f^*(x) \in R[x]$ is primitive.

(ii) *This factorization is unique in the sense that if $f(x) = dg^*(x)$, where $d \in R$ and $g^*(x) \in R[x]$ is primitive, then d and $c(f)$ are associates and $f^*(x)$ and $g^*(x)$ are associates.*

(iii) *Let $g^*(x)$, $f(x) \in R[x]$. If $g^*(x)$ is primitive and $g^*(x) \mid bf(x)$, where $b \in R$, then $g^*(x) \mid f(x)$.*

Proof.

(i) If $f(x) = a_n x^n + \cdots + a_1 x + a_0$ and $c(f)$ is a content of f, then there are factorizations $a_i = c(f)b_i$ in R for $i = 0, 1, \ldots, n$; if we define $f^*(x) = b_n x^n + \cdots + b_1 x + b_0$, then $f(x) = c(f)f^*(x)$ and $f^*(x)$ is primitive.

(ii) To prove uniqueness, it suffices to show that both $c(f)$ and d are gcd's of the coefficients $a_n, \ldots, a_1, a_0$ of $f(x)$, for then they are associates. It will then follow that $g^*(x)$ and $f^*(x)$ are associates: if $d = uc(f)$, where u is a unit, then $c(f)f^* = f = dg^* = uc(f)g^*$ and $f^* = ug^*$.

The hypothesis $f(x) = dg^*(x)$ shows that d is a common divisor of the coefficients of f. Now $c(f)$ is a gcd of these coefficients, so that $c(f) \mid d$. If $c(f)$ is a proper divisor, then $d = rc(f)$, where $r \in R$ is not a unit. Hence, $f(x) = c(f)f^*(x) = drf^*(x)$. On the other hand, $f(x) = dg^*(x)$, so that $g^*(x) = rf^*(x)$ is not primitive. This contradiction shows that d is also a gcd of the coefficients of f, and so $c(f)$ and d are associates.

(iii) Since $g^*(x) \mid bf(x)$, there is $h(x) \in R[x]$ with

$$bf(x) = g^*(x)h(x). \tag{2}$$

By part (i), we have $h(x) = c(h)h^*(x)$, where h^* is primitive. Substituting in Eq. (2),

$$bf(x) = c(h)g^*(x)h^*(x).$$

It follows that b divides each coefficient of $c(h)g^*(x)h^*(x)$; that is, b is a common divisor of these coefficients. But $g^*(x)h^*(x)$ is primitive, by Lemma 7.19, and so $c(h)$ is a content of $c(h)g^*(x)h^*(x)$, by part (ii). Thus, $c(h)$ is a gcd of the coefficients of $c(h)g^*(x)h^*(x)$, and so $b \mid c(h)$. Hence, $c(h) = ba$ for some $a \in R$, and $bf(x) = c(h)g^*(x)h^*(x) = bag^*(x)h^*(x)$. Canceling b, we have $f(x) = ag^*(x)h^*(x)$; that is, $g^*(x) \mid f(x)$. •

Theorem 7.21 (Gauss). *If R is a UFD, then $R[x]$ is also a UFD.*

Proof. We show first, by induction on $\deg(f)$, that every $f(x) \in R[x]$, neither zero nor a unit, is a product of irreducibles. If $\deg(f) = 0$, then $f(x)$ is a constant, hence lies in R. Since R is a UFD, f is a product of irreducibles. If $\deg(f) > 0$, then $f(x) = c(f)f^*(x)$, where $c(f) \in R$ and $f^*(x)$ is primitive. Now $c(f)$ is either a unit or a product of irreducibles, by the base step. If $f^*(x)$ is irreducible, we are done. Otherwise, $f^*(x) = g(x)h(x)$, where neither g nor h is a unit. Since $f^*(x)$ is primitive, however, neither g nor h is a constant; therefore, each of these has degree less than $\deg(f^*) = \deg(f)$, and so each is a product of irreducibles, by the inductive hypothesis.

Proposition 7.15 now applies: $R[x]$ is a UFD if $(p(x))$ is a prime ideal for every irreducible $p(x) \in R[x]$; that is, if $p(x) \mid f(x)g(x)$, then $p(x) \mid f(x)$ or $p(x) \mid g(x)$. Hence, let us assume that $p(x) \in R[x]$ is irreducible, with $p(x) \mid f(x)g(x)$, but with $p(x) \nmid f(x)$; we will prove that $p(x) \mid g(x)$. In this proof, $f(x)$ may be abbreviated to f.

Case (i). Suppose that $\deg(p) = 0$. Write

$$f(x) = c(f)f^*(x) \text{ and } g(x) = c(g)g^*(x),$$

where $c(f), c(g) \in R$, and $f^*(x), g^*(x)$ are primitive. Now $p \mid fg$, so that $p \mid c(f)c(g)f^*(x)g^*(x)$. Since $f^*(x)g^*(x)$ is primitive, we must have $c(f)c(g)$ an associate of $c(fg)$, by Lemma 7.20(ii). However, if $p \mid f(x)g(x)$, then p divides each coefficient of fg; that is, p is a common divisor of all the coefficients of fg, and hence $p \mid c(fg) = c(f)c(g)$ in R, which is a UFD. But Proposition 7.15 says that (p) is a prime ideal in R, and so $p \mid c(f)$ or $p \mid c(g)$. If $p \mid c(f)$, then $p(x) \mid c(f)f^*(x) = f(x)$, a contradiction. Therefore, $p \mid c(g)$ and, hence, $p(x) \mid g(x)$, as desired.

Case (ii). Suppose that $\deg(p) > 0$. Consider the ideal

$$(p, f) = \{sp + tf : s, t \in R[x]\};$$

of course, (p, f) contains $p(x)$ and $f(x)$. Choose $m(x) \in (p, f)$ of minimal degree. If $Q = \mathrm{Frac}(R)$, then the division algorithm in $Q[x]$ gives polynomials $q'(x), r'(x) \in Q[x]$ with $f(x) = m(x)q'(x) + r'(x)$, where either $r'(x) = 0$ or $\deg(r') < \deg(m)$. Clearing denominators, there are polynomials $q(x), r(x) \in R[x]$ and a constant $b \in R$ with

$$bf(x) = q(x)m(x) + r(x),$$

where $r(x) = 0$ or $\deg(r) < \deg(m)$. Since $m \in (p, f)$, which is an ideal, $r = bf - qm \in (p, f)$. Since m has minimal degree in (p, f), we must have $r = 0$; that is, $bf(x) = m(x)q(x)$, and so $bf(x) = c(m)m^*(x)q(x)$. But $m^*(x)$ is primitive, and $m^*(x) \mid bf(x)$, so that $m^*(x) \mid f(x)$, by Lemma 7.20(iii). A similar argument, replacing $f(x)$ by $p(x)$, gives $m^*(x) \mid p(x)$. Since $p(x)$ is irreducible, its only factors are units and associates. If $m^*(x)$ is an associate of $p(x)$, then $m^*(x) \mid m(x)$ implies $p(x) \mid f(x)$, contrary to the hypothesis. Hence, $m^*(x)$ must be a unit; that is, $m(x) = c(m) \in R$, and so (p, f) contains the nonzero constant $c(m)$. Now $c(m) = sp + tf$, and so

$$c(m)g = spg + tfg.$$

Since $p(x) \mid f(x)g(x)$, we have $p \mid c(m)g$. But $p(x)$ is primitive, because it is irreducible, and so $p(x) \mid g(x)$, by Lemma 7.20(iii). This completes the proof.   •

It follows from Proposition 7.17 that if R is a UFD, then gcd's exist in $R[x]$.

Corollary 7.22. *If k is a field, then $k[x_1, \ldots, x_n]$ is a UFD.*

Proof. The proof is by induction on $n \geq 1$. We proved, in Chapter 3, that the polynomial ring $k[x_1]$ in one variable is a UFD. For the inductive step, recall that $k[x_1, \ldots, x_n, x_{n+1}] = R[x_{n+1}]$, where $R = k[x_1, \ldots, x_n]$. By induction, R is a UFD, and by Theorem 7.21, so is $R[x_{n+1}]$. •

The theorem of Gauss, Theorem 3.96, can be generalized.

Corollary 7.23. *Let R be a UFD, let $Q = \text{Frac}(R)$, and let $f(x) \in R[x]$. If*

$$f(x) = G(x)H(x) \text{ in } Q[x],$$

then there is a factorization

$$f(x) = g(x)h(x) \text{ in } R[x],$$

where $\deg(g) = \deg(G)$ and $\deg(h) = \deg(H)$; in fact, $g(x)$ and $G(x)$ are associates in $Q[x]$, as are $h(x)$ and $H(x)$.

Therefore, if $f(x)$ does not factor into polynomials of smaller degree in $R[x]$, then $f(x)$ is irreducible in $Q[x]$.

Proof. Clearing denominators, there are elements $r, s \in R$ with $rG(x) \in R[x]$ and $sH(x) \in R[x]$; thus, $rsf(x) = [rG(x)][sH(x)]$ is a factorization in $R[x]$. By Lemma 7.20, there is a factorization

$$rsf(x) = c(rG)c(sH)[rG]^*(x)[sH]^*(x) \text{ in } R[x],$$

where $[rG]^*(x), [sH]^*(x) \in R[x]$ are primitive polynomials. By Lemma 7.20(ii), $c(rsf) = c(rG)c(sH)$, so that $rsf(x) = c(rsf)[rG]^*(x)[sH]^*(x)$. But $c(f) \in R$ and $c(rsf) = rsc(f)$; hence $f(x) = c(f)[rG]^*(x)[sH]^*(x)$ in $R[x]$. Therefore, there is a factorization $f(x) = g(x)h(x)$ in $R[x]$, where $g(x) = c(f)[r]^*(x)$ and $h(x) = [sH]^*(x)$. •

Irreducibility of a polynomial in several variables is more difficult to determine than irreducibility of a polynomial of one variable, but here is one criterion.

Corollary 7.24. *Let k be a field and let $f(x_1, \ldots, x_n)$ be a primitive polynomial in $R[x_n]$, where $R = k[x_1, \ldots, x_{n-1}]$. If f cannot be factored into two polynomials of lower degree in $R[x_n]$, then f is irreducible in $k[x_1, \ldots, x_n]$.*

Proof. Let us write $f(x_1, \ldots, x_n) = F(x_n)$, for we wish to view f as a polynomial in $R[x_n]$; that is, we view f as a polynomial in x_n having coefficients in $k[x_1, \ldots, x_{n-1}]$. Suppose that $F(x_n) = G(x_n)H(x_n)$; by hypothesis, the degrees of G and H (in x_n) cannot both be less than $\deg(F)$, and so one of them, say, G, has degree 0. It follows, because F is primitive, that G is a unit in $k[x_1, \ldots, x_{n-1}]$. Therefore, $f(x_1, \ldots, x_n)$ is irreducible in $R[x_n] = k[x_1, \ldots, x_n]$. •

Of course, the corollary applies to any variable x_i, not just to x_n.

Example 7.25.
We claim that $f(x, y) = x^2 + y^2 - 1 \in k[x, y]$ is irreducible, where k is a field of characteristic not 2. Write $Q = k(y) = \mathrm{Frac}(k[y])$, and view $f(x, y) \in Q[x]$. Now the quadratic $g(x) = x^2 + (y^2 - 1)$ is irreducible in $Q[x]$ if and only if it has no roots in $Q = k(y)$, and this is so, by Exercise 3.66 on page 274.

Since $k[x, y]$ is a UFD, it follows from Proposition 7.15 that $(x^2 + y^2 - 1)$ is a prime ideal, for it is generated by an irreducible polynomial.   ◄

EXERCISES

H **7.17** True or false with reasons.

    (i) If a, b, c are elements in a domain R, and if a and b are associates, then $a \mid c$ if and only if $b \mid c$.

    (ii) If a, b, c are elements in a domain R, and if a and b are associates, then $c \mid a$ if and only if $c \mid b$.

    (iii) $\mathbb{Z}$ is a UFD.

    (iv) If an element a in a UFD has two factorizations into irreducibles, $a = p_1 \cdots p_n = q_1 \cdots q_m$, then $m = n$ and $p_i = q_i$ for all i.

    (v) If an element a in a UFD has two factorizations into irreducibles, $a = p_1 \cdots p_n = q_1 \cdots q_m$, then $m = n$ and p_i and q_i are associates for all i.

    (vi) If an element a in a UFD has two factorizations into irreducibles, $a = p_1 \cdots p_n = q_1 \cdots q_m$, then $m = n$ and there is $\sigma \in S_n$ with p_i and $q_{\sigma(i)}$ associates for all i.

    (vii) If R is a PID, then there is no infinite descending chain of ideals $I_1 \supsetneq I_2 \supsetneq \cdots$.

    (viii) If R is a PID, then $R[x]$ is a PID.

    (ix) If R is a UFD, then $R[x]$ is a UFD.

    (x) If R is a PID, then $R[x]$ is a UFD.

7.18 In any commutative ring R, prove that if every pair of elements has a gcd, then every finite number of elements has a gcd.

*__7.19__ Let R be a UFD and let $Q = \mathrm{Frac}(R)$ be its fraction field. Prove that every $q \in Q$ has an expression in lowest terms; that is, $q = a/b$, where a and b are relatively prime.

*__7.20__ Let R be a UFD, and let $a, b, c \in R$. If a and b are relatively prime, and if $a \mid bc$, prove that $a \mid c$.

7.21 If R is a domain, prove that the only units in $R[x_1, \ldots, x_n]$ are units in R.

7.22 If R is a UFD and $f(x), g(x) \in R[x]$, prove that $c(fg)$ and $c(f)c(g)$ are associates.

*__7.23__   **(i)** Prove that x and y are relatively prime in $k[x, y]$, where k is a field.

    (ii) Prove that 1 is not a linear combination of x and y in $k[x, y]$.

7.24 Prove that $\mathbb{Z}[x_1, \ldots, x_n]$ is a UFD for all $n \geq 1$.

7.25 Let k be a field and let $f(x_1, \ldots, x_n) \in k[x_1, \ldots, x_n]$ be a primitive polynomial in $R[x_n]$, where $R = k[x_1, \ldots, x_{n-1}]$. If f is either quadratic or cubic in x_n, prove that f is irreducible in $k[x_1, \ldots, x_n]$ if and only if f, regarded as a polynomial in x_n, has no roots in $k(x_1, \ldots, x_{n-1})$.

7.26 Let $f(x_1, \ldots, x_n) = x_n g(x_1, \ldots, x_{n-1}) + h(x_1, \ldots, x_{n-1})$, where $(g, h) = 1$.

    (i) Prove that f is irreducible in $k[x_1, \ldots, x_n]$.

(ii)   Prove that $xy^2 + z$ is an irreducible polynomial in $k[x, y, z]$.

7.27 (***Eisenstein's criterion***) Let R be a UFD with $Q = \text{Frac}(R)$, and let $f(x) = a_0 + a_1 x + \cdots + a_n x^n \in R[x]$. Prove that if there is an irreducible element $p \in R$ with $p \mid a_i$ for all $i < n$ but with $p \nmid a_n$ and $p^2 \nmid a_0$, then $f(x)$ is irreducible in $Q[x]$.

7.28 Prove that

$$f(x, y) = xy^3 + x^2 y^2 - x^5 y + x^2 + 1$$

is an irreducible polynomial in $\mathbb{R}[x, y]$.

7.3 NOETHERIAN RINGS

One of the most important properties of $k[x_1, \ldots, x_n]$, when k is a field, is that every ideal in it can be generated by a finite number of elements. This property is intimately related to chains of ideals, which we have already met in the course of proving that PIDs are UFDs (I apologize for so many acronyms, but here comes another one).

A commutative ring has the *ascending chain condition* if every ascending chain of ideals is constant from some point on.

Definition.   A commutative ring R has **ACC**, the *ascending chain condition*, if every ascending chain of ideals

$$I_1 \subseteq I_2 \subseteq \cdots \subseteq I_n \subseteq \cdots$$

stops: there is an integer N with $I_N = I_{N+1} = I_{N+2} = \cdots$.

The proof of Lemma 7.14(ii) shows that every PID satisfies ACC.
Here is an important type of ideal.

Definition.   An ideal I in a commutative ring R is called *finitely generated* if there are finitely many elements $a_1, \ldots, a_n \in I$ with

$$I = \Big\{ \sum_i r_i a_i : r_i \in R \text{ for all } i \Big\};$$

that is, every element in I is a linear combination of the a_i's. One writes

$$I = (a_1, \ldots, a_n)$$

and calls I the *ideal generated by* $a_1, \ldots, a_n$.

A set of generators $a_1, \ldots, a_n$ of an ideal I is sometimes called a *basis* of I (even though this is a weaker notion than that of a basis of a vector space because uniqueness of expression is not assumed).

Every ideal I in a PID is generated by one element, and so I is finitely generated.

Proposition 7.26. *The following conditions are equivalent for a commutative ring* R.

(i) R *has the* ACC.

(ii) R *satisfies the* **maximum condition**: *every nonempty family* $\mathcal{F}$ *of ideals in* R *has a maximal element; that is, there is some* $I_0 \in \mathcal{F}$ *for which there is no* $J \in \mathcal{F}$ *with* $I_0 \subsetneq J$.

(iii) *Every ideal in* R *is finitely generated.*

Proof.   (i) $\Rightarrow$ (ii): Let $\mathcal{F}$ be a family of ideals in R, and assume that $\mathcal{F}$ has no maximal element. Choose $I_1 \in \mathcal{F}$. Since I_1 is not a maximal element, there is $I_2 \in \mathcal{F}$ with $I_1 \subsetneq I_2$. Now I_2 is not a maximal element in $\mathcal{F}$, and so there is $I_3 \in \mathcal{F}$ with $I_2 \subsetneq I_3$. Continuing in this way, we can construct an ascending chain of ideals in R that does not stop, contradicting the ACC.

(ii) $\Rightarrow$ (iii): Let I be an ideal in R, and define $\mathcal{F}$ to be the family of all the finitely generated ideals contained in I; of course, $\mathcal{F} \neq \varnothing$. By hypothesis, there exists a maximal element $M \in \mathcal{F}$. Now $M \subseteq I$ because $M \in \mathcal{F}$. If $M \subsetneq I$, then there is $a \in I$ with $a \notin M$. The ideal

$$J = \{m + ra : m \in M \text{ and } r \in R\} \subseteq I$$

is finitely generated, and so $J \in \mathcal{F}$; but $M \subsetneq J$, and this contradicts the maximality of M. Therefore, $M = I$, and so I is finitely generated.

(iii) $\Rightarrow$ (i): Assume that every ideal in R is finitely generated, and let

$$I_1 \subseteq I_2 \subseteq \cdots \subseteq I_n \subseteq \cdots$$

be an ascending chain of ideals in R. Lemma 7.14(i) shows that $J = \bigcup_n I_n$ is an ideal. By hypothesis, there are elements $a_i \in J$ with $J = (a_1, \ldots, a_q)$. Now a_i got into J by being in I_{n_i} for some n_i. If N is the largest n_i, then $I_{n_i} \subseteq I_N$ for all i; hence, $a_i \in I_N$ for all i, and so

$$J = (a_1, \ldots, a_q) \subseteq I_N \subseteq J.$$

It follows that if $n \geq N$, then $J = I_N \subseteq I_n \subseteq J$, so that $I_n = J$; therefore, the chain stops, and R has ACC.   •

We now give a name to a commutative ring which satisfies any of the three equivalent conditions in the proposition.

Definition.   A commutative ring R is called **noetherian**[1] if every ideal in R is finitely generated.

---

[1] This name honors Emmy Noether (1882–1935), who introduced chain conditions in 1921.

Corollary 7.27. *If I is an ideal in a nonzero noetherian ring R, then there exists a maximal ideal M in R containing I. In particular, every noetherian ring has maximal ideals.*[2]

Proof. Let $\mathcal{F}$ be the family of all those proper ideals in R which contain I; note that $\mathcal{F} \neq \varnothing$ because $I \in \mathcal{F}$. Since R is noetherian, the maximum condition gives a maximal element M in $\mathcal{F}$. We must still show that M is a maximal ideal in R (that is, that M is actually a maximal element in the larger family $\mathcal{F}'$ consisting of all the proper ideals in R). Suppose there is a proper ideal J with $M \subseteq J$. Then $I \subseteq J$, and so $J \in \mathcal{F}$; therefore, maximality of M gives $M = J$, and so M is a maximal ideal in R. •

Remark. Zorn's lemma is related to the maximum condition, statement (ii) in Proposition 7.26.

Definition. A *partially ordered set* is a set X equipped with a relation $x \preceq y$ which satisfies, for all $x, y, z \in X$,

(i) *reflexivity*: $x \preceq x$;

(ii) *antisymmetry*: if $x \preceq y$ and $y \preceq x$, then $x = y$;

(iii) *transitivity*: if $x \preceq y$ and $y \preceq z$, then $x \preceq z$.

We now generalize our earlier definition of maximal element (in a family of ideals). An element u in a partially ordered set X is called a *maximal element* if there is no $x \in X$ with $u \preceq x$ and $u \neq x$.

If A is a set, then the family $\mathcal{P}(A)$ of all the subsets of A is a partially ordered set if one defines $U \preceq V$ to mean $U \subseteq V$, where U and V are subsets of A; the family $\mathcal{P}(A)^*$, consisting of all the proper subsets of A, is also a partially ordered set (more generally, every subset of a partially ordered set is itself a partially ordered set). Another example is the real numbers $\mathbb{R}$, with $x \preceq y$ meaning $x \leq y$. There are some partially ordered sets [e.g., $\mathcal{P}(A)^*$] having many maximal elements [the complement of a point in A is a maximal element in $\mathcal{P}(A)^*$], and there are some partially ordered sets (e.g., $\mathbb{R}$) having no maximal elements. Zorn's lemma is a condition on a partially ordered set which guarantees that it has at least one maximal element.

A partially ordered set X is called a *chain* if, for every $a, b \in X$, either $a \preceq b$ or $b \preceq a$. (Since every two elements in a chain are comparable, chains are sometimes called *totally ordered sets* to contrast them with more general partially ordered sets.) We can now state Zorn's lemma.

---

[2]This corollary is true without assuming R is noetherian, but the proof of the general result needs Zorn's lemma.

Zorn's Lemma. *Let X be a nonempty partially ordered set in which every chain C has an upper bound; that is, there exists $x_0 \in X$ with $c \preceq x_0$ for every $c \in C$. Then X has a maximal element.*

It turns out that Zorn's lemma is equivalent to the ***Axiom of Choice***, which says that any cartesian product of (possibly infinitely many) nonempty sets is itself nonempty.

There is usually no need for Zorn's lemma when dealing with noetherian rings, for the maximum condition guarantees the existence of a maximal element in any nonempty family $\mathcal{F}$ of ideals. ◄

Here is one way to construct a new noetherian ring from an old one.

Corollary 7.28.   *If R is a noetherian ring and J is an ideal in R, then R/J is also noetherian.*

Proof.   If A is an ideal in R/I, then the correspondence theorem provides an ideal J in R with $J/I = A$. Since R is noetherian, the ideal J is finitely generated, say, $J = (b_1, \ldots, b_n)$, and so $A = J/I$ is generated by the cosets $b_1 + I, \ldots, b_n + I$. Therefore, every ideal A is finitely generated, and R/I is noetherian.   •

In 1890, Hilbert proved the famous Hilbert basis theorem, showing that every ideal in $\mathbb{C}[x_1, \ldots, x_n]$ is finitely generated. As we shall see, the proof is nonconstructive in the sense that it does not give an explicit set of generators of an ideal. It is reported that when P. Gordan, one of the leading algebraists of the time, first saw Hilbert's proof, he said, "Das ist nicht Mathematik. Das ist Theologie!" ("This is not mathematics. This is theology!"). On the other hand, Gordan said, in 1899 when he published a simplified proof of Hilbert's theorem, "I have convinced myself that theology also has its merits."

The following elegant proof of Hilbert's theorem is due to H. Sarges.

Lemma 7.29.   *A commutative ring R is noetherian if and only if, for every sequence $a_1, \ldots, a_n, \ldots$ of elements in R, there exists $m \geq 1$ and $r_1, \ldots, r_m \in R$ with $a_{m+1} = r_1 a_1 + \cdots + r_m a_m$.*

Proof.   Assume that R is noetherian and that $a_1, \ldots, a_n, \ldots$ is a sequence of elements in R. If $I_n = (a_1, \ldots, a_n)$, then there is an ascending chain of ideals, $I_1 \subseteq I_2 \subseteq \cdots$. By the ACC, there exists $m \geq 2$ with $I_m = I_{m+1}$. Therefore, $a_{m+1} \in I_{m+1} = I_m$, and so there are $r_i \in R$ with $a_{m+1} = r_1 a_1 + \cdots + r_m a_m$.

Conversely, suppose that R satisfies the condition on sequences of elements. If R is not noetherian, then there is an ascending chain of ideals $I_1 \subseteq I_2 \subseteq \cdots$ which does not stop. Deleting any repetitions if necessary, we may assume that $I_n \subsetneq I_{n+1}$ for all n. For each n, choose $a_{n+1} \in I_{n+1}$ with $a_{n+1} \notin I_n$. By hypothesis, there exists m and $r_i \in R$ for $i \leq m$ with $a_{m+1} = \sum_{i \leq m} r_i a_i \in I_m$. This contradiction implies that R is noetherian.   •

Theorem 7.30 (Hilbert Basis Theorem). *If R is a commutative noetherian ring, then R[x] is also noetherian.*

Proof. Assume that I is an ideal in $R[x]$ that is not finitely generated; of course, $I \neq \{0\}$. Define $f_0(x)$ to be a polynomial in I of minimal degree and define, inductively, $f_{n+1}(x)$ to be a polynomial of minimal degree in $I - (f_0, \ldots, f_n)$. Note that $f_n(x)$ exists for all $n \geq 0$; if $I - (f_0, \ldots, f_n)$ were empty, then I would be finitely generated. It is clear that

$$\deg(f_0) \leq \deg(f_1) \leq \deg(f_2) \leq \cdots.$$

Let a_n denote the leading coefficient of $f_n(x)$. Since R is noetherian, Lemma 7.29 applies to give an integer m with $a_{m+1} \in (a_0, \ldots, a_m)$; that is, there are $r_i \in R$ with $a_{m+1} = r_0 a_0 + \cdots + r_m a_m$. Define

$$f^*(x) = f_{m+1}(x) - \sum_{i=0}^{m} x^{d_{m+1} - d_i} r_i f_i(x),$$

where $d_i = \deg(f_i)$. Now $f^*(x) \in I - (f_0(x), \ldots, f_m(x))$, otherwise $f_{m+1}(x) \in (f_0(x), \ldots, f_m(x))$. It suffices to show that $\deg(f^*) < \deg(f_{m+1})$, for this contradicts $f_{m+1}(x)$ having minimal degree among polynomials in I that are not in $(f_0, \ldots, f_m)$. If $f_i(x) = a_i x^{d_i} +$ lower terms, then

$$f^*(x) = f_{m+1}(x) - \sum_{i=0}^{m} x^{d_{m+1} - d_i} r_i f_i(x)$$

$$= (a_{m+1} x^{d_{m+1}} + \text{lower terms}) - \sum_{i=0}^{m} x^{d_{m+1} - d_i} r_i (a_i x^{d_i} + \text{lower terms}).$$

The leading term being subtracted is thus $\sum_{i=0}^{m} r_i a_i x^{d_{m+1}} = a_{m+1} x^{d_{m+1}}$. •

Corollary 7.31.

  (i) *If k is a field, then $k[x_1, \ldots, x_n]$ is noetherian.*

  (ii) *The ring $\mathbb{Z}[x_1, \ldots, x_n]$ is noetherian.*

  (iii) *For any ideal I in $k[x_1, \ldots, x_n]$, where $k = \mathbb{Z}$ or k is a field, the quotient ring $k[x_1, \ldots, x_n]/I$ is noetherian.*

Proof. The proofs of the first two items are by induction on $n \geq 1$, using the theorem, while the proof of item (iii) follows from Corollary 7.28. •

It is known that if R is a noetherian ring, then the ring of formal power series $R[[x]]$ is also a noetherian ring (see Zariski-Samuel, *Commutative Algebra* II, page 138).

EXERCISES

H **7.29** True or false with reasons.

  (i) Every commutative ring is noetherian.

  (ii) Every subring of a noetherian domain is noetherian.

  (iii) If X and Y are nonempty sets, then $X \times Y \neq \varnothing$.

  (iv) Every partially ordered set has at most one maximal element.

  (v) If $\mathcal{F}$ is a nonempty family of ideals in $\mathbb{F}_2[x, y]$, then a maximal element of $\mathcal{F}$ is a maximal ideal in $\mathbb{F}_2[x, y]$.

  (vi) If R has ACC and if $J_1 \supseteq J_2 \supseteq J_3 \supseteq \cdots$ is a chain of ideals of R, then there exists M such that $J_M = J_{M+1} = J_{M+2} = \cdots$.

  (vii) If k is a field, then $k[[x]]$ is noetherian.

  (viii) If R is a commutative ring and $\varphi \colon \mathbb{Z}[x, y] \to R$ is a homomorphism, then $\ker \varphi$ is finitely generated.

7.30 Let m be a positive integer, and let X be the set of all its (positive) divisors. Prove that X is a partially ordered set if one defines $a \preceq b$ to mean $a \mid b$.

7.31 Prove that the ring $\mathcal{F}(\mathbb{R})$ of Example 3.11 on page 224 is not a noetherian ring.

7.32 Let
$$S^2 = \{(a, b, c) \in \mathbb{R}^3 \colon a^2 + b^2 + c^2 = 1\}$$

be the unit sphere in $\mathbb{R}^3$, and let

$$I = \{f(x, y, z) \in \mathbb{R}[x, y, z] \colon f(a, b, c) = 0 \text{ for all } (a, b, c) \in S^2\}.$$

Prove that I is a finitely generated ideal in $\mathbb{R}[x, y, z]$.

7.33 If R and S are noetherian rings, prove that their direct product $R \times S$ is also a noetherian ring.

7.34 If R is a ring that is also a vector space over a field k, then R is called a **k-algebra** if

$$(\alpha u)v = \alpha(uv) = u(\alpha v)$$

for all $\alpha \in k$ and $u, v \in R$. Prove that every finite-dimensional k-algebra is a noetherian ring.

7.4 VARIETIES

Analytic geometry gives pictures of equations. For example, we picture a function $f \colon \mathbb{R} \to \mathbb{R}$ as its graph, which consists of all the ordered pairs $(a, f(a))$ in the plane; that is, f is the set of all the solutions $(a, b) \in \mathbb{R}^2$ of

$$g(x, y) = y - f(x) = 0.$$

We can picture equations that are not graphs of functions. For example, the set of all the zeros of the polynomial

$$h(x, y) = x^2 + y^2 - 1$$

is the unit circle. We can also picture simultaneous solutions in $\mathbb{R}^2$ of several polynomials of two variables, and, indeed, we can picture simultaneous solutions in $\mathbb{R}^n$ of several polynomials of n variables.

Notation. Let k be a field and let k^n denote the set of all n-tuples

$$k^n = \{(a_1, \ldots, a_n) \colon a_i \in k \text{ for all } i\}.$$

The polynomial ring $k[x_1, \ldots, x_n]$ in several variables may be denoted by $k[X]$, where X is the abbreviation:

$$X = (x_1, \ldots, x_n).$$

In particular, $f(X) \in k[X]$ may abbreviate $f(x_1, \ldots, x_n) \in k[x_1, \ldots, x_n]$.

In what follows, we regard polynomials $f(x_1, \ldots, x_n) \in k[x_1, \ldots, x_n]$ as functions of n variables $k^n \to k$. Here is the precise definition.

Definition. Given a polynomial $f(X) \in k[X]$, its ***polynomial function*** $f^\flat \colon k^n \to k$ is defined in the obvious way: if $f(x_1, \ldots, x_n) = \sum_{e_1, \ldots, e_n} b_{e_1, \ldots, e_n} x_1^{e_1} \cdots x_n^{e_n}$ and $(a_1, \ldots, a_n) \in k^n$, then

$$f^\flat \colon (a_1, \ldots, a_n) \mapsto f(a_1, \ldots, a_n) = \sum_{e_1, \ldots, e_n} b_{e_1, \ldots, e_n} a_1^{e_1} \cdots a_n^{e_n}.$$

The next proposition generalizes Corollary 3.52 from one variable to several variables.

Proposition 7.32. *Let k be an infinite field and let $k[X] = k[x_1, \ldots, x_n]$. If $f(X), g(X) \in k[X]$ satisfy $f^\flat = g^\flat$, then $f(x_1, \ldots, x_n) = g(x_1, \ldots, x_n)$.*

Proof. The proof is by induction on $n \geq 1$; the base step is Corollary 3.52. For the inductive step, write

$$f(X, y) = \sum_i p_i(X) y^i \text{ and } g(X, y) = \sum_i q_i(X) y^i,$$

where X denotes $(x_1, \ldots, x_n)$ and $y = x_{n+1}$. If $f^\flat = g^\flat$, then we have $f(a, \alpha) = g(a, \alpha)$ for every $a \in k^n$ and every $\alpha \in k$. For fixed $a \in k^n$, define $F_a(y) = \sum_i p_i(a) y^i$ and $G_a(y) = \sum_i q_i(a) y^i$. Since both $F_a(y)$ and $G_a(y)$ are in $k[y]$, the base step gives $p_i(a) = q_i(a)$ for all $a \in k^n$. By the inductive hypothesis, $p_i(X) = q_i(X)$ for all i, and hence

$$f(X, y) = \sum_i p_i(X) y^i = \sum_i q_i(X) y^i = g(X, y). \quad \bullet$$

As a consequence of this last proposition, we drop the $f^\flat$ notation and identify polynomials with their polynomial functions when k is infinite. We note that algebraically closed fields are always infinite, by Exercise 3.104 on page 305. Thus, Proposition 7.32 applies whenever k is algebraically closed.

Definition. If $f(X) \in k[X] = k[x_1, \ldots, x_n]$ and $f(a) = 0$, where $a \in k^n$, then a is called a **zero** of $f(X)$. [If $f(x)$ is a polynomial in one variable, then a zero of $f(x)$ is also called a *root* of $f(x)$.]

Proposition 7.33. *If k is an algebraically closed field and $f(X) \in k[X]$ is not a constant, then $f(X)$ has a zero in k^n.*

Proof. We prove the result by induction on $n \geq 1$, where $X = (x_1, \ldots, x_n)$. The base step follows at once from our assuming that $k^1 = k$ is algebraically closed. As in the proof of Proposition 7.32, write

$$f(X, y) = \sum_i g_i(X)y^i.$$

For each $a \in k^n$, define $f_a(y) = \sum_i g_i(a)y^i$. If $f(X, y)$ has no zeros, then each $f_a(y) \in k[y]$ has no zeros, and the base step says that $f_a(y)$ is a nonzero constant for all $a \in k^n$. Thus, $g_i(a) = 0$ for all $i > 0$ and all $a \in k^n$. By Proposition 7.32, which applies because algebraically closed fields are infinite, $g_i(X) = 0$ for all $i > 0$, and so $f(X, y) = g_0(X)y^0 = g_0(X)$. By the inductive hypothesis, $g_0(X)$ is a nonzero constant, and the proof is complete. ●

We now consider solution sets of polynomials.

Definition. If $F \subseteq k[X] = k[x_1, \ldots, x_n]$ is a subset, then the **algebraic set** defined by F is

$$\mathrm{Var}(F) = \{a \in k^n : f(a) = 0 \text{ for every } f(X) \in F\};$$

thus, $\mathrm{Var}(F)^3$ consists of all those $a \in k^n$ which are zeros of every $f(X) \in F$.

Example 7.34.

(i) Here is an algebraic set defined by two equations.

$$\mathrm{Var}(x, y) = \{(a, b) \in k^2 : x = 0 \text{ or } y = 0\}.$$

Thus,

$$\mathrm{Var}(x, y) = x\text{-axis} \cup y\text{-axis}.$$

More generally, any finite union of algebraic sets is an algebraic set.

---

[3] The notation $\mathrm{Var}(F)$ arises from *variety*, which is a special kind of algebraic set to be defined later.

(ii) The *n-sphere* S^n is defined as

$$S^n = \left\{ (x_1, \ldots, x_{n+1}) \in k^{n+1} : \sum_{i=1}^{n+1} x_i^2 = 1 \right\}.$$

More generally, define a **hypersurface** in k^n to be the algebraic set defined as all the zeros of a single polynomial in $k[X]$.

(iii) Let A be an $m \times n$ matrix with entries in k. A system of m equations in n unknowns,

$$AX = B,$$

where B is an $n \times 1$ column matrix, defines an algebraic set, Var($AX = B$), which is a subset of k^n. Of course, $AX = B$ is really shorthand for a set of m linear equations in n variables, and Var($AX = B$) is usually called the **solution set** of the system $AX = B$; when this system is homogeneous, that is, when $B = 0$, then Var($AX = 0$) is a subspace of k^n, called the **solution space** of the system. ◄

The next result shows that, as far as algebraic sets are concerned, one may just as well assume the subsets F of $k[X]$ are ideals in $k[X]$.

Proposition 7.35. *Let k be a field.*

(i) *If $F \subseteq G \subseteq k[X]$, then* Var($G$) $\subseteq$ Var(F).

(ii) *If $F \subseteq k[X]$ and $I = (F)$ is the ideal generated by F, then*

$$\text{Var}(F) = \text{Var}(I).$$

(iii) *Every algebraic set can be defined by a finite number of equations.*

Proof.
(i) If $a \in$ Var(G), then $g(a) = 0$ for all $g(X) \in G$; since $F \subseteq G$, it follows, in particular, that $f(a) = 0$ for all $f(X) \in F$.
(ii) Since $F \subseteq (F) = I$, we have Var(I) $\subseteq$ Var(F), by part (i). For the reverse inclusion, let $a \in$ Var(F), so that $f(a) = 0$ for every $f(X) \in F$. If $g(X) \in I$, then $g(X) = \sum_i r_i f_i(X)$, where $r_i \in k$ and $f_i(X) \in F$; hence, $g(a) = \sum_i r_i f_i(a) = 0$ and $a \in$ Var(I).
(iii) If I is an ideal in $k[X]$, then the Hilbert basis theorem says that I is finitely generated; that is, there is a finite subset $F \subseteq I$ with Var(I) = Var(F). •

It follows that not every subset of k^n is an algebraic set. For example, if $n = 1$, then $k[x]$ is a PID. Hence, if F is a subset of $k[x]$, then $(F) = (f(x))$ for some $f(x) \in k[x]$, and so

$$\text{Var}(F) = \text{Var}((F)) = \text{Var}((f)) = \text{Var}(f).$$

But $f(x)$ has only a finite number of roots, and so $\mathrm{Var}(F)$ is finite. If k is algebraically closed, then it is an infinite field, and so most subsets of $k^1 = k$ are not algebraic sets.

In spite of our wanting to draw pictures in the plane, there is a major defect with $k = \mathbb{R}$: some polynomials have no zeros. For example, $f(x) = x^2 + 1$ has no real roots, and so $\mathrm{Var}(x^2 + 1) = \varnothing$. More generally, $g(x_1, \ldots, x_n) = x_1^2 + \cdots + x_n^2 + 1$ has no zeros in $\mathbb{R}^n$, and so $\mathrm{Var}(g) = \varnothing$. Since we are dealing with (not necessarily linear) polynomials, it is a natural assumption to want all their zeros available. For polynomials in one variable, this amounts to saying that k is algebraically closed and, in light of Proposition 7.33, we know that $\mathrm{Var}(f) \neq \varnothing$ for every nonconstant $f(X) \in k[X]$ in this case. Of course, algebraic sets are of interest for all fields k, but it makes more sense to consider the simplest case before trying to understand more complicated problems. On the other hand, many of the first results below are valid for any field k. We will state hypotheses needed for each proposition, but the reader should realize that the most important case is when k is algebraically closed.

Here are some elementary properties of Var.

Proposition 7.36. *Let k be a field.*

(i) $\mathrm{Var}(x_1, x_1 - 1) = \varnothing$ *and* $\mathrm{Var}(0) = k^n$, *where 0 is the zero polynomial.*

(ii) *If I and J are ideals in $k[X]$, then*

$$\mathrm{Var}(IJ) = \mathrm{Var}(I \cap J) = \mathrm{Var}(I) \cup \mathrm{Var}(J),$$

where $IJ = \{\sum_i f_i(X)g_i(X) : f_i(X) \in I \text{ and } g_i(X) \in J\}$.

(iii) *If $\{I_\ell : \ell \in L\}$ is a family of ideals in $k[X]$, then*

$$\mathrm{Var}\Big(\sum_\ell I_\ell\Big) = \bigcap_\ell \mathrm{Var}(I_\ell),$$

where $\sum_\ell I_\ell$ is the set of all finite sums of the form $r_{\ell_1} + \cdots + r_{\ell_q}$ with $r_{\ell_i} \in I_{\ell_i}$.

Proof.
(i) If $a = (a_1, \ldots, a_n) \in \mathrm{Var}(x_1, x_1 - 1)$, then $a_1 = 0$ and $a_1 = 1$; plainly, there are no such points a, and so $\mathrm{Var}(x_1, x_1 - 1) = \varnothing$. That $\mathrm{Var}(0) = k^n$ is clear, for every point $a \in k^n$ is a zero of the zero polynomial.
(ii) Since $IJ \subseteq I \cap J$, it follows that $\mathrm{Var}(IJ) \supseteq \mathrm{Var}(I \cap J)$; since $IJ \subseteq I$, it follows that $\mathrm{Var}(IJ) \supseteq \mathrm{Var}(I)$. Hence,

$$\mathrm{Var}(IJ) \supseteq \mathrm{Var}(I \cap J) \supseteq \mathrm{Var}(I) \cup \mathrm{Var}(J).$$

To complete the proof, it suffices to show that $\mathrm{Var}(IJ) \subseteq \mathrm{Var}(I) \cup \mathrm{Var}(J)$. If $a \notin \mathrm{Var}(I) \cup \mathrm{Var}(J)$, then there exist $f(X) \in I$ and $g(X) \in J$ with $f(a) \neq 0$ and $g(a) \neq 0$. But $f(X)g(X) \in IJ$ and $(fg)(a) = f(a)g(a) \neq 0$, because $k[X]$ is a domain. Therefore, $a \notin \mathrm{Var}(IJ)$, as desired.

(iii) For each ℓ, the inclusion $I_\ell \subseteq \sum_\ell I_\ell$ gives $\text{Var}(\sum_\ell I_\ell) \subseteq \text{Var}(I_\ell)$, and so

$$\text{Var}\Big(\sum_\ell I_\ell\Big) \subseteq \bigcap_\ell \text{Var}(I_\ell).$$

For the reverse inclusion, if $g(X) \in \sum_\ell I_\ell$, then there are finitely many ℓ with $g(X) = \sum_\ell h_\ell f_\ell$, where $h_\ell \in k[X]$ and $f_\ell(X) \in I_\ell$. Therefore, if $a \in \bigcap_\ell \text{Var}(I_\ell)$, then $f_\ell(a) = 0$ for all ℓ, and so $g(a) = 0$; that is, $a \in \text{Var}(\sum_\ell I_\ell)$. $\bullet$

Definition. A *topology* on a set X is a family $\mathcal{F}$ of subsets of X, called *closed sets*,[4] which satisfy the following axioms:

(i) $\varnothing \in \mathcal{F}$ and $X \in \mathcal{F}$;

(ii) if $F_1, F_2 \in \mathcal{F}$, then $F_1 \cup F_2 \in \mathcal{F}$; that is, the union of two closed sets is closed;

(iii) if $\{F_\ell : \ell \in L\} \subseteq \mathcal{F}$, then $\bigcap_\ell F_\ell \in \mathcal{F}$; that is, any intersection of possibly infinitely many closed sets is also closed.

A *topological space* is an ordered pair $(X, \mathcal{F})$, where X is a set and $\mathcal{F}$ is a topology on X.

Proposition 7.36 shows that the family of all algebraic sets is a topology on k^n; it is called the *Zariski topology*, and it is very useful in the deeper study of $k[X]$. The usual topology on $\mathbb{R}$ has many closed sets; for example, every closed interval is a closed set. In contrast, in the Zariski topology on $\mathbb{R}$, every proper closed set is finite.

Given an ideal I in $k[X]$, we have just defined its algebraic set $\text{Var}(I) \subseteq k^n$. We now reverse direction: to every subset $A \subseteq k^n$, we assign an ideal in $k[X]$; in particular, we assign an ideal to every algebraic set.

Definition. If $A \subseteq k^n$ is a subset, where k is a field, then its ***coordinate ring*** $k[A]$ is the commutative ring of all restrictions $f|A$ of polynomial functions $f : k^n \to k$, under pointwise operations.

The function res$: k[X] \to k[A]$, given by $f(X) \mapsto f|A$, is a ring homomorphism, and the kernel of this restriction map is an ideal in $k[X]$.

Definition. If $A \subseteq k^n$, where k is a field, define

$$\text{Id}(A) = \{f(X) \in k[X] = k[x_1, \dots, x_n] : f(a) = 0 \text{ for every } a \in A\}.$$

The Hilbert basis theorem tells us that $\text{Id}(A)$ is always a finitely generated ideal.

---

[4]One can also define a topology by specifying its *open subsets*, which are defined to be complements of closed sets.

Proposition 7.37.   *If $A \subseteq k^n$, where k is a field, then there is an isomorphism*

$$k[X]/\operatorname{Id}(A) \cong k[A].$$

Proof.   The restriction map res$: k[X] \to k[A]$ is a surjection with kernel $\operatorname{Id}(A)$, and so the result follows from the first isomorphism theorem. Note that two polynomials agreeing on A lie in the same coset of $\operatorname{Id}(A)$.   •

Although the definition of Var(F) makes sense for any subset F of $k[X]$, it is most interesting when F is an ideal. Similarly, although the definition of $\operatorname{Id}(A)$ makes sense for any subset A of k^n, it is most interesting when A is an algebraic set. After all, algebraic sets are comprised of solutions of (polynomial) equations, which is what we care about.

Proposition 7.38.   *Let k be a field.*

(i) $\operatorname{Id}(\varnothing) = k[X]$ *and, if k is algebraically closed,* $\operatorname{Id}(k^n) = \{0\}$.

(ii) *If $A \subseteq B$ are subsets of k^n, then* $\operatorname{Id}(B) \subseteq \operatorname{Id}(A)$.

(iii) *If $\{A_\ell : \ell \in L\}$ is a family of subsets of k^n, then*

$$\operatorname{Id}\left(\bigcup_\ell A_\ell\right) = \bigcap_\ell \operatorname{Id}(A_\ell).$$

Proof.
(i) If $f(X) \in \operatorname{Id}(A)$ for some subset $A \subseteq k^n$, then $f(a) = 0$ for all $a \in A$; hence, if $f(X) \notin \operatorname{Id}(A)$, then there exists $a \in A$ with $f(a) \neq 0$. In particular, if $A = \varnothing$, every $f(X) \in k[X]$ must lie in $\operatorname{Id}(\varnothing)$, for there are no elements $a \in \varnothing$. Therefore, $\operatorname{Id}(\varnothing) = k[X]$.

If $f(X) \in \operatorname{Id}(k^n)$, then $f(a) = 0$ for all $a \in k^n$; it follows from Proposition 7.32 that $f(X)$ is the zero polynomial.

(ii) If $f(X) \in \operatorname{Id}(B)$, then $f(b) = 0$ for all $b \in B$; in particular, $f(a) = 0$ for all $a \in A$, because $A \subseteq B$, and so $f(X) \in \operatorname{Id}(A)$.

(iii) Since $A_\ell \subseteq \bigcup_\ell A_\ell$, we have $\operatorname{Id}(A_\ell) \supseteq \operatorname{Id}(\bigcup_\ell A_\ell)$ for all ℓ; therefore, $\bigcap_\ell \operatorname{Id}(A_\ell) \supseteq \operatorname{Id}(\bigcup_\ell A_\ell)$. For the reverse inclusion, let $f(X) \in \bigcap_\ell \operatorname{Id}(A_\ell)$; that is, $f(a_\ell) = 0$ for all ℓ and all $a_\ell \in A_\ell$. If $b \in \bigcup_\ell A_\ell$, then $b \in A_\ell$ for some ℓ, and hence $f(b) = 0$; therefore, $f(X) \in \operatorname{Id}(\bigcup_\ell A_\ell)$.   •

One would like to have a formula for $\operatorname{Id}(A \cap B)$. Certainly, the formula $\operatorname{Id}(A \cap B) = \operatorname{Id}(A) \cup \operatorname{Id}(B)$ is not correct, for the union of two ideals is almost never an ideal (see Exercise 7.38 on page 556).

The next idea arises in characterizing those ideals of the form $\operatorname{Id}(V)$ when V is an algebraic set.

Definition. If I is an ideal in a commutative ring R, then its ***radical***, denoted by $\sqrt{I}$, is

$$\sqrt{I} = \{r \in R : r^m \in I \text{ for some integer } m \geq 1\}.$$

An ideal I is called a ***radical ideal***[5] if

$$\sqrt{I} = I.$$

Exercise 7.36 on page 556 asks you to prove that $\sqrt{I}$ is an ideal. It is easy to see that $I \subseteq \sqrt{I}$, and so an ideal I is a radical ideal if and only if $\sqrt{I} \subseteq I$. For example, every prime ideal P is a radical ideal, for if $f^n \in P$, then $f \in P$. Here is an example of an ideal that is not radical. Let $b \in k$ and let $I = ((x - b)^2)$. Now I is not a radical ideal, for $(x - b)^2 \in I$ while $x - b \notin I$.

Definition. An element a in a commutative ring R is called ***nilpotent*** if there is some $n \geq 1$ with $a^n = 0$.

Note that I is a radical ideal in a commutative ring R if and only if R/I has no nilpotent elements (of course, we mean that R/I has no *nonzero* nilpotent elements).

Here is why radical ideals are introduced.

Proposition 7.39. *If an ideal $I = \text{Id}(A)$ for some $A \subseteq k^n$, where k is a field, then it is a radical ideal. Hence, the coordinate ring $k[A]$ has no nilpotent elements.*

Proof. Since $I \subseteq \sqrt{I}$ is always true, it suffices to check the reverse inclusion. By hypothesis, $I = \text{Id}(A)$ for some $A \subseteq k^n$; hence, if $f \in \sqrt{I}$, then $f^m \in \text{Id}(A)$; that is, $f(a)^m = 0$ for all $a \in A$. But the values of $f(a)^m$ lie in the field k, and so $f(a)^m = 0$ implies $f(a) = 0$; that is, $f \in \text{Id}(A) = I$. •

Proposition 7.40.

  (i) *If I and J are ideals, then $\sqrt{I \cap J} = \sqrt{I} \cap \sqrt{J}$.*

  (ii) *If I and J are radical ideals, then $I \cap J$ is a radical ideal.*

Proof.
(i) If $f \in \sqrt{I \cap J}$, then $f^m \in I \cap J$ for some $m \geq 1$. Hence, $f^m \in I$ and $f^m \in J$, and so $f \in \sqrt{I}$ and $f \in \sqrt{J}$; that is, $f \in \sqrt{I} \cap \sqrt{J}$.

For the reverse inclusion, assume that $f \in \sqrt{I} \cap \sqrt{J}$, so that $f^m \in I$ and $f^q \in J$. We may assume that $m \geq q$, and so $f^m \in I \cap J$; that is, $f \in \sqrt{I \cap J}$.

(ii) If I and J are radical ideals, then $I = \sqrt{I}$ and $J = \sqrt{J}$ and

$$I \cap J \subseteq \sqrt{I \cap J} = \sqrt{I} \cap \sqrt{J} = I \cap J. \quad •$$

---

[5]This term is appropriate, for if $r^m \in I$, then its mth root r also lies in I.

We are now going to prove Hilbert's *Nullstellensatz*: $\sqrt{I} = \mathrm{Id}(\mathrm{Var}(I))$ for every ideal $I \subseteq \mathbb{C}[X]$; that is, a polynomial $f(X)$ vanishes on $\mathrm{Var}(I)$ if and only if $f^m \in I$ for some $m \geq 1$. This theorem is true for ideals in $k[X]$, where k is any algebraically closed field. The astute reader can adapt the proof given here for $k = \mathbb{C}$ to any uncountable algebraically closed field k. However, a new idea is needed to prove the general theorem for algebraic closures of the prime fields, for example, which are countable.

Lemma 7.41. *Let k be a field and let $\varphi \colon k[X] \to k$ be a surjective homomorphism which fixes k pointwise. If $J = \ker \varphi$, then $\mathrm{Var}(J) \neq \varnothing$.*

Proof. Now $x_i \in k[X]$ for each i; let $\varphi(x_i) = a_i \in k$, and let $a = (a_1, \ldots, a_n) \in k^n$. If $f(X) = \sum_{e_1, \ldots, e_n} c_{e_1, \ldots, e_n} x_1^{e_1} \cdots x_n^{e_n} \in k[X]$, then

$$
\begin{aligned}
\varphi(f(X)) &= \sum_{e_1, \ldots, e_n} c_{e_1, \ldots, e_n} \pi(x_1)^{e_1} \cdots \varphi(x_n)^{e_n} \\
&= \sum_{e_1, \ldots, e_n} c_{e_1, \ldots, e_n} a_1^{e_1} \cdots a_n^{e_n} \\
&= f(a_1, \ldots, a_n).
\end{aligned}
$$

Hence, $\varphi(f(X)) = f(a) = \varphi(f(a))$, because $f(a) \in k$ and φ fixes k pointwise. It follows that $f(X) - f(a) \in J$ for every $f(X)$. Now if $f(X) \in J$, then $f(a) \in J$. But $f(a) \in k$, and, since J is a proper ideal, it contains no nonzero constants. Therefore, $f(a) = 0$ and $a \in \mathrm{Var}(J)$. •

Theorem 7.42 (Weak Nullstellensatz). *If $f_1(X), \ldots, f_t(X) \in \mathbb{C}[X]$, then the ideal $I = (f_1, \ldots, f_t)$ is a proper ideal in $\mathbb{C}[X]$ if and only if*

$$
\mathrm{Var}(f_1, \ldots, f_t) \neq \varnothing.
$$

Proof. One direction is clear: if $\mathrm{Var}(I) \neq \varnothing$, then I is a proper ideal, because $\mathrm{Var}(\mathbb{C}[X]) = \varnothing$.

For the converse, suppose that I is a proper ideal in $\mathbb{C}[X]$. By Corollary 7.27, there is a maximal ideal M containing I, and so $K = \mathbb{C}[X]/M$ is a field. It is plain that the natural map $\mathbb{C}[X] \to \mathbb{C}[X]/M = K$ carries $\mathbb{C}$ to itself, so that $K/\mathbb{C}$ is an extension field; hence, K is a vector space over $\mathbb{C}$. Now $\mathbb{C}[X]$ has countable dimension, for a basis consists of all the monic monomials; it follows that $\dim_{\mathbb{C}}(K)$ is countable (possibly finite).

Suppose that K is a proper extension of $\mathbb{C}$; that is, there is some $t \in K$ with $t \notin \mathbb{C}$. Since $\mathbb{C}$ is algebraically closed, t cannot be algebraic over $\mathbb{C}$, and so it is transcendental. Consider the subset B of K,

$$
B = \{1/(t - c) \colon c \in \mathbb{C}\}.
$$

(note that $t - c \neq 0$ because $t \notin \mathbb{C}$). The set B is uncountable, for it is indexed by the uncountable set $\mathbb{C}$. We claim that B is linearly independent over $\mathbb{C}$; if so, then we will have contradicted the fact that $\dim_{\mathbb{C}}(K)$ is countable. If B is linearly dependent, there are nonzero $a_1, \ldots, a_r \in \mathbb{C}$ with $\sum_{i=1}^{r} a_i/(t - c_i) = 0$. Clearing denominators, we have $\sum_i a_i(t - c_1) \cdots \widehat{(t - c_i)} \cdots (t - c_r) = 0$ (the factor $t - c_i$ is omitted). Use this formula to define a polynomial $h(x) \in \mathbb{C}[x]$:

$$h(x) = \sum_i a_i(x - c_1) \cdots \widehat{(x - c_i)} \cdots (x - c_r).$$

Now $h(t) = 0$, so that t transcendental implies $h(x)$ is the zero polynomial. On the other hand, $h(c_1) = a_1(c_1 - c_2) \cdots (c_1 - c_r) \neq 0$, a contradiction. We conclude that $K/\mathbb{C}$ is not a proper extension; that is, $K = \mathbb{C}$. The natural map $\mathbb{C}[X] \to K = \mathbb{C}[X]/M = \mathbb{C}$ now satisfies the hypothesis of Lemma 7.41, and so $\text{Var}(M) \neq \varnothing$. But $\text{Var}(M) \subseteq \text{Var}(I)$, and this completes the proof. $\quad\bullet$

A translation of the German term *Nullstellensatz* is "Locus-of-zeros theorem"; this name comes from the following corollary.

Corollary 7.43. *For every proper ideal I in $\mathbb{C}[X]$, there is $a = (a_1, \ldots, a_n) \in \mathbb{C}^n$ with $f(a) = 0$ for all $f \in I$.*

Proof. Choose a to be any element in $\text{Var}(I)$. $\quad\bullet$

In case $\mathbb{C}[X] = \mathbb{C}[x]$ and $f(x) \in \mathbb{C}[x]$ is not a constant, then there is $a \in \mathbb{C}$ with $f(a) = 0$; that is, $f(x)$ has a complex root. Thus, the weak Nullstellensatz is a generalization to several variables of the fundamental theorem of algebra.

The following proof of Hilbert's Nullstellensatz uses the "Rabinowitch trick" of imbedding a polynomial ring in n variables into a polynomial ring in $n + 1$ variables.

Theorem 7.44 (Nullstellensatz). *If I is an ideal in $\mathbb{C}[X]$, then*

$$\text{Id}(\text{Var}(I)) = \sqrt{I}.$$

Thus, f vanishes on $\text{Var}(I)$ if and only if $f^m \in I$ for some $m \geq 1$.

Proof. The inclusion $\text{Id}(\text{Var}(I)) \geq \sqrt{I}$ is obviously true, for if $f^m(a) = 0$ for some $m \geq 1$ and all $a \in \text{Var}(I)$, then $f(a) = 0$ for all a, because $f(a) \in \mathbb{C}$.

For the converse, assume that $h \in \text{Id}(\text{Var}(I))$, where $I = (f_1, \ldots, f_t)$; that is, if $f_i(a) = 0$ for all i, where $a \in \mathbb{C}^n$, then $h(a) = 0$. We must show that some power of h lies in I. Of course, we may assume that h is not the zero polynomial. Let us regard

$$\mathbb{C}[x_1, \ldots, x_n] \subseteq \mathbb{C}[x_1, \ldots, x_n, y];$$

thus, every $f_i(x_1, \ldots, x_n)$ is regarded as a polynomial in $n + 1$ variables that does not depend on the last variable y. We claim that the polynomials

$$f_1, \ldots, f_t, 1 - yh$$

in $\mathbb{C}[x_1, \ldots, x_n, y]$ have no common zeros. If $(a_1, \ldots, a_n, b) \in \mathbb{C}^{n+1}$ is a common zero, then $a = (a_1, \ldots, a_n) \in \mathbb{C}^n$ is a common zero of $f_1, \ldots, f_t$, and so $h(a) = 0$. But now $1 - bh(a) = 1 \neq 0$. The weak Nullstellensatz now applies to show that the ideal $(f_1, \ldots, f_t, 1 - yh)$ in $\mathbb{C}[x_1, \ldots, x_n, y]$ is not a proper ideal. Therefore, there are $g_1, \ldots, g_{t+1} \in \mathbb{C}[x_1, \ldots, x_n, y]$ with

$$1 = f_1 g_1 + \cdots + f_t g_t + (1 - yh)g_{t+1}.$$

Now make the substitution $y = 1/h$, so that the last term involving g_{t+1} vanishes. Writing the polynomials $g_i(X, y)$ more explicitly

$$g_i(X, y) = \sum_{j=0}^{d_i} u_j(X)y^j$$

[so that $g_i(X, h^{-1}) = \sum_{j=0}^{d_i} u_j(X)h^{-j}$], we see that

$$h^{d_i} g_i(X, h^{-1}) \in \mathbb{C}[X].$$

Therefore, if $m = \max\{d_1, \ldots, d_t\}$, then

$$h^m = (h^m g_1)f_1 + \cdots + (h^m g_t)f_t \in I. \quad \bullet$$

Theorem 7.45. *Every maximal ideal M in $\mathbb{C}[x_1, \ldots, x_n]$ has the form*

$$M = (x_1 - a_1, \ldots, x_n - a_n),$$

where $a = (a_1, \ldots, a_n) \in \mathbb{C}^n$, and so there is a bijection between $\mathbb{C}^n$ and the maximal ideals in $\mathbb{C}[x_1, \ldots, x_n]$.

Proof. Since M is a proper ideal, we have $\mathrm{Var}(M) \neq \varnothing$, by Theorem 7.42; that is, there is $a = (a_1, \ldots, a_n) \in k^n$ with $f(a) = 0$ for all $f \in M$. Since $\mathrm{Var}(M) = \{b \in k^n : f(b) = 0 \text{ for all } f \in M\}$, we have $\{a\} \subseteq \mathrm{Var}(M)$. Therefore, Proposition 7.35 gives

$$\mathrm{Id}(\mathrm{Var}(M)) \subseteq \mathrm{Id}(\{a\}).$$

Now Theorem 7.44 gives $\mathrm{Id}(\mathrm{Var}(M)) = \sqrt{M}$. But $\sqrt{P} = P$ for every prime ideal P; as maximal ideals are prime, we have $\mathrm{Id}(\mathrm{Var}(M)) = M$. Note that $\mathrm{Id}(\{a\})$ is a proper ideal, for it does not contain any nonzero constant; and so maximality of M gives $M = \mathrm{Id}(\{a\})$. Let us compute $\mathrm{Id}(\{a\}) = \{f(X) \in \mathbb{C}[X] : f(a) = 0\}$. If, for each i, $f_i(x_1, \ldots, x_n) = x_i - a_i$, then $f_i(a) = 0$, and so $x_i - a_i \in \mathrm{Id}(\{a\})$. Hence, $(x_1 - a_1, \ldots, x_n - a_n) \subseteq \mathrm{Id}(\{a\})$. But $(x_1 - a_1, \ldots, x_n - a_n)$ is a maximal ideal, by Corollary 7.10, so that

$$(x_1 - a_1, \ldots, x_n - a_n) = \mathrm{Id}(\{a\}) = M. \quad \bullet$$

Hilbert proved the Nullstellensatz in 1893. The original proofs of the Nullstellensatz for arbitrary algebraically closed fields, in the 1920s, used "elimination theory" (see van der Waerden, *Modern Algebra*, Section 79) or the Noether normalization theorem (see Zariski-Samuel, *Commutative Algebra* II, pp. 164–167). Less computational proofs, involving *Jacobson rings*, were found around 1960, independently, by W. Krull and O. Goldman.

In spite of our calling Theorem 7.42 the weak Nullstellensatz, Theorem 7.42, the Nullstellensatz (Theorem 7.44), and Theorem 7.45 are all equivalent (see Exercise 7.43 on page 557).

We continue the study of the operators Var and Id.

Proposition 7.46. *Let k be a field.*

(i) *For every subset $A \subseteq k^n$,*

$$\mathrm{Var}(\mathrm{Id}(A)) \supseteq A.$$

(ii) *For every ideal $I \subseteq k[X]$.*

$$\mathrm{Id}(\mathrm{Var}(I)) \supseteq I.$$

(iii) *If $V \subseteq k^n$ is an algebraic set, then*

$$\mathrm{Var}(\mathrm{Id}(V)) = V.$$

Proof.
(i) This result is almost a tautology. If $a \in A$, then $f(a) = 0$ for all $f(X) \in \mathrm{Id}(A)$. But every $f(X) \in \mathrm{Id}(A)$ annihilates A, by definition of $\mathrm{Id}(A)$, and so $a \in \mathrm{Var}(\mathrm{Id}(A))$. Therefore, $\mathrm{Var}(\mathrm{Id}(A)) \supseteq A$.
(ii) Again, one merely looks at the definitions. If $f(X) \in I$, then $f(a) = 0$ for all $a \in \mathrm{Var}(I)$; hence, $f(X)$ is surely one of the polynomials annihilating $\mathrm{Var}(I)$.
(iii) If V is an algebraic set, then $V = \mathrm{Var}(J)$ for some ideal J in $k[X]$. Now

$$\mathrm{Var}(\mathrm{Id}(\mathrm{Var}(J))) \supseteq \mathrm{Var}(J),$$

by part (i). Also, part (ii) gives $\mathrm{Id}(\mathrm{Var}(J)) \supseteq J$, and applying Proposition 7.35(i) gives the reverse inclusion

$$\mathrm{Var}(\mathrm{Id}(\mathrm{Var}(J))) \subseteq \mathrm{Var}(J).$$

Therefore, $\mathrm{Var}(\mathrm{Id}(\mathrm{Var}(J))) = \mathrm{Var}(J)$; that is, $\mathrm{Var}(\mathrm{Id}(V)) = V$. $\quad \bullet$

Corollary 7.47.

(i) *If V_1 and V_2 are algebraic sets and $\mathrm{Id}(V_1) = \mathrm{Id}(V_2)$, then $V_1 = V_2$.*

(ii) *If I_1 and I_2 are radical ideals and $\mathrm{Var}(I_1) = \mathrm{Var}(I_2)$, then $I_1 = I_2$.*

Proof.
(i) If $\mathrm{Id}(V_1) = \mathrm{Id}(V_2)$, then $\mathrm{Var}(\mathrm{Id}(V_1)) = \mathrm{Var}(\mathrm{Id}(V_2))$; by Proposition 7.46(iii), we have $V_1 = V_2$.
(ii) If $\mathrm{Var}(I_1) = \mathrm{Var}(I_2)$, then $\mathrm{Id}(\mathrm{Var}(I_1)) = \mathrm{Id}(\mathrm{Var}(I_2))$. By the Nullstellensatz, $\sqrt{I_1} = \sqrt{I_2}$. Since I_1 and I_2 are radical ideals, by hypothesis, we have $I_1 = I_2$. •

The next theorem summarizes this discussion.

Theorem 7.48. *The functions $V \mapsto \mathrm{Id}(V)$ and $I \mapsto \mathrm{Var}(I)$ are inverse order-reversing bijections*

$$\{\text{algebraic sets} \subseteq \mathbb{C}^n\} \rightleftarrows \{\text{radical ideals} \subseteq \mathbb{C}[x_1, \dots, x_n]\}.$$

Proof. By Proposition 7.46, we have $\mathrm{Var}(\mathrm{Id}(V)) = V$ for every algebraic set V, while Theorem 7.44 gives $\mathrm{Id}(\mathrm{Var}(I)) = \sqrt{I}$ for every ideal I. •

Can an algebraic set be decomposed into smaller algebraic subsets?

Definition. An algebraic set V is ***irreducible*** if it is not a union of two proper algebraic subsets; that is, $V \neq W' \cup W''$, where both W' and W'' are algebraic sets that are proper subsets of V. A ***variety***[6] is an irreducible algebraic set.

Proposition 7.49. *Every algebraic set V in k^n is a union of finitely many varieties:*

$$V = V_1 \cup V_2 \cup \cdots \cup V_m.$$

Proof. Call an algebraic set $W \in k^n$ *good* if it is irreducible or a union of finitely many varieties; otherwise, call W *bad*. We must show that there are no bad algebraic sets. If W is bad, it is not irreducible, and so $W = W' \cup W''$, where both W' and W'' are proper algebraic subsets. But a union of good algebraic sets is good, and so at least one of W' and W'' is bad; say, W' is bad, and rename it $W' = W_1$. Repeat this construction for W_1 to get a bad algebraic subset W_2. It follows by induction that there exists a strictly descending sequence

$$W \supsetneq W_1 \supsetneq \cdots \supsetneq W_n \supsetneq \cdots$$

---

[6]The term *variety* arose as a translation by E. Beltami (inspired by Gauss) of the German term *Mannigfaltigkeit* used by Riemann; nowadays, this term is usually translated as *manifold*.

of bad algebraic subsets. Since the operator Id reverses inclusions, there is a strictly increasing chain of ideals

$$\text{Id}(W) \subsetneq \text{Id}(W_1) \subsetneq \cdots \subsetneq \text{Id}(W_n) \subsetneq \cdots$$

[the inclusions are strict because of Corollary 7.47(i)], and this contradicts the Hilbert basis theorem. We conclude that every variety is good. •

Varieties have a nice characterization.

Proposition 7.50. *An algebraic set V in k^n is a variety if and only if* $\text{Id}(V)$ *is a prime ideal in $k[X]$. Hence, the coordinate ring $k[V]$ of a variety V is a domain.*

Proof. Assume that V is a variety. It suffices to show that if $f_1(X)$, $f_2(X) \notin \text{Id}(V)$, then $f_1(X) f_2(X) \notin \text{Id}(V)$. Define, for $i = 1, 2$,

$$W_i = V \cap \text{Var}(f_i(X)).$$

Note that each W_i is an algebraic subset of V, for it is the intersection of two algebraic subsets; moreover, since $f_i(X) \notin \text{Id}(V)$, there is some $a_i \in V$ with $f_i(a_i) \neq 0$, and so W_i is a proper algebraic subset of V. Since V is irreducible, we cannot have $V = W_1 \cup W_2$. Thus, there is some $b \in V$ which is not in $W_1 \cup W_2$; that is, $f_1(b) \neq 0 \neq f_2(b)$. Therefore, $f_1(b) f_2(b) \neq 0$, hence $f_1(X) f_2(X) \notin \text{Id}(V)$, and so $\text{Id}(V)$ is a prime ideal.

Conversely, assume that $\text{Id}(V)$ is a prime ideal. Suppose that $V = V_1 \cup V_2$, where V_1 and V_2 are algebraic subsets. If $V_2 \subsetneq V$, then we must show that $V = V_1$. Now

$$\text{Id}(V) = \text{Id}(V_1) \cap \text{Id}(V_2) \supseteq \text{Id}(V_1) \text{Id}(V_2);$$

the equality is given by Proposition 7.38, and the inequality is given by Exercise 7.12 on page 524. Since $\text{Id}(V)$ is a prime ideal, Exercise 7.12(ii) says that $\text{Id}(V_1) \subseteq \text{Id}(V)$ or $\text{Id}(V_2) \subseteq \text{Id}(V)$. Now $V_2 \subsetneq V$ implies $\text{Id}(V_2) \supsetneq \text{Id}(V)$, and we conclude that $\text{Id}(V_1) \subseteq \text{Id}(V)$. But the reverse inequality $\text{Id}(V_1) \supseteq \text{Id}(V)$ holds as well, because $V_1 \subseteq V$, and so $\text{Id}(V_1) = \text{Id}(V)$. Therefore, $V_1 = V$, by Corollary 7.47, and so V is irreducible; that is, V is a variety. •

We now consider whether the varieties in the decomposition of an algebraic set into a union of varieties are uniquely determined. There is one obvious way to arrange nonuniqueness. Suppose that there are two prime ideals $P \subsetneq Q$ in $k[X]$ (for example, $(x) \subsetneq (x, y)$ are such prime ideals in $k[x, y]$). Now $\text{Var}(Q) \subsetneq \text{Var}(P)$, so that if $\text{Var}(P)$ is a subvariety of a variety V, say, $V = \text{Var}(P) \cup V_2 \cup \cdots \cup V_m$, then $\text{Var}(Q)$ can be one of the V_i or it can be left out.

Definition. A decomposition $V = V_1 \cup \cdots \cup V_m$ is an ***irredundant union*** if no V_i can be omitted; that is, for all i,

$$V \neq V_1 \cup \cdots \cup \widehat{V_i} \cup \cdots \cup V_m.$$

Proposition 7.51.   *Every algebraic set V is an irredundant union of varieties*

$$V = V_1 \cup \cdots \cup V_m;$$

moreover, the varieties V_i are uniquely determined by V.

Proof.   By Proposition 7.49, V is a union of finitely many varieties; say, $V = V_1 \cup \cdots \cup V_m$. If m is chosen minimal, then this union must be irredundant.

We now prove uniqueness. Suppose that $V = W_1 \cup \cdots \cup W_s$ is an irredundant union of varieties. Let $X = \{V_1, \ldots, V_m\}$ and let $Y = \{W_1, \ldots, W_s\}$; we shall show that $X = Y$. If $V_i \in X$, we have

$$V_i = V_i \cap V = \bigcup_j (V_i \cap W_j).$$

Now $V_i = V_i \cap W_j \neq \varnothing$ for some j; since V_i is irreducible, there is only one such W_j. Therefore, $V_i = V_i \cap W_j$, and so $V_i \subseteq W_j$. The same argument applied to W_j shows that there is exactly one V_ℓ with $W_j \subseteq V_\ell$. Hence,

$$V_i \subseteq W_j \subseteq V_\ell.$$

Since the union $V_1 \cup \cdots \cup V_m$ is irredundant, we must have $V_i = V_\ell$, and so $V_i = W_j = V_\ell$; that is, $V_i \in Y$ and $X \subseteq Y$. The reverse inclusion is proved in the same way. •

Definition.   An intersection $I = J_1 \cap \cdots \cap I_m$ is an ***irredundant intersection*** if no J_i can be omitted; that is, for all i,

$$I \neq J_1 \cap \cdots \cap \widehat{J_i} \cap \cdots \cap J_m.$$

Corollary 7.52.   *Every radical ideal J in $k[X]$ is an irredundant intersection of prime ideals,*

$$J = P_1 \cap \cdots \cap P_m;$$

moreover, the prime ideals P_i are uniquely determined by J.

Proof.   Since J is a radical ideal, there is a variety V with $J = \mathrm{Id}(V)$. Now V is an irredundant union of irreducible algebraic subsubsets,

$$V = V_1 \cup \cdots \cup V_m,$$

so that

$$J = \mathrm{Id}(V) = \mathrm{Id}(V_1) \cap \cdots \cap \mathrm{Id}(V_m).$$

By Proposition 7.50, V_i irreducible implies $\mathrm{Id}(V_i)$ is prime, and so J is an intersection of prime ideals. This is an irredundant intersection, for if there is ℓ with $J = \mathrm{Id}(V) = \bigcap_{j \neq \ell} \mathrm{Id}(V_j)$, then

$$V = \mathrm{Var}(\mathrm{Id}(V)) = \bigcup_{j \neq \ell} \mathrm{Var}(\mathrm{Id}(V_j)) = \bigcup_{j \neq \ell} V_j,$$

contradicting the given irredundancy of the union.

Uniqueness is proved similarly. If $J = \mathrm{Id}(W_1) \cap \cdots \cap \mathrm{Id}(W_s)$, where each $\mathrm{Id}(W_i)$ is a prime ideal (hence is a radical ideal), then each W_i is a variety. Applying Var expresses $V = \mathrm{Var}(\mathrm{Id}(V)) = \mathrm{Var}(J)$ as an irredundant union of varieties, and the uniqueness of this decomposition gives the uniqueness of the prime ideals in the intersection. •

Here are some natural problems arising as one investigates these ideas further. First, what is the dimension of a variety? There are several candidates, and it turns out that prime ideals are the key. If V is a variety, then its dimension is the length of a longest chain of prime ideals in its coordinate ring $k[V]$ (which, by the correspondence theorem, is the length of a longest chain of prime ideals above $\mathrm{Id}(V)$ in $k[X]$). Another problem involves intersections. If $\mathrm{Var}(f)$ is a curve arising from a polynomial of degree d, how many points lie in the intersection of V with a straight line? *Bézout's theorem* says there should be d points, but one must be careful. First, one must demand that the coefficient field be algebraically closed, lest $\mathrm{Var}(f) = \varnothing$ cause a problem. But there may also be multiple roots, and so some intersections may have to be counted with a certain *multiplicity* in order to have Bézout's theorem hold. Defining multiplicities for intersections of higher dimensional varieties is very subtle.

It turns out to be more convenient to work in a larger **projective space**. Recall that we gave two constructions of a projective plane over a field k. In Section 3.9, we constructed a projective plane by adjoining a line at infinity to the plane k^2, and this construction can be generalized to higher dimensions by adjoining a "hyperplane at infinity" to k^n. To distinguish the subset k^n from projective space, one calls it **affine space**, for k^n consists of the "finite points"—that is, those points not at infinity. We gave an alternate construction of a projective plane, in Example 4.26, in which points are essentially lines through the origin in k^3. Each point in this construction has **homogeneous coordinates** $[a_0, a_1, a_2]$, where $a_i \in k$ and $[a_0', a_1', a_2'] = [a_0, a_1, a_2]$ if there is a nonzero $t \in k$ with $a_i' = t a_i$ for all i. This construction is more amenable to algebraic sets. For fixed $n \geq 1$, define an equivalence relation on $k^{n+1} - \{0\}$ by

$$(a_0', \ldots, a_n') \equiv (a_0, \ldots, a_n)$$

if there is a nonzero $t \in k$ with $a_i' = t a_i$ for all i. Denote the equivalence class of $(a_0, \ldots, a_n)$ by $[a_0, \ldots, a_n]$, call it a **projective point**, and define **projective n-space** over k, denoted by $\mathbb{P}_n(k)$, to be the set of all projective points. It is now natural to define *projective algebraic sets* as zeros of a family of polynomials. For example, if $f(X) \in k[X] = k[x_0, x_1, \ldots, x_n]$, define

$$\text{Var}(f) = \{[a_0, \ldots, a_n] \in \mathbb{P}_n(k) : f([a_0, \ldots, a_n]) = 0\}.$$

There is a problem with this definition: $f(X)$ is defined on points in k^{n+1}, not on projective points; that is, we need $f(a_0, \ldots, a_n) = 0$ if and only if $f(ta_0, \ldots, ta_n) = 0$, where $t \in k$ is nonzero. A polynomial $f(x_0, \ldots, x_n)$ is called **homogeneous** of **degree** $m > 0$ if

$$f(tx_0, \ldots, tx_n) = t^m f(x_0, \ldots, x_n)$$

for all $t \in k$. For example, a monomial $cx_0^{e_0} \cdots x_n^{e_n}$ is homogeneous of degree m, where $m = e_0 + \cdots + e_n$ is its *total degree*; a polynomial $f(X) \in k[X]$ is homogeneous of degree m if $f(X) = \sum c_{e_0, \ldots, e_n} x_0^{e_0} \cdots x_n^{e_n}$, where the monomials all have total degree m. If $f(X)$ is homogeneous and $f(a_0, \ldots, a_n) = 0$, then $f(ta_0, \ldots, ta_n) = t^m f(a_0, \ldots, a_n) = 0$. Thus, when $f(X)$ is homogeneous, it makes sense to say that a projective point is a zero of f. Define a projective algebraic set as follows.

Definition. If $F \subseteq k[X] = k[x_0, \ldots, x_n]$ is a set of homogeneous polynomials, then the **projective algebraic set** defined by F is

$$\text{Var}(F) = \{[a] \in \mathbb{P}_n(k) : f([a]) = 0 \text{ for every } f(X) \in F\},$$

where $[a]$ abbreviates $[a_0, \ldots, a_n]$.

The reason for introducing projective space is that it is often the case that many separate affine cases become part of one simpler projective formula. Indeed, Bézout's theorem is an example of this phenomenon.

EXERCISES

H **7.35** Prove that if an element a in a commutative ring R is nilpotent, then $1 + a$ is a unit.

*H **7.36** If I is an ideal in a commutative ring R, prove that its radical, $\sqrt{I}$, is an ideal.

7.37 If R is a commutative ring, then its **nilradical** nil(R) is defined to be the intersection of all the prime ideals in R. Prove that nil(R) is the set of all the nilpotent elements in R:

$$\text{nil}(R) = \{r \in R : r^m = 0 \text{ for some } m \geq 1\}.$$

***7.38** If I and j are ideals in $\mathbb{C}[X]$, prove that $\text{Id}(\text{Var}(I) \cap \text{Var}(J)) = \sqrt{I + J}$.

7.39 If k is a field, a *hypersurface* is a subset of k^n of the form Var(f), where $f \in k[x_1, \ldots, x_n]$. Prove that every algebraic set Var(I) is an intersection of finitely many hypersurfaces.

7.40   (i)   Show that $x^2 + y^2$ is irreducible in $\mathbb{R}[x, y]$, and conclude that $(x^2 + y^2)$ is a prime, hence radical, ideal in $\mathbb{R}[x, y]$.

   (ii)   Prove that Var$(x^2 + y^2) = \{(0, 0)\}$.

(iii) Prove that $\mathrm{Id}(\mathrm{Var}(x^2 + y^2)) > (x^2 + y^2)$, and conclude that the radical ideal (x^2+y^2) in $\mathbb{R}[x, y]$ is not of the form $\mathrm{Id}(V)$ for some algebraic set V. Conclude that the Nullstellensatz may fail in $k[X]$ if k is not algebraically closed.

(iv) Prove that $(x^2 + y^2) = (x + iy) \cap (x - iy)$ in $\mathbb{C}[x, y]$.

(v) Prove that $\mathrm{Id}(\mathrm{Var}(x^2 + y^2)) = (x^2 + y^2)$ in $\mathbb{C}[x, y]$.

7.41 Prove that every radical ideal in $\mathbb{C}[X]$ is an irredundant intersection of prime ideals.

7.42 Prove that if $f_1, \ldots, f_t \in \mathbb{C}[X]$, then $\mathrm{Var}(f_1, \ldots, f_t) = \varnothing$ if and only if there are $h_1, \ldots, h_t \in k[X]$ such that

$$1 = \sum_{i=1}^{t} h_i(X) f_i(X).$$

*7.43 Consider the statements:

I. If I is a proper ideal in $\mathbb{C}[X]$, then $\mathrm{Var}(I) \neq \varnothing$.

II. $\mathrm{Id}(\mathrm{Var}(I)) = \sqrt{I}$.

III. Every maximal ideal in $\mathbb{C}[X]$ has the form $(x_1 - a_1, \ldots, x_n - a_n)$.

Prove **III** $\Rightarrow$ **I**. (We have proved **I** $\Rightarrow$ **II** and **II** $\Rightarrow$ **III** in the text.)

7.44 Let R be a commutative ring, and let

$$\mathrm{Spec}(R)$$

denote the set of all the prime ideals in R. If $E \subseteq \mathrm{Spec}(R)$, define the **closure** of a subset $E = \{P_\alpha : \alpha \in A\}$ of $\mathrm{Spec}(R)$ to be

$$\overline{E} = \left\{ \text{all the prime ideals } P \in R \text{ with } P_\alpha \subseteq P \text{ for all } P_\alpha \in E \right\}.$$

Prove the following:

(i) $\overline{\{0\}} = \mathrm{Spec}(R)$.

(ii) $\overline{R} = \varnothing$.

(iii) $\overline{\sum_\ell E_\ell} = \bigcap_\ell \overline{E_\ell}$.

(iv) $\overline{E \cap F} = \overline{E} \cup \overline{F}$.

Conclude that the family of all subsets of $\mathrm{Spec}(R)$ of the form $\overline{E}$ is a topology on $\mathrm{Spec}(R)$; it is called the **Zariski topology**.

7.45 Prove that an ideal P in $\mathrm{Spec}(R)$ is closed in the Zariski topology if and only if P is a maximal ideal.

7.46 If X and Y are topological spaces, then a function $g: X \to Y$ is **continuous** if, for each closed subset Q of Y, the inverse image $g^{-1}(Q)$ is a closed subset of X.

Let $f: R \to A$ be a ring homomorphism, and define $f^*: \mathrm{Spec}(A) \to \mathrm{Spec}(R)$ by $f^*(Q) = f^{-1}(Q)$, where Q is any prime ideal in A. Prove that f^* is a continuous function. [Recall Exercise 7.4 on page 523: $f^{-1}(Q)$ is a prime ideal.]

7.47 Prove that the function $\varphi: \mathbb{C}^n \to \mathrm{Spec}(\mathbb{C}[x_1, \ldots, x_n])]$, defined by

$$\varphi: (a_1, \ldots, a_n) \mapsto (x_1 - a_1, \ldots, x_n - a_n),$$

is a continous injection (where both $\mathbb{C}^n$ and $\mathrm{Spec}(\mathbb{C}[x_1, \ldots, x_n])$ are equipped with the Zariski topology; the Zariski topology on $\mathbb{C}^n$ was defined on page 545.

7.48 Prove that any descending chain

$$F_1 \supseteq F_2 \supseteq \cdots \supseteq F_m \supseteq F_{m+1} \supseteq \cdots$$

of closed sets in $\mathbb{C}^n$ stops; there is some t with $F_t = F_{t+1} = \cdots$.

7.5 GENERALIZED DIVISON ALGORITHM

Given two polynomials $f(x), g(x) \in k[x]$ with $g(x) \neq 0$, where k is a field, when is $g(x)$ a divisor of $f(x)$? The division algorithm gives unique polynomials $q(x), r(x) \in k[x]$ with

$$f(x) = q(x)g(x) + r(x),$$

where $r = 0$ or $\deg(r) < \deg(g)$, and $g \mid f$ if and only if the remainder $r = 0$. Let us look at this formula from a different point of view. To say that $g \mid f$ is to say that $f \in (g)$, the principal ideal generated by $g(x)$. Thus, the remainder r is the obstruction to f lying in this ideal; that is, $f \in (g)$ if and only if $r = 0$.

   Consider a more general problem. Given polynomials

$$f(x), g_1(x), \ldots, g_m(x) \in k[x],$$

where k is a field, when is $d(x) = \gcd\{g_1(x), \ldots, g_m(x)\}$ a divisor of f? The Euclidean algorithm finds d, and the division algorithm determines whether $d \mid f$. From another viewpoint, the two classical algorithms combine to give an algorithm determining whether $f \in (g_1, \ldots, g_m) = (d)$.

   We now ask whether there is an algorithm in $k[x_1, \ldots, x_n] = k[X]$ to determine, given $f(X), g_1(X), \ldots, g_m(X) \in k[X]$, whether $f \in (g_1, \ldots, g_m)$. A division algorithm in $k[X]$ should be an algorithm yielding

$$r(X), a_1(X), \ldots, a_m(X) \in k[X],$$

with $r(X)$ unique, such that

$$f = a_1 g_1 + \cdots + a_m g_m + r.$$

Since $(g_1, \ldots, g_m)$ consists of all the linear combinations of the g's, such a generalized division algorithm would say again that the remainder r is the obstruction: $f \in (g_1, \ldots, g_m)$ if and only if $r = 0$.

   We are going to show that both the division algorithm and the Euclidean algorithm can be extended to polynomials in several variables. Even though these results are elementary, they were discovered only recently, in 1965, by B. Buchberger. Algebra has always dealt with algorithms, but the power and beauty of the axiomatic method has dominated the subject ever since Cayley and Dedekind in the second half of the

nineteenth century. After the invention of the transistor in 1948, high-speed calculation became a reality, and old complicated algorithms, as well as new ones, could be
implemented; a higher order of computing had entered algebra. Most likely, the de-
velopment of computer science is a major reason why generalizations of the classical
algorithms, from polynomials in one variable to polynomials in several variables, are
only now being discovered. This is a dramatic illustration of the impact of external
ideas on mathematics.

Monomial Orders

The most important feature of the division algorithm in $k[x]$, where k is a field, is that
the remainder $r(x)$ has small degree. Without the inequality $\deg(r) < \deg(g)$, the
result would be virtually useless; after all, given any $Q(x) \in k[x]$, there is an equation

$$f(x) = Q(x)g(x) + [f(x) - Q(x)g(x)].$$

Now polynomials in several variables are sums of monomials $cx_1^{\alpha_1} \cdots x_n^{\alpha_n}$, where $c \in k$
and $\alpha_i \geq 0$ for all i. Here are two degrees that one can assign to a monomial.

Definition.   The ***multidegree*** of a monomial $cx_1^{\alpha_1} \cdots x_n^{\alpha_n} \in k[x_1, \ldots, x_n]$, where k is
a field, $c \in k$ is nonzero, and $\alpha_i \geq 0$ for all i, is the n-tuple $\alpha = (\alpha_1, \ldots, \alpha_n)$; its ***total
degree*** is the sum $|\alpha| = \alpha_1 + \cdots + \alpha_n$.

When dividing $f(x)$ by $g(x)$ in $k[x]$, one usually arranges the monomials in $f(x)$
in descending order, according to degree:

$$f(x) = c_n x^n + c_{n-1} x^{n-1} + \cdots + c_2 x^2 + c_1 x + c_0.$$

Consider a polynomial in several variables:

$$f(X) = f(x_1, \ldots, x_n) = \sum c_{(\alpha_1, \ldots, \alpha_n)} x_1^{\alpha_1} \cdots x_n^{\alpha_n}.$$

We will abbreviate $(\alpha_1, \ldots, \alpha_n)$ to α and $x_1^{\alpha_1} \cdots x_n^{\alpha_n}$ to X^α, so that $f(X)$ can be written
more compactly as

$$f(X) = \sum_\alpha c_\alpha X^\alpha.$$

Our aim is to arrange the monomials involved in $f(X)$ in a reasonable way, and we do
this by ordering their multidegrees.

The set $\mathbb{N}^n$, consisting of all the n-tuples $\alpha = (\alpha_1, \ldots, \alpha_n)$ of natural numbers, is
a commutative *monoid*[7] under addition:

$$(\alpha_1, \ldots, \alpha_n) + (\beta_1, \ldots, \beta_n) = (\alpha_1 + \beta_1, \ldots, \alpha_n + \beta_n).$$

---

[7]Recall that a *monoid* is a set with an associative binary operation and an identity element. Here,
the operation is +, the identity is $(0, \ldots, 0)$, and the operation is commutative: $\alpha + \beta = \beta + \alpha$.

This monoid operation is related to the multiplication of monomials:

$$X^\alpha X^\beta = X^{\alpha+\beta}.$$

Recall that a *partially ordered set* is a nonempty set X equipped with a relation $\preceq$ which is reflexive, antisymmetric, and transitive. Of course, we may write $x \prec y$ if $x \preceq y$ and $x \neq y$, and we may write $y \succeq x$ (or $y \succ x$) instead of $x \preceq y$ (or $x \prec y$).

Definition. A partially ordered set X is **well-ordered** if every nonempty subset $S \subseteq X$ contains a smallest element; that is, there exists $s_0 \in S$ with $s_0 \preceq s$ for all $s \in S$.

For example, the Least Integer Axiom says that the natural numbers $\mathbb{N}$ with the usual inequality $\leq$ is well-ordered.

Proposition 7.53. *Let X be a well-ordered set.*

(i) *X is a chain: if $x, y \in X$, then either $x \preceq y$ or $y \preceq x$.*

(ii) *Every strictly decreasing sequence is finite.*

Proof.
(i) The subset $S = \{x, y\}$ has a smallest element, which must be either x or y. In the first case, $x \preceq y$; in the second case, $y \preceq x$.
(ii) Assume that there is an infinite strictly decreasing sequence, say,

$$x_1 \succ x_2 \succ x_3 \succ \cdots .$$

Since X is well-ordered, the subset S consisting of all the x_i has a smallest element, say, x_n. But $x_{n+1} \prec x_n$, a contradiction.   •

The second property of well-ordered sets will be used in showing that an algorithm eventually stops. In the proof of the division algorithm for polynomials in one variable, for example, we associated a natural number to each step: the degree of the remainder. Moreover, if the algorithm does not stop at a given step, then the natural number associated to the next step–the degree of its remainder–is strictly smaller. Since the natural numbers, equipped with the usual inequality $\leq$, is a well-ordered set, this strictly decreasing sequence of natural numbers must be finite; that is, the algorithm must stop after a finite number of steps.

We are interested in orderings of multidegrees that are compatible with multiplication of monomials–that is, with addition in the monoid $\mathbb{N}^n$.

Definition. A *monomial order* is a well-order on $\mathbb{N}^n$ such that

$$\alpha \preceq \beta \quad \text{implies} \quad \alpha + \gamma \preceq \beta + \gamma$$

for all $\alpha, \beta, \gamma \in \mathbb{N}^n$.

A monomial order on $\mathbb{N}^n$ gives an ordering on monomials in $k[X] = k[x_1, \ldots, x_n]$: define $X^\alpha \preceq X^\beta$ if $\alpha \preceq \beta$; monomials are ordered according to their multidegrees.

Definition. If $\mathbb{N}^n$ is equipped with a monomial order, then every $f(X) \in k[X] = k[x_1, \ldots, x_n]$ can be written with its largest term first, followed by its other, smaller, terms in descending order: $f(X) = c_\alpha X^\alpha + $ lower terms. Define its *leading term*

$$\mathrm{LT}(f) = c_\alpha X^\alpha$$

and its *degree*

$$\mathrm{DEG}(f) = \alpha.$$

Call $f(X)$ *monic* if $\mathrm{LT}(f) = X^\alpha$; that is, if $c_\alpha = 1$.

There are many examples of monomial orders, but we shall give only the two most popular ones.

Definition. The *lexicographic order* on $\mathbb{N}^n$ is defined by $\alpha \preceq_{\mathrm{lex}} \beta$ in case $\alpha = \beta$ or the first nonzero coordinate in $\beta - \alpha$ is positive.[8]

The term *lexicographic* refers to the standard ordering in a dictionary. If $\alpha \prec_{\mathrm{lex}} \beta$, their first $i-1$ coordinates agree, for some $i \geq 1$: $\alpha_1 = \beta_1, \ldots, \alpha_{i-1} = \beta_{i-1}$, and there is strict inequality $\alpha_i < \beta_i$. For example, the following German words are increasing in lexicographic order (the letters are ordered $a < b < c < \cdots < z$):

ausgehen

ausladen

auslagen

auslegen

bedeuten

Proposition 7.54. *The lexicographic order $\preceq_{\mathrm{lex}}$ on $\mathbb{N}^n$ is a monomial order.*

Proof. First, we show that the lexicographic order is a partial order. The relation $\preceq_{\mathrm{lex}}$ is reflexive, for its definition shows that $\alpha \preceq_{\mathrm{lex}} \alpha$. To prove antisymmetry, assume that $\alpha \preceq_{\mathrm{lex}} \beta$ and $\beta \preceq_{\mathrm{lex}} \alpha$. If $\alpha \neq \beta$, there is a first coordinate, say the ith, where they disagree. For notation, we may assume that $\alpha_i < \beta_i$. But this contradicts $\beta \preceq_{\mathrm{lex}} \alpha$. To prove transitivity, suppose that $\alpha \prec_{\mathrm{lex}} \beta$ and $\beta \prec_{\mathrm{lex}} \gamma$ (it suffices to consider strict inequality). Now $\alpha_1 = \beta_1, \ldots, \alpha_{i-1} = \beta_{i-1}$ and $\alpha_i < \beta_i$. Let γ_p be the first coordinate with $\beta_p < \gamma_p$. If $p < i$, then

$$\gamma_1 = \beta_1 = \alpha_1, \ldots, \gamma_{p-1} = \beta_{p-1} = \alpha_{p-1}, \, \alpha_p = \beta_p < \gamma_p;$$

if $p \geq i$, then

$$\gamma_1 = \beta_1 = \alpha_1, \ldots, \gamma_{i-1} = \beta_{i-1} = \alpha_{i-1}, \, \alpha_i < \beta_i = \gamma_i.$$

---

[8]The difference $\beta - \alpha$ may not lie in $\mathbb{N}^n$, but it does lie in $\mathbb{Z}^n$.

In either case, the first nonzero coordinate of $\gamma - \alpha$ is positive; that is, $\alpha \prec_{\text{lex}} \gamma$.

Next, we show that the lexicographic order is a well-order. If S is a nonempty subset of $\mathbb{N}^n$, define

$$C_1 = \{\text{all first coordinates of } n\text{-tuples in } S\},$$

and define δ_1 to be the smallest number in C_1 (note that C_1 is a nonempty subset of the well-ordered set $\mathbb{N}$). Define

$$C_2 = \{\text{all second coordinates of } n\text{-tuples } (\delta_1, \alpha_2, \ldots, \alpha_n) \in S\}.$$

Since $C_2 \neq \varnothing$, it contains a smallest number, δ_2. Similarly, for all $i < n$, define C_{i+1} as all the $(i + 1)$th coordinates of those n-tuples in S whose first i coordinates are $(\delta_1, \delta_2, \ldots, \delta_i)$, and define δ_{i+1} to be the smallest number in C_{i+1}. By construction, the n-tuple $\delta = (\delta_1, \delta_2, \ldots, \delta_n)$ lies in S; moreover, if $\alpha = (\alpha_1, \alpha_2, \ldots, \alpha_n) \in S$, then

$$\alpha - \delta = (\alpha_1 - \delta_1, \alpha_2 - \delta_2, \ldots, \alpha_n - \delta_n)$$

has all its coordinates nonnegative. Hence, if $\alpha \neq \delta$, then its first nonzero coordinate is positive, and so $\delta \prec_{\text{lex}} \alpha$. Therefore, the lexicographic order is a well-order.

Assume that $\alpha \preceq_{\text{lex}} \beta$; we claim that

$$\alpha + \gamma \preceq_{\text{lex}} \beta + \gamma$$

for all $\gamma \in \mathbb{N}$. If $\alpha = \beta$, then $\alpha + \gamma = \beta + \gamma$. If $\alpha \prec_{\text{lex}} \beta$, then the first nonzero coordinate of $\beta - \alpha$ is positive. But

$$(\beta + \gamma) - (\alpha + \gamma) = \beta - \alpha,$$

and so $\alpha + \gamma \prec_{\text{lex}} \beta + \gamma$. Therefore, $\preceq_{\text{lex}}$ is a monomial order. •

In the lexicographic order, $x_1 \succ x_2 \succ x_3 \succ \cdots$, for

$$(1, 0, \ldots, 0) \prec (0, 1, 0, \ldots, 0) \prec (0, 0, 1, 0, \ldots, 0) \prec \cdots.$$

Any permutation of the variables $x_{\sigma(1)}, \ldots, x_{\sigma(n)}$ yields a different lexicographic order on $\mathbb{N}^n$.

Remark.   If X is any well-ordered set with order $\preceq$, then the lexicographic order on X^n can be defined by $a = (a_1, \ldots, a_n) \preceq_{\text{lex}} b = (b_1, \ldots, b_n)$ in case $a = b$ or if they first disagree in the ith coordinate and $a_i \prec b_i$. It is a simple matter to generalize Proposition 7.54 by replacing $\mathbb{N}$ with X. ◀

Definition. If X is a set and $n \geq 1$, we define a ***positive word*** on X of length n to be a function $w \colon \{1, 2, \ldots, n\} \to X$, and we denote w by

$$w = x_1 x_2 \cdots x_n,$$

where $x_i = w(i)$. Of course, w need not be injective; that is, there may be repetitions of x's. Two positive words can be multiplied: if $w' = x'_1 \ldots x'_m$, then

$$ww' = x_1 x_2 \cdots x_n x'_1 \ldots x'_m.$$

We introduce the ***empty word***, denoted by 1, as the word of length 0 such that $1w = w = w1$ for all positive words w. With these definitions, the set $\mathcal{W}(X)$, consisting of all the positive words on X, is a monoid.

Corollary 7.55. *If X is a well-ordered set, then $\mathcal{W}(X)$ is well-ordered in the lexicographic order (which we also denote by $\preceq_{\text{lex}}$).*

Proof. We will only give a careful definition of the lexicographic order here; the proof that it is a well-order is left to the reader. First, define $1 \preceq_{\text{lex}} w$ for all $w \in \mathcal{W}(X)$. Next, given words $u = x_1 \cdots x_p$ and $v = y_1 \cdots y_q$ in $\mathcal{W}(X)$, make them the same length by adjoining 1s at the end of the shorter word, and rename them u' and v' in $\mathcal{W}(X)$. If $m \geq \max\{p, q\}$, we may regard $u', v', \in X^m$, and we define $u \preceq_{\text{lex}} v$ if $u' \preceq_{\text{lex}} v'$ in X^m. (This is the word order commonly used in dictionaries, where a blank precedes any letter: for example, *muse* precedes *museum*.) •

Example 7.56.
Given a monomial order on $\mathbb{N}^n$, each polynomial $f(X) = \sum_\alpha c_\alpha X^\alpha \in k[X] = k[x_1, \ldots, x_n]$ can be written with the multidegrees of its terms in descending order: $\alpha_1 \succ \alpha_2 \succ \cdots \succ \alpha_p$. Write

$$\text{multiword}(f) = \alpha_1 \cdots \alpha_p \in \mathcal{W}(\mathbb{N}^n).$$

Let $c_\beta X^\beta$ be a nonzero term in $f(X)$, let $g(X) \in k[X]$ have $\text{DEG}(g) \prec \beta$, and write

$$f(X) = h(X) + c_\beta X^\beta + \ell(X),$$

where $h(X)$ is the sum of all terms in $f(X)$ of multidegree $\succ \beta$ and $\ell(X)$ is the sum of all terms in $f(X)$ of multidegree $\prec \beta$. We claim that

$$\text{multiword}(f(X) - c_\beta X^\beta + g(X)) \prec_{\text{lex}} \text{multiword}(f) \text{ in } \mathcal{W}(X).$$

The sum of the terms in $f(X) - c_\beta X^\beta + g(X)$ with multidegree $\succ \beta$ is $h(X)$, while the sum of the lower terms is $\ell(X) + g(X)$. But $\text{DEG}(\ell + g) \prec \beta$, by Exercise 7.51 on page 565. Therefore, the initial terms of $f(X)$ and $f(X) - c_\beta X^\beta + g(X)$ agree, while the next term of $f(X) - c_\beta X^\beta + g(X)$ has multidegree $\prec \beta$, and this proves the claim.

   If $f(X) \to f(X) - c_\beta X^\beta + g(X)$, where $c_\beta X^\beta$ is a nonzero term of $f(X)$ and $\text{DEG}(g) \prec \beta$, then $f(X) \succ_{\text{lex}} f(X) - c_\beta X^\beta + g(X)$. Since $\mathcal{W}(\mathbb{N}^n)$ is well-ordered, any sequence of steps of this form must be finite. ◄

Here is the second popular monomial order.

Definition.   The *degree-lexicographic order* on $\mathbb{N}^n$ is defined by $\alpha \preceq_{\text{dlex}} \beta$ in case $\alpha = \beta$ or

$$|\alpha| = \sum_{i=1}^{n} \alpha_i < \sum_{i=1}^{n} \beta_i = |\beta|,$$

or, if $|\alpha| = |\beta|$, then the first nonzero coordinate in $\beta - \alpha$ is positive.

In other words, given $\alpha = (\alpha_1, \ldots, \alpha_n)$ and $\beta = (\beta_1, \ldots, \beta_n)$, first check total degrees: if $|\alpha| < |\beta|$, then $\alpha \preceq_{\text{dlex}} \beta$; if there is a tie, that is, if α and β have the same total degree, then order them lexicographically. For example, $(1, 2, 3, 0) \prec_{\text{dlex}} (0, 2, 5, 0)$ and $(1, 2, 3, 4) \prec_{\text{dlex}} (1, 2, 5, 2)$.

Proposition 7.57.   *The degree-lexicographic order $\preceq_{\text{dlex}}$ is a monomial order on $\mathbb{N}^n$.*

Proof.   It is routine to show that $\preceq_{\text{dlex}}$ is a partial order on $\mathbb{N}^n$. To see that it is a well-order, let S be a nonempty subset of $\mathbb{N}^n$. The total degrees of elements in S form a nonempty subset of $\mathbb{N}$, and so there is a smallest such, say, t. The nonempty subset of all $\alpha \in S$ having total degree t has a smallest element, because the degree-lexicographic order $\preceq_{\text{dlex}}$ coincides with the lexicographic order $\preceq_{\text{lex}}$ on this subset. Therefore, there is a smallest element in S in the degree-lexicographic order.

Assume that $\alpha \preceq_{\text{dlex}} \beta$ and $\gamma \in \mathbb{N}^n$. Now $|\alpha + \gamma| = |\alpha| + |\gamma|$, so that $|\alpha| = |\beta|$ implies $|\alpha + \gamma| = |\beta + \gamma|$ and $|\alpha| < |\beta|$ implies $|\alpha + \gamma| < |\beta + \gamma|$; in the latter case, Proposition 7.54 shows that $\alpha + \gamma \preceq_{\text{dlex}} \beta + \gamma$.   •

The next proposition shows, with respect to any monomial order, that polynomials in several variables behave like polynomials in a single variable.

Proposition 7.58.   *Let $\preceq$ be a monomial order on $\mathbb{N}^n$, and let $f(X), g(X), h(X) \in k[X] = k[x_1, \ldots, x_n]$.*

(i) *If* $\text{DEG}(f) = \text{DEG}(g)$, *then* $\text{LT}(g) \mid \text{LT}(f)$.

(ii) $\text{LT}(hg) = \text{LT}(h)\text{LT}(g)$.

(iii) *If* $\text{DEG}(f) = \text{DEG}(hg)$, *then* $\text{LT}(g) \mid \text{LT}(f)$.

Proof.
(i) If $\text{DEG}(f) = \alpha = \text{DEG}(g)$, then $\text{LT}(f) = cX^\alpha$ and $\text{LT}(g) = dX^\alpha$. Hence, $\text{LT}(g) \mid \text{LT}(f)$ [and also $\text{LT}(f) \mid \text{LT}(g)$].
(ii) Let $h(X) = bX^\gamma + $ lower terms and let $h(X) = cX^\beta + $ lower terms, so that $\text{LT}(h) = cX^\gamma$ and $\text{LT}(g) = bX^\beta$. Clearly, $cbX^{\gamma+\beta}$ is a nonzero term of $h(X)g(X)$. To see that it is the leading term, let $c_\mu X^\mu$ be a term of $h(X)$ with $\mu \prec \gamma$, and let $b_\nu X^\nu$ be a term of $g(X)$ with $\nu \prec \beta$. Now $\text{DEG}(c_\mu X^\mu b_\nu X^\nu) = \mu + \nu$; since $\preceq$ is a monomial

order, we have $\mu + \nu \prec \gamma + \nu \prec \gamma + \beta$. Thus, $cbX^{\gamma+\beta}$ is the term in $h(X)g(X)$ with largest multidegree.

(iii) Since $\mathrm{DEG}(f) = \mathrm{DEG}(hg)$, part (i) gives $\mathrm{LT}(hg) \mid \mathrm{LT}(f)$ and part (ii) gives $\mathrm{LT}(h)\mathrm{LT}(g) = \mathrm{LT}(hg)$; hence, $\mathrm{LT}(g) \mid \mathrm{LT}(f)$. •

EXERCISES

7.49   (i)   Write the first 10 monic monomials in $k[x, y]$ in lexicographic order and in degree-lexicographic order.

      (ii)   Write all the monic monomials in $k[x, y, z]$ of total degree at most 2 in lexicographic order and in degree-lexicographic order.

7.50   Give an example of a well-ordered set X containing an element u having infinitely many predecessors.

*7.51   Let $\preceq$ be a monomial order on $\mathbb{N}^n$, and let $f(X), g(X) \in k[X] = k[x_1, \ldots, x_n]$ be nonzero polynomials. Prove that if $f + g \neq 0$, then

$$\mathrm{DEG}(f + g) \preceq \max\{\mathrm{DEG}(f), \mathrm{DEG}(g)\},$$

and that strict inequality can occur only if $\mathrm{DEG}(f) = \mathrm{DEG}(g)$.

Division Algorithm

We are now going to use monomial orders to give a division algorithm for polynomials in several variables.

Definition.   Let $\preceq$ be a monomial order on $\mathbb{N}^n$ and let $f(X), g(X) \in k[X] = k[x_1, \ldots, x_n]$. If there is a nonzero term $c_\beta X^\beta$ in $f(X)$ with $\mathrm{LT}(g) \mid c_\beta X^\beta$ and

$$h(X) = f(X) - \frac{c_\beta X^\beta}{\mathrm{LT}(g)} g(X),$$

then *reduction* $f \xrightarrow{g} h$ is the replacement of f by h.

Reduction is precisely the usual step involved in long division of polynomials of one variable; if $f \xrightarrow{g} h$, then we have used g to eliminate a term from f, yielding h. Of course, a special case of reduction is when $c_\beta X^\beta = \mathrm{LT}(f)$.

Proposition 7.59.   *Let $\preceq$ be a monomial order on $\mathbb{N}^n$, let $f(X), g(X) \in k[X] = k[x_1, \ldots, x_n]$, and assume that $f \xrightarrow{g} h$; that is, there is a nonzero term $c_\beta X^\beta$ in $f(X)$ with $\mathrm{LT}(g) \mid c_\beta X^\beta$ and $h(X) = f(X) - \frac{c_\beta X^\beta}{\mathrm{LT}(g)} g(X)$.*

If $\beta = \mathrm{DEG}(f)$, then either

$$h(X) = 0 \quad or \quad \mathrm{DEG}(h) \prec \mathrm{DEG}(f);$$

if $\beta \prec \mathrm{DEG}(f)$, then $\mathrm{DEG}(h) = \mathrm{DEG}(f)$. In either case,

$$\mathrm{DEG}\left(\frac{c_\beta X^\beta}{\mathrm{LT}(g)} g(X)\right) \preceq \mathrm{DEG}(f).$$

Proof. Let us write

$$f(X) = \mathrm{LT}(f) + c_\kappa X^\kappa + \text{lower terms};$$

since $c_\beta X^\beta$ is a term of $f(X)$, we have $\beta \preceq \mathrm{DEG}(f)$. If $\mathrm{LT}(g) = a_\gamma X^\gamma$, so that $\mathrm{DEG}(g) = \gamma$, let us write

$$g(X) = a_\gamma X^\gamma + a_\lambda X^\lambda + \text{lower terms}.$$

Hence,

$$\begin{aligned}
h(X) &= f(X) - \frac{c_\beta X^\beta}{\mathrm{LT}(g)} g(X) \\
&= f(X) - \frac{c_\beta X^\beta}{\mathrm{LT}(g)}\left[\mathrm{LT}(g) + a_\lambda X^\lambda + \cdots\right] \\
&= \left[f(X) - c_\beta X^\beta\right] - \frac{c_\beta X^\beta}{\mathrm{LT}(g)}\left[a_\lambda X^\lambda + \cdots\right].
\end{aligned}$$

Now $\mathrm{LT}(g) \mid c_\beta X^\beta$ says that $\beta - \gamma \in \mathbb{N}^n$. We claim that

$$\mathrm{DEG}\left(-\frac{c_\beta X^\beta}{\mathrm{LT}(g)}\left[a_\lambda X^\lambda + \cdots\right]\right) = \lambda + \beta - \gamma \prec \beta.$$

The inequality holds, for $\lambda \prec \gamma$ implies $\lambda + (\beta - \gamma) \prec \gamma + (\beta - \gamma) = \beta$. To see that $\lambda + \beta - \gamma$ is the degree, it suffices to show that $\lambda + \beta - \gamma = \mathrm{DEG}\left(-\frac{c_\beta X^\beta}{\mathrm{LT}(g)} a_\lambda X^\lambda\right)$ is the largest multidegree occurring in $-\frac{c_\beta X^\beta}{\mathrm{LT}(g)}\left[a_\lambda X^\lambda + \cdots\right]$. But if $a_\eta X^\eta$ is a lower term in $g(X)$, i.e., $\eta \prec \lambda$, then $\preceq$ being a monomial order gives $\eta + (\beta - \gamma) \prec \lambda + (\beta - \gamma)$, as desired.

If $h(X) \neq 0$, then Exercise 7.51 on page 565 gives

$$\mathrm{DEG}(h) \preceq \max\left\{\mathrm{DEG}\left(f(X) - c_\beta X^\beta\right), \mathrm{DEG}\left(-\frac{c_\beta X^\beta}{\mathrm{LT}(g)}\left[a_\lambda X^\lambda + \cdots\right]\right)\right\}.$$

Now if $\beta = \mathrm{DEG}(f)$, then $c_\beta X^\beta = \mathrm{LT}(f)$,

$$f(X) - c_\beta X^\beta = f(X) - \mathrm{LT}(f) = c_\kappa X^\kappa + \text{lower terms},$$

and, hence, $\mathrm{DEG}(f(X) - c_\beta X^\beta) = \kappa \prec \mathrm{DEG}(f)$. Therefore, $\mathrm{DEG}(h) \prec \mathrm{DEG}(f)$ in this case. If $\beta \prec \mathrm{DEG}(f)$, then $\mathrm{DEG}(f(X) - c_\beta X^\beta) = \mathrm{DEG}(f)$, while $\mathrm{DEG}\left(-\frac{c_\beta X^\beta}{\mathrm{LT}(g)}[a_\lambda X^\lambda + \cdots]\right) \prec \beta \prec \mathrm{DEG}(f)$, and so $\mathrm{DEG}(h) = \mathrm{DEG}(f)$ in this case.

The last inequality is clear, for

$$\frac{c_\beta X^\beta}{\mathrm{LT}(g)} g(X) = c_\beta X^\beta + \frac{c_\beta X^\beta}{\mathrm{LT}(g)}[a_\lambda X^\lambda + \cdots].$$

Since $\mathrm{DEG}\left(-\frac{c_\beta X^\beta}{\mathrm{LT}(g)}[a_\lambda X^\lambda + \cdots]\right) \prec \beta$, we see that

$$\mathrm{DEG}\left(\frac{c_\beta X^\beta}{\mathrm{LT}(g)} g(X)\right) = \beta \preceq \mathrm{DEG}(f). \quad \bullet$$

Definition. Let $\{g_1, \ldots, g_m\}$, where $g_i = g_i(X) \in k[X]$. A polynomial $r(X)$ is *reduced mod* $\{g_1, \ldots, g_m\}$ if either $r(X) = 0$ or no $\mathrm{LT}(g_i)$ divides any nonzero term of $r(X)$.

Here is the division algorithm for polynomials in several variables. Because the algorithm requires the "divisor polynomials" $\{g_1, \ldots, g_m\}$ to be used in a specific order (after all, an algorithm must give explicit directions), we will be using an m-tuple of polynomials instead of a subset of polynomials. We use the notation $[g_1, \ldots, g_m]$ for the m-tuple whose ith entry is g_i, because the usual notation $(g_1, \ldots, g_m)$ would be confused with the notation for the ideal $(g_1, \ldots, g_m)$ generated by the g_i.

Theorem 7.60 (Division Algorithm in $k[x_1, \ldots, x_n]$). *Let $\preceq$ be a monomial order on $\mathbb{N}^n$, and let $k[X] = k[x_1, \ldots, x_n]$. If $f(X) \in k[X]$ and $G = [g_1(X), \ldots, g_m(X)]$ is an m-tuple of polynomials in $k[X]$, then there is an algorithm giving polynomials $r(X), a_1(X), \ldots, a_m(X) \in k[X]$ with*

$$f = a_1 g_1 + \cdots + a_m g_m + r,$$

where r is reduced $\mathrm{mod}\{g_1, \ldots, g_m\}$, *and* $\mathrm{DEG}(a_i g_i) \preceq \mathrm{DEG}(f)$ *for all* i.

Proof. Once a monomial order is chosen, so that leading terms are defined, the algorithm is a straightforward generalization of the division algorithm in one variable. First, reduce $\mathrm{mod} g_1$ as many times as possible, then reduce mod g_2, then reduce mod g_1 again, etc. Here is a pseudocode describing the algorithm more precisely.

Input: $f(X) = \sum_\beta c_\beta X^\beta$, $[g_1, \ldots, g_m]$
Output: $r, a_1, \ldots, a_m$
$r := f$; $a_i := 0$
WHILE f is not reduced mod $\{g_1, \ldots, g_m\}$ DO
    select smallest i with $\mathrm{LT}(g_i) \mid c_\beta X^\beta$ for some β
    $f - [c_\beta X^\beta / \mathrm{LT}(g_i)] g_i := f$
    $a_i + [c_\beta X^\beta / \mathrm{LT}(g_i)] := a_i$
END WHILE

At each step $h_j \xrightarrow{g_i} h_{j+1}$ of the algorithm, multiword(h_j) $\succ_{\text{lex}}$ multiword(h_{j+1}) in $\mathcal{W}(\mathbb{N}^n)$, by Example 7.56, and so the algorithm does stop, because $\preceq_{\text{lex}}$ is a well-order on $\mathcal{W}(\mathbb{N}^n)$. Obviously, the output $r(X)$ is reduced mod $\{g_1, \ldots, g_m\}$, for if it has a term divisible by some $\mathrm{LT}(g_i)$, then one further reduction is possible.

Finally, each term of $a_i(X)$ has the form $c_\beta X^\beta / \mathrm{LT}(g_i)$ for some intermediate output $h(X)$ (as one sees in the pseudocode). It now follows from Proposition 7.59 that either $a_i g_i = 0$ or $\mathrm{DEG}(a_i g_i) \prec \mathrm{DEG}(f)$.   •

Definition. Given a monomial order on $\mathbb{N}^n$, a polynomial $f(X) \in k[X]$, and an m-tuple $G = [g_1, \ldots, g_m]$, we call the output $r(X)$ of the division algorithm the *remainder of f mod G*.

The remainder r of f mod G is reduced mod $\{g_1, \ldots, g_m\}$, and $f - r \in I = (g_1, \ldots, g_m)$. The algorithm requires that G be an m-tuple, because of the command

$$\text{select smallest } i \text{ with } \mathrm{LT}(g_i) \mid c_\beta X^\beta \text{ for some } \beta$$

specifying the order of reductions. The next example shows that the remainder may depend not only on the set of polynomials $\{g_1, \ldots, g_m\}$ but also on the ordering of the coordinates in the m-tuple $G = [g_1, \ldots, g_m]$. That is, if $\sigma \in S_m$ is a permutation and $G_\sigma = [g_{\sigma(1)}, \ldots, g_{\sigma(m)}]$, then the remainder r_σ of f mod G_σ may not be the same as the remainder r of f mod G. Even worse, it is possible that $r \neq 0$ and $r_\sigma = 0$, so that the remainder mod G is not the obstruction to f being in the ideal $(g_1, \ldots, g_m)$.

Example 7.61.
Let $f(x, y, z) = x^2 y^2 + xy$, and let $G = [g_1, g_2, g_3]$, where

$$g_1 = y^2 + z^2$$
$$g_2 = x^2 y + yz$$
$$g_3 = z^3 + xy.$$

We use the degree-lexicographic order on $\mathbb{N}^3$. Now $y^2 = \mathrm{LT}(g_1) \mid \mathrm{LT}(f) = x^2 y^2$, and so $f \xrightarrow{g_1} h$, where $h = f - \frac{x^2 y^2}{y^2}(y^2 + z^2) = -x^2 z^2 + xy$. The polynomial $-x^2 z^2 + xy$

is reduced mod G, because neither $-x^2z^2$ nor xy is divisible by any of the leading terms $\text{LT}(g_1) = y^2$, $\text{LT}(g_2) = x^2y$, or $\text{LT}(g_3) = z^3$.

On the other hand, let us apply the division algorithm using the 3-tuple $G' = [g_2, g_1, g_3]$. The first reduction gives $f \xrightarrow{g_2} h'$, where

$$h' = f - \frac{x^2y^2}{x^2y}(x^2y + yz) = -y^2z + xy.$$

Now h' is not reduced, and reducing mod g_1 gives

$$h' - \frac{-y^2z}{y^2}(y^2 + z^2) = z^3 + xy.$$

But $z^3 + xy = g_3$, and so $z^3 + xy \xrightarrow{g_3} 0$.

Thus, the remainder depends on the ordering of the divisor polynomials g_i in the m-tuple.

For a simpler example of different remainders (but with neither remainder being 0), see Exercise 7.52. ◄

The dependence of the remainder on the order of the g_i in the m-tuple $G = [g_1, \ldots, g_m]$ will be treated in the next subsection.

EXERCISES

*7.52 Let $G = [x - y, x - z]$ and $G' = [x - z, x - y]$. Show that the remainder of x mod G (degree-lexicographic order) is distinct from the remainder of x mod G'.

7.53 Use the degree-lexicographic order in this exercise.
  (i) Find the remainder of $x^7y^2 + x^3y^2 - y + 1$ mod $[xy^2 - x, x - y^3]$.
  (ii) Find the remainder of $x^7y^2 + x^3y^2 - y + 1$ mod $[x - y^3, xy^2 - x]$.

7.54 Use the degree-lexicographic order in this exercise.
  (i) Find the remainder of $x^2y + xy^2 + y^2$ mod $[y^2 - 1, xy - 1]$.
  (ii) Find the remainder of $x^2y + xy^2 + y^2$ mod $[xy - 1, y^2 - 1]$.

*7.55 Let $c_\alpha X^\alpha$ be a nonzero monomial, and let $f(X), g(X) \in k[X]$ be polynomials none of whose terms is divisible by $c_\alpha X^\alpha$. Prove that none of the terms of $f(X) - g(X)$ is divisible by $c_\alpha X^\alpha$.

*7.56 A *monomial ideal* in $k[X]$ is an ideal I generated by monomials: $I = (X^{\alpha(1)}, \ldots, X^{\alpha(q)})$.
  (i) Prove that $f(X) \in I$ if and only if each term of $f(X)$ is divisible by some $X^{\alpha(i)}$.
  (ii) Prove that if $G = [g_1, \ldots, g_m]$ and r is reduced mod G, then r does not lie in the monomial ideal $(\text{LT}(g_1), \ldots, \text{LT}(g_m))$.

7.6 GRÖBNER BASES

We will assume in this section that $\mathbb{N}^n$ is equipped with some monomial order (the reader may use the degree-lexicographic order), so that $\mathrm{LT}(f)$ is defined and the division algorithm makes sense.

We have seen that the remainder of f mod $[g_1, \ldots, g_m]$ obtained from the division algorithm can depend on the order in which the g_i are listed. A *Gröbner basis* $\{g_1, \ldots, g_m\}$ of the ideal $I = (g_1, \ldots, g_m)$ is a basis such that, for any of the m-tuples G formed from the g_i, the remainder of f mod G is always the obstruction to whether f lies in I; this will be a consequence of the definition (which is given to make make sure Gröbner bases are sets and not m-tuples).

Definition.   A *set* of polynomials $\{g_1, \ldots, g_m\}$ is a ***Gröbner basis***[9] of the ideal $I = (g_1, \ldots, g_m)$ if, for each nonzero $f \in I$, $\mathrm{LT}(g_i) \mid \mathrm{LT}(f)$ for some g_i.

Example 7.61 on page 568 shows that

$$\{y^2 + z^2, x^2y + yz, z^3 + xy\}$$

is not a Gröbner basis of the ideal $I = (y^2 + z^2, x^2y + yz, z^3 + xy)$.

Proposition 7.62.   *A set $\{g_1, \ldots, g_m\}$ of polynomials is a Gröbner basis of $I = (g_1, \ldots, g_m)$ if and only if, for each m-tuple $G_\sigma = [g_{\sigma(1)}, \ldots, g_{\sigma(m)}]$ (where $\sigma \in S_m$), every $f \in I$ has remainder 0 mod G_σ.*

Proof.   Assume there is some permutation $\sigma \in S_m$ and some $f \in I$ whose remainder mod G_σ is not 0. Among all such polynomials, choose f of minimal Degree. Since $\{g_1, \ldots, g_m\}$ is a Gröbner basis, $\mathrm{LT}(g_i) \mid \mathrm{LT}(f)$ for some i; select the smallest $\sigma(i)$ for which there is a reduction $f \overset{g_{\sigma(i)}}{\to} h$, and note that $h \in I$. Since $\mathrm{DEG}(h) \prec \mathrm{DEG}(f)$, by Proposition 7.59, the division algorithm gives a sequence of reductions $h = h_0 \to h_1 \to h_2 \to \cdots \to h_p = 0$. But the division algorithm for f adjoins $f \to h$ at the front, showing that 0 is the remainder of f mod G_σ, a contradiction.

Conversely, let $\{g_1, \ldots, g_m\}$ be a Gröbner basis of $I = (g_1, \ldots, g_m)$. If there is a nonzero $f \in I$ with $\mathrm{LT}(g_i) \nmid \mathrm{LT}(f)$ for every i, then in any reduction $f \overset{g_i}{\to} h$, we have $\mathrm{LT}(h) = \mathrm{LT}(f)$. Hence, if $G = [g_1, \ldots, g_m]$, the division algorithm mod G gives reductions $f \to h_1 \to h_2 \to \cdots \to h_p = r$ in which $\mathrm{LT}(r) = \mathrm{LT}(f)$. Therefore, $r \neq 0$; that is, the remainder of f mod G is not zero, and this is a contradiction.   •

---

[9]It was B. Buchberger who, in his dissertation, proved the main properties of Gröbner bases. He named these bases to honor his thesis advisor, W. Gröbner.

Corollary 7.63. *Let $I = (g_1, \ldots, g_m)$ be an ideal, let $\{g_1, \ldots, g_m\}$ be a Gröbner basis of I, and let $G = [g_1, \ldots, g_m]$ be any m-tuple formed from the g_i. If $f(X) \in k[X]$, then there is a unique $r(X) \in k[X]$, which is reduced mod $\{g_1, \ldots, g_m\}$, such that $f - r \in I$; in fact, r is the remainder of f mod G.*

Proof. The division algorithm gives a polynomial r reduced mod $\{g_1, \ldots, g_m\}$ and polynomials $a_1, \ldots, a_m$ with $f = a_1 g_1 + \cdots + a_m g_m + r$; clearly, $f - r = a_1 g_1 + \cdots + a_m g_m \in I$.

To prove uniqueness, suppose that r and r' are reduced mod $\{g_1, \ldots, g_m\}$ and that $f - r$ and $f - r'$ lie in I, so that $(f - r') - (f - r) = r - r' \in I$. Since r and r' are reduced mod $\{g_1, \ldots, g_m\}$, none of their terms is divisible by any $\mathrm{LT}(g_i)$. If $r - r' \neq 0$, then Exercise 7.55 on page 569 says that no term of $r - r'$ is divisible by any $\mathrm{LT}(g_i)$; in particular, $\mathrm{LT}(r - r')$ is not divisible by any $\mathrm{LT}(g_i)$, and this contradicts Proposition 7.62. Therefore, $r = r'$. •

The next corollary shows that Gröbner bases resolve the problem of different remainders in the division algorithm arising from different m-tuples.

Corollary 7.64. *Let $I = (g_1, \ldots, g_m)$ be an ideal, let $\{g_1, \ldots, g_m\}$ be a Gröbner basis of I, and let G be the m-tuple $G = [g_1, \ldots, g_m]$.*

(i) *If $f(X) \in k[X]$ and $G_\sigma = [g_{\sigma(1)}, \ldots, g_{\sigma(m)}]$, where $\sigma \in S_m$ is a permutation, then the remainder of f mod G is equal to the remainder of f mod G_σ.*

(ii) *A polynomial $f \in I$ if and only if f has remainder 0 mod G.*

Proof.
(i) If r is the remainder of f mod G, then Corollary 7.63 says that r is the unique polynomial, reduced mod $\{g_1, \ldots, g_m\}$, with $f - r \in I$; similarly, the remainder r_σ of f mod G_σ is the unique polynomial, reduced mod $\{g_1, \ldots, g_m\}$, with $f - r_\sigma \in I$. The uniqueness assertion in Corollary 7.63 gives $r = r_\sigma$.
(ii) Proposition 7.62 shows that if $f \in I$, then its remainder is 0. For the converse, if r is the remainder of f mod G, then $f = q + r$, where $q \in I$. Hence, if $r = 0$, then $f \in I$. •

There are several obvious questions. Do Gröbner bases exist and, if they do, are they unique? Given an ideal I in $k[X]$, is there an algorithm to find a Gröbner basis of I?

The notion of *S-polynomial* will allow us to recognize a Gröbner basis, but we first introduce some notation.

Definition. If $\alpha = (\alpha_1, \ldots, \alpha_n)$ and $\beta = (\beta_1, \ldots, \beta_n)$ are in $\mathbb{N}^n$, define

$$\alpha \vee \beta = \mu,$$

where $\mu = (\mu_1, \ldots, \mu_n)$ is given by $\mu_i = \max\{\alpha_i, \beta_i\}$.

Note that $X^{\alpha \vee \beta}$ is the least common multiple of the monomials X^α and X^β.

Definition.   Let $f(X), g(X) \in k[X]$, where $\text{LT}(f) = a_\alpha X^\alpha$ and $\text{LT}(g) = b_\beta X^\beta$. Define

$$L(f, g) = X^{\alpha \vee \beta}.$$

The *S-polynomial* $S(f, g)$ is defined by

$$
\begin{aligned}
S(f, g) &= \frac{L(f, g)}{\text{LT}(f)} f - \frac{L(f, g)}{\text{LT}(g)} g \\
&= a_\alpha^{-1} X^{(\alpha \vee \beta) - \alpha} f(X) - b_\beta^{-1} X^{(\alpha \vee \beta) - \beta} g(X).
\end{aligned}
$$

Note that $S(f, g) = -S(g, f)$.

Example 7.65.
We show that if $f = X^\alpha$ and $g = X^\beta$ are monomials, then $S(f, g) = 0$. Since f and g are monomials, we have $\text{LT}(f) = f$ and $\text{LT}(g) = g$. Hence,

$$S(f, g) = \frac{L(f, g)}{\text{LT}(f)} f - \frac{L(f, g)}{\text{LT}(g)} g = \frac{X^{\alpha \vee \beta}}{f} f - \frac{X^{\alpha \vee \beta}}{g} g = 0. \quad \blacktriangleleft$$

The following technical lemma indicates why S-polynomials are relevant.

Lemma 7.66.   *Given* $g_1(X), \ldots, g_\ell(X) \in k[X]$ *and monomials* $c_j X^{\alpha(j)}$, *let* $h(X) = \sum_{j=1}^{\ell} c_j X^{\alpha(j)} g_j(X)$.

Let δ be a multidegree. If $\text{DEG}(h) \prec \delta$ *and* $\text{DEG}(c_j X^{\alpha(j)} g_j(X)) = \delta$ *for all* $j < \ell$, *then there are* $d_j \in k$ *with*

$$h(X) = \sum_j d_j X^{\delta - \mu(j)} S(g_j, g_{j+1}),$$

where $\mu(j) = \text{DEG}(g_j) \vee \text{DEG}(g_{j+1})$, *and for all* $j < \ell$,

$$\text{DEG}\left(X^{\delta - \mu(j)} S(g_j, g_{j+1})\right) \prec \delta.$$

Remark.   The lemma says that if $\text{DEG}(\sum_j a_j g_j) \prec \delta$, where the a_j are monomials, while $\text{DEG}(a_j g_j) = \delta$ for all j, then h can be rewritten as a linear combination of S-polynomials, with monomial coefficents, each of whose terms has multidegree strictly less than δ. $\blacktriangleleft$

Proof. Let $\text{LT}(g_j) = b_j X^{\beta(j)}$, so that $\text{LT}(c_j X^{\alpha(j)} g_j(X)) = c_j b_j X^{\delta}$. The coefficient of X^{δ} in $h(X)$ is thus $\sum_j c_j b_j$. Since $\text{DEG}(h) \prec \delta$, we must have $\sum_j c_j b_j = 0$. Define monic polynomials

$$u_j(X) = b_j^{-1} X^{\alpha(j)} g_j(X).$$

There is a telescoping sum

$$h(X) = \sum_{j=1}^{\ell} c_j X^{\alpha(j)} g_j(X)$$

$$= \sum_{j=1}^{\ell} c_j b_j u_j$$

$$= c_1 b_1 (u_1 - u_2) + (c_1 b_1 + c_2 b_2)(u_2 - u_3) + \cdots$$

$$+ (c_1 b_1 + \cdots + c_{\ell-1} b_{\ell-1})(u_{\ell-1} - u_{\ell})$$

$$+ (c_1 b_1 + \cdots + c_{\ell} b_{\ell}) u_{\ell}.$$

Now the last term $(c_1 b_1 + \cdots + c_{\ell} b_{\ell}) u_{\ell} = 0$ because $\sum_j c_j b_j = 0$. We have $\alpha(j) + \beta(j) = \delta$, since $\text{DEG}(c_j X^{\alpha(j)} g_j(X)) = \delta$, so that $X^{\beta(j)} \mid X^{\delta}$ for all j. Hence, for all $j < \ell$, we have $\text{lcm}\{X^{\beta(j)}, X^{\beta(j+1)}\} = X^{\beta(j) \vee \beta(j+1)} \mid X^{\delta}$; that is, if we write $\mu(j) = \beta(j) \vee \beta(j+1)$, then $\delta - \mu(j) \in \mathbb{N}^n$. But

$$X^{\delta - \mu(j)} S(g_j, g_{j+1}) = X^{\delta - \mu(j)} \left(\frac{X^{\mu(j)}}{\text{LT}(g_j)} g_j(X) - \frac{X^{\mu(j)}}{\text{LT}(g_{j+1})} g_{j+1}(X) \right)$$

$$= \frac{X^{\delta}}{\text{LT}(g_j)} g_j(X) - \frac{X^{\delta}}{\text{LT}(g_{j+1})} g_{j+1}(X)$$

$$= b_j^{-1} X^{\alpha(j)} g_j - b_{j+1}^{-1} X^{\alpha(j+1)} g_{j+1}$$

$$= u_j - u_{j+1}.$$

Substituting this equation into the telescoping sum gives a sum of the desired form, where $d_j = c_1 b_1 + \cdots + c_j b_j$:

$$h(X) = c_1 b_1 X^{\delta - \mu(1)} S(g_1, g_2) + (c_1 b_1 + c_2 b_2) X^{\delta - \mu(2)} S(g_2, g_3) + \cdots$$

$$+ (c_1 b_1 + \cdots + c_{\ell-1} b_{\ell-1}) X^{\delta - \mu(\ell-1)} S(g_{\ell-1}, g_{\ell}).$$

Finally, since both u_j and u_{j+1} are monic with leading term of multidegree δ, we have $\text{DEG}(u_j - u_{j+1}) \prec \delta$. But we have shown that $u_j - u_{j+1} = X^{\delta - \mu(j)} S(g_j, g_{j+1})$, and so $\text{DEG}(X^{\delta - \mu(j)} S(g_j, g_{j+1})) \prec \delta$, as desired. •

Let $I = (g_1, \ldots, g_m)$. By Proposition 7.62, $\{g_1, \ldots, g_m\}$ is a Gröbner basis of the ideal I if every $f \in I$ has remainder $0 \bmod G$ (where G is any m-tuple formed by ordering the g_i). The importance of the next theorem lies in its showing that it is necessary to compute the remainders of only finitely many polynomials, namely, the S-polynomials, to determine whether $\{g_1, \ldots, g_m\}$ is a Gröbner basis.

Theorem 7.67 (Buchberger).   *A set $\{g_1, \ldots, g_m\}$ is a Gröbner basis of $I = (g_1, \ldots, g_m)$ if and only if $S(g_p, g_q)$ has remainder 0 mod G for all p, q, where $G = [g_1, \ldots, g_m]$.*

Proof.   Clearly, $S(g_p, g_q)$, being a linear combination of g_p and g_q, lies in I. Hence, if $G = \{g_1, \ldots, g_m\}$ is a Gröbner basis, then $S(g_p, g_q)$ has remainder 0 mod G, by Proposition 7.62.

Conversely, assume that $S(g_p, g_q)$ has remainder 0 mod G for all p, q; we must show that every $f \in I$ has remainder 0 mod G. By Proposition 7.62, it suffices to show that if $f \in I$, then $\mathrm{LT}(g_i) \mid \mathrm{LT}(f)$ for some i. Since $f \in I = (g_1, \ldots, g_m)$, we may write $f = \sum_i h_i g_i$, and so

$$\mathrm{DEG}(f) \preceq \max_i \{\mathrm{DEG}(h_i g_i)\}.$$

If $\mathrm{DEG}(f) = \mathrm{DEG}(h_i g_i)$ for some i, then Proposition 7.58 gives $\mathrm{LT}(g_i) \mid \mathrm{LT}(f)$, a contradiction. Therefore, we may assume strict inequality: $\mathrm{DEG}(f) \prec \max_i \{\mathrm{DEG}(h_i g_i)\}$.

The polynomial f may be written as a linear combination of the g_i in many ways. Of all the expressions of the form $f = \sum_i h_i g_i$, choose one in which $\delta = \max_i \{\mathrm{DEG}(h_i g_i)\}$ is minimal (which is possible because $\preceq$ is a well-order). We are done if $\mathrm{DEG}(f) = \delta$, as we have seen above; therefore, we may assume that there is strict inequality: $\mathrm{DEG}(f) \prec \delta$. Write

$$f = \sum_{j,\, \mathrm{DEG}(h_j g_j) = \delta} h_j g_j + \sum_{\ell,\, \mathrm{DEG}(h_\ell g_\ell) \prec \delta} h_\ell g_\ell. \tag{1}$$

If $\mathrm{DEG}(\sum_j h_j g_j) = \delta$, then $\mathrm{DEG}(f) = \delta$, a contradiction; hence, $\mathrm{DEG}(\sum_j h_j g_j) \prec \delta$. But the coefficient of X^δ in this sum is obtained from its leading terms, so that

$$\mathrm{DEG}\left(\sum_j \mathrm{LT}(h_j) g_j\right) \prec \delta.$$

Now $\sum_j \mathrm{LT}(h_j) g_j$ is a polynomial satisfying the hypotheses of Lemma 7.66, and so there are constants d_j and multidegrees $\mu(j)$ so that

$$\sum_j \mathrm{LT}(h_j) g_j = \sum_j d_j X^{\delta - \mu(j)} S(g_j, g_{j+1}), \tag{2}$$

where $\mathrm{DEG}\left(X^{\delta - \mu(j)} S(g_j, g_{j+1})\right) \prec \delta$.[10]

---

[10]The reader may wonder why we consider all S-polynomials $S(g_p, g_q)$ instead of only those of the form $S(g_i, g_{i+1})$. The answer is that the remainder condition is applied only to those $h_j g_j$ for which $\mathrm{DEG}(h_j g_j) = \delta$, and so the indices viewed as i's need not be consecutive.

Since each $S(g_j, g_{j+1})$ has remainder 0 mod G, the division algorithm gives $a_{ji}(X) \in k[X]$ with

$$S(g_j, g_{j+1}) = \sum_i a_{ji} g_i,$$

where $\mathrm{DEG}(a_{ji} g_i) \preceq \mathrm{DEG}(S(g_j, g_{j+1}))$ for all j, i. It follows that

$$X^{\delta - \mu(j)} S(g_j, g_{j+1}) = \sum_i X^{\delta - \mu(j)} a_{ji} g_i.$$

Therefore, Lemma 7.66 gives

$$\mathrm{DEG}(X^{\delta - \mu(j)} a_{ji}) \preceq \mathrm{DEG}(X^{\delta - \mu(j)} S(g_j, g_{j+1})) \prec \delta. \tag{3}$$

Substituting into Eq. (2), we have

$$\sum_j \mathrm{LT}(h_j) g_j = \sum_j d_j X^{\delta - \mu(j)} S(g_j, g_{j+1})$$
$$= \sum_j d_j \left(\sum_i X^{\delta - \mu(j)} a_{ji} g_i \right)$$
$$= \sum_i \left(\sum_j d_j X^{\delta - \mu(j)} a_{ji} \right) g_i.$$

If we denote $\sum_j d_j X^{\delta - \mu(j)} a_{ji}$ by h_i', then

$$\sum_j \mathrm{LT}(h_j) g_j = \sum_i h_i' g_i, \tag{4}$$

where, by Eq. (3), $\mathrm{DEG}(h_i' g_i) \prec \delta$ for all i.

Finally, we substitute the expression in Eq. (4) into Eq. (1):

$$f = \sum_{\substack{j \\ \mathrm{DEG}(h_j g_j) = \delta}} h_j g_j \quad + \sum_{\substack{\ell \\ \mathrm{DEG}(h_\ell g_\ell) \prec \delta}} h_\ell g_\ell$$
$$= \sum_{\substack{j \\ \mathrm{DEG}(h_j g_j) = \delta}} \mathrm{LT}(h_j) g_j \quad + \sum_{\substack{j \\ \mathrm{DEG}(h_j g_j) = \delta}} [h_j - \mathrm{LT}(h_j)] g_j \quad + \sum_{\substack{\ell \\ \mathrm{DEG}(h_\ell g_\ell) \prec \delta}} h_\ell g_\ell$$
$$= \sum_i h_i' g_i \quad + \sum_{\substack{j \\ \mathrm{DEG}(h_j g_j) = \delta}} [h_j - \mathrm{LT}(h_j)] g_j \quad + \sum_{\substack{\ell \\ \mathrm{DEG}(h_\ell g_\ell) \prec \delta}} h_\ell g_\ell.$$

We have rewritten f as a linear combination of the g_i in which each term has multidegree strictly smaller than δ, contradicting the minimality of δ. This completes the proof. •

Corollary 7.68. *If $I = (f_1, \ldots, f_s)$ is a monomial ideal in $k[X]$; that is, each f_i is a monomial, then $\{f_1, \ldots, f_s\}$ is a Gröbner basis of I.*

Proof. By Example 7.65, the S-polynomial of any pair of monomials is 0. •

Here is the main result.

Theorem 7.69 (Buchberger's Algorithm).   *Every ideal $I = (f_1, \ldots, f_s)$ in $k[X]$ has a Gröbner basis[11] which can be computed by an algorithm.*

Proof. Here is a pseudocode for an algorithm.

> Input: $B = \{f_1, \ldots, f_s\}$ $G = [f_1, \ldots, f_s]$
> Output: a Gröbner basis $B = \{g_1, \ldots, g_m\}$
> containing $\{f_1, \ldots, f_s\}$
> $B := \{f_1, \ldots, f_s\}$; $G := [f_1, \ldots, f_s]$
> REPEAT
> $B' := B$; $G' := G$
>   FOR each pair g, g' with $g \neq g' \in B'$ DO
> $r :=$ remainder of $S(g, g')$ mod G'
>   IF $r \neq 0$ THEN
> $B := B \cup \{r\}$; $G' := [g_1, \ldots, g_m, r]$
>   END IF
>   END FOR
> UNTIL $B = B'$

Now each loop of the algorithm enlarges a subset $B \subseteq I = (g_1, \ldots, g_m)$ by adjoining the remainder mod G of one of its S-polynomials $S(g, g')$. As $g, g' \in I$, the remainder r of $S(g, g')$ lies in I, and so the larger set $B \cup \{r\}$ is contained in I.

The only obstruction to the algorithm stopping at some B' is if some $S(g, g')$ does not have remainder 0 mod G'. Thus, if the algorithm stops, then Theorem 7.67 shows that B' is a Gröbner basis.

To see that the algorithm does stop, suppose a loop starts with B' and ends with B. Since $B' \subseteq B$, we have an inclusion of monomial ideals

$$\big(\mathrm{LT}(g') \colon g' \in B'\big) \subseteq \big(\mathrm{LT}(g) \colon g \in B\big).$$

We claim that if $B' \subsetneq B$, then there is also a strict inclusion of ideals. Suppose that r is a (nonzero) remainder of some S-polynomial mod B', and that $B = B' \cup \{r\}$. By definition, the remainder r is reduced mod G', and so no term of r is divisible by $\mathrm{LT}(g')$ for any $g' \in B'$; in particular, $\mathrm{LT}(r)$ is not divisible by any $\mathrm{LT}(g')$. Hence,

---

[11] A nonconstructive proof of the existence of a Gröbner basis can be given using the proof of the Hilbert basis theorem; for example, see Section 2.5 of the book of Cox, Little, and O'Shea (they give a constructive proof in Section 2.7).

$\mathrm{LT}(r) \notin (\mathrm{LT}(g') : g' \in B')$, by Exercise 7.56 on page 569. On the other hand, we do have $\mathrm{LT}(r) \in (\mathrm{LT}(g) : g \in B)$. Therefore, if the algorithm does not stop, there is an infinite strictly ascending chain of ideals in $k[X]$, and this contradicts the Hilbert basis theorem, for $k[X]$ has ACC. •

Example 7.70.
The reader may show that $B' = \{y^2 + z^2, x^2y + yz, z^3 + xy\}$ is not a Gröbner basis because $S(y^2 + z^2, x^2y + yz) = x^2z^2 - y^2z$ does not have remainder 0 mod G'. However, adjoining $x^2z^2 - y^2z$ does give a Gröbner basis B because all S-polynomials in B have remainder 0 mod B'. ◄

Theoretically, Buchberger's algorithm computes a Gröbner basis, but the question arises how practical it is. In very many cases, it does compute in a reasonable amount of time; on the other hand, there are also many examples in which it takes a very long time to produce its output. The efficiency of Buchberger's algorithm is discussed in Section 2.9 of the book by Cox, Little, and O'Shea.

Corollary 7.71.

(i) If $I = (f_1, \ldots, f_t)$ is an ideal in $k[X]$, then there is an algorithm to determine whether a polynomial $h(X) \in k[X]$ lies in I.

(ii) If $I = (f_1, \ldots, f_t)$ and $I' = (f_1', \ldots, f_s')$ are ideals in $k[X]$, then there is an algorithm to determine whether $I = I'$.

Proof.
(i) Use Buchberger's algorithm to find a Gröbner basis B of I, and then use the division algorithm to compute the remainder of h mod G (where G is any m-tuple arising from ordering the polynomials in B). By Corollary 7.64(ii), $h \in I$ if and only if $r = 0$.
(ii) Use Buchberger's algorithm to find Gröbner bases $\{g_1, \ldots, g_m\}$ and $\{g_1', \ldots, g_p'\}$ of I and I', respectively. By part (i), there is an algorithm to determine whether each $g_j' \in I$, and $I' \subseteq I$ if each $g_j' \in I$. Similarly, there is an algorithm to determine the reverse inclusion, and so there is an algorithm to determine whether $I = I'$. •

One must be careful here. Corollary 7.71 does not begin by saying "If I is an ideal in $k[X]$;" instead, it specifies a basis: $I = (f_1, \ldots, f_t)$. The reason, of course, is that Buchberger's algorithm requires a basis as input. For example, if $J = (h_1, \ldots, h_s)$, then the algorithm cannot be used to check whether a polynomial $f(X)$ lies in the radical $\sqrt{J}$ unless one has a basis of $\sqrt{J}$. (There do exist algorithms giving bases of $\sqrt{(f_1, \ldots, f_t)}$; see the book by Becker and Weispfenning.)

A Gröbner basis $B = \{g_1, \ldots, g_m\}$ can be too large. For example, it follows from Proposition 7.62 that if $f \in I$, then $B \cup \{f\}$ is also a Gröbner basis of I; thus, one may seek Gröbner bases that are, in some sense, minimal.

Definition.   A basis $\{g_1, \ldots, g_m\}$ of an ideal I is **reduced** if

(i) each g_i is monic;

(ii) each g_i is reduced mod $\{g_1, \ldots, \widehat{g_i}, \ldots, g_m\}$.

Exercise 7.63 on page 580 gives an algorithm for computing a reduced basis for every ideal $(f_1, \ldots, f_t)$. When combined with the algorithm in Exercise 7.65 on page 581, it shrinks a Gröbner basis to a *reduced* Gröbner basis. It can be proved that a reduced Gröbner basis of an ideal is unique.

In the special case when each $f_i(X)$ is linear, that is,

$$f_i(X) = a_{i1}x_1 + \cdots + a_{in}x_n.$$

then the common zeros $\mathrm{Var}(f_1, \ldots, f_t)$ are the solutions of a homogeneous system of t equations in n unknowns. If $A = [a_{ij}]$ is the $t \times n$ matrix of coefficients, then it can be shown that the reduced Gröbner basis corresponds to the row reduced echelon form for the matrix A (see Section 10.5 in the book of Becker and Weispfenning).

Another special case occurs when $f_1, \ldots, f_t$ are polynomials in one variable. The reduced Gröbner basis obtained from $\{f_1, \ldots, f_t\}$ turns out to be their gcd, and so the Euclidean algorithm has been generalized to polynomials in several variables.

We end this chapter by showing how to find a basis of an intersection of ideals. Given a system of polynomial equations in several variables, one way to find solutions is to eliminate variables (van der Waerden, *Modern Algebra* II, Chapter XI). Given an ideal $I \subseteq k[X]$, we are led to an ideal in a subset of the indeterminates, which is essentially the intersection of $\mathrm{Var}(I)$ with a lower-dimensional plane.

Definition.   Let k be a field and let $I \subseteq k[X, Y]$ be an ideal, where $k[X, Y]$ is the polynomial ring in disjoint sets of variables $X \cup Y$. The **elimination ideal** is

$$I_X = I \cap k[X].$$

For example, if $I = (x^2, xy)$, then a Gröbner basis is $\{x^2, xy\}$ (by Corollary 7.68, because its generators are monomials), and $I_x = (x^2) \subseteq k[x]$, while $I_y = \{0\}$.

Proposition 7.72.   *Let k be a field and let $k[X] = k[x_1, \ldots, x_n]$ have a monomial order for which $x_1 \succ x_2 \succ \cdots \succ x_n$ (for example, the lexicographic order) and, for fixed $p > 1$, let $Y = x_p, \ldots, x_n$. If $I \subseteq k[X]$ has a Gröbner basis $G = \{g_1, \ldots, g_m\}$, then $G \cap I_Y$ is a Gröbner basis for the elimination ideal $I_Y = I \cap k[x_p, \ldots, x_n]$.*

Proof.   Recall that $\{g_1, \ldots, g_m\}$ being a Gröbner basis of $I = (g_1, \ldots, g_m)$ means that for each nonzero $f \in I$, there is g_i with $\mathrm{LT}(g_i) \mid \mathrm{LT}(f)$. Let $f(x_p, \ldots, x_n) \in I_Y$ be nonzero. Since $I_Y \subseteq I$, there is some $g_i(X)$ with $\mathrm{LT}(g_i) \mid \mathrm{LT}(f)$; hence, $\mathrm{LT}(g_i)$ involves only the "later" variables $x_p, \ldots, x_n$. Let $\mathrm{DEG}(\mathrm{LT}(g_i)) = \beta$. If g_i has a term

$c_\alpha X^\alpha$ involving "early" variables x_i with $i < p$, then $\alpha \succ \beta$, because $x_1 \succ \cdots \succ$ $x_p \succ \cdots \succ x_n$. This is a contradiction, for β, the degree of the leading term of g_i, is greater than the degree of any other term of g_i. It follows that $g_i \in k[x_p, \ldots, x_n]$. Exercise 7.62 on page 580 now shows that $G \cap k[x_p, \ldots, x_n]$ is a Gröbner basis for $I_Y = I \cap k[x_p, \ldots, x_n]$. •

We can now give Gröbner bases of intersections of ideals.

Proposition 7.73. *Let k be a field, and let $I_1, \ldots, I_t$ be ideals in $k[X]$, where $X = x_1, \ldots, x_n$.*

(i) *Consider the polynomial ring $k[X, y_1, \ldots, y_t]$ in $n + t$ indeterminates. If J is the ideal in $k[X, y_1, \ldots, y_t]$ generated by $1 - (y_1 + \cdots + y_t)$ and $y_j I_j$, for all j, then $\bigcap_{j=1}^t I_j = J_X$.*

(ii) *Given Gröbner bases of $I_1, \ldots, I_t$, a Gröbner basis of $\bigcap_{j=1}^t I_j$ can be computed.*

Proof.
(i) If $f = f(X) \in J_X = J \cap k[X]$, then $f \in J$, and so there is an equation

$$f(X) = g(X, Y)\left(1 - \sum y_j\right) + \sum_j h_j(X, y_1, \ldots, y_t) y_j q_j(X),$$

where $g, h_j \in k[X, Y]$ and $q_j \in I_j$. Setting $y_j = 1$ and the other y's equal to 0 gives $f = h_j(X, 0, \ldots, 1, \ldots, 0) q_j(X)$. Note that $h_j(X, 0, \ldots, 1, \ldots, 0) \in k[X]$, and so $f \in I_j$. As j was arbitrary, we have $f \in \bigcap I_j$, and so $J_X \subseteq \bigcap I_j$.

For the reverse inclusion, if $f \in \bigcap I_j$, then the equation

$$f = f\left(1 - \sum y_j\right) + \sum_j y_j f$$

shows that $f \in J_X$, as desired.
(ii) This follows from part (i) and Proposition 7.72 if we use a monomial order in which all the variables in X precede the variables in Y. •

Example 7.74.
Consider the ideal $I = (x) \cap (x^2, xy, y^2) \subseteq k[x, y]$, where k is a field. Even though it is not difficult to find a basis of I by hand, we shall use Gröbner bases to illustrate Proposition 7.73. Let u and v be new variables, and define

$$J = (1 - u - v, ux, vx^2, vxy, vy^2) \subseteq k[x, y, u, v].$$

The first step is to find a Gröbner basis of J; we use the lexicographic monomial order with $x \prec y \prec u \prec v$. Since the S-polynomial of two monomials is 0, Buchberger's algorithm quickly gives a Gröbner basis[12] G of J:

$$G = \{v + u - 1, x^2, yx, ux, uy^2 - y^2\}.$$

It follows from Proposition 7.72 that a Gröbner basis of I is $G \cap k[x, y]$: all those elements of G that do not involve the variables u and v. Thus,

$$I = (x) \cap (x^2, xy, y^2) = (x^2, xy). \quad \blacktriangleleft$$

EXERCISES

Use the degree-lexicographic monomial order in the following exercises.

7.57 Let $I = (y - x^2, z - x^3)$.

    (i) Order $x \prec y \prec z$, and let $\preceq_{\text{lex}}$ be the corresponding monomial order on $\mathbb{N}^3$. Prove that $[y - x^2, z - x^3]$ is not a Gröbner basis of I.

    (ii) Order $y \prec z \prec x$, and let $\preceq_{\text{lex}}$ be the corresponding monomial order on $\mathbb{N}^3$. Prove that $[y - x^2, z - x^3]$ is a Gröbner basis of I.

7.58 Find a Gröbner basis of $I = (x^2 - 1, xy^2 - x)$.

7.59 Find a Gröbner basis of $I = (x^2 + y, x^4 + 2x^2y + y^2 + 3)$.

7.60 Find a Gröbner basis of $I = (xz, xy - z, yz - x)$. Does $x^3 + x + 1$ lie in I?

7.61 Find a Gröbner basis of $I = (x^2 - y, y^2 - x, x^2y^2 - xy)$. Does $x^4 + x + 1$ lie in I?

*7.62 Let I be an ideal in $k[X]$, where k is a field and $k[X]$ has a monomial order. Prove that if a set of polynomials $\{g_1, \ldots, g_m\} \subseteq I$ has the property that, for each nonzero $f \in I$, there is some g_i with $\text{LT}(g_i) \mid \text{LT}(f)$, then $I = (g_1, \ldots, g_m)$. Conclude, in the definition of Gröbner basis, that one need not assume that I is generated by $g_1, \ldots, g_m$.

*7.63 Show that the following pseudocode gives a reduced basis Q of an ideal $I = (f_1, \ldots, f_t)$.

```
Input: P = [f_1, ..., f_t]
Output: Q = [q_1, ..., q_s]
Q := P
WHILE there is q ∈ Q which is
        not reduced mod Q − {q} DO
    select q ∈ Q which is not reduced mod Q − {q}
    Q := Q − {q}
    h := the remainder of q mod Q
    IF h ≠ 0 THEN
        Q := Q ∪ {h}
    END IF
END WHILE
make all q ∈ Q monic
```

7.64 If G is a Gröbner basis of an ideal I, and if Q is the basis of I obtained from the algorithm in Exercise 7.63, prove that Q is also a Gröbner basis of I.

---

[12]This is actually the reduced Gröbner basis given by Exercise 7.65 on page 581.

*7.65 Show that the following pseudocode replaces a Gröbner basis G with a reduced Gröbner basis H.

Input: $G = \{g_1, \ldots, g_m\}$
Output: H
$H := \varnothing; \quad F := G$
WHILE $F \neq \varnothing$ DO
   select f' from F
   $F := F - \{f'\}$
   IF $\mathrm{LT}(f) \nmid \mathrm{LT}(f')$ for all $f \in F$ AND
       $\mathrm{LT}(h) \nmid \mathrm{LT}(f')$ for all $h \in H$ THEN
     $H := H \cup \{f'\}$
   END IF
END WHILE
apply the algorithm in Exercise 7.63 to H

A

Inequalities

We prove some elementary properties about inequalities of real numbers, but we first record some properties of the set P of all positive real numbers.

(i) P is closed under addition and multiplication; that is, if a, b are in P, then $a + b$ is in P and ab is in P.

(ii) There is a **trichotomy**: if a is any real number, then exactly one of the following is true:

$$a \text{ is in } P; \quad a = 0; \quad -a \text{ is in } P.$$

Definition. For any two real numbers a and A, define $a < A$ (also written $A > a$) to mean that $A - a$ is in P. We write $a \leq A$ to mean either $a < A$ or $a = A$.

Proposition A.1. *For all $a, b, c \in \mathbb{R}$,*

(i) *$a \leq a$;*

(ii) *if $a \leq b$ and $b \leq c$, then $a \leq c$;*

(iii) *if $a \leq b$ and $b \leq a$, then $a = b$;*

(iv) *either $a < b$, $a = b$, or $b < a$.*

Proof.
(i) We have $a \leq a$ because $a - a = 0$.
(ii) If $a \leq b$, then $b - a$ is in P or $b = a$; if $b \leq c$, then $c - b$ is in P or $c = b$. There are four cases. If $b - a$ is in P and $c - b$ is in P, then $(b - a) + (c - b) = c - a$ is in P and $a \leq c$. If $b - a$ is in P and $c = b$, then $c - a$ is in P and $a \leq c$. The other two cases are similar.
(iii) Assume that $a \leq b$ and $b \leq a$. As in part (i), there are four easy cases. For example, if $b - a$ is in P and $a - b$ is in P, then $(b - a) + (a - b) = 0$ is in P,

a contradiction, so this case cannot occur. If $b - a$ is in P and $b = a$, then $b - a = b - b = 0$ is in P, another contradiction; similarly, $a - b$ is in P and $a = b$ cannot occur. The only remaining possibility is $a = b$.

(iv) Either $a - b$ is in P, $a - b = 0$, or $-(a - b) = b - a$ is in P; that is, either $b \leq a$, $a = b$, or $a \leq b$. ●

Notice that if $a < b$ and $b < c$, then $a < c$ [for $c - a = (c - b) + (b - a)$ is a sum of two numbers in P and, hence, lies in P]. One often abbreviates these two inequalities as $a < b < c$. The reader may check that if $a \leq b \leq c$, then $a \leq c$, with $a < c$ if either inequality $a \leq b$ or $b \leq c$ is strict.

Proposition A.2. *Assume that b and B are real numbers with $b < B$.*

(i) *If m is positive, then $mb < mB$; if m is negative, then $mb > mB$.*

(ii) *For any number N, positive, negative, or zero, we have*

$$N + b < N + B \qquad and \qquad N - b > N - B.$$

(iii) *Let a and A be positive numbers. If $a < A$, then $1/a > 1/A$, and, conversely, if $1/A < 1/a$, then $A > a$.*

Proof.   (i) By hypothesis, $B - b > 0$. If $m > 0$, then the product of positive numbers being positive implies that $m(B - b) = mB - mb$ is positive; that is, $mb < mB$. If $m < 0$, then the product $m(B - b) = mB - mb$ is negative; that is, $mB < mb$.

(ii) The difference $(N + B) - (N + b)$ is positive, for it equals $B - b$. For the other inequality, $(N - b) - (N - B) = -b + B$ is positive, and, hence, $N - b > N - B$.

(iii) If $a < A$, then $A - a$ is positive. Hence, $1/a - 1/A = (A - a)/Aa$ is positive, being the product of the positive numbers $A - a$ and $1/Aa$ (by hypothesis, both A and a are positive). Therefore, $1/a > 1/A$. Conversely, if $1/A < 1/a$, then part (i) gives $a = Aa(1/A) < Aa(1/a) = A$; that is, $A > a$. ●

For example, since $2 < 3$, we have $-3 < -2$ and $\frac{1}{3} < \frac{1}{2}$. One should always look at several particular cases of a formula (even though the validity of these few cases does not prove the truth of the formula), for it helps one have a better feeling about what is being asserted.

B

Pseudocodes

An *algorithm* solving a problem is a set of directions which gives the correct answer after a finite number of steps, never at any stage leaving the user in doubt as to what to do next. The division algorithm is an algorithm in this sense: one starts with a and b and ends with q and r. We are now going to treat algorithms more formally, using ***pseudocodes***, which are general directions that can easily be translated into a programming language. The basic building blocks of a pseudocode are *assignments*, *looping structures*, and *branching structures*.

An ***assignment*** is an instruction written in the form

$$\langle \text{variable} \rangle := \langle \text{expression} \rangle.$$

This instruction evaluates the expression on the right, using any stored values for the variables appearing in it; this value is then stored on the left. Thus, the assignment replaces the variable on the left by the new value on the right.

Example B.1.
Consider the following pseudocode for the division algorithm.

```
1: Input: b ≥ a > 0
2: Output: q, r
3: q := 0;   r := b
4: WHILE r ≥ a DO
5:    r := r − a
6:    q := q + 1
7: END WHILE
```

The meaning of the first two lines is clear; line 3 has two assignments giving initial values to the variables q and r. Let us explain the ***looping structure*** WHILE...DO before considering assignments 5 and 6. The general form is

$$\text{WHILE } \langle\text{condition}\rangle \text{ DO}$$
$$\langle\text{action}\rangle.$$

Here, *action* means a sequence of instructions. The loop repeats the action as long as the condition holds, but it stops either when the condition is no longer valid or when it is told to end. In the example above, one begins with $r = b$ and $q = 0$; since $b \geq a$, the condition holds, and so assignment 5 replaces $r = b$ by $r = b - a$. Similarly, assignment 6 replaces $q = 0$ by $q = 1$. If $r = b - a \geq 0$, this loop repeats this action using the new values of r and q just obtained.

This pseudocode is not a substitute for a proof of the existence of a quotient and a remainder. Had we begun with it, we would still have been obliged to prove two things: first, that the loop does stop eventually; second, that the outputs q and r satisfy the desiderata of the division algorithm, namely, $b = qa + r$ and $0 \leq r < a$. ◄

Example B.2.
Another popular looping structure is REPEAT, written

$$\text{REPEAT } \langle\text{action}\rangle \text{ UNTIL } \langle\text{condition}\rangle.$$

In WHILE, the condition tells when to proceed, whereas in REPEAT, the condition tells when to stop. Another difference is that WHILE may not do a single step, for the condition is checked before acting; REPEAT always does at least one step, for it checks that the condition holds only after it acts.

For example, consider Newton's method for finding a real root of a polynomial $f(x)$. Recall that one begins with a guess a_0 for a root of $f(x)$ and, inductively, defines

$$a_{n+1} = a_n - \frac{f(a_n)}{f'(a_n)}.$$

If the sequence $\{a_n\}$ converges (and it may not), then its limit is a root of $f(x)$. The following pseudocode finds a real root of $f(x) = x^3 + x^2 - 36$ with error at most .0001.

```
Input: positive a
Output: a, y, y'
REPEAT
    y := a³ + a² − 36
    y' := 3a² + 2a
    a := a − y/y'
UNTIL y < .0001  ◄
```

Example B.3.

Here is an example of the *repetition structure* FOR, written

$$\text{FOR each } k \text{ in } K \text{ DO } \langle \text{action} \rangle.$$

Here, a (finite) set $K = \{k_1, \ldots, k_n\}$ is given, and the action consists in performing the action on k_1, then on k_2, through k_n.

For example,

$$\text{FOR each } n \text{ with } 0 \le n \le 41 \text{ DO}$$
$$\quad f := n^2 - n + 41$$
$$\text{END FOR} \quad \blacktriangleleft$$

Example B.4.

An example of a *branching structure* is

$$\text{IF } \langle \text{condition} \rangle \text{ THEN } \langle \text{action \#1} \rangle \text{ ELSE } \langle \text{action \#2} \rangle.$$

When this structure is reached and the condition holds, then action #1 is taken (only once), but if the condition does not hold when this structure is reached, then action #2 is taken (only once). One can omit ELSE $\langle$action #2$\rangle$, in which case the directions are

$$\text{IF } \langle \text{condition} \rangle \text{ THEN } \langle \text{action \#1} \rangle \text{ ELSE } \text{ do nothing.} \quad \blacktriangleleft$$

Here is a pseudocode implementing the Euclidean algorithm.

$$\text{Input: } a, b$$
$$\text{Output: } d$$
$$d := b; \quad s := a$$
$$\text{WHILE } s > 0 \text{ DO}$$
$$\quad \text{rem} := \text{remainder after dividing } d \text{ by } s$$
$$\quad d := s$$
$$\quad s := \text{rem}$$
$$\text{END WHILE}$$

Hints for Selected Exercises

Hint 1.1 (i) True. (ii) True. (iii) False. (iv) True. (v) True. (vi) True. (vii) True. (viii) False.

Hint 1.2(ii) Either prove this by induction, or use part (i).

Hint 1.3 This may be rephrased to say that there is an integer q_n with $10^n = 9q_n + 1$.

Hint 1.8 The sum is n^2.

Hint 1.9 The sum $1 + \sum_{j=1}^{n} j!j = (n+1)!$.

Hint 1.10(ii) One must pay attention to hypotheses. Consider $a^3 + b^3$ if b is negative.

Hint 1.11 There are $n+1$ squares on the diagonal, and the triangular areas on either side have area $\sum_{i=1}^{n} i$.

Hint 1.12(i) Compute the area R of the rectangle in two ways.

Hint 1.12(ii) As indicated in Figure 1.3, a rectangle with height $n+1$ and base $\sum_{i=1}^{n} i^k$ can be subdivided so that the shaded staircase has area $\sum_{i=1}^{n} i^{k+1}$, while the area above it is

$$1^k + (1^k + 2^k) + (1^k + 2^k + 3^k) + \cdots + (1^k + 2^k + \cdots + n^k).$$

Hint 1.12(iii) Write $\sum_{i=1}^{n} (\sum_{p=1}^{i} p) = \frac{1}{2} \sum_{i=1}^{n} i^2 + \frac{1}{2} \sum_{i=1}^{n} i$ in Alhazen's formula, and then solve for $\sum_{i=1}^{n} i^2$ in terms of the rest.

Hint 1.13(i) In the inductive step, use $n \geq 10$ implies $n \geq 4$.

Hint 1.13(ii) In the inductive step, use $n \geq 17$ implies $n \geq 7$.

Hint 1.14 You may assume that $\sum_{n=0}^{\infty} ar^n = a/(1-r)$ if $0 \leq r < 1$.

Hint 1.15 The base step is the product rule for derivatives.

Hint 1.16 The inequality $1 + x > 0$ allows one to use Proposition A.2.

Hint 1.17 Model your solution on the proof of Proposition 1.14. Replace "even" by "multiple of 3" and "odd" by "not a multiple of 3."

Hint 1.18 What is the appropriate form of induction to use?

Hint 1.19 Use Theorem 1.15 and geometric series.

Hint 1.20 For the inductive step, try adding and subtracting the same term.

Hint 1.21 If $2 \leq a \leq n+1$, then a is a divisor of $a + (n+1)!$. Most proofs do not use induction!

Hint 1.25 Use the Inequality of the Means.

Hint 1.26(i) Use *Heron's formula*: if a triangle has area A and sides of lengths a, b, c, then $A^2 = s(s-a)(s-b)(s-c)$, where $s = \frac{1}{2}(a+b+c)$.

Hint 1.26(ii) Use Heron's formula and the Inequality of the Means.

Hint 1.27 If $p \geq q > 0$ and $p' \geq q' > 0$, then $pp' \geq qq'$.

Hint 1.28 (i) True. (ii) False. (iii) True. (iv) True. (v) True. (vi) False. (vii) True.

Hint 1.29 Check that the properties of addition and multiplication used in the proof for real numbers also hold for complex numbers.

Hint 1.31 Consider $f(x) = (1+x)^n$ when $x = 1$.

Hint 1.32(i) Consider $f(x) = (1+x)^n$ when $x = -1$.

Hint 1.33 Take the derivative of $f(x) = (1+x)^n$.

Hint 1.35(i) Use the triangle inequality and induction on n.

Hint 1.35(ii) Use the following properties of the dot product: if $u, v \in \mathbb{C}$, then $|u|^2 = u \cdot u$ and $u \cdot v = |u||v| \cos \theta$, where θ is the angle between u and v.

Hint 1.37 Only odd powers of i are imaginary.

Hint 1.38(ii) Compare with part (i).

Hint 1.40 How many selections of 5 numbers are there?

Hint 1.42 Even though there is a strong resemblance, there is no routine derivation of the Leibniz formula from the binomial theorem (there is a derivation using a trick of hypergeometric series).

Hint 1.45(i) The polar coordinates of $(8, 15)$ are $(17, 62°)$, and $\sin 31° \approx .515$ and $\cos 31° \approx .857$.

Hint 1.45(ii) $\sin 15.5° \approx .267$ and $\cos 15.5° \approx .967$.

Hint 1.46 (i) False. (ii) True. (iii) True. (iv) False. (v) True. (vi) True. (vii) False. (viii) True. (ix) True. (x) False.

Hint 1.47 Use the portion of the full division algorithm that has already been proved.

Hint 1.49 $19 \mid f_7$, but 7 is not the smallest k.

Hint 1.51 Use Corollary 1.37.

Hint 1.52 Write m in base 2.

Hint 1.54(i) Assume $\sqrt{n} = a/b$, where a/b is in lowest terms, and adapt the proof of Proposition 1.43.

Hint 1.54(ii) Assume that $\sqrt[3]{2}$ can be written as a fraction in lowest terms.

Hint 1.58 If $ar + bm = 1$ and $sr' + tm = 1$, consider $(ar + bm)(sr' + tm)$.

Hint 1.59 If $2s + 3t = 1$, then $2(s + 3) + 3(t - 2) = 1$.

Hint 1.60 Use Corollary 1.40.

Hint 1.61 If $b \geq a$, then any common divisor of a and b is also a common divisor of a and $b - a$.

Hint 1.62 Show that if k is a common divisor of ab and ac, then $k \mid a(b, c)$.

Hint 1.64 Use the idea in antanairesis.

Hint 1.68 (i) False. (ii) True. (iii) True. (iv) True. (v) True.

Hint 1.70(ii) Use Corollary 1.53.

Hint 1.71 The sets of prime divisors of a and b are disjoint.

Hint 1.72 Assume otherwise, cross multiply, and use Euclid's lemma.

Hint 1.76(i) If neither a nor b is 0, show that $ab/(a, b)$ is a common multiple of a and b that divides every common multiple c of a and b.

Hint 1.77 (i) True. (ii) False. (iii) False. (iv) False. (v) False. (vi) False. (vii) True. (viii) False.

Hint 1.79 Cast out 9's.

Hint 1.80 $10 \equiv -1 \bmod 11$.

Hint 1.81 $100 = 2 \cdot 49 + 2$.

Hint 1.85 Use the fact, proved in Example 1.61, that if a is a perfect square, then $a^2 \equiv 0$, 1, or 4 mod 8.

Hint 1.86 If the last digit of a^2 is 5, then $a^2 \equiv 5 \bmod 10$; if the last two digits of a^2 are 35, then $a^2 \equiv 35 \bmod 100$.

Hint 1.88 Use Euclid's lemma.

Hint 1.90 By Exercise 1.60 on page 54, we have $21 \mid (x^2 - 1)$ if and only if $3 \mid (x^2 - 1)$ and $7 \mid (x^2 - 1)$.

Hint 1.92 Use the proof of the Chinese remainder theorem. The answer is 199.

Hint 1.94(i) Consider the parity of a and of b.

Hint 1.97 Try -4 coconuts.

Hint 1.98 Easter always falls on Sunday. (There is a Jewish variation of this problem, for Yom Kippur must fall on either Monday, Wednesday, Thursday, or Saturday; secular variants can involve Thanksgiving Day, which always falls on a Thursday, or Election Day, which always falls on a Tuesday.)

Hint 1.99 The year $y = 1900$ was not a leap year.

Hint 1.100 On what day did March 1, 1896, fall?

Hint 1.101(iii) Either use congruences or scan the 14 possible calendars: there are 7 possible common years and 7 possible leap years, for January 1 can fall on any of the 7 days of the week.

Hint 1.102 1900 was not a leap year in America.

Hint 2.1 (i) True. (ii) False. (iii) True. (iv) True. (v) True. (vi) False. (vii) False. (viii) False. (ix) True.

Hint 2.4(iv) Show that each of $A + (B + C)$ and $(A + B) + C$ is described by Figure 2.7.

Hint 2.5 One of the axioms constraining the $\in$ relation is that the statement

$$a \in x \in a$$

is always false.

Hint 2.6(i) You may use the facts: (1) lines ℓ_1 and ℓ_2 having slopes m_1 and m_2, respectively, are perpendicular if and only if $m_2 m_2 = -1$; (2) the midpoint of the line segment having endpoints (a, b) and (c, d) is $(\frac{1}{2}(a + c), \frac{1}{2}(b + d))$.

Hint 2.7(i) Use Proposition 2.2.

Hint 2.8 Does g have an inverse?

Hint 2.10 Either find an inverse or show that f is injective and surjective.

Hint 2.11 It isn't.

Hint 2.12 If f is a bijection, there are m distinct elements $f(x_1), \ldots, f(x_m)$ in Y, and so $m \le n$; using the bijection f^{-1} in place of f gives the reverse inequality $n \le m$.

Hint 2.13(i) If $A \subseteq X$ and $|A| = n = |X|$, then $A = X$; after all, how many elements are in X but not in A?

Hint 2.15(i) Compute composites.

Hint 2.20(i) What is y?

Hint 2.21 (i) False. (ii) True. (iii) True. (iv) False. (v) False. (vi) True. (vii) False. (viii) True. (ix) False. (x) False.

Hint 2.23 Use the complete factorizations of σ and of σ'.

Hint 2.24(i) There are r cycle notations for any r-cycle.

Hint 2.25(i) If $\alpha = (i_0 \ldots i_{r-1})$, show that $\alpha^k(i_0) = i_k$ for $k < r$.

Hint 2.25(ii) Use Proposition 2.24.

Hint 2.27 Use induction on $j - i$.

Hint 2.29(i) If $\alpha = (a_1 \, a_2 \cdots a_k)(b_1 \, b_2 \cdots b_k) \cdots (c_1 \, c_2 \cdots c_k)$ is a product of disjoint k-cycles involving all the numbers between 1 and n, show that $\alpha = \beta^k$, where $\beta = (a_1 \, b_1 \cdots z_1 \, a_2 \, b_2 \cdots z_2 \, \ldots a_k \, b_k \cdots z_k)$.

Hint 2.30(i) First show that $\beta \alpha^k = \alpha^k \beta$ by induction on k.

Hint 2.32 Let $\tau = (1 \; 2)$, and define $f \colon A_n \to O_n$, where A_n is the set of all even permutations in S_n and O_n is the set of all odd permutations, by

$$f \colon \alpha \mapsto \tau \alpha.$$

Show that f is a bijection, and conclude that $|A_n| = |O_n|$.

Hint 2.35 No.

Hint 2.36 (i) False. (ii) False. (iii) True. (iv) False. (v) False. (vi) True. (vii) False. (viii) False. (ix) False. (x) True.

Hint 2.39(i) There are 25 elements of order 2 in S_5 and 75 in S_6.

Hint 2.39(ii) You may express your answer as a sum not in closed form.

Hint 2.40 Clearly, $(y^t)^d = 1$. Use Lemma 2.53 to show that no smaller power of y^t is equal to 1.

Hint 2.43(i) Use induction on $k \geq 1$.

Hint 2.45 Consider the function $f \colon G \to G$ defined by $f(x) = x^2$.

Hint 2.46 Pair each element with its inverse.

Hint 2.47 No general formula is known for arbitrary n.

Hint 2.52 (i) True. (ii) True. (iii) False. (iv) False. (v) True. (vi) False. (vii) True. (viii) False. (ix) True. (x) True. (xi) False.

Hint 2.55 Let G be the four-group **V**.

Hint 2.57 Consider $|H \cap K|$.

Hint 2.58 Can an infinite group have only finitely many cyclic subgroups?

Hint 2.61 If $G \neq ST$, find disjoint subsets of G having $|S|$ and $|T|$ elements, respectively.

Hint 2.62(ii) Use induction on the number of distinct subgroups S_i.

Hint 2.63(ii) Consider $aH \mapsto Ha^{-1}$.

Hint 2.64 (i) True. (ii) False. (iii) True. (iv) True. (v) False. (vi) True. (vii) False. (viii) True. (ix) True. (x) True.

Hint 2.65 If $\alpha \in S_X$, define $\varphi(\alpha) = f \circ \alpha \circ f^{-1}$. In particular, show that if $|X| = 3$, then φ takes a cycle involving symbols 1, 2, 3 into a cycle involving a, b, c, as in Example 2.88.

Hint 2.75 Use a conjugation.

Hint 2.77(i) Consider

$$\varphi : A = \begin{bmatrix} \cos\alpha & -\sin\alpha \\ \sin\alpha & \cos\alpha \end{bmatrix} \mapsto (\cos\alpha, \sin\alpha).$$

Hint 2.78 List the prime numbers $p_0 = 2$, $p_1 = 3$, $p_2 = 5, \ldots$, and define

$$\varphi(e_0 + e_1 x + e_2 x^2 + \cdots + e_n x^n) = p_0^{e_0} \cdots p_n^{e_n}.$$

Hint 2.82 Show that squaring is an injective function $G \to G$, and use Exercise 2.13 on page 105.

Hint 2.83 Take $G = S_3$, $H = \langle (1\ 2) \rangle$, and $g = (2\ 3)$.

Hint 2.84 Show that if A is a matrix which is not a scalar matrix, then there is some nonsingular matrix that does not commute with A. (A proof of this for $n \times n$ matrices given in Proposition 4.86.)

Hint 2.85(iii) Consider cases $A^i A^j$, $A^i B A^j$, $B A^i A^j$, and $(B A^i)(B A^j)$.

Hint 2.86(i) Note that $A^2 = -I = B^2$.

Hint 2.87 Use Exercise 2.69 on page 169.

Hint 2.89(ii) Use Proposition 2.97(ii).

Hint 2.90(iii) See Example 2.48(iv).

Hint 2.91 The vertices $X = \{v_0, \ldots, v_{n-1}\}$ of π_n are permuted by every isometry $\varphi \in \Sigma(\pi_n)$.

Hint 2.95 (i) False. (ii) False. (iii) True. (iv) True. (v) False. (vi) True. (vii) True. (viii) False. (ix) False. (x) True.

Hint 2.97(iii) Define $f: H \times K \to H$ by $f: (h, k) \mapsto h$.

Hint 2.98 If $G/Z(G)$ is cyclic, use a generator to construct an element outside of $Z(G)$ which commutes with each element of G.

Hint 2.99 $|G| = |G/H||H|$.

Hint 2.100 Use induction on $n \geq 1$, where $X = \{a_1, \ldots, a_n\}$. The inductive step should consider the quotient group $G/\langle a_{n+1}\rangle$.

Hint 2.105 If $H \leq G$ and $|H| = |K|$, what happens to elements of H in G/K?

Hint 2.106(i) Use the fact that $H \subseteq HK$ and $K \subseteq HK$.

Hint 2.111(ii) Use Exercise 2.110.

Hint 2.112 Use Wilson's theorem.

Hint 2.114 (i) False. (ii) True. (iii) False. (iv) False. (v) False. (vi) False. (vii) True. (viii) True. (ix) False. (x) True. (xi) True. (xii) False. (xiii) True. (xiv) False.

Hint 2.117 Use Cauchy's theorem.

Hint 2.120 Use Proposition 2.135.

Hint 2.121(i) Recall that A_4 has no element of order 6.

Hint 2.121(ii) Each element $x \in D_{12}$ has a unique factorization of the form $x = b^i a$, where $b^6 = 1$ and $a^2 = 1$.

Hint 2.122(ii) Use the second isomorphism theorem.

Hint 2.123 You may use the fact that the only nonabelian groups of order 8 are D_8 and $\mathbf{Q}$.

Hint 2.124(i) There are 8 permutations in S_4 commuting with (1 2)(3 4), and only 4 of them are even.

Hint 2.125(i) If $\alpha = (1\ 2\ 3\ 4\ 5)$, then $|C_{S_5}(\alpha)| = 5$ because $24 = 120/|C_{S_5}(\alpha)|$; hence $C_{S_5}(\alpha) = \langle \alpha \rangle$. What is $C_{A_5}(\alpha)$?

Hint 2.127(i) Show that (1 2 3) and $(i\ j\ k)$ are conjugate as in the proof of Lemma 2.155.

Hint 2.129 Use Proposition 2.33, checking the various cycle structures one at a time.

Hint 2.131 Use Proposition 2.97(ii).

Hint 2.133(i) Kernels are normal subgroups.

Hint 2.133(ii) Use part (i).

Hint 2.134 Show that G has a subgroup H of order p, and use the representation of G on the cosets of H.

Hint 2.135 If H is a second such subgroup, then H is normal in S_n and hence $H \cap A_n$ is normal in A_n.

Hint 2.137 (i) False. (ii) False. (iii) False. (iv) True. (v) True. (vi) True.

Hint 2.138 The parity of n is relevant.

Hint 2.141(i) The group $G = D_{10}$ is acting. Use Example 2.64 to assign to each symmetry a permutation of the vertices, and then show that

$$P_G(x_1, \ldots, x_5) = \tfrac{1}{10}(x_1^5 + 4x_5 + 5x_1 x_2^2)$$

and

$$P_G(q, \ldots, q) = \tfrac{1}{10}(q^5 + 4q + 5q^3).$$

Hint 2.141(ii) The group $G = D_{12}$ is acting. Use Example 2.64 to assign to each symmetry a permutation of the vertices, and then show that

$$P_G(x_1, \ldots, x_6) = \tfrac{1}{12}(x_1^6 + 2x_6 + 2x_3^2 + 3x_2^3 + 4x_2^3)$$

and so

$$P_G(q, \ldots, q) = \tfrac{1}{12}(q^6 + 2q + 5q^2 + 4q^3).$$

Hint 3.1 (i) False. (ii) True. (iii) False (iv) True. (v) True. (vi) False. (vii) False. (viii) True.

Hint 3.7(i) You may use some standard facts of set theory:

$$U \cap (V \cup W) = (U \cap V) \cup (U \cap W);$$

if V' denotes the complement of V, then

$$U - V = U \cap V';$$

the de Morgan laws (Exercise 2.3 on page 104):

$$(U \cap V)' = U' \cup V' \quad \text{and} \quad (U \cup V)' = U' \cap V'.$$

Hint 3.11(i) If $zw = 0$ and $z = a + ib \neq 0$, then $z\bar{z} = a^2 + b^2 \neq 0$, and

$$\left(\frac{\bar{z}}{z\bar{z}}\right) z = 1.$$

Hint 3.13 Every subring R of $\mathbb{Z}$ contains 1.

Hint 3.14 Use Theorem 1.69.

Hint 3.15(i) Yes.

Hint 3.15(ii) No.

Hint 3.17 (i) True. (ii) True. (iii) False. (iv) False. (v) True. (vi) True. (vii) False.

Hint 3.20 If $R^\times$ denotes the set of nonzero elements of R, prove that multiplication by r is an injection $R^\times \to R^\times$, where $r \in R^\times$.

Hint 3.21 Use Corollary 1.23.

Hint 3.28(i) See Example 2.48(iv).

Hint 3.29 (i) True. (ii) True. (iii) False. (iv) True. (v) True. (vi) True. (vii) False.

Hint 3.30 If x^{-1} exists, what is its degree?

Hint 3.32(i) Compute degrees.

Hint 3.33 Use Fermat's theorem.

Hint 3.34(i) Compare the binomial expansions of $(1+x)^{pm}$ and of $(1+x^m)^p$ in $\mathbb{F}_p[x]$.

Hint 3.36 This is not a hard exercise, but it is a long one.

Hint 3.38(ii) The condition is that there should be a polynomial $g(x) = \sum a_n x^n$ with $f(x) = g(x^p)$; that is, $f(x) = \sum b_n x^{np}$, where $b_n^p = a_n$ for all n.

Hint 3.39(i) The proof for polynomials, Proposition 3.25, works here.

Hint 3.40(i) If R is a domain and $\sigma, \tau \in R[[x]]$ are nonzero, prove that $\mathrm{ord}(\sigma\tau) = \mathrm{ord}(\sigma) + \mathrm{ord}(\tau)$, and hence $\sigma\tau$ has an order.

Hint 3.41 (i) True. (ii) False. (iii) False. (iv) True. (v) False. (vi) False. (vii) True. (viii) True. (ix) True. (x) True.

Hint 3.43(ii) First prove that $1 + 1 = 0$, and then show that the nonzero elements form a cyclic group of order 3 under multiplication.

Hint 3.48 Use the previous exercise to prove that φ is a homomorphism.

Hint 3.51(i) Define $\Phi\colon \mathrm{Frac}(A) \to \mathrm{Frac}(R)$ by $[a, b] \mapsto [\varphi(a), \varphi(b)]$.

Hint 3.54(i) Show that (r, s) is a unit in $R \times S$ if and only if r is a unit in R and s is a unit in S.

Hint 3.54(ii) See Theorem 2.128.

Hint 3.55(ii) Define $\varphi\colon F \to \mathbb{C}$ by $\varphi(A) = a + ib$.

Hint 3.56 (i) True. (ii) False. (iii) False. (iv) True. (v) False. (vi) True. (vii) True. (viii) True. (ix) True. (x) True.

Hint 3.57(ii) Use Corollary 3.52.

Hint 3.58 The answer is $x - 2$.

Hint 3.60 Use Frac(R).

Hint 3.63 Use Frac(R).

Hint 3.64 See Exercise 1.58 on page 54.

Hint 3.66 Mimic the proof of Proposition 1.43 which shows that $\sqrt{2}$ is irrational.

Hint 3.67 Use Exercise 3.37 on page 243.

Hint 3.69(ii) The general proof can be generalized from a proof of the special case of polynomials.

Hint 3.71 There are $q, r \in R$ with $b^i = qb^{i+1} + r$.

Hint 3.72 Use Exercise 3.40 on page 243.

Hint 3.73(i) Example 3.39.

Hint 3.74 See Exercise 1.76 on page 59.

Hint 3.77 See Proposition 1.34.

Hint 3.78(i) Use a degree argument.

Hint 3.79 Show that $\sqrt{x} + 1$ is not a polynomial.

Hint 3.81 Let k be a field and let R be the subring of $k[x]$ consisting of all polynomials having no linear term; that is, $f(x) \in R$ if and only if

$$f(x) = s_0 + s_2 x^2 + s_3 x^3 + \cdots .$$

Show that x^5 and x^6 have no gcd.

Hint 3.82 (i) False. (ii) True. (iii) False. (iv) True. (v) False. (vi) True.

Hint 3.84(i) See Exercise 3.67 on page 274 and Corollary 3.75.

Hint 3.85(i) Use Theorem 3.50.

Hint 3.85(ii) Set $x = a/b$ if $b \neq 0$.

Hint 3.86 (i) True. (ii) False. (iii) True. (iv) False. (v) True. (vi) False. (vii) False. (viii) True. (ix) False. (x) True. (xi) False. (xii) True. (xiii) False. (xiv) True.

Hint 3.87 (i) irreducible. (ii) . (iii) irreducible. (iv) irreducible. (v) irreducible. (vi) irreducible. (vii) irreducible. Show that $f(x)$ has no roots in $\mathbb{Q}$ and that a factorization of $f(x)$ as a product of quadratics would force impossible restrictions on coefficients. (viii) irreducible. Show that $f(x)$ has no rational roots and that a factorization of $f(x)$ as a product of quadratics would force impossible restrictions on coefficients. (ix) irreducible. (x) irreducible.

5

Hint 3.89 The irreducible quintics in $\mathbb{F}_2[x]$ are:

$$x^5 + x^3 + x^2 + x + 1 \qquad x^5 + x^4 + x^2 + x + 1$$
$$x^5 + x^4 + x^3 + x + 1 \qquad x^5 + x^4 + x^3 + x^2 + 1$$
$$x^5 + x^3 + 1 \qquad x^5 + x^2 + 1.$$

Hint 3.90(i) Use the Eisenstein criterion.

Hint 3.91 $f(x) \mapsto f^*(x)$, which reverses coefficients, is not a well-defined function $k[x] \to k[x]$.

Hint 3.92 (i) True. (ii) True. (iii) True. (iv) False. (v) False. (vi) True. (vii) True. (viii) True. (ix) False. (x) False. (xi) True. (xii) True. (xiii) False. (xiv) True. (xv) True.

Hint 3.94(i) Adapt the proof of Theorem 1.73.

Hint 3.94(ii) See the proof of Theorem 2.128.

Hint 3.95 See Exercise 3.84 on page 281.

Hint 3.97(i) Use Exercise 2.13 on page 105.

Hint 3.98 Use Exercise 2.61 on page 158.

Hint 3.99 Show that $\mathbb{F}_p^\times \cong \langle -1 \rangle \times H$, where H is a group of odd order m, say, and observe that either 2 or -2 lies in H because

$$\mathbb{F}_2 \times \mathbb{I}_m = (\{1\} \times H) \cup (\{-1\} \times H).$$

Finally, use Exercise 2.82 on page 170.

Hint 3.100(ii) Equate like coefficients after expanding the right-hand side.

Hint 3.100(iii) In the first case, set $a = 0$ and use b to factor $x^4 + 1$. If $a \neq 0$, then $d = b$ and $b^2 = 1$ (so that $b = \pm 1$); now use a to factor $x^4 + 1$.

Hint 3.100(iv) Use Exercise 3.99 on page 304.

Hint 3.103 See Example 4.127 on page 419.

Hint 3.104(ii) Use the existence of a field with exactly p^n elements.

Hint 3.105 If E has characteristic p, then every nonzero element in E has order p.

Hint 4.1 (i) True. (ii) True. (iii) False. (iv) False. (v) True. (vi) True. (vii) False. (viii) False. (ix) True. (x) True.

Hint 4.4 If $u, v \in V$, evaluate $-[(-v) + (-u)]$ in two ways.

Hint 4.8(i) When are two polynomials equal?

Hint 4.9 The slope of a vector $v = (a, b)$ is $m = b/a$.

Hint 4.10(ii) Rewrite the vectors u, v, and n using coordinates in $\mathbb{R}^3$.

Hint 4.12(ii) If A is skew symmetric, then all its diagonal entries are 0.

Hint 4.13 Use Theorem 3.83.

Hint 4.14 Prove that $(e_i, e_j) = \delta_{ij}$ for all i, j, where δ_{ij} is the Kronecker delta.

Hint 4.15 Given A, prove that there is some m such that $I, A, A^2, \ldots, A^m$ is a linearly dependent list.

Hint 4.17 Prove that if $v_1 + U, \ldots, v_r + U$ is a basis of V/U, then the list $v_1, \ldots, v_r$ is linearly independent.

Hint 4.19(ii) Take a basis of $U \cap U'$ and extend it to bases of U and of U'.

Hint 4.23 (i) False. (ii) True. (iii) False. (iv) True. (v) True. (vi) True.

Hint 4.24(ii) Let A be the matrix whose rows are the given vectors, and see whether $\text{rank}(A) = m$.

Hint 4.25 If A is the matrix whose rows are v_1, v_2, v_3, is $\text{rank}(A) = 3$?

Hint 4.27 If $\gamma \in k^m$, prove that $A\gamma$ is a linear combination of the columns of A.

Hint 4.29(ii) Let A be Gaussian equivalent to an echelon matrix U, so that there is a nonsingular matrix P with $PA = U$. Prove that β lies in the row space $Row(A)$ if and only if $P\beta \in Row(U)$.

Hint 4.30(ii) If $E_p \cdots E_1 A = I$, then $A^{-1} = E_1^{-1} \cdots E_p^{-1}$. Conclude that the elementary row operations which change A into I also change I into A^{-1}. The answer is

$$A^{-1} = \tfrac{1}{4} \begin{bmatrix} 1 & -3 & -1 \\ 1 & 1 & -1 \\ -1 & 3 & 5 \end{bmatrix}.$$

Hint 4.31(ii) Use Corollary 4.40.

Hint 4.32 (i) False. (ii) False. (iii) True. (iv) False. (v) True. (vi) True. (vii) True. (viii) False. (ix) True. (x) True.

Hint 4.38(ii) Here is the statement. If $f \colon V \to W$ is a linear transformation with $\ker f = U$, then U is a subspace of V and there is an isomorphism $\varphi \colon V/U \to \text{im } f$, namely, $\varphi(v + U) = f(v)$.

Hint 4.40 Use Theorem 4.62.

Hint 4.46 (i) False. (ii) True. (iii) False. (iv) False. (v) False. (vi) True. (vii) False. (viii) True. (ix) False. (x) True.

Hint 4.49 See the elementary row operations on page 345.

Hint 4.52(ii) $0 = 1 - \omega^n = (1 - \omega)(1 + \omega + \omega^2 + \cdots + \omega^{n-1})$.

Hint 4.54(i) Define $T^{\#}(e_i) = \bar{a}_{1i}e_1 + \cdots + \bar{a}_{ni}e_n$.

Hint 4.54(iv) Adapt the proof of Theorem 4.104.

Hint 4.60 If $B_i = P_i A_i P_i^{-1}$ for all i, then $(P_1 \oplus \cdots \oplus P_t)^{-1} = P_1^{-1} \oplus \cdots \oplus P_t^{-1}$.

Hint 4.62 Assume first that $c \in k$.

Hint 4.65 Recall the power series $1/x = 1 - x + x^2 - x^3 + \cdots$, where x is a nonzero real.

Hint 4.70 C is the disjoint union $\bigcup_{c \in C} B_t(c)$.

Hint 4.73 If $C = \{(a, a) \in \mathbb{F}^2\}$, then C corrects up to 1 error. If $w = (a, b)$, where $a \neq b$, then $w \notin C$ and $\delta\big(w, (a, a)\big) = 1 = \delta\big(w, (b, b)\big)$.

Hint 4.76(i) Use the factorization $x^{15} - 1 = \Phi_1(x)\Phi_3(x)\Phi_5(x)\Phi_{15}(x)$ in $\mathbb{Z}[x]$, where $\Phi_d(x)$ is the dth cyclotomic polynomial.

Hint 4.76(ii) Try $g(x) = x^4 + x + 1$.

Hint 4.76(iii) Show that ζ^2 is a root of $g(x)$.

Hint 4.77 Use Table 4.1 in Example 4.127. The answer is

$$c = (\zeta^3, \zeta, 1 + \zeta^3, \zeta^3 + \zeta, 0, \zeta^3, 1).$$

Hint 5.1(ii) As a practical matter, given a monic polynomial in $\mathbb{Q}[x]$, one should first use Theorem 3.90 to see whether it has any rational (necessarily integral) roots.

Hint 5.6 Apply complex conjugation to the equation $f(u) = 0$.

Hint 5.8 $r = \cos 3\theta = \cos 3(\theta + 120°) = \cos 3(\theta + 240°)$.

Hint 5.9(i) By definition, $\cosh \theta = \frac{1}{2}(e^\theta + e^{-\theta})$. Expand and then simplify

$$4[\tfrac{1}{2}(e^\theta + e^{-\theta})]^3 - 3[\tfrac{1}{2}(e^\theta + e^{-\theta})]$$

to obtain $\frac{1}{2}(e^{3\theta} + e^{-3\theta})$.

Hint 5.9(ii) By definition, $\sinh \theta = \frac{1}{2}(e^\theta - e^{-\theta})$.

Hint 5.10 The roots are -4 and $2 \pm \sqrt{-3}$.

Hint 5.11 The roots are 17 and $\frac{1}{2}(-1 \pm \sqrt{-3})$.

Hint 5.12(i) The roots appear in unrecognizable form.

Hint 5.12(ii) The roots are 4 and $-2 \pm \sqrt{3}$.

Hint 5.13 The roots are 2 and $-1 \pm \sqrt{3}$.

Hint 5.14 This is a tedious calculation. The roots are $-3, -1, 2 \pm \sqrt{6}$.

Hint 5.15 (i) False. (ii) False. (iii) False. (iv) False. (v) False. (vi) True. (vii) False. (viii) True. (ix) False.

Hint 5.20 No.

Hint 5.21(ii) Use Proposition 1.39.

Hint 5.21(iii) Use Exercise 2.13 on page 105 to prove the Frobenius $F: k \rightarrow k$ is surjective when k is finite.

Hint 5.22(ii) Use Proposition 3.116.

Hint 5.22(iii) Prove that if $\sigma \in G$, then σ is completely determined by $\sigma(\alpha)$, which is a root of the irreducible polynomial of α.

Hint 5.22(iv) Prove that F has order $\geq n$.

Hint 5.25 Observe that $x^{30} - 1 = (x^6 - 1)^5$ in $\mathbb{F}_5[x]$.

Hint 5.29(i) If α is a real root of $f(x)$, then $\mathbb{Q}(\alpha)$ is not the splitting field of $f(x)$.

Hint 5.29(ii) Use part (i).

Hint 5.29(iii) Try $g(x) = 3x^3 - 3x + 1$.

Hint 5.30(ii) Use Exercise 3.67 on page 274.

Hint 5.32(i) Consider $f(x) = x^p - t \in \mathbb{F}_p(t)[x]$.

Hint 6.1 (i) False. (ii) False. (iii) True. (iv) True. (v) True. (vi) False. (vii) False. (viii) False. (ix) False. (x) True.

Hint 6.6(ii) Use part (i).

Hint 6.6(iii) Use part (i).

Hint 6.8 There are 14 groups.

Hint 6.10 If B is a direct sum of k copies of a cyclic group of order p^n, then how many elements of order p^n are in B?

Hint 6.11(iii) If A and B are nonzero subgroups of $\mathbb{Q}$, then $A \cap B \neq \{0\}$.

Hint 6.12(i) Use the proof of the basis theorem (Theorem 6.11).

Hint 6.13 If F is a direct sum of m infinite cyclic groups, prove that $F/2F$ is an m-dimensional vector space over $\mathbb{F}_2$.

Hint 6.20 (i) False. (ii) True. (iii) True. (iv) True. (v) False. (vi) True. (vii) False. (viii) True. (ix) False. (x) True.

Hint 6.22 Consider $S_3 \times S_3$.

Hint 6.24 If $g \in G$, then $g P g^{-1}$ is a Sylow p-subgroup of K, and so it is conjugate to P in K.

Hint 6.25 See Exercise 3.28 on page 235.

Hint 6.26 It suffices to find a subgroup of S_6 of order 16. Consider the disjoint union $\{1, 2, 3, 4, 5, 6\} = \{1, 2, 3, 4\} \cup \{5, 6\}$, and use Exercise 2.106 on page 191.

Hint 6.27 Use the fact that any other Sylow p-subgroup of G is conjugate to P.

Hint 6.28 Compute the order of the subgroup generated by the Sylow subgroups.

Hint 6.29(i) Show that p divides neither $[G/H : HP/H]$ nor $[H : H \cap P]$.

Hint 6.29(ii) Choose a subgroup H of S_4 with $H \cong S_3$, and find a Sylow 3-subgroup P of S_4 with $H \cap P = \{1\}$.

Hint 6.31 Some of these are not tricky.

Hint 6.32 Adapt the proof of the primary decomposition.

Hint 6.34 By Cauchy's theorem, G must contain an element a of order p, and $\langle a \rangle \lhd G$ because it has index 2.

Hint 6.35(i) Every independent subset can be extended to a basis.

Hint 6.35(ii) The group $\mathrm{GL}(r, k)$ acts on the set X of all linearly independent r-lists in $(\mathbb{F}_q)^n$, and use the proof of Theorem 6.30.

Hint 6.38(i) Use determinant.

Hint 6.38(ii) Use part (i) and Theorem 6.30.

Hint 6.38(iii) Prove that the matrices $A = \left[\begin{smallmatrix} 0 & 1 \\ -1 & 0 \end{smallmatrix} \right]$ and $B = \left[\begin{smallmatrix} 0 & 2 \\ 2 & 0 \end{smallmatrix} \right]$ generate a subgroup of order 8 which is isomorphic to **Q**.

Hint 6.39 (i) False. (ii) False. (iii) False. (iv) False. (v) True. (vi) False. (vii) True. (viii) True.

Hint 6.48(i) If $\rho(z) = e^{i\theta} \bar{z}$, define $R(z) = e^{i\alpha} z$, where $\alpha = \frac{1}{2}(2\pi - \theta)$.

Hint 6.51 Prove that every $g \in G$ has a unique expression $g = a^i b^j$, where $i \in \{0, 1\}$ and $j \in \mathbb{Z}$.

Hint 7.1 (i) True. (ii) True. (iii) False. (iv) False. (v) True. (vi) False. (vii) False. (viii) True. (ix) True. (x) True.

Hint 7.3 When is a Boolean ring a domain?

Hint 7.5(ii) Let $f : \mathbb{Z} \to \mathbb{I}_4$ be the natural map, and take $Q = \{0\}$.

Hint 7.15(ii) For surjectivity, if I and J are coprime, there are $a \in I$ and $b \in J$ with $1 = a + b$. If $r, r' \in R$, prove that $(d + I, d + J) = (r + I, r' + J) \in R/I \times R/J$, where $d = r'a + rb$.

Hint 7.16(iii) You may assume that every nonunit in a commutative ring lies in some maximal ideal (this result is proved using Zorn's lemma).

Hint 7.17 (i) True. (ii) True. (iii) True. (iv) False. (v) False. (vi) True. (vii) False. (viii) False. (ix) True. (x) True.

Hint 7.29 (i) False. (ii) False. (iii) True. (iv) False. (v) False. (vi) False. (vii) True. (viii) True.

Hint 7.35 Use Exercise 1.2(i) on page 15.

Hint 7.36 If $f^r \in I$ and $g^s \in I$, prove that $(f + g)^{r+s} \in I$.

Bibliography

Albert, A. A., *Introduction to Algebraic Theories*, University of Chicago Press, 1941.

Artin, M., *Algebra*, Prentice Hall, Upper Saddle River, NJ, 1991.

Baker, A., *Transcendental Number Theory*, Cambridge University Press, 1979.

Becker, T., and Weispfenning, V., *Gröbner Bases: a Computational Approach to Commutative Algebra*, Springer-Verlag, New York, 1993.

Berlekamp, E. R., Conway, J. H., and Guy, R. K., *Winning Ways for Your Mathematical Plays*, Academic Press, Orlando, FL, 1982.

Biggs, N. L., *Discrete Mathematics*, Oxford University Press, 1989.

Birkhoff, G., and Mac Lane, S., *A Survey of Modern Algebra*, 4th ed., Macmillan, New York, 1977.

Blake, I. F., and Mullin, R. C., *The Mathematical Theory of Coding*, Academic Press, New York, 1975.

Burn, R. P., *Groups: A Path to Geometry*, Cambridge University Press, Cambridge, 1985.

Burnside, W., *The Theory of Groups of Finite Order*, Cambridge University Press, 1911.

Cajori, F., *A History of Mathematical Notation*, Open Court, 1928; Dover reprint, 1993.

Carmichael, R., *An Introduction to the Theory of Groups of Finite Order*, Ginn, Boston, 1937.

Cox, D., Little, J., and O'Shea, D., *Ideals, Varieties, and Algorithms*, 3d ed., Springer-Verlag, New York, 1992.

Curtis, C., *Linear Algebra; An Introductory Approach*, Springer-Verlag, New York, 1984.

Dornhoff, L. L., and Hohn, F. E., *Applied Modern Algebra*, Macmillan, New York, 1978.

Eisenbud, D., *Commutative Algebra with a View Toward Algebraic Geometry*, Springer-Verlag, New York, 1995.

Fröhlich, A., and Taylor, M. J., *Algebraic Number Theory*, Cambridge University Press, Cambridge, 1991.

Gorenstein, D., Lyons, R., and Solomon, R., *The Classification of the Finite Simple Groups*, Math. Surveys and Monographs, Volume 40, American Mathematical Society, Providence, 1994.

Hadlock, C., *Field Theory and its Classical Problems*, Carus Mathematical Monographs No. 19, Mathematical Association of America, Washington, 1978.

Herstein, I. N., *Topics in Algebra*, 2d ed., Wiley, New York, 1975.

Hoffman, D. G., Leonard, D. A., Lindner, C. C., Phelps, K. T., Rodger, C. A., and Wall, J. R., *Coding Theory: The Essentials*, Marcel Dekker, New York, 1991.

Jacobson, N., *Basic Algebra* I, Freeman, San Francisco, 1974.

————, *Basic Algebra* II, Freeman, San Francisco, 1980.

Kaplansky, I., *Fields and Rings*, 2d ed., University of Chicago, 1974.

Laywine, C. F., and Mullen, G. L., *Discrete Mathematics Using Latin Squares*, Wiley, New York, 1998.

Leon, Steven J., *Linear Algebra with Applications*, 6th ed., Prentice Hall, Upper Saddle River, 2002.

Li, C. C., *An Introduction to Experimental Statistics*, McGraw-Hill, New York, 1964.

Lidl, R., and Niederreiter, H., *Introduction to Finite Fields and Their Applications*, Cambridge University Press, 1986.

Ling, S., and Xing, C., *Coding Theory; A First Course*, Cambridge University Press, 2004.

Martin, G. E., *Transformation Geometry: An Introduction to Symmetry*, Springer-Verlag, New York, 1982.

McCoy, N. H., and Janusz, G. J., *Introduction to Modern Algebra*, 5th ed., Wm. C. Brown Publishers, Dubuque, 1992.

Nagpaul, S. R., and Jain, S. K., *Topics in Applied Abstract Algebra*, Brooks/Cole, Belmont, 2005.

Niven, I., and Zuckerman, H. S., *An Introduction to the Theory of Numbers*, Wiley, New York, 1972.

Pollard, H., *The Theory of Algebraic Numbers*, Carus Mathematical Monographs No. 9, Mathematical Association of America, Washington, 1950.

Rotman, J. J., *Advanced Modern Algebra*, Prentice Hall, Upper Saddle River, 2002.

———, *Galois Theory*, 2d ed., Springer-Verlag, New York, 1998.

———, *An Introduction to the Theory of Groups*, 4th ed., Springer-Verlag, New York, 1995.

———, *Journey into Mathematics*, Prentice Hall, Upper Saddle River, 1998.

Ryser, H. J., *Combinatorial Mathematics*, Carus Mathematical Monographs No.14, Mathematical Association of America, Washington, 1963.

Stark, H. M., *An Introduction to Number Theory*, Markham, Chicago, 1970.

Stillwell, J., *Mathematics and Its History*, Springer-Verlag, New York, 1989.

Suzuki, M., *Group Theory I*, Springer-Verlag, New York, 1982.

Thompson, T. M., *From Error-Correcting Codes Through Sphere Packings to Simple Groups*, Carus Mathematical Monographs No. 21, Mathematical Association of America, Washington, 1983.

Tignol, J.-P., *Galois' Theory of Equations*, World Scientific, Singapore, 1988.

Trappe, W., and Washington, L. C., *Introduction to Cryptography with Coding Theory*, Prentice Hall, Upper Saddle River, 2002.

Tucker, A., *Applied Combinatorics*, 2d ed., Wiley, New York, 1984.

Uspensky, J. V., and Heaslet, M. A., *Elementary Number Theory*, McGraw-Hill, New York, 1939.

van der Waerden, B. L., *Geometry and Algebra in Ancient Civilizations*, Springer-Verlag, New York, 1983.

———, *A History of Algebra*, Springer-Verlag, New York, 1985.

———, *Modern Algebra*, 4th ed., Ungar, New York, 1966.

———, *Science Awakening*, Wiley, New York, 1963.

Weyl, H., *Symmetry*, Princeton University Press, 1952.

Zariski, O., and Samuel, P., *Commutative Algebra*, volume II, von Nostrand, Princeton, 1960.

Index